WATER RESOURCES

ENGINEERING

VOLUME I

Proceedings of the First International Conference
Sponsored by the
Water Resources Engineering
Division of the American Society of Civil Engineers

in cooperation with
American Institute of Hydrology
American Society of Civil Engineers:
  Environmental Engineering Division
  Water Resources Planning and Management Division
  Waterway, Port, Coastal, and Ocean Division
American Water Resources Association
Canadian Society for Civil Engineering
International Association for Hydraulic Research
U. S. Geological Survey
U. S. Bureau of Reclamation
Natural Resources Conservation Service
Society of Range Conservation
Soil and Water Conservation Society
Texas Water Development Board
Texas Natural Resource Conservation Commission

San Antonio, Texas
August 14-18, 1995

Edited by William H. Espey, Jr. and Phil G. Combs

Published by the
American Society of Civil Engineers
345 East 47th Street
New York, New York 10017-2398

ABSTRACT

These proceedings, Water Resources Engineering, contains papers that were presented at the First International Conference on Water Resources Engineering in San Antonio, Texas, August 14-18, 1995. Beginning in 1995, the annual conferences of the Hydraulic and Irrigation and Drainage Divisions have combined to become the International Water Resources Engineering Conference. The objective of the Conference is to provide an annual forum for discussion and exchange of information on the broad spectrum of important issues in Water Resources Engineering. The themes discussed in this proceedings are: 1) new methodology of hydraulic engineering; 2) new directions for hydraulic research; 3) disaster hazard reduction; 4) groundwater; 5) sedimentation; 6) hydraulic data acquisition and display; 6) innovative hydraulic structures; 7) wetlands mitigation/sedimentation; 8) environmental hydraulics; 9) stream restoration; 10) optimization of water resource systems; 11) water quality enhancement technology; 12) irrigation and marginal reclaimed water; 13) 3D free surface modeling; 14) flow forecasting under uncertainties; 15) irrigation efficiency and uniformity; and 16) theory, numerical modeling, and numerical analysis of flow in steep channels.

ISBN 0-7844-0108-X
Manufactured in the United States of America.

PREFACE

Beginning in 1995, the annual conferences of the Hydraulic and Irrigation and Drainage Divisions have combined to become the International Water Resources Engineering Conference. The 1995 First International Conference on Water Resources Engineering, in conjunction with the International Groundwater Management Symposium and the Watershed Management Symposium, was held at the Hyatt Regency San Antonio on the Riverwalk, San Antonio, Texas, August 14-18, 1995. In addition, the conference included Texas Water '95, which focused on Texas water at the turn of the century. The theme for the Conference and the Symposiums was "Water: A Resource in Crisis".

The emphasis was on practical state-of-the-art water resources engineering among practicing professionals. Topics that were discussed are as follows:

- Canal Automation
- Developing Country Technology Transfer
- Hydromechanics
- Wetlands Mitigation/Sedimentation
- Hydrologic Extremes and Risk Analyses
- Theory, Numerical Modeling, and Numerical Analysis of Flow in Steep Channels
- Overtopping Protection for Dams
- Non-Point Source Pollution
- Sedimentation and Scour
- Oil Spill Modeling/Analysis
- Stream Restoration
- Surface Water Hydrology
- Remote Sensing/GIS
- Urban Hydrology
- Stochastic Hydrology
- Hydraulic Structures
- Tidal and Coastal Hydraulics
- Optimization of Water Resources Systems
- Irrigation with Marginal Reclaimed Water
- 3D Free Surface Modeling
- Irrigation Efficiency and Uniformity
- Farm Irrigation System
- Flow Forecasting Under Uncertainties
- Water Quality Enhancement Technology
- Environmental Hydraulics

Panel discussions on *Establishing Research Priorities in Developing Countries* and *3-D Free Surface Modeling* were also held. A pre-conference technical tour of *Edwards Hydrogeology and Seco Creek National Watershed Demonstration Project* was held Saturday, August 12. A technical tour of selected features of the San Antonio River Authority's Project on Flood Drainage Tunnel Protection Works for the City of San Antonio was also conducted.

Over 600 papers were selected to be presented at the Conference. Papers were received through the Call for Papers and accepted based on the reviews of the Conference Committee. All papers are eligible for discussion in the appropriate journal and are also eligible for ASCE awards.

We wish to thank all the authors, moderators and panelists for their participation and cooperation in helping make the Conference a successful one.

William H. Espey, Jr.
Phil G. Combs
Editors

CONTENTS

SESSION C-1
CANAL AUTOMATION: PART I
Moderator: ALBERT J. CLEMMENS, U.S. Department of Agriculture/Agricultural Research Service, U.S. Water Conservation Laboratory, Phoenix, Arizona

SECTION C-2
RED RIVER: PART I
Moderator: C. FRED PINKARD, JR., U.S. Army Corps of Engineers, Vicksburg, MS

SECTION C-3
DEVELOPING COUNTRY TECHNOLOGY TRANSFER: PART I
Moderator: CARL NORDIN, Colorado State University, Ft. Collins, CO

SESSION C-4
HYDROMECHANICS
Moderator: EDWARD R. HOLLEY, University of Texas, Austin, Texas

SESSION C-5
WETLANDS MITIGATION/SEDIMENTATION
Moderator: BRAD R. HALL, U.S. Army Waterway Experiment Station, Vicksburg, MS

SESSION C-6
HYDROLOGIC EXTREMES AND RISK ANALYSES: PART I
Moderator: JEFF BRADLEY, WEST Inc., Seattle, WA

SESSION C-7
THEORY, NUMERICAL MODELING, AND NUMERICAL ANALYSIS OF FLOW IN STEEP CHANNELS: PART I
Moderator: CHINTU LAI, National Taiwan University, Taipei, Taiwan

SESSION C-8
OVERTOPPING PROTECTION FOR DAMS: PART I
Moderator: MORRIS SKINNER, Colorado State University, Ft. Collins, CO

SESSION C-9
CANAL AUTOMATION: PART II
Moderator: ALBERT J. CLEMMENS, U.S. Department of Agriculture/Agricultural Research Service, U.S. Water Conservation Laboratory, Phoenix, Arizona

SESSION C-10
RED RIVER: PART II
Moderator: BEN LITTLEPAGE, Red River Waterway Commission, Natchitoches, LA

SESSION C-11
CHEMICAL FATE AND TRANSPORT: PART I
Moderator: POOJITHA D. YAPA, Clarkson Univeristy, Potsdam, NY

SESSION C-12
DEVELOPING COUNTRY TECHNOLOGY TRANSFER: PART II
Moderator: DANIEL GUNARATHAM, The World Bank, Washington D.C.

SESSION C-13
WETLAND HYDROLOGY
Moderator: ANANTA K. NATH, South Florida Management District, Naples, FL

SESSION C-14
HYDROLOGIC EXTREMES AND RISK ANALYSIS: PART II
Moderator: CHARLES S. MELCHING, USGS Water Resources Division, Urbana, IL

SESSION C-15
THEORY AND NUMERICAL MODELING OF FLOW IN STEEP CHANNELS: PART II
Moderator: RICHARD STOCKSTILL, U.S. Army Waterways Experiment Station, Vicksburg, MS

SESSION C-16
OVERTOPPING PROTECTION FOR DAMS: PART II
Moderator: JAMES RUFF, Colorado State University, Ft. Collins, CO

SESSION C-17
CANAL AUTOMATION: PART III
Moderator: CHARLES M. BURT, Irrigation Training and Research Center, California Polytechnic State University, San Luis Obispo, California

SESSION C-18
NON-POINT SOURCE POLLUTION: PART I
Moderator: LESLIE SHOEMAKER, Tetra Tech, Inc., Fairfax, VA

SESSION C-19
CLIMATE CHANGE
Moderator: SHAFIQUL ISLAM, University of Cincinnati, Cincinnati, OH

SESSION C-20
COMPUTERIZATION OF HYDRAULICS: PART I
Moderator: JEFF HOLLAND, U.S. Army Waterways Experiment Station, Vicksburg, MS

SESSION C-21
SEDIMENTATION AND SCOUR: PART I
Moderator: MARCELO GARCIA, University of Illinois at Urbana Champaign, Urbana, IL

SESSION C-22
CHEMICAL FATE AND TRANSPORT: PART II
Moderator: E.Z. HOSSEINIPOUR, Law Engineering and Environmental Services, Kennesaw, GA

SESSION C-23
RED RIVER: PART III
Moderator: DAVID BIEDENHARN, Lower Mississippi Valley Division, U.S. Army Corps of Engineers, Vicksburg, MS

SESSION C-24
CANAL AUTOMATION PART IV
Moderator: DAVID C. ROGERS, U.S. Bureau of Reclamation, Denver, Colorado

SESSION C-25
SEDIMENTATION AND SCOUR: PART II
Moderator: RONALD COPELAND, U.S. Army Waterway Experiment Station, Vicksburg, MS

SESSION C-26
CHEMICAL FATE AND TRANSPORT: PART III
Moderator: LISA ROIG, U.S. Army Waterways Experiment Station, Vickburg, MS

SESSION C-27
FISH BYPASS
Moderator: JOHN GEORGE, U.S. Army Waterways Experiment Station, Vicksburg, MS

SESSION C-28
COMPUTERIZATION OF HYDRAULICS: PART II
Moderator: EDWARD R. HOLLEY, University of Texas, Austin, Texas

SESSION C-29
DEVELOPING COUNTRY TECHNOLOGY TRANSFER: PART III
Moderator: MICHAEL A. STEVENS, Consultant, Boulder, CO

SESSION C-30
CANAL AUTOMATION: PART V
Moderator: ROBERT S. GOOCH, Salt River Project, Phoenix, Arizona

SESSION C-31
SEDIMENT AND SCOUR: PART III
Moderator: LYLE ZEVENBURGER, Ayres and Associates, Ft. Collins, CO

SESSION C-32
OIL SPILL MODELING/ANALYSIS
Moderator: POOJITHA YAPA, Clarkson University, Postsdam, NY

SESSION C-33
STREAM RESTORATION: PART I
Moderator: DOUG SHIELDS, U.S. Department of Agriculture National Sedimentation Laboratory, Oxford, MS

SESSION C-34
COMPUTERIZATION OF HYDRAULICS: PART III
Moderator: JEFF BRADLEY, WEST, Inc., Seattle, WA

SESSION C-35
NEEDS, GOALS AND FEASIBILITY STUDIES FOR DESIGN AND OPERATION OF PRECIPITATION ENHANCEMENT PROJECTS
Moderator: CONRAD KEYES, JR., Director, Water Resources Special Project, Boyle Engineering Corporation, El Paso, TX

SESSION C-36
MANUAL 57: MANAGEMENT OPERATION AND MAINTENANCE OF IRRIGATION AND DRAINAGE SYSTEMS
Moderator: JAMES B. ROBERTSON, U.S. Bureau of Reclamation, Denver, Colorado

SESSION C-37
DEVELOPING COUNTRY TECHNOLOGY TRANSFER ESTABLISHING RESEARCH PRIORITIES IN DEVELOPING COUNTRIES, PANEL DISCUSSION: PART 4
Moderator: MAHMOUD ABU-ZEID, Ch. Water Research Center, Cairo, Egypt

SESSION C-38
SEDIMENTATION AND SCOUR: PART IV
Moderator: DAVID WILLIAMS, WEST, Inc., Carlsbad, CA

SESSION C-39
STREAM RESTORATION: PART II
Moderator: CARLOS ALONZO, U.S. Department of Agriculture/National, Sedimentation Laboratory, Oxford, MS

SESSION C-40
WATER SUPPLY
Moderator: MARSHALL ENGLISH, Oregon State University, Covallis, Oregon

SESSION C-41
FLOODING: PART I
SOUTHEAST UNITED STATES FLOODING 1994
Moderator: BERT HOLLER, South Atlantic Division, U.S. Army Corps of Engineers, Atlanta, GA

SESSION C-42
BRIDGE SCOUR: PART I
Moderator: PEGGY JOHNSON, University of Maryland, College Park, MD, and Carl Nordin, Colorado State University, Ft. Collins, CO

SESSION C-43
STREAM RESTORATION: PART III
Moderator: ANDREW BROOKES, National Rivers Authority, United Kingdom

SESSION C-44
SURFACE WATER HYDROLOGY—GENERAL: PART I
Moderator: VICTOR ZITTA, Mississippi State University, Starksville, MS

SESSION C-45
REMOTE SENSING/GIS: PART I
Moderator: PAUL A. DEBARRY, R.K.R. Hess Associates, Pennsylvania

SESSION C-46
FLOODING: PART II
MIDWEST UNITED STATES FLOOD 1993
Moderator: RAY KOPSKY, U.S. Army Corps of Engineers, St. Louis, MO

SESSION C-47
BRIDGE SCOUR: PART II
Moderators: JEFF BRADLEY, West Consultants, Seattle, WA, and DAVID FROEHLICH, Lexington, KY

SESSION C-48
SURFACE WATER HYDROLOGY—GENERAL: PART II
Moderator: ELLIOTT SIVERSTON, Greiner, Inc., Tampa, FL

SESSION C-49
REMOTE SENSING/GIS: PART II
Moderator: RAFAEL G. QUIMPO, University of Pittsburg, Pittsburg, PA

SESSION C-50
HYDRAULIC MEASUREMENTS: PART I
Moderator: GREGORY GARTRELL, Contra Costsa Water District, Concord, CA

SESSION C-51
NEW IRRIGATION AND DRAINAGE PRACTICES FOR MARGINAL QUALITY WATERS: PART I
Moderator: JONANNES C. GUITJENS, University of Nevada, Reno, Nevada

SESSION C-52
FLOODING: PART III
MIDWEST UNITED STATES FLOOD 1993
Moderator: RAY KOPSKY, U.S. Army Corps of Engineer, St. Louis, MO

SESSION C-53
BRIDGE SCOUR: PART III
BRIDGE SCOUR IN BEDROCK AND COARSE BED MATERIAL
Moderators: JERRY R. RICHARDSON, University of Missouri at Kansas City, MO and STEVE R. ABT, Colorado State University, Ft. Collins, CO

SESSION C-54
SURFACE WATER HYDROLOGY—GENERAL: PART III
Moderator: ARLEN FELDMAN, Hydrologic Engineering Center, U.S. Army Corps of Engineers, Davis, CA

SESSION C-55
NEW IRRIGATION AND DRAINAGE PRACTICES FOR MARGINAL QUALITY WATERS: PART II
Moderator: GLEN D. SANDERS, U.S. Bureau of Reclamation, Denver, Colorado

SESSION C-56
FLOOD: PART IV
NATIONAL FLOOD POLICY DISCUSSION
Moderator: BRIG. GEN. GERALD GALLOWAY, Academic Dean at West Point Academy, West Point, NY

SESSION C-57
OVERTOPPING PROTECTION FOR DAMS: PART III
Moderator: RODNEY WITTLER, U.S. Bureau of Reclamation, Denver, CO

SESSION C-58
WATER CONSERVATION AND REUSE
Moderator: BILL S. LANDIN, Boyle Engineering, Albuquerque, NM

SESSION C-59
NEW IRRIGATION AND DRAINAGE PRACTICES FOR MARGINAL QUALITY WATERS: PART III
Moderator: JOHNANNES C. GUITJENS, University of Nevada, Reno, Nevada

SESSION C-60
BRIDGE SCOUR: PART IV
Moderators: DAVE MUELLER, U.S. Geological Survey, Louisville, KY and D. MAX SHEPPARD, University of Florida, Gainesville, FL

SESSION C-61
URBAN HYDROLOGY: PART I
Moderator: ARLEN FELDMAN, Hydrologic Engineering Center, U.S. Army Corps of Engineers, Davis, CA

SESSION C-62
STOCHASTIC HYDROLOGY: PART I
Moderator: JOHN J. INGRAM, National Oceanic Atmospheric Administration/ National Weather Service, Office of Hydrology, Silver Spring, MD

SESSION C-63
HYDRAULIC STRUCTURES: PART I
Moderator: PHIL BURGI, U.S. Bureau of Reclamations, Denver, CO

SESSION C-64
TIDAL AND COASTAL HYDRAULICS
Moderator: P.D. SCARLATOS, Florida Atlantic University, Boca Raton, FLorida

SESSION C-65
OPERATIONAL EXPERIENCES WITH AUTOMATED WEATHER STATIONS
Moderator: RONALD L. ELLIOTT, Oklahoma State University, Stillwater, Oklahoma

SESSION C-66
BRIDGE SCOUR: PART V
BRIDGE SCOUR COUNTERMEASURES
Moderators: GEORGE PAGAN, Federal Highway Administration, Washington, D.C. and STANLEY R. DAVIS, Maryland SHD, Baltimore, MD

SESSION C-67
RESERVOIR SEDIMENTATION/FERC: PART I
Moderator: SHOU-SHAN FAN, Office of Hydropower Licensing, FERC, Washington, D.C.

SESSION C-68
URBAN HYDROLOGY: PART II
Moderator: VICTOR ZITTA, Mississippi State University, Starkville, MS

SESSION C-69
OPTIMIZATION OF WATER RESOURCES SYSTEMS
Moderator: KEVIN LANSEY, University of Arizona, Tucson, AZ

SESSION C-70
HYDRAULIC STRUCTURES: PART II
Moderator: JOHN GEORGE, U.S. Army Waterways, Experiment Station, Vicksburg, MS

SESSION C-71
BRIDGE SCOUR: PART II
Moderator: J. STERLING JONES, Federal Highway Administration, McLean, VA and A. C. PAROLA, University of Louisville, Louisville, KY

SESSION C-72
URBAN HYDROLOGY: PART III
Moderator: DONALD FREVERT, U.S. Bureau of Reclamation, Denver, CO

SESSION C-73
IRRIGATION WITH MARGINAL RECLAIMED WATER
Moderator: J.C. GUITJENS, University of Nevada, Reno, NV

SESSION C-74
HYDRAULIC STRUCTURES: PART III
Moderator: JACOB ODGAARD, University of Iowa, Iowa City, Iowa

*Manuscript not received at time of printing.

SESSION C-75
3D FREE SURFACE MODELING: PART I
Moderator: BILLY JOHNSON, U.S. Army Waterways Experiment Station, Vicksburg, MS

SESSION C-76
AUSTIN/URBAN HYDROLOGY
Moderator: ARLEN D. FELDMAN, Hydrologic Engineering Center, U.S. Army Corps of Engineers Davis CA

SESSION C-77
EROSION CONTROL: PART I
Moderator: CHESTER WATSON, Department of Civil Engineering, Colorado State University, Ft. Collins, CO

SESSION C-78
TRANSBOUNDARY WATER USE
Moderator: STEPHEN DRAPER, Atlanta, GA

SESSION C-79
WATERSHED MANAGEMENT
Moderator: ARLEN D. FELDMAN, Hydrologic Engineering Center, U.S. Army Corps of Engineers Davis CA

SESSION C-80
NON-POINT SOURCE POLLUTION: PART II
Moderator: LESLIE SHOEMAKER, Tetra Tech, Inc., Fairfax, VA

SESSION C-81
HYDRAULIC STRUCTURES: PART IV
Moderator: MORRIS SKINNER, Colorado State University, Ft. Collins, CO

SESSION C-82
RESERVOIR SEDIMENTATION/FERC: PART II
Moderator: SHOU-SHAN FAN, Office of Hydropower Licensing, FERC, Washington, D.C.

SESSION C-83
EROSION CONTROL: PART II
Moderator: CHARLES D. LITTLE, U.S. Army Corps of Engineers, Vicksburg, MS

SESSION C-84
WATER QUALITY ENHANCEMENT TECHNOLOGY: PART I
Moderator: JEB GULK, University of Minnesota, Minneapolis, MN

SESSION C-85
STORMWATER MANAGEMENT: PART I
Moderator: JOSE SALAS, Colorado State University, Ft. Collins, CO

SESSION C-86
HYDRAULIC STRUCTURES: PART V
Moderator: JEFF BRADLEY, WEST, Inc., Seattle, WA

SESSION C-87
ESTUARIES AND RIVERS: PART I
Moderator: HOWARD CHANG, San Diego State University, San Diego, CA

SECTION C-88
IRRIGATION EFFICIENCY AND UNIFORMITY
Moderator: CHARLES M. BURT, Irrigation Training and Research Center, California, Polytechnic State University, San Luis Obispo, California

SESSION C-89
3D FREE SURFACE MODELING: PART II
Moderator: SAM S.Y. WANG, University of Mississippi, Oxford, MS

SESSION C-90
ENVIRONMENTAL HYDRAULICS
Moderator: NANI BHOMWIK, Hydrology Division, Illinois State Water Survey, Champaign, IL

SESSION C-91
EROSION CONTROL: PART III
Moderator: NOLAN RAPHELT, U.S. Army Waterways Experiment Station, Vicksburg, MS

SESSION C-92
FARM IRRIGATION SYSTEM I
Moderator: THEODOR S. STRELKOFF, U.S. Department of Agriculture/Agricultural Research Service, U.S. Water Conservation Laboratory, Phoenix, Arizona

SESSION C-93
STORMWATER MANAGEMENT: PART II
Moderator: MICHAEL PORTS, Parsons Brinckerhoff, Baltimore, Maryland

SESSION C-94
ESTUARIES AND RIVERS: PART II
Moderator: W.H. MCANALLY, U.S. Army Waterways Experiment Station, Vicksburg, MS

SESSION C-95
STOCHASTIC HYDROLOGY: PART II
Moderator: JOHN J. INGRAM, National Oceanic Atmospheric Administration/National Weather Service, Office of Hydrology, Silver Spring, MD

SESSION C-96
NUMERICAL METHODS IN HYDRAULICS AND HYDROLOGY
Moderator: RONALD COPELAND, U.S. Army Waterways Experiment Station, Vicksburg, MS

Computer Aided Design of Riprap Revetments
MARTIN J. TEAL, and DAVID T. WILLIAMS, WEST Consultants, Inc., Carlsbad, CA 1673

SESSION C-97
RUNOFF CHARACTERISTICS OF WATERSHEDS
Moderator: VICTOR ZITTA, Mississippi State University, Starkville, MS

Unifying Unit Hydrograph Methods
ALLEN T. HJELMFELT, JR., U.S.Department of Agriculture/Agricultural Research Services, Cropping Systems and Water Quality Research Unit, University Missouri, Columbia, MO 1678
Clark's Unit Hydrograph Method—50th Anniversary
DANIEL W. KULL, Department of Civil and Environmental Engineering, University of California, Davis, Hydrologic Engineering Center U.S. Army Corps of Engineers, Davis, CA., ARLEN D. FELDMAN, Hydrologic Engineering Center, U.S. Army Corps of Engineers, Davis, CA 1683
Time of Concentration
ARTHUR C. MILLER, DENNIS L. JOHNSON, GERT ARON, Penn State University, University Park, PA 1688
NRCS Unit Hydrographs
D. E. WOODWARD, NRCS, Washington, D.C., WILLIAM MERKEL, NRCS-SNTC, Ft. Worth, TX, JOSEPH SHERIDAN, Agricultural Research Service, Southeast Watershed Laboratory, Tifton, GA 1693

SESSION C-98
FLOW FORECASTING UNDER UNCERTAINTIES
Moderator: JOSE D. SALAS, Colorado State University, Ft. Collins, CO

Flow Forecasting Under Uncertainties
CHAO-LIN CHIU, Department of Civil and Environmental Engineering, School of Engineering, University of Pittsburg, Pittsburg, PA, RANDY D. CRISSMAN, New York Power Authority, Niagara Falls, NY 1698
Chaotic Phenomena of Rainfall and Runoff in Tou-Chien Creek
SU-CHIN CHEN, and MEI-CHING HUANG, Department of Soil and Water Conservation, National Chung Hsing University, Taichung, Taiwan 1703
Water Quality Management Planning in Hungary: A Case Study
LAURENS VAN DER TAK, CH2M HILL International, Budapest, Hungary, and Herndon, VA, BROOKS NEWBRY, CH2M HILL, Krakow, Poland, and Denver, CO., PETER BINNEY, CH2M HILL, Denver, CO 1708

SESSION C-99
FARM IRRIGATION SYSTEMS II
Moderator: STUART W. STYLES, Irrigation Training and Research Center, California Polytechnic State University, San Luis Obispo, California

Jacksonville Wastewater Land Treatment Application Selection and Irrigation System Design
EDWARD N. ANTOUN, PAUL H. POWERS, WILLIAM S. MCCOY, and RICHARD J. THEISS, Malcolm Pirnie, Inc., Newport News, VA 1713

SESSION C-100
WATER QUALITY ENHANCEMENT TECHNOLOGY: PART II
Moderator: A.R. RAO, School of Civil Engineering, Purdue University, West Lafayette, IN.

SESSION C-101
ESTUARIES AND RIVERS: PART III
Moderator: BRIEN R. WINKLEY, Vicksburg, MS

SESSION C-102
HYDRAULIC MEASURMENTS: PART II
Moderator: GREGORY GARTRELL, Contra Costsa Water District, Concord, CA

Introduction to Canal Control Algorithm Needs

A.J. Clemmens, M. ASCE, C.M. Burt, M. ASCE, D.C. Rogers, M. ASCE[1]

Abstract

This paper provides an introduction to a series of papers by the ASCE Task Committee on *Canal Automation Algorithms* to be presented in this proceedings. The purpose of the task committee is to promote the development of improved canal automation algorithms. The committee has had active participation from both North America and Western Europe. The committee pursued three activities; Classification of Algorithms, Canal Characteristics, and Test Cases. The task committee results on these topics are presented in 8 additional papers.

Introduction

The general trend in U.S. and overseas projects is towards modernization and rehabilitation of existing projects rather than new construction. Meeting expected project benefits may require upgrading the capabilities of the system, rather than simply rebuilding the original facilities. Thus there is increasing interest in improving the flexibility of water delivery to users. Canal automation is one method for achieving improved flexibility on some projects.

In 1987, several committees within the Irrigation and Drainage Division of ASCE presented a symposium on *Planning, Operation, Rehabilitation and Automation of Irrigation Water Delivery Systems* to summarize the state-of-the-art in providing water delivery flexibility and service (Zimbelman 1987). At that time, application of canal automation was limited to a few sites and research on canal automation was limited by the ability to conveniently simulate unsteady flow in canals. A task committee was formed in 1989 to promote the advancement of user-friendly unsteady-flow software as a tool for canal automation research. This resulted in a symposium on *Irrigation Canal System Hydraulic Modeling* (In Ritter 1991), which summarized the

---

[1] Respectively, Research Hydraulic Engineer, U.S. Water Conservation Laboratory, USDA/ARS, 4331 E. Broadway, Phoenix, AZ 85040; Professor, California Polytechnic State University, Irrigation Training and Research Center, San Luis Obispo, CA, 93407; Hydraulic Engineer, U.S. Bureau of Reclamation, P.O. Box 25007, D-8560, Denver, CO 80225.

progress on canal modeling by the task committee and others.

In 1993, a task committee was formed to clarify the state-of-the-art in canal automation research and to promote the development of improved canal automation algorithms. The results of this task committee on *Canal Automation Algorithms* are reported in a series of papers at this conference, with this paper providing the introduction.

Canal Control Algorithms

Numerous papers have been written over the past two decades on canal automation -- here, canal *automation* means the automatic control of canal gates and not remote-manual (supervisory) control. The fact remains that very few canals in the world operate under true automatic control. Those that are automated generally fall into one of the following categories; 1) mechanical devices for control of water levels immediately adjacent to a controlled gate, 2) very simple microprocessor-based feedback control limited in scope, or 3) control based on detailed engineering knowledge of canal hydraulic properties (which requires continual refinement).

On the other hand, the technical papers dealing with canal control cover a wide variety of methods, few of which have been implemented in a significant application. It is very difficult to get an accurate picture of the merits of one or another canal control scheme, since each researcher has independently chosen the canal conditions studied and developed his own criteria for evaluation, testing, and rating of the software and results. Some have been tested on unsteady-flow simulation models, some on a model canal at Cal Poly (Parrish & Burt, pg 481-486 in Ritter, 1991) while other have only been tested with linear approximations to the unsteady flow equations (i.e., those used to develop the controller).

It is well known that the properties of the canal on which the control method is applied has a significant influence on the performance of the controller. Thus a particular controller may work well on one canal and not another. No systematic studies have been performed to determine the conditions under which these various control algorithms are useful.

Finally, the language of control engineering may be foreign to many practicing civil engineers, making it difficult for them to discern the differences in control algorithms and to determine which type of algorithm is best suited for a particular canal.

Task Committee Objectives and Activities

The overall purpose of the task committee is to provide information and guidelines which the "consumers" of canal control automation software can use in making their decisions regarding the most applicable form to apply. The task committee is composed of active members from the U.S., Canada, Mexico, The Netherlands, France and Spain.

To address the difficulties described in the previous section, the task committee divided its activities into three areas:

- Classification of Canal Control Algorithms (proposed and in use)
- Canal Characteristics (and their influence on the suitability of various control algorithms
- Canal Automation Test Cases (and associated performance measures)

Classification

A survey was made regarding control algorithms and methods that have been proposed or are currently in use. Approximately 45 algorithms or methods were identified, very few of which have been implemented. This activity focused on the logic used to recommend control actions and not the method of physical implementation. The group did not deal with *system* issues, such as the mechanical means of moving gates, communication from sensors and to gate actuators, or whether the logic was performed mechanically, with electrical circuits, or by microprocessor or computer.

Three papers dealing with classification are included in this proceedings. The first deals with classification categories and categorization of the algorithms. The second presents a summary of the properties and characteristics of the various algorithms. The third presents a summary of control algorithms currently in use. While significant progress has been made on understanding the various control schemes, more research is needed to determine the effective differences -- those differences which influence the algorithms' real performance. In addition, the general area of tuning the algorithms has not been fully addressed. Tuning is the process of adjusting mechanical settings or electrical components, or, with digital logic, determining numerical values for algorithm coefficients.

Canal Characteristics

Canals vary widely in their ability to respond to varied water demands and control measures. This group initiated a general study to examine canal properties and characteristics which influence the potential applicability of various types of control algorithms. Unsteady-flow simulation models were used to study canal response to various demands and control measures. The St. Venant equations were put into dimensionless form to reduce the number of variables which characterize a particular canal. Several preliminary studies were performed to give the committee an idea on what properties and conditions were of interest. Considerably more work is needed in this area before these results can be applied to the selection of control algorithms.

One study was performed to determine the response of canal pools under what could be called perfect feedback with no anticipation of demands. In this scenario, a withdrawal is made at the downstream end of the pool and at the same time the same change in discharge is introduced at the upstream end. Unsteady-flow simulation is used to determine the response in the pool's water level. This is intended to define the limits under which feedback alone can be used to provide adequate control, as opposed to control with anticipated or prescheduled changes in demand in addition to feedback.

A second study has been undertaken to model the propagation of waves in irrigation pools. While such processes are well understood, the influence of the travel and deformation of the wave on a canal's controllability is not well known. Further work is needed in both studies to determine how these properties relate to specific control methods (as opposed to the ideal controllers studied here). Three papers are presented here dealing with non-dimensional approaches and the two studies outlined above.

Test Cases

The development of test cases is primarily useful in formal comparisons of control algorithms. Most control algorithms in use have generally been developed for the specific canal on which they are currently applied. Furthermore, in the literature control algorithms are typically applied to some arbitrary canal, without investigating the significance of its properties.

The task committee chose two real canals with very different properties to propose as test cases. The first is a very steep canal with very rapid response times, but very little storage. The second is a larger canal with relatively slower response. These two canals represent somewhat extreme conditions. The specific physical dimensions have been altered slightly to make the theoretical studies more convenient. Several test scenarios have been proposed for the two canals, including both scheduled (known ahead of time) and unscheduled (and unknown) changes in demand.

The task committee has also proposed a set of criteria to be used in presenting the results of the various control algorithms. There is still debate on how the results should be presented and interpreted. Control algorithms are typically designed and tuned on the basis of different criteria; thus they tend to optimize some performance criteria at the expense of others. The potential users of these algorithms should understand which are the most appropriate for their desired operations; then controllers with the desired characteristics can be chosen for a particular canal. Two papers are presented, dealing with the presentation of controller performance and the test cases.

Publications

The following additional task committee papers (with appropriate references) are included in this proceedings:

Classification of Control Algorithms
by P.O. Malaterre, D.C. Rogers, J. Schuurmans

Properties and Characteristics of Canal Control Algorithms
by V. Ruiz, J.Schuurmans, J. Rodellar

Canal Control Algorithms Currently In Use
by D.C. Rogers, J. Goussard

Dimensionless Characterization of Canal Pools
by T.S. Strelkoff, A.J. Clemmens, R.S. Gooch

Response of Ideally Controlled Canals to Downstream Withdrawals
by C.M. Burt, R.S. Gooch T.S. Strelkoff, J.L. Deltour

Propagation of Upstream Control Measures along a Canal Pool
by J.L. Deltour, J.P. Baume, T.S. Strelkoff

Guidelines for Presentation of Canal Control Algorithms
by W.Schuurmans, B. Grawitz, A.J. Clemmens

Test Cases for Canal Control Algorithms with Examples
by T. Kacerek, A.J. Clemmens, F. Sanfilippo

References

Ritter, W.F. (ed) 1991. *Irrigation and Drainage.* Proceedings of the 1991 National Conference, Irrigation and Drainage Division, ASCE, 821 p.

Zimbelman, D.D. (ed.) 1987 *Planning, Operation, Rehabilitation and Automation of Irrigation Water Delivery Systems.* Symposium Proceedings, Irrigation and Drainage Division, ASCE, 381 p.

Active Committee Members

A. J. Clemmens, Chair	U.S.
C. M. Burt, Vice Chair	U.S.
R. S. Gooch, Secretary	U.S.
D. C. Rogers, Control Member	U.S.

J. P. Baume	France	J. B. Parrish, III	U.S.
B. Deleon	Mexico	J. Rodellar	Spain
J.-L. Deltour	France	F. Sanfilippo	France
J. Goussard	France	J. Schuurmans	Netherlands
B. Grawitz	France	W. Schuurmans	Netherlands
T. Kacerek	U.S.	T. S. Strelkoff	U.S.
P. Kosuth	France	M. Reddy	U.S.
P. O. Malaterre	France	V. Ruiz	Mexico
D. Manz	Canada	E. Zagona	U.S.
G. Merkley	U.S.		

Classification of Canal Control Algorithms

Pierre-Olivier Malaterre[1], David C. Rogers[2], Jan Schuurmans[3]

Abstract

Different control algorithms have been developed and applied in the world for the regulation of irrigation canals. Each of them can be characterized according to several criterias among which: the considered variables, the logic of control, and the design technique. The following text presents definitions of these terms, and a classification of the algorithms detailed in the literature.

1. Introduction

A control system is an elementary system (algorithm + hardware) in charge of operating canal cross structures, based on information from the canal system. This information may include measured variables, operating conditions (e.g. predicted withdrawals) and objectives (e.g. hydraulic targets). Boundaries of the control system are outputs of the sensors placed on the canal system, and inputs of the actuators controlling the cross structures. The following text presents definitions and a classification of canal control algorithms developed or used in the world. Hardware aspects will be presented in a separate paper.

2. Definitions

Several criterias can be used to define control algorithms. The three essential ones are: considered variables, logic of control, and design technique.

2.1. Considered variables

Variable location is given in reference to a pool and not to a structure (e.g. upstream end, intermediate or downstream end of a pool). This avoids confusion in the case of a multivariable control algorithm, where a variable can be controlled

[1] Cemagref, BP 5095, 34033 Montpellier Cedex 1, France
[2] Bureau of Reclamation, PO Box 25007, D-8560, Denver, Colorado 80225, USA
[3] Delft University of Technology, PO Box 5048, 2600 GA Delft, The Netherlands

by both upstream and downstream structures. The location of controlled variables in a pool is indicative of hydraulic behavior (e.g. available storage volume) and civil engineering constraints (e.g. bank slopes). Three types of variables are considered in control algorithms:

Controlled variables are target variables controlled by the control algorithm. Examples are water level at the upstream end of a pool (Yu), water level at the downstream end of a pool (Yd), flow rate at a structure (Q), volume of water in a pool (V), and weighted water level (e.g. α Yu + β Yd). Controlled variables are not necessarily directly measurable.

Measured variables, also called inputs of the control algorithm, are the variables measured on the canal system. Examples are water level at the upstream end of a pool (Yu), water level at the downstream end of a pool (Yd), water level at an intermediate point of a pool (Yin), flow rate at a structure (Q), and setting of a structure (G).

Control action variables, also called outputs of the control algorithm, are variables issued from the control algorithm and supplied to the cross structures' actuators. They are either gate positions (G) or flow rates (Q). In this latter case, another algorithm transforms the flow rate into a gate position. This algorithm is important from hydraulic and control points of view, and is considered as a separate control algorithm.

Remark: **I/O structure** is the number of inputs and outputs considered by the control algorithm. A control algorithm is said nImO when it has n inputs (measured variables) and m outputs (control action variables). Special names are given in specific cases: SISO (Single Input, Single Output, if n = m = 1), MISO (Multiple Inputs, Single Output, if n > 1 and m = 1), and MIMO (Multiple Inputs, Multiple Outputs, if n and m > 1).

2.2. Logic of control

The logic of control refers to the type and direction of the links between controlled variables and control action variables.

Type: the control algorithm uses either feedback control (FB, also called closed-loop control), feedforward control (FF, also called open-loop control) or a combination (FB + FF). In a feedback control algorithm, the controlled variables are measured, or directly obtained from measurements. Any deviation from the targets is fed back into the control algorithm in order to produce a corrective action. In a feedforward control algorithm, the control action variables are computed from targeted variables, perturbation estimations and process modelisation. Feedforward control usually improves control performance when few unknown perturbations occur in the canal system.

Direction: a structure can be operated to control a variable located further downstream, which is called downstream control. All variables (discharge, level or volume) can be controlled with downstream control. A structure can also be operated to control a variable located further upstream, which is called upstream control. Only levels or volumes can be controlled with upstream control, when flow conditions are subcritical and under the limitations of the backwater effects.

2.3. Design technique

The design technique is the algorithm or methodology used within the control algorithm in order to generate the control action variables, from the measured variables.

Main design techniques examples are three position, heuristic, PID, pole placement, predictive control, optimal control, fuzzy control, neural network, backward simulation, linear optimization, and non-linear optimization.

Additional components: a main technique can benefit from additional components that may improve control algorithm performance by accounting for canal system features. Examples are filter, decoupler, observer, Smith predictor, and autoadaptative tuning.

3. Classification

Canal control algorithms detailed in the literature are classified in Table 1. They are listed according to their main design technique (e.g. heuristic or PID). Complete references can be found in Zimbelman 1987, Goussard 1993 or Malaterre 1994.

4. References

Goussard J. 1993. Automation of canal irrigation systems. International Commission of Irrigation and Drainage, Working Group on Construction, Rehabilitation and Modernisation of Irrigation Project, ICID, 103 p.

Malaterre P.O. 1994. Modélisation, analyse et commande optimale LQR d'un canal d'irrigation. Ph.D., ENGREF - Cemagref - LAAS CNRS, 255 references, 200 p.

Zimbelman D.D. 1987. Planning, operation, rehabilitation and automation of irrigation water delivery systems. Proceedings of a symposium ASCE, Portland, Oregon, USA, 28-30 July 1987, 377 p.

Table 1. Classification of canal control algorithms.

IDENTIFICATION		CHARACTERIZATION							APPLICATIONS OR TESTS
Name	Developer	Considered variables			I/O	Logic of control		Design Technique	
		controlled	measured	ctrl. act.	Struct.	Type	Direct.		
DACL	USWC Lab	y_{dn}	y_{dn}	G	SISO	FB	up	3 position	
LittleMan	USBR and others	y_{up} or dn	y_{up} or dn	G	SISO	FB	dn or up	3 position	Several in USA
CARDD	Burt & Parrish	y_{dn}	3-5 y_{in}	G	3-5ISO	FB	dn	Heuristic	CalPoly scale canal
RTUQ	Rogers	Q	y_{up} y_{dn} & G	G	3ISO	FB	dn	Heuristic	Dolores Project
	Zimbelman	y_{dn}	y_{dn}	G	SISO	FB	dn	Heuristic	Model
	CARA	Q & y	Q & y	Q	SISO	FB + FF	dn	Heuristic + PID	Several in France
	Najim	y	y	Q	?	FB + FF		Variable structure	
AMIL, AVIS, AVIO	GEC Alsthom	y_{up} or dn	y_{dn} or up	G	SISO	FB	up or dn	P	Several countries
Danaïdean system		y_{dn}	y_{dn}	G	SISO	FB	up	P	Several in USA
Mixed Gates	GEC Alsthom	$f(y_{up},y_{dn})$	y_{up} & y_{dn}	G	2ISO	FB	up + dn	P	Several countries
BIVAL	Sogreah	$f(y_{up},y_{dn})$	y_{up} & y_{dn}	G	2ISO	FB	dn	PI	Mali, Mexico, etc.
Dynamic Regulation	SCP - Gersar	V	y_{up} y_{dn} & G	Q	3ISO	FB + FF	dn	PI	France, Marocco
	IMTA - Cemagref	y_{dn}	y_{dn}	Q	SISO	FB + FF	dn	PI	Begonia (Mexico)
	UMA Engineering	y_{dn}	y_{dn}	G	SISO	FB	up	PI	Imperial Valley
ELFLO / P+PR	Buyalski, Serfozo	y_{dn}	y_{dn} & G	G	2ISO	FB	dn / up	PI + filter	Several in USA
	Sogreah	y_{dn}	y_{dn}	G	SISO	FB + FF	dn	PI + filter	Kirkuk (Iraq)
	IMTA - Cemagref	y_{dn}	Q_{dn} y_{up} y_{dn}	Q_{up}	3ISO	FB + FF	dn	PID + pole placement	Yaqui (Mexico)
ELFLO + Decoupler	Schuurmans	y_{dn}	y_{dn} & G	G	2-3ISO	FB	dn	PI + filter + decoupler	CalPoly scale canal
PIR	SCP - Gersar	y_{dn}	y_{dn}	Q	2ISO	FB	dn	PI + Smith predictor	France (SCP)

	CACG Cemagref	Q_{dn}	Q_{dn} & Q_{in}	Q_{up}	3ISO	FB + FF	dn	Pole placement	Several in France
CARAMBA	De Leon	Q_{dn}	Q_{dn}	Q_{up}	SISO	FB + FF	dn	Pole placement	Model
	Sawadogo	Q_{dn}	Q_{dn}	Q_{up}	SISO	FB + FF	dn	GPC	Non linear model
	Rodellar, Gomez	Q_{dn}	Q_{dn}	Q_{up}	SISO	FB	dn	Predictive control	Non linear model
	Zagona & Clough	Q & y	Q & y	G	MIMO	FB + FF	dn	Predictive control	Model
	Corriga	y	y_{up} & y_{dn}	G	MIMO	FB	up + dn	LQR + observer	Non linear model
	Davis U.	Q & y	y_{up} & y_{dn}	Q & G	MIMO	FB	up + dn	LQR + observer	Non linear model
PILOTE	Cemagref	Q & y	y_{up} y_{dn} & Q	Q_{up} & G	MIMO	FB + FF	up + dn	LQR + observer	Non linear model
	Reddy	Q & y	y_{up} & y_{dn}	G	MIMO	FB	up + dn	LQR + observer	Non linear model
FKBC	BRL - Gersar	Q_{dn}	Q_{dn}	Q_{up}	SISO	FB + FF	dn	Fuzzy control	T2 (Morocco)
ANN	Schaalje & Manz	y	y	G	MIMO	FB	up + dn	Neural network	Model
	Toudeft	Q_{dn}	Q_{dn}	Q_{up}	SISO	FB	dn	Neural network	Model
ACS	CAP, USBR	Q & y	-	G	MIMO	FF	dn	Model inversion	Central Arizona P.
CLIS	Liu	Q & y_{dn}	y	G	MIMO	FB + FF	dn	Model inversion	Non linear model
Controlled Volumes	CSWP	V	y or Q	G	MIMO	FB + FF	dn	Model inversion	Calif. Aqueduct
Gate Stroking	Wylie, Falvey	Q & y_{dn}	-	G	MIMO	FF	dn	Model inversion	CAP (USA)
	O'Laughlin	Q & y_{dn}	-	Q & G	MIMO	FF	dn	Model inversion	Scale Model
SIMBAK	Chevereau	Q_{dn}	-	Q_{up}	SISO	FF	dn	Model inversion	Non linear model
DYN[2]	Filipovic	V	y	Q	MIMO	FB + FF	dn	Linear optimization	Yugoslavia
	Sabet	V	-	Q	MIMO	FF	dn	Linear optimization	CSWP
	Cemagref	Q & y	-	Q & G	MIMO	FF	dn	Non-linear optimisation	Wateringues
NLP	Lin & Manz	Q & y	-	Q & G	MIMO	FF	dn	Non-linear optimisation	Model
	Tomicic	Q & y	-	Q & y	MIMO	FF	dn	Non-linear optimisation	Model

Properties and Characteristics of Canal Control Algorithms

V. Ruiz[1], J. Schuurmans[2] and J. Rodellar[3]

Abstract

Different feedback control algorithms, based on control theory, have been applied to solve the canal operation problem. The control algorithms and design methods used go from the Proportional-Integral-Derivative controller to the Optimal control. This paper presents the properties and characteristics of some control algorithms to illustrate their performance and limitations in the canal operations control.

Introduction

Providing flexibility in terms of water quantity and timing to achieve improved crop yield and water use efficiency is of utmost importance. Control theory offers different tools to analyze the canal dynamics and to design the regulator capable of obtaining the desired canal operation performance.

The following sections presents the methodology used in control theory and its relation to canal operation, and describe the properties and characteristics of some feedback control algorithms and design methods used in canal operation control.

Control Theory Methodology

Control theory offers analytical and design tools, in the time-space and frequency-space domains, to determine the dynamics process properties and to select the control algorithm and its design parameters required for the desired process performance.

---

[1]Riego y Drenaje, IMTA, P. Cuauhnáhuac 8532 , Jiutepec, Mor. 62550, Mexico
[2]Delft University of Technology, P.O. Box 5048, 2600 GA Delft, The Netherlands
[3]Dep. de Mat. Apli., Uni. Pol. de Catalunya, Gran Capitá s/n, 08034 Barcelona, Spain

The control model developed using the analytical tools, specifies the simplified dynamics relation between the control and controlled process variables. The controlled variable are process outputs that must be held at pre-set values to obtain the desired process performance. The process input variables may be divided in control variables and perturbations. The control variables are modified to obtain the desired values of the controlled variables. The perturbations can not be modified, as may or may not be measured. The process control model may be a differential (continuous-time systems) or difference (sampled systems) equation in the time-space domain or a transfer function in the frequency-space domain with polynomials in the Laplace variable "s" (continuous-time systems) or the Z variable "z" (sampled systems). The control model parameters can be estimated by parametric (direct determination of the parameters) or nonparametric (transient or correlation analysis) identification methods (Söderström and Stoica, 1989).

For pool operation methods and control concepts, Buyaslki et al. (1991) specified the control and controlled variables used in canal control. The operation method determines the canal controlled variables: upstream depth, downstream depth or the volume in each pool. The control concept specifies gate openings and the flow rates at the control structures upstream and/or downstream of each pool as control variables. The use of flow rate as a control variable reduces the interactions (cross coupling effects) between controlled pools in a canal. The control model used in the canal pool control goes from a first order difference or differential equation with delay, or its transfer function, to a set of first order coupled differential or difference equations (state-space representation). The perturbations in the canal operation are the outflows.

The dynamics properties of the process determine the regulator or control algorithm that can be used, since some regulators are unable to stabilize processes with time-varying delay, a characteristic often found in canals. The control algorithms calculate the control variables required to make the controlled variables equal to their reference profile, considering the known and unknown perturbations. The control algorithm may be either an open or closed loop. In open-loop control, the control variables are independent of the controlled variables. The closed-loop control is characterized by the presence of a link or a feedback from the controlled variables to the control variables of the process, where the control variables are functions of the controlled ones. As no field measurements are used in an open-loop control, it is unable to compensate for perturbations and canal dynamics variations. A closed-loop control system has some advantages. It can reduce the effect of noise and perturbations on the process performance, increase the process response speed, eliminate the steady state error, and reduce the sensitivity to process variations. The main disadvantages of the feedback or closed-loop control system are that an originally stable process may become unstable, and special analysis, to determine the domain of stable operation of the regulator-process couple, is indispensable.

Feedback Control Algorithms

The feedback control algorithms may be designed using a model or not.

Non-Model-Based Regulators

Proportional-Integral-Derivative Regulator (PID)

The PID regulator determines the control variables evolution, u(t), with the following expression:

$$u(t) = K_P \left(e(t) + \frac{1}{T_I} \int_{t_o}^{t} e(\tau)\partial\tau + T_D \frac{\partial e(t)}{\partial t} \right) \quad (1)$$

where e(t)=y*(t)-y(t) is the difference between the reference and the controlled variable, K_P is the proportional gain, T_I is the integral time, T_D is the derivative time and u(t) is the control variable. The regulator parameters T_I, K_P and T_D are obtained using the tuning rules of Ziegler and Nichols (Aström and Wittenmark, 1989).

Model-Based Regulators

Pole Placement Design Method (Aström and Wittenmark, 1989)

The control variable evolution is obtained from:

$$u(t) = \frac{1}{S(s)} \left(T(s)y^*(t) - R(s)y(t) \right) \quad (2)$$

where R, S and T are polynomials in s (Laplace variable) or z (Z transform variable).The pole placement method offers some guidelines on how to determines the R, S and T polynomials, (order and parameters), to obtain the desired relation between the reference and the controlled variable.

Predictive Control (Clarke et al, 1987)

The objective of the predictive control law is to drive the future controlled variable, y(t+j), "close" to its reference profile, y*(t+j), bearing in mind the control activity required to do so. This is done using a receding-horizon approach for which at each sample-time, the future controlled variable predictions are generated. The control variable vector is determined from the minimization of a quadratic function assuming that after some "control horizon" further increments in control are zero and the first element of the control variable vector is asserted and the appropriate data vector shifted so that the calculation can be repeated at the next sample instant. The cost function is :

$$J(N_1,N_2) = E\left\{ \sum_{j=N_1}^{N_2} \left[y(t+j) - y^*(t+j) \right]^2 + \sum_{j=1}^{N_2} \lambda(j) \left[\Delta u(t+j-1) \right] \right\} \quad (3)$$

where N1, N2 are the minimum and maximum costing horizon, λ(j) is a control weighting sequence and Δu(t+j-1) = u(t+j-1) - u(t+j-2) is the control variable increment.

State-Space Linear Optimal Control

The optimal control, u(k), is calculated from the minimization of a quadratic cost function of the states and the control signals. The discrete-time cost function is :

$$J = \sum_{k=1}^{K_\infty} \left[x(k)^T \, Q_{p \times p} \, x(k) + u(k)^T \, R_{m \times m} \, u(k) \right] \qquad (4)$$

with $k = 0, \ldots, K_\infty$; Q = p x p state-cost weighting matrix; R = m x m control-cost weighting matrix. Q and R are positive-defined symmetric matrices for $\le k \le K_\infty$ and x(k) = p x 1 the state vector. The control variable vector has the form, $u(k) = -K\,x(k)$, where K = m x p controller-gain matrix. The control model used is a state space representation (Aström and Wittenmark, 1989).

Two other model-based controls used in canal operation are the robust and the adaptive control. The adaptive control is the combination of any one of the previously mentioned control algorithms with a parameter identification algorithm. The robust control tools give additional design guidelines for the previously described control algorithms, to stabilize the process-regulator system at all the specified operation points.

Properties and Characteristics of the Control Algorithms used in Canal Operation

The PID algorithm is the classical algorithm used in the industry. It is also the simplest. The cross-coupling effects between controlled canal pools complicate the parameters selection, making this a trial-and-error task. The PID integral action eliminates the steady state error between the controlled variables and their reference profiles. The performance of the PID regulator is poor for processes with delays, as canals. To improve its performance under these conditions, the PID regulator has been combined with a Smith predictor. The Smith predictor uses a model of the process dynamics and the time delay to predict the process controlled variables. The canal time-varying dynamics and delays make the prediction of future controlled variables difficult.

The pole placement design method is simple to use, to understand, and to implement. Various constraints, such as high-loop gain at certain frequencies (integral action at $\omega=0$) and low-loop gain at other ones are easily introduced. This design method leads to a regulator that includes a feedback term from the measured controlled variable and a feedforward term from the reference signal. For time-varying systems, such as canals, special attention must be given to the selection of the desired closed-loop model and the observer polynomial.

Predictive control is a suitable control algorithm for stable open-loop time-varying processes, dynamics and delay, as seen in canals. Control and controlled variables may be constrained. However, there is no general stability theorem to explain the stability of the closed-loop system, nor can the desired closed-loop specifications be directly incorporated in the design parameters.

The state space optimal control involves numerous calculations for highly dimensioned systems (number of gates to be controlled), a problem overcome with the new hardware and software available. When state variables are not measurable, they must be estimated. When the separation theorem is applicable, the optimal control can be divided in the state estimator, which gives the best estimates of the states from the measured outputs, and the linear-feedback law of the estimated states. In the case of Gaussian disturbances, a Kalman filter, or optimal estimator, may be used to minimize the variance of the estimator error. The use of estimated states reduces the stability robustness of the optimal control. Modifying the weighting matrices, it is possible to introduce some constraints on inputs and outputs and to improve the stability robustness when the estimated states are used. The closed-loop system specifications must be translated into the cost function. This is usually an iterative procedure, which requires the use of a good interactive computer program.

Conclusions and Comments

Canal dynamics presents some characteristics (large time-varying delay, non-linear dynamics, high cross coupling effects between the canal pools) for which control theory has only a partial solution. Research in analysis and design may provide the necessary tools to tackle the canal control problems.

The most severe limitation in the application of control theory to canals is the control model selection. As control models incorporate more information on system dynamics and reduce undesirable process dynamics (nominal phase behavior, cross coupling), the performance of any regulator and its capacity to decrease and incorporate the outflows variations will improve.

As long as it is not possible to find an ad hoc control model and the performance obtained with non-model-based regulators is insufficient, the techniques of fuzzy logic, expert systems and neuronal networks can be used with them to organize the available process information to improve process control.

References

Aström, J. and B. Wittenmark (1989). *Computer Controlled Systems Theory and Design*. Prentice Hall, USA.

Buyalski, C. P., D. G. Ehler, H. T. Falvey, D. C. Rogers and E. A. Serfozo (1991). *Canal Systems Automation Manual*. Bureau of Reclamation. Denver, USA.

Clarke, D. W., C. Mohtadi and P. S. Tuffs (1987). Generalized Predictive Control-Part I. The Basic Algorithm. *Automatica*, 23 (2), 137-148.

Söderström, T. and P. Stoica (1989). *System Identification*. Prentice Hall, USA.

CANAL CONTROL ALGORITHMS CURRENTLY IN USE

David C. Rogers,[1] Member, ASCE, and Jean Goussard[2]

Abstract

Many canal control methods and algorithms have been developed, but only some of them are being used on operating canal projects. As a part of the ASCE task committee on *Canal Automation Algorithms*, this paper discusses field application of automatic control algorithms. Based on available data, brief information on algorithm implementation is presented.

Introduction

Canal automation has been evolving for several decades now, to the point where most new canal designs and canal modernization projects include some level of automation. Numerous canal control algorithms have been developed, but how many of these algorithms have been implemented in the field? The practical implications, successes, and failures of control algorithms may be more important than theoretical performance. Canal control algorithms currently in use are summarized in the sections below, categorized as implicit algorithms in self-regulating gates, local automatic feedback controllers, and supervisory control algorithms. Many of these algorithms are described in the references (Buyalski et al 1991, Goussard 1993, and Zimbelman 1987).

Implicit Algorithms Integrated in Self-regulating Gate Design

Although they do not execute an algorithm in the customary sense, these hydro-mechanical control devices are used successfully on many canal projects:

- ***Constant upstream level gates (AMIL gates)*** - The first operational AMIL gates were installed in Algeria (Oued Rhiou area) in 1937 for automatic upstream control

---

[1] Hydraulic Engineer, U.S.Bureau of Reclamation, P.O.Box 25007, Denver, Colorado 80225

[2] Irrigation and Water Supply Engineering Advisor, 28 rue Gay-Lussac, 38100 Grenoble, France

of a main canal (10 m^3/s max.). Most of those gates are still working. In 1950, nearly 1000 such gates had been installed, mainly in North Africa. Among recent significant references are the North Jazirah 1 & 2 projects, Iraq (1987-1989, 30 m^3/s max., 30 gates) and the Selangor project, Malaysia (1989, 20 m^3/s max., three gates). AMIL gates also are used to control levels in drainage systems (Disney World, Florida).

• ***Constant downstream level gates (AVIS and AVIO gates)*** - This gate type was developed and first applied in the late 1940's (over 400 gates installed before 1951, mainly in France and Algeria). Hundreds have been installed throughout the world since then, for downstream control of level-top canals. A recent reference (1989) is the Sidorejo area of the Kedung Ombo project, Indonesia, with four AVIS gates on the main canal (9.5 m^3/s max.) and four AVIO gates on turnouts to secondary canals. Significant also is the fact that the flow (40 m^3/s max.) at the head of the Canal de Provence system, France, is automatically controlled according to downstream demand through two AVIS gates in parallel.

• ***Mixed gates*** - Mixed gates are used for related level control of reservoir pools, which is their basic operating mode, and for mixed control. One of the earliest applications (1955-1961) has been the control of the reservoir pools forming the two main branches of the Bas-Rhône Canal in France (respectively 61.5 and 13.5 m^3/s) through 7 mixed gates, for the purpose of compensating for the mismatch between the pumped head supply and the lateral on-demand deliveries. The most recent reference (1993) is Canal T2, ORMVA Haouz, Morocco, with two mixed gates controlling a 20 km reservoir reach linking an upstream feeder section (53 km, 12 m^3/s max.) under upstream control to a downstream section of 20 km under downstream control.

• ***Danaidean system*** - Some applications of this system to canal control can be found in several European, Asian, and American countries. For example, in the USA it has been applied to upstream level control with maximum flows ranging from 0.5 m^3/s (Tranquility I.D., California) to 30 m^3/s (Imperial I.D., California). Though simple and efficient, the system has not been widely used because of the bulky additional structures required to provide buoyant counterweights.

Local Automatic Feedback Controllers

a. **Three-position controllers.**

• ***Little-Man*** - The first automatic gate controller in the USA was an electro-mechanical, three-position (floating, set-operate-time, set-rest-time) controller called the Little-Man, installed in 1952 by the Bureau of Reclamation (USBR). Little-Man controllers have been used to maintain a target water level adjacent to (either upstream or downstream from) the controlled gate. They are most effective in applications with a single structure or a few isolated structures, because instability can develop when three-position controllers are installed on a series of check structures. Among numerous installations are Friant-Kern Canal in California (243 km, 113 m^3/s max., 13 check structures) and the Columbia Basin project in Washington (several branch canals with a total of some 50 pools extending over 385 km, with maximum flows ranging from 16 to 144 m^3/s). More recently, the Little-Man algorithm has been programmed into microprocessor canal controllers on projects such as Government Highline Canal in Colorado.

• ***Colvin*** - The Colvin controller adds a rate control mode to the three-position mode to improve performance. Developed and improved by the USBR from 1971 to 1980, Colvin controllers have been applied to upstream control of diversion dam gates (North Poudre supply canal diversion dam, Colorado, and San Juan-Chama Project, New Mexico) and to downstream control of turnout or outlet gates (Loveland turnout from Hansen Feeder Canal, Colorado, and Flatiron afterbay outlet, Colorado).

b. **PI and PID controllers.**

A number of analog and microprocessor-based controllers integrating PI or PID (Proportional-Integral-Derivative) algorithms have been developed and applied for the last two decades. They differ not only in hardware but also in their internal control logic and in their application.

Control of a distant downstream level:

• ***ELFLO*** - In the early 1970's, the USBR developed the analog controller EL-FLO plus Reset from the results of a previous research program on a controller suitable for distant downstream level control (HyFLO, then EL-FLO). ELFLO (Electronic Filter Level Offset) controllers were first installed on Corning Canal, California (1974, 14 m^3/s max., 34 km, 12 check structures) and on Coalinga Canal, California (three check structures). Automatic control was implemented on these projects because flow changes were straining the capabilities of manual gate control. Control performance is good at low flows but degrades as canal flow approaches design capacity; at high flows, canal operators switch controllers into an upstream mode. In recent years, microprocessor PID controllers have replaced ELFLO in USBR applications.

• ***Sogreah PID*** - This PID controller by Sogreah, France, was installed in the 1970's to control the level at the downstream end of the 37 km head reach (278 m^3/s max.) of the Kirkuk-Adhaim main canal in Iraq; similar controllers are currently being installed on the Cupatitzio-Tepalcatepec Project in Mexico (five check structures on a secondary canal of the right bank system and a dam outlet to the left bank system).

• ***IMTA PID*** - PI and PID controllers have been developed by IMTA, Mexico, in collaboration with CEMAGREF, France, and recently installed on Mexican projects under modernization, e.g. La Begoña main canal (20 km, 10 m^3/s, five gate structures), and the Canal Alto of Rio Yadui I.D. (120 km, 110 m^3/s, 15 structures).

Control of a constant level close upstream or downstream:

• ***P+PR*** - The P+PR (Proportional plus Proportional Reset) algorithm is essentially the same as ELFLO, except applied in an upstream (supply-oriented) mode. P+PR has been implemented at the Yuma Desalting Plant Bypass Drain Canal (Arizona), Umatilla Basin (Washington), Closed Basin Canal (Colorado), and Dolores Project (Colorado). In each of these applications, controllers are installed at several canal check structures in series to route flow changes downstream through the canal while maintaining water level upstream from each check.

• ***UMA*** - UMA Engineering, Canada, in collaboration with Armtec, has developed a system combining drop-leaf gates and programmable local controllers

(Modicon or TeleSafe). This system is installed on the St. Mary River I.D. main canal, Alberta, Canada (280 km, 91 m^3/s max., upstream level control at check structures, indirect flow control via downstream level control at outlet gates) and at the South San Joaquin I.D. main canal, California (1989, 40 km, 26 m^3/s max., upstream control of 10 check structures, flow control through downstream level control for two check structures).

• ***Related level control*** - To our knowledge, the only controllers using this logic are those installed in the 1970's by Sogreah, France, to control the two reservoir-reaches (22 km each, 232 and 130 m^3/s respectively) of the Kirkuk-Adhaim main canal in Iraq, to maintain a constant difference between the level just upstream from each regulator and the level at the far end of the pool downstream from the same regulator.

• ***Constant volume (BIVAL) control*** - This logic, developed by Sogreah, has been applied to two reaches (62 km each, 75 m^3/s) of the Sahel canal in the Fala de Modolo system in Mali, since 1983. As only infrequent gate adjustments were required, the concerned regulators are operated manually from level readings and using charts. An automated BIVAL control system is currently under implementation on the right bank of the Cupatitzio-Tepalcatepec Project, Mexico.

• ***PIR control*** - The PIR algorithm has been developed by Société du Canal de Provence in 1992-1993 and has been satisfactorily controlling a branch of the Canal de Provence system since 1994. For this first operational application, the software has been integrated into the Dynamic Regulation system and no specific hardware has yet been developed or selected for a possible canalside PIR controller.

c. **Heuristic controllers.**

• ***RTUQ*** - The RTUQ (Remote Terminal Unit flow control) algorithm was developed by USBR in 1992. The algorithm uses a feedback loop to maintain a target flow through a gated check structure. RTUQ is being used at the Dolores Project, Colorado (125 km, 11 m^3/s max., 60 check gate structures) as part of a supervisory control system. The algorithm has performed well when all data is in order. Enhancements have been added to improve stability when downstream water levels are in the transition zone between free gate flow and submerged gate flow.

Supervisory (or Centralized) Control Algorithms

• ***Dynamic Regulation*** - Dynamic Regulation was developed by Société du Canal de Provence, France, for application on the Canal de Provence system (1971, 105 km of main and branch canals, 130 km of pressure pipes and tunnels, 40 m^3/s max., 33 regulating gates, 24 emergency gates, four pumping stations, and two in-line hydro plants). The system has shown a high degree of efficiency and reliability. Dynamic Regulation also has been successfully applied to complex systems in Greece (Athens water supply), Macedonian Republic (Stretzevo Irrigation Project), and Morocco (Rocade Canal, 127 km, 20 m^3/s, seven check structures, two main turnouts, 15 RTU's).

• ***ACS*** - ACS (Aqueduct Control Software) was developed by USBR in the 1980's to control the Central Arizona Project canal system (540 km of open canals, inverted siphons, and tunnels, 85 m^3/s max., 36 check structures, 14 in-line pumping plants). The project is demand-oriented, delivering water for irrigation and municipal use without any wasteways to spill excess water. ACS solves canal hydraulics using model inversion (backwards simulation) to control water volumes throughout the canal system while minimizing pump starts. ACS executes on a master station computer and sends control schedules to microprocessor RTU equipment at pumping plants, check structures, and turnouts.

• ***Controlled Volume*** - This control method was developed in the 1970's for the California Aqueduct (710 km, 290 m^3/s max., 242 turnouts, and some 90 aqueduct pools with tunnels, siphons, 66 gated check structures, and 27 pumping and power plants). The centralized algorithm controls pool volumes to satisfy scheduled water demand while minimizing pumping power costs and avoiding overloading of the power supply network. Although the system was designed to respond to delivery changes on relatively short notice, farmers (30% of the yearly deliveries) reproach the system some lack of flexibility.

• ***CACG*** - This method was developed from the mid-1960's and continuously improved since, by Compagnie d'Aménagement des Coteaux de Gascongne, France, for central management of flows and reservoirs in a system of rivers. The main reference is the Neste system, France, for which the method was devised (17 rivers totaling 1300 km, four in-line dams, and a 29 km, 14 m^3/s feeder canal). The objectives of the project--satisfying user demand, maintaining minimum flows required for water quality, and improving the conveyance efficiency (now about 90%) by reducing operational losses--are considered to be fully met.

• ***FKBC (Fuzzy Knowledge-Based Controller)*** - This controller has been developed very recently by BRL-Ingénierie (a division of Compagnie Nationale d'Aménagement de la Région du Bas-Rhône et du Languedoc, France). A first FKBC was installed and put into operation at the beginning of 1995, as a part of the supervisory control system of Canal T2 in Morocco (see Mixed Gates above). FKBC determines the optimal flow setpoint for the two radial head gates, based on demand forecasts, current system-wide status, a rule base, and a database.

Conclusions

Although this paper summarizes canal control algorithm implementation, it is not an all-inclusive compilation. The authors welcome additional information on canal projects where control algorithms currently are being used.

References

Buyalski, C.P., Ehler, D.G., Falvey, H.T., Rogers, D.C., and Serfozo, E.A. (1991). *Canal systems automation manual, vol.I.* Bureau of Reclamation, Denver, Colo.

Goussard, J. (1993). *Automation of canal irrigation systems.* ICID, New Delhi, India.

Zimbelman, D.D., ed. (1987). *Planning, operation, rehabilitation, and automation of irrigation water delivery systems.* ASCE, New York, N.Y.

DIMENSIONLESS CHARACTERIZATION OF CANAL POOLS

T.S. Strelkoff, M. ASCE, A.J. Clemmens, M. ASCE, R.S. Gooch, M. ASCE[1]

Abstract
The response of canals to control measures can be studied in a systematic program of simulations covering the practical range of interest of each variable. The large number of variables is reduced by putting the governing equations in nondimensional form, with normal depth at design discharge the principal reference variable. Units of Manning *n* and associated coefficients in various dimensional systems are derived. The concept of hypothetical dimensioned canal is introduced to give a physical meaning to dimensionless results. Adaptation of standard, dimensioned simulation models to dimensionless studies is explored. A steady-state example is given.

Introduction and governing equations
For a given offtake schedule and given control measure the fluctuations in depths and discharges in a canal pool depend on length, slope, roughness, cross section, initial discharge, and amount of check-up at the downstream end. To reduce the ponderousness of a general program of numerical simulations (*e.g.*, for discerning the influence of pool properties on controllability) without reducing its scope, all governing equations and initial and boundary conditions can be utilized in dimensionless form, *i.e.*, with each of the variables in the original, dimensioned equations divided by a constant reference value with the same dimensions. One of the possible systems of reference variables is presented here.

Conservative Saint Venant equations with Manning roughness are given by

$$\frac{\partial Q}{\partial x} + \frac{\partial A}{\partial t} + q_o = 0$$

$$\frac{\partial Q}{\partial t} + \frac{\partial}{\partial x}\left[\frac{Q^2}{A}\right] + u_o q_o + Ag\left[\frac{\partial h}{\partial x} + \frac{Q^2 n^2}{c_u^{\,2} A^2 R^{4/3}}\right] = 0 \qquad \textbf{(1)}$$

---

[1] Respectively, Research Hydraulic Engineer (and University of Arizona Research Professor), Supervisory Research Hydraulic Engineer, U.S. Water Conservation Laboratory, USDA/ARS, 4331 E. Broadway, Phoenix, Arizona 85040, and Senior Engineer, Water Engineering, Salt River Project, P.O. Box 52025, Phoenix, Arizona 85072-2025

in which x=distance along the channel; t=time; Q=discharge; A=cross-sectional area; q_o=distributed lateral outflow per unit length of channel; u_o=longitudinal component of velocity of lateral outflow; h=water-surface elevation; n=Manning n, in units $m^{1/6}$; c_u=units coefficient allowing the same numerical value of Manning n to be used in any dimensional system (see Appendix); R=hydraulic radius (cross-sectional area / wetted perimeter); g=ratio of weight to mass.

Reduction to dimensionless form

All transverse dimensions of the channel flow are referred to a reference transverse length Y_R, not yet defined, and all longitudinal canal dimensions to a reference longitudinal length, X_R, thus introducing

$$h^* = \frac{h}{Y_R}; \quad R^* = \frac{R}{Y_R}; \quad A^* = \frac{A}{Y_R^2}; \quad x^* = \frac{x}{X_R} \tag{2}$$

Similarly, dimensionless discharges are defined through a reference Q_R, velocities through a reference velocity V_R, and times through a reference time T_R,

$$Q^* = \frac{Q}{Q_R}; \quad q_o^* = \frac{q_o X_R}{Q_R}; \quad u_o^* = \frac{u_o}{V_R}; \quad t^* = \frac{t}{T_R} \tag{3}$$

A *normal* discharge Q_N, characteristic of the canal pool, is now introduced, for example, the design discharge. The reference depth Y_R is then specified as normal depth y_N at that discharge, at some reference section. The aspect ratio a of the channel is defined as the ratio of average breadth, $\overline{B}_N$, to depth under these normal conditions, *i.e.*, as the dimensionless normal area A_N^*,

$$a = \frac{\overline{B}_N}{y_N} = \frac{A_N/y_N}{Y_R} = A_N^* \tag{4}$$

The *reference* discharge is defined as the ratio of normal discharge to aspect ratio,

$$Q_R = \frac{Q_N}{a} \tag{5}$$

This definition of reference discharge insures that dimensionless velocities will approach unity under normal-depth conditions.

For further simplification, it is also specified that

$$\frac{X_R Y_R^2}{T_R Q_R} = 1; \quad X_R = \frac{Y_R}{S_{0_R}}; \quad V_R = \frac{Q_R}{Y_R^2} \tag{6}$$

in which S_{0R} is the bottom slope at the reference section; it follows, too, that the characteristic reference velocity is the normal velocity, $V_R = Q_N/A_N$.

Now, all of the reference variables are defined, and eqs. 1 assume the form

$$\frac{\partial Q^*}{\partial x^*} + \frac{\partial A^*}{\partial t^*} + q_o^* = 0$$

$$F_R^2 \left[\frac{\partial Q^*}{\partial t^*} + \frac{\partial}{\partial x^*} \left(\frac{Q^{*2}}{A^*} \right) + u_o^* q_o^* \right] + A^* \left(\frac{\partial h^*}{\partial x^*} + R_N^{*4/3} \frac{Q^{*2} n^{*2}}{A^{*2} R^{*4/3}} \right) = 0 \qquad \textbf{(7)}$$

in which n^* is the ratio of local Manning n to Manning n in the reference section, n_R; for a pool with constant roughness, $n^*=1$. The term F_R^2 is a kind of Froude number, related to the normal-depth Froude number of the design inflow, F_N,

$$F_R^2 = \frac{Q_R^2}{g\,Y_R^5} = \frac{Q_N^2 B_N}{a^2 g A_N^3} \left(\frac{A_N}{Y_R^2} \right)^3 \left(\frac{Y_R}{B_N} \right) = F_N^2 \left(\frac{A_N^*}{B_N^*} \right) = F_N^2 d_N^* \qquad \textbf{(8)}$$

in which A_N, B_N, and d_N are cross-sectional area, top width, and hydraulic depth, respectively, under design normal conditions; d_N^* is a shape factor characterizing the reference flow cross section -- in a rectangular channel, it is simply unity.

The term R_N^* is also a shape factor. Indeed,

$$R_N^{*4/3} = \frac{1}{S_{0_R}} \frac{Q_R^2 n_R^2}{c_u^2 Y_R^{16/3}} = \frac{Q_N^2 n_R^2/c_u^2}{a^2 S_{0_R} A_N^2 R_N^{4/3}} \left(\frac{A_N^2}{Y_R^4} \right) \left(\frac{R_N}{Y_R} \right)^{4/3} = \left(\frac{R_N}{Y_R} \right)^{4/3} \qquad \textbf{(9)}$$

Clearly, R_N^* is the dimensionless hydraulic radius of the channel at uniform flow at the normal discharge and characterizes the shape of the channel from the point of view of flow resistance. In a broad rectangular channel, it approaches unity.

For channels with variable slope and roughness, the dimensionless normal-depth condition at any location is given by

$$R_N^{*4/3} \frac{Q^{*2} n^{*2}}{A^{*2} R^{*4/3} S_0^*} = 1 \qquad \textbf{(10)}$$

in which S_0^* is the ratio of local bottom slope to reference bottom slope, unity in a pool of constant slope. The dimensionless critical depth-discharge relation is

$$\frac{Q^{*2} B^*}{g^* A^{*3}} = 1 \qquad \textit{with} \qquad g^* = \frac{1}{F_R^2} \qquad \textbf{(11)}$$

For a given shape of canal cross section, eqs. 7 contain but one variable governing parameter, F_R. In other words, for all canals of given shape subject to the same relative initial and boundary conditions, all possible solutions form a one-parameter family. Specifically, these solutions depend on the dimensionless length of channel L^*, on the shape of the inflow hydrographs, and on the operation of the gates relative to the characteristic normal depth and reference time.

Application of dimensionless equations

A generalized study of canal behavior can be performed directly in dimensionless mode, with the equations, input conditions, and solutions all in dimensionless form. Canal shape would be specified by b^*, base width relative to normal depth, and side slopes s. With a dimensionless normal depth of unity, the shape factors A_N^*, R_N^* and d_N^* would follow. Pool length would be specified as L^*. In a uniform pool, n^* and S_0^* would both be unity. Dimensionless design inflow Q_N^* would be a (A_N^*); its Froude number at normal depth F_N must be specified (eq. 8 then leads to F_R). Any other dimensionless discharges would also be given in reference to Q_R. Gate openings and amount of checkup would be specified relative to normal depth and gate-schedule times relative to T_R. Offtake-flow velocities (longitudinal components), u_o^*, would be input as a fraction of characteristic, normal velocity, typically, zero.

Dimensionless steady nonuniform flows exhibit a particularly simple form. In Fig. 1, initial, steady-state profiles of *all* checked up canal pools are described by some portion of the single generalized backwater curve shown. In this case, variation of normal Froude number over a customary range and variation of channel cross section and side slopes over their customary ranges have little effect.

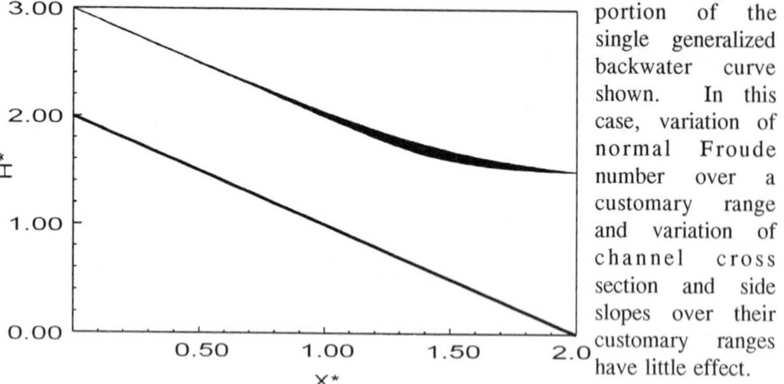

Figure 1. Dimensionless Steady State Backwater Curves: $0.1 \leq F_N \leq 0.7$; $0.5 \leq b^* \leq 3$; $1 \leq s \leq 2$

Interpretation of dimensionless variables -- hypothetical dimensioned channel

To assist in the interpretation of systematically varied, dimensionless input and output, the concept of hypothetical dimensioned channel is introduced. The physical significance of the dimensionless variables depends upon the geometrical scale of the canal flow at hand. The scale is completely determined once the characteristic normal depth, y_N, and the reference Manning roughness n_R have been specified.

Indeed, with the dimensionless base width b^* and side slopes s of a symmetrical trapezoidal cross section and length L^* given, as well as normal Froude number F_N

and some dimensionless time in the problem, say, gate rise time t_r^*, corresponding dimensioned values of A_N, d_N, Q_N, R_N, S_{OR}, L, and t_r can be successively determined.

For example, in a channel of trapezoidal cross section with dimensionless bottom width $b^*=1$, side slopes $1.25H/V$, Manning $n=0.02$, and a design flow normal depth of 2 m, a given choice of F_N corresponds to the following bottom slope S_{OR}, design discharge Q_N, normal velocity V_N, and reference distance X_R, time T_R, discharge Q_R:

F_N	S_{OR}	Q_N(cms)	V_N (mps)	X_R (m)	T_R(min)	Q_R (cms)
0.10	0.00005	3.20	0.36	43400	34 hr	1.42
0.30	0.00041	9.59	1.07	4830	75.5	4.26
0.50	0.00115	16.0	1.78	1740	16.3	7.10
0.70	0.00226	22.4	2.49	887	5.94	9.95

Execution of dimensionless studies using dimensional models

The dimensionless eqs. 7 can be brought into congruence with eqs. 1 by substituting

$$F_R^2 = \frac{1}{g^*}; \qquad R_N^{*4/3} = \frac{1}{c_u^{*2}} \tag{12}$$

If industrial programs for simulating unsteady flow in open channels allow user entry of numerical values for g and c_u, they can be used not only in different dimensional systems (metric, English), but also for nondimensional studies. With eq. 12 in force, and with the understanding that the unstarred variables of eq. 1 are in fact dimensionless, these standard equations could be used for a dimensionless study.

Appendix

Inasmuch as the Manning n is a measure of the absolute roughness of the canal wall, its units must be expressible in terms of length alone. Since resistance, under the high-Reynolds number, high-roughness conditions contemplated in the Manning formula, depends on relative roughness, $R^{1/6}/n$, the units of n must be in $L^{1/6}$. In the original (metric) definition, R is in m, so n must be in $m^{1/6}$. Historically, and by current convention, the numerical value of n is taken the same, whether the units of the remaining variables in the formula are metric or English. The units of n, then, must also remain the same, *i.e.*, $m^{1/6}$.

The coefficient c_u, in the original definition of n has the value of 1.000 $m^{1/2}/sec$; in the English system, its value is $3.2808^{1/3}$ in units of $ft^{1/2}/sec\ (m/ft)^{1/6}$, the mixture of length types stemming from the above convention. In consequence, the units of c_u/n are simply $m^{1/3}/sec$ in the metric system and $ft^{1/3}/sec$ in the English system.

Red River Waterway Project: General Design

C. Fred Pinkard, Jr., P.E. *

Project History and Status

The Red River Waterway Project was authorized in 1968 with the primary purpose of providing a 2.7 meter (9 feet) deep by 61 meter (200 feet) wide navigation channel from the Mississippi River to Shreveport, Louisiana. A waterway project location map is provided as Figure 1. The construction of a system of five locks and dams in conjunction with a program of channel realignment and bank stabilization is required to provide and maintain the authorized project channel. At present, all five locks and dams are complete and in operation. Lock and Dam No. 1 (Lindy C. Boggs Lock and Dam) was put into

Figure 1. Red River Location Map

* Hydraulic Engineer, U.S. Army Corps of Engineers, Vicksburg District, 2101 North Frontage Road, Vicksburg, MS 39180

operation during the fall of 1984. Lock and Dam No. 2 (John H. Overton Lock and Dam) became operational during the fall of 1987. Lock and Dam No. 3 has been in operation since December 1991. Lock and Dam No. 4 and Lock and Dam No. 5 (Joe D. Waggonner, Jr. Lock and Dam) were constructed concurrently and were put into operation during December 1994. The channel realignment and bank stabilization program has progressed with practically all work complete and developing through Pool 4. Some capout of revetments continues in Pool 4 while revetment and capout construction continues in Pool 5.

Locks and Dams

On the Red River, each of the five locks and dams includes a lock chamber that is 25.6 meters (84 feet) wide with a useable length of 208.8 meters (685 feet). This size chamber will allow for the lockage of six barges, each barge being 10.7 meters (35 feet) wide and 59.4 meters (195 feet) long, and a tow boat in a single lift. The total lift provided by the five locks is 43 meters (141 feet). Lock and Dam No. 1 provides the greatest lift since stages downstream of this structure are dependent upon backwater from the Mississippi River. The maximum lift at the locks and dams range from 7.3 meters (24 feet) to 11 meters (36 feet). For flows that exceed the 10-year frequency flood, channel velocities become sufficiently high to restrict commercial navigation. Therefore, the lock walls are designed so that the lock is operational only up to the 10-year frequency flood event. Each lock and dam also includes a gated dam that maintains the minimum pool during low water periods and passes flood flows. Lock and Dam No. 1 was designed to pass the project design flood (100-year frequency post project flood) with only 0.3 meters (1 foot) of swellhead. A gated dam with eleven tainter gates was required to meet this criteria. For the other four locks and dams, gate optimization studies were conducted. These studies compared the cost of each additional tainter gate to the cost associated with inundating additional upstream lands. These studies resulted in the upstream four locks and dams having fewer tainter gates than Lock and Dam No. 1. Unlike Lock and Dam No. 1, these four structures include either an uncontrolled or hinge crest gated overflow section or both. A photograph of Lock and Dam No. 3 is included as Figure 2. In selecting the sites for the locks and dams and minimum pool levels, consideration is given to impacts associated with increased water levels. These impacts include inundating lands not normally flooded under pre-project conditions, increased groundwater levels associated with holding a minimum pool that is higher than normal low water pre-project stages, and the permanent ponding of water against existing levees.

Figure 2. Lock and Dam No. 3

Channel Realignment

Many bends on the Red River are too tight to accommodate commercial navigation by the Red River Waterway channel design tow (3 barges in tandem). Therefore, these bends are realigned via a channel cutoff across the neck of the old bendway. A photograph of a typical channel realignment constructed on the Red River is included as Figure 3. On the Red River, the pilot channel concept for channel realignment is utilized. This technique includes excavating a pilot channel that is smaller than the desired channel section and allowing the natural erosive action of the river to develop the pilot channels to their ultimate section. On the Red River, pilot channels as small as 24.4 meters (80 feet) wide have been excavated that quickly developed to the typical channel width of 137.2 meters (450 feet) to 182.9 meters (600 feet). A non-overtopping earth closure dam is provided across the upstream end of the old bendways that are at least 1.6 kilometers (1 mile) long. This closure dam prevents sediment laden river flow from entering these old bendways and thus helps preserve the bendways for environmental and recreational purposes. A non-overtopping closure dam also helps facilitate the development of the pilot channel by forcing all of the channel developing flow through the pilot cut. On the shorter bendways that are not being preserved, only a low stone closure is provided across the upstream end of the old bendway. This low closure helps facilitate development of the pilot channel during periods of low flow. During periods of higher flow, the stone closure is overtopped and divided flow occurs. This condition results in sediment deposition in the old bendway.

Figure 3. Darrow Realignment

An added benefit associated with the channel realignment program is a reduction in flood stages which results from a shorter, more efficient channel. The channel realignments on the Red River results in an approximate 18 percent reduction in length of the river downstream of Shreveport. During the May 1990 flood on the Red River, an event that exceeded the 100-year frequency flood at Shreveport, the peak stage was approximately one to two feet lower than it would have been prior to construction of the project realignments.

Bank Stabilization

The type revetments currently used on the Red River are trenchfill, stonefill, and timber pile revetments. At those sites at which the desired bankline is located landward of the existing bankline, trenchfill revetment is used. This type revetment includes the excavation of a trench along the desired channel alignment and filling the trench with stone. As the bankline continues to erode, the trench is undermined and the stone in the trench launches down the face of the bank. This process stabilizes the bank, thus locking the channel alignment into place. Trenchfill revetments have proved very effective on high energy rivers like the Red River that primarily traverse easily erodible soils.

At those sites at which the desired bankline is located riverward of the existing bankline, stonefill revetment or timber pile revetment is used. These type revetments protect the bank by inducing sediment deposition behind the revetment and thus building the bankline out toward the revetment. If the river is shallow along the desired alignment, stonefill revetment is used. This revetment includes placing stone in a peak section or a section with a flat crown along the desired alignment. In the deeper river locations, timber pile revetment is used. Timber pile revetment includes driving timber piles along the desired alignment and placing stone around the toe of the piling. These revetments are used in the deeper sections of river because they are cheaper to construct than are massive stonefill revetments. Once sediment deposition has occurred behind the stonefill or the timber pile revetment, the revetment is raised or "capped-out" by placing additional stone on top of the deposited sediment. Capouts utilize the sediment deposition behind the revetment to reduce the stone required to raise the revetment to its desired level. This method of construction is less expensive than initially constructing the revetment to its ultimate elevation. Due to the heavy suspended sediment load carried by the Red River, sufficient deposition usually occurs behind the revetments to allow the capout to be constructed after one or two highwater seasons.

Channel Control Structures

On the downstream end of revetments, especially those in the upper reach of each pool where channel depth is most critical, kicker dikes are provided. This type dike is an extension of the revetment and forces the channel crossing from the revetment to the revetment on the opposite bank. Providing these structures helps maintain navigable depths within the channel crossing, thus reducing maintenance dredging. In the very upstream most reach of the pools, additional contraction structures (ACS) are provided. ACS are stone dikes that extend from the convex bank to develop and maintain the channel against a revetment on the opposite bank. By constricting the channel, a narrower, deeper channel that accommodates commercial navigation is maintained.

Summary

In its natural state, the Red River is wide and shallow with numerous shifting sand bars. In this condition, the Red River does not accommodate any level of commercial navigation except during unreliable highwater periods. However, through the construction of a system of locks and dams, bank stabilization, and channel realignment, the Red River is being developed to provide for navigation from the Mississippi River to Shreveport, Louisiana. Through the use of sound engineering design, dependable navigation can be provided far into the 21st century.

LOCKS AND DAMS - RED RIVER WATERWAY

Ralph R. Robertson, Jr., M. ASCE*

Introduction

The Red River Waterway Project, Mississippi River to Shreveport, Louisiana, consists of a navigation channel 2.7 meters (9 feet) deep by 61.0 meters (200 feet) wide, five locks and dams providing a total lift of 43.0 meters (141 feet), and a number of channel realignments and revetments for a length of 379.8 kilometers (236 miles).

The required lift of 43.0 meters (141 feet) is provided by five locks with sidewall port filling and emptying systems. The navigation pools are maintained by gated dams adjacent to each lock. The lift provided by each structure varies from 7.3 to 11.0 meters (24 to 36 feet).

The sidewall port filling and emptying system is well suited to the Red River since relatively low lifts are required. The system is designed to produce hawser forces of less than about 4500 kilograms (five tons). These locks are 25.6 meters (84 feet) wide by 239.3 meters (785 feet) long and can accomodate a tow 208.8 meters (685 feet) long consisting of six barges in a two wide by three long configuration and the towboat.

The dams were originally designed to minimize the effect on pre-project flood elevations. The most downstream lock and dam consists of 11 tainter gates. This configuration results in swellhead of 0.3 meters (1 foot) for the project design flood, the 100-year event. The design process for the remaining dams included a gate optimization study to take advantage of the flood elevation reduction resulting from the channel realignments. Consequently, the other dams have fewer gates. The stilling basin is designed to provide

----

*Hydraulic Engineer, Hydraulics Branch, Corps of Engineers, Vicksburg District, 2101 N. Frontage Road, Vicksburg, MS 39180

submergence of 85% of the conjugate depth of the entering flow and consists of a concrete slab with two rows of baffle blocks and a sloping end sill. The stilling basin and associated downstream riprap are designed to maintain the integrity of the structure in the event of a navigation accident or gate mis-operation.

The design process for the locks and dams included physical and numerical model studies conducted at the Waterways Experiment Station (WES) in Vicksburg, MS. The spillway model was used to evaluate the stilling basin, riprap and other structural features of the project. The navigation model was used to evaluate navigation conditions in the vicinity of the structure. The moveable bed model was used to identify depositional trends and methods to manage the sediment in the vicinity of the structure. A numerical model was used to evaluate depositional trends of the suspended sediment and current direction and velocity in the vicinity of the structure.

As construction of the Waterway progressed upstream, each lock and dam presented interesting and challenging problems as the river reacted to the changes being imposed upon it. Innovation and creativity were often required to solve some of these problems.

Lindy C. Boggs Lock and Dam

The first in the series of locks and dams of the Red River Waterway is Lindy C. Boggs Lock and Dam. The navigation pool is maintained at elevation 12.2 m NGVD (40 ft) by a tainter gated dam with 11 gates. The upstream and downstream lock approaches are separated from the active flow portion of the river up to a specific stage by an earthen embankment with a concrete I-wall. Each approach has a floating guidewall 208.8 meters (685 feet) in length to assist towboats entering and leaving the lock.

Shortly after the structure was placed in operation, the I-wall at the lower lock approach was overtopped which resulted in major deposition in the lower lock approach. This deposition had severe impacts to the operation of the lower miter gates since the material deposited against the gates and fell into the lock chamber when the gates were opened. It was determined that the I-wall separating the lock approach from the active flow portion of the river should be to a higher elevation. A vertical extension of the concrete I-wall was constructed of treated timbers supported by steel "H" beams. This timber wall extends 274.3 meters (900 feet) downstream of the miter gates and has been very successful in reducing the deposition that occurs in the lower approach. Deposition still occurs, but in much smaller amounts and is not against the miter gates.

John H. Overton Lock and Dam

The second in the series of locks and dams of the Red River Waterway is John H. Overton Lock and Dam. The navigation pool is maintained at elevation 19.5 m NGVD (64 ft) by a tainter gated dam with five gates and an uncontrolled weir 57.9 meters (190 feet) in length with a crest at elevation 20.1 m NGVD (66 ft).

John H. Overton Lock and Dam was under construction when Lindy C. Boggs Lock and Dam was placed in operation. The original design of the lower lock approach, as did the remaining locks and dams, consisted of a landside guidewall about 198.1 meters (650 feet) in length with a short rock dike that would be overtopped with relatively small flows. Our experience with the first lock and dam indicated that the downstream lock approach should be totally separated from the active flow portion of the river. Since the project was under construction and significant work had been done on the lower guidewall, it was determined to not be feasible to modify the contract to place the guidewall on the riverside. Our options to modify the approach were very limited since the modification had to be quickly designed and constructed. The selected modification was a rock dike with a reach the same elevation as the lock walls and a section 3.0 meters (10 feet) above the lower pool elevation. This configuration is designed to provide a slack water area for the lower approach and allow some flow near the surface to enter the approach to reduce the eddies that would occur.

After being placed in operation, the navigation conditions in the upstream lock approach were difficult at medium to high flows. The upstream approach consists of a ported guidewall 213.4 meters (700 feet) in length on the riverside of the approach. During medium to high flows, the flow in the lock approach was concentrated at the lower ports. At times the flow would pin the tow against the wall resulting in great difficulty approaching or leaving the lock. It was apparent to correct the situation that the flow through the ports had to be redistributed and the lateral flow distribution in this reach of the river had to be altered to reduce the percentage of total flow that entered the lock approach. The physical models were incorporated into the study to help find a solution. It was determined that reducing the flow entering the approach did in fact reduce the force exerted on tows at the downstream end of the guidewall. We then used the navigation model to evaluate a system of submerged dikes upstream from the guidewall as a means of forcing some of the flow away from this side of the river. The model showed that two dikes with their crest 4.3 meters (14 feet) below the normal pool level improved the navigation conditions in the approach. The dikes were constructed and a change in the flow characteristics of the river was noticed almost immediately, even during low flow periods. The project personnel reported a dramatic decrease in the amount of debris that collected in front of the upper miter gates. Also, during the subsequent high water season, pilots using the lock indicated they had a much easier time entering and leaving the lock. However, the flow distribution along the guidewall was still not acceptable. Prototype data including discharge measurements of the flow through the ports indicated that 62% of the flow that entered the portion of the river bounded by the guidewall passed through the downstream 25% of the wall. To restrict the flow through this portion of the wall, concrete blocks were placed in the three full-sized ports at the downstream end of the wall that reduced the flow area of these ports by about 50%. Subsequent prototype data indicated that about 38% of the flow that entered this portion of the river passed through the downstream 25% of the wall. Reports of the navigation conditions in the lock approach have been favorable and no problems with the blocks in place have surfaced.

Lock and Dam No. 3

The third in the series of locks and dams of the Red River Waterway is Lock and Dam No. 3. The normal navigation pool is maintained at elevation 29.0 m NGVD (95 ft) by a tainter gated dam with six gates and an uncontrolled weir 96.0 meters (315 feet) in length with a crest at elevation 29.6 m NGVD (97 ft).

Considering the deposition problems experienced in the lower lock approach of the first two locks and dams, the lower guidewall was moved to the riverside of the approach at the remaining structures to separate the lock approach from the active flow portion of the river. However, deposition immediately downstream of the lower miter gates continued to be a concern. We installed pipes 7.6 centimeters (3 inches) in diameter on 0.9 meter (3 feet) centers in the lower miter gate sill as a means to provide essentially constant flow to prevent deposition immediately downstream of the miter gates. The reports from the project personnel have been very positive. They report little if any material in this vicinity. As a result, similar pipes were incorporated into the lower sill at the remaining locks and dams.

Pool 3 is unique in that it is operated as a hinge pool. During low flows, a constant pool elevation of 29.0 m NGVD (95 ft) is maintained. As the flow increases, the water surface at the lock and dam is drawn down to elevation 27.1 m NGVD (89 ft). The pool will remain at this level as the flow increases until the tailwater begins to control the pool level. The hinge pool operation reduces the area inundated by the pool if held at a constant level.

Lock and Dam No. 4 and Joe D. Waggonner, Jr. Lock and Dam

The fourth and fifth in the series of locks and dams of the Red River Waterway are Lock and Dam No. 4 and Joe D. Waggonner, Jr. Lock and Dam. The navigation pools are maintained at elevation 36.6 m NGVD (120 ft) and 44.2 m NGVD (145 ft), respectively by a dam with five tainter gates, a hinge crest gate 30.5 meters (100 feet) in length and an uncontrolled weir 45.7 meters (150 feet) in length with a crest at an elevation 0.6 meter (2 feet) above the normal pool.

These two locks and dams include the hinged crest gate as a means to enhance the water quality in Pools 3 and 4 when dissolved oxygen levels are the most crucial, during extreme low flow periods. The hinged crest gate draws water from the surface of the pool and discharges it onto a baffled chute which increases the level of dissolved oxygen from the turbulence on the chute.

Summary

Lock and Dam No.4 and Joe D. Waggonner, Jr. Lock and Dam were placed in operation in December 1994. The other three have been in operation for several years. The modifications and design changes implemented during construction of the Waterway are functioning as intended and indications are that they will continue to fulfill their function as part of the Red River Waterway for many years to come.

GEOMORPHIC HISTORY OF THE RED RIVER OF LOUISIANA AND TEXAS

Chester C. Watson,[1] F. ASCE and Phil G. Combs, [2] M. ASCE

Abstract

The history of the Red River has been prepared based on a review of early accounts, pertinent recent literature, and hydrographic surveys beginning with the 1886 survey. A series of events on the river, including the removal of the Great Raft during the 1870s and 1880s, removal of the rapids at Alexandria, Louisiana, reservoir construction, flood control levees and improved conveyance have been constructed. The navigation project is presently under construction. The effect of these projects on the Red River is reviewed.

Introduction

The Red River initiates in eastern New Mexico following a generally southeastern course across the Texas panhandle, then flows along the Texas-Oklahoma boundary, and through southwestern Arkansas and Louisiana to the confluence with the Atchafalaya River. The lower Red River, downstream of Denison Dam and to the mouth, occupies a valley that is filled with alluvium that ranges in depth of from 75 to 150 feet. The alluvium grades vertically from red clay at the surface through red silt, brown and gray sand, to gravel at the base of the deposit (Newcome, 1960). Valley width varies from about 8 to 12 miles in width. In general, borings and bed samples indicate that the river flows in relatively fine alluvium along the entire lower valley, with the capacity to freely transport the bed and bank material. However, local deposits of gravel or bedrock outcrop could exist.

Major impacts on the Red River include construction of the Denison dam, removal of the Great Raft, changes in the Atchafalaya River, effect of geology and active tectonics, the effect of diversions, and the removal of the Rapides. A summary of the Red River characteristics during the 1930s was presented by the USACE in 1936. The 1936 report conclusions are that the great variability of discharge and high sediment loads made improvements in navigation difficult to achieve and that large floods made bank stability almost impossible (U.S. Congress, 1936). At that time the river banks and bed were unstable from Shreveport to the mouth. George Price, Chief, Hydrologic Investigations Sections, New Orleans District (USACE, 1964) is quoted, "The bed of the Red River from Fulton to Alexandria is subject to rapid and erratic scour and fill, amounting to 8 to 12 feet above and below the average bed, and as much as 10 feet in a single week." Presently, the U.S. Army, Corps of Engineers (USACE), Vicksburg District is constructing a major navigation project on the lower Red River, which provides commercial navigation from the mouth upstream to Shreveport, Louisiana. Significant channel stabilization and lock and dam construction has been completed, which dramatically changes the character of the Red River. Each of the major historical impacts prior to the current navigation project will be discussed in the following paragraphs.

---

[1] Assoc. Prof. of Civil Engr., Colorado State Univ., Ft. Collins, CO 80523

[2] Act Chief of Channel Stab. Branch, Vicksburg District, USACE, Vicksburg, MS 39180

Impacts

Fisk (1940) recognized that the lower Red River had, through geologic time, formed a low alluvial fan, which was the result of deposition along the many previous courses of the river. Abington (1973) concludes that the Red River had changed course about 400 to 500 years ago. The new course provided a shorter route to the Mississippi River valley by about 32 miles, and caused the river to incise. Abington (1973) believes that this incision was retarded by bedrock in the vicinity of Alexandria, and that bedrock was the cause of the Rapides recognized by the early settlers at Alexandria.

Geologic evidence suggests that active tectonics has affected the Red River. Murray (1948) states that anomalies associated with the Sabine Uplift in the river floodplain indicate that uplift of positive structural anomalies has continued into recent times. Russ (1973) concludes that continuing uplift of the Sabine structure has affected the width of the floodplain, warped terrace surfaces, and has caused entrenchment of parts of the modern Red River.

A major event in the history of the Red River was the removal of the Great Raft. Capt. Henry Miller Shreve began removing the raft in the early 1830s, however, new jams were removed for another 40 years. This great log jam began to form in the old courses of the Red River (Bayou Boeuf and Bayou Teche), and the upstream most extent of the raft was near the present site of Alexandria in the latter part of the 15th century. According to Veatch (1906), the raft started as a series of log jams, which retarded flow and may have caused flooding of timbered bottom land that, in turn, would cause more timber debris to be accumulated. Through time the raft would grow upstream, and with time, the downstream debris would decay and be carried downstream. Modern examples of this process have been documented.

As the raft moved upstream, lakes would form at tributaries and at natural drains. By 1835, the Red River was a series of inter-connected lakes extending to within 3 miles of the Arkansas-Louisiana boundary. Following removal of the raft, the lakes drained and the river began to incise. Veatch (1906) states that the river bed had lowered over 15 feet between 1873 and 1892 at a point 15 miles upstream Shreveport, and had lowered 3 feet at Shreveport. Removal of the Great Raft was a dramatic event in the history of the Red River, and was strong influence upon the modern geomorphology of the river.

Watson (1982) reported that in about the 15th century, the Red River entered the Mississippi River at Turnbull Bend. The bend then migrated, intersecting a prior course of the Red River that flowed into the Gulf of Mexico through the Atchafalaya River. Following this capture of the Red River, the Red River and some of the Mississippi River flood discharge followed the Atchafalaya course. Manipulation of the Turnbull Bend-Old River area now insures that the Red River will discharge only into the Atchafalaya River. During the period 1839 through 1842, burning of the timber raft in the Atchafalaya River cleared that waterway. The Simmesport gauge is about 5 miles downstream of the Red River confluence, and during the period 1890 through 1976, the stage for a discharge of 100,000 cfs has dropped more than 25 feet. The long term effect of this lowering on the Red River is not known.

The Rapides at Alexandria was a 2-mile-long shallow reach that was a major obstacle t navigation. The Union gunboats and troop transports encountered difficulty in moving downstrea following defeat at the Battle of Mansfield during the Civil War. Col. Joseph Bailey used timbe barrages to narrow and deepen the flow for the fleet (Herndon, 1995). In 1892-93 the USAC excavated more than 270,000 cubic feet of bedrock from the Rapides reach. Abington (197 believes that the excavation caused degradation at Alexandria and a decline in sinuosity upstream c this point as a result of the increase in channel gradient.

The Denison Dam was completed in 1943 at the upstream extent of the reach unde investigation. Williams and Wolman (1984) present data indicating the effect of the dam o discharge. Average daily discharge was decreased 35%, and average annual peak discharge wa decreased 68%. Sediment concentrations were decreased by 50% for the same discharge at distance of 58 miles downstream. In 1969, maximum degradation below Denison Dam was 9 fee at a location about 10 miles downstream of the dam. At greater distances the amount of degradatio decreased.

<u>Effects</u>

Prior to the present navigation project, which incorporates numerous sites of ban stabilization and a series of navigation locks to stabilize and regulate the depth and slope of the Re River, the major impacts recorded in history have acted to destabilize the river and to cause incisio as a result of increasing channel gradient, and decreasing sediment supply from upstrean Comparison of the gauge elevation for a specific discharge for a single location can give insight int the history of channel response. The specific gauge diagrams were compared for the Alexandri gauge, the Shreveport gauge, and the Index gauge.

For 80,000 cfs at Alexandria, data from 1870 to about 1930 was limited, however, the dat suggests a trend to aggradation. Following 1930, data is available on an annual basis, and a clea degradation trend is present. The river stage for 80,000 cfs drops in excess of 10 feet during th period of 1930 to 1990. Degradation at the Alexandria gauge may be related to flood control cuto that occurred during that period, however, the timing of the 1930 change in aggradation t degradation may also be related to upstream sediment supply related to severe channel incision in th Alexandria to Shreveport reach during the period 1890 to approximately 1942. During the perio from 1906 to 1932, the river stage for 80,000 cfs at Shreveport dropped approximately 14 feet, an the degradation continued as a slower pace until approximately 1942. The abrupt incision from 190 to 1932 may have over supplied the downstream reach, below Alexandria, and delayed the incisior Records at the Index gauge begin in 1936 and indicate a continuing trend of degradation. During th period from 1936 to 1986, the gauge height decreased approximately 3 feet, and this may be th direct result of the Denison Dam closure.

Because most of the historical impacts were generally downstream of Shreveport, Deniso Dam being the exception, the changes in width and depth of the river between Shreveport and Fulto are a significant indicator of the effect of historical events. Between 1886 and 1938, the averag width a bankfull increased from 509 feet to 1251 feet; for the same period the depth increased fror 37 feet to 40 feet. This trend continued to 1980, with width increasing to 1386 feet and the dept emaining about the same at 39 feet. From 1886 to 1980, the slope increased from 0.000087 to .000135. Closer examination of data compiled from historical survey indicates that the changes elated to the clearing of the Great Raft were much more significant at Shreveport than at Fulton. or example the average width change near Shreveport for the period 1886 to 1990 was by a factor f 5.5, and was only 1.7 near Fulton. The data suggests that a hinge points exists between Fulton and hreveport, with the downstream reach being dominated by Raft and Rapides removal, and the pstream reach dominated by the construction of Denison Dam.

onclusion

The Red River is a product of the geologic history and man's involvement in influencing the alance sediment supply and the capacity of the river to transport the river bed and bank materials. he previous statement is not unique to the Red River, and is true of rivers in general; however, the ed River provides a unique opportunity to trace the effects extreme and well-documented impacts f man's activities.

cknowledgment

Much of the background information for this paper was prepared under contract DACW38-6-D-0062 from the Vicksburg District, USACE to Water Engineering and Technology, Inc. ermission was granted by the Chief, Corps of Engineers to publish this information.

eferences

bington, O. D., 1973, Changing meander morphology and hydraulics, Red River: Unpublished Ph.D. dissertation, Louisiana State Univ., 42 p.

isk, H. N., 1940, Geology of Avoyelles and Rapides Parishes: Louisiana Geol. Survey, Bull. 18, 240 p.

arvey, M. D., 1987, Geomorphic and hydraulic analysis of the Red River from Shreveport, Louisiana, to Denison Dam, Texas, Prepared for U.S. Army Engineer Waterways Experiment Station, Research Contract DACW38-86-D-0062.

erndon, J. E., 1995, Personal communication.

uss, D. P., 1975, The Quaternary geomorphology of the lower Red River Valley, Louisiana: Unpublished Ph.D. dissertation, Pennsylvania State Univ., 207 p.

J.S. Army Corps of Engineers, 1964, Red River channel scour and fill: Unpublished memorandum in files: New Orleans District, 28, Aug. 1964.

J.S. Congress, 1936, Red River, Louisiana, Arkansas, Oklahoma, and Texas, House Document No. 378, U. S. Government Printing Office, Washington, DC.

Veatch, A. C., 1906, Geology of the underground water resources of northern Louisiana and southern Arkansas: U.S. Geol. Survey Prof. Paper 46, 394 p.

Watson, C. C., 1982, An assessment of the lower Mississippi River below Natchez, Mississippi: Unpublished Ph.D. dissertation, Colorado State Univ., 162 p.

Williams, G. P. and Wolman, M. G., 1984, Downstream effects of dams on alluvial rivers: U.S. Geol. Survey Prof. Paper 1286, 83 p.

El Niño Effect on Colombian Hydrology Practice

Ricardo A. Smith and Oscar J. Mesa[1]

Abstract

El Niño-Southern Oscillation Phenomena (ENSO) has a direct effect on Colombian Climate. In the periods of time when the ENSO occurs, Colombian weather is dryer than normal, and in some cases it could be very critical. The occurrence of ENSO at the end of 1991 and during 1992 took by surprise the Colombian electric generation system and in march of 1992 the system started a electricity supply rationing that lasted for one year (until march of 1993). The social and economical consequences for Colombia were enormous, and even the president asked for an investigation to define responsibilities.

The expansion (timing and sequencing of new projects) and operation of Colombian Electric Sector is done using hydrology models developed in the early 60's and are based on the Matalas multivariate model (a multivariate autoregressive model of order one) or in simple autoregressive univariated models. Prior to the ENSO occurrence of 1991-1992 it was almost impossible to propose a research for consideration of the Colombian Electric System institutions. They were sure that the models developed in the 60's and in use by the system did not need to be reviewed. After the 1991-1992 ENSO occurrence and the government actions to define responsibilities, there is a new attitude in the Colombian Electric Sector. They has started to support and finance research, specially related to hydrology predictions involving the ENSO phenomena. They are very interested that research results to be included in the actual modeling practices. They even are financing a project to analyze the practical potentialities of new developments in water resources, operational research, hydrology, system modeling, etc.

---

[1] Water Resources Graduate Program, Universidad Nacional de Colombia, Facultad de Minas, Apartado Aéreo 1027, Medellín, Colombia.

An analysis of this transformation of the Colombian Electric Sector, as an example of successful causes of transferring research results to practice, is presented. The new research projects financed by the Colombian Electric Sector are discussed with its practical consequences. Some conclusions and recommendations are also presented.

Introduction

Colombian interconnected electric generation system has a installed capacity of about 10.050 Mega-watts (MW), of which 8.060 MW are from hydropower (80.2%) and 1.990 MW are from thermopower plants (19.8%). The system has 35 reservoirs with and aggregated storage capacity of 6.676 hm^3 that is equivalent to 13.859 Giga-watts hour (GWh). The system also has 18 thermopower plants of which 28.6% are coal-vapor systems, 45.7% are gas and/or fuel oil vapor systems and 25.7% are gas turbine systems. The system has 14.471 Km of transmission lines.

During the period of 1960 to 1965 several hydrologic methodologies were developed and used for the operation and for the expansion planning of the interconnected system. At the time the system was managed as a centralized decision government owned system. The operational planning and the expansion planning was done for years by a government agency called Electric Interconnected System (ISA). The regional electric enterprises, owners of most of the electric generation plants, just followed the central government instructions, and did not developed any analysis capacity to perform their own operational or planning decision policies.

After the initial period of methodology development the Colombia Electric Sector (CES) did not make almost any new development in the hydrologic modelling area for years. The hydrologic models developed in the 60's, based in the Matalas multivariate model (a multivariate autoregressive model of order one), or in simple autoregressive models, were used for synthetic streamflow generation until very recently.

At the end of 1991 and during 1992 occurred a El Niño - Southern Oscillation (ENSO) event that had tremendous consequences over Colombian weather. During the periods of time when ENSO occurs weather in Colombia is dryer than normal, and in some cases it could be very critical. Records of low streamflows are associated with ENSO occurrences (Mesa et al, 1994). The ENSO occurrence of 1991-1992 took by surprise the CES and in March of 1992 an electric rationing was imposed to the whole country that lasted for one year. The social and economic consequences for Colombia were enormous. Because of the rationing, the president of Colombia, for the first time in Colombia history, created a central commission to define the causes of the rationing and to define responsibilities. From the inquires of this commission and from other investigations and researches it was clear that the CES was well behind in hydrologic analysis methodologies and that actions should be taken to correct this situation. The non availability of modern methodologies was no excuse to justify the situation.

This critical situation gave an alert to the CES that felt that applications, research and analysis of all kinds should be constantly developed. After the 1991-1992 ENSO occurrence the CES was willing to analyze and develop new methodologies, and to transfer research results to practice (Mesa et al, 1995).

Another situation in the CES that helped to change attitude in the public utilities about research and transfer research results to practice was a structural change in the CES. Recently (from 1992) the CES has been moving from a centralized operational scheme to a market style operational scheme. In the new scheme, the different electric generation companies (private or public), will make offer bids to a central dispatch office which decides who is dispatched using a minimum cost criteria. The regional public electric generation enterprises are now competing in an electric market between them and with new private electric generation companies. The regional public electric generation companies, that for years has been doing what ISA told them, feel now that they need to develop analysis capacity that will allow them to use their own methodologies. This situation helped also to increase the needs for methodology developments and for transferring research results to practice.

The El Niño - Souther Oscillation (ENSO) Phenomena

The El Niño is a Peruvian name associated with the abnormal warming of the normally cold waters of the southamerican Pacific ocean coast. This temperature changes and the changes in the atmospheric pressure in the tropical Pacific ocean are two related aspects of the same phenomena that has important consequences in the earth climate and that is known as the ENSO phenomena.

The main effect of the ENSO over Colombian climate is associated with the displacement of the convection center from its normal position to the south-east, being attracted by the abnormal Pacific ocean warm waters in front of Peru and Ecuador coasts. This situation significantly reduces rainfall amounts in the north west central region and occasionally an increase in rainfall in the south-west region (Mesa et al, 1994).

The sea warming during the ENSO occurrence weakness the temperature gradients not only in the west - east Pacific ocean regions, but also between the region in front of the Peruvian coasts and the region in front of Colombia. This weakness affects the Walker circulation and the flow that cross over Ecuador and penetrates Colombia from the Pacific ocean. In the absence of this flow the convergence diminishes and also the humidity, convection and rainfall. This involves a change in the atmospheric pressure. A clear proof of this change is the movement of the convection center mentioned above that in normal conditions is over Colombia.

Some research has shown (Mesa et al, 1994) that there is an important relationship between river streamflows at the west and central Andean mountain chains in Colombia and some ENSO indicators such as the Pacific ocean sea surface level temperature and the souther oscillation index. Correlation coefficients grater than 0.5 has been found. Although there is a clear relationship between ENSO and Colombian Hydrology, no research was done in this respect prior to the 1991-1992 ENSO occurrence. After the 1991-1992 ENSO occurrence the CES has done research to study the ENSO relationship to Colombian hydrology, to develop and implement streamflows prediction models that include ENSO effects over Colombian hydrology, and to review and develop streamflow synthetic generation models that also include the ENSO effects (Smith et al, 1995). Some of this developments are now being used in the operational planning of the CES.

New Electric Market

Since 1992 CES is moving from a centralized government owned electric generation market to a energy generation open free market. This change in government policy looks to encourage efficiency in public utilities and private participation. The different electric generation companies will make offer bids to a central dispatch office which will decide who is dispatched using minimum cost criteria. The new electric generation market suppose to be in operation by year 1998.

For years Colombian electric generation market was managed by a central government agency (ISA) and the different generation companies (all public regional companies) just followed the operational instructions of ISA. The regional companies did not have the needs to develop any analysis or research capability. With the new electric generation market, the regional companies and the new private enterprises need to develop analysis and research capability that will improve the market knowledge and its decision making procedures. These companies are already hiring engineers with knowledge on aspects related to the new energy generation environment, and developing methodologies and transferring research results to practice in those topics. Research projects in areas of economic regulation, dispatch modeling, energy integrated models, and others has been done. Some of this developments are now being used by several electric generation companies.

Another aspect that contributed to the research development and in the transfer of research results to practice is that electric generation expansion planning is now done considering all energy alternatives. Few years back the CES expansion planning was done as an independent energy sector with no relation with the other energy sectors (gas, coal, solar energy, oil, others). Now the planning of the CES is done integrated with all energy sectors inside a national energy planning scheme. This new way of planning has opening interesting areas for research in Colombia.

Conclusions

- Several recent critical situations has occurred in Colombia that has encouraged the development of new methodologies and the transfer of research results to practice. The 1991-1992 ENSO occurrence with severe consequences over Colombian economy and the new electric generation market scheme are the main situations that has triggered important research activities in Colombia.

- The actual general situation of the Colombian Electric Sector has forced to change the research attitude of this sector from a conservative attitude to a more liberal one. They want to develop or to transfer methodologies and results specially related with the energy open free market and with the modelling of different aspects of that market. The CES even has approved a proposal to the National University of Colombia so that some researchers in that university will study, analyze and recommend methodologies and developments of potential use in the new electric energy generation environment.

- There are many research opportunities in Colombia in many different areas. The electric generation companies are supporting research activities and are willing to transfer research results to practice. They are quite interested in developments that have been applied in other regions of the world with similar conditions than the Colombian case.

References

- Mesa O., Salazar J.E., Carvajal L.F., Poveda G., and Velez J.I., 1994. Methodologies for Hydrologic Predictions (in Spanish). Empresas Públicas de Medellín, Research Report, 1994, Medellín, Colombia.

- Mesa O., Smith R.A., Salazar J.E., Carvajal L.F. and Velasquez J.D., 1995. Streamflow prediction models for three rivers in Colombia(in Spanish). Interconexión Eléctrica S.A., ISA, Research Report, 1995, Medellín, Colombia.

- Smith R.A., Mesa O., Jimenez C. and Peña G.E., 1995. Proposal to develop streamflow synthetic generation models for the Colombian Electric Sector (in Spanish). Interconexión Eléctrica S.A., ISA, Research Report, 1995, Medellín, Colombia.

River Basin Management in India

L. Douglas James[1]

Abstract

In India, where water resources development is a state responsibility, the Central Water Commission (CWC) became concerned that planning was not meeting long-term needs at the river-basin scale. The CWC worked with national and expatriate experts to establish a course that imparts a broad perspective in Integrated River Basin Planning and Management. The course went through several rounds and is now conducted by a Central Training Unit (CTU) in Pune. The notes are being prepared as a monograph.

Introduction

India has a hot, dry climate and a large population. Vast numbers of people draw on a limited water resource to grow food and crops to sell to have cash. Over time, an economy industrializes, and cities dispersed throughout the country will need more water. Since water is limited by nature and nearly fully allocated by government, the required dependable supplies of high quality water have to be obtained by increasing water use efficiency.

This pressure has lead India to adopt integrated river basin planning and management (IRBP&M) as a national policy; however water resources are a state responsibility. The constitution limits the Government of India to ensuring technical standards, expediting financing, and resolving water disputes. Thus, the obtained funding from USAID and the World Bank to launch an effort by Harza Engineering and Utah State University to develop a CTU to train IRBP&M

---

[1]Member, Professor, Utah Water Research Laboratory, Utah State University, Logan, UT 84322-8200

skills to state agency personnel. This paper outlines the concept and evaluates the progress of the work.

The IRBP&M Concept

The National Water Policy of the Government of India (1987) proclaims: "Water is a prime natural resource, a basic human need and a precious national asset." India, like most countries, has developed it water resources by project-by-project planning (PPP). New needs are defined, and new projects are built and operated independently by separate agencies. India followed PPP to construct vast numbers of water projects ranging from ponds to some of the largest in the world. This irrigation program accomplished the seemingly impossible job of feeding and raising the living standards of 800,000,000 people.

However as PPP approaches the hydrologic limit to available water, the projects become larger. A crowded country lacks land, and a poor country lacks finances for many vast reservoirs and long canals. The alternative is to deliver more water from given facilities (add water delivery efficiency), obtain more value from given deliveries (add economic and social efficiency), and be more benign and sustainable (add environmental efficiency).

Opportunities to Increase Efficiency

Greater efficiencies can be achieved by connecting projects, operating them as river basin systems, and supplementing structures with complementary educational and regulatory programs to adjust water and land uses. Planners use principles set forth by (White, 1969), organized in James and Lee (1970), and officialized by the U.S. Water Resources Council (1980) to integrate system construction, operations, and management. They craft institutions; evaluate total worth to society; and seek equity among regions and social classes. The planning process takes advantage of 3Ds:

1. Diversity. Some rivers have high flows when others are low, aquifers fill and deplete on long time scales, and some demands are large when others are small.
2. Dependency. Water is a flowing resource. It is never truly consumed as it travels from use to use. Each user depends on the management by prior users.
3. Dynamics. Present decisions affect future opportunities. Managers can act early to mitigate problems and schedule ahead to reduce costs.

Scope and Methodology

IRBP&M introduces a new vocabulary. Projects include all structural and nonstructural elements used in water resources development. Programs are groups of projects managed together. Planning analyzes interfaces among projects so that programs perform better. Development takes advantage of inter-project dynamics to craft programs for changing times. IRBP&M plans facilities and conjunctive ground-surface water use to deal with floods and droughts. It plans operation and maintenance to take advantage of system economies. It addresses large-scale issues such as where best to grow what crops and where to foster urban-industrial growth.

In the process, planners must subdivide basins to capture spatial differences, temporal changes, and internal interactions. All three are important. First, water deliveries, drainage provisions, and measures for flood and water pollution control are best varied from one part of a basin to another. Second, different economic, environmental, and social conditions change at different rates in different places. Third, the interactions are tied to phenomena parameterized with local characteristics.

With a good representation of the interconnected system, river basin authorities can reduce losses and make the water saved more productive. They can draw from gains to compensate losers. They can reduce storage needs by coordinating reservoir operations and ground water use, keep water systems functioning as elements are replaced and plan replacements that make projects work as systems, reduce large-scale and long-term adverse social and environmental impacts, and manage upstream water so that river flows better match, in quantity and quality, downstream needs. Future management must have an element of people accommodating to nature with nonstructural approaches (demand management) rather than always trying to conquer nature with structural measures (supply management).

Issues in Implementation

The initiation of IRBP&M faces important issues. It needs authority to make needed changes, access to reliable information, engineering expertise to find workable system modifications, and finances. For planning the transfer of water from times and places with excess to times and places with shortage, a basin is subdivided into units where times of excess and shortage can be identified with water balance accounting. Opportunities for water trades, needed canal capacities, and needed reservoir storage are assessed by comparing gains with sacrifices in taking advantage of opportunities. The practical challenge is to recognize the impossibility being comprehensive coverage and reducing the work to doable steps to manage continuing investment in economic growth. IRBP&M uses nonstructural measures to modify demands and applies principles from public administration to improve program management. Planners must work as advisors who make thoughtful assessments without telling people how to manage their lives. They characterize problems, make recommendations, and facilitate consensus.

The Training Structure

The course is taught to about 20 trainees at a time. A full course covers eight prerequisite topics (to bring people from widely varied backgrounds to a common level), eleven synthesis courses (to integrate engineering, economic, ecologic, and social perspectives into long-term basin wide plans), a case study (for water management in a selected basin), self-directed projects to let the training officers develop topics of personal interest, and a two-week tour of the United States (to provide demonstration of how river-basin planning is conducted here).

The prerequisite topics are basics in computer usage, hydrology, agricultural science, probability and statistics, sociology, environmental science, economics, and modeling and optimization. The synthesis topics introduce IRBP&M concepts, information and display systems, irrigation systems, river basin hydrology, water resource engineering, social and environmental analysis, systems engineering, economic analysis, system operations, and applications.

Program History

An operating CTU was established on the campus of the Central Water and Power Research Station near Pune in 1989. At first, introductory (3-month) and advanced (9-month) courses were taught with the prerequisite topics covered by faculty from Indian universities and the synthesis topics covered by faculty from the U.S. The shorter courses presented concepts to senior people and the longer courses trained junior people in methodology. Over time, the primary teaching burden has shifted to CTU staff who took the earlier courses and came to the U.S. for an in depth experience called "training of the trainers."

Assessment and Conclusions

The conversion from PPP to IRBP&M switches planning from immediate to long-term goals in the spirit presented by Fisher and Ury (1983). People are asked what they want their communities to be in 2050 and experts discuss how water resources planning can contribute to getting there. Success depends not on sophisticated tools but on holistic coordination of structural measures for water control with nonstructural measures to reduce waste and inefficiency.

In essence, IRBP&M is an exercise in creative thinking. Practitioners learn concepts, teach others, and spread enthusiasm to overcome three common misconceptions. IRBP&M has no comprehensive model and depends less on proficiency in running models than on skill in assembling facts and reasoning logically. IRBP&M never "arrives" but forever encounters new conditions, gains new information, and improves by adjustment. Finally, IRBP&M is not done by finding a world class expert but by uniting teams from many disciplines, governmental roles, and water users. It seeks harmony among people and of people with nature, and the work underway at the CTU can become an example to the world.

Acknowledgment

The author expresses gratitude to the project sponsors in India, USAID, and the World Bank, to the faculty in both countries whose names would fill the space allocated to this paper, and especially to the inspiration of Dr. Warren Hall. We are all proud to see the CTU working under the leadership of Mr. Rajan Nair its Director.

References

Fisher, R., and Ury, W. (1983), *Getting to Yes*, The Penguin Group, New York, N.Y.

Government of India, Ministry of Water Resources (1987), "National Water Policy." New Delhi, India.

James, L. D. and Lee, R. R., (1970), *Economics of Water Resources Planning*, McGraw Hill, New York, N.Y.

U.S. Water Resources Council, Part II, (Apr. 14, 1980) "Proposed Rules; Principles, Standards, and Procedures for Planning Water and Related Land Resources," *Federal Register*, pp. 25302-25348.

White, G. F., (1969), *Strategies of American Water Management*, Univ. of Michigan Press, Ann Arbor, Mi.

A critical review of the Pangue Dam project EIA (Biobío River, Chile)

Claudio I. Meier[1], Student Member ASCE

Abstract

In 1992, I was asked to revise the Environmental Impact Assessment (EIA) for the Pangue project, a large dam currently under construction in the upper reaches of the Biobío, a river of enormous social and economical importance. The report was carried out by an American consulting firm for ENDESA, the utility that owns Pangue.

The EIA conclusions were found to be incomplete, invalid, and highly biased, i.e., supportive of the project a priori. Moreover, the assessment was based on scant and inadequate data (with the exception of hydrological series), and many essential studies (such as fish migration patterns; the downstream routing of scheduled releases; its effects on the fluvial and riparian communities, on water quality, on irrigation diversions; etc.) were not carried out. In conclusion, the report was more a propagandistic justification for the project than an EIA of its potential effects.

This case-study raises serious ethical issues regarding the objectivity and quality of some consulting work done in less-developed countries by profit-seeking private firms. It also shows how untruthful are ENDESA's claims that Pangue is being built following the principles of the sustainable development concept.

---

[1] Instructor Professor, Civil Engineering Dept., Universidad de Concepción, Chile. Present address: Civil Engineering Dept., Colorado State University, Fort Collins, CO 80523

Introduction

Of the 950,000 inhabitants that live in the 24,300 km^2 Biobío basin, about 550,000 drink potable water obtained directly from the river, whilst the untreated domestic sewage of some 620,000 people is dumped into its main course or tributaries. The river is the basic resource for Chile's second largest industrial pole: iron, steel, chemical, petrochemical, oil refining, pulp and paper, textile, leather, fish meal, etc., all these industries use Biobío's waters, diverting 8.3 m^3/s, and dumping back 6.2 m^3/s of mostly untreated liquid wastes. Across the basin, 192 m^3/s are diverted for irrigation. The installed hydroelectric capacity is 1050 MW, but should reach almost 4000 MW in the next two decades.

Chile's fast growth has caused electricity consumption to soar at sustained yearly rates of more than 10%. The government's response has been to rely on private utilities to build and operate powerplants, in order to keep up with the increased demand. ENDESA, the largest utility in the country, has planned to build five large dams along the main course of the Biobío. Pangue, currently under construction, is first in the series.

In 1992, Chile's House of Representatives requested Universidad de Concepción to comment on Pangue's EIA, carried out for ENDESA by an American consulting firm (E&E,Inc and Agrotec Ltda., 1991). My specific task was to revise possible effects on the physical environment (hydrology, morphology, ecology, and water quality) and on downstream water users. The resulting document(Centro EULA, 1992)concluded that the EIA was of poor quality and that a new assessment should be carried out. Meier (1993a) analysed the most probable impacts on the natural environment, proposed mitigation measures to counter the more obvious effects, and recommended performing a series of specific environmental studies before attempting to carry out a new EIA. A later paper (Meier, 1993b) written at the request of the Regional Justice Court analysed the probable effects on downstream users (mainly on irrigation diversions).

This paper describes the Pangue dam project and lists its most probable environmental impacts. Mitigation measures are proposed. The original EIA is criticised by comparing its main characteristics with the expected standards for this kind of report. Finally, the different studies to be carried out before performing a serious EIA are listed.

Pangue Dam and its Impacts

The following information about the project is taken from Pangue SA(1992) and E&E,Inc and Agrotec Ltda(1991). The 113 m height and 450 m long concrete dam will be located in a narrow scenic gorge on the main course of the Biobío. It will impound a 540 ha reservoir, with a volume of $175x10^6$ m^3. Regulated storage is $6x10^6$ m^3. The subterranean plant will house two identical Francis-type turbines, with a total output of 450 MW at a discharge of 500 m^3. It will operate as hydropeaking, with daily generation at full capacity for 3 hours. This $470x10^6$ dollar project, mostly financed by the World Bank, will generate 9% of Chile's demand when operative, in 1998.

Pangue's most direct effect is on the river hydrology. Mean annual flow is 290 m^3/s; the driest month is March with 88 m^3/s. Frequency analysis was performed in order to obtain a critical scenario: a value of 57 m^3/s, corresponding to a ten year return period was adopted as 'worst-case' flow (compare to the EIA's 99 m^3/s). The flow duration curve confirms the lower value: there are on average 5 days per year with flows under 60 m^3/s.

The peaking operation will cause highly adverse impacts: daily channel dewatering episodes during the low-flow periods and violent fluctuations in stage. Under the utility proposed scheme, the river would be dewatered for flows below 167 m^3/s (137 days in an average year), and these episodes would last up to 21 hours. For an alternative operation, tending towards the 'run-of-the-river' type, the Biobío would run dry for flows under 112.5 m^3/s (102 days on average), for periods of up to 13 hours. Thus, Pangue cannot work as a run-of-the-river facility, no matter how it is operated. This is due to a turbine design that does not allow flexibility in matching the incoming hydrograph on an hourly basis. This seemingly irrational design is justified if one considers that the next planned dam, Ralco, will lie just upstream and has inter-annual regulation capacity.

The deposition of most of the sediment load in the impoundment could have important effects. The decrease in bed-material load and increase in transport capacity could cause channel widening in some places. Degradation will not occur in this strongly armoured cobble bed. The trapping of a large part of its heavy glacial washload will result in enhanced clarity of the Biobío's waters; this can in turn have vast effects on the river ecology.

Pangue's operation could also impact other traditional uses of the river, such as irrigation, potable water, and dilution of sewage, as well as instream uses, like fishing, rafting, etc. The violent flow fluctuations will be dangerous to recreational users and will also affect water diversions. To predict their effects, the releases were routed down the channel. The fluctuations were found to be still significant at a town 55 km downstream, where 52.5 m^3/s are diverted, and untreated sewage is dumped into the river. Two very large paper pulp mills are located only 50 km from here. This raises serious concerns regarding water quality (mainly with regard to drinking water).

In order to mitigate some of Pangue's adverse effects, it was proposed to implement a minimum instream flow for the Biobío River, downstream of the dam. As such a flow would have to be released without generating (due to the large size of turbines), it was also recommended to install multiple unequally-sized turbines. Finally, variable depth intakes (to minimise water quality changes) and fish passage facilities were prescribed.

The EIA for Pangue Dam

The EIA paid for by ENDESA was mostly based on existing -and insufficient- information. No studies of the aquatic environment were carried out; some were proposed for completion after the dam construction. An EIA must be based on adequate environmental studies. In this case, some research should have been carried out -before building the dam, and before attempting to write an EIA- about the following topics: habitat needs of relevant species present in the system; water quality changes due to reduced flows; impacts of fluctuations on diversion intakes; ecological effects on riparian and fluvial communities; fish migration; channel survey and hydraulic study (for routing and habitat modelling); morphological effects; etc.

The impounded sector upstream of the dam was considered to be the main impact area, with all downstream effects arbitrarily labeled as 'negligible'. Thus, the report only assessed impacts in the 540 ha reservoir!

There was no discussion of project design and/or operation alternatives. There seemed to be no chance of altering the original choices. Anyhow, all impacts were mild according to the report. From its very beginning it was clearly supportive of the project. An EIA is not supposed to take stand regarding whether a project is good or not, it only must assess its impacts.

Notwithstanding the fact that Pangue is only the first in a series of five very large dams, no mention was ever made of the possible interactions. Of course, a cumulative assessment was needed in this case.

The EIA based all computations (including an application of a two-dimensional reservoir water quality model) on a 'critical' flow that was a monthly average over a 30 year period. This is a serious mistake, as all EIAs must consider some kind of worst-case scenario.

Based on this case, I would argue that private companies willing to develop projects should pay for, but should not be allowed to directly hire consultants. The latter should be responsible not to the corporation paying their bill, but to the society that will be affected by their work. In this regard, I find it extremely unethical to profit from less-developed countries, where research funds are very scarce, and the social impacts of these mega-projects can be devastating.

References

Centro EULA, 1992. Analysis of Pangue Project's EIA. 92 p. Carried out at the request of the House of Representatives. Universidad de Concepción, Chile. (In Spanish)

Ecology & Environment, Inc. and Agrotec Ltda. 1991. Assessment of Pangue Project's Relevant Environmental Impacts. Final Report prepared for Pangue S.A. Electrical Company. Santiago. (In Spanish)

Meier C.I., 1993a. Probable impacts of the Pangue dam on the natural environment. Pp. 347-383 **In**: Faranda F. and O.Parra (Eds): Ecology and Water Quality of the Biobío System. Monografía Científica 12, Centro Eula, Universidad de Concepción, Chile . (In Spanish)

Meier C.I., 1993b. Possible effects of Pangue dam on irrigation rights. Technical report written for the Regional Justice Court. Concepción, Chile. (In Spanish)

Pangue S.A., 1992. Environmental Aspects of Pangue Project. Pangue S.A. Electrical Company, Santiago.

THE CANAL DE PROVENCE DYNAMIC REGULATION SYSTEM

A SAFE AND SUITABLE PROCESS FOR OPERATING WATER CONVEYANCE STRUCTURES

Jean-Luc DELTOUR[1]

Situated in the South of France, the Société du Canal de Provence et d'Aménagement de la Région Provençale (Canal de Provence and Provençal Region Development Company) was created to contribute to the economic development of Provence and the Côte d'Azur regions in the South of France. As such it studies, executes and manages several multipurpose hydraulic schemes. These schemes supply 70,000 ha, more than 100 communities, 500 companies and many individuals in sub-urban areas with raw water.

The main structure is the Canal de Provence which carries the Verdon River water to eastern and coastal Provence. So that all the users can have all the water they require at all times, whilst optimizing management of the resources, the company developed a remote management system called "dynamic regulation" in 1970. This system provided and still provides automatic and permanent control of canal facilities and safety systems.

Since then the Société du Canal de Provence has devoted a great deal of time and effort to research and development. Whilst retaining its basic principles, dynamic regulation has benefited from the progress made in simulating hydraulic conveyance systems and in the techniques of automatic control.

----

(1) Water control engineer, Société du Canal de Provence (SCP-member company of GERSAR), BP 100, 13603 Aix-en-Provence, Cedex 1 - FRANCE.

1 - The Canal de Provence

Canal de Provence facilities can be classified into two groups :

- The main structures comprising the main canal and four branch canals. These operate through gravity flow and transport a flow decreasing from 40 to 10m3/s. There are 62 km of open canal and 140 km of pressurized galleries.
- The supply networks include :
  . 440 km of piping with between 500 and 1300 mm diameter and 3200 km of secondary piping with less than 500 mm diameter.

2 - Dynamic regulation

The automatic and centralized system developed by S.C.P. controls the movements of water in the conveyance structures. It was first implemented in 1971 in answer to certain technical constraints associated with the operation of open canals.

2.1 Hydraulic constraints

Supplying water on demand leads to wide fluctuations in flow which are difficult to predict. As opposed to pressurized pipes, open canals introduce the principle of long response times and are very inflexible in the absence of a storage volume.

Since upstream control alone could not be envisaged (forecasting is too difficult), and downstream control was too costly (banks have to be heightened to make them horizontal), it was decided to associate these two methods in one.

Dynamic regulation implements the principle of control by anticipation based on the forecasting of needs and continuous correction of canal settings in order to adapt the status of the canal to the real conditions of water usage.

2.2 Basic principles of dynamic regulation

Dynamic regulation or "controlled volume" regulation is based on the management of volumes of water transiting through open surface canals. It is applied on all the company facilities whatever their hydraulic operating regime. As a result all water demand can be satisfied and operating costs optimized at all points on the system and at all times by minimizing pumping energy costs and maximizing income from turbining.
The different tasks performed by real-time dynamic regulation software are :

a) Pre-processing

- measurement acquisition (every minute),
- data checking,
- storage of data in a real-time data base.

b) The operations required at the check gates are determined using three components :

- demand forecasting : every hour or every two hours, the programme assesses the offtaking flows and from this data it deduces the forecast flow which will be required at each check gate moving upstream,
- comparison of actual canal status with a reference status : every fifteen minutes, the programme defines a reference volume for the canal based on the forecasts of flows and optimizing of energy utilization. A correction is calculated by comparing this volume with the volume measured on the canal,
- taking into account adjustments made on other check gates : the corrective flows calculated at the check gates are cumulated upstream without warting that this having an effect on the canal levels.

c) Post-processing

- checking of adjustments' validity,
- dispatch of control orders,
- verification of execution,
- in case part of the equipment does not operate, degraded operating procedures are automatically tripped and alarms dispatched to O & M employees.

2.3 Comments on the dynamic regulation concept

After 25 years in service, the main strengths of the dynamic regulation programme can be summarized. They are :

a) - Coordination of canal settings : the changes in the flow which results from successive adjustments on the canal are cumulated from downstream to upstream and taken into account before they can have a measurable effect on the reach and reservoir volumes. This increases the system's response time.

b) -The simplicity and reliability of the software : the canal simulation model included in the programme does not completely integrate the Saint Venant equations and uses pre-established tables for :

Propagation times which are used to read off the adjustments needed based on advanced information of changes at canal offtakes.

Depth/Flow/Volume curves which are used to assess the volumes present in the canal based on the flow setting at the regulators and the measurement of canal level downstream of a pool.

These models are tested during the design with models which resolve all the Saint Venant equation.

- The relative inaccuracy of the simplified simulation is compensated for by its extreme reliability and by the closed loop corrections which allow easy and fast resetting based on the measured values. The correction module, which was organized around a proportional integral controller operating in relation to the volume, has been improved using the pole placement technique.

- The same simplicity characterizes the static simulation used to forecast irrigation water consumption. However, this forecast is updated every two hours. The accuracy achieved for mean daily volumes is thus ±5%.

c) - Taking into account all the operating problems : the software was originally developed to Canal de Provence specifications by a multi-disciplinary team of computer, telemetering, hydraulics engineers and the operating personnel. Little by little it has been adapted to meet the changing requirements of the operator.

Similarly, when being installed on the other structures, a study is first carried out during which the operating instructions are defined. Société du Canal de Provence then supervises installation of the equipment while checking that the different elements are compatible, and finally trains the operators and helps them with system takeover.

d) - The importance of the degraded operating procedures included in the programme : They enable most current dysfunctions to be managed without major problems and, above all, maintain a normal and automatic service if a defect occurs on one part of the installations. Experience has shown that the most frequent failures, which are usually minor, occur locally. It is therefore essential to be able to incorporate them into the automatic regulation process.

3. Equipment

3.1 Data processing

Two VAX series 3000 micro-computers are used with 12 Megabyte memory operating under the VMS operating system. Each one has two 150 Megabyte disks and a 290 Megabyte magnetic tape store. The centre is equipped with two Tektronik XN 11 19" colour graphic terminals and two VT 420 alphanumeric terminals. Two other terminals are used to operate the central computers remotely. Nine PC or VAX supervisors located in regional operating centres are connected to the central computers and manage their own telemeasurement network.

3.2 - Transmission system

There are 2000 km of telephone line, 350 km of which are buried along the canal, for transmitting data. 170 teletransmission stations code, dispatch and receive measurements and orders.

3.3 - Measurements and orders

1300 measurements are made directly or calculated based on other measured parameters. These cover mainly level, pressure, flow, gate position and water physical and chemical characteristics. 7800 signals are relayed. These are the logic variables which describe the operating status of the equipment.

The system generates and transmits 230 on/off instructions and 30 continuously controlled adjustments.

CONCLUSIONS

After 25 years' operation, the Canal de Provence system of "dynamic regulation" has given entire satisfaction both from the standpoints of safety and of quality of the water supply service.

The experience accumulated by the Société du Canal de Provence in this field has been applied to other major projects which, through their diversity, have amply demonstrated the flexibility of this process :

- Canal de Rocade in Marrakech : supply of gravity irrigation networks, urban supply, pressure network supply.
- Athens water conveyance canal to supply the city.
- Flumendosa project in Sardania : multipurpose project integrating optimized use of several dams.
- King Abdullah Canal in Jordan which proposes a system which includes manual operation of gates.

Other projects are currently in hand.

Apart from providing simply regulation at structures, Dynamic Regulation equipment has formed the basis of an entire centralized technical management system which is developing and which shows great promise in the field of hydraulic Operation and Maintenance of structures.

Dispersion of Unsteady Line Source in Turbulent Shear Flow

Kyung Soo Jun,[1] Associate Member, ASCE

Abstract

A numerical model for unsteady dispersion of a horizontal line source in turbulent shear flow is developed. A fractional step finite difference method is used which splits the unsteady two-dimensional advective diffusion equation into the longitudinal advection and the vertical diffusion equations, and solves them alternately for half time intervals by the Holly-Preissmann scheme and the Crank-Nicholson scheme, respectively. The developed numerical model is verified using analytic solutions for the case of constant velocity and diffusivity. The speed of vertical mixing varies approximately as the square root of the friction factor. The source position, which gives the most rapid mixing, lies above the mid-depth and moves toward the water surface as the friction factor increases.

Introduction

The initial period of mixing of pollutants discharged into turbulent shear flow can be described by the advective diffusion equation. An important engineering problem is to predict how rapidly the pollutants are mixed over the cross section, and this also needs analyses through the advective diffusion equation. Many researches have been performed on the dispersion of steady line source in turbulent shear flow (McNulty and Wood, 1984; Nokes et al., 1984). However, those for unsteady source are rare.

In this study a numerical model for the unsteady dispersion in turbulent shear flow is developed. Sensitivities of the dispersion process to the friction factor and the initial source position are analyzed.

[1]Assistant Prof., Dept. of Civ. Engrg., Sung Kyun Kwan Univ., Suwon, Korea 440-330.

Mathematical Formulation

The advective diffusion equation for an unsteady line source in turbulent shear flow in a wide open channel can be written as

$$\frac{\partial C}{\partial t} + u\frac{\partial C}{\partial x} = \frac{\partial}{\partial z}(\varepsilon_z \frac{\partial C}{\partial z}) \tag{1}$$

where $C = C(x,z,t)$ = mass concentration; $u = u(z)$ = longitudinal flow velocity; $\varepsilon_z = \varepsilon_z(z)$ = vertical turbulent diffusion coefficient; z = vertical coordinate; x = longitudinal coordinate. and t = time. Diffusive transport in the x direction is assumed to be negligibly small compared with the advective transport. If a logarithmic velocity distribution is assumed, u can be written as

$$u(z) = U + \frac{u_*}{\kappa}[1 + \ln(\frac{z}{d})] \tag{2}$$

where κ = von Karman constant; d = water depth; U = mean flow velocity; and u_* = shear velocity. From the Reynolds analogy the vertical turbulent diffusion coefficient can be expressed as

$$\varepsilon(z) = \kappa u_* z(1 - \frac{z}{d}) \tag{3}$$

Boundary conditions at the bottom and on the water surface are given by no-flux conditions as

$$\frac{\partial C}{\partial z}(x,0,t) = \frac{\partial C}{\partial z}(x,d,t) = 0 \tag{4-5}$$

Upstream boundary condition in the x direction and initial conditions for the entire region of interset must also be given.

To normalize the governing equation and the initial and boundary conditions, the following dimensionless variables are introduced:

$$z' = \frac{z}{d},\ x' = \frac{x}{d},\ t' = \frac{tu_*}{d},\ u' = \frac{u}{u_*},\ \varepsilon' = \frac{\varepsilon}{u_* d},\ C' = \frac{C}{C_{ref}} \tag{6-11}$$

where C_{ref} is some reference concentration.

The normalized advective diffusion equation in terms of dimensiooless variables defined above is

$$\frac{\partial C'}{\partial t'} + u'\frac{\partial C'}{\partial x'} = \frac{\partial}{\partial z'}(\varepsilon_z' \frac{\partial C'}{\partial z'}) \tag{12}$$

$$u' = \sqrt{\frac{8}{f}} + \frac{1}{\kappa}(1 + \ln z') \tag{13}$$

$$\varepsilon_z' = \kappa z'(1 - z') \tag{14}$$

where the friction factor, f is defined as

$$f = 8(\frac{u_*}{U})^2 \tag{15}$$

Normalized boundary conditions in the z direction are

$$\frac{\partial C'}{\partial z'}(x',0,t') = \frac{\partial C'}{\partial z'}(x',1,t') = 0 \tag{16-17}$$

Upstream boundary conditions and the initial conditions can also be normalized using the above-defined dimensionless variables.

Numerical Model

The governing equation (12) is a partial differential equation, which is parabolic in the z direction, and hyperbolic in the x direction. These two types of equations have different characteristics from each other; and thus, appropriate numerical methods for each equation are quite different. Therefore, a fractional steps technique is used where the advective diffusion equation is time-split to separate the physical processes involved, i.e., the longitudinal advection and the vertical diffusion. Eq. (12) can be written as the set of equations (the symbol, ' is omitted hereafter for convenience)

$$\frac{\partial C}{\partial t} + 2u\frac{\partial C}{\partial x} = 0, \qquad n\Delta t \le t \le (n+\frac{1}{2})\Delta t \tag{18}$$

$$\frac{\partial C}{\partial t} = \frac{\partial}{\partial z}(2\varepsilon_z\frac{\partial C}{\partial z}), \qquad (n+\frac{1}{2})\Delta t \le t \le (n+1)\Delta t \tag{19}$$

which may be solved separately in two half-time steps to get from time level n to n+1. Note that Eqs. (18) and (19) sum to give Eq. (12) over one time step. Eq. (18) contains twice the x-direction advective effect concentrated over half a time step; similarly, Eq. (19) incorporates twice the z-direction diffusive effect applied over the second half of the step.

Advantage of the fractional step method is that the most suitable numerical schemes for each equation can be applied to accurately simulate corresponding physical process. In particular, usual Eulerian schemes for the advection equation often accompany with numerical diffusion or oscillations (Noye, 1987) while they are widely used to solve the diffusion equation. In this study, the two-point fourth-order Holly-Preissmann scheme (Holly and Preissmann, 1977) is used to solve the longitudinal advection equation (18), and the Crank-Nicholson scheme for the vertical diffusion equation (19). Since the Crank-Nicholson scheme for vertical diffusion calculation and the Holly-Preissmann scheme (Toda and Holly, 1987) for longitudinal advection calculation are both unconditionally stable, the fractional step method described above has unconditional stability.

Test of the Model

The numerical model was tested by comparing the results computed for uniform velocity and diffusivity with the analytic solutions for steady and unsteady problems. The comparison showed good agreements between the numerical and the analytic solutions. Fig. 1 illustrates the comparison of vertical concentration profiles at two downstream positions, where a steady source is given at (x,z) = (0,0.5).

Simulation Results and Analysis

Dispersion of an instantaneous plane source was simulated, which is discharged at time $t = 0$, at the position $x = 0$. The source strength is assumed uniform in vertical direction. The initial condition can be written as

$$C^0_{i,j} = \frac{m}{\Delta x} \qquad (i = 0) \qquad \text{for all } j \tag{20a}$$

$$= 0 \qquad (i = \text{elsewhere}) \qquad \text{for all } j \tag{20b}$$

where m is the mass per unit area normal to the longitudinal direction.

A series of numerical simulations was carried out for various friction factors with $\Delta x = 1.0$, $\Delta z = 0.01$, $\Delta t = 0.01$ and $m = 1.0$. Fig. 2 illustrates the longitudinal distributions of vertical average of concentrations computed by the model. One can see that the degree of mixing at a certain downstream location is larger for larger friction factor as it is expected. Also observed is that at the same dimensionless time $t = 200$, the distribution for $f = 0.04$ is almost identical to that for $f = 0.01$. However, the longitudinal distance traveled by the pollutant mass for $f = 0.01$ is about twice as long as it is for $f = 0.04$. From comparisons of concentration distributions for various friction factors, it was found that the downstream distance required to reach a certain degree of mixing is inversely proportional to the square root of the friction factor. In other words, the speed of mixing varies directly as the square root of the friction factor.

The effect of vertical source position was investigated by comparing vertical concentration profiles at downstream locations for various vertical positions of steady line source. Fig. 3 represents the best source position for each friction factor which requires the shortest distance for complete vertical mixing. It is seen that the optimum source position is located slightly above the mid-depth and moves upwards as the friction factor increases.

Acknowledgement

This study is supported by the Korea Science and Engineering Foundation.

References

Holly, F.M. and Preissmann, A.(1977). "Accurate calculation of transport in two dimensions." *J. Hyd. Div.*, ASCE, 103(11), 1259-1277.

McNulty, A.J. and Wood, I.R.(1984). "A new approach to predicting the dispersion of a continuous pollutant source." *J. Hyd. Res.*, 22, 147-159.

Nokes, R.I., McNulty, A.J., and Wood, I.R.(1984). "Turbulent dispersion from a steady two-dimensional horizontal source." *J. Fluid Mech.*, 149, 23-34.

Noye, J.(1987). "Numerical methods for solving the transport equation." *Numerical Modelling: Applications to Marine Systems*, ed. J. Noye, Elsevier, Amsterdam, 195-229.

Toda, K. and Holly, F.M.(1987). "Hybrid numerical method for linear advection-diffusion." *Microsoftware for Engineers*, 3(4), 199-205.

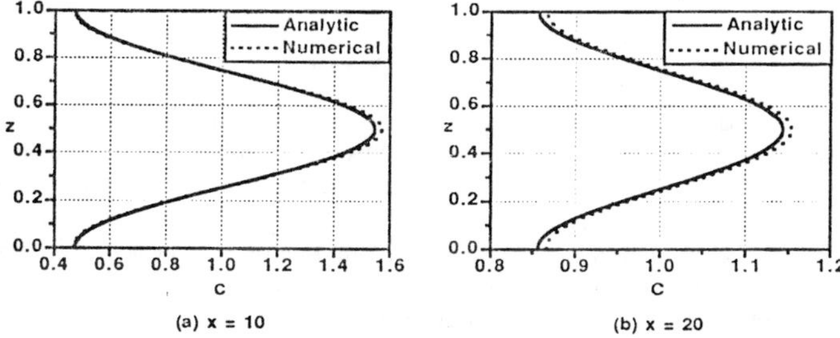

Fig. 1. Vertical Concentration Distributions Computed for Steady Line Source

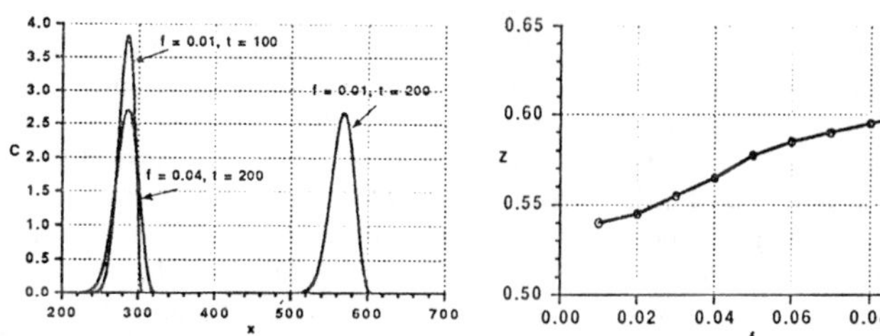

Fig. 2. Depth-Averaged Concentration Distributions Computed for Instantaneous Plane Source

Fig. 3. Optimum source Position for Various Friction Factors

TURBULENCE CHARACTERISTICS OF FLOW OVER A COBBLE BED

Fabián López[1], Chad Dunn[1] and Marcelo García[2], A.M. ASCE

Abstract

Experiments are carried out on uniform, turbulent open-channel flow over large roughness elements (cobbles) using recently available technology (Acoustic Doppler Velocimetry) in order to characterize the turbulence structure and the role played by coherent events, for a given range of Reynolds and Froude numbers. Large gradients of vertical turbulent kinetic energy were observed offering a plausible explanation for the lack of fit of the log law. Conditional quadrant-technique shows the flow depth divided into two regions of marked different structure with ejections and sweeps making the major contributions to the Reynolds stress above and within the cobbles, respectively. One of the main effects of the roughness elements has been found to be in the way by which mean-flow energy is transferred towards the turbulent fluctuations in different flow regions.

1. Introduction

It has long been recognized that turbulent flows in mountain rivers, with roughness elements comparable in size to the flow depth and steep gradients, exhibit resistance coefficients which are quite different from those predicted by typical resistance relationships for open-channel flows (Bathurst, 1985). Field measurements in the last decade have also shown that the vertical velocity distribution deviates substantially from the logarithmic shape, and under certain conditions a two-zone velocity profile with a characteristic S-shape seems to dominate (Bathurst, 1994). Several resistance equations have been proposed based on field and laboratory observations and some characteristics of the velocity profile have been investigated theoretically as well (e.g. Wiberg and Smith, 1991). However, little attention has been paid to the particular turbulence structure of these kind of flows, mostly due to the lack of detailed measurements under different flow conditions.

[1] Research Assistant, [2] Assistant Professor. Department of Civil Engineering, University of Illinois at Urbana Champaign. 205 N. Mathews, Urbana, IL 61801. USA.

The role played by coherent structures in smooth-wall boundary layers has received a lot of attention. However, in spite of being typical of natural environments, less attention has been paid to the corresponding research on rough wall flows. It is therefore natural that far less is known about the role played by coherent structures in flows with large-scale roughness, like mountain streams or vegetated open-channel flows. Grass (1971) showed that, in the outer region, both fluid ejections and sweeps make strong, intermittent contributions to the Reynolds stresses and turbulence production, irrespective of wall roughness. On the other hand, Nakagawa and Nezu (1977) and Raupach (1981) found, by conditional measurements, that sweeps are more important than ejections in maintaining the Reynolds stresses close to rough surfaces.

The present work aims at helping characterize the turbulence structure and the role of coherent structures for the case of a cobble-bed open channel flow. Comparisons are made when appropriate with results from measurements in smooth-wall conditions.

2. Experiments

The experiments were performed in a laboratory tilting flume, 12.2 m. long and 0.9 m. wide. The test section was located about 6 m. downstream from the entrance. Discharge measurements were conducted using two parallel weighting tanks of 20,000 pounds capacity each. Velocity measurements were taken by means of an Acoustic Doppler Velocimeter, of 25 Hz maximum sampling rate, which allowed the acquisition of the three cartesian velocity components. The cobbles used were natural rocks with an average diameter of 0.115 m. and a standard deviation of 0.023 m., and were set in the closest possible arrangement. Five experiments were conducted, all in nearly uniform flow conditions, with longitudinal slopes ranging from 0.0091 up to 0.025 and flow depths ranging from 0.24 m to 0.27 m.

4. Experimental Results

Velocity Statistics

The measured vertical profiles of mean streamwise velocity deviate from the semi-logarithmic law, taking an S-shaped distribution as observed by others in laboratory as well as in field measurements (Bathurst, 1994). The existence of an inflection point was detected near the averaged top of the cobbles, i.e. showing a strongly localized sheared velocity. While normalized values of streamwise turbulent intensities for typical smooth (and rough) boundary layer flows exhibit a peak of the order of 0.17, present results for cobble-bed flows in general show much higher values throughout the whole flow depth. These high ratios clearly indicate that as the channel bottom is approached the mean velocity decreases at a higher rate than the averaged turbulent kinetic energy.

Turbulent Kinetic Energy Balance

One necessary condition for the existence of an equilibrium layer, and hence for the log law to apply, is that the net rate of energy loss by turbulent diffusive transport is negligible compared to the production of turbulence (Townsend, 1961). In other words, in the inertial region one expects the vertical gradient of the term $<0.5\,q^2 v'>$ to vanish compared to the production term, where $q^2 = u'^2+v'^2+w'^2$. The profile of the measured vertical flux of turbulent kinetic energy is illustrated in Figure 1 , whereas Figure 2 shows the relative magnitude of diffusive transport and production of turbulent energy for run Grav4.

Coherent Structures

Following conditional quadrant techniques (Raupach, 1981), the fractional contribution to Reynolds stresses from each quadrant, as well as the fractional time, was computed and is illustrated in Figure 3 both for flow within the cobbles and for flow atop of them for run Grav5. Above the roughness elements the coherent structure of the turbulence resemblances the one corresponding to the typical outer layer (although as shown no equilibrium layer

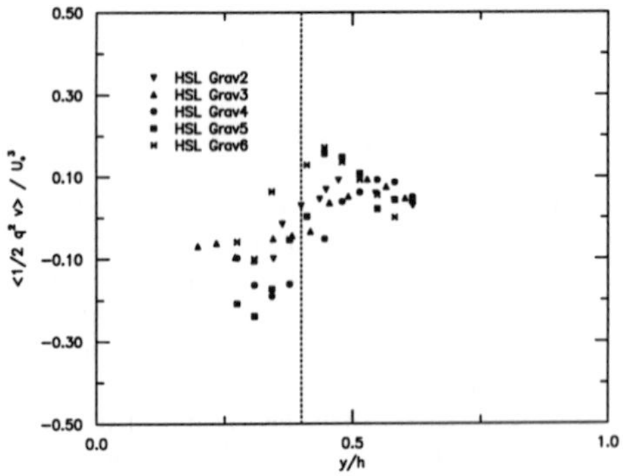

Figure 1 Variation of the Vertical Flux of Turbulent Kinetic Energy.

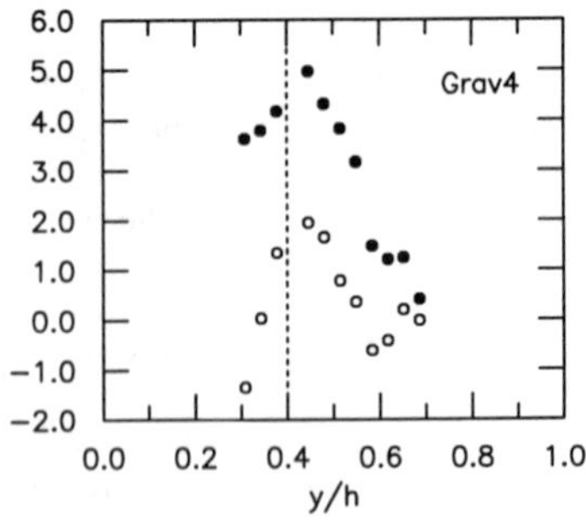

Figure 2 Variation of the Diffusive Transport (○) and Production of Turbulent Kinetic Energy (•) using flow depth and shear velocity as scaling variables.

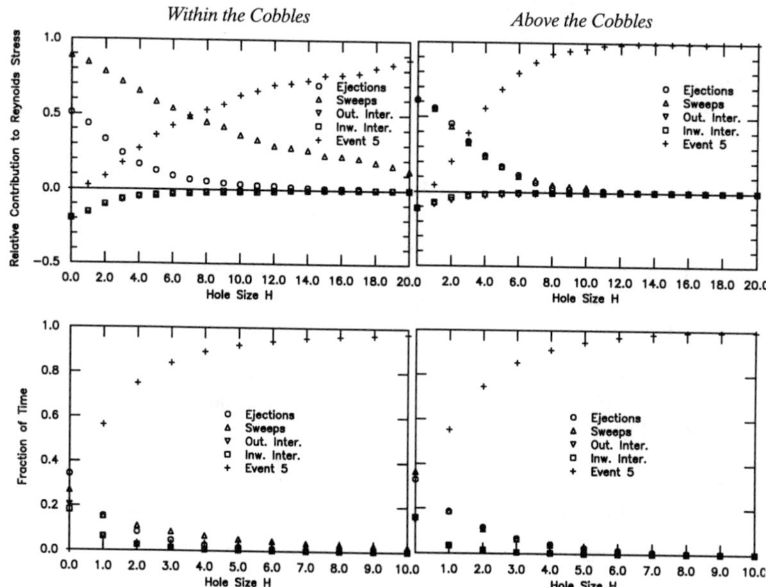

Figure 3 Fractional Contribution to Reynolds Stress and Fraction of Time corresponding to each Quadrant

exists), in the sense that ejections dominate the Reynolds stresses contributions over the one due to sweeps. On the other hand, below the averaged top of the cobbles the situation reverts, and sweeps dominate the mentioned contribution independently of hole size. It is interesting to note that the structure of odd quadrant events does not seem to be affected by the presence of the cobbles, and the absolute contribution to Reynolds stress is higher from outward than from inward interactions irrespectively of distance from the bed. Regarding the fractional time corresponding to each event, it was observed that above the cobbles, like in the smooth bed case, the percentage corresponding to sweeps is higher than the one corresponding to ejections, and this trends reverts for the flow within the roughness elements.

4. Summary and Conclusions

Normalized values of turbulent intensities show a peak near the top of the cobbles, and remain almost constant in the near region below that level. While the mean velocity decreases towards the bed within the cobbles, two observed factors are believed to contribute in keeping the local energy of the fluctuations at a relative high level: the mechanism that accounts for the downward transport of energy (Figure 1 and 2), which seems to be still efficient in this region, and the production term, which does not show a rapid decay towards the bed (Figure 2). It is believed that the production of energy due to wakes may also play an important role in maintaining high levels of almost constant energy. Conditional quadrant results show the flow depth subdivided into two subregions with different turbulence structure. In the region above the cobbles (AC) large gradients of vertical turbulent kinetic energy fluxes exist, which inhibits the development of a dynamical equilibrium region, constituting a plausible explanation for the lack of applicability of a logarithmic law. In the region below the cobbles (BC) the vertical flux of turbulent kinetic energy points downward, and correspondingly sweeps contribute the most to the Reynolds stresses. Hence, in region AC short ejection-dominated periods of high-intensity coexist with longer sweep-dominated periods of lower-intensity, while the opposite is true for region BC. Noting further that odd quadrant-events represent negative production (energy flux from the turbulence to the mean flow) and that even quadrant-events imply a positive production (energy form mean flow to turbulence) one may conclude by saying that one of the main effects of large-scale roughness is to affect the way by which energy is transferred from the mean flow toward the turbulent fluctuations, while the backscatter flux remains unaffected. The implications of this finding to turbulent transport processes, as well as the effect of roughness concentration upon the herein obtained results, need the assessment of further investigations.

5. Acknowledgements

The support of the Fluid, Hydraulic, and Particulate System Program of the National Science Foundation (Grant CTS 92−0211) is gratefully acknowledged

6. References

BATHURST , J.C, 1985, *Flow Resistance Estimation in Mountain Rivers.* Journal of Hydraulic Engineering. ASCE. Vol **111**, pp. 625-641.

BATHURST , J.C, 1994, *At-A-Site Mountain River Flow Resistance Variation.* ASCE 1994 Hydr. Eng. Conference. Buffalo, New York.

GRASS A.J. , 1971, *Structural features of turbulent flow over smooth and rough boundaries*. J. Fluid Mech. **50**, pp 233−255.

NAKAGAWA H. and NEZU I., 1977, *Prediction of the Contributions to the Reynolds Stress from Bursting Events in Open-Channel Flows.* JFM Vol. **80**, part 1, pp. 99-128.

RAUPACH M.R., 1981, *Conditional Statistics of Reynolds Stress in Rough-Wall and Smooth-Wall Turbulent Boundary Layers.* JFM, Vol. **108**, pp. 363-382.

TOWNSEND A.A, 1961, *Equilibrium Layers and Wall Turbulence.* JFM, Vol. **11**, pp. 97-120.

WIBERG P.L. and SMITH J.D., 1991, *Velocity Distribution and Bed Roughness in High-Gradient Streams.* Water Resources Research. Vol. **27** pp. 825-838.

A Spectral Domain Decomposition Method for Computational Fluid Dynamics

J. Keskar[1] and **D. A. Lyn**[1]

Abstract

Spectral computational methods are capable of attaining higher resolution than comparable finite-difference or finite-element schemes for numerical simulation of flows, but traditional global methods have been restricted to relatively simple geometries. Incorporating domain decomposition into spectral codes has widened their range of applicability. This paper describes a modification of the method of Ku et al. (1989), and its application to two incompressible laminar benchmark flows: the flow in a regularized lid-driven cavity and the flow over a backward-facing step.

1. Introduction

Numerical diffusion due to inadequate spatial resolution poses a major problem in simulating flows. For incompressible flows, spectral methods, in which velocity and pressure fields are expanded in high-order spectral basis functions, are superior in this regard to finite-difference or finite-element methods using a comparable number of nodes (Canuto et al., 1988). This is attributed to the exponential convergence of the former compared to the algebraic convergence of the latter. Traditionally, they have been limited in their range of applicability because of their global nature, which usually requires simple geometries. Recently, concepts of domain decomposition have been combined with spectral methods thereby obtaining greater flexibility in treating complicated geometries, while retaining high spatial resolution.

In the present work, the pseudospectral matrix element (PSME) method of Ku et al. (1989) (hereafter referred to as K89), based on a Chebyshev collocation scheme, is examined. Its attraction lies in its exact (to machine accuracy) satisfaction of the incompressibility constraint at all nodes in the flow domain including the boundary points. In its original form, the PSME method was found limited in the type of boundary conditions on the velocity field that could

[1] School of Civil Engineering, Purdue Univ., W. Lafayette, IN47907

be conveniently imposed, and a modification was implemented to overcome this limitation. Two benchmark steady laminar two-dimensional incompressible flow cases are considered: the regularized lid-driven cavity at Reynold number, $Re = 1000$, and the backward-facing step flow at $Re = 109$. The steady-flow solution is obtained by solving the unsteady problem to a steady-state.

2. The pseudospectral matrix element (PSME) method

2.1 The governing equations

In Cartesian tensor notation, the governing equations for unsteady two-dimensional laminar incompressible flow in a domain are expressed in dimensionless form as

$$\frac{\partial u_i}{\partial x_i} = 0, \tag{1}$$

$$\frac{\partial u_i}{\partial t} + \frac{\partial p}{\partial x_i} = F_i. \tag{2}$$

Here, u_i is the velocity component in the x_i direction, t is the time variable, p is the static pressure, and

$$F_i = -u_j \frac{\partial u_i}{\partial x_j} + \left(\frac{1}{Re}\right) \frac{\partial^2 u_i}{\partial x_j^2}. \tag{3}$$

Eqn. 1 is the continuity equation expressing the incompressibility constraint and Eqn. 2 are momentum equations in the two coordinate directions. Appropriate initial and boundary conditions (Gresho, 1991) must also be imposed.

2.2 The projection method for the treatment of pressure

Pressure in an incompressible flow has no evolution equation, but rather acts as a constraint on the velocity field. A direct solution of Eqn. 1 is possible, but most approaches have chosen to solve a derived Pressure Poisson Equation. The PSME method is based on a projection method (Chorin, 1968), which may be viewed as a predictor-corrector scheme. In the first step, an intermediate predicted velocity field, $\overline{u}_i$, is obtained in the flow interior by explicit integration of Eqn. 2 *without* the pressure-gradient term, namely,

$$\overline{u}_i^{n+1} = u_i^n + \Delta t\, F_i^n. \tag{4}$$

The corrector step is made through a relationship between p and u_i,

$$u_i^{n+1} = \overline{u}_i^{n+1} - \Delta t \frac{\partial p}{\partial x_i}, \tag{5}$$

which is combined with the continuity equation at the $(n+1)^{\text{th}}$-time level to yield,

$$\frac{\partial^2 p^{n+1}}{\partial x_i^2} = \left(\frac{1}{\Delta t}\right) \frac{\partial \overline{u}_i^{n+1}}{\partial x_i}. \tag{6}$$

This Poisson equation for p has a forcing function proportional to the divergence of the predicted velocity field.

Eqn. 6 is imposed at all nodes, including boundary nodes, but Eqn. 4 is used only in the interior, resulting in a velocity field which is everywhere divergence-free. When only Dirichlet velocity conditions are imposed, as in the lid-driven cavity, the PSME method imposes only velocity boundary conditions because, on the boundary, the pressure gradient terms can be eliminated by invoking Eqn. 5. When more complicated boundary conditions are imposed, e.g., at an open-flow boundary, the elimination of the pressure terms in favor of the imposed velocity boundary condition becomes difficult or impossible. In contrast to the PSME method in K89, $\overline{u}_i^{n+1}$ on the boundary are, in the present work, explicitly considered as unknown variables to be determined simultaneously with p. Thus, at each boundary point, three unknowns (in two dimensions, $\overline{u}^{n+1}$, $\overline{v}^{n+1}$, and p) must be solved for with three equations (two conditions on velocity and the continuity equation or possibly a pressure boundary condition). For Dirichlet velocity conditions are imposed, this reduces to the original PSME method, but more general velocity boundary conditions can be more conveniently handled.

A weakness of the PSME method is the presence of spurious modes in the pressure solution (Canuto et al., 1988). In the present work, the steady-state pressure field is obtained in a post-processing step by a solution of the Eqn. 6 with boundary conditions derived from the normal momentum equation (Gresho, 1991).

2.3 Domain decomposition

The basic concept of domain decomposition, familiar from finite-element and finite-difference methods, consists of decomposing the original flow region into smaller domains, on each of which a pseudospectral technique can be applied. The treatment of conditions at the interface between domains must be considered. The PSME method elects to match explicitly only the function value, while continuity of the normal derivative is satisfied only implicitly (i.e., continuity of the normal derivative is approached as the number of nodes is increased). Since it does not require explicit continuity of derivatives across domain boundaries, the PSME method uses the additional degree of freedom to enforce the divergence-free condition at each interface point. K89 described an implementation of the PSME method for parallel machines, in which the solution for overlapping 'macro'-elements consisting of more than one element are obtained by an iterative technique. In the present work, the focus has so far been on a non-iterative direct solution on a non-parallel machine (a IBM RS/6000 workstation) of the algebraic system resulting from the discretization of Eqn. 6. For two-dimensional problems, this is sufficient, and avoids any specification of convergence criteria.

3. The flow cases

3.1 The lid-driven cavity for $Re = 1000$.

The lid-driven cavity provides a convenient and clean benchmark problem because it involves only Dirichlet velocity conditions. The flow domain and boundary is shown in Fig. 1. On all boundaries, the velocities are zero, except the lid where the regularized condition, $u(x, y = 1) = 16x^2(1-x)^2$, $v(x, y = 1) = 0$, is imposed. The streamlines as well as the pressure field from a computation with four equal-sized domains, each with 13×13 nodes, for flow at $Re = 1000$ are shown in Fig. 1 together with the individual domain boundaries. The solution of both fields are notably smooth at domain boundaries. Good agreement is found with the results of Phillips and Roberts (1993) who used a single domain spectral method.

3.2 The backward-facing step for $Re = 109$

The backward-facing step exhibits additional features, which make it an interesting case for a spectral domain decomposition method. The relatively complex geometry and the separated shear layer at the step with the associated sharp poses velocity gradients in the flow interior poses problems for traditional spectral methods, and its downstream open-flow boundary may require a complicated boundary condition. The flow domain and boundary is shown in Fig. 2, with the steplocated at $(x, y) = (0, 0)$. On all solid boundaries, the velocities are zero, while at the inflow ($x = -2$, $0 \le y \le 2$), a fully developed parabolic profile, $u(x = 0, y) = 2y(1 - y/2)$, $v = 0$, is imposed, and at the outflow ($x = 20$, $-1 \le y \le 2$), the so-called traction-free conditions, $-p + (2/Re)\partial u/\partial x = 0$, $\partial v/\partial x = 0$, are imposed. A total of 12 variably sized domains, two upstream of the step, each with 8×8 nodes, were used. The streamlines and pressure contours are shown in Fig. 2. Again, the solutions are smooth across domain boundaries. A reattachment length slightly larger than 5 was found, agreeing with the result of Patera (1984) who used a spectral-element approach.

References

1. Canuto, C., Hussaini, M. Y., Quarteroni, A., and Zang, T. A. (1988). *Spectral Methods in Fluid Dynamics*, Springer-Verlag, New York.
2. Chorin, A. J. (1967). *Math. Comp*, **22**, pp. 745–762.
3. Gresho, P. M. (1991). *Ann. Rev. Fluid Mech.*, **23**, pp. 413–453.
4. Ku, H. C., Hirsh, R. C., Taylor, T. D. and Rosenberg, A. P. (1989). *J. Comp. Phys.*, **83**, pp. 260–291.
5. Patera, A. T. (1984). *J. Comp. Phys*, **54**, pp. 468–488.
6. Phillips, T. N. and Roberts, G. W. (1993). *J. Comp. Phys.*, **105**, pp. 150–164.

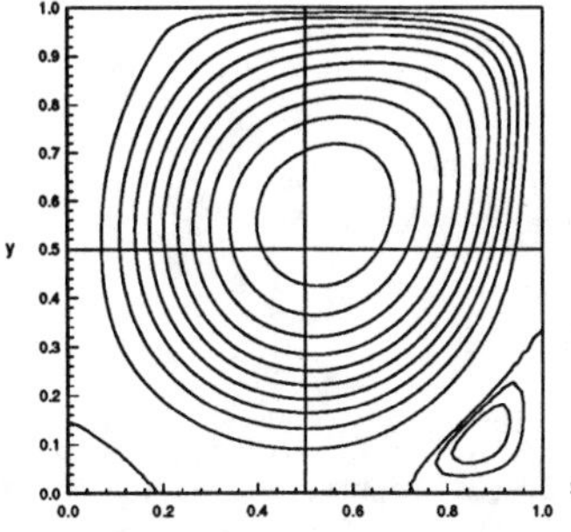

Fig.1a. Lid Driven Cavity : Streamfunction contours

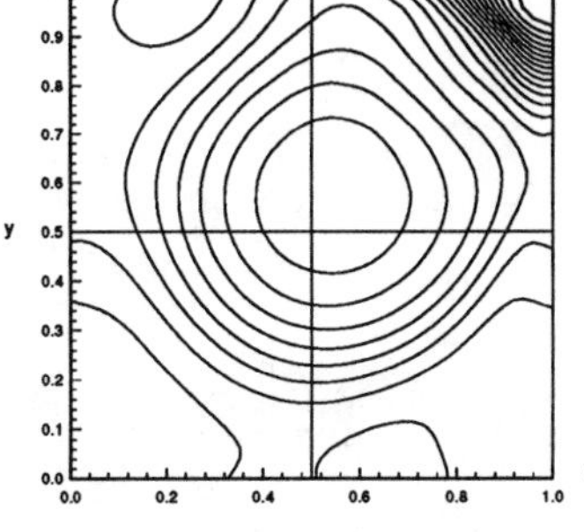

Fig.1b. Lid Driven Cavity : Pressure contours

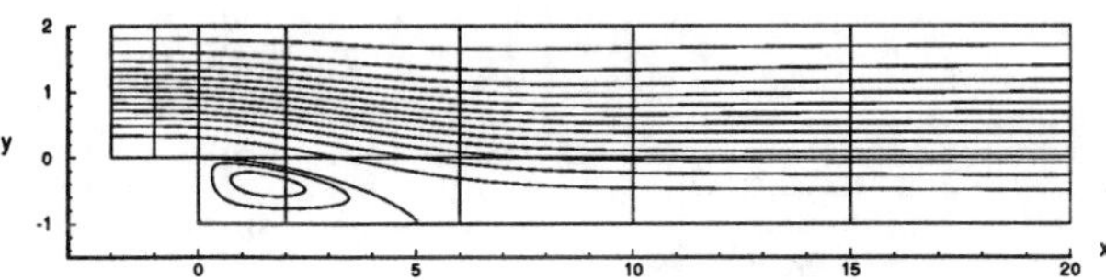

Fig.2a. Backward Facing Step : Streamfunction contours

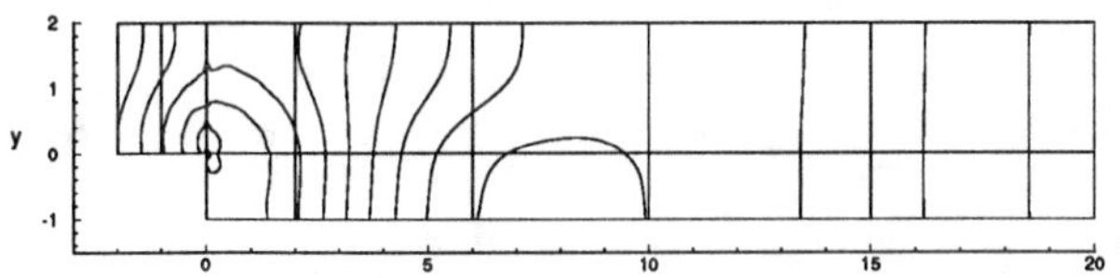

Fig.2b. Backward Facing Step : Pressure Contours

Manning's Equation and the Internal Combustion Engine

Douglas J. Trieste, PH [1]

Abstract

Both the Manning equation and the internal combustion engine were developed in the late 1800's. The same basic designs are still in use today with an industry developed around each. Were the designs that sound when developed over 100 years ago? Or, have we adopted accepted designs, improved on them, but, not considered other designs. This paper is not intended to answer these questions, but, to offer food for thought.

Discussion

The internal combustion engine was invented in 1859. It consists of an engine block, pistons, intake and exhaust valves, carburetor, crank shaft, flywheel, etc. It was based on the combustion of a mixture of fuel and air expanding in a cylinder, moving a piston, and turning a crankshaft. It's main nemises were friction and heat loss. Since that time, there have been many refinements and improvements, but, the basic design remains the same. Most improvements have come about by variation on a theme. There has been honing and refining, but, it is still the same basic design. Is there no other way to make an internal combustion engine, or are the concepts and principles used in 1859 still the best that we can do in today's world?

Manning's equation for open channel flow was developed in 1889. The general Manning equation is:

$$Q=1.49AR^{2/3}S^{1/2}/n \quad \text{(1)}$$

in which Q = the discharge (ft^3/s); R = the hydraulic radius (ft); S = the energy gradient, and n = Manning's roughness coefficient.

Manning's equation was based on data from flume studies, and developed for uniform flow conditions in which the water-surface slope, friction slope, and energy gradient are parallel to the streambed, and the cross-sectional area, hydraulic radius, and depth remain constant throughout the reach. Today, the Manning equation is probably the most popular for practical open-channel flow computations, including hydraulic computer models. It is easy to use, gives results that range from reasonable to accurate in many situations, and is accepted by the industry. It has served us well for many years and is to be commended.

---

[1] Hydraulic Engineer, Bureau of Reclamation, PO Box 25007, Denver, CO 80225

However, the results from the Manning equation are essentially at the mercy of n-values. And the selection of appropriate n values is as much an art as a science. Many sources offering guidance are available on n selection. Some of these include Barnes (1967), Benson and Dalrymple (1967), Chow (1959), Limerinos (1970), and, Jarrett (1985). But, due to the variability found in nature, it is difficult, if not impossible, to accurately estimate n in complex hydraulic situations.

Manning's equation is commonly used in natural channels for conditions that are not consistent with that from which it was developed. These conditions include non-uniform reaches, unsteady flow, irregular shaped channels, turbulence, steep channels, sediment and debris transport, moveable beds, etc. It is assumed that the equation is valid in these conditions, and the energy gradient adjusted via roughness coefficients (n-values) to make the equation as accurate as possible. As a result, much research has been performed on n-values.

Most improvements pertaining to the Manning equation have come about by variation on a theme - the original design of the Manning equation remains an industry standard. Only the "theme" (n-values) is changed to improve its performance. We work on making "Volkswagon improvements" on n-values - honing, shaping, defining, etc. But, even the famous and ever popular Volkswagon Bug was eventually discontinued for new and different models that combine and integrate all that has been learned and developed. Is it best to keep refining what we have? Or, would we be ahead to develop new equations that would eventually give better results?

<u>Can no better equation than the Manning equation be developed, or are the concepts and principles used in 1889 still the best possible today?</u>

It is interesting to wonder that if the Manning equation, or, piston-based internal combustion engine as we know it were never developed, then what would we use today?

For both the internal combustion engine and the Manning equation, a basic design was adopted and an industry developed around it. Were these designs that sound when first developed that they have stayed with us for so many years. If so, that is remarkable and a tribute to the developers. Or, have we not looked beyond the original designs, but, simply adopted and built on "accepted" technology?

Is it possible to replace the Manning equation with something new and different that draws upon all the knowledge that we learned since it's development? The Manning equation is at the mercy of n-values which are a black box (Trieste and Jarrett, 1987) in many situations. The equation itself is rarely challenged, but, n-values are continually debated. Could there be a better approach?

Is it time to develop new concepts in engines to better meet future needs such mechanical efficiency, simplicity, fuel type and consumption, pollution, and costs? And, is it time to develop new open-channel flow equations to better solve continual nemesis in computation such as non-uniform channels, unsteady flow, large floods, high-gradient channels, unstable beds, sediment and debris transport, supercritical/subcritical flow regimes, etc.?

This paper in no way intends to discount the Manning equation, or internal combustion engine, but, to provide food for thought on improvement of old designs, versus development of new designs.

References

1. Barnes, H.H., Jr., Roughness Characteristics of Natural Channels, U.S. Geological Survey Water-Supply Paper 1849, 1967.

2. Benson, M.A. and Tate Dalrymple, General Field and Office Procedures for Indirect Discharge Measurements, U.S. Geological Survey, Techniques of Water-Resources Investigations, Book 3, Chapter A-1, 1967.

3. Chow, V.T., Open Channel Hydraulics, New York, McGraw-Hill, 1959.

4. Jarrett, R.D., Determination of Roughness Coefficients for Streams in Colorado, U.S. Geological Survey Water Resources Investigations Report 85-400

5. Limerinos, J.T., Determination of the Manning Coefficient from Measured Bed Roughness in Natural Channels, U.S. Geological Survey Water-Supply Paper 1898-B, 1970.

6. Trieste and Jarrett, ASCE Proceeding of a Conference, Irrigations Systems for the 21st Century, Portland, Oregon, July 28-30, 1987.

FLOW RESISTANCE: FRICTION OR ENERGY

BY

Ben Chie Yen,[1] Fellow ASCE

Abstract

In hydraulic analysis of flows, particularly for flows in open channels, often a flow resistance coefficient or slope is required. However, in many applications it is not clear what type of resistance coefficient is used in the Manning, Darcy-Weisbach or other equations. At times some people argue whether it is a friction or energy coefficient. In this paper the resistance coefficients are defined according to the energy and momentum concepts respectively, and the general relationship between them are derived.

Introduction

In hydraulics literature both friction and energy slopes are frequently mentioned. They are often used interchangeably. However, from the fluid mechanics point of view they are distinctively different. Momentum is a vector quantity whereas energy is a scalar quantity. This problem is far from purely an academic interest because a clear understanding is important in the selection of proper values in solving problems. In some computer-based channel-flow models in which the values of the resistance coefficient are determined through calibration, the calibrated values are not really the resistance coefficient but a lumped factor in accordance with the flow equations adopted and the reliability of the data used. From the momentum viewpoint, the resistance is the resultant of the forces acting against the flow on the boundary of a control volume or a reach. Conversely, from the energy viewpoint, the resistance reflects the energy lost inside the control volume or

[1] Prof. Civil Eng. Univ. of Illinois at Urbana-Champaign, Urbana IL 61801

reach. Conventional approach in hydraulic analyses is to express the flow resistance coefficient in the Manning n, Weisbach f, or Chezy C form, without specifying whether it's a momentum or energy coefficient,

$$n = \frac{K_n}{V} R^{2/3} S^{1/2} \qquad f = \frac{8g}{V^2} RS \qquad c = \frac{V}{\sqrt{RS}} \tag{1}$$

in which K_n = 1 $m^{1/2}$/sec, 1.486 $ft^{1/3}$ -$m^{1/6}$/sec or $\sqrt{g}$ (Yen, 1993); V = cross-sectional average velocity; R = hydraulic radius; S = slope; and g = gravitational acceleration. The above equations provide the relationship between the slope and the resistance coefficients n, f, or C,

$$\frac{\sqrt{gRS}}{V} = \sqrt{\frac{f}{8}} = \frac{n}{R^{1/6}} \frac{\sqrt{g}}{K_n} = \frac{\sqrt{g}}{C} \tag{2}$$

Obviously, if the friction and energy-loss slopes are different, the resistance coefficient, be that f, n, or C, would also have two different, momentum and energy versions. The main purpose of this paper is to explore the difference and relationship between the friction and energy slopes from the basic momentum and energy equations.

Momentum and Energy Equations

The complete and exact momentum and energy equations in one-dimensional form can be derived from the Navier-Stokes equation averaged over turbulence time scale and integrated over the flow cross section or control column (Yen, 1973). By using a gravity-oriented coordinate system with x horizontal along the channel longitudinal direction and coordinate y and flow depth Y both vertical, the one-dimensional x-direction momentum equation for a homogeneous incompressible fluid is

$$\frac{1}{gA}\frac{\partial Q}{\partial t} + \frac{1}{gA}\frac{\partial}{\partial x}\left(\frac{\beta}{A}Q^2\right) + \frac{\partial}{\partial x}(KY) + (K - K')\frac{Y}{A}\frac{\partial A}{\partial x}$$

$$= S_o - S_f + \frac{1}{\gamma A}\frac{\partial T}{\partial x} + \frac{1}{gA}\int_\sigma \bar{U}_x \bar{q}\, d\sigma \tag{3}$$

in which t = time; A = cross-sectional flow area; Q = discharge through A; Y = flow depth; S_o = $-\partial y_b/\partial x$ is the channel slope along x direction; y_b = channel bed elevation; q = time rate of lateral flow per unit length of σ which is the perimeter bounding the cross section A; U_x = x-component of velocity of the lateral flow when joining the channel flow. The friction slope is defined as

$$S_f = \frac{-1}{\gamma A}\int_\sigma [\bar{\tau}_j]_\sigma N_j d\sigma \tag{4}$$

where N = directional normal of a surface, positive outward; i, j = orthogonal coordinate directions, i,j = 1, 2, 3 for x, y and z directions, respectively; γ =

specific weight of the fluid; the stress τ_{ij} averaged over turbulence time and the force T due to internal stresses are

$$\tau_{ij} = \mu\left(\frac{\partial \bar{u}_i}{\partial x_j} + \frac{\partial \bar{u}_j}{\partial x_i}\right) - \rho\overline{u_i' u_j'} \quad (5) \qquad T = \int_A \left[2\mu\left(\frac{\partial \bar{u}_x}{\partial x}\right) - \rho\overline{u_x'^2}\right] dA \quad (6)$$

where μ = dynamic viscosity of the fluids; u_i = instantaneous local (point) velocity component along x_i direction; u_i' = turbulent fluctuation with respect to $\bar{u}_i$; and the velocity and pressure distribution correction factors are

$$\beta = \frac{A}{Q^2}\int_A \bar{u}_x^2 dA \quad (7)$$

$$K = \frac{1}{\gamma AY}\int_A \bar{P}dA \qquad K' = \left(\gamma Y\frac{\partial A}{\partial x}\right)^{-1}\int_\sigma\left(\bar{P}\frac{\partial \bar{r}}{\partial x} + \frac{\overline{p'\partial 2r'}}{\partial x}\right)d\sigma \quad (8)$$

where $P = p + \gamma\,(y - y_b)$ is the local piezometric pressure with respect to channel bed; $p = \tau_{ii}/3$ is the local pressure intensity; and r = normal displacement of σ with respect to space or time projected on a plane parallel to A, positive outward.

The corresponding one-dimensional energy equation is

$$\frac{\partial H_B}{\partial t} + (H_c - H_B)\frac{1}{A}\frac{\partial Q}{\partial x} + \frac{Q}{A}\frac{\partial H_c}{\partial x} - W - \frac{1}{A}\int_\sigma (H_L - H_B)\bar{q}\; d\sigma + \frac{Q}{A}S_e = \zeta\frac{\partial Y}{\partial t} \quad (9)$$

in which the true cross-sectional averaged total head of the flow, H_B, is

$$H_B = \beta_B\frac{V^2}{2g} + KY + y_b \quad (10)$$

$$H_c = \alpha\frac{V^2}{2g} + \eta Y + y_b \quad (11)$$

$$\beta_B = \frac{1}{V^2 A}\int_A \bar{u}_i\bar{u}_i\, dA \quad (12) \qquad \alpha = \frac{1}{V^2 Q}\int_A \bar{u}_i\bar{u}_i\bar{u}_x\, dA \quad (13)$$

$$\eta = \frac{1}{\gamma QY}\int_Z \bar{P}\bar{u}\, dA \quad (14) \qquad \zeta\frac{\partial Y}{\partial t} = \frac{1}{\gamma A}\int_A \frac{\partial \bar{p}}{\partial t} dA \quad (15)$$

$$W = \frac{1}{\gamma A}\left(\frac{\partial}{\partial x}\int_A \bar{u}_i\,\bar{\tau}_{ix}\, dA + \int_\sigma \left[\bar{u}_x\,\bar{\tau}_{ij}\right]_\sigma N_j\, d\sigma\right) \quad (16)$$

H_L is the total head of the lateral flow q; and the energy loss slope S_e is

$$S_e = \frac{1}{\gamma AV}\int_A \bar{\tau}_{ij}\frac{\partial \bar{u}_i}{\partial x_j} dA \quad (17)$$

Different Slopes in Open-Channel Flow

As pointed out by Yen (1973) the following slopes often appear in open channel analysis:

(a) channel slope $S_o = -\partial y_b/\partial x$;
(b) water surface slope with respect to the horizontal x-direction, $S_w = -\partial(Y + y_b)/\partial x$;
(c) water surface slope with respect to channel bed, $S_w' = -\partial Y/\partial x$;
(d) $S_H = -\partial H/\partial x$, the slope of conventionally used approximately total head

$$H = H_p + y_b + \frac{V^2}{2g} \tag{15}$$

where H_p is the cross-sectional average piezometric head with respect to the channel bed;
(e) The slope of actual total head, $-\partial H_B/\partial x$, in which H_B is given as Eq. 10;
(f) The piezometric-head slope with respect to the channel bed, $\partial H_p/\partial x$;
(g) The hydraulic gradient, $-(\partial H_p/\partial x) + S_o$ which is the gradient of H_p with respect to a horizontal reference plane;
(h) The friction slope, S_f, which accounts for only the shear resistance from the wetted perimeter as defined in Eq. 4;
(i) The energy slope, or more precisely, the gradient of the dissipated mean-motion (over turbulence) energy, S_e, as defined in Eq. 17.

Relationship Between Friction and Energy Slopes

The relationship between any of the eight slopes, (b) through (i), can be obtained with the help of Eqs. 3 and 9. Particularly, for the relationship between S_e and S_f, by noting that

$$\frac{\partial H_c}{\partial x} = \frac{\partial}{\partial x}\left(\frac{\alpha V^2}{2g}\right) + \eta\,\frac{\partial Y}{\partial x} + Y\,\frac{\partial \eta}{\partial x} - S_o \tag{19}$$

one obtains

$$\begin{aligned} S_e = S_f &+ \frac{\partial}{\partial x}\left[(2\beta - \alpha)\frac{V^2}{2g}\right] + \left[(K - K')Y - \frac{\beta V^2}{g}\right]\frac{1}{A}\frac{\partial A}{\partial x} + (K - \eta)\frac{\partial Y}{\partial x} + Y\frac{\partial K}{\partial x} \\ &- Y\frac{\partial \eta}{\partial x} + \left[(\alpha - \beta_B)\frac{V^2}{2g} + (\eta - K)Y\right]\frac{1}{Q}\frac{\partial Q}{\partial x} + \frac{A}{Q}W - \frac{1}{\gamma A}\frac{\partial T}{\partial x} + \frac{1}{gA}\frac{\partial Q}{\partial t} \\ &+ \frac{\zeta}{V}\frac{\partial Y}{\partial t} - \frac{1}{V}\frac{\partial H_B}{\partial t} + \int_\sigma \left[\frac{1}{Q}(H_L + H_B) - \frac{\bar{U}_x}{gA}\right]\bar{q}\, d\sigma \end{aligned} \tag{20}$$

Likewise, from $S_w = -\partial(Y + y_b)/\partial x$ and Eq. 3, the water surface slope is

$$S_w = S_f + \frac{1}{gA}\frac{\partial}{\partial x}\left(\beta\frac{Q^2}{A}\right) + (K - K')\frac{Y}{A}\frac{\partial A}{\partial x} + (K - 1)\frac{\partial Y}{\partial x} + Y\frac{\partial K}{\partial x}$$

$$- \frac{1}{\gamma A}\frac{\partial T}{\partial x} + \frac{1}{gA}\frac{\partial Q}{\partial t} - \frac{1}{gA}\int_\sigma \bar{U}_x \bar{q}\, d\sigma \qquad (21)$$

and from Eqs. 8 and 3,

$$S_H = S_f + \frac{\partial}{\partial x}(KY - H_p) + \frac{\partial}{\partial x}\left[(2\beta - 1)\frac{V^2}{2g}\right] + \left[(K - K')Y + \frac{\beta V^2}{g}\right]\frac{1}{A}\frac{\partial A}{\partial x}$$

$$- \frac{1}{\gamma A}\frac{\partial T}{\partial x} + \frac{1}{gA}\frac{\partial Q}{\partial t} - \frac{1}{gA}\int_\sigma \bar{U}_x \bar{q}\, d\sigma \qquad (22)$$

According to Eqs. 20, 21, 22, or similar relationships for other slopes, in general all the nine slopes discussed previously are not equal. Only for steady uniform flow in a prismatic channel without lateral flow that $S_f = S_e = S_o = S_w = S_H = -\partial H_B/\partial x$. Yen et al. (1972) provided examples of quantitative differences of the momentum, energy and nominal resistance coefficients in the Weisbach form, f_f, f_e and f_H. For a flow with lateral inflow such as rainfall, $f_e > f_H > f_f$. The differences between the slopes or coefficients increase with increasing flow nonuniformity and unsteadiness. The explanation of the effect of boundary friction, form drag, wave, and flow unsteadiness on resistance can be found in Rouse (1965) and Yen (1991).

References

Rouse, H., "Critical analysis of open-channel resistance," *J. Hydraulics Div.*, ASCE, *91* (HY4): 1-25, July 1965.

Yen, B.C., "Open-channel flow equations revisited," *J. Eng. Mech. Div.*, ASCE, *99*, (EM5): 979-1009, Oct, 1973.

Yen, B.C., "Hydraulic resistance in open channels," In: *Channel Flow Resistance: Centennial of Manning's Formula*, ed. by B.C. Yen, 1-135, *Water Resources Publications*, Highlands Ranch, CO, 1991.

Yen, B.C., Wenzel, H. G. Jr., and Yoon, Y.N., "Resistance Coefficients for steady spatially varied flow," *J. Hydraulics Div.*, ASCE, *98* (HY8): 1395-1410, 1972.

SEDIMENTATION IN THE CACHE RIVER WETLANDS

Misganaw Demissie[1], M. ASCE, and Richard A. Cahill[2]

ABSTRACT

The Cache River wetlands, located in the Cache River basin in southern Illinois, are small remnants of a vast wetland system that used to occupy the Cache River valley before commercial logging and agricultural developments significantly altered the area. In the Lower Cache River basin, a wetland system called "Buttonland Swamp" is under great stress because of increased sediment inflow from tributary streams. The Cache River wetlands act as sedimentation basins that trap significant amounts of sediment inflow from tributary streams.

The rates of sedimentation in the wetlands were investigated by using two different methods to provide a better understanding of the amount and areal distribution of the sediment. The first method is based on monitoring sediment flowing into and out of the wetlands. The second method is based on radiometric dating techniques of sediment cores collected at selected points within the study area.

INTRODUCTION

The Lower Cache River wetlands are small remnants of an extensive wetland system in the Cache River valley. The wetlands are bottomland forests that are frequently flooded by tributary streams that drain upland watersheds. The Cache River basin is located in the extreme southern part of Illinois, just north of the confluence of the Ohio and Mississippi Rivers (figure 1). The total drainage area of the watershed is 1909 km^2. Since the construction of the Post Creek Cutoff in 1915, the Cache River basin has been divided into the Upper Cache and Lower Cache River subwatersheds, as shown in figure 1. The Upper Cache River watershed consists of the eastern part of the Cache River basin with a drainage area of 953 km^2; it drains directly to the Ohio River at River Mile 957.8 through the Post Creek Cutoff. The Lower Cache River watershed consists of the western part of the Cache River basin

---

[1] Principal Scientist and Director, Office of Sediment & Wetland Studies, Illinois State Water Survey, 2204 Griffith Drive, Champaign, IL 61820-7495.

[2] Geochemist, Illinois State Geological Survey, 615 E. Peabody, Champaign, IL.

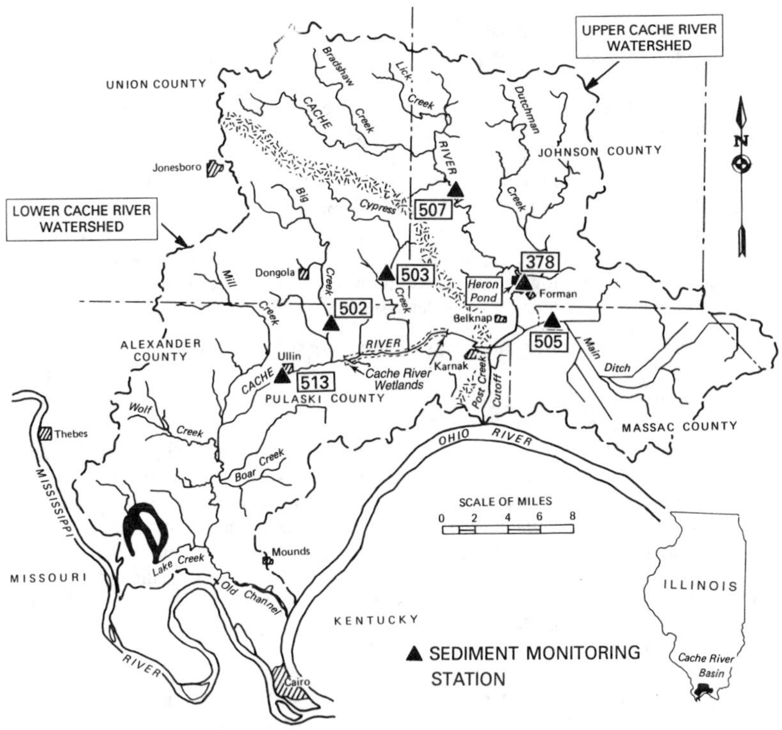

Figure 1. Cache River basin in Illinois

with a drainage area of 927 km^2; it drains to the Mississippi River at River Mile 13.2 through the diversion channel at the downstream end of the river.

The Upper and Lower Cache River basins are separated by the Cache River levee along the western bank of the Post Creek Cutoff near Karnak. This levee was built across the old Cache River channel and forces drainage from the Upper Cache River to flow directly to the Ohio River through the Post Creek Cutoff; drainage from the Lower Cache basin flows to the west to empty into the Mississippi River. Two 4-foot culverts in this levee allow some water from the Lower Cache River basin to discharge into the Post Creek Cutoff.

Two different approaches were integrated to gain a detailed understanding of the sedimentation problem in the Lower Cache River. The two approaches are the hydrologic budget and geochemical analysis. Neither approach by itself was capable of providing all the information needed to quantify the sedimentation pattern in the complex ecosystem found in the Lower Cache River. Very often, the geochemical approach relies on a limited number of sediment cores (usually only one core) to determine the sedimentation rate; the results are then extrapolated over unreasonably large areas, often incorrectly assuming a uniform sedimentation rate. On the other hand, the hydrologic approach is designed to determine average sedimentation rates based on sediment budgets for large areas; those values are then applied uniformly to all the different units within the ecosystem. Application of the two approaches independently could result in contradicting concepts and conclusions as discussed by Christophersen and Neal (1990). This study demonstrates how the two approaches should be used to complement each other and provide a more complete picture of the sedimentation process in a complex environment. More detailed discussion on the Cache River basin hydrology and sedimentation studies is presented in several reports from the Illinois State Water Survey (Demissie et al., 1990; 1992;).

SEDIMENTATION RATE BASED ON SEDIMENT BUDGET

Sediment monitoring stations were located at strategic sites to provide the best data on which to construct the sediment budget for the study area. The sediment monitoring stations in the entire Cache River basin, in both the upper and lower subwatersheds, are shown in figure 1. In the Lower Cache River, two stations were located on the two major tributaries draining into the Buttonland Swamp area to monitor the inflow of sediment, and the third was located on the main stem of the Lower Cache River at Ullin to monitor sediment outflow from the Buttonland Swamp area. This arrangement made it possible to estimate the amount of sediment entering and leaving this wetland complex, and thus develop a sediment budget. The sediment budget for the area was determined after calculating the sediment yields at the monitoring stations and extrapolating the results to the ungaged area.

Sediment yield data collected at the three Lower Cache sediment monitoring stations were used in developing a sediment budget for the Buttonland Swamp area to determine the amount of sediment accumulated. Sediment yield values calculated for Big Creek and Cypress Creek were used for those watersheds not monitored to generate the total sediment inflow into the Buttonland Swamp area. The sediment outflow from the area was calculated from outflow measured in the Cache River at Ullin and was adjusted for outflow through the culverts at the Cache River levee on the east end of the Buttonland Swamp area. The sediment budget for the Lower Cache River wetland area is then determined by computing the total sediment inflow from the tributaries and subtracting the outflows from the wetland area.

The results of the sediment budget calculations are given in table 1. Calculations were performed for three water years where data were collected at all stations. For Water Year 1986, sediment inflow into the Buttonland Swamp area was calculated to be 157,300 tons, but only 25,700 tons left the area. Therefore, 131,600 tons of sediment were trapped in the area, indicating an 84 percent trap efficiency. In Water Year 1987, a dry year, sediment inflow into the area was only 40,000 tons and the outflow was 8,700 tons, indicating 78 percent trap efficiency. In Water Year 1988, another dry year, sediment inflow was 49,400 tons and outflow was 15,300 tons, indicating 69 percent trap efficiency. Therefore, the results from the three years of data collection indicate that 69 to 84 percent of the total amount of sediment that enters the Buttonland Swamp area is trapped in the area. Over the three-year monitoring period, 80 percent of the inflowing sediment is trapped. These very high sediment trapping efficiencies, similar to those for man-made reservoirs, account for the high sedimentation rate within the Buttonland Swamp area.

Table 1. Sediment Budget and Sedimentation Rates for the Cache River Wetlands

	Water Year			
	1986	*1987*	*1988*	*Average*
Sediment inflow from tributaries (tons × 10^3)	157.3	40.0	49.4	82.2
Sediment outflow at Cache River at Ullin (tons × 10^3)	23.3	7.9	14.0	15.1
Adjusted sediment outflow from the Lower Cache River (tons × 10^3)	25.7	8.7	15.3	16.6
Sediment trapped in the Buttonland Swamp area (tons × 10^3)	131.6	31.3	34.1	65.7
Trap efficiency (percent)	84	78	69	80
Sedimentation rate (cm)	1.3	0.33	0.33	0.66

The sediment thickness or the rate of vertical deposition is calculated by dividing the sediment volume by the area of sediment deposition. The calculations were performed by assuming an average sediment density of 800 kg/m^3 and the sediment depositional area includes all stream channels, backwaters, sloughs, and wetlands adjacent to the river. These areas are generally flooded every year, and most of them support some form of wetland vegetation. The most probable depositional area is expected to be below 330 ft msl and has an area of 1,200 hectares. The sedimentation rate based on the most probable depositional area ranges from a low of 0.33 cm in 1987 and 1988 to a high of 1.32 cm in 1986. The rates are equivalent to an average of 0.66 cm over the three-year period. One major factor that should be noted is that 1987 and 1988 were dry years and only 1986 was a near-normal year.

SEDIMENTATION RATE BASED ON RADIOMETRIC ANALYSIS

The sedimentation rate calculations based on sediment load monitoring at tributary streams inflowing into an area and at the outlets provide the overall average sedimentation rate for the entire area for the period of data collection. However, it is well known that the sedimentation rate will vary both in time and space. Certain areas such as backwaters and side channels experience more sedimentation than areas such as floodplain fringes and constricted stream channels. Therefore, to obtain a more detailed resolution of the temporal and spatial variation of the sedimentation rate, sediment core samples were collected at selected areas for analysis using a radiometric dating technique.

Cesium-137 (^{137}Cs) is among the radioisotopes distributed globally as a result of atmospheric testing of nuclear weapons in the atmosphere. The initial input of ^{137}Cs in measurable quantities represents the period from 1952-1954. The peak period of fallout was 1963. By measuring the activity of ^{137}Cs in discrete intervals of sediment, the pattern observed from atmospheric testing is often replicated in a sediment column. The onset of measurable activity in the core and peak activity are assumed to be 1954 and 1963, respectively.

Sediment cores were collected from the channel and floodplain deposits of the Cache River in the vicinity of Buttonland Swamp. Nine core samples were collected in July 1988 and one core sample in September 1986. The sedimentation rates determined based on ^{137}Cs analysis at the ten locations where core samples were collected are summarized in table 2. The annual sedimentation rate varies from a low of 0.3 cm at site A to a high of 2.8 cm at site F. In general, the side channels to the Cache River had the highest rate of sedimentation followed by backwater sloughs, ponds, and wetland meadows. The samples collected from the floodplain showed the lowest rate of sedimentation, and the sites directly connected to the Cache River channel have the highest sedimentation rates.

Table 2. Sedimentation Rates Based on ^{137}Cs Analysis

Core	*Location type*	*Sedimentaion rate (cm/yr)*	*Core*	*Location type*	*Sedimentation rate (cm/yr)*
A	Floodplain	0.3	F	Side Channel	2.8
B	Floodplain	0.3	G	Backwater Slough	0.5
C	Pond	1.4	H	Backwater Slough	1.6
D	Pond	1.4	I	Wetland Meadow	1.6
E	Side Channel	1.1	J	Wetland Meadow	0.5

CONCLUSIONS

Sedimentation rates determined by using the two techniques (sediment budget and radiometric dating) were found to be consistent with each other. The radiometric technique provides site-specific information and the spatial variation of the sedimentation rate, while the sediment budget technique provides the total amount of sediment being trapped in the area and the average rate of sediment accumulation over an assumed area of sediment deposition. Even though sedimentation rates vary widely from one location to another, the radiometric technique provides quick, reliable sedimentation rate values if an adequate number of samples are collected and analyzed with the proper interpretation of the results as they relate to the physical settings of the areas of sediment deposition.

REFERENCES

Christophersen, N., and C. Neal. 1990. Linking Hydrological, Geochemical, and Soil Chemical Processes on the Catchment Scale: An Interplay between Modeling and Field Work. *Water Resources Research* 26(12):3077-3086.

Demissie, M., T.W. Soong, R.L. Allgire, L.L. Keefer, and P.B. Makowski. 1990. *Cache River Basin: Hydrology, Hydraulics, and Sediment Transport, Vol. 1: Background, Data Collection and Analysis*. Illinois State Water Survey Contract Report 484.

Demissie, M., W.P. Fitzpatrick, and R.A. Cahill. 1992. *Sedimentation in the Cache River Wetlands: Comparison of Two Methods*. Illinois State Water Survey Miscellaneous Publication 129, Champaign, IL.

Constructed Wetlands for Sediment and Non-point Source Pollution Control at US Army Corps of Engineers Projects

Charles W. Downer[1] Member and Tommy E. Myers[2]

Abstract

Two wetlands, recently constructed at United States Army Corps of Engineers (USACE) reservoirs, were chosen as demonstration sites for sediment control and non-point source pollution abatement, and monitored for their ability to remove suspended sediments, nutrients and herbicides from storm runoff. Intensive sampling of selected storm events was conducted during the period from spring of 1992 to the fall of 1993. Monitoring efforts indicate that while the wetlands can remove suspended particulate matter from storm flows, removal of dissolved constituents is less effective. The short detention times of the wetlands is thought to be a primary cause for the lack of treatment effectiveness.

Introduction

The intended uses of some USACE reservoirs are threatened by declining water quality due to non-point source pollution (NPSP). Baker (1992) points out in a recent study that three-forths of lake water quality impairment is due to NPSP. While reduction of NPSP is probably best accomplished by source control, the USACE has limited ability to affect off-site land usage and must look for ways to control NPSP on its' own property. Conventional water treatment methods don't work well for NPSP because flows and constituent loadings vary widely. However, wetlands have the ability to spread and detain

---

[1]Research Hydraulic Engineer, Hydraulics Laboratory, US Army Engineer Waterways Experiment Station, 3909 Halls Ferry Rd., Vicksburg, MS 39180.

[2]Environmental Engineer, Environmental Laboratory, US Army Engineer Waterways Experiment Station, 3909 Halls Ferry Rd., Vicksburg, MS 39180.

flows from sporadic high intensity storm events and provide a variety of potential treatment mechanisms to treat NPSP. Wetlands therefore have the potential to be used for treatment of NPSP at USACE projects where large tracts of land are typically available around and within the reservoir flood pool.

Study Sites

Spring Creek Wetland, Bowman-Haley Reservoir

Bowman-Haley Reservoir is a 708 hectare flood control and water supply reservoir located along the North Dakota/South Dakota border, near Bowman, North Dakota. The initial good water quality of the reservoir has declined due to non-point sources of suspended sediments and excess nutrients. The resulting turbidity, excess algae growth and low winter dissolved oxygen concentrations are well documented (USACE 1984).

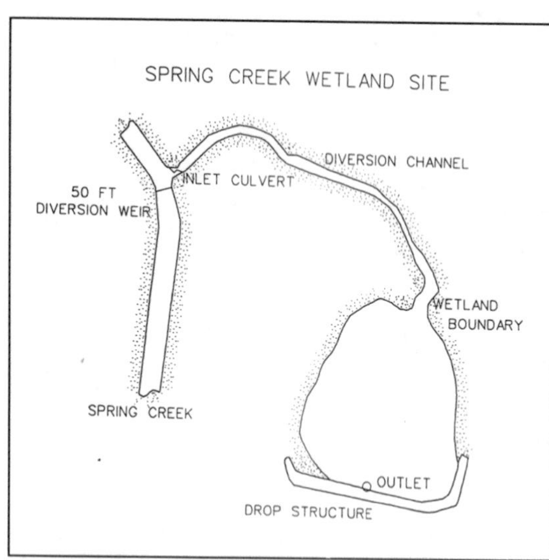

Figure 1. Spring Creek Wetland Site

In an effort to improve water quality in the reservoir the Soil Conservation Service (SCS) initiated a plan to reduce NPSP loadings to the reservoir, primarily by source reduction through good agricultural land management. In addition to source reduction and other efforts, a 9.31 hectare emergent marsh was constructed in 1991 on Spring Creek, one of three major tributaries into Bowman-Haley Reservoir. The marsh is connected to the creek by a diversion canal. Flows into the wetland are controlled by a gated inlet culvert. Flows out of the wetland are controlled by a drop structure. The water level in Spring Creek is controlled by a diversion weir located just downstream of the wetland inlet. Stop logs, located in the center of the diversion weir, allow the water level in the creek to be drawn down to drain the wetland at the end of each growing season. Flows from Spring Creek are the only source of water to the wetland except direct precipitation. Though constructed on USACE lands, Ducks Unlimited provided funding and engineering services for the project; and the wetlands are operated to maximize waterfowl use by capturing spring runoff and detaining it until late summer, when the wetland is drained.

Range Creek Wetland 1, Ray Roberts Reservoir

Ray Roberts Lake is a 11,880 hectare flood control and water supply reservoir located in Denton County, Texas about 50 km north of Dallas. Ray Roberts lake was constructed in 1987 in a largely agricultural watershed. Five wetlands were constructed in 1991 along Range Creek, a major tributary into the reservoir. Wetland 1, a 2 hectare wetland located nearest the reservoir, from this complex was chosen for the study. Flows to this wetland come from an upstream wetland/reservoir used primarily to supply water to the downstream wetland. Inflows from the upstream wetland and outlflows are controlled by gated culverts and emergency spillways. The wetland is designed so that at high creek levels, flood flows spill in over the the dikes. Because the wetland is located within the flood pool of the reservoir, it can and does, become inudated by the reservoir at high stages. The principle purpose for the construction of these wetlands was for wildlife habitat, and Wetland 1 is operated for moist soil management where the wetland is filled in the spring, drawn down in the summer to allow the growth of moist soil

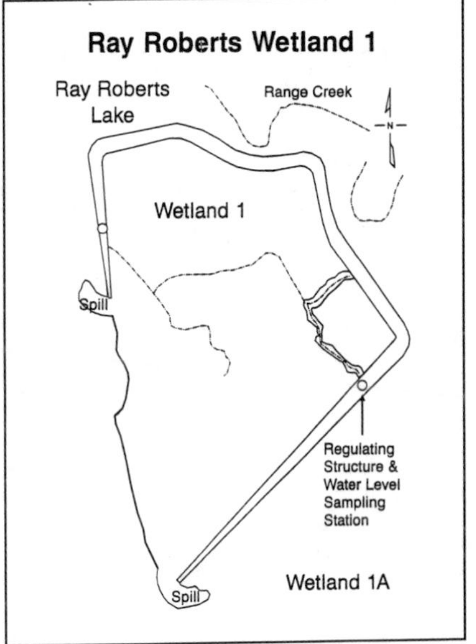

Figure 2. Range Creek Wetland 1

vegetation, and then flooded in the fall to provide a food source for overstopping and wintering waterfowl.

Monitoring

Monitoring of the sites began in the spring of 1992 and continued until fall 1993. Two years of data were collected at the Bowman Haley site. Because of flooding that occured in the spring of 1992, data from the Ray Roberts site were collected in 1993 only. Automated water level and sampling stations were located at the inlet and outlet of each wetland. Sampling efforts were aimed at capturing large storm events.

Because the wetlands were located in primarily agricultural watersheds, analysis of water samples focused on non-point source pollutants generally associated with agricultural runoff: suspended sediments, nutrients and herbicides. Sediment accretion and mass accumulation were measured at several sites within each of the wetlands.

Results and Analysis

Bowman-Haley Reservoir

Because wetland inflows and outflows were controlled, mass treatment efficiencies for total suspended sediments (TSS) and nutrients could be computed. Removal of TSS was good, 82% for 1992 and 61% in 1993. Removal of total phosphorus (TP) was also good, 37% for 1992 and 33% for 1993. Removal of nitrogen as total Kjeldahl nitrogen (TKN) or nitrite plus nitrate (NO_2+NO_3) was poor. Results for nitrogen sampling may have been affected by an application of fertilizer on the inlet canal side slopes. The concentration of herbicides was too low to calculate mass balances.

The affect of the wetland on Spring Creek varied according to creek flows and wetland operation. In 1992, when creek flows were low and the wetland was dry going into the summer sampling period, the wetland was capable of treating 26 percent of the total creek flow. In 1993, when creek flows were higher and the wetland was filled by early spring snow melt, the wetland was capable of treating only 2 percent of the total creek flow. A dye study indicated that at normal flows, the hydraulic retention time (HRT) of the wetland is about 5 days.

Sediment sampling indicated that the wetlands retained 9.9 mm of sediments over the two year study, with 3 mm, 1.31 kg/m^2, of sediments retained in 1993.

Sediments were largely inorganic, 7.4% combustible material. A large portion of this sediment is thought to originate from the erosion of the inlet canal and dikes.

Ray Roberts

Three major storm events were sampled at the Range Creek Wetland 1A during the spring of 1993. All three storm events caused the dikes of the wetland to be overtopped so that flows and mass treatment efficiencies could not be calculated. Qualitatively, the wetland had limited effect on removal of TSS, nutrients and atrazine. Because of hydraulic overloading, the wetland detention time was inadequate for removal of constituents during the monitored events. A dye study indicated that the HRT of the wetland was about 5 hours when flow was confined to the inlet and outlet culverts. The actual HRT during the storm events is unknown, but probably shorter.

Sediment sampling indicated that the wetlands accumulated 2.2 mm, 1.37 kg/m^2, of sediments over the four month sampling period. Yearly sediment accumulation of sediments is probably about twice this amount. The percent organic material was nearly 19%.

Conclusion

The ability of the wetlands to remove sediments and nutrients from inflows to the wetlands appears to be closely tied to the hydrologic design and operation of the wetlands. At Bowman-Haley, the construction and operation of the wetland for waterfowl nesting habitat greatly decreases the amount of water that the wetland can potentially treat. At Ray Roberts, the treatment capacity of the wetlands is hindered by the hydraulic overloading of the system. In order to effectively use wetlands for treatment of NPSP, the wetlands should be designed and operated specifically for this purpose. Obtaining a sufficient hydraulic retention time for the removal of NPSP may require large areas of land to be devoted to wetland treatment.

References

Baker, L. A. 1992. "Introduction to Non-point source Pollution in the United States and Prospects for Wetland Use," Ecological Engineering, Vol 1, Nos. ½, pp 1-26.

USACE 1994. Reservoir Water Quality Data Report, Missouri River Basin, Bowman-Haley Reservoir, North Dakota. USACE Omaha District, Omaha, NE.

Modeling of Sedimentation Processes in a Bottomland Hardwood Wetland

Brad R. Hall[1]
John Engel[2]

The US Army Engineer Waterways Experiment Station has monitored several physical and biological parameters on the Rex Hancock Swamp on the Cache River, Arkansas for several years for ecosystem modeling purposes. As part of these efforts, measurements of suspended sediment grain size, concentration, and deposition quantities within the wetland system were obtained. Daily suspended sediment loads on the Cache River were also sampled at the upstream and downstream limits of the wetland system to identify boundary conditions and a overall sediment budget for the wetland. These data were used for development and testing of a TABS-MD numerical model of the wetland system. A portion of the field sampling and numerical modeling results are reported in this paper.

Field Measurement Efforts

The B-transect, which extends from the Cache River bankline to the adjacent upland forest habitat, was previously established in the study area for ecological studies. Water surface elevation, water temperature, and suspended sediment concentrations were measured at three locations on the B-transect from January through June 1992. This time period encompasses the normal high water season for the Cache River bottomland hardwood wetland. Water samples for suspended sediment concentration were obtained with ISCO automatic water samples. Water surface elevation was measured with Lundahl ultrasonic

---

1. Research Hydraulic Engineer, Hydraulics Laboratory, US Army Engineer Waterways Experiment Station, Vicksburg, MS 39180

2. Graduate Student, Department of Civil Engineering, University of Nebraska, Lincoln, NE 68501

distance recorders. These instruments were found to be very reliable for this application. The instruments and field sampling results are described in greater detail in Abraham and Hall, 1994. Suspended sediment concentration was also measured on a daily basis on the Cache River at the Patterson, AR gage by the United States Geological Survey using standard depth integrated sampling techniques. Laboratory analysis for suspended sediment concentration was performed by both the U.S. Geological Survey and the U.S. Army Engineer Waterways Experiment Station.

Numerical Modeling Approach

The TABS-MD hydrodynamic and sediment transport model (Thomas and McAnally, 1991) was used for simulating overbank sediment concentration and deposition. The model was chosen for the ability to simulate wetting and drying of the overbank areas, as well as model capabilities for sediment transport calculations.

A finite element mesh was developed to simulate the transient hydrodynamic and sediment transport within the Cache River study area. The mesh extends from the Cache River at the Patterson, AR gage at the upstream limit to the Cache River at the Cotton Plant, AR gage at the downstream limit. The mesh covers the entire floodplain of the study area, and is shown in Figure 1. The location of the B-transect within the grid is also indicated on this figure. The mesh contains a total of 3041 elements, and a total of 7790 corner and mid-side nodes.

A TABS hydrodynamic simulation of the for the entire 1992 calendar year was developed to assess inundation hydroperiod of the wetland system. Hydraulic roughness coefficients (Manning's n values) of 0.040 for the channel elements and 0.500 for the overbank vegetation elements were used. These Manning's n values are in general agreement with the hydraulic roughness values for dense vegetation proposed by Hall and Freeman, 1994.

The primary sediment modeling parameters are the particle fall velocity, shear stress criteria for erosion and deposition, as well as turbulent diffusion coefficients used in the conservation of mass equations for sediment. Boundary conditions for the sediment simulations consist of the inflowing sediment load recorded at the Cache River at the Patterson, AR stream gage.

Numerical Modeling Results

The TABS model was used to simulate the distribution of suspended sediment on the floodplain during the winter-spring 1992 high water season. The period was characterized by several flooding and drying sequences over the simulation period. Sedimentation model adjustment parameters include erosive and depositional shear stress criteria, particle fall velocity, and turbulent diffusion coefficients. All parameters were set at physically realistic values determined from the field measurements. Numerical model results visualized using the FAST-TABS model interface (Lin, et. al., 1992) reveal an interesting downvalley succession of depositional splay deposits controlled by localized overbank flooding. This observation indicates that overbank deposition is hydraulically controlled by either local channel capacity under conditions of limited backwater effects and by high backwater conditions from downstream stage control.

An example of computed overbank deposition quantities for the high water season of the 1992 water year along the B-transect is given on Figure 2. The quantities are reasonable in comparison with earlier maximum overbank deposition depths of 2.5 cm reported by Kleiss, 1993. Computed suspended sediment concentrations indicate a reduction in concentration with increasing distance from the river, which is consistent with a depositional regime and in agreement with the floodplain deposition and development theory described and quantified by Pizzuto, 1986.

Acknowledgments

The tests described and the resulting data presented herein were obtained from research conducted under the Wetlands Research Program of the US Army Corps of Engineers by the US Army Engineer Waterways Experiment Station. Permission was granted by the Chief of Engineers to publish this information.

Appendix I - Conversion Factors, Units of Measurement

To Convert	To	Multiply By
Foot	Meter	0.3048
Cubic foot	Cubic meter	0.02832
Mile (U.S. statute)	Kilometer	1.609
Pound (mass)	Kilogram	0.4536

References

Abraham, D. D. and B. R. Hall, 1994. Ultrasonic sensors viable option for water surface measurements, Wetlands Research Program Bulletin, U.S. Army Engineer Waterways Experiment Station, Volume 4, Number 2, Vicksburg, MS.

Hall, B. R. and G. F. Freeman, 1994. Study of hydraulic roughness in wetland vegetation takes new look at Manning's n, Wetlands Research Program Bulletin, U.S. Army Engineer Waterways Experiment Station, Volume 4, Number 1, Vicksburg, MS.

Kleiss, B. A. 1993. Cache River Arkansas: Studying a bottomland hardwood wetland ecosystem. Wetlands Research Program Bulletin, U.S. Army Engineer Waterways Experiment Station, Volume 3, Number 1, Vicksburg, MS.

Pizzuto, J.E., 1990. Sediment diffusion during overbank flows. Sedimentology 34: 301-317.

Lin, H. C., N. L. Jones, and D. R. Richards, 1992. Real Time simulation and visualization of 2-D surface Water Flow, Proceedings of the ASCE Hydraulic Engineering Sessions at Water Forum '92, Baltimore MD. 335-340.

Thomas, W. A. and W. H. McAnally, Jr., 1991. Users manual for the generalized computer program system: open-channel flow and sedimentation, TABS-2" U.S. Army Engineer Waterways Experiment Station, Vicksburg, MS.

Patterson, AR --->
stream gage

N
O
R ↑
T
H

Mesh Scale
1:100000

B Transect ------>

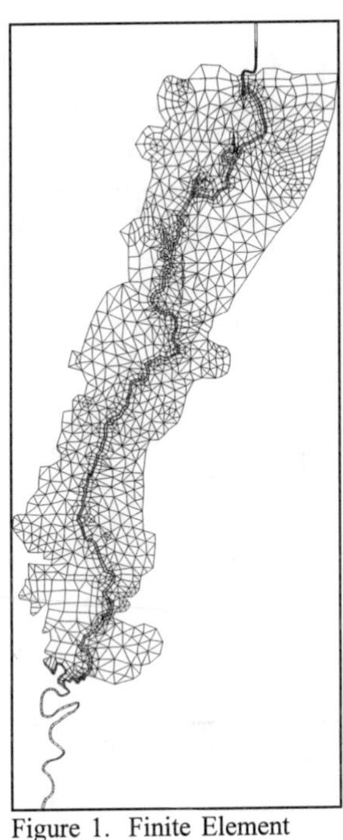

Cotton Plant, --->
AR stream gage

Figure 1. Finite Element Mesh of the Cache River Study Area

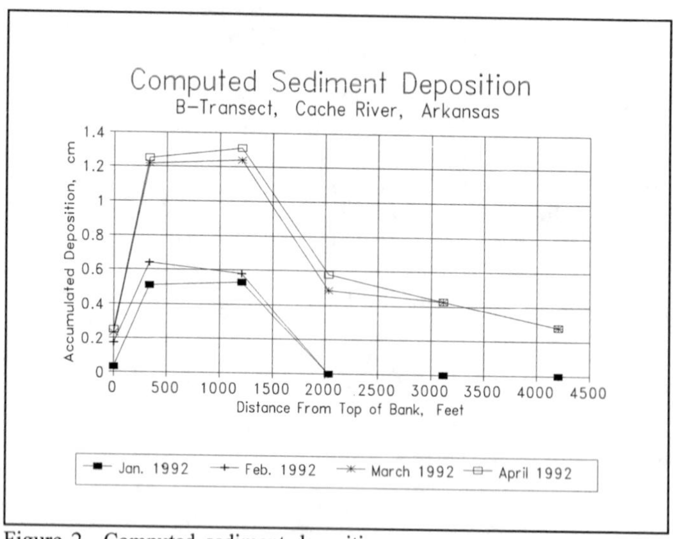

Figure 2. Computed sediment deposition

TURBULENT OPEN-CHANNEL FLOW THROUGH SIMULATED VEGETATION

Fabián López[1] , Chad Dunn[1], and Marcelo García[2], A.M. ASCE

Abstract

The present work concerns experiments conducted under uniform flow conditions in a laboratory flume. Acoustic-Doppler Velocimetry and Hot-Film Anemometry are combined to measure the turbulence structure within elements and in the free surface reagion of the flow at several locations within the grid. Mean values and standard deviations for the three velocity components are computed, together with vertical, longitudinal and spanwise profiles of turbulent kinetic energy, Reynolds' stresses, power spectra etc. Conditional-quadrant technique computations show the structure of sweeps and ejections at different locations in the flow field. Analysis of the results obtained provide valuable information concerning the overall dynamics of these kind of flows. The existence of a short-circuited cascade process is validated.

1. Introduction

In the last few years, there has been an increasing need for the understanding of transport processes in plant environments, specifically for the mechanisms governing momentum, heat, and mass exchange. Clarification of the dynamics of such processes is essential for a great variety of applications in biology, hydraulics, hydrology, agriculture, etc. These strong and diverse motivations have led to several laboratory and field experiments mainly for the case of air flow above plant canopies.

In atmospheric boundary layers over plant canopies two distinct regions may be distinguished, namely a surface layer above the vegetation and a region close to and within the canopy, where the scaling laws of the former do not apply. In the first region a logarithmic velocity profile seems to apply downward to a height $(d+z_o)$, where z_o represents the roughness length and d the so called displacement height (Kaimal and Finnigan, 1994). From the ground to a distance of about three times the canopy height however, the character of the turbulence is directly affected by the presence of the vegetation (Raupach, 1981). This latter

[1] Research Assistant, [2] Assistant Professor. Department of Civil Engineering, University of Illinois at Urbana Champaign. 205 N. Mathews, Urbana, IL 61801. USA.

region is commonly referred to as "roughness sublayer". Measurements in this second region pose difficulties related to very high turbulence intensities and the need for some kind of spatial average. Summarizing, one may affirm that the essential differences between turbulence above and within the canopy result from the multiple sources and sinks of momentum and scalars, spread throughout the canopy.

Although some similarities of the above described canopy flow may be applied to vegetated open-channels, the existence of a free surface constitutes an extra constraint for the development of an equilibrium layer, i.e for the existence of a logarithmic velocity profile above non-emergent plants. Moreover, the role played by turbulent coherent structures needs further investigations.

The present work aims at helping characterize the structure of turbulence and the associated coherent structures by means of three-dimensional measurements conducted in a laboratory channel with simulated vegetation, using recently available Acoustic Doppler Velocimetry combined with traditional Hot-Film Anemometry. It is believed that a comprehensive description and understanding of the flow structure in this particular type of flow is essential for any attempt to study the related transport processes.

2. *Experiments*

The experiments were carried out in a tilting flume, 12.2 m. long and 0.9 m. wide. The test section was located about 6 m. downstream from the entrance. Discharge measurements were conducted using two parallel weighting tanks of 20,000 pounds capacity each. Velocity measurements were taken by means of an Acoustic Doppler Velocimeter (ADV), of 25 Hz maximum sampling rate, which allowed the acquisition of the three velocity components, and a traditional one-dimensional Hot-Film Sensor (HFS).

The vegetation was simulated using flexible elements consisting of commercially available plastic straws, with a circular cross-section of one quarter inch in diameter, and lengths varying from 7.5 to 8.5 inches. The "plants" were placed on a board lying at the bottom of the channel in a staggered arrangement with 3.00 inches grid-space.

Two set of experiments were conducted (Veg1 and Veg2), where vertical velocity profiles were measured, both using ADV and HFS, at three different locations in a cross-section: behind a plant, 0.75 inches to the right and 2.25 inches to the left, looking in the direction of the mean flow. These three locations will be designated as P1, P2 and P3, respectively. The longitudinal slope was about 0.009 and the flow depth 0.2 m.

3. *Results*

Figure 1 shows the measured mean velocity profile (U) and streamwise turbulent intensities (u_{rms}), for locations P1, P2 and P3. The horizontal dashed line represents the average location of the top of the plants. A highly sheared velocity profile is observed atop of the plants, where turbulent intensities attain their maximum values. The measured correlation coefficient for Reynolds stresses, R, shows an almost constant value of -0.30 within the plants irrespective of lateral location. Since this coefficient represents some measure of the efficiency of turbulence in transferring momentum relative to the absolute amount of energy available, the almost constant value of R within plants suggests that the ability of downward momentum transfer decreases at the same rate as the energy of the flow fluctuations.

Of particular interest is how turbulence spectra measured within the vegetation compares to typical shapes in open channel flows. Figure 2 illustrates streamwise, vertical and spanwise spectra of velocity fluctuations for location P1, together with their corresponding counterparts for smooth-wall flows. Some important aspects are worth highlightening: (1) the similar shape of the streamwise spectra in both cases, which suggests that the transfer of energy down the cascade due to vortex stretching is the principal mechanism; (2) the higher and almost constant level of energy in the vertical fluctuations compared to the smooth-wall case, which is herein attributed to the relative importance of direct feeding due to wake and waving production and to the "spectral short-cut" (Kaimal and Finnigan, 1994); and (3) the peak existing in the spanwise spectra, associated with the observed lateral waving of the plants.

Coherent Structures

A remarkable different turbulence structure was found for the flow behind the plants and between them. Figure 3 shows fractional contributions to the Reynolds stress from each of the quadrants, and the fraction of the total time in which each event is present. Event 5 represents the contribution from a hyperbolic hole of size H (Nezu and Nakagawa, 1977). As can be observed, behind the plants an almost symmetrical structure predominates irrespectively of hole size H, with highly intermittent events, each contributing about 20% (for sweeps this value is closer to 60%) at values of H as high as 20. Regarding the fractional time corresponding to each quadrant, a symmetrical behavior is also observed, with all events being present the same percentage of the total time independent of the value of H. On the other hand, between plants the structure of turbulence resemblances the typical one corresponding to flow close to roughness elements (Grass, 1971; Raupach, 1981, García et al., 1995), with the predominant contribution made by sweep events. Here a highly asymmetric structure is observed, in the sense that for $H \geq 9.00$ only sweeps contribute to the Reynolds

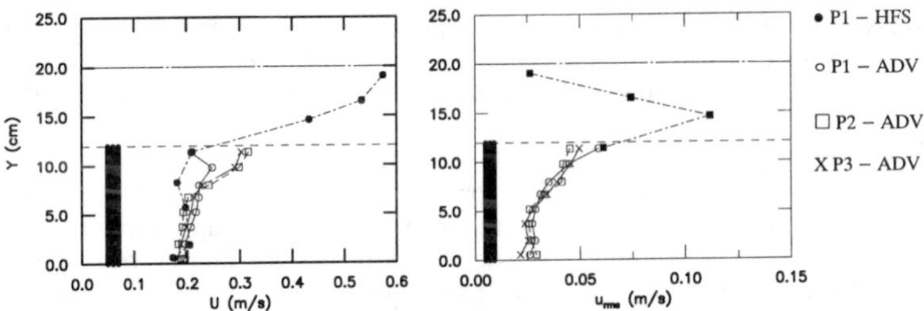

Figure 1. Mean Velocity and Streamwise Intensity Profiles

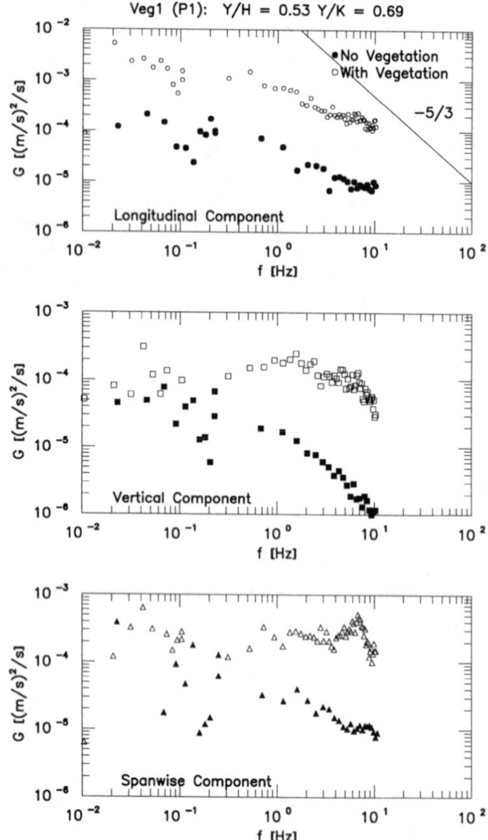

Figure 2. Spectra of Each Velocity Component at Position P1

stresses. Regarding the fractional time, the asymmetry is also evident, with sweeps and ejections each being present about 35% of the total time, whereas outward and inward interactions contribute to the Reynolds stresses only 18% of the time.

4. Conclusions

Experimental results show the existence of a highly sheared mean velocity profile at the top of the roughness elements, coinciding with turbulence intensity peaks. The turbulent kinetic energy flux points downwards, with the energy decreasing at almost the same rate as the ability of downward momentum transfer. A highly three-dimensional structure was found for the flow within the plants, with short periods of high-intensity sweeps dominating throughout the flow

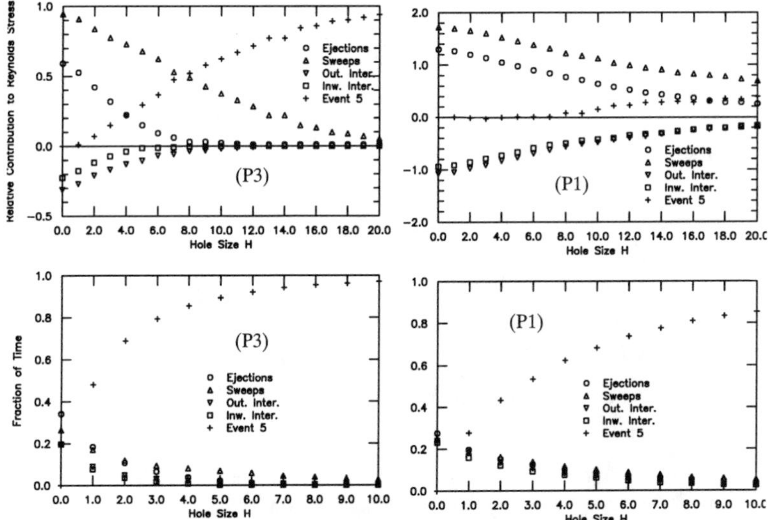

Figure 3. Fractional Contribution to Reynolds Stress and Fraction of Time corresponding to each Quadrant ($y/h = 0.39$).

field. A trend to a symmetric distribution of the Reynolds stress contribution from the quadrant-events was observed to prevail just behind the elements. Further investigations are needed in order to extend present results and relate them to the associated transport processes.

5. Acknowledgements

The support of the US Army Corps of Engineers, Waterways Experiment Station (DACW39-94-K-0010) is gratefully acknowledged.

6. References

LOPEZ F., DUNN C and GARCIA M.H., 1995, *Turbulence Characteristics of Flow over Cobble-Bed* ASCE Hydr. Eng. Conference. San Antonio, Texas.

GRASS A.J. , 1971, *Structural features of turbulent flow over smooth and rough boundaries.* J. Fluid Mech. **50**, pp 233–255.

KAIMAL J.C. and FINNIGAN J.J., 1994, *Atmospheric Boundary Layer Flows.* Oxford University Press.

NAKAGAWA H. and NEZU I., 1977, *Prediction of the Contributions to the Reynolds Stress from Bursting Events in Open-Channel Flows.* JFM Vol. **80**, part 1, pp. 99-128.

RAUPACH M.R., 1981, *Conditional Statistics of Reynolds Stress in Rough-Wall and Smooth-Wall Turbulent Boundary Layers.* JFM, Vol. **108**, pp. 363-382.

SIMULATION OF SUSPENDED SEDIMENT TRANSPORT IN VEGETATED OPEN CHANNEL FLOWS WITH A k-ε TURBULENCE MODEL

Fabián López[1] and Marcelo García[2], A.M. ASCE

Abstract

A k-ε turbulence model is employed, coupled with special boundary conditions for dissipation at the free surface, to numerically simulate uniform open-channel flows through vegetation. For the calibration of some parameters involved in the model, predicted vertical profiles of spatially averaged mean velocities, standard deviations of the velocity components, and eddy viscosities are compared with experimental observations. The computed velocity and turbulence fields are then used to solve the vertical sediment diffusion equation and hence compute the equilibrium vertical distribution of suspended sediment. Numerical integration of suspended sediment and velocity profiles allows for the estimation of sediment transport rates, and thus sediment transport capacities under different flow conditions and vegetation characteristics. Numerical results are compared against computations for non-vegetated channels.

1. Introduction

Historically the engineering practice of vegetated open channels has been limited to the determination of flow resistance factors, usually in the form of Manning's *n* (Chow, 1959). In the last few years the increasing need for understanding transport processes in plant environments, specifically for the mechanisms governing momentum, heat and mass exchange, has motivated the realization of various laboratory and field measurements as well as the development of mathematical models for the numerical description and study of such processes, mainly in atmospheric canopy flows (Raupach and Thom, 1981). Several attempts have been made to estimate vertical profiles of mean velocities within vegetation using simple turbulence closure schemes (e.g. Plate and Quraishi, 1965), while other works used two-equations (Tsujimoto et al., 1991) and even higher order closure schemes (Wilson and Shaw, 1977). Despite some objections against flux-gradient approaches in canopy boundary layers (Finnigan,

[1] Research Assistant, [2] Assistant Professor. Department of Civil Engineering, University of Illinois at Urbana Champaign. 205 N. Mathews, Urbana, IL 61801. USA.

1985), first order closures using the k-ε model have shown to reproduce reasonably well the available experimental observations of turbulence descriptors in vegetated open channels (Tsujimoto et al., 1991).

The relative importance of wetlands, compared to other landscape components, in trapping sediments and toxicants (heavy metals, pesticides, etc.) clearly shows the engineering need for the understanding and modeling of transport processes in natural environments. However, to date practically no physically based model exists in hydraulic engineering for the evaluation of sediment transport and retention capabilities of vegetated waterways.

The present work reports numerical results of a two-equation turbulence closure model employed to compute the structure of steady, uniform flows in vegetated open channels and to estimate suspended sediment transport capacities.

2. Numerical Model

One-dimensional steady, uniform flow over vegetated bed was simulated using a modified form of the standard k-ε model in order to account for the presence of vegetation. The existence of vegetation creates a multi-connected, highly three-dimensional flow region, and thus a spatial averaging procedure is required in order to adequately model the problem (Raupach and Shaw, 1982). The following set of equations is obtained:

Continuity eqn.

$$\frac{\partial U}{\partial x} = 0 \tag{1}$$

x-momentum eqn.

$$0 = g\varrho S_o + \frac{\partial}{\partial z}\left[(\nu_t + \nu)\frac{\partial(\varrho U)}{\partial z}\right] - \varrho f_x \tag{2}$$

Turbulent kinetic energy eqn.

$$0 = \frac{\partial}{\partial z}\left[(\frac{\nu_t}{\sigma_k} + \nu)\frac{\partial k}{\partial z}\right] + P_k - \varepsilon + C_{fk} f_x U \tag{3}$$

Dissipation eqn.

$$0 = \frac{\partial}{\partial z}\left[(\frac{\nu_t}{\sigma_\varepsilon} + \nu)\frac{\partial \varepsilon}{\partial z}\right] + \frac{\varepsilon}{k}\left[C_1 P_k + C_{f\varepsilon} f_x U - C_2 \varepsilon\right] \tag{4}$$

where

$$P_k = \nu_t\, 2\, (\frac{\partial U}{\partial z})^2 \qquad f_x = 0.50 C_D a U \sqrt{U^2} \qquad \nu_t = \frac{C_\mu k^2}{\varepsilon} \tag{5a,b,c}$$

and (x,z) represent the streamwise and vertical coordinates, respectively; U is the turbulence-averaged streamwise velocity component averaged also over space; ν is the kinematic viscosity of water; ν_t is the kinematic eddy viscosity; f_x represents the drag force exerted by the vegetation per unit water volume which models the spatial average of gradients of deviations from the spatial mean pressure; C_D is the drag coefficient; a represents the vegetation density (front area per unit volume); k is the turbulent kinetic energy averaged over space; ε is the turbulent dissipation rate averaged over space; P_k stands for production of turbulent energy; S_o

represents the longitudinal slope; ϱ is the water density and g the gravitational acceleration. According to the standard form of the k-ε model: $C_\mu = 0.09$, $C_1 = 1.44$, $C_2 = 1.92$, $\sigma_k = 1.0$ and $\sigma_\varepsilon = 1.3$. The two additional parameters C_{fk} and $C_{f\varepsilon}$ were estimated from a best fit to experimental results. The above set of equations was solved using an equation solver developed by Svensson (1986). Additional assumptions involve the use of the log-law as a bridge function for streamwise momentum near the bottom, and the damping of the turbulence near the free surface according to Celik and Rodi (1984).

Once the turbulence characteristics were computed, the results were employed in order to solve the equation for the vertical diffusion of suspended sediment under equilibrium conditions:

$$0 = \frac{\partial}{\partial z}\left[\frac{\nu_t}{\sigma_c}\frac{\partial C}{\partial z}\right] - w_s\, C \tag{6}$$

where C is the mean sediment concentration averaged over turbulence and space; σ_c the Schmidt number and w_s the terminal fall velocity of the sediment. The bottom boundary condition for this equation was the sediment entrainment function of Garcia and Parker (1991).

3. Results

In order to check the model a first set of runs were performed to simulate turbulence characteristics and suspended sediment transport capacities as well as vertical concentration profiles under equilibrium conditions in open channel flows without vegetation. Results were compared against laboratory measurements and the Rousean distribution for suspended load. These results suggested the value of $\sigma_c = 1.0$.

Afterwards a set of numerical simulations of vegetated open channel flows were conducted for different plant densities, and the results were compared against experimental observations (López et al., 1995) in order to calibrate the parameters C_{fk} and $C_{f\varepsilon}$. A best fit was achieved for $C_{fk} = 0.15$ and $C_{f\varepsilon} = 0.30$. Figure 1. illustrates the computed normalized vertical profile of kinematic eddy diffusivity compared to the parabolic distribution, using $(ghSo)^{0.5}$ as the characteristic velocity. With the computed ν_t the sediment diffusion equation was solved numerically for a mean sediment diameter of 8.2 10^{-5} m (w_s= 5.8 10^{-3} m/s). Figure 2. depicts a computed vertical profile of relative concentration compared to the Rousean distribution for the same total shear velocity, i.e. $(ghSo)^{0.5}$. Suspended transport capacities in equilibrium conditions, qs, were estimated by integrating the product UC along the vertical. Figure 3. shows the ratio qs/qso, for a given specific water discharge of qw= 73 l/s and So = 1 E−03, as a function of vegetation density a, where qso denotes the suspended load for the same qw and slope in an open channel without vegetation. Lastly, Figure 4. illustrates the variation of Manning's n as a function of vegetation density, for the case of hydraulically smooth conditions at the bottom.

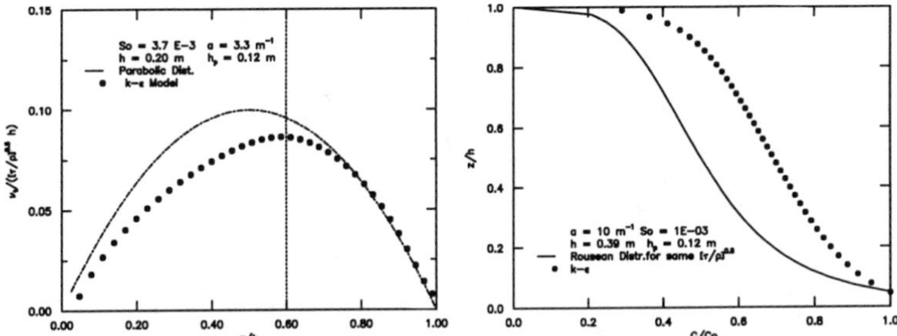

Figure 1. Normalized Vertical Profile of Kinematic Eddy Viscosity compared to the Parabolic Shape.

Figure 2. Vertical Profile of Relative Suspended Sediment Concentration compared to the Rousean Distribution

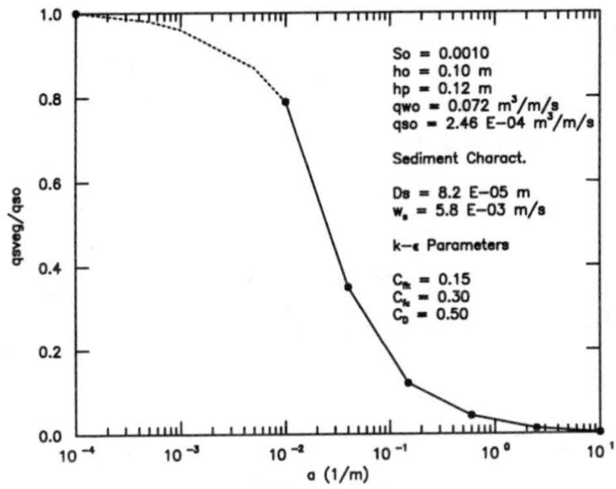

Figure 3. Relative Transport Capacity as function of Vegetation Density

4. Conclusions

The computed eddy viscosities in the presence of vegetation show similar values as the parabolic distribution although the maximum is at the top of the plants. The suspended sediment distribution is more uniform in the vertical than the one predicted by the Rousean distribution without vegetation. Numerical simulations confirm the ability of vegetated waterways to reduce sediment transport capacity. Resistance coefficients in terms of Manning's n show good agreement with standard recommended values (Chow, 1959).

5. Acknowledgements

The support of the US Army Corps of Engineers, Waterways Experiment Station (DACW39-94-K-0010) is gratefully acknowledged.

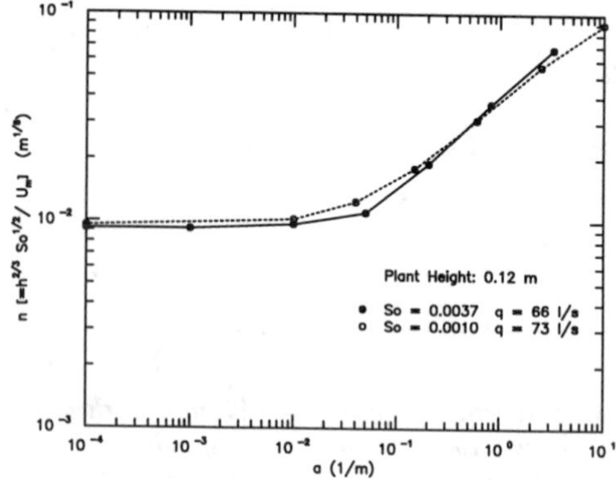

Figure 4. Variation of Manning's n with Vegetation Density

6. References

CELIK I. and RODI W., 1984, *Simulation of Free-Surface Effects in Turbulent Channel Flows.* PhysicoChem. Hydrod. **5**: 217-227.

CHOW V.T, 1959, *Open-Channel Hydraulics.* McGraw Hill, Inc.

FINNIGAN J.J, 1985, *Turbulent Transport in Flexible Plant Canopies.* Proc. Forest Env. Meas. Conf. Oak Bridge, TN, USA.

LOPEZ F., DUNN C and GARCIA M.H., 1995, *Turbulence Open Channel Flow Through Simulated Vegetation* ASCE Hydr. Eng. Conference. San Antonio, Texas.

GARCIA M.H. and PARKER G., 1991, *Entrainment of Bed Sediment into Suspension*. Hydr. Engrg. **117** (4), 414-435.

PLATE and QURAISHI, 1965, *Modeling Velocity Distributions Inside and Above Tall Crops.* J. Appl. Meteorol. **4**: 400-408.

RAUPACH, M.R. and THOM A.S, 1981, *Turbulence in and above Plant Canopies.* Ann. Rev. Fluid Mech. **13**: 97-129.

RAUPACH, M.R. and SHAW R.H., 1982, *Averaging Procedures for Flow Within Vegetation Canopies.* Bound. Layer Meteorol.**22**: 79-90.

SVENSSON U., 1986, *PROBE.* The Swedish Meteorol. and Hydr. Inst. S-601 76. Norrköping, Sweden.

TSUJIMOTO T., SHIMIZU Y. and NAKAGAWA H., 1991, *Concentration Distribution of Suspended Sediment in Vegetated Sand Bed Channel.* Int. Symp. Transp. of Susp. Sed. and Math. Mod. Florence, Italy.

WILSON N.R. and SHAW R.H., 1977, *A Higher Order Closure Model for Canopy Flow.* J. Appl. Meteorol. **16**: 1197-1205.

RAINFALL INTENSITY EQUATIONS FOR DURATIONS FROM ONE TO 24 HOURS

by David C. Froehlich[1] and Mohammad Tufail[2]

INTRODUCTION

A *design storm* is a precipitation pattern used by a hydrologic model of stormwater runoff [Chow and others (1988)]. Depending the type of hydrologic model, a design storm can consist of a precipitation depth at a single geographic location, a design hyetograph prescribing the time distribution of rainfall during a storm, or an isohyetal map specifying the spatial pattern of the precipitation. Typically design storms are constructed from rainfall intensity-duration-frequency (IDF) relations [Chow and others (1988); Pilgrim and Cordery (1993); Wenzel (1982)] that apply to the geographical location. Graphical IDF relations have been prepared from recorded rainfall for some locations in the United States [see, for example, "Rainfall" (1955)], and equations expressing rainfall IDF relations have been developed for many cities [Chow (1962), Wenzel (1982). However, at most locations rainfall intensity-duration relations will not be available and will need to be created using isopluvial maps prepared for a large geographic region.

A method is presented here for rapidly obtaining rainfall intensity equation parameters for intermediate durations from one to 24 hours for any location in the central and eastern United States, Alaska, Hawaii, Puerto Rico, and the U.S. Virgin Islands using readily available isopluvial maps prepared by the U.S. National Weather Service (NWS). Four types of commonly used intensity-duration equations are evaluated. However, a simple single-parameter equation is found to provide a good fit for all of the regions covered and is suggested for use. The method provides an easy, rapid, and accurate way of obtaining rainfall IDF equations that will be useful in designing hydraulic structures.

NATIONAL WEATHER SERVICE ISOPLUVIAL MAPS

Estimates of total point precipitation (that is, precipitation occurring at a single geographic location) for durations ranging from one hour to 24 hours can be obtained from isopluvial maps prepared by NWS, formerly known as the U.S. Weather Bureau. Isopluvial maps for the contiguous U.S., Alaska, Hawaii, Puerto Rico, and U.S. Virgin

[1]Consulting Engineer, 4612 Thornwood Circle, Lexington, Kentucky 40515-6129.
[2]Graduate Student, University of Kentucky, Department of Civil Engineering, 161 CE/TRANS Building, Lexington, Kentucky 40506-0281.

Islands appear in *TP-40* (*Technical Paper No. 40*) [Hershfield (1961)], *TP-47* [Miller (1963)], *TP-43* ["Rainfall frequency atlas" (1962)], and *TP-42* ["Generalized estimates" (1961)], respectively. Although *TP-40* includes isopluvial maps that cover the entire contiguous U.S., more up-to-date maps for durations from one hour to 24 hours have been prepared for the 11 states west of the 105th meridian [Miller and others (1973)]. Therefore, the procedure presented here for estimating rainfall IDF equation parameters does not apply to the 11 western states. Additionally, 1-hour duration isopluvial maps in *TP-40* for the eastern and central U.S. have been superseded by maps presented in a National Oceanic and Atmospheric Administration Technical Memorandum known as *HYDRO-35* [Frederick and others (1977)].

Other Durations

For a given location, rainfall depths for durations of 1, 2, 3, 6, 12, and 24 hours having return periods of 2, 5, 10, 25, 50, and 100 years can be found using a standard interpolation between the 1-hour and 24-hour isopluvial values as

$$P_{t,T} = P_{1,T} + f_t \times \left(P_{24,T} - P_{1,T}\right) \tag{1}$$

where $P_{t,T}$ = t-hour duration rainfall depth having a T-year return period, $P_{1,T}$ = 1-hour, T-year rainfall depth, $P_{24,T}$ = 24-hour, T-year rainfall depth, and $_tf$ = t-hour rainfall duration factor that applies to rainfall of all return periods. The factor f_t is nearly constant in the regions studied [Hershfield (1961)] and was scaled from the standard NWS 1-hour to 24-hour duration-interpolation diagrams used in *TP-40*, *TP-42*, *TP-43*, and *TP-47*.

The ratio of t-hour to 1-hour rainfall depth with a T-year return period found by dividing equation 1 by $P_{1,T}$ is

$$\frac{P_{t,T}}{P_{1,T}} = 1 + f_t \times \left(\frac{P_{24,T}}{P_{1,T}} - 1\right) \tag{2}$$

Using equation 2 and the standard factors f_t, six t-hour rainfall depth ratios (that is, $P_{t,T}/P_{1,T}$ for t = 1, 2, 3, 6, 12, and 24 hours) can be found as a function of the 24-hour to 1-hour rainfall depth ratio $P_{24,T}/P_{1,T}$. The corresponding dimensionless average rainfall intensity for a t-hour duration is

$$i^* \equiv \frac{i}{i_1} = \frac{P_{t,T}}{P_{1,T}} \times \frac{1}{t} \tag{3}$$

Chen (1983) carries out a similar analysis of data from isopluvial maps in *TP-40*, including rainfall intensities for durations of 5, 10, 15, and 30 minutes, found using standard NWS rainfall depth ratios that are the same for all values of the 24-hour to a 1-hour rainfall depth ratio. As a result, Chen's (1983) relative intensity-duration relations [see Chen's (1983) Figure 2] are not smooth for the range of 24-hour to 1-hour rainfall depths generally found in the United States, displaying noticeable kinks at the 60-minute point. The slope discontinuity at 60 minutes adversely affects the accuracy of a fitted rainfall intensity-duration equation for durations from one to 24 hours.

RAINFALL INTENSITY-DURATION EQUATIONS

Four basic forms of rainfall intensity-duration equations used at various locations throughout the United States [Chow (1962)] are summarized in Table 1. These equations are empirical and show that rainfall intensity decreases with rainfall duration for a given return period.

TABLE 1. Rainfall Intensity-Duration Equations

Equation type[a]	Equation form	Equation parameters
I	$i = a_1/(t + b_1)$	a_1, b_1
II	$i = a_2/t^{c_2}$	a_2, b_2
III	$i = a_3/(t + b_3)^{c_3}$	a_3, b_3, c_3
IV	$i = a_4/(t^{c_4} + b_4)$	a_4, b_4, c_4

[a]Chow (1962).

Each of the four types of equations was made dimensionless by dividing by the 1-hour rainfall intensity, i_1, for a specific return period. For example, the dimensionless form of a Type I equation is

$$i^* = \frac{a_1}{t + b} \tag{4}$$

where $a_1^* = a_1/i_1$, and t = rainfall duration in hours. If rainfall intensity is in units per *hour*, then $i_1 = P_{1,\,T}$ and $a_1 = a_1^* \times P_{1,T}$. Parameters of each of the four dimensionless nonlinear intensity-duration equations were found from an intensive pattern search over the entire feasible range of each parameter by minimizing the error sum-of-squares for the given data (that is, the dimensionless rainfall intensities for durations of 1, 2, 3, 6, 12, and 24 hours given by the ratio $P_{24,\,T}/P_{1,\,T}$)

$$S(\theta) = \sum_{=}^{N} (\hat{i}_j^* - i_j^*)^2 \tag{5}$$

where θ = vector of equation parameters, $\hat{i}_j^*$ = jth estimate of dimensionless rainfall intensity given by an intensity-duration equation, $i_j^* \equiv i_j/i_1$ = jth dimensionless rainfall intensity obtained from the ratio $P_{24,\,T}/P_{1,\,T}$ found from the NWS isopluvial maps, and N = number of rainfall intensity observations (N = 6 for all evaluations). Froehlich (1993) uses a similar procedure to obtain optimal intensity equation parameters for short-duration rainfall of one-hour or less. Chen (1983) analyzes only the Type III equation and minimizes an error function based on a logarithmic transformation of the Type III expression.

The rainfall depth ratios depend on only the ratio of 24-hour to 1-hour rainfall depths. Accurate expressions for each fitted parameter based on regression analysis of the 26 optimal estimates for values of $P_{24,\,T}/P_{1,\,T}$ ranging from 1.5 to 4.0 are given in Table 2.

From the error sums-of-squares of each equation, the two three-parameter equations (Type III and Type IV) were found to be are generally superior, as expected, and provide about the same quality fits to the data. The Type I equation yields significantly inferior fits. However, the two-parameter Type II equation provides nearly as good a fit as the Type III and Type IV expressions. Additionally, the a_2^* parameter is nearly constant at 1.

Equation Type II was simplified by replacing a_2^* with the constant 1. The resulting dimensionless *simple Type II* equation provides nearly as good a fit to the range of data as the Type II expression and is given by

$$i^* = \frac{1}{t^{c_{2s}}} \tag{6}$$

TABLE 2. Intensity-Duration Relation Parameter Equations

Equation type	Parameter equations[a]	Coef. of determ.[b]
Type I	$a_1^* = 1.134 - 0.2220y_T + 0.1852y_T^2$	0.999
	$b_1 = 0.1975 - 0.2946y_T + 0.2089y_T^2$	0.999
Type II	$a_2^* = 1.012 - 0.01555y_T + 0.002858y_T^2$	0.999
	$c_2 = 1.144 - 0.1992y_T + 0.01152y_T^2$	0.998
Simple Type II	$a_{1s}^* = 1$	--[c]
	$c_{2s} = 1.092 - 0.1450y_T$	0.993
Type III	$a_3^* = 1.069 - 0.1950y_T + 0.04356y_T^2$	0.995
	$b_3 = 0.4026 - 0.5033y_T + 0.09998y_T^2$	0.995
	$c_3 = 1.185 - 0.3012y_T + 0.03343y_T^2$	0.997
Type IV	$a_4^* = 1.352 - 0.4793y_T + 0.09820y_T^2$	0.999
	$b_4 = 0.3548 - 0.4821y_T + 0.09904y_T^2$	0.999
	$c_4 = 1.313 - 0.4234y_T + 0.05730y_T^2$	0.999

[a] $y_T = P_{24\text{-hr}, T\text{-yr}}/P_{1\text{-hr}, T\text{-yr}}$
[b] From regression analysis of optimal parameter estimates for 26 evenly spaced values of y_T from 1.5 to 4.0.
[c] Not calculated.

The parameter c_{2s} is nearly a linear function of the ratio $P_{24,T}/P_{1,T}$ (the linear relation is given in Table 2). Considering the approximate nature of the six rainfall ratios used to fit the dimensionless intensity-duration equations, use of the simple Type II expression rather than the more accurate Type III or Type IV equation for intermediate-duration rainfall seems well justified.

SUMMARY AND CONCLUSIONS

A method for finding rainfall intensity-duration equation parameters for intermediate durations between one and 24 hours for any location in the central and eastern United States, Alaska, Hawaii, Puerto Rico and the U.S. Virgin Islands using readily available isopluvial maps prepared by the National Weather Service was presented. Optimal parameter values were found for four types of commonly used intensity-duration equations, and equations that provide close fits were found for each parameter. A simple intensity-duration equation form requiring only a single parameter was found to provide good estimates for all 24-hour to 1-hour depth ratios that might be encountered in the mapped regions and is suggested for use. The procedure modifies, expands, and simplifies the analysis of

Chen (1983), and will be useful for developing intermediate-duration design storms needed to evaluate hydraulic structures when more accurate rainfall intensity-duration-frequency relations are not available.

APPENDIX -- REFERENCES

Chen, C. L. (1983). "Rainfall intensity-duration-frequency formulas." *Journal of Hydraulic Engineering*, 109(12), 1603-1621.

Chow, V. T. (1962). "Hydrologic determination of waterway areas for the design of drainage structures in small drainage basins." *Bulletin No. 462*, Engineering Experiment Station, University of Illinois, Urbana, Illinois.

Chow, V. T., Maidment, D. R., and Mays, L. W. (1988). *Applied hydrology*. McGraw-Hill, New York, New York.

Frederick, R. H., Myers, V. A., and Auciello, E. P. (1977). "Five- to 60-minute precipitation frequency for the eastern and central United States." *NOAA Technical Memorandum NWS HYDRO-35*, National Weather Service, Silver Spring, Maryland.

Froehlich, D. C. (1993). "Short-duration-rainfall intensity equations for drainage design." *Journal of Irrigation and Drainage Engineering*, 119(5), 814-828.

"Generalized estimates of probable maximum precipitation and rainfall-frequency data for Puerto Rico and Virgin Islands for areas to 400 square miles, durations to 24 hours, and return periods from 1 to 100 years." (1961). *Technical Paper No. 42*, U.S. Weather Bureau, Washington, D.C.

Hershfield, D. M. (1961). "Rainfall frequency atlas of the United States for durations from 30 minutes to 24 hours and return periods from 1 to 100 years." *Technical Paper No. 40*, U.S. Weather Bureau, Washington, D.C.

Miller, J. F. (1963). "Probable maximum precipitation and rainfall-frequency data for Alaska for areas to 400 square miles, durations to 24 hours, and return periods from 1 to 100 years." *Technical Paper No. 47*, U.S. Weather Bureau, Washington, D.C.

Miller, J. F., Frederick, R. H., and Tracey, R. J. (1973). "Precipitation-frequency atlas of the United States: Volumes 1 to 10." *NOAA Atlas 2*, National Weather Service, Silver Spring, Maryland.

Pilgrim, D. H., and Cordery, Ian (1993) "Flood runoff." *Handbook of hydrology*, D. R. Maidment (ed.), McGraw-Hill, New York, New York, 9.1-9.42.

"Rainfall-frequency atlas of the Hawaiian Islands for areas to 200 square miles, durations to 24 hours, and return periods from 1 to 100 years" *(1962). Technical Paper No. 43*, U.S. Weather Bureau, Washington, D.C.

"Rainfall intensity-duration-frequency curves for selected stations in the United States, Alaska, Hawaiian Islands, and Puerto Rico." (1955). *Technical Paper No. 25*, U.S. Weather Bureau, Washington, D.C.

Wenzel, H. G. (1982). "Rainfall for urban stormwater design." *Urban stormwater hydrology; Water Resources Monograph 7*, D. F. Kibler (ed.), American Geophysical Union, Washington, D.C.

Outlier Detection in Annual Maximum Flow Series

A.R. Rao[1] and D. McCormick[1]

Abstract

In using the water resources council method (Water Resources Council, 1981), it is recommended that adjustments must be made for outliers. Outliers are extreme data which significantly deviate from the general trend of the data.

In the present paper, annual maximum flow data from the Wabash River basin in Indiana are analyzed. The first part of the study demonstrates that the water resources council method of identifying outliers does not identify even very large annual maximum flows as outliers.

In the second part of the study, an alternate outlier detection method, based on the least median of squares (LMS) criterion, is used.

Introduction

Traditional methods of analyzing flood data can lead to difficulties in identifying outlier values. Flow values which deviate greatly from the mean flow are considered as potential outliers and may have a dominant effect on the predicted flows. The purpose of this study is to examine the method recommended by the U.S. Water Resource Council and the Least Mean Squares Method to identify the outliers.

Two common distributions applied to flood frequency analysis are the Gumbel and the log-Pearson Type III distribution. The latter distribution is the basis for Water Resource Council (WRC) Method of flood frequency analysis.

Outliers are identified by the WRC Method and by a robust method called the Least Median Squared (LMS) Method. The Gumbel distribution is used as the basis for LMS Method.

---

[1] School of Civil Engineering, Purdue University, W. Lafayette, IN 47907

Data

The annual maximum flow data used in this study were from United States Geological Survey (USGS) gaging stations in Indiana in the Ohio River Basin (24 stations); Upper Mississippi River Basin (3 stations); tributaries to Lake Erie (2 stations); tributary to Lake Michigan (1 station). The watershed areas corresponding to these stations range from 01 mi^2 to 789 mi^2. Periods of record range from 25 years to 68 years. An effort was made to select data in which the flows were not significantly influenced by controlled releases from reservoirs.

Outlier Detection by Water Resources Council Method

The Water Resource Council Method (WRCM) is a method that was introduced in order to promote uniformity of flood frequency analyses. It is a commonly used method. WRCM is based on a Log-Pearson III (LPIII) distribution.

The first step in an LPIII analysis is to take the logarithms of the flow values. The mean $\bar{L}$, standard deviation s_L of the logarithms of the flows are computed. The thresholds L_L and L_H are computed by using the equations given below.

$$L_L = \bar{L} - K_n s_L \qquad L_H = \bar{L} + K_n s_L$$

K_n is dependent on sample size and is given in WRC (1963). If a value is higher or lower than L_H or L_L respectively then it is called an outlier.

Outlier Detection by the Least Median Square (LMS) Method

The general equation used for parameter estimation in this study is of the form in Equation 1.

$$Y_i = \beta_0 + \beta_1 X_1 + \varepsilon_i \tag{1}$$

In the above equation Y_i is the response variable; β_0 and β_1 are the parameters to be estimated; X_i is the independent variable; and ε_i is the deviation from the predicted value. The estimated value from the model is given by Equation 2.

$$\hat{Y}_i = b_0 + b_1 X_1 \tag{2}$$

where $\hat{Y}_i$ is the estimated response variable; b_0 and b_1 are the estimates of the parameters β_0 and β_1. The residual value, the difference between the predicted and observed values, is then defined as in Equation 3.

$$e_i = Y_i - \hat{Y}_i \tag{3}$$

In the *least median of squares* (LMS) method, b_0 and b_1 in Equation 2 are determined such that the median value of the squared residual is minimized:

$$\text{Minimize } \left[\text{med } (e_i)\right]$$

A *robust method* is one which will not overemphasize any particular portion of the data range (such as the higher values of the dependent variable). The LMS method is a robust method. A *scale estimate* is used by the LMS method to define how well the data are fitted by the straight line. The initial scale estimate, s^0 is given by Equation 4

$$s^0 = 1.486\ (1 + \frac{5}{n-2})\ \sqrt{MED(e_i^2)} \tag{4}$$

Each observation is then assigned a weight, corresponding to whether it is within a reasonable range of the initial scale estimate.

$$w_i = \begin{cases} 1 \text{ if } |e_i| / s^0 \leq 2.5 \\ 0 \text{ otherwise} \end{cases}$$

The final scale estimate used in the LMS method is then:

$$\sigma^* = \sqrt{\frac{\sum_{i=1}^{n} (w_i e_i^2)}{\sum_{i=1}^{n} w_i - 2}} \tag{5}$$

Outliers may be detected from a plot of e_i vs. $\hat{Y}_i$, also known as a *residual plot.* The criterion for whether an observation is an outlier will be whether it has a residual value greater than a multiple of the final scale estimate. In this study, the distance chosen to detect an outlier was, $2.5\ \sigma^*$.

Results

Results from WRC Method

The data were tested for outliers by the Water Resources Council method. No station had any upper outliers according to the WRC Method. However, many of the stations had low outliers.

The WRCM appears to have a very high threshold. For example, in one of the stations the mean value of 9796 cfs and a standard deviation of 5166 cfs for observed flows. The mean plus two standard deviations is 20128 cfs. There were two observed flows higher than 20128 cfs. In this case the high outlier threshold by the WRCM is 48640 cfs, which corresponds to the mean plus seven and half times the standard deviation, which is extremely high.

Results from LMS Method

The data were analyzed by the LMS method by using the Gumbel distribution. The Extreme Value Type I, or Gumbel, distribution may be expressed as in Equation 6.

$$Q_T = b_0 + b_1 Y_T \tag{6}$$

Q_T is the flow corresponding to the return period T, and Y_T is the reduced variate.

$$Y_T = -\ln\left[\ln\left(\frac{T}{T-1}\right)\right] \tag{7}$$

The return period T is estimated by the California formula.

The LMS solution is highly dependent on the spacing of the independent variable. Since the differences in the values of Y_T decrease as the rank of the value increases, there is a heavier concentration of data points in the lower Y_T range. This concentration can result in a solution that will best fit the lower to middle values of Y_T, while leaving values in the upper ranks outside of the $2.5\sigma^*$ range. This problem was overcome by dropping data points corresponding to the lowest values of Y_T for each station until the number of values outside of the $2.5\sigma^*$ range is approximately the same as the number of values above the upper threshold. This forces the median to account for the upper portion of the data, rather than lower values.

A typical result from a station with the highest 20 observations is given in Figure 1. The β_0, β_1 and σ^* values for this case are 266, 162 and 12.5. A plot of e_i vs $\hat{Y}_i$ is given in Fig. 1. The largest observed flow is an outlier in this case, as it is outside the $\pm 2.5^*$ band. Any observations below the upper threshold which are outside the $2.5\sigma^*$ band would not be classified as outliers, but simply poorly fitted. Further details of the method are found in McCormick and Rao (1995).

Conclusions

Based on the results of the present study the following conclusions may be presented.

(1) The high threshold in the water resources council method is so high as to indicate no high outliers in the data.

(2) The Least Median Squares method can be successfully used to identify outliers in flood data.

References

McCormick, D.L. and A.R. Rao, "Outlier detection in Indiana flood data", Tech. Rept. CE-EHE-95-4, School of Civil Engineering, Purdue University, W. Lafayette, IN 47907, 1995.

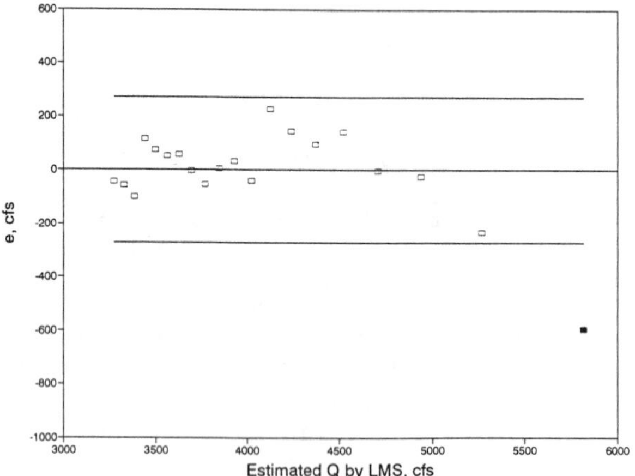

Figure 1. Sample Residual Plot using LMS Method

Estimating Extreme Discharges in a Colombian Region

Ricardo A. Smith, María Victoria Vélez, Carlos A. Pérez,
Hernán D. Bolaños y Carlos J. Franco[1]

Abstract

Regionalization procedures are recommended that allows to estimate extreme discharges where its needed. The main proposed regionalization proposed procedure regionalize probability distribution functions (pdf) that are used to estimate the extreme events associated to a given return period. The basic statistical characteristics are related to morphometric watershed characteristics using regression and correlation analysis. Once that the statistical characteristics are estimated using the regional equations, the parameters of the pdf can be estimated. To define the pdf to be used several distributions were tested in the sites with discharge information. The selected distribution was then used with the regional procedure described above. Due that in most cases streamflow information is limited at the gaging stations (around 15 years of data), two parameters pdf were preferred. Nevertheless attempts to use three parameter distributions were made using regional relations of third order moments with basin areas to estimate skewness coefficients. The proposed procedures were used in Antioquia, one of the biggest Colombian states located in the west central region. The procedures were applied over the complete region and also in several subregions. The two best fit distributions are the two parameter log normal and the Gumbel. When subregions are used there is a tendency to one of these distributions to best fit all gaging stations in the subregions. The Log Pearson three parameter distribution was also used estimating its parameters with the regional procedure described above. It did not show a clear advantage over the two parameter distributions. Some conclusions and recommendations are presented.

---

[1] Water Resources Graduate Program, Universidad Nacional de Colombia, Facultad de Minas, Apartado Aéreo 1027, Medellín, Colombia.

Introduction

The State of Antioquia is located to the central north west of Colombia covering an important part of the Andes mountain chain. It has an area of 63612 Km^2. There are important regions of Antioquia where there is no information about maximum instantaneous discharges and mean minimum discharges (Universidad Nacional, 1988).

In the areas with no information, this problem could be solved if regionalization procedures are used where some statistics, or probability distribution parameters, could be related to morphometric watershed characteristics. These regionalization procedures could be used to estimate probability distribution functions in locations with no information and in turn to define the discharges associated with different return periods (Velez et al, 1993).

To determine the best probability distribution function to be regionalized, goodness of fit test were performed using the stations with available records. This procedure allowed to identify regions where some distributions fitted best than others.

The main morphometric characteristics used in this study were the watershed area, the length of the main stream and the main stream slope.

Methodology

Initially a selection procedure of available runoff records was performed. Stations with more than 10 years of records and with good quality of data were selected. The quality of the data was defined considering stations not affected by upstream developments and located in homogeneous watersheds. In this way 74 stations for minimum discharges and 77 stations for maximum discharges were selected.

At each selected station several probability distribution function were fitted and analyzed. For the maximum instantaneous case the two parameter Log-Normal and the Gumbel distributions were fitted. For the case of the mean minimum discharges the two parameter Log-Normal, Gumbel and Weibull distributions were used. Moment and maximum likelihood estimation procedures of probability distribution functions were used for all the above distributions (Kite, 1977; Salas et al, 1993). Smirnov-Kolmogorov , Cramer-Von Mises and graphical goodness of fit test were performed to define the best fit distribution (Salas et al, 1993). With the best fit distributions for all the stations several region were identified where some particular distribution fitted best for maximum or minimum discharges.

In the case of minimum discharges Antioquia was divided in six regions. In general the best fit distribution in this case was the Weibull distribution. Nevertheless, the differences with the other distributions was not significant. So, in some cases the Weibull distribution was replaced by one of the other distributions to avoid the use of a three parameter distribution with short records length. Finally the two parameter Log Normal (LNII) and the Gumbel distribution, with parameters estimated using the method of moments (GMOM), were recommended for the different regions.

For the case of the maximum instantaneous discharges Antioquia was divided in four regions. In this case only two parameter distribution were used to avoid problems with the estimation of three parameter distributions with short records. The finally selected distributions were the Gumbel distribution by the method of moments (GMOM) and the two parameter Log Normal (LNII) distributions.

To define the regression equations that relates the mean and standard deviation of the maximum and minimum discharges with the watershed morphometric characteristics, step wise regression procedures were used.

Results

Using the goodness of fit tests different regions were defined for Antioquia where a specific probability distribution function fit best. The regional regression equations were defined for the whole state and for the different regions. Some of the results are:

- For maximum discharges and considering the whole State as one single region the regional equations found were:

  $$\text{Mean} = 10^{-4.518} * A^{0.551} * S^{0.181} * P^{1.512} \text{, with } R = 0.932$$

  $$\text{Standard Deviation} = 10^{-4.194} * A^{0.436} * S^{0.252} * P^{1.352} \text{, with } R = 0.826$$

  When only A and P are used in the regional regression equations the results are about the same with $R = 0.928$ for the mean and $R = 0.813$ for the standard deviation.

  In the above equations A is the watershed area (A) in [Km^2], L is the length of the main stream in [Km]; S is the slope in [%], P is the precipitation (P) in [mm/year], and R is the correlation coefficient. The mean and the standard deviation are in [m^3/s] units.

- Regional equations were also defined for maximum discharges in the different regions. For example, for the Cauca river region (16 stations were included) the results were:

  $$\text{Mean} = 10^{-3.457} * A^{0.560} * P^{1.246} \text{, with } R = 0.973$$

  $$\text{Standard Deviation} = 10^{-3.267} * S^{0.383} * P^{0.963} \text{, with } R = 0.980$$

  For the Medellín - Porce - Nechí region (23 stations) the results were:

$\text{Mean} = 10^{0.129} * L^{1.259}$, with $R = 0.935$

$\text{Standard Deviation} = 10^{0.132} * L^{0.949}$, with $R = 0.833$

- For minimum discharges and considering the whole State as one single region the regional equations found were:

  $\text{Mean} = 10^{-5.281} * A^{0.903} * P^{1.086}$, with $R = 0.958$

  $\text{Standard Deviation} = 10^{-.0437} * A^{0.683} * L^{0.516} * S^{0.322} * P^{0.868}$, with $R = 0.964$

- Regional equations were also defined for maximum discharges in the different regions. For example, for the Cauca and Porce river region (32 stations were included) the results were:

  $\text{Mean} = 10^{-5.269} * A^{0.948} * P^{1.046}$, with $R = 0.993$

  $\text{Standard Deviation} = 10^{-5.293} * A^{0.985} * P^{0.875}$, with $R = 0.980$

Conclusions

- The results obtained with the regional equations are encouraging. In all cases significant correlation was obtained, and in most cases correlation values were extremely high.

- The proposed procedure really means a regionalization of the probability distribution functions, and not just of a specific statistic

- The predominant probability distribution function for maximum discharges in Antioquia is the Log Normal type II distribution. Parameter estimation for this distribution and for the Gumbel distribution (in the areas where Gumbel fit best) could be done using the regional equations for the mean and for the standard deviation developed for the different regions of Antioquia.

- The watershed area and the precipitation are the two most important variables in all the regression models.

- In the case of minimum discharges the Gumbel distribution trend to fit best to the stations with longer records. On the contrary, the Log Normal distribution type II trends to fit best to the stations with shorter records.

- Comparing the sizes of the used watersheds, it can be concluded that Weibull distribution fit best to small watersheds.

- The test statistics computed for the Smirnov-Kolmogorov and Cramer-Von Mises goodness of fit tests, for the different probability distribution functions used in the case of minimum discharges, are about the same for all distributions. In the case of maximum discharges there are significant differences.

- The Log Pearson three parameter distribution was also used estimating its parameters with the regional procedure described above. In this case a regional relation of third moments with basin areas was used to estimate the skewness coefficient. It did not show a clear advantage over the two parameter distributions.

References

Kite G.W., 1977. Frequency and risk analyses in hydrology. Water Resources Publications, Fort Collins, Colorado.

Salas, J., Smith R., Tabios G. and Heo J., 1993. Statistical computer techniques in hydrology and water resources. Colorado State University, Fort Collins, Colorado.

Universidad Nacional de Colombia, 1988. Regional models for maximum instantaneous, mean and minimum discharges in 180 watersheds of a Colombian Region (in Spanish). Research Report, Medellín, Colombia.

Vélez M.V., Smith R.A., Mesa O., and Vélez J.I., 1993. Hydrology methodologies with scarce or no available information (in Spanish). Water Resources Graduate Program, Universidad Nacional de Colombia, Medellín, Colombia.

An Evaluation of Frequency Distributions for Flood Hazard Analysis

Wilbert O. Thomas, Jr.[1], Minoru Kuriki[2] and Tadashi Suetsugi[2]

Abstract

Many different frequency distributions and fitting methods are used to determine the magnitude and frequency of floods and rainfall. Ten different combinations of frequency distributions and fitting methods are evaluated by summarizing the differences in the 0.002 exceedance probability quantile (500-year event), presenting graphical displays of the 10 estimates of the 0.002 quantile, and performing statistical tests to determine if differences are statistically significant. This evaluation indicated there are some statistically significant differences among the methods but, from an engineering standpoint, these differences may not be significant.

Introduction

Accurate estimates of the magnitude and frequency of floods and rainfall are needed for planning and designing flood-control structures and transportation facilities and for floodplain management. The U.S. Geological Survey (USGS) and the Public Works Research Institute (PWRI) of Japan have undertaken a joint research project on flood hazard analysis, a major part of which involves the evaluation of commonly-used frequency distributions and fitting methods for determining the magnitude and frequency of floods and rainfall. These methods are compared and evaluated using flood and rainfall data from both the United States of America (USA) and Japan. The objective of this analysis is to quantify and characterize the differences among the various frequency methods.

Data Base

A data base of flood and rainfall data from both the USA and Japan were assembled. Annual maximum peak discharges were compiled for 47 streamflow stations in 47 different States throughout the USA with record lengths greater than 50 years and drainage areas ranging from 520 to 14,800 square kilometers (200 to 5,700 square miles). Annual peak discharges were compiled for 43 streamflow stations in

----

[1]U.S. Geological Survey, Reston, Virginia 22092 USA

[2]Public Works Research Institute, Tsukuba, Ibaraki, Japan

43 different Perfectures throughout Japan with record lengths greater than 20 years and drainage areas ranging from 100 to 2,500 square kilometers (40 to 1,000 square miles). Rainfall data (annual maximum daily values) for 47 stations in 47 States in the USA and 47 stations in 47 Prefectures in Japan were also assembled. All rainfall records exceeded 50 years in length. A distribution of record lengths for all streamflow and rainfall stations is given in table 1. A total of 184 streamflow and rainfall stations were used in the analysis.

Table 1. Distribution of records length for streamflow and rainfall stations

Period of record (years)	USA		Japan	
	Flood	Rainfall	Flood	Rainfall
20-29			13	
30-39			16	
40-49	0	0	12	0
50-59	3	7	2	1
60-69	16	12	0	0
70-79	11	0		0
80-89	12	11		1
90-99	5	12		10
100-	0	5		35

Distributions and Fitting Methods

The four frequency distributions selected for evaluation are the most commonly-used distributions for extremes of both rainfall and floods throughout the world (Cunnane, 1989). The frequency distributions are the three-parameter lognormal (3LN), the Pearson Type III (PIII), the Generalized Extreme Value (GEV) and the log-Pearson Type III (LPIII). The GEV distribution includes the Extreme Value Types I, II and III as special cases.

Three methods for estimating the parameters of the four frequency distributions were evaluated. The three methods are the method of moments (MM), maximum likelihood estimation (MLE) and L moments (LM). The method of moments uses the ordinary sample moments (mean, standard deviation and skewness) to estimate the parameters of the distribution. In maximum likelihood estimation, the parameters that maximize the likelihood function are used. In the L-moment method, linear combinations of the ranked observed data (L moments) are used to estimate the distribution parameters.

Software for applying the L-moment method to the LPIII distribution is not available and, additionally, the method of moments was not applied to the PIII distribution simply for lack of time. Therefore, 10 different sets of frequency estimates were obtained for the four distributions and the three fitting methods.

The number of estimates available for analysis for the MLE method is less than 184 because it was not always possible to obtain a reasonable MLE solution. The number of reasonable estimates was 183 for the 3LN, 126 for the PIII, 178 for the GEV, and 151 for the LPIII. A PIII, LPIII or GEV MLE solution was considered unreasonable if estimates of selected quantiles were more than twice or less than half of the 3LN estimate for the same station.

Quantiles with annual exceedance probabilities of 0.10, 0.02, 0.01, 0.005, and 0.002 were computed and used as a basis for the evaluation. Software for performing the computations came from a variety of sources: L moments from Hosking (1990, 1991), and Hosking and Wallis (1993), method of moments from

Interagency Advisory Committee on Water Data (1982) and Pilon and others (1985), and maximum likelihood estimation from Pilon and others (1985) and Stedinger and others (1988).

Graphical Presentation

Estimates of the 0.002 quantile varied the most and are used to characterize the differences among the frequency distributions and fitting methods. The following symbols were adopted to represent the 10 different estimates of the 0.002 quantile or 500-year event:

NLM500 - 3LN distribution, L moments
PLM500 - PIII distribution, L moments
GLM500 - GEV distribution, L moments
NMO500 - 3LN distribution, method of moments
LMO500 - LPIII distribtion, method of moments
GMO500 - GEV distribution, method of moments
NML500 - 3LN distribution, maximum likelihood
PML500 - PIII distribution, maximum likelihood
LML500 - LPIII distribution, maximum likelihood
GML500 - GEV distribution, maximum likelihood

The 10 estimates of the 0.002 quantile for flood data in the USA are illustrated in box plots in figure 1. Pertinent features of the box plots are as follows: the solid box represents the range of data between the lower and upper quartiles (25th and 75th

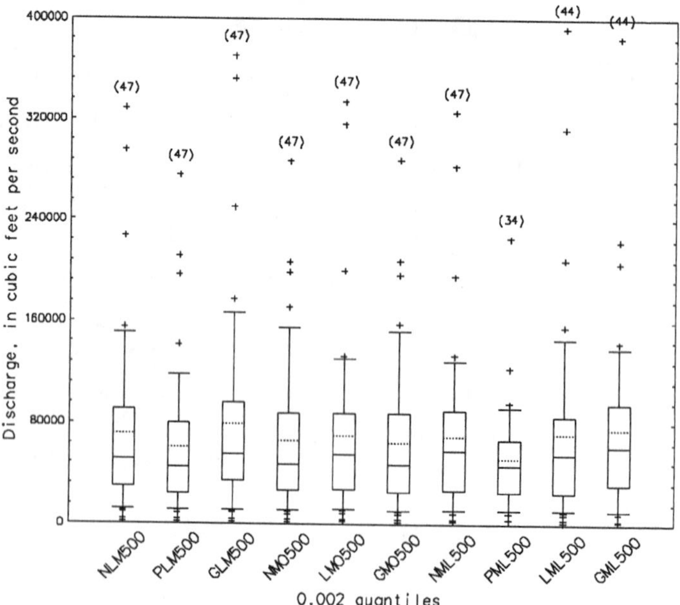

Figure 1. Summary of 0.002 quantiles for flood data in the USA.

percentiles, respectively), the dotted line within the box is the mean, the solid line within the box is the median, the horizontal lines at the end of the vertical lines represent the 10th and 90th percentiles, and the + values represent extreme values outside of the 10th and 90th percentiles. The numbers in parentheses (47) represent the number of observations in that sample.

The data in figure 1 show that the mean (dotted line in box) and standard deviation of the estimates for the 10 methods are reasonably similar. The GEV distribution using L moments and maximum likelihood estimation (GLM500 and GML500, respectively) give the highest estimates on average. Likewise, the PIII distribution using L moments and maximum likelihood estimation (PLM500 and PML500, respectively) give the lowest estimates on the average.

Statistical tests

Paired t tests were used to determine if significant differences existed between the other 9 estimates of the 0.002 quantile and the LPIII MM estimates (LMO500). The LPIII MM was selected as the base method because of its recommended use within the USA for frequency analysis of flood data (Interagency Advisory Committee on Water Data, 1982). There is no implication here that the LPIII MM method is necessarily the best method.

The null hypothesis or the hypothesis to be tested is that the means of the 0.002 quantiles, estimated by the other methods and the LPIII MM, are equal or that the difference in mean quantiles ($\overline{D}$) is zero. The test statistic is $t = (\overline{D})/(S_D/\sqrt{N})$ where S_D is the standard deviation of the differences D and N is the number of differences. Results from the t tests for the 0.02 and 0.002 quantiles for a 1 percent level of significance are summarized in table 2.

Table 2. Paired t test results for differences in the 0.02 and 0.002 quantiles between the other methods and log-Pearson Type III method of moments (1 percent level of significance)

Method	USA				Japan			
	Flood		Rainfall		Flood		Rainfall	
	.02	.002	.02	.002	.02	.002	.02	.002
3LN LM	N	N	N	N	N	N	N	N
PIII LM	N	Y-	Y-	Y-	Y-	Y-	Y-	Y-
GEV LM	N	Y+	Y+	Y+	Y-	N	Y+	Y+
3LN MM	N	N	N	Y-	Y-	Y-	Y-	Y-
GEV MM	N	N	Y-	Y-	Y-	Y-	Y-	Y-
3LN ML	N	N	N	Y-	N	N	N	N
PIII ML	Y-	Y-	Y	Y-	N	N	Y-	Y-
LPIII ML	N	N	N	N	N	N	N	N
GEV ML	Y+	Y+	N	N	N	N	Y+	Y+

N no significant difference + Other method higher than LPIII MM

Y significant difference - Other method lower than LPIII MM

The results in table 2 indicate that, in general, there are no significant differences (at the 1 percent significance level) in the 0.02 and 0.002 quantiles for the following three methods, 3LN LM and ML, LPIII ML, and the base method LPIII MM. When the GEV LM and ML are statistically different from the LPIII MM estimates, the GEV estimates are generally higher. When the PIII LM and ML, 3LN MM, and GEV MM estimates are statistically different from the LPIII MM estimates, these estimates are lower. Note that the results for the 0.02 and 0.002 quantiles are nearly identical implying the other 9 methods give similar estimates for the 0.02 and 0.002 quantiles as does the log-Pearson Type III method of moments.

Concluding Remarks

Quantiles from some of the methods tend to be statistically different from quantiles based on the log-Pearson Type III method of moments used for flood-frequency analysis in the USA (Interagency Advisory Committee on Water Data, 1982). However, from an engineering standpoint, estimates of the various quantiles did not vary that much with average differences usually less than 20 percent. Additional study is needed to determine if significant differences exist among other combinations of frequency methods. If significant differences exist, then the statistical, hydrological or meteorological factors contributing to these differences should be identified.

The maximum likelihood method did not always provide reasonable estimates of the various quantiles for certain distributions. Additional research is needed to develop improved maximum likelihood estimation techniques for the Pearson Type III and log-Pearson Type III distributions. The frequency estimates computed in this study are archived in data bases in the Office of Surface Water, USGS, Reston, Virginia and in the Urban River Division, PWRI, Tsukuba, Japan. Further comparisons and analyses of the data are needed before recommending frequency distributions and/or fitting methods for flood and rainfall data in Japan or the USA.

References

Cunnane, C., 1989, Statisitcal distributions for flood frequency analysis: World Meteorological Organization Operational Hydrology Report No. 33, Geneva, Switzerland, 115 p.

Hosking, J. R. M., 1990, L-moments: Analysis and estimation of distributions using linear combinations of order statistics: Journal of Royal Statistical Society B, 52, no. 1, p. 105-124.

Hosking, J. R. M., 1991, Fortran routines for use with the method of L-moments, Version 2: Research Report RC-17097, IBM Research Division, T. J. Watson Research Center, Yorktown Heights, NY, 17 P.

Hosking, J. R. M., and Wallis, J. R., 1993, Some statistics useful in regional frequency analysis: Water Resources Research, Vol. 29, No. 2, pp. 217-281.

Interagency Advisory Committee on Water Data, 1982, Guidelines for determining flood flow frequency: Bulletin 17B of the Hydrology Subcommittee, Office of Water Data Coordination, U.S. Geological Survey, Reston, Va., 183 p.

Pilon, P. J., Condie, R., and Harvey, K. D., 1985, Consolidated Frequency Analysis (CFA) User Manual for Version 1: Water Resources Branch, Environment Canada, Ottawa, Ontario, 148 p.

Stedinger, J. R., Surani, R., and Therivel, R., 1988, MAX Users Guide: A program for flood frequency analysis using systematic-record, historical, botanical, physical palehydrologic and regional hydrologic information using maximum likelihood techniques: Department of Environmental Engineering, Cornell University, Ithaca, NY, 51 p.

Great Lakes Water Level Extremes and Risk Assessment

Deborah H. Lee[1], P.E., M. ASCE, and Anne H. Clites[2]

Abstract

A method for developing Great Lakes probabilistic monthly water level forecasts, adapted from the Extended Streamflow Prediction technique, is presented. The method is applied retrospectively to quantify the risk of flooding at Milwaukee, Wisconsin and assess the need to enhance flood control measures during a period of record high water levels. The results show that during the 24 months following October, 1986 when Lake Michigan reached an all time high, the risk of exceeding the maximum flood protection level of 177.76 m was less than 2%. This information would have reduced the uncertainty regarding the need to prepare a high lake level protection plan for downtown Milwaukee. Further refinements to the probabilistic forecast technique are discussed.

Introduction

The adverse consequences of extreme Great Lakes water level fluctuations on public and private interests are well documented. During low water level periods, such as those experienced in the 1920's, 1930's, and 1960's, commercial navigation suffers from loss of adequate navigation depths and reduced cargo capacity, hydropower generation is reduced, water intakes are exposed, and recreational use of the lakes is impaired. During high water level periods, such as those experienced in the 1950's, 1970's, and most recently in 1985 and 1986, riparians suffer property damage due to flooding and increased erosion, metropolitan sewer outfalls are submerged, and recreational use of beaches and marinas is impaired.

During periods of extreme water levels, governments (local, state, and federal) and commercial and private interests are faced with making decisions

---

[1]Hydrologist and [2]Physical Scientist, Great Lakes Environmental Research Laboratory, 2205 Commonwealth Blvd., Ann Arbor, MI, 48105-1593.
GLERL contribution no. 948.

regarding what actions, if any, they should take to avoid or mitigate damages and losses. They must weigh the risks (costs) of taking action versus no action. They must also decide when to take action. Because many measures take time to implement (i.e., construction of shore protection) or to become effective (i.e., deviations from lake regulation plans), decisions must be made well in advance of reaching critical water levels. Also, they must make these decisions with little certainty of future water levels. Recent analyses (Croley and Lee 1993, Lee 1992) of state-of-the-art deterministic Great Lakes water supply and lake level forecasts have shown that these forecasts are only marginally better than climatology for a 1-month outlook, and that their skill declines to the same or worse than climatology for a 6-month outlook. Probabilities of exceedance or non-exceedance are not explicitly given. How, then, can interests affected by fluctuating Great Lakes water levels make decisions that depend on knowledge of future water levels? In addition, how do they measure the risks associated with their decision? As suggested by Croley and Lee (1993), the answer lies in the use of probabilistic forecasts. This type of forecast was first applied to water resources by Day (1985) for streamflow forecasting. He presented the Extended Streamflow Prediction (ESP) procedure as an objective means for long-range hydrologic forecasting and the assessment of forecast uncertainty.

The ESP technique is adapted here to illustrate the use of Great Lakes probabilistic water level forecasts to assess the risk of flooding at Milwaukee, Wisconsin and the need to implement flood control projects if lake levels were to continue to rise above the October, 1986 record. Refinements to the forecast technique are suggested.

Adaptation of the ESP Forecast Approach

The ESP procedure (Day 1985) uses conceptual hydrologic models to produce alternative future scenarios of stream flows using periods of historical meteorological data and current watershed conditions as initial conditions. Frequency analysis is then performed on the streamflow scenarios to derive a probabilistic forecast. Similarly, for the following case study, alternative Great Lakes water level scenarios were produced by routing scenarios of historical water supplies through a hydrologic response model of the Great Lakes, using recorded lake levels and outflows for initial conditions.

The alternative water supply scenarios were routed through a hydrologic response model obtained from Environment Canada. This model embodies the Lake Superior regulation plan and middle lakes stage-discharge relationships. The hydraulic conditions of the system (lake outlet conditions, diversion rates, and ice and weed retardation) represent those of the present system. Initial conditions of lake levels and outflows were taken from forecast summary sheets from the U.S. Army Corps of Engineers.

For each month of an extended (multi-month) outlook, the water levels resulting from the routed alternative water supply scenarios were ranked, and probabilities were assigned using the empirical distribution (Linsley, et al. 1982):

$$p_i(x) = \frac{m_i}{n_i + 1} \tag{1}$$

where p_i is the probability of water level x being equaled or exceeded in a given month i, and n_i is the number of historical water supply scenarios available for that month. The variable m_i is the rank assigned to the water levels, sorted in descending order. The highest water level has $m_i = 1$; the lowest, $m_i = n_i$. The empirical distribution for each month of the outlook was then interpolated to determine water levels associated with even intervals of exceedance probabilities, and the results plotted to produce a probabilistic forecast.

Risk Assessment: A Case Study

Throughout 1986, Lake Michigan set record high monthly mean water levels. As recorded at Milwaukee, Wisconsin, the maximum monthly lake level of 177.36 m occurred in October of that year. In November, concerned that the trend of rising lake levels would continue, the Milwaukee County Board of Supervisors requested the Southeastern Wisconsin Regional Planning Commission to prepare a prospectus for a possible study of the impacts of high Lake Michigan water levels on the area surrounding the Milwaukee Harbor (downtown Milwaukee). Potential problems from rising lake levels included flooding in the Menomonee River Valley and other riverine areas along the Milwaukee Harbor estuary; the flow of inner harbor estuary waters back over diversion gates into intercepting sewers, and through sewer surcharging, into basements; impaired flows from storm sewers and industrial and other clearwater discharge pipes; very high groundwater levels affecting the infiltration and inflow of clear waters into sewers, utility tunnels, and basements; the flooding of transportation facilities; and overland flooding of major utility installations such as the Jones Island sewage treatment plant.

The completed prospectus (Southeastern Wisconsin Regional Planning Commission, 1987) concluded that as long as lake levels did not exceed 177.76 m, damages from direct overland flooding problems would be localized and no large harborwide flood control projects, such as elevating the breakwater, needed to be considered. However, if lake levels were to continue to rise, the preparation of a contingency plan for flood protection was recommended. The prospectus estimated the cost of preparing this plan at $253,200.00. The Milwaukee County Board of Supervisors was faced with the decision as to whether the plan should be prepared. They could have used their knowledge of the probability that the level would exceed 177.76 m in the 12 to 24 months following October, 1986 to assess the imminent risk of flooding and aid them in making their decision.

The adapted ESP method was applied to estimate these risks. Alternative water supplies for 24-month periods, beginning in November for 1940 through 1985 [a relatively wet climate period as identified by Quinn (1981)] were routed using the Great Lakes hydrologic response model and recorded initial conditions of lake levels and outflows for November, 1986. Probabilities of exceedance were computed based on the resulting lake level scenarios; the results are shown in Figure 1.

These results show that during the next 24 months, the probability of Lake Michigan's level exceeding 177.36 m was 21%, 22%, and 19% for the upcoming months of June, July, and August, respectively. For the remainder of the 24 months, the probability of exceedance was substantially less. The probability of exceeding 177.76 m was much less than 2%. The 2% exceedance line peaked in July, 1987 at 177.5 m, 26 cm below the critical flood level. Thus, the Milwaukee Board of Supervisors could have made the decision with a high level of confidence at a time of record high levels, that a contingency flood plan was not needed .

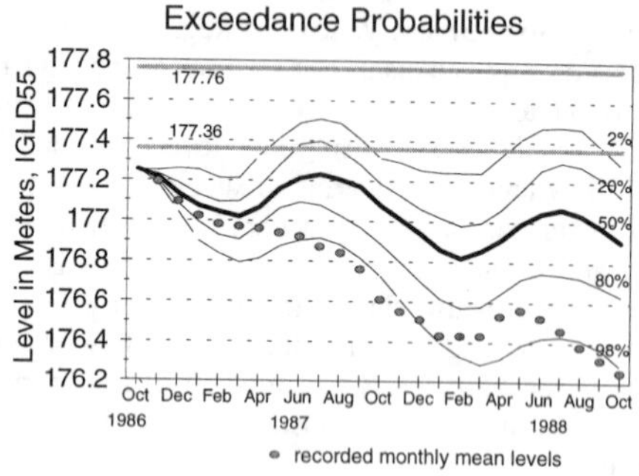

Figure 1. Probabilistic forecast of Lake Michigan water levels from October, 1986 to October, 1988.

The prospectus commissioned in November, 1986 by the Board of Commissioners to review the need for a flood protection plan was completed 1 year later in December, 1987. The prospectus recommended the development of the flood protection plan if levels were to continue rising. Ironically, Lake Michigan levels began to decline in response to below average precipitation in the winter and spring of 1987, and by December, 1987, the lake was at an elevation of 176.51 m, 0.85 m below the October 1986 high, and 1.25 m below the proposed flood level. This was an unprecedented drop in lake levels. Figure 1 shows that the actual 1987 and 1988 lake levels at times fell near the 2% non-exceedance (98% chance exceedance) line of expected probabilities. This shows that the adapted ESP approach also provides a good estimate of the probable range (with 96% probability of occurrence) of expected lake levels.

Conclusion

We have adapted a simple technique to produce Great Lakes probabilistic water level forecasts for risk assessment and have demonstrated its use in assessing the need for flood protection at Milwaukee, Wisconsin during a period of record high lake levels. The technique has many other potential applications, including lake regulation and use by hydropower and commercial navigation interests. The forecast method is simple and practical, however, several improvements to the method are suggested. First, physically-based hydrology and lake evaporation models should be used to generate the alternate water supply scenarios, driven by historical or stochastically generated meteorology. A second suggestion is to develop a more sophisticated approach to identifying climate regimes and trends (decadal to interdecadal) and to relate this to the selection or stochastic generation of alternate water supply scenarios. A final suggestion is that various distribution functions be explored and compared to the results obtained using the empirical function given in equation 1. A formal assessment of probabilistic water level forecast skill should also be made.

References

Croley, T. E, II, and D. H. Lee. (1993). "Evaluation of Great Lakes net basin supply forecasts." *Water Resources Bulletin*, 29(2), 267-282.

Day, G. N. (1985). "Extended streamflow forecasting using NWSRFS." *Journal of Water Resources Planning and Management*, 111(2), 157-170.

Lee, D. H. (1992). "An evaluation of deterministic Great Lakes water supply and lake level forecasts." Open file report, Great Lakes Environmental Research Laboratory, Ann Arbor, Michigan.

Linsley, R. K., M. A. Kohler, and J. L. H. Paulhus. (1982). *Hydrology for Engineers*. 3rd Edition. McGraw-Hill, Inc., New York, N.Y.

Quinn, F. H. (1981). Secular changes in annual and seasonal Great Lakes precipitation, 1854-1979, and their implications for Great Lakes water resource studies. *Water Resources Research*, 17, 1619-1624.

Southeastern Wisconsin Regional Planning Commission. (1987). *Milwaukee High Lake Level Impact Study Prospectus*. Waukesha, Wisconsin.

A Study on the Characteristics of Shallow-Water Flow in a Steep Channel

Chang-Shian Chen[1], Chyan-Deng Jan[2] and Kun-Chih Lin[3]

Abstract

Velocity measurements were conducted in very steep open-channel flows over smooth and rough beds. The channel slopes were changed from 1.2° to 14.0° by steps while the flow depths were retained to be 15 mm. As a result, the flows was all supercritical and the Froude number varied from 2.0 to 7.2. The velocity-distribution laws, such as logarithmic law, Coles' velocity-defect law, power law and Hama's velocity-defect law, were examined in more detail.

Introduction

The velocity profile and the resistance law in steep open-channel flow are necessary in dealing with the problems of soil erosion and sediment transport on steep channels. A large number of velocity-distribution formulas have been developed for uniform flows over a very wide hydraulically smooth bed with a mild slope. Four well-known formulas of them are briefly reviewed as follows.

(1) The logarithmic law (or called the law of the wall) is written as

$$\frac{u}{u_*} = \frac{1}{\kappa}\ln\left(\frac{u_* y}{\nu}\right) + A \tag{1}$$

where u is the mean point velocity at the vertical distance y from the bottom; u_* is the friction velocity; κ is the universal von Kármán constant; ν is the

---

[1]Associate Professor, Department of Hydraulic Engineering, Feng Chia University, Taichung Taiwan 40227, R.O.C.

[2]Associate Professor, Department of Hydraulics and Ocean Engineering, National Cheng Kung University, Tainan, Taiwan 70101, R.O.C.

[3]Graduate Student, Department of Hydraulic Engineering, Feng Chia University, Taichung Taiwan 40227, R.O.C.

kinematic viscosity; and A is the integration constant. Both κ and A are to be established by experimentation, and their values are usually taken as $\kappa \cong 0.40$ and $A \cong 5.5$, with large variations for A.

(2) Coles' velocity-defect law: Recently studies established that the entire velocity profile for fully turbulent open-channel flow could be written as (Coles, 1956, Zezu and Rodi, 1986)

$$\frac{U_{max} - u}{u_*} = -\frac{1}{\kappa}\ln\left(\frac{y}{\delta}\right) + \frac{2\Pi}{\kappa}\cos^2\left(\frac{\pi y}{2\delta}\right) \qquad (2)$$

in which U_{max} is the maximum profile velocity observed at $y=\delta$, where δ is the thickness of turbulent boundary layer and it is slightly smaller than the water depth h. For very shallow water flow, $\delta=h$. Π is the Coles' wake parameter and was given as 0.55 with $\kappa=0.4$ by Coles (1956) for flows over a smooth bed.

(3) Power law ：A simple power law of velocity distribution is given as

$$\frac{u}{U_{max}} = \left(\frac{y}{\delta}\right)^{\frac{1}{N}} \qquad (3)$$

where the exponent N varies slightly with the Reynolds number. For velocity distribution in smooth pipes, N varies from 6 to 10 when Reynolds number changes from 4,000 to 3,240,000, with the flow depth in the above equation replaced by hydraulic radius (Schlichting, 1979, p.600).

(4) Hama's velocity-defect law: According to Sarma et al. (1983), Hama proposed a parabolic velocity distribution for the region, $y/h>0.15$, in the form

$$\frac{U_{max} - u}{u_*} = C\left(1 - \frac{y}{\delta}\right)^2 \qquad (4)$$

where C is an empirical coefficient to be determined. According to Sarma et al. (1983), it was given as 9.6 and 6.3, by Hama and Bazin, respectively.

Based on the assumption of that the velocity laws established from the flows over mild slopes are still valid for flows over steep slopes. This paper is aimed to clarify the values of the von Kármán constant κ and the integral constant A in the log-law velocity distribution, the Coles' wake parameter Π, the exponent N in the power law, and the coefficient C in the Hama's velocity-defect law for very steep open-channel flows over smooth and rough beds.

Experiments

In order to investigate the velocity distribution of steep open-channel flow over smooth and rough beds, experiments were conducted in a laboratory. They were carried out in an indoor glass-walled rectangular channel having dimensions of 6.0 m in length, 0.3 m in width and 0.3 m in depth. The channel slope could be tilted from -3° to 21°. The channel was connected at its upstream end to a head supply tank. The water was supplied to the channel through the head tank by a centrifugal pump from the a large main water tank in the laboratory. The water flowing through the channel was allowed to fall into a rectangular notch and then fall back to the main tank. A settling chamber and screens were installed at the entrance of the channel in order to prevent the occurrence of large-scale disturbances and to achieve uniform entrance flow. The velocities were measured at the mid-vertical of the flow cross section with a Prandtl-type pitot-static tube. In order to assure the establishment of the fully developed flow the measuring section was chosen near the downstream end of the channel. The mean water depth was adjusted to be 15 mm for all experiments and thus the aspect ratio was retained to be 20. The inclinations of the channel bed were changed between 1.2° to 14.0° by steps in the present study. As a result, the flows were all supercritical and the Froude number varied from 2.0 to 7.2. Roll waves were observed since the flows were supercritical and they resulted in the slightly variations of water depth. Velocity measurements were conducted in the steep open-channel flows over smooth and rough beds. A varnished wooden board was placed on the bottom of the experimental tilting channel to form a smooth bed and a piece of A-80 emery cloth (having mean emery size of 0.18 mm) was closely placed on the smooth wooden board to form a rough bed. Some details of the experiments are summarized in Table 1.

Results and Conclusions

Velocity profiles in the very steep open-channel flows over smooth and rough beds were measured, as indicated in Figs. 1 and 2. They were then compared with four kinds of velocity-distribution laws, such as logarithmic law, Coles' velocity-defect law, power law and Hama's velocity-defect law, with the help of curve fitting techniques. Through these comparisons, the von Kármán constant κ in the log law was found to be 0.39 and 0.41 for flows over smooth and rough beds, respectively, which was close to the universal value of κ (0.40) for flows over mild-slope channels, as shown in Fig. 3. On the other hand, as indicated in Fig. 4, the integral constant A of the log-law distribution became smaller with an increase of the Froude number. The value of A became negative when the Froude number was larger than about 3. Fig. 4 also indicated that the value of A for smooth bed was larger than that over rough bed. The Coles' wake parameter Π was found to be about 0.07 in supercritical flows, close to 0.08 obtained by Cardoso (1989) in subcritical flows. The exponent N in the power law was 7 for flows over the smooth bed, and was 6.4 over the rough bed. The coefficient C in the Hama's velocity-defect law was 6.2 and 5.8 for flows over the smooth and rough beds, respectively. Both N and C seem independent of the Froude number, but smaller with the increase of bed roughness.

References

1. Cardoso, A. H., Graf, W. H., and Gust, G. (1989). "Uniform flow in a smooth open channel." J. Hydr. Res., IAHR, 27(5), 603-615.
2. Coles, D. (1956). "The law of the wake in the turbulent boundary layer." J. Fluid Mech., 1. 191-226.
3. Kirkgöz, M. S. (1989)."Turbulent velocity profile for smooth and rough open channel flow." J. Hydr. Engrg., ASCE, 115(11), 1543-1561.
4. Nezu, I., and Rodi, W. (1986). "Open channel flow measurements with a laser doppler anemometer." J. Hydr. Engrg., ASCE, 112(5), 335-355.
5. Sarma, K.V.N. et al. (1983). "Velocity distribution in smooth rectangular open channels." J. Hydr. Engrg., ASCE, 109(2) 270-289.
6. Schlichting, H. (1979). "Boundary layer theory." 7 th edition, McGraw-Hill, New York, N.Y.

Table l. Some details of experiments, in which θ is the bed inclination angle, h is the mean water depth, U is the depth averaged velocity, F_r is the Froude number defined as $F_r = U/\sqrt{gh}$, g is the gravitational acceleration .

Bed Type	Test No.	θ (O)	h mm	U (m/s)	F_r	Bed Type	Test No.	θ (O)	h mm	U (m/s)	F_r
Smooth	A1	1.2	15	0.965	2.52	Rough	B1	1.2	15	0.784	2.04
Smooth	A2	1.6	15	1.100	2.87	Rough	B2	1.6	15	0.964	2.51
Smooth	A3	2	15	1.224	3.19	Rough	B3	2	15	1.053	2.74
Smooth	A4	3	15	1.399	3.65	Rough	B4	3	15	1.204	3.14
Smooth	A5	4	15	1.683	4.39	Rough	B5	4	15	1.405	3.66
Smooth	A6	5	15	1.806	4.71	Rough	B6	5	15	1.575	4.11
Smooth	A7	6	15	1.979	5.16	Rough	B7	6	15	1.659	4.33
Smooth	A8	7	15	2.186	5.70	Rough	B8	7	15	1.785	4.65
Smooth	A9	8	15	2.267	5.91	Rough	B9	8	15	1.929	5.03
Smooth	A10	9	15	2.367	6.17	Rough	B10	9	15	2.018	5.26
Smooth	A11	10	15	2.503	6.53	Rough	B11	10	15	2.096	5.46
Smooth	A12	11	15	2.522	6.58	Rough	B12	11	15	2.190	5.71
Smooth	A13	12	15	2.596	6.77	Rough	B13	12	15	2.234	5.83
Smooth	A14	13	15	2.655	6.92	Rough	B14	13	15	2.240	5.84
Smooth	A15	14	15	2.759	7.20	Rough	B15	14	15	2.399	6.25

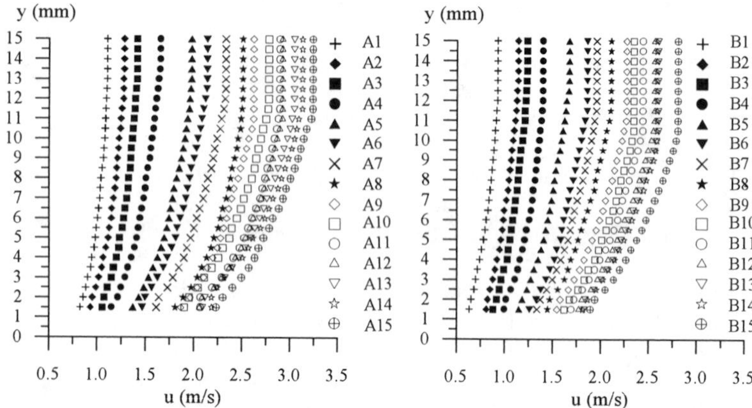

Fig. 1. Velocity Distributions for Smooth Bed.

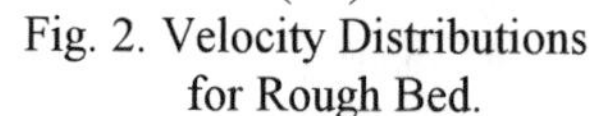

Fig. 2. Velocity Distributions for Rough Bed.

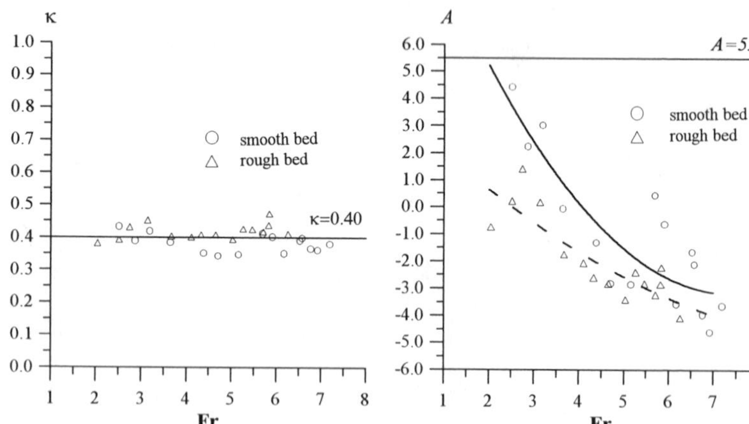

Fig. 3. Variation of the von Kármán k with the Froude Number.

Fig. 4. Variation of the Integration Constant A with the Froude Number.

ROBUSTNESS OF DE SAINT VENANT EQUATIONS FOR SIMULATING UNSTEADY FLOWS

Robert A. Baltzer[1], Raymond W. Schaffranek[1], and Chintu Lai[2], Members,ASCE

ABSTRACT: Long-wave motion in open channels can be expressed mathematically by the one-dimensional de Saint Venant equations describing conservation of fluid mass and momentum. Numerical simulation models, based on either depth/velocity or water-level/discharge dependent-variable formulations of these equations, are typically used to simulate unsteady open-channel flow. However, the implications and significance of selecting either dependent-variable form—on model development, discretization and numerical solution processes, and ultimately on the range-of-application and simulation utility of resulting models—are not well known. Results obtained from a set of numerical experiments employing two models—one based on depth/velocity and the other on water-level/discharge equation formulations—reveal the sensitivity of the two equation sets to various channel properties and dynamic flow conditions. In particular, the effects of channel gradient, channel width-to-depth ratio, flow-resistance coefficient, and flow unsteadiness are analyzed and discussed.

INTRODUCTION

Engineering practice, today, is confronted by a broad array of unsteady, open-channel flow problems resulting from a spectrum of superposed wave types occurring in a variety of channel geometries. Insight and knowledge gained about the hydraulic properties of such systems are often necessary to address environmental concerns. Wave-generated flows in waterways are recognized as unsteady open-channel flows and, as such, can be simulated in the space/time domain by numerical models based, typically, on either a depth/velocity or water-level/discharge dependent-variable formulation of the de Saint Venant equations. During the past four decades, these two forms have been the "starting point" for many investigators developing numerical simulation models. Yet, the implications and significance of selecting a particular dependent-variable form of the equations upon model development, numerical treatment, and ultimately upon model implementation and appropriateness for a broad range of prototype applications are not well known.

In an ongoing study, the appropriateness and range-of-applicability of the two equation forms are being evaluated using two different models—a method-of-characteristics model with dependent variables depth, h, and velocity, u, and a four-point implicit finite-difference model with dependent variables water level, Z, and discharge, Q. Preliminary results and discussions presented in this paper represent initial findings in which only the

[1]Hydrologist, U.S. Geological Survey, National Center, MS 430, Reston, VA 22092.

[2]Visiting Professor, Dept. of Civil Engineering, National Taiwan University, Taipei, Taiwan.

inherent form of the equations is evaluated using the applicable model in a set of numerical experiments. Follow-on efforts will utilize these same numerical models to evaluate the alternate equation forms and thereby identify possible solution dependencies. Sets of numerical experiments are being conducted in rectangular channels having different bottom gradients and widths and by employing varied flow-resistance coefficients. Unsteady flows defined by discharge hydrographs representing floods of various rates-of-rise and fall are imposed as boundary conditions. Channel properties and hydrodynamic conditions were selected to exercise both models for a broad range of sub-critical flows.

EQUATION SETS

A variety of differential-equation sets can be derived for depicting unsteady open-channel flow (Cunge et al. 1980; Lai 1986). Two sets, one using h and u and the other using Z and Q as dependent variables, in common use in numerical models are:

$$\frac{\partial h}{\partial t} + u\frac{\partial h}{\partial x} + H\frac{\partial u}{\partial x} + \frac{u}{B}A_x^h - \frac{q}{B} = 0, \tag{1}$$

$$\frac{\partial u}{\partial t} + u\frac{\partial u}{\partial x} + g\frac{\partial h}{\partial x} - gS_b + gS_f + \frac{q(u-u')}{A} - \xi\frac{V_w^2}{H}\cos\alpha = 0, \tag{2}$$

and

$$\frac{\partial Z}{\partial t} + \frac{1}{B}\frac{\partial Q}{\partial x} - \frac{q}{B} = 0, \tag{3}$$

$$\frac{\partial Q}{\partial t} + \frac{Q}{A}\frac{\partial Q}{\partial x} + Q\frac{\partial}{\partial x}\left(\frac{Q}{A}\right) + gA\frac{\partial Z}{\partial x} + gAS_f - qu' - \xi BV_w^2\cos\alpha = 0. \tag{4}$$

In these equations, water is assumed to be of constant homogeneous density, internal pressure is considered hydrostatic, the channel bottom is taken to be rigid or relatively stable and fixed with respect to time, and velocity is assumed to be uniform over the cross-sectional area. In both equation sets, distance, x, along the longitudinal axis of the channel and time, t, are the independent variables. Other terms in the equations represent the nonprismatic-channel factor, $A_x^h = \frac{\partial A}{\partial x}|_h$ (in which A is cross-sectional area); channel top width, B; hydraulic depth of the cross section, H ($\approx A/B$); lateral flow discharge per unit length of channel, q; x-component of lateral-flow velocity, u'; gravitational acceleration, g; bed slope, S_b; friction slope, S_f; wind-stress coefficient, ξ; and wind velocity (occurring at angle α with the x-axis of the channel), V_w. These two sets of partial-differential equations are mathematically equivalent if, and only if, all functions and variables are at least once differentiable.

MODEL DESCRIPTIONS

A multimode method-of-characteristics (MMOC) model based on dimensionless characteristics equations derived from Eqs. 1 and 2 and an implicit, finite-difference model, referred to as BRANCH, using Eqs. 3 and 4 were employed in the numerical experiments. The MMOC model (Lai 1988) combines the implicit, temporal reachback, spatial reachback, and classical schemes of the specified-time-interval MOC schemes into a single, multimode scheme. The computational algorithm is designed so that for a given time and location the model automatically selects the appropriate numerical mode for the characteristic of concern. The BRANCH model (Schaffranek et al. 1981; Schaffranek 1987) uses a four-point, implicit, finite-difference approximation of the unsteady-flow equations. In the four-point, Preissmann (1961) or box, scheme, the finite-difference approximation of the spatial derivatives can be varied by weighting factor, θ, from box-centered ($\theta = 0.5$) to fully forward ($\theta = 1$), which is the range for which its stability has been proven (Fread 1974). In the BRANCH numerical experiments, $\theta = 0.6$ was used, unless otherwise indicated. Both models use a Manning formulation of the S_f term.

EXPERIMENTAL CHANNEL AND FLOW CONDITIONS

Results presented in this paper were obtained from numerical experiments conducted in hypothetical, uniform, rectangular channels 8000 m in length, having bottom gradients (S_b) of 0.0005, 0.0020, and 0.0050, and widths (B) of 6 and 24 m. Manning-type flow-resistance coefficients (η) of 0.025 and 0.050 were assigned to each channel. The set of hydraulic conditions represented by these 12 combinations of geometric and flow-resistance parameters were purposely chosen to evaluate the sensitivity and range-of-applicability of the two dependent-variable formulations embodied in the models for a broad range of open-channel conditions.

Unsteady flows were generated by imposing a triangular-shaped hydrograph upon an initial, steady-state uniform-flow discharge of 18 m^3s^{-1} at the channel entrance. The results, presented herein, were generated using a $Q(x,t)$ hydrograph with the following flow characteristics, where $x = 0$ is the channel entrance and t is time in seconds: $Q(0,t) = Q_s$, $(0 \leq t \leq 240)$; $Q(0,t) = Q_s + r_1(t - 240)$, $(240 < t \leq 1440)$; $Q(0,t) = Q_{\max} + r_2(t - 1440)$, $(1440 < t \leq 3840)$; and $Q(0,t) = Q_s$, $(t > 3840)$. Here, $Q_s = 18$ m^3s^{-1}, $Q_{\max} = 54$ m^3s^{-1}, $r_1 = 0.030$ m^3s^{-2}, and $r_2 = -0.015$ m^3s^{-2}. In summary, the hydrograph at the channel entrance consists of a two-fold increase in discharge in 20 minutes that completely subsides in 40 minutes, after which the initial steady-state discharge resumes. Uniform flow depth was employed as a boundary condition at the downstream end of the channel in all experiments.

EXPERIMENTAL RESULTS

For inter-model comparison, consistency in simulation design and model setup are important for proper interpretation and evaluation of results. Initially, all numerical experiments with both models were to be conducted using a one-minute time step. However, due to the inherent properties of MMOC, a time step appropriate to the prevailing hydrodynamic conditions was required. Thus, whereas for eight of the 12 MMOC simulations a 10 s time step was used, a reduced time step of 5 s was required for the $B = 24$ m, $S_b = 0.0050$, and $\eta = 0.05$ condition. A larger time step of 30 s was used for both the 6 and 24 m channels with $S_b = 0.0020$ and $\eta = 0.025$ and a 60 s time step was used for the $B = 6$ m, $S_b = 0.0005$, and $\eta = 0.025$ condition. All 12 BRANCH model simulations were conducted using a one-minute time step. The spatial-derivative weighting factor, θ, was increased to 0.65 in one case and to 0.70 in two other cases, to minimize some observed secondary oscillations, or computational mode, evident in the computed results following the flood recession.

The 8000 m channel was subdivided into 400 m segments for both model discretizations and computed results were obtained at 1600 and 3200 m downstream of the channel entrance for each of the 12 combinations of channel and flow-resistance properties. Computed discharges from the BRANCH and MMOC models at these locations, for each of the three channel gradients, are grouped into four graphs according to channel width and flow-resistance coefficient as illustrated in Figs. 1 and 2, respectively. Computed depths (hydrographs not shown) were likewise observed to be very comparable at both locations for each dependent-variable formulation represented by the two models. Differences of

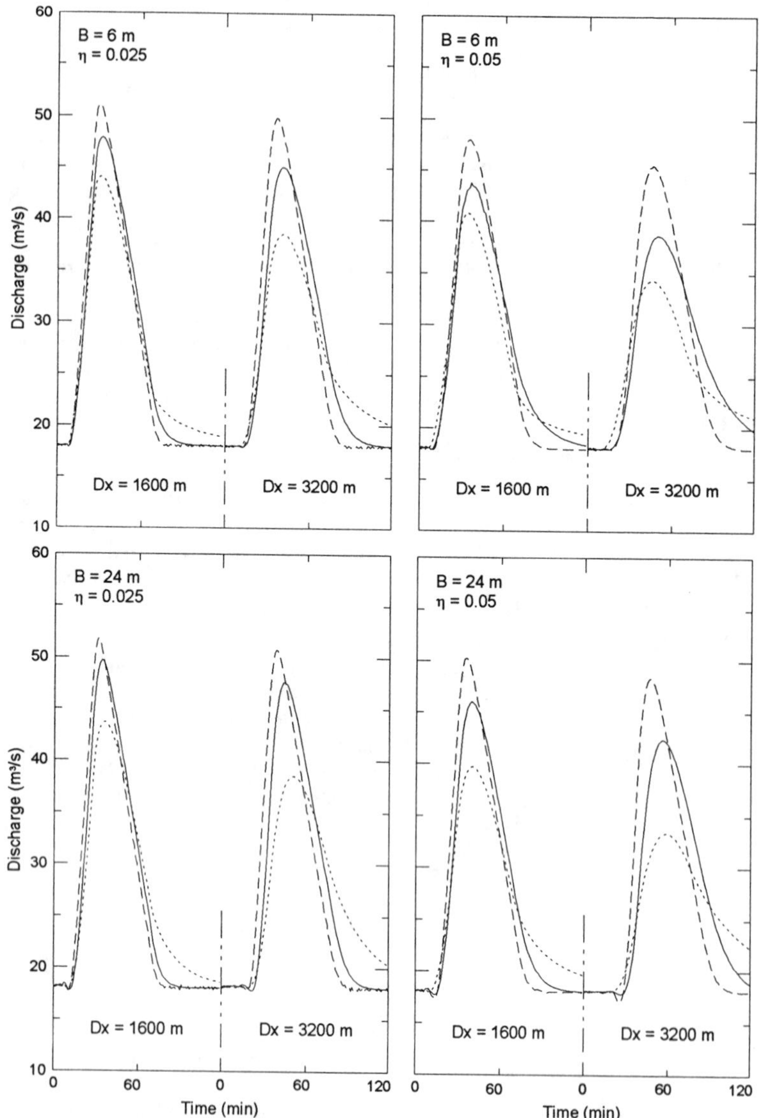

Fig. 1. Discharges Computed Using Resistance-Coefficient Values of 0.025 and 0.050 and Obtained from BRANCH Model at 1600 and 3200 m Locations in 6 and 24 m Wide Rectangular Channels with Bottom Gradients of 0.0005 (Dotted Line), 0.0020 (Solid Line), and 0.0050 (Dashed Line).

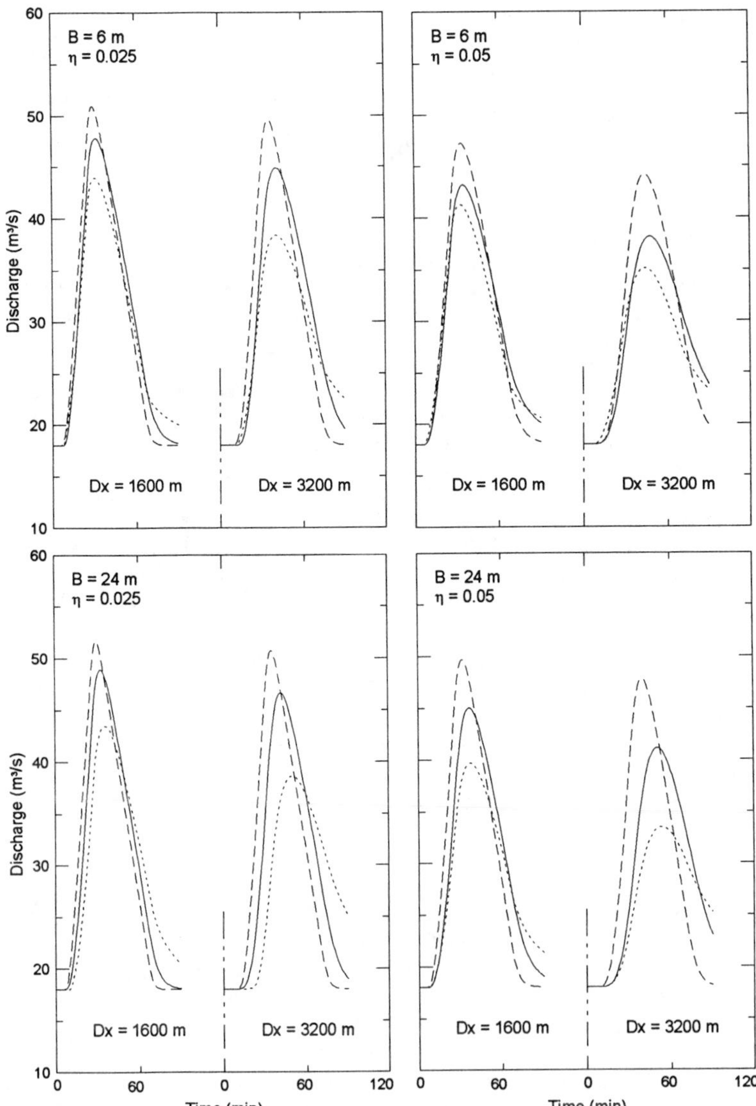

Fig. 2. Discharges Computed Using Resistance-Coefficient Values of 0.025 and 0.050 and Obtained from MMOC Model at 1600 and 3200 m Locations in 6 and 24 m Wide Rectangular Channels with Bottom Gradients of 0.0005 (Dotted Line), 0.0020 (Solid Line), and 0.0050 (Dashed Line).

0.02 m or less in peak depths computed by the two models were noted for 17 of the 24 observations. Times of occurrence of the peak depths and maximum discharges also compare favorably, with differences of two minutes or less for 14 of the 24 observations. However, a subtle increase in travel-time differences between the two models is noted for the wide-channel experiments with a tendency for diverging travel times with distance propagated downstream. For the imposed dynamic flows and channel properties, maximum Froude numbers ranged from 0.16 to 0.87 and maximum width-to-depth ratios ranged from 1.38 to 27.9.

SUMMARY AND CONCLUSIONS

A study is being conducted to evaluate the significance and range-of-application implications of using unsteady-flow simulation models based on either depth/velocity or water-level/discharge formulations of the de Saint Venant equations. Preliminary results of numerical experiments conducted over a broad range of hydrodynamic conditions in hypothetical rectangular channels using two models, each embodying one of the dependent-variable formulations, are presented. Despite dissimilar solution methods for treating the two different formulations, computed results from the models compare very favorably. These results reveal no evidence to support a conclusion that either formulation is better suited for simulation of high Froude number, sub-critical flows in high-gradient open channels of similar characteristics. However, observed secondary differences in the experimental results, most evident in the propagation rates, cannot be attributed, as yet, to formulation differences, to discretization or numerical solution processes, or to other undetermined factors. To provide more complete and definitive answers, numerical simulations in which both dependent variables are treated using the same solution technique are planned as a continuation of this effort. Additionally, hypothetical non-rectangular channels are being investigated.

REFERENCES

Cunge, J.A., Holly, F.M., and Verwey, A. (1980). *Practical Aspects of Computational River Hydraulics*, Pitman Publishing Ltd., London, Eng.

Fread, D.L. (1974). "Numerical properties of implicit four-point finite difference equations of unsteady flow." *Technical Memorandum NWS HYDRO-18*, U. S. National Oceanic and Atmospheric Admin., Silver Spring, Md.

Lai, C. (1986). "Numerical modeling of unsteady open-channel flow." *Advances in Hydroscience*, V.T. Chow and B.C. Yen, eds., Vol. 14, Academic Press, Orlando, Fla., 163-333.

Lai, C. (1988). "Comprehensive method of characteristics models for flow simulation." *J. Hydr. Engrg.*, ASCE, 114(9), New York, N.Y., 1074-1097.

Preissmann, A. (1961). "Propagation of translatory waves in channels and rivers." *Proc., First Congress of French Assoc. for Computation*, Grenoble, France, 433-442.

Schaffranek, R.W. (1987). "Flow model for open-channel reach or network." *Professional Paper 1384*, U.S. Geological Survey, Denver, Colo.

Schaffranek, R.W., Baltzer, R.A., and Goldberg, D.E. (1981). "A model for simulation of flow in singular and interconnected channels." *Book 7, Chap. C3, Techniques of Water Resources Investigations*, U.S. Geological Survey, Denver, Colo.

2D MODELING OF CLASS B BRIDGE FLOW IN STEEP CHANNELS

By Richard L. Stockstill,[1] M. ASCE

ABSTRACT

Bridge piers located in flood control channels are classified by the relation of flow depth through the bridge section to critical depth upstream, between, and downstream of the piers. The term Class B flow is applied to conditions in which subcritical flow approaches the bridge, passes through critical depth at a point along the piers, and then jumps to subcritical flow or remains supercritical depending on the downstream conditions. The far field flow upstream of the bridge may be subcritical or supercritical. The flows studied in this paper are of the Class B type in hydraulically steep channels. The bridge pier constriction chokes the flow producing a hydraulic jump upstream of the piers. The flow downstream of the piers is supercritical. The numerical flow model HIVEL2D is used to simulate the rapidly varied flow. HIVEL2D is a depth-averaged, two-dimensional (2D) flow model designed specifically for flow fields that contain supercritical and subcritical regimes as well as the transitions between the regimes. Simulation results are compared with published laboratory data.

INTRODUCTION

Analysis of flow conditions in the vicinity of bridge crossings is complicated by the fact that the bridge piers serve as flow obstructions. Bridge pier scour during flood events is a major engineering concern; however, in concrete-lined high-velocity channels, scour is not a problem but pier effects on the flow are significant. Flood control channel design is primarily interested in sizing channel sidewall heights. An accurate prediction of the flow depth is imperative in successful design of flood control channels. Flow depth prediction is complicated by bridge piers which can influence the flow depth both upstream and downstream of the bridge crossing.

Bridge piers located in high-velocity channels where the channel slope produces

[1]Research Hydraulic Engineer, Hydraulics Laboratory, US Army Engineer Waterways Experiment Station, 3909 Halls Ferry Road, Vicksburg, MS 39180-6199.

a normal depth less than critical depth can choke the flow resulting in backwater effects from the bridge. Choking situations in steep channels produce a hydraulic jump upstream of the bridge. Choking in high-velocity channels results in flow depths that are significantly higher than those in the absence of choke. Downstream of the bridge, supercritical flow through the bridge piers produces standing waves generated at the pier tails.

In the analysis of high-velocity channel flow at bridge piers, the design engineer must be able to predict whether the flow obstructions will choke the flow and if they do, determine the upstream depth of flow. The flow depth downstream of the bridge must also be determined and in supercritical flow, the height of standing waves must be predicted. Accurate determination of flow depths are required for sidewall height and minimum bridge soffit elevation determinations.

A one-dimensional analysis is adequate when the flow conditions are similar to those assumed in the model development. These assumptions commonly include uniform (or known) velocity distribution across the channel. However, is cases where the velocity distribution is not uniform (and not known a priori), 2D modeling of the flow is more appropriate. Two-dimensional modeling lends itself to the simulation of skewed bridges, bridges in or immediately downstream of channel bends where the velocity is not uniformly distributed, and to complex pier configurations and bridge abutments.

This paper describes the application of HIVEL2D (Stockstill and Berger 1994) to the simulation of flow at a Class B bridge crossing a steep channel. HIVEL2D is a 2D depth-averaged flow model for high-velocity channels. The model is a Petrov-Galerkin finite element representation of the shallow water equations. The simulation results are compared with laboratory data reported by Stonestreet (1990).

MODEL TESTING

Tests were conducted to evaluate the usefulness of HIVEL2D in simulating Class B bridge flow. The hydraulic conditions chosen for testing were one of those tested in a laboratory flume by Stonestreet (1990). The tests were conducted in a 60 ft (18.29 m)-long by 4 ft (1.22 m)-wide rectangular tilting flume. The pier configuration consisted of four 1 inch (0.025 m)-wide by 1 ft (0.305 m)-long piers each having a semicircular nose and tail. The piers were spaced evenly across the flume width. The flume discharge was 4.75 cfs (0.135 cms) and the slope was set at 0.0032. Although this is a simple geometric design for a 2D analysis, the test case demonstrates HIVEL2D's ability to capture the choking and hydraulic jump upstream and the standing waves downstream of the piers.

Computations were made using a 486 personal computer. The finite element mesh consisted of 773 nodes and 720 elements. Lateral resolution across the flume varied from 4 elements at the upstream and downstream ends to 28 elements in the vicinity of the piers. The numerical model was time stepped to steady state from the initial conditions which were those given for normal flow conditions in the absence of any piers. Both inflow and outflow boundaries are supercritical and the boundary conditions were set appropriately. At the inflow boundary, a unit discharge of 1.1875 cfs/ft (0.1103 cms/ft) and flow depth of 0.32 ft (0.0975 m) were specified; whereas, no boundary conditions were specified at the outflow boundary.

Water-surface profiles along the right wall and centerline are provided on Figs. 1 and 2, respectively. The profiles illustrate that the flow is subcritical upstream of the piers, accelerates through critical as it flows between the piers, and is supercritical downstream of the piers. The hydraulic jump was computed to be located approximately 17 ft (5.18 m) upstream of the piers; whereas, Stonestreet noted an undular jump occurring about 10 ft (3.05 m) upstream of the piers. Generally, the comparison between flume data and model results is good. Examination of the results with regard to design considerations mentioned previously indicates the model is a beneficial design/evaluation tool. The model captured the choke and resulting hydraulic jump upstream of the bridge and the flow depth upstream and downstream of the bridge though the computed upstream water surface elevation is slightly conservative. The model accurately reproduced the water surface elevation at the pier nose which is essential in determining the minimum bridge soffit elevation. Although local differences along the profile are troubling, the overall energy losses through the bridge compare quite well. Simulation results indicate the flow accelerating more rapidly than the flume results show. That is model results show the flow passing through critical upstream of that observed in the flume. This difference is probably due to the hydrostatic pressure distribution assumption enforced in the numerical model (see Berger 1994). Fig. 3 is a map of computed depth contours. The contours show the oblique standing waves generated at the pier tails.

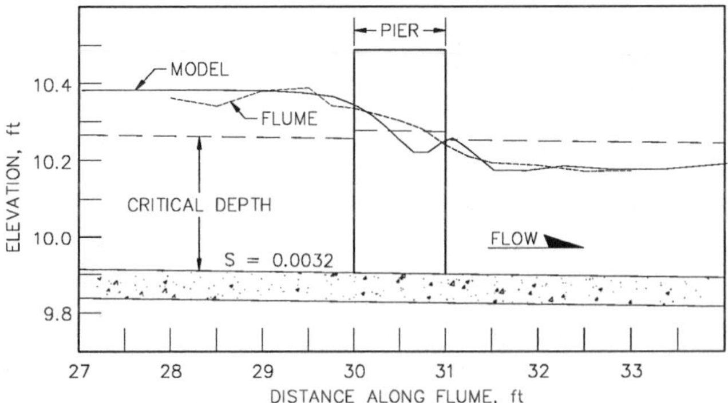

FIG. 1. Water-Surface Profiles along Right Wall

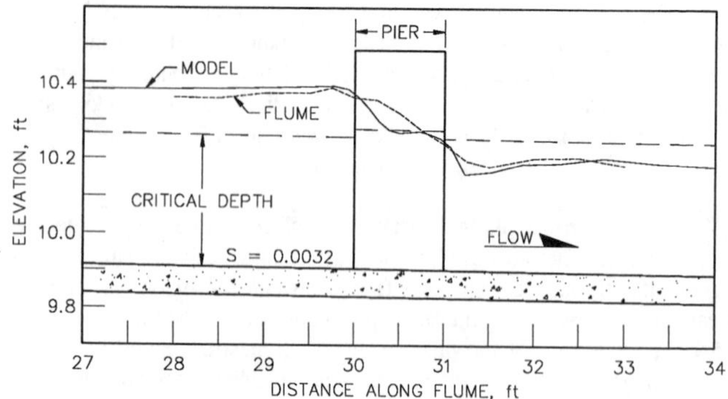

FIG. 2. Water-Surface Profiles along Centerline

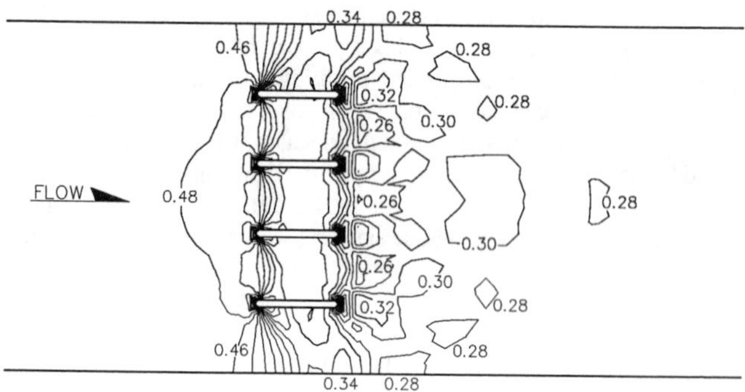

FIG. 3 Depth Contours, ft

SUMMARY AND CONCLUSIONS

Bridge piers located in high-velocity channels are sources of flow disturbances downstream and possibly upstream of the bridge crossing. If the piers result in a choked flow condition, a hydraulic jump is formed upstream of the structure. The hydraulic design engineer is required to determine the wave heights downstream of the bridge and whether a choke situation exists at the design flow rate. Furthermore, if a choke occurs (Class B bridge) the resulting flow depth must be computed for sidewall height design.

HIVEL2D is a viable means of simulating flow conditions at Class B bridges in high-velocity channels. This conclusion is substantiated by the results presented herein. The model reproduced the choke condition, gave a conservative estimate for the bridge soffit elevation and matched the water surface elevations of the supercritical flow downstream of the bridge. The model is most useful when applied to geometrically complex designs consisting of nonuniform velocity distribution in the approach flow to the bridge. Such cases exist at skewed bridges and bridges having complex pier configurations associated with multiple bridge crossings or when the bridge is located in or immediately downstream of a channel bend.

ACKNOWLEDGEMENT

The tests described and the resulting data presented, unless otherwise noted, were obtained from research funded by the Corps of Engineer's Repair, Evaluation, Maintenance, and Rehabilitation (REMR) research program. Permission was granted by the Chief of Engineers to publish this information.

APPENDIX I. CONVERSION FACTORS, NON-SI TO SI UNITS

Non-SI units of measurements used in this paper can be converted to SI units as follows:

Multiply	By	To Obtain
cubic feet	0.02831685	cubic meters
feet	0.3048	meters

APPENDIX II. REFERENCES

Berger, R. C. (1994). "Strengths and weaknesses of shallow water equations in steep open channel flow," *Hydraulic Engineering, Proc. 1994 National Hydraulics Conference*, ASCE, G. V. Cotroneo and Rumer, R. R. eds., Buffalo, New York, 1257-1262.

Stockstill, R. L., and Berger, R. C. (1994). "HIVEL2D: a two-dimensional flow model for high-velocity channels." *Technical Report No. REMR-HY-12*, U.S. Army Engr. Waterways Experiment Station, Vicksburg, Mississippi.

Stonestreet, S. E. (1990). "Critical flow through bridge piers in supercritical lined channels," thesis presented to the California State University, at Long Beach, California, in partial fulfillment of the requirements for the degree of Master of Science.

NUMERICAL OSCILLATION ANALYSES OF RAPIDLY–VARIED UNSTEADY CHANNEL FLOWS

Chin-Lien Yen,[1] F. ASCE, Chintu Lai,[2] M. ASCE,
and Wen-Cheng Lee[3]

Abstract

As a part of comprehensive numerical stability studies on rapidly-varied unsteady open-channel flow models, a series of numerical investigations on several dimensionless parameters have been conducted for identification of the most sensitive one in terms of numerical oscillations. First, five parameters, or "numbers" have been organized; these include: (a) the steepness number, or the modeling slope, $S_m = S_o/(h_o/\Delta x)$, a ratio of physical slope S_o to computational slope $h_o/\Delta x$; (b) the Courant number, $C_r = \lambda/r = (dx/dt)$ / $(\Delta x/\Delta t)$, a ratio of physical celerity λ to computational celerity r; (c) the B.C. number, $P_{bc} = \Delta h/h_p$, an indicator for boundary-value variation in terms of flow depth h; (d) the nonprismatic factor, $P_a = (A_x^h L_o)/A$, (A_x^h = change of cross-sectional area A in x-direction with h held constant), reflecting cross-sectional variation; and (e) the Froude number. Two numerical models, one using the four-point implicit finite difference method (IFDM), and the other using the multimode method of characteristics (MMOC), have been selected for the study. Using the Froude number as the base parameter going across the other four, the experimental results have indicated that the effect of cross-sectional variation, P_a, has the most sensitive response to numerical oscillations. Based on this finding, further experiments designed to highlight the P_a effects are carried out with the IFDM model, which has demonstrated that two finite difference forms corresponding to water-depth gradient, $\frac{\partial h}{\partial x}$, and the velocity-head gradient, $\frac{\partial}{\partial x}(\frac{u^2}{2g})$, are most apt to oscillate numerically with changes in parameter values. These findings are useful in the further task of model improvement.

Introduction

As a part of comprehensive and continuing studies concerning numerical stability of rapidly-varied unsteady open-channel flow modeling, a series of numerical investigations have been conducted with respect to numerical oscillations. Such oscillations, when occuring in nonlinear cases, are sometimes referred to as spurious oscillations, parasitic waves, or Gibb's effects. [For example, see Sod (1985)] These numerical instabilities

[1] Professor, Dept. of Civil Engrg. and Hydraulic Research Laboratory, National Taiwan University, Taipei, Taiwan.

[2] Visiting Professor, Dept. of Civil Engrg. and Hydraulic Research Laboratory, National Taiwan University, Taipei, Taiwan.

[3] Graduate student and Research assistant, Dept. of Civil Engrg. and Hydraulic Research Laboratory, National Taiwan University, Taipei, Taiwan.

or difficulties, often encountered in discontinuous or high-gradient flows, are usually not the same as other numerical instability problems normally analyzed or treated by more or less regular numerical analysis techniques such as von Neumann's Fourier series analysis. Their cause, behavior, or remedy are generally remain mysterious, difficult to analyze, or simply intractable. Currently, sets of carefully designed and well controlled numerical experiments seem to be the best and most rational approach to address this type of problems.

The purpose of this paper is to investigate factors causing numerical oscillations in simulation of rapidly-varied flood flows and to identify parameters that have relatively high sensitivity in regards to the oscillation. The location of such parameters should enable the model developer to devise a counter measure with which to reduce numerical oscillations.

In the following paragraphs, five dimensionless parameters are organized for the proposed stability analyses. Two numerical models, one using the four-point implicit finite difference method (IFDM), and the other relying on the multimode method of characteristics (MMOC), are used for this study. With the focus on sensitive parameters, further numerical experiments are used to uncover the most oscillation-prone terms.

Equations for Unsteady Open-Channel Flows

The governing PDEs have a variety of forms (Lai, 1986). The following two forms, one using discharge, Q, and depth h, as the two dependent variables,

$$B\frac{\partial h}{\partial t} + \frac{\partial Q}{\partial x} = 0 \tag{1}$$

$$\frac{\partial Q}{\partial t} + \frac{\partial}{\partial x}\left(\frac{Q^2}{A}\right) + gA\frac{\partial h}{\partial x} = gAS_o - gAS_f \tag{2}$$

and the other using velocity, u, and depth, h, as the two dependent variables,

$$\frac{\partial h}{\partial t} + u\frac{\partial h}{\partial x} + H\frac{\partial u}{\partial x} + \frac{u}{B}A_x^h = 0 \tag{3}$$

$$\frac{\partial u}{\partial t} + u\frac{\partial u}{\partial x} + g\frac{\partial h}{\partial x} = g\left(S_o - S_f\right) \tag{4}$$

in which A = cross-sectional area of flow; B = top width; S_f = friction slope; S_o = bed slope; g = acceleration of gravity; $H = A/B$ = hydraulic depth; and $A_x^h \equiv \left.\frac{\partial A}{\partial x}\right|_h$ = rate of change of A in x-direction with h held constant, or non-prismatic channel factor. Both sets of equations use x(=distance) and t(=time) as the independent variables, and consist of the equation of continuity (the first one) and the equation of motion (the second one).

Parameters for Sensitivity Analysis

Five dimensionless parameters, or "numbers" have been organized for sensitivity analysis with respect to numerical oscillation. These numbers except one, represent some modeling characteristics associated with both the physics of steep channel flows and the numerical mathematics of flow computation. They are: (a) the Steepness number, or the Modeling slope, S_m, —a ratio of physical slope S_o to computational slope $h_o/\Delta x$ (where h_o is some characteristic flow depth), i.e., $S_m = S_o\Delta x/h_o$; (b) the Courant number, C_r, —a ratio of physical disturbance-wave celerity $\lambda(=dx/dt)$ to computational information-transmission speed $r(=\Delta x/\Delta t)$, i.e., $C_r = \lambda/r$, (Note that $\Delta x/\Delta t$ is a commonly referred computational celerity and not always the maximum information-transmission speed); (c) the B.C. number, P_{bc}, —an indicator for the effect of model boundary-value variation in terms of flow depth, typically of the form $(h_b - h_p)/h_p$, where h_b and h_p are, respectively, the new boundary flow depth and the normal depth equivalent to the peak discharge. The parameter value is often influential in defining the water surface profile; (d) the Nonprismatic factor reflecting cross-sectional variation, $P_a = (A_x^h L_o)/A$, where L_o = some characteristic distance, which reduces to a top-width variation $\Delta B_r = (\Delta B/\Delta x)L_o/B$ in case of a rectangular cross section; and (e) the Froude number, $F = u/c$, ($c = \sqrt{gh}$ = celerity of gravity wave in shallow water), the only pure physical quantity among the five numbers, is used as the base parameter appearing concurrently with each of the other four.

Numerical Models for Experiments

To avoid a possible bias resulting from depending on only one model, two numerical models are used in parallel for numerical experiments for sensitivity analysis. These are a four-point implicit finite-difference model (IFDM) using (1) and (2) as the governing equations (Yen and Hus, 1984), and a multimode method of characteristics model (MMOC) with (3) and (4) for its governing equations (Lai, 1988). The IFDM model expresses (1) and (2) directly in finite-difference forms for numerical solution, whereas the MMOC model first converts (3) and (4) into the characteristics forms, then express these characteristic equations in difference forms. The simulation capabilities of the two models have been tested by a series of numerical experiments based on the aforementioned parameters.

Numerical Experiments for Sensitivity Analysis

Effects of numerical purturbation induced by variables and parameters in the models were investigated by controlled numerical experiments through a systematic use of the five numbers mentioned earlier. By using the Froude number as the base parameter for unified analyses, each of the other four was numerically tested for its sensitivity with respect to numerical oscillations associated with flow-depth computation.

Before proceeding with display of the experimental results, the definition of the maximum oscillation, $(O_s)_{max}$, used in this study, is briefly described: During the process of flow computation, if a wavy undulation in computed depth hydrograph at a node point j is detected, the following two equations are used to smooth the disturbance,

$$h_{j+\frac{1}{2}} = 0.5(h_j + h_{j+1}) \quad , \quad (h_u)_j = 0.5(h_{j-\frac{1}{2}} + h_{j+\frac{1}{2}}) \tag{5}$$

in wfich h_j and h_u are, respectively, the computed and smoothed flow depths. If the undulation persists, repeat the same process [using (5)] until the depth hydrograph becomes sufficiently smooth. The measure of oscillation is then defined by

$$O_s = \left| \frac{h_j - h_u}{h_u} \right| \tag{6}$$

A larger value of O_s indicates a more severe case of numerical oscillation. The maximum O_s value along the reach (i.e., $j = 0, 1, ...J$), at any time step is marked as $(O_s)_{max}$.

The experimental results indicate:

1. Effect of the distance increment, Δx, via analysis of S_m: The maximum value of $(O_s)_{max}$, increases as parameter S_m increases. The change of S_m has a greater influence on the $(O_s)_{max}$ in subcritical flow than in supercritical flow. Between the two models, the MMOC has less influence on the $(O_s)_{max}$ in subcritical flow, and the IFDM in supercritical flow.

2. Effect of the time increment, Δt, via analysis of C_r: In unsteady open-channel flow, the hydrodynamic Courant number, C_r, can be expressed by the transport Courant number P_t and the Froude number F, i.e.,

$$C_r = \frac{\lambda}{r} = \frac{|u \pm c|}{\Delta x/\Delta t} = \left| P_t \left(1 \pm F^{-1}\right) \right| \;\; ; \;\; P_t = \frac{u}{\Delta x/\Delta t} \tag{7}$$

Varying the C_r value in subcritical flow gives little influence on the $(O_s)_{max}$ for both methods, with MMOC being the lesser of the two. In supercritical flow, on the other hand, the IFDM renders less influence on the $(O_s)_{max}$.

3. Effect of the downstream boundary value variation P_d: (Effective only in subcritical flow). The $(O_s)_{max}$ value increase as the P_d increases for both methods, with the MMOC showing less disturbance.

4. Effect of cross-sectional variation, P_a(or ΔB_r): The $(O_s)_{max}$ increases as the nonprismatic factor P_a(or ΔB_r in case of a rectangular channel) increases. The order of disturbance is found larger with this parameter than with the preceding three, which means that P_a(or ΔB_r) has the highest sensitivity among all the parameters tested.

Summarizing the results of the above numerical experiments, the effect of cross-sectional variation, P_a(or ΔB_r), has the most sensitive response to numerical oscillations among the four parameters investigated. As a distant second comes the B.C. number pertaining to the downstream-end depth variation (P_d).

Numerical Experiments for Development of Disturbance

Having successfully identified the most sensitive parameter, further numerical experiments have been conducted to study the disturbance inception and amplification under the most severe conditions. First, a set of rectangular channels with various degrees of channel constriction (see Fig. 1) is designed, which furnishes various ΔB_r values. The experiments run on this set of channels using the IFDM model, reveal that two terms in the model, i.e., the finite-difference forms of the longitudinal flow-depth gradient, $\frac{\partial h}{\partial x}$, and the velocity-head gradient, $\frac{\partial}{\partial x}(\frac{u^2}{2g})$, are most apt to oscillate numerically by controlled change of parameters. These findings are valuable in that they enable or facilitate model developers to identify origins of numerical troubles and thereby to seek means for improvements. A similar kind of numerical experiments is planned for the MMOC model.

Summary and Conclusions

Focusing on numerical oscillations arising in numerical stability studies of rapidly-varied unsteady flow modeling, five numbers or dimensionless parameters have been organized for sensitivity analysis. They are the steepness number, or the modeling slope, $S_m = S_o/(h_o/\Delta x)$; the Courant number, $C_r = (dx/dt) \ / \ (\Delta x/\Delta t)$; the B.C. number, $P_{bc} = (h_b - h_p)/h_p$; nonprismatic factor, $P_a = (A_x^h L_o)/A$; and the Froude number, $F = u/c$. Through a series of numerical experiments using two models, the IFDM model and the MMOC model, the effect of cross-sectional variation, represented by parameter P_a, is found to be most sensitive to the numerical oscillation. Based on this finding, the next step of numerical experiments has been designed (cf. Fig. 1) and two terms, the depth gradient, $\frac{\partial h}{\partial x}$, and the velocity-head gradient, $\frac{\partial}{\partial x}(\frac{u^2}{2g})$, in their finite-difference forms, are found most apt to oscillate in the IFDM model.

Acknowledgement: Support for this research by the National Science Council, Taiwan, ROC, under grant NSC 81-0414-P-002-07-B, is gratefully acknowledged. The computing facilities provided by the Hydraulic Research Laboratory for the this research are highly appreciated.

References

Lai, C. (1986). "Numerical Modeling of Unsteady Open-Channel Flow." *in Advances in Hydroscience*, V.T. Chow and B.C. Yen, ed., v.14, Orlando, FL, Academic Press, 161-333.

Lai, C. (1988). "Comprehensive Method of Characteristics Model for Flow Simulation." *J. Hydr. Engrg.*, ASCE, 114(9), 1074-1097.

Yen, C. L., and Hsu, M. H., (1984). "Numerical Stability of Unsteady Flow Simulation in Open Channel," Proc., *4th Congress*, APD-IAHR, Chiangmai, Thailand, 815-829.

Sod, G. A., (1985). "Numerical Methods in Fluid Dynamics," Cambridge University Press.

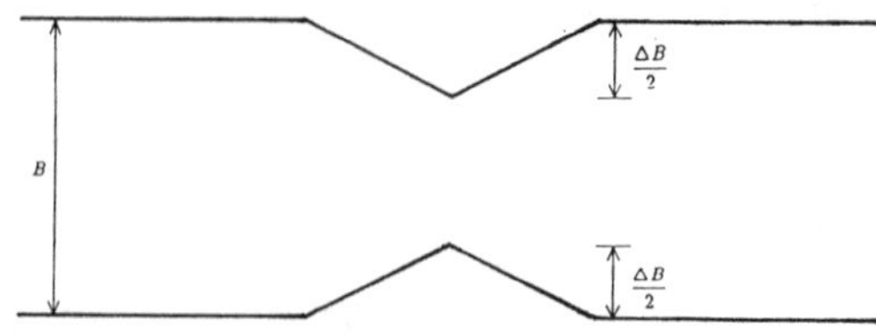

Figure 1. Definition Sketch for Variable Channel Constriction

RAILROAD CANYON DAM OVERTOPPING PROTECTION

by
Gordon Marsh[1], Tom MacDonald[1],
John Hoagland[2], Noel Wong[1]

Abstract

Railroad Canyon Dam, completed in 1928, is a 75-foot high, 375-foot long concrete arch dam, plus associated thrust blocks and wingwalls, and is located on the San Jacinto River, just upstream from Lake Elsinore, in Riverside County, California. The San Jacinto River is a tributary of the Santa Ana River. The dam impounds an 11,900 acre-ft. reservoir, which is surrounded by luxury homes. The drainage area above the dam is 700 square miles. The existing spillway was designed to pass about 35,000 cfs. Due to inadequate spillway capacity and recent changes in data and procedures for calculating Probable Maximum Precipitation (PMP), the dam would be overtopped during a theoretical Probable Maximum Flood (PMF). The California Division of Safety of Dams (DSOD) is concerned that erosion at the abutments and toe of the dam could cause it to fail during overtopping. Thus, modifications to increase spillway capacity and to provide protection of the abutments are required.

Approach

Studies were carried out to identify key issues to consider in selecting a preferred modification plan. The presence of the luxury homes around the reservoir proved to be a unique challenge for the modification of the spillway. Four key issues were identified:

---

1 Woodward-Clyde Consultants, 500 12th Street, Suite 100, Oakland, California 94607.

2 Elsinore Valley Municipal Water District, 31315 Chaney Street, Lake Elsinore, California 92530.

1) Could a larger spillway reduce reservoir surcharge to avoid flooding lakefront properties on a cost effective basis?
2) What was the cost of larger spillways?
3) What was the frequency of flooding of lakefront properties?
4) What were the downstream impacts of larger spillways?

The first step of the studies was to develop the PMP and PMF data. This was complicated by the fact that the National Oceanic and Atmospheric Administration (NOAA) was in the process of revising the hydrometeorological data to publish the HMR 58, which will replace HMR 36. The initial studies used interim and draft PMP data provided by the NOAA, estimates of the aerial and temporal distribution of the PMP, and watershed characteristics for future development conditions. These studies were done in collaboration with the Corps of Engineers, who were conducting concurrent studies for their billion dollar Santa Ana River Project. Initial results yielded PMF values ranging up to 300,000 cfs. The PMF was routed through the different conceptual spillway configurations. Flood frequency studies were also conducted to establish the probability of flooding of lakefront properties for the different conceptual spillway configurations.

Changes in downstream channel flow conditions were evaluated. This was complicated by variations in the water surface elevations in Lake Elsinore under various size floods. Lake Elsinore has a small outlet channel, therefore, most flows that enter Lake Elsinore go into storage. The National Weater Service (NWS) DAMBRK flood routing model was used to calculate the water levels between Railroad Canyon Dam and Lake Elsinore. Only small changes in flow depths occurred for small floods (<100 year return frequency).

Five spillway configurations were evaluated. The spillway configuration with the least modification, with the largest reservoir surcharge, and lowest cost, required overtopping of the entire existing dam and adding extensive downstream slope protection. The second through fifth configurations developed ever larger spillways, to result in lower reservoir surcharge levels, with increasing areas of downstream slope protection. This data is presented in Figure 1. The costs of the different configurations ranged from $5 million to $20 million.

Solution

The final inflow PMF was established at 167,000 cfs after further studies by the Corps, NOAA and Woodward-Clyde. This value was accepted by the DSOD. The configuration shown in Figures 2 and 3 was selected for design. It is the lowest cost, does not significantly change upstream reservoir surcharge levels or downstream flow conditions and provides 100-year flood protection for lakefront property. It requires removing the bridge and piers at the existing spillway, removing the top 16 feet of the arch dam, flanking the existing spillway, raising the thrust blocks, and raising and extending the wing walls.

The primary design issues included the structural analysis of the modified arch dam for the new PMF reservoir and tailwater levels, and for an updated Magnitude 6.5 earthquake on the nearby Elsinore Fault. A 3-dimensional finite element analysis was performed and included a response spectra analysis for the seismic case. The arch dam was found to be stable. The thrust blocks, to be raised by adding upstream walls, were checked for stability, again primarily for the new PMF level and seismic loadings, and were found to require five 48-strand and four 48-strand post-tensioned rock anchors in the left and right thrust blocks, respectively. The modification of the thrust blocks was complicated by the fact that the original thrust block design was a classic "V" shaped structure. However, the lift lines in the "thrust blocks" were in such poor condition that in 1965 the "V's" were filled in with concrete to create large single blocks and the concrete was compressed together vertically with post-tensioned cables. The wing walls are to be raised and extended, and stabilized with a total of six 27- strand post tensioned rock anchors.

The downstream slopes are to be protected with 24-inch thick reinforced concrete slabs with rock anchors. A training wall is to be provided on each abutment. The protective channel apron is to be extended with 5-foot thick 1,000 psi concrete. The slope protection slabs and apron will have an extensive drain system.

Other improvements include three new intake slide gates (with submerged hydraulic operators) and trashracks, two new 48-inch butterfly valves (with hydraulic operators), a new water supply outlet system (with a hydraulically operated Polyjet valve), a new control building for the electrical and hydraulic control system, and miscellaneous road and waterline relocation work.

The reservoir cannot be drawn down significantly during construction due to agreements with the property owners around the reservoir. Consequently, there will be substantial underwater construction work to install the new upstream gates and trashracks. Due to the compact nature of the work there were no significant environmental issues involved in the project.

Bids were received in early 1995 and the low bid was $5.36 million. Construction is expected to be completed in late 1995. This will complete the mitigation of a dam safety deficiency at Railroad Canyon Dam, which was first identified in 1983. The selected solution represents a balanced compromise of dam safety remediation, cost of construction, and avoidance of additional flooding of lakefront property.

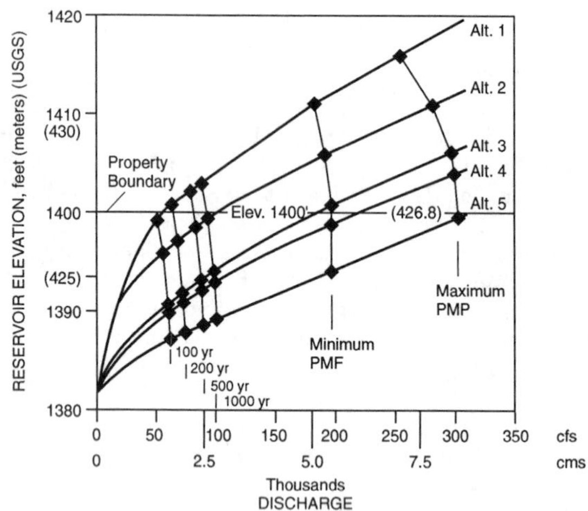

Figure 1. ELEVATION-DISCHARGE CURVES FOR ALTERNATIVE CONCEPTS

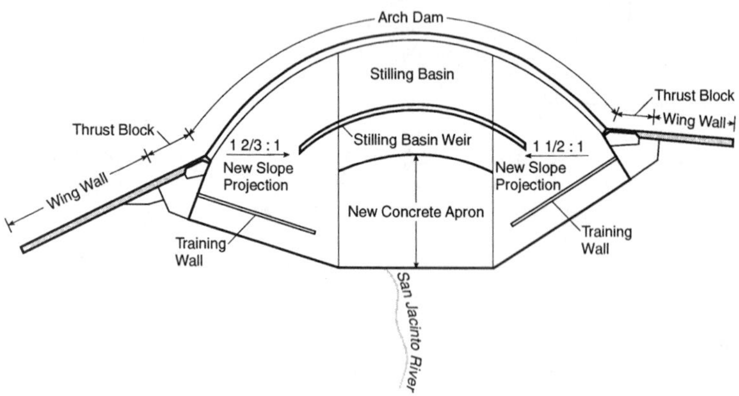

Figure 2. PLAN

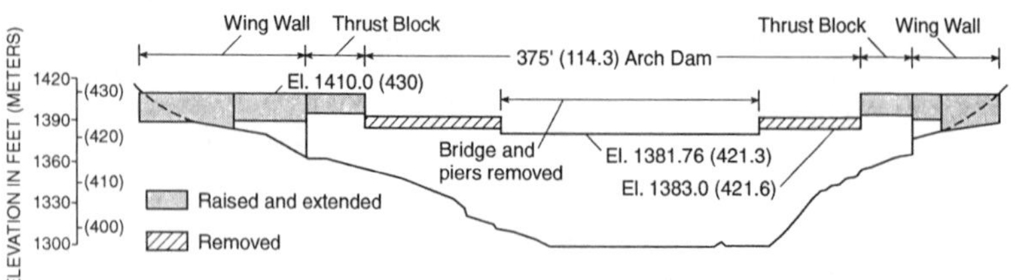

Figure 3. DEVELOPED ELEVATION (LOOKING UPSTREAM)

Abutment Erosion Control System at Boundary Dam

Ram P. Sharma[1], Harry E. Jackson[2],
Walter L. Davis[3], and Donald Gwilym[4]

Introduction

Boundary Dam is a major feature of Seattle City Light's 1000 MW Boundary Hydroelectric Project. It was constructed between 1963 and 1967 on the Pend Oreille River in northeastern Washington State, about one mile south of the Canadian border. The dam is a thin concrete arch structure with a maximum height of 104 m (340 ft) and a crest length of 155 m (508 ft). The total crest length including the two ogee crested spillways and a skimmer chute is 226 m (740 ft). Photo No. 1 shows an overview of Boundary Dam.

The dam site is a symmetrical, deep canyon with steep abutments. The foundation rock is primarily interbedded limestone and dolomite, which strikes about normal to the river and dips from 40 to 50 degrees upstream (south). It is relatively hard but has wide variations in physical characteristics.

The stability of the dam has been the subject of numerous discussions and studies over the life of the project because of several unusual characteristics, including its shape and thinness. A Board of Consultants was retained by Seattle in 1990 to review the safety of the dam at the request of the Federal Energy Regulatory Commission. Subsequent studies have confirmed that the dam can structurally withstand all credible loading conditions, including water loads imposed by overtopping during the probable maximum flood (PMF).

The currently established PMF for the 65,000 sq km (25,000 sq mi) drainage area results in a peak inflow of 14,000 cms (495,000 cfs). The dam's two spillways and seven mid-level sluice gates have a combined capacity of 10,200 cms (360,000 cfs).

---

[1] M, ASCE, Chief, Dams & Structures Dept., Morrison Knudsen Corp., San Francisco,CA

[2] M, ASCE, Principal Civil Engineer, Morrison Knudsen Corp., San Francisco, CA.

[3] Supervising Civil Engineer, Seattle City Light, Seattle,WA.

[4] M, ASCE, Associate Civil Engineer, Seattle City Light, Seattle, WA.

Photo No. 1 - Overview of Boundary Dam

Reservoir and spillway flood routing of the PMF shows that the maximum water surface elevation would be 8.8 feet above the top of the dam parapet. The overtopping duration would be 21 days. Because of concern regarding the overtopping depth and duration, the Board of Consultants has recommended that a erosion protection system be constructed to protect the dam abutment rock from overtopping flows.

Design Studies

Topographic Mapping - Supplemental topographic mapping was developed to better define the topography downstream of the dam to assist in the geologic mapping and for use in hydraulic model construction and in the final design. The work consisted of mapping of the downstream area developed from aerial photographic surveys, oblique angle aerial photographs of the dam and abutment area, and river cross sections in the tailwater pool downstream of the dam developed from bathometric survey data (underwater measurements).

Geologic Mapping - Additional geotechnical field investigations were carried out to map and evaluate the integrity of the abutment rock surfaces and to delineate the relative susceptibility of the rock abutments at Boundary Dam to erosion by dam overflow discharge. Mapping was conducted along regularly spaced intervals in the overtopping impact area and structures were mapped where encountered. The data collected consisted of intact rock strength (measured using a Schmidt Hammer); rock description including color, rock type, degree of weathering, visible bedding, joints (orientation, spacing, condition and infilling); and photographs of important features.

The actual mapping was conducted along the rock slopes by geologists trained in rock climbing techniques. The steepness of the terrain in the area to be mapped made for a very challenging assignment for the two geologists collecting the field data. Photo No. 2 shows one of the geologists performing the mapping activities on the left abutment by slowly descending the cliff face along a double rope system.

Photo No. 2 - Geologic Mapping Activity

Rock classification data and significant structural data were collected and compiled and the data was presented on photo mosaic maps created during the exploration program for use by the design engineer.

Hydraulic Model Studies - The final design of the abutment erosion control system will be based on the results of the hydraulic model studies conducted to simulate PMF overflow conditions. The studies were conducted at the R. L. Albrook Hydraulic Laboratory at Washington State University, Pullman, Washington. The work involved the design and fabrication of a hydraulic model of the arch dam including the discharge facilities at a scale of 1:72, performance of tests, and test data analysis to obtain: (1) path of the trajectory of the overtopping water and the zone of jet impingement on the abutment area; (2) hydrodynamic forces at the area of jet impingement on the abutment areas; (3) requirements for aeration of the under nappe using splitter piers or walls; and (4) maximum reservoir water and tailwater surface elevations.

The studies were conducted under a variety of crest overtopping conditions. During the tests, the seven sluices, the two spillways, and the skimmer structure operated at full capacity. Discharge through the power intake was not considered since it is likely that the power facilities will not operate under PMF conditions. Test conditions included the existing (no modification)condition, existing crest parapet modifications and two types of nappe aeration devices.

Photo No. 3 shows the flow over the model of the dam and left abutment during one of the tests with a discharge of 14,000 cms (495,000 cfs).

Photo No. 3 - Hydraulic Model Test, Q=14,000 cms

Final Determination of the Probable Maximum Flood - The currently established probable maximum flood (PMF) at Boundary Dam was developed in 1973 in connection with the evaluation of the existing spillway adequacy. In September 1994 a review study of the 1973 PMF was completed. This review concluded that the currently established PMF may be somewhat overly conservative, the peak discharge being perhaps 12 percent higher. This evaluation is based on recently published estimates of PMP and the PMF could also be affected by new, recently available meteorological information and current operating procedures for upstream reservoirs.

Since the magnitude of the PMF will have a significant effect upon the final design of the erosion protection system, a new detailed determination of the PMF is being performed using the updated hydrometeorological data and by refining the watershed simulation model as appropriate. The study is being performed in close consultation with the Corps of Engineers who developed and use the Stream Flow Synthesis and Reservoir Regulation (SSARR) model. New PMP development procedures presented in the recently published Hydrometeorological Report, HMR 57 for the Northwest States will be employed. After completion of the study and review and acceptance of the updated PMF by the appropriate regulating agencies, the final design of the Boundary Dam erosion control system will be completed.

Final Design of the Erosion Control System

The design concept for the abutment erosion control system is based upon studies of a number of modification alternatives which led to the conclusion that the reservoir outflow conditions during the occurrence of the PMF would cause dam overtopping of varying amounts depending upon the mitigation measures adopted. Mitigation of the potential erosion on the abutment rock could be accomplished by: (1) reducing the overflow discharge concentration on the abutment and protecting the rock surfaces against erosion; (2) Preventing overflow on the abutment by raising the dam crest parapet wall; or (3) providing an auxiliary spillway with adequate discharge capacity to pass additional flood flow to either reduce or prevent dam overtopping.

Alternative No. 1 was selected for implementation because of the much lower cost compared to the other two alternatives.

The final design consists of providing erosion protection of the rock abutment surfaces subjected to flow over the parapets. The protection will consist of (1) reinforced concrete slabs on the relatively flat areas of the rock abutment and (2) shotcrete reinforced with welded wire mesh on the steeply sloping portions above the tailwater level. Rockbolts will be provided to tie the concrete or shotcrete to the underlying rock, to strengthen the near surface of the rock, and to hold individual rock masses in place if necessary. Drains will be provided in the shotcrete to prevent build-up of pore pressures under the shotcrete that could create a potential for displacement of the shotcrete mass. Prior to shotcreting, removal of semi-detached individual rock masses and some minor shaping of the abutment side walls may be desirable to improve hydraulic conditions and to reduce impact. Figure 1 shows a typical cross section of the proposed erosion protection in a critical area of the left abutment.

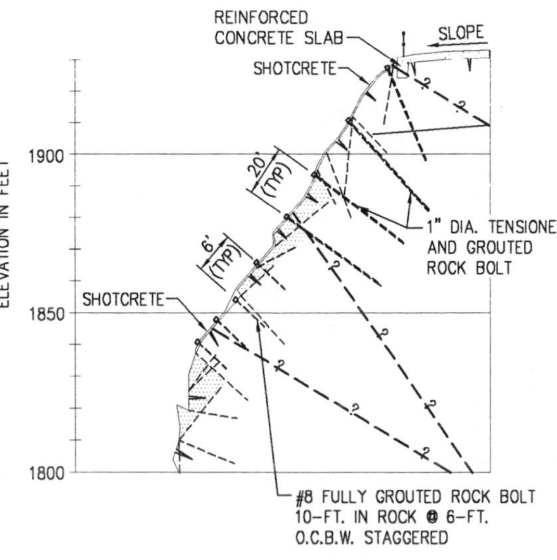

Figure 1 - Erosion Protection Typical Section

Depending upon the magnitude of the updated PMF, a portion of the existing upstream and downstream crest parapets may be removed in the central portion of the dam. The opening created by such parapet removal will thus be above the tailwater pool and the concentrated flow will fall into the pool without affecting the abutment rock. Consequently, the flow overtopping the remaining portions of the parapets above the abutments will be reduced, thereby lessening the impact on the abutment rock.

Aeration of the overtopping nappe will be accomplished by the use of splitter piers or walls.

Reclamation Experience With Overtopping Protection

William Fiedler[1], Member, ASCE; Peter Grey[1]; and Thomas Hepler[1], Member, ASCE

Abstract

Overtopping protection for dams can take various forms, depending upon dam type and size, foundation properties, hydraulic requirements, construction materials, and other factors. The Bureau of Reclamation has recent design and/or construction experience with several types of overtopping protection, including conventional concrete on the abutments of a concrete multiple-dome dam, a roller-compacted concrete (RCC) cap for a hydraulic fill/rockfill dam, a continuously reinforced concrete (CRC) slab for the downstream face of a zoned earthfill dam, and concrete caps for a cellular cofferdam. Details and design considerations for these overtopping protection methods are provided.

Introduction

Many dams in the Western United States were constructed during the first half of this century. Hydrologic data at the sites were limited and flood forecasting was poor, compared to present forecasting methods. Major flooding during spring 1964 in western Montana overtopped concrete and embankment dams causing extensive damages and several dam failures. The flooding was many times larger than any maximum flood then predicted for that region of the country.

New prediction methods were developed using these and other more recent flooding events in the western states. The new methods have been applied to existing dams under the Safety of Dams Program, with many found to be subject

---

[1] Civil Engineer, Bureau of Reclamation, P.O. Box 25007, Denver CO 80225-0007

to overtopping. Overtopping protection is now being provided at some of these dams.

Concrete Overtopping Protection for Concrete Dams

Gibson Dam, a massive concrete arch dam located on the Sun River near Augusta, Montana, was overtopped by up to 1.0 m (3.2 feet) for 20 hours during the major flooding in 1964. The dam remained in place during the flood; however, the abutments suffered extensive erosion. Between 1980 and 1982, conventional concrete was used to cover those areas of the abutments that would be expected to erode during future overtopping. The Gibson Dam overtopping experience was influential in determining the remedial actions required at other concrete dams subject to overtopping, including Stony Gorge Dam in California, Santa Cruz Dam in New Mexico, and Coolidge Dam in Arizona.

Coolidge Dam is located on the Gila River within the San Carlos Indian Reservation, east of Phoenix. The dam, completed in 1929, was the first multiple-dome dam built in the United States. It rises 76 m (250 feet) above bedrock and is 280 m (920 feet) long, including the spillways on each abutment. The three inclined domes are supported by two massive buttresses 55 m (180 feet) apart, by the rock foundation between the buttresses, and by a smaller buttress located on the upper portion of each abutment. The spillways consist of an uncontrolled crest, a curved transition section, and a chute. Discharges in excess of 566 m^3/s (20,000 ft^3/s) in either spillway will overtop the chute walls.

The probable maximum flood (PMF) for Coolidge Dam has a peak inflow of 22,700 m^3/s (800,000 ft^3/s) and a 17-day volume of 3.4×10^9 m^3 (2,754,000 acre-feet). Routing the PMF results in dam overtopping for 70 hours with a maximum depth of 6.7 m (22 feet). Hydraulic model studies indicate that dam overtopping begins when releases reach 4250 m^3/s (150,000 ft^3/s). The dam abutments are sedimentary sandstone and quartzite, and would be expected to erode by water flowing over the spillway walls and the dam crest, resulting in dam failure.

A concrete overlay was constructed on the downstream abutments for protection from flood flows overtopping the spillway walls and the dam crest. The concrete overlay consisted of CRC slabs tied to the abutments by grouted anchor bars. A network of drain pipes was installed between the foundation rock and the slabs to relieve any seepage pressures. Although the minimum concrete thickness for stability was 0.8 m (2.5 feet), the slabs had to be shaped for good hydraulic performance, requiring actual thicknesses greater than 3 m (10 feet) in some places. All work was completed in December 1994.

RCC Overtopping Protection for Embankment Dams

RCC overtopping protection will be provided at Black Rock Dam, located on the Zuni River within the Zuni Indian Reservation, south of Gallup, New Mexico. The dam was constructed between 1904 and 1908, and consists of hydraulic fill upstream and rockfill downstream. The dam has a crest length of 160 m (525 feet) and a structural height of 34 m (110 feet). The downstream face of the dam is sloped at 1.25(H):1(V), and is protected with hand-placed, stepped basalt blocks. Flood routings indicated the dam would be overtopped during the PMF (peak inflow - 3210 m^3/s (113,400 ft^3/s), 190-hour volume - 183 x 10^6 m^3 (148,300 acre-feet)) for over 24 hours at depths up to 3.0 m (9.9 feet). It was concluded that the dam would fail under these conditions.

RCC overtopping protection will be provided on the dam crest, downstream face, abutments, and along the toe. The RCC on the downstream face of the dam will be placed in 2.7-m- (9-foot) wide lifts, with formed steps. The distinctive appearance of the dam, which is on the State Historical Register, will be preserved by replacement of the basalt blocks on the modified dam face. In addition, the dam crest will be lowered 1.2 m (4 feet) to reduce the maximum water surface elevation by 0.2 m (0.7 foot). Although lowering the dam crest will increase the frequency of dam overtopping, it will minimize the required raise for the dikes on both abutments. Under the planned modifications, Black Rock Dam will be overtopped by 4.0 m (13.2 feet) during the PMF.

The stability of the RCC overlay was analyzed at the maximum section for several loading conditions, including: 1) the maximum overtopping condition, and 2) the postovertopping condition when the tailwater has receded. The stability analysis included dead loads, tractive forces of overtopping flows, uplift forces, and horizontal hydrostatic loads. For the postovertopping condition, it was assumed that full uplift due to the maximum tailwater still existed beneath the overlay. The stability of the RCC overlay was analyzed for resistance against sliding, overturning, and buckling failures.

Drainage systems will be provided for the stepped RCC overlay on the downstream face, and for the variable thickness RCC blanket overlays at the abutments and toe. The volume of seepage through the RCC and into the foundation was determined by computing the flow through cracks, with assumed crack spacings and widths. The drainage system will consist of various-diameter polyethylene pipes.

A hydraulic model study was used to determine the wall height and layout of a guide wall on the right abutment. The guide wall is needed to contain flow within the RCC apron and minimize the potential for erosion. A second model study was used to determine wall heights for the spillway chute and the requirements for a guide wall along the left side of the spillway outlet channel. As a result of the model study, a deflector wall will also be provided just downstream from the spillway chute to spread out the spillway flows and minimize the extent of the RCC erosion protection. The contract for the modifications at Black Rock Dam is scheduled to be awarded in September 1995.

CRC Overtopping Protection for Embankment Dams

A smooth CRC slab has been proposed for overtopping protection of A. R. Bowman Dam, a 75-m- (245-foot) high-embankment dam located on the Crooked River near Prineville, Oregon. Passage of the PMF results in overtopping of the dam for about 4.5 days with a maximum depth of 6.1 m (20 feet). The 0.3-m- (1-foot) thick slab would cover the crest and entire downstream face of the embankment, from the left abutment contact to the spillway wall on the right abutment. The slab would be restrained by subsurface blocks at the crest and toe by a gravity wall along the spillway, and by rock anchors along the left abutment. The slab would be thicker in the transition zone between the upper 2:1 and lower 4:1 slopes for protection from potential flood debris impacts.

The ability to control cracks and offsets is critical to the performance of the CRC slab during overtopping. The slab would be slip-formed in alternating panels from toe to crest to minimize shrinkage cracking, with the final closure panel scheduled to minimize temperature stresses within the completed slab. A single mat of steel reinforcement would be provided to prevent offsets and to limit crack widths to 0.1 mm (0.003 inch) at 32 °F and crack spacings to 0.6 m (2 feet). The seepage volume through the slab during the PMF was estimated assuming laminar flow through the cracks to determine potential uplift pressures. Subsurface drainage would be provided by a 1-m- (3-foot) thick gravel blanket on the embankment slopes beneath the slab, with a row of drain outlets located above the toe of the hydraulic jump.

Flow conditions from the dam crest to the stilling basin were investigated using a hydraulic model. The overtopping flow converges from a crest length of 256 m (840 feet) to a basin width of 116 m (380 feet). Water surface profiles were developed, and maximum pressure fluctuations were measured within the hydraulic jump. Additional studies will be performed during final design to further address the stability of the CRC treatment during overtopping and subsequent drawdown.

Overtopping Protection for Cellular Cofferdams

Dam safety modifications to Theodore Roosevelt Dam, located in Arizona on the Salt River, required the construction of cellular cofferdams on each abutment to permit replacement of both spillways. Each cofferdam included three cells and two connecting arcs of interlocking sheet piles filled with sand. Each cell was 11.5 m (37.6 feet) in diameter and extended up to 11.3 m (37 feet) above the irregular bedrock foundation. Concrete gravity walls provided closure at each end between the existing dam and the abutment contact.

Overtopping protection was provided by capping the cells and connecting arcs with a 0.3-m- (1-foot) thick, reinforced concrete slab, and by constructing a concrete buttress along the downstream toe within 1.5 m (5 feet) of the sheet piles. The concrete cap prevents loss of the fill material during overtopping, and the buttress seals the foundation contact and provides additional sliding resistance for the higher reservoir loads. Weep holes were cut on 1-m (3-foot) centers in the downstream face for drainage of the cells, with each 25 mm (1 inch) hole fitted with a cone-shaped piece of wire mesh to prevent loss of the sand fill.

The left abutment cofferdam was overtopped for 42 hours in January 1993, with a maximum overtopping depth of 0.3 m (1.1 feet). Overtopping flow was estimated to be up to 20 m^3/s (700 ft^3/s) along the 70-m- (230-foot) long crest. Inspection of the cofferdam following this event revealed only minor deflections, at the top of the structure, between 3 and 21 mm (0.01 and 0.07 foot) downstream, and radial bulging near the lower quarter-point of height between 12 and 46 mm (0.04 and 0.15 feet). The concrete cap, buttress, and wire mesh were effective in preventing any significant loss of fill material.

Conclusions

The Bureau of Reclamation has employed various types of overtopping protection for both concrete and embankment dams, and for temporary cofferdams. Overtopping protection has become an efficient means of safely passing floods.

Response of Ideally Controlled Canals to Downstream Withdrawals

C.M. Burt, M. ASCE, R.S. Gooch, M. ASCE,
T.S. Strelkoff, M. ASCE, and J.L. Deltour[1]

Abstract

In recognition of the different degrees of success achieved by a given downstream controller in different canals, the effect of pool characteristics on controllability of water levels is investigated. To eliminate the properties of the controller-algorithm from consideration, control is assumed ideal, *i.e.*, simultaneous, exact upstream replacement of withdrawals from a downstream turnout. The maximum drawdown at the turnout is then, for any given withdrawal fraction, a function primarily of pool geometry and dynamics. This is viewed, first, for a pool of specified dimensions in terms of consecutive steady states. Then, the transitory depth variations of unsteady flow are considered, through an example from a general nondimensional study. When complete, the results will allow prediction of maximum drawdown at the turnout and, so, indicate circumstances in which even ideal control is insufficient, and *anticipatory* control measures become necessary.

Introduction

Most research on downstream control of canals has dealt with the problem of selecting and calibrating a suitable algorithm for dictating gate movements aimed at achieving rapid and stable recovery of a downstream water level, following any deviation from a desired target depth. This paper addresses another, equally important factor in achieving the desired control -- the influence of geometric and dynamic canal characteristics on controllability. This paper does not provide any formula for predicting controllability. It does not even *address* issues of stability. We view only that aspect of controllability which describes the changes in water level and how long it takes to restore original conditions.

With all of the many types of local, independent downstream control (Burt and

---

[1]Respectively, Professor and Director, Irrigation Training and Research Center, Cal Poly State University, San Luis Obispo, CA 93407; Senior Engineer, Water Engineering, Salt River Project, P.O. Box 52025, Phoenix, Arizona 85072-2025; Research Hydraulic Engineer (and Research Professor, Univ. of Arizona), U.S. Water Conservation Laboratory, USDA/ARS, 4331 E. Broadway, Phoenix, AZ 85040; and Hydraulic Engineer and Water Control Specialist, GERSAR - Societe du Canal de Provence, Le Tholonet B.P. 100, Aix-en-Provence CEDEX 1, France.

Plusquellec, 1990), a gate at the upstream end of the pool is modulated in an effort to control the water level at a designated downstream point, as offtakes and other gates are exercised. That controlled point can be at the upstream end of the pool, leading to a constant depth there (*e.g.*, *AVIS* gate), in the interior of the pool, or at its downstream end. Unfortunately, no satisfactory general guidelines regarding pool characteristics in downstream control have emerged, beyond the recognition that longer and steeper pools are more difficult to control than short pools of small slope.

In the present study, neither the control algorithm nor gate characteristics are under consideration. For even if a controller were *ideal*, *i.e.*, capable of inducing immediate and exact replacement of offtake withdrawals, the depth at the point of withdrawal would still decrease until the replacements arrive in sufficient quantity. After the wave from upstream is first noted at the downstream end of the pool, only gradually, over a period of time, does the depth there revert to its original level.

It is evident, then, that regardless of the algorithm, the canal must possess certain physical characteristics to realize adequate downstream control without anticipation[2]. It follows that, before implementation of an imperfect canal-control algorithm for downstream control (and all algorithms are imperfect), the canal pools should first be studied to determine if their response to *ideal* control is acceptable, in terms of water-level deviation from the target depth. The phenomena related to withdrawal replacement and water-level recovery are controlled by gravity, resistance, pressure, and inertial forces, the proportions varying with specific geometric and dynamic circumstances: reach length, slope, roughness, cross section, initial discharge, degree of check-up at the downstream end, and withdrawal rate.

Steady-state characteristic curves for a pool

Figure 1, drawn from Deltour (1992), graphically displays ultimate pool-volume responses to control measures. These curves characterize the full range of *steady states* possible in a pool of given dimensions. Developed from multiple computer simulations of backwater curves in a single, specific pool, they shows how the volume in transit within the pool changes with flow rate and downstream depth. The dashed curves are lines of constant upstream depth; the solid curves are contours of downstream depth. The heavy solid curve on the right represents normal depth. If depths between normal and critical were included, they would occupy a narrow region to the right of the normal-depth curve (Deltour 1992). Both depth changes and volume changes arising from discharge changes are of interest. The volume changes, along with the discharges, relate to the time necessary to effect the change in storage required to complete the process of adjustment to new flow rates. The diagram provides valuable insight into the controllability of the given pool.

For example, with the control point at the downstream end of the pool, and the water

[2]A control measure can be instituted in *anticipation* of a demand, if its timing and magnitude are known, a separate problem discussed in Bautista, Clemmens, and Strelkoff, 1995.

depth there held, say, to 2.75m, all possible states of steady flow with depths greater than normal in the canal pool are represented by the heavy gray curve in the figure. Then, if the flow were to change, say, from 14 to 20 *cms*, as shown by the heavy arrow ***A***, the upstream depth will change from 1.89m to 2.20m; furthermore, a volume of 21000 m^3 must be added to the pool before a new steady state can be reached.

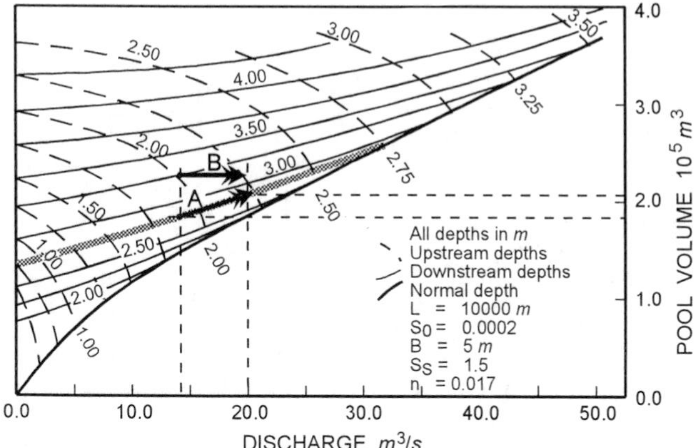

Figure 1. Steady State Characteristic Curves for a specified pool. Stored volume as a function of flow rate and water level

In terms of initial and final steady states, when wave action has died down and a new equilibrium state has evolved, Figure 1 can also be used to chart the changes resulting from ideal control based on a downstream control point. With inflow and outflow rates always identical, the volume in the pool remains constant. Thus, for example, if the discharge was 14 m^3/s initially, with a downstream depth of 3.25m (and 2.02m upstream), and then we simultaneously increase the flow at both ends of the reach to 20 m^3/s, the new downstream water level must be 3.10m, and the upstream depth, 2.28m. This constant-volume change is shown by the heavy arrow ***B*** on Figure 1.

This approach was also used in Deltour (1992) to estimate the minimum depth reached in unsteady-flow simulations, before correction by the controller, increasing the volume in the pool. The adequacy of this estimate was also tested in a field study (Sanfilippo, 1994) on the MSIDD WM lateral canal, with favorable results. The minimum reached during the simulation was lower than the graphics of the type of Figure 1 would predict, because the reaction of the controller was not immediate, and the volume decreased a little because of the increase in offtake withdrawal.

Nondimensional unsteady response curves

The curves of Figure 1 were prepared for a specific canal pool. Furthermore, they show possible steady states in that pool and do not formally consider the transient depth changes. So, a general program of unsteady-flow simulations was undertaken with the ideal-control scenario, to quantify the influence of pool geometry and dynamics on maximum, unsteady drawdown. To reduce the number of variables to consider and hence reduce the computational and presentational effort without losing generality, dimensionless variables were used[3]. This allows, in principle, the generation of a pattern of solutions blanketing the practical range of interest.

In the specified scenario, given fractions R_Q of the initial flow rate are withdrawn suddenly and for an indefinite period from an offtake just upstream from the downstream gate (see Figure 2). Simultaneously the same fraction is added upstream; it is assumed that the upstream gate is somehow controlled to produce this increase. The downstream gate remains at its original setting, which yielded the initial checked-up depth at the downstream end of the pool. Downstream from the downstream gate, an indefinite length of additional canal of the same cross section, slope and roughness as the given pool is assumed. As a result, the depth on the downstream side remains at normal for whatever discharge is passed. The offtake discharge remains constant after the augmentation. However, in contrast to the preceding constant-volume example (arrow B, Figure 1), with given gate discharge, now, the discharge through the downstream gate varies in accord with variations in depths and discharge coefficient. Gate width is assumed equal to the base width of the canal trapezoidal cross section, resulting in a small decrease in water surface elevation as the flow enters the control structure.

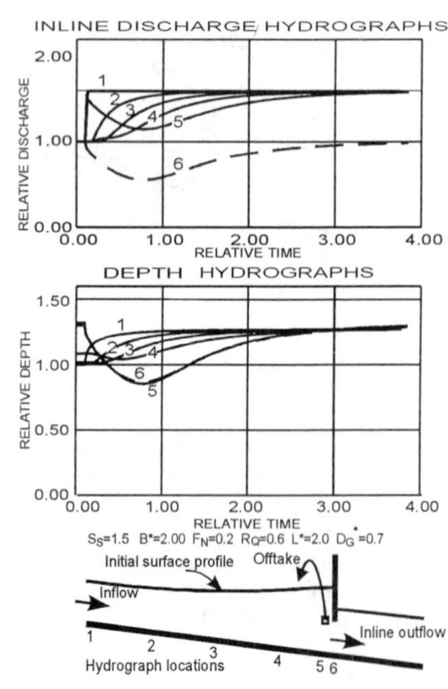

Figure 2. Ideal control -- no anticipation of demand

[3] In dimensionless terms, typically designated by a *, all depths and other transverse lengths (breadth, hydraulic radius, etc.) are expressed in ratio to a reference -- normal depth in the canal at the initial flow; all lengths are expressed in ratio to a reference length equal to normal depth divided by bottom slope; all times, to the time to traverse the characteristic length at normal velocity. The reference discharge is the inital flow rate divided by the aspect ratio of the normal-flow cross section (*aspect ratio* is defined as the ratio of average breadth to depth under normal conditions, A_N/y_N^2. The relative measure of dynamic effects is provided by the normal Froude number, F_N, of the initial flow. The procedures are described in a companion paper, by Strelkoff, Clemmens, and Gooch (1995).

Typical discharge and depth hydrographs are shown in Figure 2, for withdrawal fraction R_Q=0.6, pool length L^*=1.6, relative check-up $y_D{}^*$=1.3, and initial normal Froude number, F_N=0.2. The relative base width in the canal, B^*=2, side slopes, S_S=1.5; the relative gate opening needed to achieve the given check-up, $D_G{}^*$=0.7. The numbers 1-6 represent hydrograph locations: at the upstream end, at the quarter, half, and three-quarter points, and at pool end on either side of the offtake. Station 6, in the gate structure downstream from the offtake, is shown dot-dash. Noteworthy is the extreme length of time required to restore original conditions at the gate.

Figure 3 is an example of generalized dimensionless graphs quantifying pool response in the chosen scenario. With relative drawdown defined as the maximum reduction in downstream depth divided by initial checked-up value, the curves show for a 60% withdrawal fraction, the drawdown as a function of checked-up depth relative to normal depth, relative pool length, and initial Froude number. Complementary graphs would show the effects of different offtake ratios, relative bottom widths, and canal side slopes. From such curves can be established whether even with an ideal controller the drawdown is tolerable, or whether anticipatory control is required.

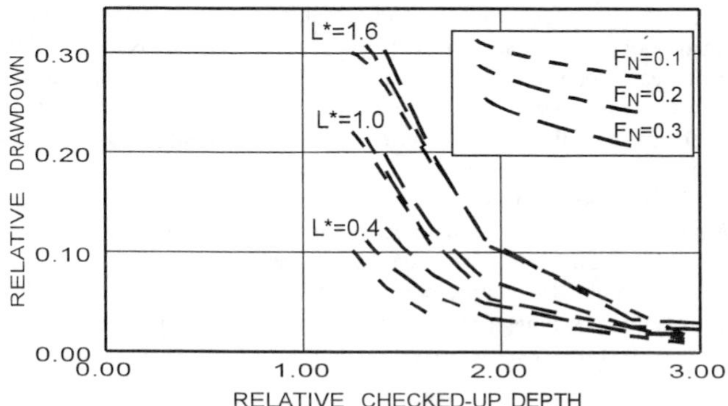

Figure 3. Maximum drawdown at downstream offtake under ideal control SS=1.5, B^*=2.0, F_N=0.1 -- 0.3, R_Q=0.6, L^*=0.4 -- 1.6

References

- Bautista, E., Clemmens, A.J., and Strelkoff, T.S. 1995. "Inverse Computational Methods for Open-Channel Flow Control," *Proceedings of the First International Conference on Water Resources Engineering,* ASCE, San Antonio, Texas, August 14-18.
- Burt, C.M. and H. Plusquellec. 1990. Water Delivery Control. Chapter 11 in Management of Farm Irrigation Systems (Hoffman, Howell, and Solomon, ed.). American Society of Ag. Engr., St. Joseph, MI.
- Deltour, J.L. 1992. *Application de lautomatique numerique a la regulation des canaux*, Phd *These*, Institute de Mecanique de Grenoble.
- San Filippo, P. 1994. *Application du PIR au cas des canaux a forte pente*, DEA *Memoire*, Universitaté Claude Bernard.
- Strelkoff, T.S., Clemmens, A.J., and Gooch, R.S. 1995. "Dimensionless Characterization of Canal Pools," *Proceedings of the First International Conference on Water Resources Engineering,* ASCE, San Antonio, Texas, August 14-18.

Propagation of Upstream Control Measures Along a Canal Pool

T.S. Strelkoff, M. ASCE, J.L. Deltour, J.P. Baume[1]

Abstract

A step increase in inflow to a canal pool is routed to its downstream boundary by numerically solving the Saint Venant equations. The wave front deforms as it propagates down the pool, evolving to an evermore-gradually rising form in response to the canal-pool characteristics: length, slope, roughness, cross section, initial flow rate, downstream boundary condition, and degree of checkup. Propagation times of different wave components are obtained for a wide range of pool conditions. The study is performed in dimensionless terms to get the maximum amount of information with minimum effort of calculation and display. The downstream boundary conditions investigated comprised: fixed reservoir elevation, long-crested (essentially constant-head) weirs, and submerged, undershot gates with normal depth downstream, as well as a wide-open gate, such that the downstream boundary condition in the pool was simply normal depth at the local discharge. The stage-discharge relation defining the downstream boundary condition is seen to have a major influence on the delay in arrival of the bulk of the wave.

Introduction

In order to provide the required amount of water at delivery-canal turnouts in a timely manner, some control measures must be applied upstream from the turnouts as demands change. Whether automatically or manually, pumps must be turned on and off; canal gates need to be raised or lowered. Determination of the appropriate control measures is complicated by the fact that the response of the canal at the turnouts is not instantaneous. A substantial time delay typically exists between the implementation of a control measure at the upstream end of a canal pool and its arrival at a downstream point.

[1] Respectively, Research Hydraulic Engineer (and University of Arizona Research Professor), U.S. Water Conservation Laboratory, USDA/ARS, 4331 E. Broadway, Phoenix, Arizona 85040; Water Control Engineer, Société du Canal de Provence (SCP-Member company of GERSAR), BP 100, 13603 Aix-en-Provence, Cedex 1, France; and Hydraulic Engineer, CEMAGREF, 361 rue J-F Breton, BP 5095, 34033 Montpellier, Cedex 1, France

Furthermore, the wave profile slumps as it propagates, so that different portions of the wave arrive at different times, with the delay between the first harbinger of an sudden upstream change and the substantial bulk of the wave ever increasing. If the pool were a prism of indefinite length, the wave profiles would eventually assume the fixed form of the well known monoclinal rising wave (Henderson, 1966, p.372), and all portions would propagate at the same, kinematic-shock, speed. But for a significant length of time, the form of the wave initiated upstream as a step increase would be gradually evolving, leading to the observed delays in arrival between the various wave components. Furthermore, nonprismatic canal flows preclude the evolution to a profile of fixed shape, and a general scheme of solution of the Saint Venant governing equations becomes necessary in order to predict the time delays.

Procedure

A simulation model capable of accepting input and calculating output in dimensionless terms (Strelkoff, Clemmens, and Gooch, 1995) was recast as a computer subroutine. The simulations of the model are based on a network of characteristic curves stemming from the characteristic form of the Saint Venant equations. This subroutine was called repeatedly by a main program, which systematically changed the dimensionless canal-pool geometry and dynamics in a predetermined pattern. At the conclusion of each simulation, pertinent input and output variables (geometry, initial Froude number, delay times) were automatically entered into a text file, one line per simulation. An auxiliary program would then read this text file, extract the desired data, and plot it. Additional simulations were performed with an implicit finite-difference model for corroboration.

Results

Figure 1 shows the evolution of discharge hydrographs in three canal pools, similar but for the downstream structure, which in (a) is a normal-depth stage-discharge relation, (b) is a long-crested weir (duck bill with a length about 21 times the initial normal depth, y_N) of such height that the water depth upstream from the weir equals y_N, and with a head on it of about 0.1 y_N, and (c) is a reservoir, held to an elevation of y_N above downstream channel invert. Thus, in every case, the initial dimensionless downstream depth in the pool is $Y_D^*=1.0$. The upstream step increase in discharge in each case is 10% of the initial flow. Further, in each case, the pool cross section is rectangular, and for the sake of definiteness, can be thought of as $2m$ wide and $1m$ deep (when flowing at normal depth at the initial discharge, 1.9 *cms*), 2300 *m* long, set on a bottom slope $S_0=0.00044$, and with Manning $n=0.014$. Other dimensions are of course possible and still yield the same dimensionless parameters as the given example. In the example case, the time reference value is 41 minutes.

Hydrographs are shown at the 0, 1/4, 1/2, 3/4, and full-length points. At once evident is that the toe of the wave arrives at the same time in all cases. The bulk of the wave on the other hand, when the downstream boundary condition is a normal-depth stage-discharge relation, arrives substantially delayed, as compared to the other

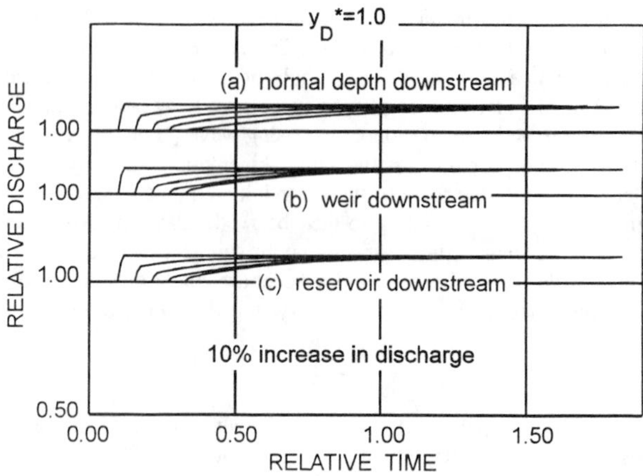

Figure 1. Discharge hydrographs in a canal pool at the upstream end, 1/4, 1/2, 3/4, and full-length points, with a step increase in discharge at the upstream end and initial downstream depth at normal.

downstream boundaries. The long weir and reservoir give essentially the same results, not surprising since the intent of the long crest is to fix the water-surface elevation independent of the discharge.

Figure 2 shows similar tendencies for the same pool, but with a checked-up initial

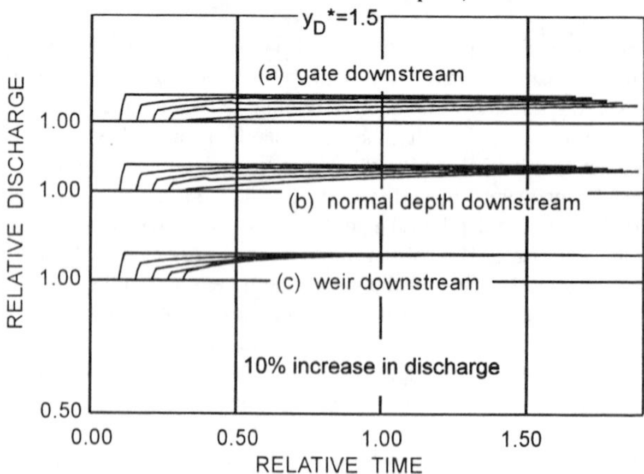

Figure 2. Discharge hydrographs in a canal pool at the upstream end, 1/4, 1/2, 3/4, and full-length points, with a step increase in discharge at the upstream end and initial downstream depth at 1.5 times normal.

flow, provided in (a) by a submerged radial gate, in (b), by a normal-depth stage discharge relation with increased roughness downstream from the nominal end of the pool, and in (c) by the long weir. In each case, the water level was checked up, initially, to a depth about 1.5 y_N.

Again, the leading component of the disturbance arrives at the same time in each case. The bulk of the wave, with nearly the full increase, however, arrives latest with the gate, just a little sooner with the normal-depth stage-discharge relation, and much sooner with the constant downstream depth (long-crested weir).

Dimensionless delay curves

A program of variation of pool and flow properties was initiated with the three downstream boundary conditions: submerged gate, long-crested weir, and reservoir. Simulations were performed by the characteristics based model in the first two cases, and with the CEMAGREF SIC implicit finite-difference model in the third case for comparison. In this initial study, cross sections were limited to rectangular, canal widths to 2 times initial normal depth, a range of dimensionless lengths from 0.3 to 2.0 (see Strelkoff, Clemmens, and Gooch, 1995, for dimensional significance of these values), checked up depths, 1.0 and 1.25 times normal depth, and Froude numbers of the initial flow equal to 0.1, 0.3, and 0.5. The step increase in discharge was held to 10% of the initial flow. Relative times of arrival of 10% and 85% of the original upstream step increase are shown in the Figures 3 and 4.

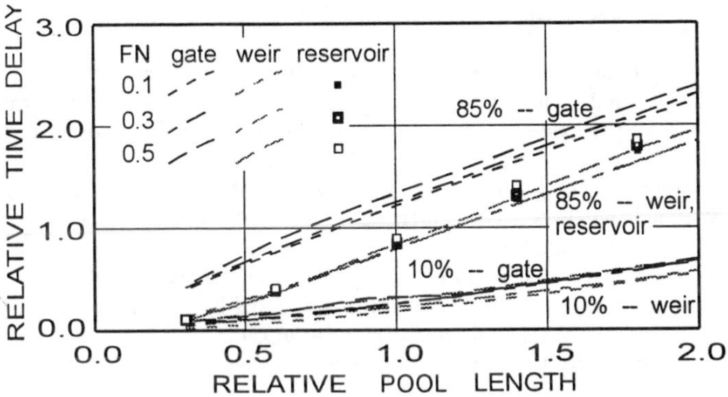

Figure 3. Delay of wave components. $y_D^*=1.0$, $F_N=0.1, 0.3, 0.5$

These curves represent the first steps in the development of more general curves which could be used at the initial stage of a project to estimate the delays required to bring conditions at the downstream end of a pool in line with upstream changes; *i.e.*, to be able effectively to use the modified discharge at the downstream end of the pool. The more general curves should show the effect of larger increases in discharge, nonzero side slopes, and other relative canal widths.

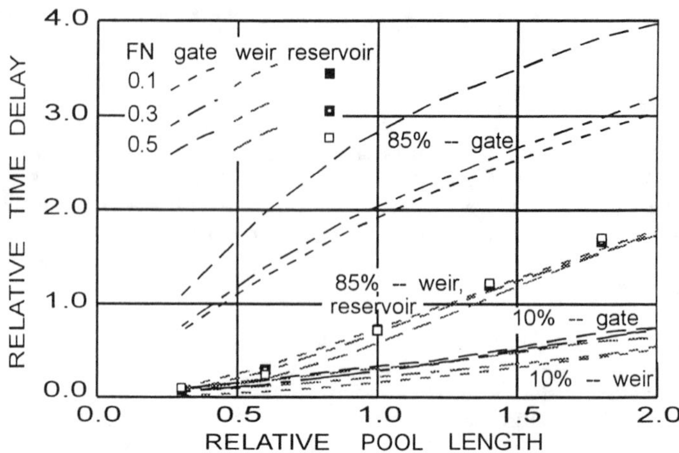

Figure 4. Delay of wave components. $y_D^*=1.25$, $F_N=0.1, 0.3, 0.5$

As noted earlier in connection with the sample hydrographs, the delay in arrival of the initial stages of the release wave (the 10% curves), is somewhat dependent on initial Froude number, but little dependent on the degree of check-up or the kind of control structure downstream.

The 85% (global) delay time varies significantly with checked depth, somewhat less so with Froude number, and is greatly dependent on the nature of the downstream boundary. With the long-crested weir or reservoir downstream, the delay of this component is decreased with an increase in checked-up depth. With the submerged gate, the larger the checked-up depth, the greater is the delay. With the gate, increasing Froude numbers result in increased delay; with the long-crested weir, the opposite trend is noted, except with $y_D^*=1$. At a dimensionless pool length of unity, the global delay is reduced by 30% if the downstream checked depth is close to normal depth, and 60 - 75%, if the downstream checked depth is close to 1.25 y_N. The delay increases because with increasing checkup, the head loss introduced by the gate increases the variation in level as the wave arrives.

This confirms one of the advantages of introducing check structures to control the level on canals operating under upstream control (advance scheduling). The time lag introduced by the canal pools is indeed an important constraint with this type of operation.

REFERENCES

- Henderson, F.M. 1966 *Open Channel Flow*, Macmillan Company, New York
- Strelkoff, T.S., Clemmens, A.J., and Gooch, R.S. 1995. "Dimensionless Characterization of Canal Pools," *Proceedings of the First International Conference on Water Resources Engineering*, ASCE, San Antonio, Texas, August 14-18.

Guidelines for Presentation of Canal Control Algorithms

W. Schuurmans, B. Grawitz, A.J. Clemmens, ASCE[1]

Abstract

This paper is one in a series of papers by the ASCE Task Committee on *Canal Automation Algorithms* to be presented in this proceedings. The purpose of the task committee is to promote the development of improved canal automation algorithms. One group within the task committee developed test cases to be used to test and compare different algorithms. This paper presents the criteria with which to judge the performance of various controllers and the format in which to provide performance results.

Introduction

Many canal control algorithms are presented in the literature on canals, under conditions specified by the algorithm developer. The properties of the canal and the severity of the test have a significant influence on the reported performance of the proposed controller. Thus it is very difficult to judge the general suitability of these controllers from the given examples. Standardized test cases on canals with well studied properties are important for the presentation and comparison of control algorithms. In a companion paper, the details on two test canals are presented (Kacerek et al, 1995). In this paper, we present the suggested criteria with which to judge controller performance and the recommended reporting format.

Test Cases

The task committee chose two real canals with very different properties to propose as test cases. The first is a very steep canal with very rapid response times, but very little storage. The second is a larger canal with relatively slower response. These two canals represent somewhat extreme conditions. The specific physical dimensions have been altered slightly to make the theoretical studies more convenient. Several

---

[1] Respectively, Hydraulic Engineer, Delft Hydraulics, Delft, The Netherlands; Hydraulic Engineer, Society Canal de Provence, Aix en Provence, France; Research Hydraulic Engineer, U.S. Water Conservation Laboratory, USDA/ARS, 4331 E. Broadway, Phoenix, AZ 85040.

test scenarios have been proposed for the two canals, including both pre-scheduled (anticipated or known ahead of time) and unscheduled (and unknown) changes in demand. (See Kacerek et al, 1995). Each test consists of two part, each 12 hours long. Tests are run on the canal as tuned and for an untuned case where canal and gate properties are changed. Testing should be performed on a fully non-linear unsteady-flow simulation program.

Performance Indicators

Control algorithms are typically designed and tuned on the basis of different criteria; thus they tend to optimize some performance criteria at the expense of others. The potential users of these algorithms should understand which criteria are the most appropriate for their desired operations; then controllers with the desired characteristics can be chosen for their particular canal.

The task committee recommends presentation of the following performance indicators.

The maximum absolute error, MAE

$$MAE = \frac{\max(|y_t - y_{target}|)}{y_{target}} \quad (1)$$

Integral of absolute magnitude of error, IAE

$$IAE = \frac{\frac{\Delta t}{T}\sum_{t=0}^{T} |y_t - y_{target}|}{y_{target}} \quad (2)$$

for a constant time step over the 12 or 24 hour interval specified.

The steady state error, StE is defined as the maximum of the average error over the last two hours of each 12 hour test. The assumption is that conditions will be stable during this time.

$$StE = \frac{\max(\ (\overline{y_{t=10,12}} - y_{target}),\ (\overline{y_{t=22,24}} - y_{target}))}{y_{target}} \quad (3)$$

Integrated average absolute gate movement, IAW,

$$IAW = \frac{\Delta t}{T}\sum_{t=0}^{T} |w_t - w_{t-1}| \tag{4}$$

Integrated average absolute discharge change, IAQ,

$$IAQ = \frac{\Delta t}{T}\sum_{t=0}^{T} |Q_t - Q_{t-1}| \tag{5}$$

For the above equations,

y_t = observed (computed from simulation) water level at time t
y_{target} = target water level
Δt = regulation time step
T = time period for test (12 or 24 hours)
$y_{t=10,12}$ = average water depth between 10 and 12 hours
w_t = check gate position at time t
Q_t = check gate discharge at time t

Several issues remain unresolved. First, IAW and IAQ have units and thus may depend on the specific gate relationships used in the simulation model. It may be possible to express these in terms of the maximum design flow rate and the associated gate opening. However, this information was not provided with the test cases. The $\Delta t/T$ term also may not be necessary.

Next, we specify that the largest value of these indicators from the different pools should be reported. This may favor some controllers over others. Local controllers try to minimize this criteria in each pool, while central controllers use criteria that averages the performance over all the pools (e.g., square root of sum of the squares of performance values over all pools).

Other performance indicators can also be reported, but are optional for the current tests.

- Standard deviation and coefficient of variation of deviation from target (standard deviation divided by average target depth); for period of increase, period of decrease, and total duration of test.

- Standard deviation and coefficient of variation of offtake discharges.

- MAE, IAE, StE for offtake discharges.

•Maximum water-level draw-down rate (over 30 min period).

•Time to steady state -- i.e., the time to return to within 10% of maximum deviation from target level, for each increase and decrease.

•Whether the downstream canal reach overtopped or spilled, and the amount of spill.

•Effectiveness of delivered volume (gravity offtakes only)-- defined below

$$D_{EF_{vol}} = \frac{\sum_t Q_e}{\sum_t Q_d} \tag{6}$$

where $Q_e = 0$ for $Q_o < 0.9Q_d$,
$Q_e = Q_o$ for $0.9Q_d \le Q_o \le 1.1Q_d$ and
$Q_e = 1.1Q_d$ for $Q_o > 1.1Q_d$.
Q_d = desired discharge
Q_e = observed discharge

•Number of pump shut downs and restarts caused by low water level (not relevant for gravity offtakes).

Guidelines for canal control algorithm presentation

1 Classify algorithm according to classification guidelines in Malaterre et al (1995).
- Name of Algorithm
- Status (research, off-the-shelf available, project-specific)
- Name of developer
- control logic
- input/output structure
- design method
- considered variables

2 Provide mathematical description of control algorithm
- input variables, output variables, control strategy
- additional information needed to implement the controller in practice, alarm devices

3 Describe tuning method used
- trial and error, simulation, or real life system
- formulae or tools needed
- tuning effort and knowledge required

4 Present test results for tuned conditions

- specification of test canal
- specification of simulation model used
- model details: Courant number, time step for simulation and regulation, etc.
- main performance indicators MAE; IAE; IAW; IAQ; STE
- graphs (time versus control variable)
- secondary performance indicators

5 Present test results for untuned conditions (Robustness -- i.e. how much performance deteriorated when conditions change)

- (same as above)

6 Describe existing applications and known limitations

- steep/mild/flat canals
- long delay: with or without storage in pool or reservoir
- suitable for gravity offtakes and/or pump offtakes

7 Define hardware requirements

- sensors: type, number, accuracy
- communication: frequency, number
- Remote Terminal Units's & Central Terminal Unit's: number, type, capabilities needed
- gates: speed, frequency, minimum possible movement

References

Kacerek, T., Clemmens, A.J. and Sanfilippo, F. 1995. Test Cases for Canal Control Algorithms with Examples, *Proc. First International Conference on Water Resources Engineering*, ASCE, San Antonio, TX

Malaterre,P.O., Rogers, D.C. and Schuurmans, J. 1995. Classification of Control Algorithms, *Proc. First International Conference on Water Resources Engineering*, ASCE, San Antonio, TX

Test Cases for Canal Control Algorithms with Examples

Timothy F. Kacerek[1], Albert Clemmens[2], M. ASCE, and Franck Sanfilippo[3]

Abstract

Standard test cases for canal control algorithms have been developed. These test cases are based on two real canals. Two flow schedules are applied to each test canal, resulting in four test cases. After all four tests are run and the control algorithm tuned, the same four test cases are run again with small changes in canal roughness and gate discharge. These test cases were applied to several different types of canal control algorithms, and some of the test results and performance indicators are discussed.

Introduction

In order to compare the myriad of different canal control algorithms, standardized test cases are needed. One could develop an infinite variety of test cases, but in order to make the testing process feasible, the number of test cases must be limited. Additionally, the test cases should be somewhat realistic, yet simple enough to apply to most algorithms. Finally, the test cases should be challenging enough to thoroughly test the algorithm, but also be constrained to a reasonable time period so the test will produce a manageable amount of data.

Scope of Test Cases

---

[1] Water Control Manager, Central Arizona Project, 23636 N. 7th Street, Phoenix, Arizona 85024

[2] Research Hydraulic Engineer, USDA/ARS Water Conservation Lab, 4331 E. Broadway Rd., Phoenix, Arizona 85040-8832

[3] Hydraulic Engineer, GERSAR, Le Tholonet, Boîte Postale 100, 13603 Aix en Provence CEDEX 1, France

Originally, a wide range and number of test cases were considered. After much thought and discussion, two test schedules were selected to be applied to two test canals, resulting in a total of four test cases. It is recognized that these test cases do not represent all canal control problems. However, they do represent actual canals and somewhat realistic extremes in physical systems. Conditions and parameters were simplified and made into whole numbers to simplify data input, but the general conditions of the test canals are similar to the real canals. Also, the test cases were checked for reasonableness using CANALCAD and a simple PID control strategy.

The test cases presented in this paper are to be applied to the canal control algorithm to be tested, then evaluated using an unsteady-state flow simulation model. This model must be fully non-linear and based on the St. Venant equations.

The testing and evaluation process will consist of the canal control algorithm run interactively with an unsteady-state simulation model. The developer will tune the algorithm's control parameters and tuning coefficients until the algorithm can produce optimal results for the test case. In the real world, however, it is difficult to determine the true conditions of any canal. Furthermore, canal conditions may change over time: the canal roughness may change due to biological growth, the cross section may change due to scouring or siltation, etc. Thus, to rigorously test any algorithm, the testing process must also evaluate how the algorithm can cope with unknown canal characteristics. To evaluate the algorithm's robustness under these unknown conditions, the four test cases discussed above shall be applied first under tuned conditions, then again under untuned conditions. The entire testing process will include eight tests overall.

Test Canal Conditions

The test canal conditions that are uniform over the length of the two canals is given in Table 1. Test Canal 1 is based on lateral canal WM within the Maricopa Stanfield Irrigation and Drainage District, in Maricopa, Arizona. (Clemmens, et al., 1994) The canal is very steep with a high Froude number and little storage, and the target water levels are checked-up above normal depth (creating a backwater condition) in each pool. Test Canal 2 is based on the Corning Canal located in Red Bluff, California, which is much flatter and has significant storage in all pools. (Buyalski, et al., 1979) However, at maximum flow, the target depth is at the normal depth for each pool.

Condition	Test Canal 1	Test Canal 2
Bottom Slope	0.002	0.0001
Manning n	0.014	0.02
Side Slopes	1.5	1.5
Drop at each gate	1.0 m	0.2 m
y_{target}/y_{normal}	1.45	1.0

Table 1. Conditions for Test Canals

The offtakes are located 5 meters from the downstream end of each pool. Gravity offtakes are to be modelled, where the discharge through the offtake, Q_o, is described by

$$Q_0 \sim \sqrt{y - y_0}$$

where

y is the water depth in the canal at the offtake

y_o is the water depth on the downstream side of the offtake, to be set at half the pool's target depth ($y_o = y_{target}/2$).

Test Canal Geometry and Schedules

Because of space limitations, test canal geometry and test schedules are not included in this paper, but can be obtained by contacting any of the authors. It is assumed that the algorithm to be tested has the capability of handling both scheduled and unscheduled changes in flow. For each canal, the first test has a relatively small scheduled flow change at 2 hours, followed by a small unscheduled change at 14 hours. The second test for each canal represents multiple changes, with test 2-2 including dramatic changes in flow rate.

Tuning

Controller parameters or tuning coefficients may be developed by whatever methods are necessary (simulation, equations, mathematical analysis, etc.) for the conditions listed in Table 1. Additional testing should be done to demonstrate the sensitivity of the algorithm to changes in canal conditions over time. The control algorithm should continue to function over a range of conditions without requiring retuning. To satisfy this requirement, the tests shall be run with parameters as tuned (as listed in Table 1), then run again with concurrent changes in the canal roughness (Manning n coefficient), checkgate discharge, and offtake discharge as listed below.

Canal roughness. The Manning n coefficient shall be changed from 0.014 to 0.018 in Test Canal 1, and from

EVALUATOR: LIU ALGORITHM: CLIS							
Regulation Time Step = 5 Minutes, 15 Minutes							
TESTS		MAE (%)	IAE (%)	STE (%) 10-12	STE (%) 22-24	IAW	IAQ
CANAL 1	Test 1 Tuned	19.4	1.35	1.0	.8	.10	.36
	Test 1 Untuned	23.1	1.55	.9	3.2	.11	.35
	Test 2 Tuned	35.1	3.00	1.8	3.6	.30	.75
	Test 2 Untuned	39.5	3.10	1.4	3.1	.40	.78
CANAL 2	Test 1 Tuned	7.1	.85	.8	.9	.52	8.22
	Test 1 Untuned	7.5	1.65	1.0	1.2	1.85	16.18
	Test 2 Tuned	21.5	3.20	.2	1.6	.98	13.47
	Test 2 Untuned	22.9	6.80	2.2	8.1	2.62	17.68

EVALUATOR: SANFILIPPO ALGORITHM: PIR							
Regulation Time Step = 6 Minutes							
TESTS		MAE (%)	IAE (%)	STE (%) 10-12	STE (%) 22-24	IAW	IAQ
CANAL 1	Test 1 Tuned	22.1	3.53	5.5	4.5	.45	1.27
	Test 1 Untuned	25.8	4.06	4.2	5.2	.74	1.87
	Test 2 Tuned	62.6	9.00	3.0	32.3	3.26	9.48
	Test 2 Untuned	(Supercritical Flow Attained)					
CANAL 2	Test 1 Tuned	8.0	1.35	.5	.6	1.31	18.25
	Test 1 Untuned	14.6	2.00	1.7	1.6	4.24	32.25
	Test 2 Tuned	16.8	5.01	5.5	15.6	1.89	33.17
	Test 2 Untuned	16.8	5.34	4.1	14.1	4.39	37.78

Figure 1. Examples of Test Results

0.020 to 0.026 in Test Canal 2.

Checkgate discharge. The checkgate discharges for the untuned case should be 10% less than under the tuned case. This can be accomplished several ways, depending on how the controller functions (e.g. gate coefficient, gate width, assumed discharge, etc.). Any method of adjustment is acceptable as long as it reflects the condition that the controller does not know the correct tuned response.

Offtake discharge. The actual offtake discharge changes shall be 5% higher than as scheduled.

The combination of the tuned and untuned tests applied to the four test cases results in eight tests overall, as shown in Figure 1.

Test Results and Performance Indicators

The test results shall be presented in a standard format, which will facilitate understanding of the controller's behavior, and comparison to other algorithms. Standard performance indicators have been defined, which are described and presented in the paper entitled *Guidelines for Presentation of Canal Control Algorithms*. (Schuurmans, et al., 1995) Some of the performance indicators include:

MAE, maximum absolute error
IAE, Integral of absolute error
StE, Steady-state error
IAW, Integrated average absolute gate movement
IAQ, Integrated average absolute discharge change

The term $\Delta t/T$ has been removed from IAW and IAQ of the original definitions in order to make them independent of the time parameters.

References

Buyalski, C.P. and Serfozo, E.A. 1979. Electronic Filter Level Offset (EL-FLO) Plus Reset Equipment for Automatic Downstream Control of Canals. Report REC-ERC-79-3, U.S. Bureau of Reclamation, Denver, CO, 145 p.

Clemmens, A.J., Sloan, G. and Schuurmans, J. 1994. Canal control needs: example. *J. of Irrig. and Drain. Engin.*, ASCE 120(6):1067:1085

Schuurmans, W., Grawitz, B. and Clemmens, A.J. 1995. Guidelines for presentation of canal control algorithms, Proceedings, The First International Conference on Water Resources Engineering, ASCE (this volume).

Red River Waterway: A Sedimentation Challenge

C. Fred Pinkard, Jr., P.E.*

Abstract

The lower Red River from the Mississippi River to the Caddo-Bossier Port near Shreveport, Louisiana was opened to commercial navigation during early January 1995. The waterway project included the construction of locks and dams, channel realignment, bank stabilization, and channel control dikes. The Red River has one of the highest sediment loads per unit area of drainage basin of all navigable rivers within the United States. Given this high sediment load, the design, construction, and operation of waterway project features must provide for scour and sediment deposition impacts. The responsibility of the design engineer is to provide a navigable waterway that maintains the best possible balance between a channel with adequate sediment transport capacity to minimize excessive sediment deposition while at the same time limit channel velocities so as not to create a hazard for tows navigating the river. The identification of sediment problems and the development of ways in which to most efficiently manage sediment deposition is crucial in the effective maintenance of the waterway project.

General Sediment Conditions

The Red River is a heavily sediment laden stream with an average annual suspended sediment load of approximately 29 million metric tons (32 million tons) at Shreveport, Louisiana and approximately 33.5 million metric tons (37 million tons) at Alexandria, Louisiana. This suspended sediment load basically consists of 25% fine sand and 75% silt. Previous studies have indicated that the bed load on the lower Red River is less than 10% of the total load. The bed is primarily composed of fine to medium sand. The bed material grain size distribution

* Hydraulic Engineer, U.S. Army Corps of Engineers, Vicksburg District, 2101 North Frontage Road, Vicksburg, MS 39180

at Shreveport (Post Project River Mile 228.4) consists of approximately 5% coarse sand, 52% medium sand, 39% fine sand, and 4% very fine sand. Like typical rivers, the grain size on the Red River becomes finer downstream. At Alexandria (Post Project River Mile 88.6), the grain size distribution includes approximately 2% coarse sand, 18% medium sand, 65% fine sand, and 15% very fine sand. The Red River is a high energy system characterized by high channel velocities. During highwater, mean channel velocities often approach 2.1 meters per second (7 feet per second) with maximum velocities exceeding 3.0 meters per second (10 feet per second). The banks of the lower Red River generally consists of fine sand and silt. This combination of high channel velocities with easily erodible banks results in very active bank caving. The loss of 61 meters (200 feet) of bankline during a single highwater event is not uncommon. The primary source of the sediment transported on the Red River comes from erosion of unrevetted banks, especially those upstream of Shreveport. The sediment contribution from tributaries is minimal.

Channel Control Features

The Red River Waterway Project includes the construction and maintenance of a stabilized, rectified channel alignment that consists of a series of navigable bendways. Providing such a channel requires bank stabilization of the concave bank in bends, occasional channel control from the convex bank, and channel control on both banks in crossings. Sedimentation played an important role in the design of the waterway project. The project design included utilizing to the maximum extent possible, the natural erosive action and the large sediment transport capacity of the river to develop project features. The pilot channel concept of channel realignment was used. Trenchfill revetments were constructed along all reaches that would allow the stabilized bankline to lie riverward of the existing bankline. Both of these project features result in construction cost savings but require development by the river. Also, the high suspended sediment load of the river is utilized with the capout of stonefill and timber pile revetments. This construction procedure results in a less costly revetment than initially constructing the revetment to its ultimate height. In general, the channel throughout the Red River Waterway has adequately developed in all reaches in which full construction of the project stabilization and rectification features have been completed and sufficient time for channel development has been provided. In those reaches in which project structures are not complete, some dredging of the channel is now being required. As the structures are completed, additional channel development will occur.

The Vicksburg District is continuing to monitor channel conditions and is committed to providing additional channel control at isolated problem sites.

Channel crossings are naturally common sediment deposition problem locations. Project design studies determined that in order to maintain navigation depths, channel widths must be limited to 182.9 meters (600 feet) in crossings in the lower reaches of the pools where depth is not critical and to 137.2 meters (450 feet) in crossings in the upper reaches of the pools where depth is critical. To provide these limiting channel crossing widths, kicker dikes are provided on the downstream end of revetments. These dikes are an extension of the revetments and reduce sediment deposition in the crossings by forcing the flow to the revetment on the opposite bank. Once raised to their ultimate height, kicker dikes on the Red River have proved very effective in maintaining navigation depths in channel crossings. Figure 1 is an aerial photograph of a typical kicker dike constructed on the Red River.

Figure 1. Typical Kicker Dike

In the very upper end of the pools where channel depths are most critical, navigation depths are provided by the construction of additional contraction structures (ACS). These structures are stone dikes that extend from the convex bank to contract the channel. The contracted channel creates a deeper channel due to bed scour. Figure 2 is an aerial photograph of the Hog Lake ACS which are located in the very upper end of Pool 1. Since these ACS were constructed in 1987, the channel has developed against the left descending bank revetment and no dredging has been required at this site.

Figure 2. Hog Lake ACS

Locks and Dams

Each of the five locks and dams required to provide navigation to Shreveport, Louisiana are now in operation. After Lock and Dam No. 1 (Lindy C. Boggs L&D) was opened during the fall of 1984, significant sediment deposition problems developed. The problem areas were (1) in the upstream lock approach, (2) along the riverside lockwall, (3) in the downstream lock approach channel, and (4) in the lock chamber. This sediment deposition resulted in damage to the lower miter gates. Repair of this damage closed the river to navigation for approximately 3 months. Analysis indicated that areas of channel expansion and flow separation created slack water and eddies at the lock and dam. Two-dimensional numerical model studies were used to evaluate the problems and aid in the determination of appropriate solutions. The results of these studies indicated that structural measures were required to either reduce the amount of sediment deposition or at least relocate the deposition into more manageable areas of the channel. These measures included constructing dikes in the upstream approach channel, raising the wall that separates the downstream lock approach and dam outlet channels, and laying back the right descending bank downstream of the dam. After these measures were constructed, periodic dredging has been required at Lock and Dam No. 1, but not to the extent previously required. Also, the sediment deposition that is occurring is limited to areas that can be easily dredged.

Lock and Dam No. 2 (John H. Overton L&D) was opened during the fall of 1987. As a result of the sediment problems experienced at Lock and Dam No. 1, some sediment deposition was anticipated at Lock and Dam No. 2.

However, several design features at Lock and Dam No. 2 which differ from Lock and Dam No. 1 indicated that a reduction in deposition problems could be expected. These features included a cross section at the structure more representative of the natural river section, no separation between the lock and dam, and fixed guidewalls instead of floating guidewalls. Both physical and numerical model studies were conducted to determine if additional structural measures were needed. As a result, a stone sediment control dike extending downstream from the riverside lockwall and a narrowing of the approach channels were incorporated. Subsequent to opening the lock and dam, sediment deposition occurred in the vicinity of the upstream miter gates and the downstream lock approach channel. While some sediment deposition does occur in the downstream lock approach channel, it has been limited to an acceptable level. To reduce the deposition at the miter gates, a high velocity scour jet system was installed during July 1988. This system has been very successful in limiting sediment deposition.

Based on the experience gained at Lock and Dam Nos. 1 and 2 and on physical and numerical model studies, the design of Lock and Dam Nos. 3, 4, and 5 (Joe D. Waggonner, Jr. L&D) included structural modifications aimed at reducing sediment deposition. At all three structures, the downstream guidewall was moved from the landside of the lock to the riverside, uncontrolled pipes were provided through the downstream lock sill, and, like Lock and Dam No. 2, a channel cross section that closely approximates the natural river section was provided. At Lock and Dam Nos. 4 and 5, more elaborate scour jet systems were included at both the upstream and downstream miter gates. Since Lock and Dam No. 3 was opened during December 1991, some dredging of the downstream lock approach channel and very limited removal of sediment in the upper approach has been required. Lock and Dam Nos. 4 and 5 were constructed concurrently and opened during late December 1994. Highwater experienced during January and February 1995 has resulted in some sediment deposition in the downstream lock approaches. The District will monitor these channels closely and conduct any necessary dredging as required.

Summary

Due to the nature of the sediment and the hydraulic characteristics of the Red River, sediment deposition and scour does occur. Sediment conditions have been given serious consideration in the design of the waterway project features. Areas of excessive sediment deposition must be minimized along the channel as well as at the locks and dams in order to limit the need for costly maintenance dredging.

Channel Realignment on the Red River Waterway

C. Fred Pinkard, Jr., P.E. *

Abstract

As frequently the case with natural alluvial rivers that are being developed for commercial navigation, the Red River experiences active bank caving and contains bendways that are too sharp to accommodate commercial navigation. The bank caving in bends and the tight bendways that restrict navigation can be eliminated by realigning the channel through a bendway cutoff.

Channel Realignment Features

On the Red River, the pilot channel concept of channel realignment was used. This construction procedure includes excavating a pilot channel of smaller section than the desired river section and allowing the natural erosive action of the river to develop the channel to its ultimate size. This method of channel realignment reduces project cost by significantly reducing channel excavation. At each realignment, a trenchfill revetment was constructed along the desired concave bankline, parallel to the pilot channel. As the pilot channel develops, the trench that contains the revetment is undermined and the stone in the trench launches down the channel bank. This occurrence stabilizes the bank and locks the channel along the desired alignment.

Project design criteria includes the preservation of the old bendways that are at least 1.6 kilometers (1 mile) long for recreational and environmental use. These bendways are preserved by constructing a non-overtopping earthen closure dam across the upstream end of the old bendway. A closure dam prevents the sediment laden flow from entering the old bendway from upstream and also helps

* Hydraulic Engineer, U.S. Army Corps of Engineers, Vicksburg District, 2101 North Frontage Road, Vicksburg, MS, 39180

facilitate the development of the pilot channel by forcing all of the river flow through it. The non-overtopping closure dams are constructed to an elevation equal to the post project 100-year frequency flood elevation plus 0.9 meters (3 feet). Once constructed, these closure dams are seeded with grass to prevent localized erosion from rainfall impact and runoff. Whenever practical, positive closure is made by tying the closure dam into natural high ground or an existing levee. Figure 1 is an aerial photograph of a typical channel realignment that includes a non-overtopping closure dam.

Figure 1. Phillip Bayou Realignment

The old bendways less than 1.6 kilometers (1 mile) long are not being preserved. A lower stone closure is provided across the upstream end of these bendways. The height of each stone closure is dependent upon its location within the pool but is sufficiently high to insure channel control. The stone closures force all the channel developing flow during low water periods down the pilot channels. During periods of high flow, the stone closures are over-topped and suspended sediment laden river flow enters the old bendway. The sediment that enters the old bendway from upstream is deposited within the bendway. Continued sediment deposition over time results in the ultimate filling of the old bendway. Kateland Cutoff was constructed on the Red River in 1972 with the purpose of alleviating active bank caving along the concave bank of the existing bendway. This channel realignment did not include a non-overtopping closure dam. An aerial photograph of Kateland Cutoff as taken in

November 1990 is provided as Figure 2. This photograph shows that due to sediment deposition, all that remains of the old bendway is a meander scar.

Figure 2. Kateland Cutoff

The downstream end of the old bendways are left open to the river to allow fish migration and recreational access into the old bendway and to allow an interchange with river water. However, undesirable sediment deposition that threatens to close off river access occurs in the downstream end of the old bendways. The Vicksburg District is currently analyzing the severity of this sediment deposition and will determine the best alternative for maintaining access.

Pilot Channel Design and Development

The U.S. Army Corps of Engineers gained valuable experience in the design and development of pilot channels during the 1930's on the Mississippi River and during the 1950's and 1960's on the Arkansas River. Since many reaches on the Red River are considered to be hydraulically similar to those on the Arkansas River, the design knowledge gained on the Arkansas River was utilized in developing the design criteria for pilot channels on the Red River. The tractive force ratio theory of pilot channel development was used to size the pilot channels. A comparison of the tractive force of the pilot channel to that of the old bendway provides a seemingly reliable indication of whether the site conditions are favorable for pilot channel development. Tractive force is proportional to the product of hydraulic radius and slope. Since sediment transport capacity is a function of the tractive force, the higher the tractive force of the pilot channel to that of the old bendway, the more favorable are the conditions for pilot channel development. Experience indicates that a tractive force ratio of at least 1.5 is needed for development of pilot channels that traverse sand. For pilot channels cut through material more resistant to erosion, a higher tractive force ratio is required. A tractive force ratio of 2.0 or greater is needed for these pilot channels to adequately develop. Box cut pilot channels as narrow as 24.4 meters (80 feet) have been constructed on the Red River that quickly developed to the ultimate channel section of 137.2 meters (450 feet) to 182.9 meters (600 feet). For channel realignments whose pilot channels are cut through materials more resistant to erosion or does not have a large slope advantage (slope advantage comes from realigning a long natural bendway with a short pilot channel), the conditions for development are less favorable. In these instances, a wider pilot channel cut is required. On the Red River, pilot channels as wide as 61 meters (200 feet) have been excavated.

For the Red River Waterway, 36 channel realignments that included the excavation of a pilot channel were constructed. The project construction sequence included finishing the channel realignments as far in advance of raising the pools at the locks and dams as possible. This sequence allowed for increased pilot channel development while channel velocities were higher. Of these 36 pilot channels, 33 adequately developed naturally within a relatively short time period. Once the realignment features are complete and the pilot channel is opened, adequate development typically occurs within a couple of highwater seasons. The remaining 3 pilot channels were Once More and Bijou Realignments in Pool 1 and Grand Bend Realignment in Pool 2. All 3 of these realignments are located downstream of Alexandria within a reach of the river that prior to construction of the waterway project was impacted by backwater from the Mississippi River. Therefore, these pilot channels were cut through back swamp deposits of stiff clay. Shortly after opening the pilot channels, typical high water occurred on the Red River. The slowly developing pilot channels constricted the channel and significantly increased local channel velocities. During this time, contractors working on the waterway project were shipping construction materials by barge. The increased channel velocities resulted in the pilot channels being hazardous to navigate. Therefore, all 3 of these pilot channels were widened by mechanical dredge.

Flood Stage Reduction

Along the lower 450 kilometers (280 miles) of the Red River from the Mississippi River to Shreveport, Louisiana, channel realignments were required to provide a commercially navigable channel. The construction of these cutoffs and the five locks and dams resulted in an approximate 80.5 kilometer (50 mile) reduction in channel length. This 18 percent reduction in association with the other project features provides a shorter, more efficient channel that produces a lowering of flood stages. A comparison of the pre-waterway project and the post project flowlines indicates that for the project design flood (post project 100-year frequency event), the post project stage at Alexandria is approximately .3 meters (1.0 foot) lower than the pre-project stage. At Shreveport the reduction increases to approximately .45 meters (1.5 feet). Therefore, the channel realignment program of the waterway project provides the added benefit of flood stage reduction.

Summary

Channel realignment is a primary feature of the Red River Waterway Project. The channel realignments constructed throughout the project reach, provide a commercially navigable channel, indirect bank stabilization, and a reduction in flood stages. With proper design and adequate development time, the pilot channel concept as used on the Red River has proved to be an effective, cost efficient method of channel realignment. In general, development of the pilot channels constructed on the Red River has been good with 92 percent of these channels adequately developing without additional widening by mechanical means.

FLUME STUDIES ON THE EROSION OF COHESIVE SEDIMENTS

K.E. Dennett[1], S.M.ASCE, T.W. Sturm[2], M.ASCE,
A. Amirtharajah[3], M.ASCE, and T. Mahmood[4], S.M.ASCE

Abstract

The erosion of pure kaolinite and bottom sediment sampled from the Calcasieu River in Louisiana was studied under conditions of uniform flow in a straight, recirculating, tilting flume. Sediment samples were resuspended in tap water and allowed to settle quiescently for 24 hours. Typically, the rate of erosion was initially rapid and then gradually tapered to zero, indicating that the sediment samples were partially-consolidated. The initial rate of erosion and the total mass of sediment eroded increased linearly as excess shear stress increased. In some tests on kaolinite, the sediment pH was varied. At lower pH values, the particles were flocculated and behaved as partially-consolidated samples, while for pH $\geq$ 7, the particles were dispersed and erosion occurred at a slower but steady rate.

Introduction

Cohesive sediments are composed primarily of clay mineral particles (e.g. illite, kaolinite, and montmorillonite) along with silt-sized particles, organic debris, microorganisms, and small amounts of fine sand (Parker, 1994; Hayter and Mehta, 1986). Clay mineral particles are surrounded by an electrical double layer resulting in electrostatic (repulsive) forces and are also subject to van der Waals (attractive) forces. Recently, it has become evident that in addition to these forces, it is necessary to include short-range forces such as Born repulsion and hydration forces (hydrophobic and hydrophilic) to explain erosion events at the microscopic level (Raveendran and Amirtharajah, 1995). The net surface forces are several orders of magnitude larger than gravitational forces and, therefore, control the physico-chemical properties of fine sediments (Hayter and Mehta, 1986). Cohesion of particles is caused by net attractive inter-particle forces.

[1,4]Research Assts., [2]Assoc. Prof., [3]Prof., School of Civil and Environmental Engineering, Georgia Institute of Technology, Atlanta, GA 30332.

Because the size of the particles in cohesive sediments is so small, these particles have a large surface area per unit mass. This property gives cohesive sediments considerable capacity to adsorb contaminants like heavy metals, pesticides, and radioisotopes where they tend to accumulate without degradation (Hayter and Mehta, 1986; Sigg *et al.*, 1987). Numerical modeling of the transport of cohesive sediments is important for determining the fate of adsorbed contaminants. A critical, related research need is to accurately predict the resuspension rate at which fine cohesive sediments move into the water column. It is well known that the resuspension of cohesive sediment is dependent on bed shear stress, τ_o, in excess of a critical value, τ_c, but the empirical constants used in sediment transport models to predict erosion are dependent on clay mineralogy, water content, total salt concentration, ionic species in the water, pH, and temperature (Mehta *et al.*, 1989). Furthermore, the empirical constants are estimated from a variety of ad hoc laboratory procedures and tend to be site-specific.

Objectives

This on-going research is attempting to link the behavior of clay particles on a microscopic level to the hydrodynamic conditions which cause erosion on a macroscopic level. Specific objectives of the research are to investigate the effects of pH and natural organic matter on the erosion and transport of cohesive sediments and relate the results to the microscopic, interparticle forces such as electrostatic, van der Waals, Born repulsion, and hydration forces.

Methodology

The behavior of kaolinite clay and a sediment collected from the Calcasieu River near a contaminated industrial site in Lake Charles, Louisiana, has been studied. Experiments were performed under conditions of uniform flow in a recirculating, tilting flume which was 20 m long and 38 cm wide. The surface of the flume bed was covered with a uniform, fine gravel having a mean size of 3.5 mm. The flow was recirculated using a variable-speed, progressing-cavity pump. Experiments were performed at bed slopes of 0.002 and 0.003. The bed shear stress was varied from 1.0 to 2.4 N/m^2 by changing the flume discharge and bed slope. Flow depths varied from 4.2 to 9.1 cm, and mean velocities ranged between 0.33 and 0.54 m/s. Bed shear stresses were calculated from a uniform-flow formula after correlation with shear stresses determined from detailed velocity profiles measured near the bed at the location of the test section. The test section was located far enough downstream of the flume entrance for full development of the boundary layer as determined by velocity profile measurements.

Two different types of experiments were performed depending on the method of sample preparation. In the first method, a slurry of the cohesive sediment was poured over fine gravel in a sample tray in the flume bed. The total mass of cohesive sediment detached by the flow from the fine gravel was monitored. For the second method, the cohesive sediment was resuspended in tap water and then allowed to settle for 24 hours into a sample tray that was 10 cm wide, 30 cm long, and 2.5 cm deep. The water content of the settled samples was typically about 60%, and the sediment pH values varied between 4.5 and 5.5. The full sample tray was then placed into the flume bed. The rate of erosion and the total mass of sediment eroded were monitored. In order to evaluate the effect of sediment pH on erosion, additional kaolinite samples were prepared at sediment pH values varying from 3 to 8.

The concentration of eroding sediment was monitored during a test using a Chemtrac Particle Monitor. This instrument monitored the fluctuations in light intensity transmitted through a sample stream flowing in transparent flexible tubing. The variation in voltage output of the particle monitor was calibrated with particle concentration for the two sediments which were studied.

Results

During the first type of experiment, detachment of the cohesive river sediment from the fine gravel grains occurred rapidly over the initial one minute of the experiments and then gradually declined for approximately three minutes which indicated a limiting quantity of detachable mass for a given shear stress. The total mass of sediment detached varied linearly with shear stress as shown in Figure 1.

For the second type of experiment with pure cohesive sediment, the rate of erosion was initially rapid and then gradually tapered off to zero after about 25 to 30 minutes. The concentration profile for a typical erosion event is shown in Figure 2. Both the rate of erosion and the total mass eroded increased linearly with increasing shear stress for the Calcasieu River sediment and kaolinite. Using a plot of the initial erosion rates versus shear stress, the value of the critical shear stress, τ_c, was estimated by extrapolating to an erosion rate of zero. Initial erosion rates are shown in Figure 3 as a function of excess shear stress, $(\tau_o-\tau_c)$. Kaolinite had a higher critical shear stress than the Calcasieu River sediment, but showed a more rapid increase in initial erosion rate with increasing values of excess shear stress.

The physical characteristics of the two sediments were different. An X-ray diffraction analysis of the clay mineralogy of the Calcasieu River sediment indicated that it was a mixed-layer clay composed primarily of smectite (60-70%) and illite (30-40%). The kaolinite particles had a very uniform size distribution with a mean size of about 1.5 μm, while the Calcasieu River sediment had a much broader size distribution ranging from dispersed colloidal particles up to very fine sand.

When the pH of the kaolinite samples was varied, those samples at lower pH conditions (pH 3 and pH 5) were flocculated with a mean particle size of 5.25 μm; the higher pH conditions (pH 7 and pH 8) resulted in a dispersed suspension with a mean particle size of 0.75 μm. The cumulative mass curve in Figure 4 for pH 3 exhibits the same behavior as the tests which are included in Figure 3 with a rapid initial erosion rate that gradually declined to zero. This behavior is characteristic of partially-consolidated sediments. At higher pH conditions, the results in Figure 4 show that erosion occurred continuously at a steady rate.

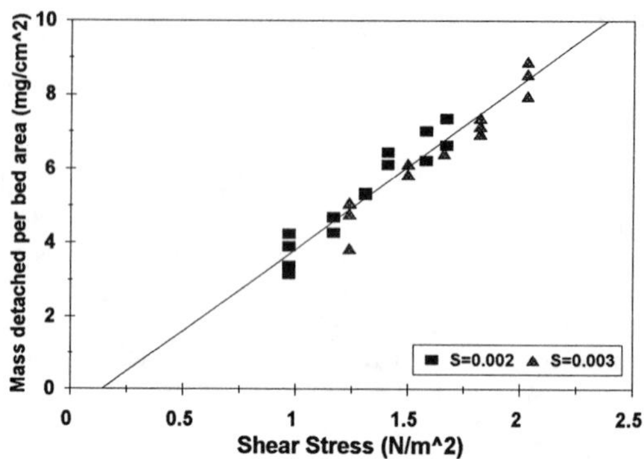

Figure 1. Detachment of Calcasieu River sediment from fine gravel.

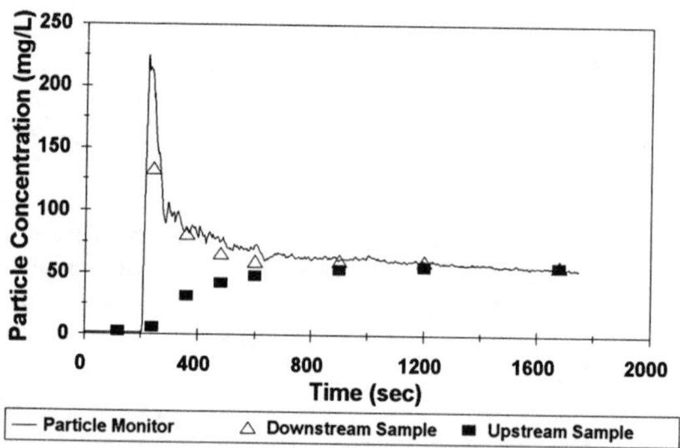

Figure 2. Variation of kaolinite concentration during an erosion event based on particle monitor response and grab samples.

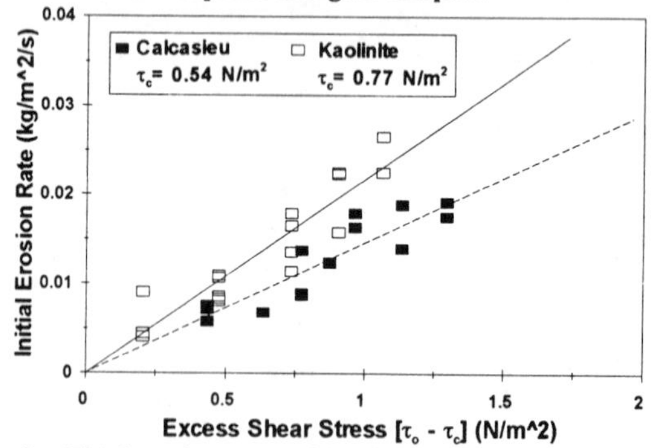

Figure 3. Initial erosion rates of kaolinite and Calcasieu River sediment.

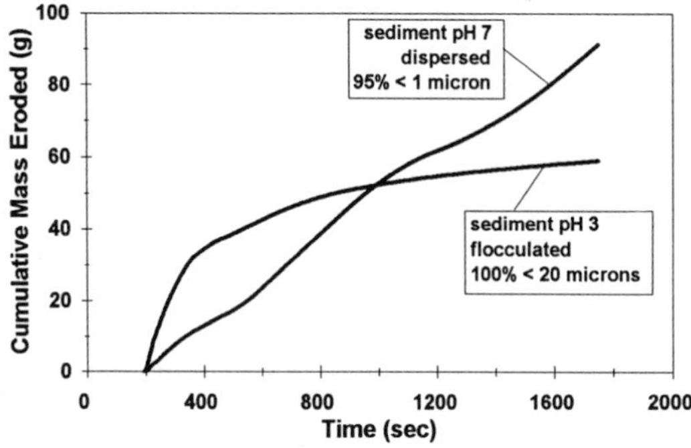

Figure 4. Variation in cumulative mass of kaolinite eroded with sediment pH (τ_o= 0.97 N/m^2).

Conclusions

For the two sediments which were studied, the initial erosion rates increased linearly with increasing shear stress. The critical shear stress of the kaolinite was higher than for the river sediment, and the initial erosion rate of the kaolinite increased more rapidly with increasing excess shear stress. Lower sediment pH conditions enhanced flocculation among kaolinite particles which increased cohesion and resistance to erosion, while higher sediment pH conditions caused dispersion and reduced resistance to erosion. It is expected that microscopic interparticle force calculations can be used to interpret these results in continuing research.

Acknowledgement

This research was partially funded by the EPA Hazardous Substance Research Center, South and Southwest.

References

Hayter, E.J., and Mehta, A.J., 1986. Modelling Cohesive Sediment Transport in Estuarial Waters. *Appl. Math. Modelling*, Vol. 10, 294-303.

Mehta, A.J., Hayter, E.J., Parker, W.G., Krone, R.B., and Teeter, A.M., 1989. Cohesive Sediment Transport. I. Process Description. *J. Hyd. Eng.*, ASCE, Vol. 115, 1076-1093.

Parker, W.R., 1994. Cohesive Sediments - Scientific Background. *Coastal, Estuarial, and Harbour Engineer's Reference Book*, edited by M.B. Abbott and W.A. Price Chapman and Hall, London, pp. 571-576.

Raveendran, P., and Amirtharajah, A., 1995. The Role of Short Range Forces in Particle Detachment. *J. Env. Eng.*, ASCE, accepted for publication.

Sigg, L., Sturm, M., and Distler, D. 1987. Vertical transport of heavy metals by settling particles in Lake Zurich. *Limnol. Ocean.* Vol. 32, 112-130.

Limitations on Use of the HELP Model Version 3

William E. Fleenor,[1] Student Member, ASCE and
Ian P. King,[2] Associate Member, ASCE

Abstract

In underground storage or disposal of wastes, understanding the movement of water through the waste matrix is critical to predicting contaminant transport and designing for its collection and treatment. This paper examined the ability of an Environmental Protection Agency sponsored model to predict vertical moisture transport in order to determine its limitations. The empirical vertical moisture transport of the EPA model was isolated for three climate conditions varying from arid to humid. A direct comparison was made with a numerical solution of the physical equation of unsaturated flow. Although the model performed well under temperate climate conditions, the results demonstrated significant errors with the model's predictions under arid and semi-arid conditions. Understanding these limitations can prevent misapplication of the model, allowing confident use where conditions indicate and demonstrating where more rigid analysis is required.

Introduction

The U.S. Environmental Protection Agency (EPA) model Hydrologic Evaluation of Landfill Performance, Version 3, (HELP) was tested to evaluate its simulation of vertical moisture transport (Schroeder et al. 1984a, 1984b, 1992, 1994a, 1994b). Concerns are raised by the results of a comparison of an earlier version of HELP with a one-dimensional finite difference model (UNSAT1D) performed for the Electrical Power Research Institute (EPRI) by Batelle, Pacific Northwest Laboratories (EPRI 1984). The EPRI report found the earlier version of HELP to overestimate downward moisture flux in a landfill with an unvegetated surface under arid or semi-arid climates.

We compared a beta release of Version 3 of HELP (results were subsequently verified with the release version) with a two-dimensional finite element unsaturated ground water flow model, RMA42. HELP is a water budget model using Darcy's Law for vertical flow with empirical calculations of unsaturated hydraulic conductivity by Campbell. RMA42 (and UNSAT1D) solved the Richard's equation with unsaturated hydraulic conductivity calculations from a more current relationship by van Genuchten (Sisson et al 1980).

[1] Ph. D. Candidate, Civil & Envir. Engrg. Dept., Univ. of California, Davis, CA 95616
[2] Professor, Civil & Envir. Engrg. Dept., Univ. of California, Davis, CA 95616

Comparison of results with unvegetated soil surface demonstrated that HELP simulated vertical water fluxes under humid climate conditions reasonably well. As conditions became more arid, empirical assumptions used in HELP increasingly limited its ability to predict rational design values of vertical transport. Analysis with vegetated surfaces indicated that the predictions became increasingly erroneous.

Models

HELP Version 1 (Schroeder et al. 1984a, 1984b) was developed to "facilitate rapid, economical estimation of the amount of surface runoff, subsurface drainage, and leachate...". Version 1 advances the Hydrologic Simulation Model for Estimating Percolation at Solid Waste Disposal Sites (HSSWDS), the first model developed for predicting percolation through landfills. HELP and HSSWDS are sponsored by EPA and developed by the Waterways Experiment Station, Corp of Engineers (WES). Version 1 of HELP incorporates most of the runoff, evaporation, and transpiration routines of the Chemical Runoff and Erosion from Agriculture Management Systems (CREAMS) model from the U. S. Department of Agriculture (USDA). Runoff and infiltration calculations rely primarily on the Hydrology Section of the National Engineering Handbook, using the Soil Conservation Service (SCS) runoff curve number method. HSSWDS modeled only the cover layer, while Version 1 incorporates lateral drainage and liner leakage and simulates the entire landfill.

Version 2 of HELP (Schroeder et al. 1992) adds a synthetic weather generator (WGEN) developed by the USDA Agriculture Research Service. WGEN can simulate up to twenty years of climate input. Both a five year (1974-78) climatological database and manual input option are retained. WGEN calculates daily values of the maximum temperature, minimum temperature, and solar radiation values, for any climate input option chosen. A vegetative growth model from the Simulator for Water Resources in Rural Basins (SWRRB) is used to calculate leaf area indices. Refinements were made to the unsaturated hydraulic conductivity and the lateral drainage flow algorithms. Interface improvements were made to the ease of use of the model.

Version 3 of HELP (Schroeder et al. 1994a, 1994b) contains additional refinements making it easier to use and improving many transport algorithms. Snow melt calculations are performed with an energy-based model. Calculations of evapotranspiration are made with a Penman model. Increases in the number of layers modeled and the available default definitions for these layers are added. Default layer types have been increased to include geomembranes, geosynthetic drainage nets, and compacted soils. Leakage through membranes and recirculation of leachate is included. Runoff calculations have been improved to incorporate the effects of surface slopes and lengths and of frozen soil surfaces. Unsaturated vertical moisture flux is modified to improve moisture storage calculations. HELP is made more user interactive and includes on-line help at all steps.

Version 3 remains a quasi two-dimensional (the lateral drainage element can be used as needed), deterministic water budget that includes a variety of physics based calculations, empirical estimates, and simplifying approximations.

The RMA42 code is a two-dimensional finite element solver of the Richard's equation for two-dimensional, unsaturated ground water flow (Bear 1972). The Richard's equation, defining unsaturated ground water flow, in two-dimensional form is represented by:

$$\frac{\partial}{\partial x}\left[K_x(\psi)\frac{\partial \psi}{\partial x}\right]+\frac{\partial}{\partial z}\left[K_z(\psi)\left[\frac{\partial \psi}{\partial z}+1\right]\right]=\Theta(\psi)\frac{\partial \psi}{\partial t} \qquad (1)$$

As is evident by the equation, both gravity and capillary forces are taken into consideration, so water is able to be moved up or down depending on the magnitudes of the two opposing driving forces (Freeze and Cherry 1979).

A system is approximated as a continuous series of discrete elements with approximation functions defined in the distribution for each element. The Galerkin method of weighted residuals is used to minimize errors in an integral sense over all the elements. This produces a finite set of equations describing the system. This equation is non-linear in both hydraulic conductivity, $K(\Psi)$, and moisture content, $\theta(\Psi)$, so a Newton-Raphson iteration scheme is used to derive a set of linear equations.

Procedure

A simple landfill was modeled with 60.96 cm (2 ft) of cover soil over a 30.48 cm (1 ft) clay barrier layer and a 914.40 cm (30 ft) waste layer profile. The material characteristics of the waste layer were chosen to represent fly ash landfill. Three climatic test cases of humid, semi-arid, and arid conditions were represented by Cincinnati, OH, Brownsville, TX, and Phoenix, AZ respectively (Fig. 1). Simulations were made for a two year period (1974-75) using the historical climatological database in HELP. Vegetative growth was simulated with HELP.

Identical initial and boundary conditions were applied to both models. Initial moisture content was set at field capacity in each layer except the root zone (cover layer), which is set to the average of field capacity and wilting point.

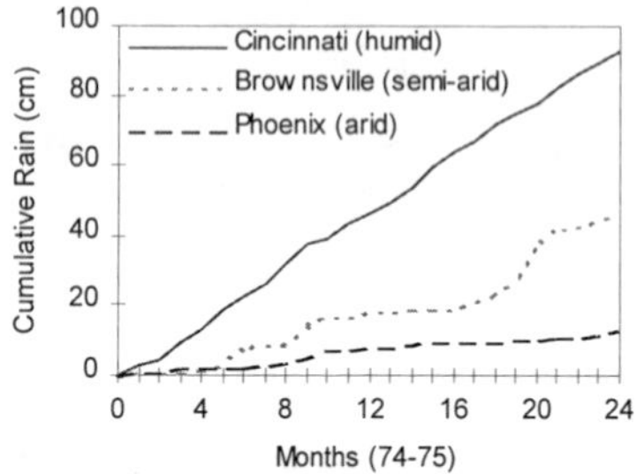

Fig. 1. Precipitation of test cases used.

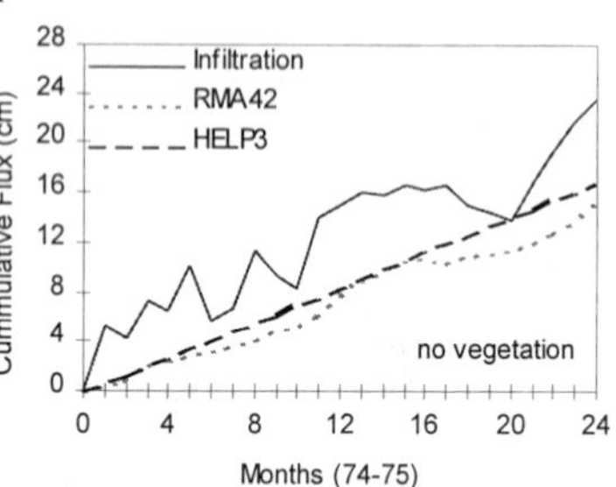

Fig. 2. Barrier Flux, (humid).

Runoff and infiltration in RMA42 were forced to the values predicted in the HELP simulations. Using the same net infiltration values in both models isolated and provided a direct comparison of vertical moisture transport calculations. Differences could be expected since RMA42 used a solution of the Richard's equation accounting for capillary forces not considered in the empirical HELP model.

HELP predicted greater downward moisture transport for all simulations. Consideration of capillary forces by RMA42 in the Richard's equation demonstrated both retardation and upward movement of moisture not possible in HELP.

Results

Flux through the barrier layer was generally over-predicted by HELP in increasingly greater amounts as the climate became more arid. This supports the conclusions of the EPRI report. Under humid conditions and without surface vegetation, HELP estimations compared closely with predictions of RMA42 (Fig. 2). Arid climate simulations show HELP continuing to predict positive downward flux through the barrier even when the net infiltration was upward out of the landfill (Fig. 3). The humid climate with vegetation cover produced highly cyclic values of infiltration to which the HELP simulation failed to react (Fig. 4). RMA42 closely responded to the input variations of these infiltration processes.

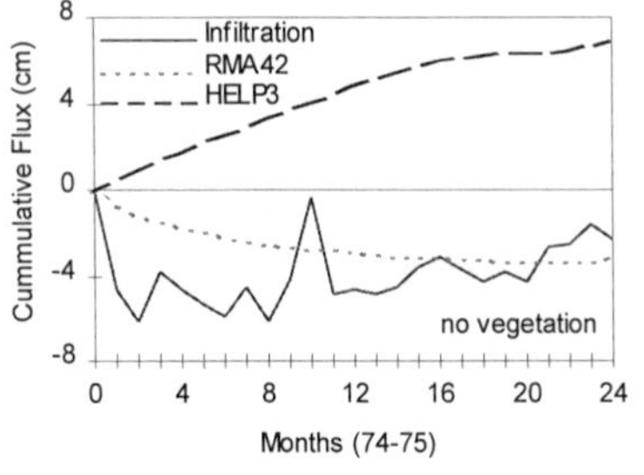

Fig 3. Barrier Flux, (arid)

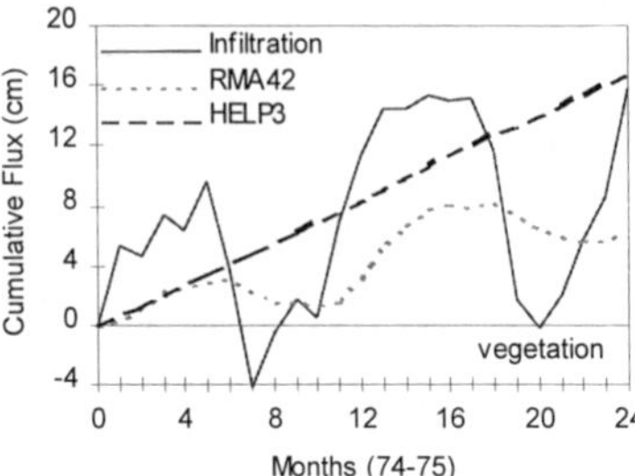

Fig. 4. Barrier Flux, (humid)

For the arid climate with vegetation, the prediction of barrier flux by HELP ceased after six months (Fig. 5). This reflected drainage from the moisture of the initial conditions and rainfall early in the simulation, but without additional precipitation the barrier flux was negligible. RMA42 code predicted a gradually tapering flux up through the barrier layer. This upward flux, caused by capillary forces, is not accounted for in the HELP model. HELP only has provisions to remove moisture from the soil root zone. Once moisture is transported below this level it is transported downward until moisture content is reduced to field capacity.

Flux through the bottom of the waste layer also demonstrates the deficiency of the HELP code and the insensitivity to the climate (Fig. 6). The RMA42 code bottom flux predictions, although very minimal, are also fairly constant with climate because bottom flux is a function of the boundary condition applied to the bottom of the landfill profile. The lower boundary condition was set low enough to provide drainage if the head at the bottom of the waste layer did not decrease and could overcome capillary forces.

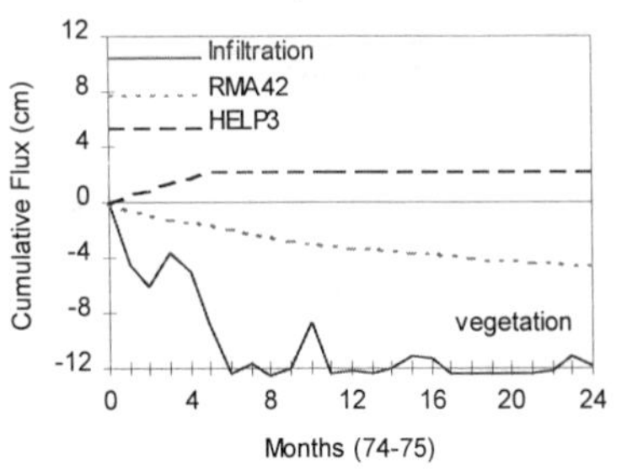

Fig. 5. Barrier Flux, (arid)

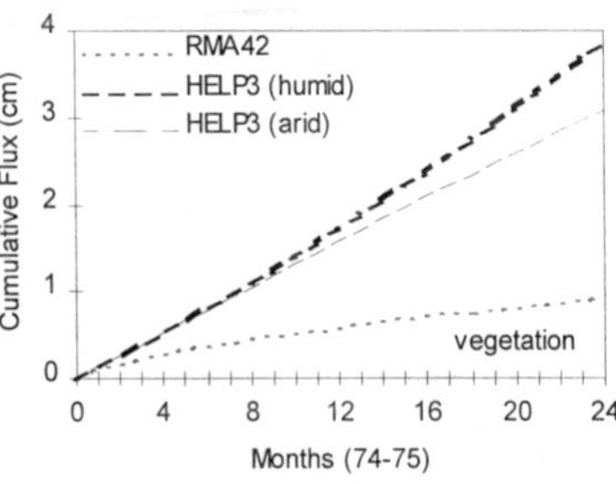

Fig. 6. Bottom Flux Comparison

Conclusions

Without specific modification to the empirical HELP code to account for capillary forces and the removal of water below the soil root zone, HELP will continue to over-predict downward vertical moisture fluxes. This accelerated downward movement of moisture will cause associated difficulties in the infiltration and runoff calculations, since both are dependent functions of surface soil moisture content.

Extensive modifications have been made over the years to portions of the code that are very sensitive to values of soil moisture content near the surface. Without attention to the accuracy of vertical moisture calculations downward and upward through the landfill, improvements to relationships dependent on surface soil moisture are of limited use.

Meanwhile, a complete understanding of these limitations can allow confident use of HELP where conditions permit and demonstrate where more rigid analysis is warranted.

Nomenclature

Ψ,	pressure head (negative capillary pressure)
$K_x(\Psi), K_y(\Psi)$,	non-linear hydraulic conductivity
$\theta(\Psi)$,	non-linear moisture content
x, z,	horizontal and vertical Cartesian coordinates
t,	time

References

Bear, Jacob, *Dynamics of Fluids in Porous Media* (1972), Dover Publications, Inc., New York, New York

Comparison of Two Groundwater Models - UNSAT1D and HELP (1984), Palo Alto, California, Electric Power and Research Institute, CS-3659.

Freeze, R.A., and Cherry, J.A. (1979), *Groundwater*, Prentice-Hall, Inc., Englewood Cliffs, New Jersey.

Schroeder, P.R., Morgan, J.M., Waliski, T.M., and Gibson, A.C. (1984), *The Hydrologic Evaluation of Landfill Performance (HELP) Model. Vol I. User's Guide for Version I.* PB85-100840. U.S. Environmental Protection Agency, Office of Solid Waste and Emergency Response, Washington, D.C.

Schroeder, P.R., Morgan, J.M., Waliski, T.M., and Gibson, A.C. (1984), *The Hydrologic Evaluation of Landfill Performance (HELP) Model. Vol II. Documentation for Version I,* PB85-100832. U.S. Environmental Protection Agency, Office of Solid Waste and Emergency Response, Washington, D.C.

Schroeder, P. R. (1992), *Interim Guide for HELP Version 2 for Experienced Users.* personal correspondence.

Schroeder, P.R., Lloyd, C.M., and Zappi, P.A. (1994), *The Hydrologic Evaluation of Landfill Performance (HELP) Model User's Guide for Version 3*, EPA/600/R-94/168a, U.S. Environmental Protection AgencyRisk Reduction Engineering Laboratory, Cincinnati, OH.

Schroeder, P.R., Dozier, T.S., Zappi, P.A., McEnroe, B.M., Sjostrom, J.W., and Peyton, R.L. (1994), *The Hydrologic Evaluation of Landfill Performance (HELP) Model: Engineering Documentation for Version 3*, EPA/600/R-94/168b, U.S. Environmental Protection AgencyRisk Reduction Engineering Laboratory, Cincinnati, OH.

Sisson, J.B., Furgeson, A.H., and M. Th. van Genuchten (1980), "Simple Methods for Predicting Drainage from Field Plots", *Soil Sci. Soc. Amer. J.*, Vol. 44, pp. 1147-1152.

Water Quality Modeling in San Diego Bay

Parmeshwar L. Shrestha,[1] Associate Member, ASCE

Abstract

A two-dimensional, vertically-averaged finite element model for water quality was applied to San Diego Bay. Model results were compared to results obtained from a quasi two-dimensional model. Investigations were also carried out to determine the sediment and sediment-borne contaminant transport within the estuary.

Introduction

Sediments and pollutants present in wastewater discharges pose a potential threat to the health and diversity of aquatic life in surface water ecosystems. High contaminant concentrations impair the suitability of the water for public use. An increasing awareness of the environment and the need to preserve it is being reflected by the introduction of more stringent water quality standards. To define effective conservation actions that will achieve these goals, it is necessary to predict the quantity, fate, and impacts of sediments and pollutants within the water body after these pollutants are introduced from adjacent lands.

The focus of this paper is to present an integrated modeling approach that couples models for hydrodynamics, water quality, and sediment-toxicant transport, within the same structured framework. Preliminary simulations were carried out to investigate the response of San Diego Bay to changes in live stream discharges of reclaimed wastewater from the Otay River. Water quality constituents simulated were temperature, nitrate-nitrogen ($N\text{-}NO_3$), algae, dissolved oxygen (DO), sediment, and an associated toxicant.

[1] Assistant Professor, Department of Civil and Structural Engineering,
The Hong Kong University of Science and Technology,
Clearwater Bay, Kowloon, Hong Kong.

Modeling Strategy

The bathymetry of San Diego Bay was first digitized and the data accessed by an interactive computer-aided graphics pre-processor, RMAGEN (King, 1994a), to generate a two-dimensional finite element grid of the bay. The network geometry, initial and boundary conditions were input to the hydrodynamic model RMA-2V (Norton et al, 1973; King, 1990). Finite element network refinement, and calibration and verification of the hydrodynamic model were carried out previously by Shrestha (1994). Output from the hydrodynamic model, the network configuration, initial and boundary conditions were then input separately to two models, the water quality model, RMA-4Q (King, 1994), and the sediment-toxicant transport model, TSEDH (Shrestha, 1991). These models utilize the same finite element grid structure as the hydrodynamic model. Both RMA-4Q and TSEDH solve the two-dimensional advection-dispersion equation for mass conservation of constituent, with appropriate source and sink terms. Output from the models include the spatial and temporal values of water column concentrations of constituents, the total mass of eroded or deposited sediment, the toxicant mass associated with the bottom sediments, and changes in bed elevation. Temporal variation of the above variables can be depicted by means of color contours or as time series plots using RMAPLT (King, 1994b).

Hydrodynamic Simulation

Hydrodynamic simulations were carried out for four different discharges from the Otay River, e.g., 0, 0.3 m^3 s^{-1}, 1.1 m^3 s^{-1}, and 2.5 m^3 s^{-1}. Each simulation was carried out for 96 hours with 1-hour time steps. The initial water surface elevation was fixed at MLLW. A tidal boundary condition was imposed at the inlet to the bay. Hydrodynamic simulation results were depicted using RMAPLT. Results of the simulation showed that the flow in San Diego Bay was primarily influenced by the tide. Flow circulation was observed in the shallow areas of South Bay, and Otay River flows did not greatly influence the circulation patterns in this area, even for the 2.5 m^3 s^{-1} discharge. It was apparent that water quality constituents associated with Otay River inflows would not be transported over any appreciable area in the South Bay.

Water Quality Simulation

The hydrodynamic simulation results corresponding to the Otay River discharge of 2.5 m^3 s^{-1} was input to RMA-4Q to investigate the variation, with time and space, of specific water quality constituents such as temperature, N-NO_3, Algae, and DO. RMA-4Q can simulate up to 15 different quality constituents at one time. Constituent reactions and interrelationships, and source and sink terms are based upon QUAL2E (Brown and Barnwell, 1987) formulations. The initial and boundary conditions for the simulation are shown in Table 1. These corresponded to those used by Bale and Orlob (1991), who carried out water quality simulations in San Diego

Bay using the RMA/UC Davis Water Quality Model (Link-Node Model), which is a quasi two-dimensional model that employs a Link-Node (L-N) configuration.

Water quality simulation results were analyzed to obtain the maximum concentrations (maximum daily values) of the four water quality constituents stated earlier, at nodes located along the central transect of the bay. Figure 1 shows a comparison between RMA-4Q model predictions and the results obtained by Bale and Orlob (1991). Node numbers from the finite element mesh used by RMA-4Q are shown with the corresponding node numbers from the Link-Node model. Data points refer to the results from the Link-Node model and lines represent the RMA-4Q model output. Results from the two models were fairly consistent. Daily maximum temperatures increased approximately 5 °C in the South Bay compared to near the ocean boundary. The impact of the Otay River discharge on the temperature of South Bay was minimal. The temperature at node 691 (closest to the Otay River discharge boundary) was about one degree less than the maximum temperature in South Bay. N-NO_3 concentrations dropped rapidly from the boundary value of 3.6 mg L^{-1} to 2.7 mg L^{-1} at node 691, followed by another decrease as dilution increased. For the major part of the bay, the concentration of N-NO_3 remained more or less constant. Algae concentrations increased near the vicinity of the discharge because of high N-NO_3 levels there, and due to the flow hydrodynamics near the source, as a result of tidal action and low Otay River inflows. DO levels in South Bay increased from 5 mg L^{-1} in the discharge waters to approximately 7.5 mg L^{-1}, probably due to the high concentration of algae near the discharge source. The DO values progressively increased along the central transect towards the tidal inlet, to a value of 9.0 mg L^{-1} at the ocean boundary.

Table 1. Initial and Boundary Conditions

	Temperature (°C)	N-NO_3 (mg L^{-1})	Algae (mg L^{-1})	Dissolved Oxygen (mg L^{-1})
Initial	15	0.40	0.40	9.0
Boundary				
Otay River	20	3.6	0.50	5.0
Ocean	15	0.40	0.40	9.0

Sediment-Toxicant Simulation

Cohesive sediments and a hypothetical toxicant was simulated for 96 hours. The same hydrodynamic model results, as used for the water quality simulation, was input to TSEDH. For this study, adsorption of dissolved pollutants onto sediment surfaces was regarded as the primary mechanism for pollutant uptake from the water column. Instantaneous local equilibrium, and complete reversibility of the adsorption/desorption mechanism was also assumed. Initial conditions in the bay were set to 10 mg L^{-1} of suspended sediment and 1.0 μg L^{-1} of a toxicant. The wastewater discharge from the Otay River carried a sediment load of 100 mg L^{-1} and a toxicant concentration of 10 μg L^{-1}. The distribution coefficient was assumed to be 20 m^3 kg^{-1}. Erosion was assumed to occur when the bed shear stress exceeded a critical shear stress of 0.06 N m^{-2}, else deposition occurred.

Simulation results showed that the concentration of suspended sediment and toxicant was a maximum near the discharge source. The suspended sediment concentration near node 682 ranged from 85 mg L^{-1} to about 95 mg L^{-1}; while at mid-bay the concentration range was between 9 mg L^{-1} and 10 mg L^{-1}. Toxicant concentrations remained unchanged in much of the bay at initial levels of 1 μg L^{-1}. Since the hydrodynamics of South Bay is tidally influenced, and the Otay River flows are small, there is not much advective-dispersive transport towards the central bay. These results are in general agreement to those obtained by Bale and Orlob (1991) for water quality constituent simulation. Sediments deposited near the vicinity of the Otay River discharge were subject to cycles of erosion and deposition as a result of tidal incursion and excursion. With lower river inflows, it is likely that the sediment depositional pattern and the suspended sediments would remain in close proximity to the Otay River discharge point.

Conclusions

The integrated modeling approach, as proposed here, is essential not only to assess the importance and interaction of the various processes, but also to further the state of the art in water quality modeling. Physically-based mathematical representations of pollutant concentrations and interactions will provide a means to examine and develop alternative management strategies for waste load allocation. Such models will play a crucial role in assessing the necessary linkage between the physicochemical processes, and surface water function. Without this linkage, it is not possible to follow the fate of contaminants from source to ultimate receptor. Future research efforts should concentrate on assessing the response of the bay to alternative scenarios of discharges not only from the Otay River but from the watershed as well.

References

1. Bale, A.E., and G.T. Orlob, 1991. "Assessing Water Quality Enhancement Alternatives in San Diego Bay - Development of a Decision Support System," G.T. Orlob and Associates, Suisan, California.
2. Brown, L.C., and T.O. Barnwell, Jr. 1987. "The Enhanced Stream Water Quality Modules QUAL2E and QUAL2E-UNCAS: Documentation and User Manual," EPA/600/3-87/007, USEPA, Athens, Georgia.

3. King, I.P., 1994a. "RMAGEN - A Program for Generation of Finite Element Networks - User Instructions," Version 3.3a, Resource Management Associates, Lafayette, California.
4. ________, 1994b. "RMAPLT - A Program for Displaying Results from RMA Finite Element Programs - User Instructions," Version 1.1a, Resource Management Associates, Lafayette, California.
5. ________, 1990. "Program Documentation RMA-2V - Two-dimensional Finite Element Model for Flow in Estuaries and Streams," Version 4.3, Resource Management Associates, Lafayette, California.
6. _________,1994. "RMA-4Q - A Two-dimensional Water Quality Model," Resource Management Associates, Lafayette, California.
7. Norton, W.R., King, I.P., and G.T. Orlob, 1973. "A Finite Element Model for Lower Granite Reservoir," Water Resources Engineers, Walnut Creek, CA.
8. Shrestha, P.L., 1991, "Multiphase Distribution of Cohesive Sediments and Associated Toxic Heavy Metals in Surface Water Systems," Ph.D. Dissertation, University of California, Davis.
9. Shrestha, P.L., 1994. "Hydrodynamic Modeling of San Diego Bay," Proc. ASCE Conf. Hyd. Div., pp. 140-144.

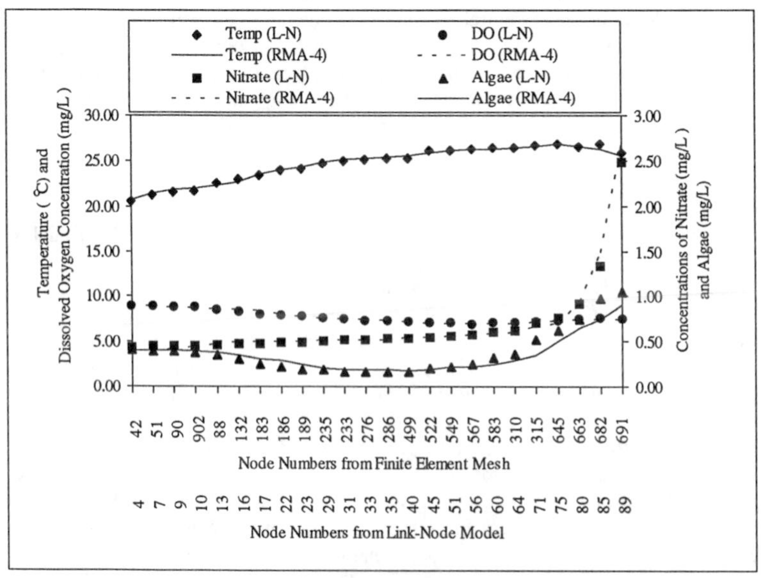

Figure 1. Comparison of Water Quality Constituent Concentrations Obtained from Simulations Using RMA-4Q and RMA/UC Davis (Link-Node) Water Quality Models

Calibration and Verification of QUAL2E Water Quality Model in Sub-Tropical Canals

Vassilios A. Tsihrintzis, Hector R. Fuentes and Leonardo Rodriguez[1]

Abstract

The water quality model QUAL2E is calibrated and verified in the Caloosahatchee River in Central Florida using flow rates and water quality data for the months of June and July 1980. Kinetic rates and coefficients obtained in a previous study on the Tampa Bypass Canal in Tampa, Florida, were used as a first step in calibration, and were adjusted to obtain better model predictions. Fairly good agreement for the sets of tested parameters was obtained.

Introduction

Water quality models have been in existence since 1925. Streeter and Phelps (1925) developed the first water quality model. Bingham et al. (1984) used QUAL2E and demonstrated that ignoring algal effects may result in incorrect conclusions when evaluating waste load allocation permits. The effects of nitrification on streams and swamps has been documented by Courchaine (1968) and Dierberg and Brezonik (1982). As a result, state of the art water quality models must have the capability to simulate nitrogen, phosphorus and algal production.

Ongoing efforts to calibrate and verify QUAL2E for a slow velocity water body in Central Florida, the Caloosahatchee River, using existing flow and water quality data obtained from the South Florida Water Management District (SFWMD) and other local agencies are presented in this paper. The coefficients obtained from the calibration study of the Tampa Bypass Canal in Tampa, Florida, are used as a first step in calibration (Tsihrintzis et al. 1995). The calibration of the model includes dissolved oxygen, the nitrogen series, the phosphorus series and chlorophyll-a.

---

[1] Respectively, Assistant Professor, Associate Professor and Graduate Student, Department of Civil & Environmental Engineering and Drinking Water Research Center, Florida International University, University Park, VH 160, Miami, FL 33199, USA.

Methodology

The Caloosahatchee River is a natural river whose fresh water portion has been dredged to convey flood water to the Gulf of Mexico from Lake Okeechobee (Camp Dresser & McKee Inc. 1991; Miller et al. 1982). It extends 45 miles from structure S-77 to S-79 (Figure 1). The canal cross-sections are typically trapezoidal, and average 20 to 30 feet in depth. The width ranges within 150 to 450 feet. Several tributary canals drain to the Caloosahatchee River, as shown in Figure 1. The portion of the Caloosahatchee River modeled is the 11 mile stretch between Lake Hicpochee and S-78 (Figure 1).

The enhanced stream water quality model QUAL2E was selected as the model to use in the study because the flow conditions are relatively steady between regulatory releases of water from Lake Okeechobee. This model is a steady-state model used for modeling conventional pollutants in streams, and is primarily used as a Level I planning tool. The conceptual representation used is a stream reach divided into computational elements. Each computational element is treated as a completely mixed reactor for which constituent mass, heat and hydraulic balance is calculated. The model can simulate addition and loss of mass from a computational element via sources and sinks by solving the mass balance equation in a implicit backward difference method.

The study portion of the river was divided into five reaches containing computational elements of 0.2 miles long each. Inflow of water and nutrients from tributary canals were considered as point loads. Flow velocities encountered in the reaches vary from 0.01 ft/s to 0.1 ft/s. The Army Corps of Engineers Water Surface Profile Program HEC-2 was used to develop power equations required for the hydraulic calculations in QUAL2E. The predictive capabilities of QUAL2E for dissolved oxygen, nitrogen, phosphorus, and chlorophyll-a were investigated by adjusting rates and coefficients within the expected range presented by Brown and Barnwell (1987) and Tsihrintzis at al. (1995). This investigation was done by perturbing the midpoint value of the acceptable range of each of the rates and coefficients by 10%, 25% and 50% in the positive and negative directions.

Results

During verification, best predictions were obtained for dissolved oxygen concentrations. The model predicted dissolved oxygen within 4% at most of the sampling stations along the river (Figure 2). The model also predicted Chlorophyll-a concentrations within approximately 8% of the observed values (Figure 3). Larger prediction errors were observed for the nitrogen and phosphorus series, which may be attributed to insufficient water quality and quantity data for most of the tributary canals draining into the Caloosahatchee River. Although the predictions for nitrogen and phosphorus were not as good as those for dissolved oxygen and chlorophyll-a, the model predicted the same trend to that of the measured data.

This similarity indicates a relationship between rates and coefficients used by QUAL2E and the flow conditions of the river which is now being investigated.

Conclusions

The results indicate that dissolved oxygen and chlorophyll-a are predicted with fairly good accuracy by QUAL2E. The study also indicates that lack of additional water quality and quantity data for the tributary canals draining into the Caloosahatchee River hinders the ability of the model to predict the nitrogen and phosphorus series concentration. On the other hand, after a careful review of the river flows and their associated nutrient concentrations, the data indicate a correlation between flows and the rates and coefficients used by QUAL2E which is being explored to improve the predictive capability of the model.

Acknowledgements

Useful data, maps and engineering plans for this study were obtained from Dr. Vinio Floris, Supervising Professional-Civil Engineer and Guy Germain, Staff Environmental Specialist both with the South Florida Water Management District; Kevin Petrus, Environmental Specialist with the Florida Department of Environmental Protection; Richard Boler of the Hillsborough County Environmental Protection Commission; and Brian Blake of the U.S. Army Corps of Engineers.

Appendix. References

Bingham, D.R., Lin, C. and Hoag, R.S. (1984). "Nitrogen cycle and algal growth modeling." *Journal of the Water Pollution Control Federation*, 56, 1118-1122.

Brown, L.C. and Barnwell, T.O. *(1987). The Enhanced Stream Water Quality Models QUAL2E and QUAL2E-UNCAS: Documentation and Users Manual*, US Environmental Protection Agency, Georgia.

Camp Dresser & McKee Inc. (1991). *Caloosahatchee River Basin Assessment: Phase 1, Task 3, Model Selection*, Report presented to the South Florida Water Management District, West Palm Beach, Florida.

Courchaine, R.J. (1968). "Significance of nitrification in stream analysis-effects on the oxygen balance." *Journal of the Water Pollution Control Federation*, 40, 835-847.

Dierberg, F.E. and Brezonik, P.L. (1982). "Nitrifying population densities and inhibition of ammonium oxidation in natural and sewage-enriched cypress swamps." *Water Research*, 16, 123-126.

Miller, T.H., Federico, A.C. and Milleson, J.F. (1982). *A Survey of Water Quality Characteristics and Chlorophyll-a Concentrations in the Caloosahatchee River System, Florida*, South Florida Water Management District, West Palm Beach, Florida.

Streeter, H.W. and Phelps, E.B. (1925). *A Study of the Pollution and Natural Purification of the Ohio River*, Treasury Department United States Public Health Service, Washington Government Printing Office.

Tsihrintzis, V.A., Fuentes, H.R. and Rodriguez, L. (1995). "Modeling water quality in low velocity sub-tropical streams." *Proceedings of Water Pollution 95, 3rd International Conference on Water Pollution Modelling, Measuring and Prediction, April 25-28, Porto Carras, Greece.*

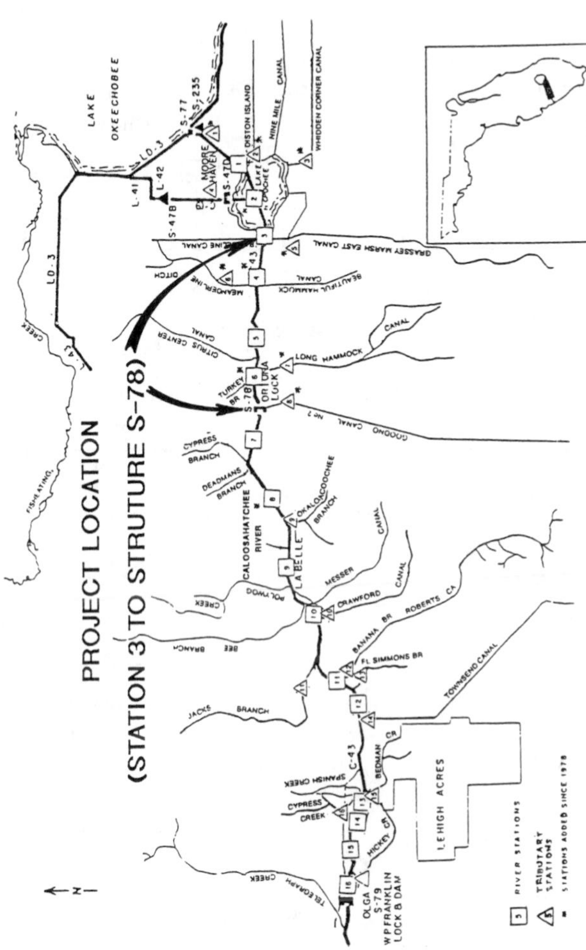

Figure 1. Location of the Study Reach on the Caloosahatchee River

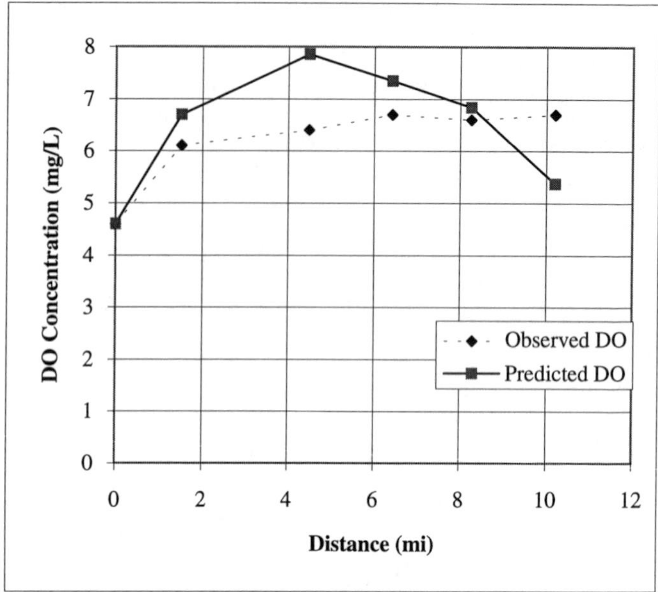

Figure 2. Verification Model Predictions for Dissolved Oxygen

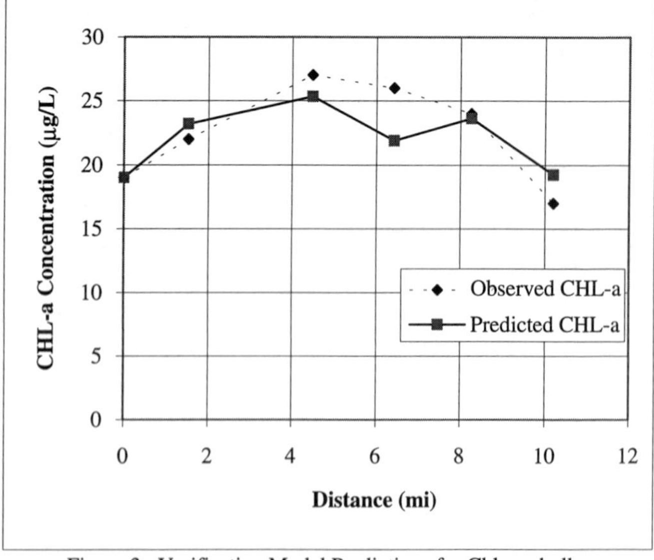

Figure 3. Verification Model Predictions for Chlorophyll-a

Pollutant Transport Beneath Porous Stream Beds

D. Zhou[1] and C. Mendoza[2]

Abstract

The presence of porous streambeds generates a significant interaction between the main stream flow above it and the flow through the porous substratum. This coupling affects both flows, and specially the pollutant transport in the region beneath the porous interface. The mechanism of such complicated interaction on the contaminant transport is explored and its magnitude examined using a simple model.

Introduction

It is now clear from the studies of Zhou and Mendoza (1993) and Mendoza and Zhou (1993) that the coupling of the main flow and the porous bed is significant, and important consequences must therefore be recognized for the transport of solute over and through the bed and for stream ecology. The flow zone adjacent to the riverbed is affected by the porous boundary; in it, strong gradients of physical, chemical, and biological variables occur. The present investigation, however, only concentrates on how the hydrodynamic interaction across the interface affects the pollutant transport process in the flow through the porous bed just beneath the interface.

---

[1]Center for Computational Hydroscience and Engineering, 102 Carrier Hall, University of Mississippi, University, MS 38677.

[2]Dept. of Civ. Engrg. and Engrg. Mech., 610 S. W. Mudd Bldg., Columbia University, New York, NY 10027.

Hydrodynamic Interaction

In the study of the hydrodynamic interaction across a flat porous bed, an incompressible free-surface shear flow coupled with the flow through porous substrate is considered in a three-dimensional Cartesian coordinate system (see Fig. 1). The turbulence is assumed to be statistically homogeneous in planes parallel to the (x_1, x_2) plane (pervious bottom surface), and the mean velocity is assumed to exist only in the x_1-direction. The mean velocity, shear stress, and the statistical quantities of the turbulence vary in the x_3-direction which is perpendicular to the (x_1, x_2) plane. The flow in the channel is fully developed and uniform.

From the analyses by Zhou and Mendoza (1993) and Mendoza and Zhou (1993) it was found that: (1) there is a mean slip velocity at the interface; (2) the bulk velocity through the bed is not uniform and changes drastically in x_3-direction as is depicted in Fig. 1 (the flow is not statistically homogeneous in that direction); (3) because of the turbulent free surface flow above the stream bed, high velocities which render Darcy's law invalid, develop in the porous medium region just beneath the interface; (4) the equations describing the macroscale vertical variation of the flow velocity in the substratum, as well as the boundary at the pervious interface is

$$U = U_1 + U_2(x_3) = U_1 + U_0 \exp(M_0 x_3) \tag{1}$$

in which U is the bulk flow velocity; U_1 and U_2 are the uniform and nonuniform parts of U, respectively; U_1 is the Darcy velocity, U_0 is the maximum value of U_2, and M_0 is a constant dependent on the properties of the porous material. The interface slip velocity $U_s = U_1 + U_0$. The comparison with the data of Shimizu et al. (1990) revealed that for a given porous medium, M_0 is a constant (Zhou and Mendoza 1993; Mendoza and Zhou 1993).

Effects on Pollutant Transport

The general diffusion equation for pollutant concentration C takes the form (Fischer et al. 1979)

$$\frac{\partial C}{\partial t} + \mathbf{U} \bullet \nabla C = D_p \nabla^2 C \tag{2}$$

where t is time, $\mathbf{U}$ is the advective velocity vector of the flow, D_p takes the value of the molecular diffusivity D_0 in the shear flow above the interface; while in the porous region below the interface, $D_p = D_0 \epsilon B_t$ is the effective molecular diffusivity taking into account the effect of the porosity ϵ and the porous medium tortuosity B_t. For granular materials the tortuosity is usually approximated as $B_t = \epsilon^{-1/3}$ (Millington and Quirk 1961), so that $D_p \approx D_0 \epsilon^{2/3}$.

To illustrate the effect, we consider a line source of contaminant as the initial condition, that is

$$C(x_i, t=0) = \begin{cases} 0, & x_1 > 0 \\ C_0, & x_1 < 0 \end{cases} \tag{3}$$

The corresponding analysis for flows with and without interface hydrodynamic interaction, are presented below.

Case A: Decoupled flow.

For such a flow, the bulk velocity profile is uniform along the x_3 direction and is equal to the Darcy value. As a result, the concentration is uniform along the x_2- and x_3-directions, and Eq. (2) reduces to

$$\frac{\partial C}{\partial t} + U_1 \frac{\partial C}{\partial x_1} = D_p \frac{\partial^2 C}{\partial x_1^2} \tag{4}$$

Its solution is

$$C(x_i, t) = \frac{C_0}{2}\left[1 - \text{erf}\left(\frac{x_1 - U_1 t}{\sqrt{4 D_p t}}\right)\right] \tag{5}$$

Case B: Coupled flow.

For such a flow, the bulk velocity profile is no longer uniform; therefore, the problem falls into the category of shear flow dispersion, even if for simplicity the effect of the penetrated turbulence is neglected. Employing the depth-averaged method and adopting the assumptions introduced by Taylor (1953), Eq. (2) can then be transformed into the depth-averaged one-dimensional dispersion equation

$$\frac{\partial C_m}{\partial t} + U_m \frac{\partial C_m}{\partial x_1} = K_p \frac{\partial^2 C_m}{\partial x_1^2} \tag{6}$$

where all variables with subscript m denote their depth-averaged values. Equation (6), however, is only valid for $t > 0.4h^2/K_p$ (Fischer et al. 1979), with h being the flow depth beneath the interface. In Eq. (6), K_p is given by

$$K_p = \frac{-1}{hD_p} \int_{-h}^{0} u' \int_{-h}^{x_3} \int_{-h}^{x_3} u' dx_3 dx_3 \tag{7}$$

in which u' is the difference between the local velocity and the depth-averaged one. Even though the solution to Eq. (6) is similar in form to Eq. (5), that is

$$C_m(x_i, t) = \frac{C_0}{2}\left[1 - \text{erf}\left(\frac{x_1 - U_m t}{\sqrt{4 K_p t}}\right)\right] \tag{8}$$

the physical process and the physical meaning of the variables are quite different, specially, their time scales. Figure 2 shows a comparison of the concentration as a function of the moving coordinate $x_1 - U_m t$ (traveling with the depth-averaged mean advective velocity) with time as a parameter for decoupled and coupled models. In computing K_p, the following typical values of the Mississippi River (Thibodeaux 1979) $\epsilon = 0.4$, $U_1 = 150cm/s$, $h = 10cm$ are taken. The values of $U_0 = 25.67cm/s$ and $M_0 = 1.02cm^{-1}$ used are those computed by Zhou and Mendoza (1993) and Mendoza and Zhou (1993) from the experimental data of Shimizu et al. (1990). The calculation indicates that K_p is more than a million times the magnitude of D_p, a situation similar to that encountered for the shear dispersion phenomenon in free surface flows (Fischer et al. 1979). Besides the difference in the mean advective velocities between the two models, the diffusion process of the coupled model is much faster than that of the uncoupled one, as clearly indicated by the time scales in Figure 2.

Conclusions

To reflect the effects of the hydrodynamic interaction existing at a porous streambed, interface, the model for pollutant transport beneath it must be modified accordingly. For the longitudinal mixing, instead of using a molecular diffusion based model, a depth-averaged shear dispersion model should be used. The model proposed in this study is based on the results of the previous studies of the hydrodynamic interaction and is applicable to stream conditions.

References

Fischer, H. B. et al. (1979). *Mixing in Inland and Coastal waters*, Academic Press, Orlando, FL.

Mendoza, C., and Zhou, D. (1993). "Turbulent channel flow over porous beds," *Advances in Hydro-science and Eng. — Proc. 1st Int. Conf. on Hydro-science and Eng.*, Washington, D. C., Sam S. Y. Wang edited, Vol. 1, Part A, 221-228.

Millington, R. J., and Quirk, J. P. (1961). "Permeability of porous solids," *Trans. Faraday Soc.*, 57, 1200.

Shimizu, Y., Tsujimoto, T., and Nakagawa, H. (1990). "Experiment and macroscopic modelling of flow in highly permeable porous medium under free-surface flow," *J. Hydrosc. and Hydr. Eng.*, 8(1), 69-78.

Taylor, G. I. (1953). "Dispersion of soluble matter in solvent flowing slowly through a tube," *Proc. R. Soc. London Ser. A*, 219, 186-203.

Thibodeaux, L. J. (1979). *Chemodynamics*, John Wiley and Sons, New York.

Zhou, D., and Mendoza, C. (1993). "Flow through porous bed of turbulent stream," *J. Engrg. Mech.*, ASCE, 119(2), 365-383.

Appendix — Figures

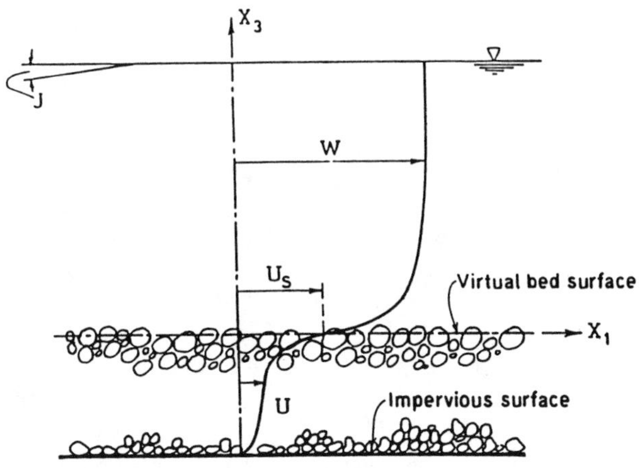

Figure 1 Turbulent flow above and through the substrate

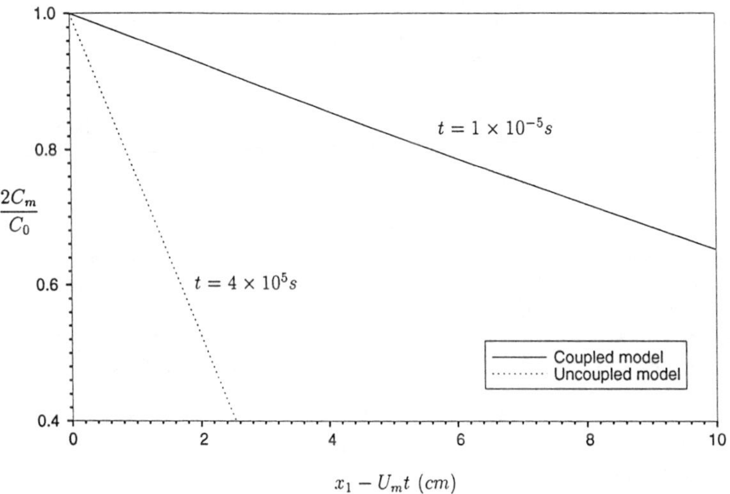

Figure 2. Comparison of diffusion process between coupled and uncoupled models

A Perspective on Sediment Research in China

Carl F. Nordin, Jr.[1], Member ASCE

Abstract

During the past decade, the writer has reviewed sediment problems associated with several important water-control projects in the People's Republic of China. A large amount of the recent sediment research in China has been directed to solving practical problems associated with these projects. Important advances have been made in the design and operation of reservoirs to preserve long-term storage, in the use of mathematical and physical models, in characterizing the properties of hyperconcentrated flows, and in developing materials that resist sediment abrasion.

Introduction

Two major water control projects are presently under construction in the People's Republic of China, the Yangtze River Three Gorges Project, and the Xiaolangdi Project on the Yellow River. For both projects, managing the sediment problems was a major concern, and much of the recent sediment research in China has been driven by practical and theoretical considerations that were involved in the design of these two projects. The research was carried out by a large number of government-sponsored research institutions and universities involving thousands of individuals. During the past decade, I had the opportunity to review some of this research. In this paper, I have outlined some of the common sediment problems of the two projects and some of the research and developments that contribute to the solutions to the problems.

Although the Yangtze and Yellow Rivers are quite different in their hydrologic characteristics and in the nature of their sediment loads, the two projects share many

[1] Professor, Civil Engineering Department, Colorado State University, Fort Collins, CO. 80523.

common problems. Both reservoirs are small, relative to the mean annual flows and flood volumes, both are designed primarily to reduce the hazards of catastrophic flooding along the alluvial reaches downstream of the reservoirs, and both are designed and will be operated in such a way as to preserve long-term storage in the reservoirs.

Three Gorges Project

The Three Gorges Project (TGP) is a multi-purpose project designed to generate power, improve navigation, and reduce the hazards of flooding. The project is located at Sandouping, about 40 km upstream of the existing low-head dam at Gezhouba. The river upstream of TGP drains approximately 10^6 km^2, or about 60% of the total basin area. The mean annual flow is about 14,300 m^3/s, with an annual volume of 450 x 10^9 m^3, the annual sediment discharge is 450 x 10^6 tonnes. The main element of the project is a concrete gravity dam with crest elevation of 185 m and length of 1983 m. The maximum dam height is 175 m. The reservoir upstream of the dam will extend almost 700 km to the City of Chongqing. The sediment problems associated with this project are conventional sediment problems - reservoir sedimentation and depletion of storage, sediment encroachment at the structure, aggradation upstream of the reservoir in the backwater region, and possible degradation and channel modifications downstream of the structure.

Xiaolangdi Project

Xiaolangdi is also a multi-purpose project, designed to generate power, reduce the hazards flooding along the lower Yellow River, and improve water supply for irrigation and other uses. The project is locate about 130 km downstream of Sanmenxia reservoir and about 40 km north of the city of Luoyang. The main structure is a rockfill dam 167 m high with a crest elevation of 281 m and a crest length of 1350 m. The normal pool level is 275 m. The layout is considerably more complex that TGP. The sediment problems in Xiaolangdi Project are similar to those in TGP.

Preserving Long-Term Storage

Flood control is the main purpose of both projects, so it is important to design and operate the reservoirs in such a way that some increment of long-term storage for flood control can be preserved indefinitely. In order to preserve the long-term capacity of reservoirs, three conditions should be met. First, the reservoir volume should be small relative to the mean annual flow. This means that a fairly large volume of flow has to be spilled each year, and this excess flow can be used to flush sediment through the reservoir. Second, the river should be able to transport its sediment load at a flatter slope than exists at present. Usually, this means that the reservoir is in a gorge section or in a reach of the river controlled by bedrock. Third, the reservoir width should be roughly comparable to the width of an equilibrium channel carrying the same sediment load. All three of these conditions are met for TGP and Xiaolangdi.

In practice, the operation of a reservoir to preserve long-term storage may be rather complex, but the concept is fairly simple. The Chinese describe it simply as "impounding the clear water and discharging the turbid flow" (Qian, 1987). During the flood season, the water level at the dam is kept at a low elevation, the flood control level (FCL). During the initial years of operation, the normal flood flows and some of the sediment load are passed through the reservoir, while some of the sediments, mostly the coarser fractions, are deposited upstream of the dam in the backwater region. As the flood passes, the receding flows with lower sediment concentrations are impounded as the water level is raised and maintained at the normal pool level (NPL). The natural channel though the reservoir has an excess capacity and excess slope to carry its sediment load. Eventually, a new alluvial channel will develop through the deposited sediments. Alluvial channels are self-forming. The new channel will develop an equilibrium slope smaller than the original channel slope that will allow the prevailing flow through the flood season to just transport the annual sediment load through the reservoir. After that equilibrium is reached, there will be no net deposition or erosion through the reservoir. The volume of storage occupied by the sediment deposit is lost, but any storage remaining between the sediments and the NPL can be preserved indefinitely.

The water surface profile upstream of the dam and the upstream extent of the deposited sediments are determined by the water level of the dam during the flood season, so the most important design considerations in managing the sediment are selecting the flood control level and ensuring that the low level outlets have sufficient capacity to flush the annual sediment load with the excess flows that have to be spilled during the flood season. The calculation of the equilibrium slope is the critical element in estimating the amount of storage that can be preserved indefinitely. The calculation is straightforward, but it requires a good bit of information about the cross sections and roughness coefficients along the reservoir. For TGP, the equilibrium slopes and deposition patterns in the reservoir for various combinations of pool elevations were calculated using a one-dimensional non-equilibrium mathematical model (Han and He, 1990). For Xiaolangdi, the equilibrium slopes were calculated using empirical transport relations developed mostly from observations at Sanmenxia (Zhang, 1994).

The average slope through the Three Gorges Reservoir area is about 20 cm/km, the equilibrium slope is about 6.5 to 7 cm/km, or about one-third of the original. The initial storage is 40 billion cubic meters, the long-term storage that can be preserved is about 23 billion cubic meters. For Xiaolangdi, the initial slope is around 110 cm/km, and the final equilibrium slope is about 26 cm/km. The initial storage is 12.7×10^9 m^3, the final preserved storage is about 5.1×10^9 m^3. For TGP the time to equilibrium is about 120 years, while for Xiaolangdi, the time to equilibrium is on the order of 12 to 15 years.

The concept of preserving long-term storage has interesting implications. It is a commitment for both projects to the notion that the major design function of the reservoirs is to manage the sediment problems and it is a clear recognition that sediment problems have to be managed over the long term. Also, it implies a willingness to forgo immediate short-term economic gains from power generation for the long-term benefits of flood control that will accrue to future generations. This last issue is contentious and highly controversial, especially among the consultants and international lending agencies, who are interested mostly in an immediate return on their investment.

Physical and Mathematical Models

One-dimensional mathematical models were used extensively in the design of TGP, but for Xiaolangdi, calculations for equilibrium slope were carried out using an empirical transport function (Zhang, 1994), and in general, hydrologic routing models with empirical transport functions are used more often for Yellow River design problems than the more theoretically based mathematical models. In part this is due to the difficulties of characterizing the properties of the hyperconcentrated flows, and in part it reflects the difficulties of collecting suitable field data to calibrate, verify, and implement the mathematical models.

Physical model studies were used extensively for both projects. Usually, the output from the mathematical model was used as the boundary condition for the physical model. The modeling techniques are well developed mostly in connection with previous studies for Ghezhouba or Sanmenxia projects, and model results have been verified against post-project evaluations so the predictions are used with considerable confidence. It is fairly common to carry out physical model studies at two separate institutions using different scales and usually different lightweight materials to simulate the sediment. In all cases where I have reviewed model results from two different institution of the same structural features, the results have been comparable, which adds considerably to our confidence in the results.

Yellow River Problems

Most of the sediment problems of the Yellow River are associated with flood flows of short duration carrying extremely high sediment concentrations, usually in excess of 400 kg/m^3, called hyperconcentrated flows. Such flows are generally non-Newtonian, and neither the characteristics of the flows or their sediment transporting characteristics can be described by the conventional flow and transport equation used for the TGP design. A tremendous amount of research has been carried out on hyperconcentrated flows, and quite a bit is known about their properties, but so far as I know, there is still no general agreement among the Chinese experts as to the most effective way to introduce these properties into mathematical models and there is no consensus on a suitable model or models. A large amount of research in China is now underway to resolve some of the issues.

The high concentrations of sediments in the Yellow River flows cause substantial damage to turbines, pumps, gates, valves, and other components of hydraulic structures and machinery. Damage is so extensive that turbines are usually shut down for a few months during the flood season. Concerns for sediment damage have given rise to extensive research in two general areas: first, the optimum arrangements for inlets and sluicing tunnels, and second, the development and testing of abrasive-resistant materials for gate sills, turbine runner and wicket gates, tunnel linings and other structural component that are known from experience to sustain high damage from the hyperconcentrated flows. The design of Xiaoloangdi calls for continuous power generation so special attention was directed towards designing the turbines to minimize sediment damage and to establishing an efficient method for the replacement and repair of the turbines. So far as I know, though, the first turbines at least will be purchased abroad. Although the science of developing abrasion-resistant materials is well-advanced, apparently the technology to manufacture units from these advanced materials does not yet exist.

Conclusions

Much of the sediment research in China is directed to solving practical problems associated with major water control projects. The research has lead to some important advances in the design and operation of reservoirs to preserve long-term storage, in the combined use of mathematical and physical models, and in characterizing the properties of hyperconcentrated flows of the Yellow River. However, there is still no general agreement on the type of mathematical model best suited for the Yellow River flows.

References

Han Qiwei and He Mingmin (1990). "A New Mathematical Model for Reservoir Sedimentation and Fluvial Process." Int. Jour. of Sediment Research, 5(2), 43-84.

Lin Bingnan (1994). "Some Facts and Issues About TGP." Int. Jour. of Sediment Research, 9(3), 75-84.

Qian Ning, Zhang Ren, and Chen Zhicong (1987). "On Some Sedimentation Problems of the Three Gorges Project." Int. Jour. of Sediment Research, No. 1, 5-38.

Zhang Ren (1994). "Regulation of the Lower Yellow River and the Xiaolangdi Project." Int. Jour. of Sediment Research, 9(2), 1-17.

Transfer of Knowledge to Overseas Practice

Michael A. Stevens[1] and Susan Scott-Stevens[1]

Abstract

The transfer of knowledge to engineers in Third-World Nations is fraught with pitfalls, some almost insurmountable. In addition to "naive theories", which exist for engineers within any culture, special problems confront the practitioner working in other cultures. For example, fear of "losing face" can prevent engineers from changing to better methods. Respect for elders and their "old" ways also hinders change in places. Hardware, such as roller-compacted concrete dams, labyrinth weirs, and computer programs, are more likely to be accepted than software—ideas and design criteria are examples. Some organizations, usually universities and federal government agencies, have much more respect in many cultures than in others. That respect can be utilized to accomplish change. Often, individual consultants are viewed somewhat less favorably than institutions. Examples of these and other problems encountered by the authors are recollected in this paper.

Introduction

Technical knowledge can be categorized as being of two types, labeled *hardware* and *software*. *Hardware* technology is defined as those "things" or material objects associated with technology transfer; for example, computers, automobiles, hydroelectric generators, and dams. *Software* knowledge is the invisible component of technology. Software may express itself, for example, as the decision procedure used to determine the depth at which to place the bridge piers in a sandbed river (after Bulfin 1972). In this paper, the focus is with transferring software, which can be very difficult.

Transfer of software knowledge to others, compatriot or foreign, is best done within a structure or method that is known to work -- at least most of the time. Experiences with learning at North American universities seem superficially to serve as a model for transferring technology: Why not? It worked for us. In fact, it is not nearly so successful as we age. As young students, and with grades as incentive, we were prone and encouraged to accept the professor's lectures and text books as "truth". The first writer had only one class in nine years of university training at two universities in which the teacher asked, and sometimes demanded, students to question what he said or what was in the refereed journals. As professionals, it seems as if everyone questions what we say or do, and not so pleasantly at times. Why is this so?

---

[1] Consultants, P.O. Box 3263, Boulder, Colorado, USA, 80307.

Naive Theory

The fundamental presumption of the concept of "naive theories" (after Resnick 1983) as they pertain to the transfer of technology (and to one's own learning) is that much of what we believe to be true about how the world "should" and "does" operate is firmly fixed in our minds long before we ever walk into our first science class. In short, we learn a "science" of the world that is more determined by cultural factors than so-called "scientific ones". Moreover, this same "science" also encompasses such things as the value we attach to knowledge itself, as well as to the premium we place on the acquisition of new knowledge. In more specific terms, one tends to take any new piece of technical information and unconsciously compares it to what one already knows. If the new information agrees with our "naive theories", it is most likely accepted and the transfer is realized. If there is conflict between new and old, it is the new and not the old, which is most likely to be rejected. The reason for acceptance or rejection is apt to go unquestioned.

As students in homeland universities, most of us were unquestioning, but with good reason. Our teachers were smart and certainly more knowledgeable than we were. Then, as practicing professionals, our experiences began to accumulate and our data base expanded by many means: work, conversations with peers, reading, and perhaps formal and informal study. We became more critical and prone to reject things as time went by because there were more comparisons to pass before acceptance. Some rejections seemed warranted because all new information is not correct.

For engineering, knowledge can be divided into three categories: a theoretical base, experience, and a data base. Each individual is limited in knowledge. Nobody has "it all". Furthermore, people collect erroneous and inaccurate information in their knowledge base or give mistaken interpretations to correct knowledge. Consequently, our knowledge base remains incomplete and most likely flawed. That is, we are "naive" with respect to some or many facets of knowledge in our profession. What many of us have done is to replace the "naive theories" of childhood and university with naive theories of a professional nature [knowledge], validated seemingly by experience.

To correct one's naive theories, one must improve his or her theoretical base by formal or informal study. One way is to buy the next-level text book in one's subject of interest. With an extended theoretical base, one can give more interpretations to any piece of information. For example, the motion of water can be viewed in more ways with hydrodynamic theory than with hydraulics theory. The former encompasses the latter. Experiences and data bases can be expanded from the writings and conversations with other professionals as well as by experimentation. With less theory and no computers prior to the late 1950's, engineers wrote more about their experiences and experiments. Their keen observations are still of great value.

Cultures

Transfer of software technology to other countries and cultures, generally speaking, tends to fail if the knowledge is perceived as irrelevant, if it is imposed from without, or if it threatens the recipients in some fundamental way (Spicer 1952). Birth control, an example of both hardware [the devices themselves] and software [the economic and social concepts underlying such a program] technology, threatens people in many countries, is not successfully imposed from without, and is often perceived as irrelevant to the existing social order. High-achieving foreign students in the USA indicate that they will not be using all they have learned when they return to their homelands. Why? Because the knowledge received from "outside" is often perceived as irrelevant at home, or because the possession of 'greater knowledge' is perceived as threatening by professionals who have stayed at home. Conversely, overseas experience is often considered irrelevant in the U.S.

Culture and its institutions directly affects the process of the transfer of technology in a variety of ways. Culture designates what we pay attention to and what we ignore. Furthermore, all people within a culture are not the same. There are a great number of so-called 'normal' people whose endeavors define and maintain any culture, maintainers of the status quo. Then there are 'fringe' people in a culture, outside the norm, who make contributions and changes far out of proportion to their numbers. Engineers usually do not make up much of the percentage of the total 'fringe' population, but have within their profession their own 'fringe' people.

In any one culture, it is critical to understand engineers. Cultural norms dictate who becomes engineers and what they do *as* engineers. For example, a society that gives great prestige (and commensurate income) to engineers will have a great number of successful engineers. In many places, the prestige awarded engineers comes as a result of management skills and not technical knowledge. In countries where government service is highly prized, engineers congregate in government departments and agencies where they are likely to be managers and not technicians. An engineer who works at the transfer of technology is likely to find himself or herself in a country where neither foreign or national engineers are esteemed. An expatriate may find himself or herself unwanted but tolerated, because he or she comes "with the money". Under such conditions it can be extremely difficult to transfer knowledge - it may be perceived as being imposed from without, irrelevant, or threatening to those in charge of the status quo.

Theories surrounding transfer of technology across cultures can be reviewed more extensively elsewhere (*see* for example, Scott-Stevens 1987). Subsequently, we would like to present some issues and anecdotes about what can happen in efforts to transfer technology.

What Did and Didn't Work and Why

Large projects fuel transfer and change in technology. Resources, people, and money are available and a need exists, usually in the form of the question: "Is there a better way?" Large engineering endeavors in foreign countries (and at home as well) can result in significant expansion of knowledge for all involved and spinoff for the national staff. Young, national engineers travel to new countries for education, while local senior engineers form their own companies to provide the same kind of service, perhaps at first on a smaller scale. Foreign engineers, who are almost always senior, expand their knowledge base by being in a new country with enough funds to make advances in their field of work. More often than not, there is a transfer of technology component to a project in which engineers are expected to exchange knowledge with their counterparts or to learn together. Unfortunately, this transfer component is often presumed to occur without effort and thus is not funded.

Large U. S. government agencies have prestige and technical knowledge desired by many developing nations. In civil engineering, the U. S. Bureau of Reclamation, the U.S. Geological Survey, and the Corps of Engineers are foremost. The Bureau's *Design of Small Dams* and other books are to be found around the world; the Corps HEC computer programs likewise. On the other hand, most individuals do not carry the prestige of the agencies. An exception seems to be university professors who write technical text books.

The success of an individual endeavoring to transfer knowledge is often proportional to how much that person is willing to learn. An example of how naive theory gets in the way of knowledge transfer is as follows. A foreign expert arrives to review the laboratory analysis of hundreds of soil samples for engineering properties, and to pass on some of his knowledge. That person rejects all results that indicate a specific gravity of approximately 2.5 and accepts those that have values around 2.7. Why? Possibly because there is a "naive theory" that his or her knowledge base is sufficient in this matter. The person has taken the new information and compared it to his or her knowledge base and rejected that which did not fit. The expert says, "The lab must be wrong!" It would have been prudent to do a few checks first, because it was the correct information which was rejected.

As mentioned earlier, many engineers are not motivated to learn new technology. They want to become managers, directing the efforts of others and the allocation of money. Transfers are irrelevant to them, or the prestige is not with technology but perhaps with government service. In such an environment, engineers congregate in government agencies and departments. The transfer of technology under these conditions is best accomplished in the private sector where people are *doing* the engineering.

With a foreign colleague, the senior author once visited a newly constructed barrage. It was different than any he had seen before. The senior author asked why it was made in such a manner, considering the loads and stresses. His colleague replied that an elderly professor had been given the task of design. The professor's younger assistants knew of the added expenses of the unusual design, but in their culture, they would not question or challenge such an esteemed teacher. Criticizing and questioning their seniors or "betters" was contrary to their value system. In such cases, there is no opportunity or need for the transfer of technology.

Another experience regarding rejection is more difficult to comprehend. The acoustic doppler current profiler (ADCP) is an instrument that can change how engineers measure and view the theory of water movement. It maps the spacial and temporal velocity field (three components) beneath the boat on which it is mounted and tracks the boat's position relative to where the instrument is turned on. After being set up, the instrument is divinely simple to operate. It comes with software and graphics so that one can see the output immediately. It is high tech, yet it is like operating a video player. A discharge measurement can be made at a cross section in an estuary or river in the time it takes to motor across. The number of measurements that can be made in a period of time can be increased two orders of magnitude. In estuaries, one can measure the tidal flows from minute to minute. Yet, our experience has been that people who could benefit from such an instrument and have the power to make the purchase have declined with hardly more than a superficial look. Why? Expense? Or is it perceived as unnecessary; that the old ways are good enough?

One extreme example of rejection encompasses all three engineering categories of knowledge: theory, experience, and data base. The senior author worked in a lagoon on a tropical island for a total of 11 weeks, measuring a vast array of variables for the calibration of a 3-dimensional hydrodynamic mathematical model: tidal levels, velocity fields, discharge, temperature, salinity, sediment concentrations, nutrients, mud properties, wind, rainfall, and bathemetry. From past experience it was known that there were remarkable dynamics going on at times. Hardly a day went by when the staff was not surprised by unforeseen or unpredicted events. Experts with world-wide experience who visited the project were at times amazed. It was a disappointing aspect of the project that, except for two people, no professional would come to work in the field, possibly because the living conditions were rather primitive, but equally as likely because of being anchored to their naive theories. For example, the math modelers said more than once, "Your data are wrong!" The search for new knowledge was made available to all, but nobody came.

Suggestions

Whenever we can, we query and test participants in "transfer" programs before we begin the transfer. We are interested in how the participants learn in order to utilize the most appropriate type of presentation, as well as what impediments there might be to learning. We then prepare the materials in accordance to what we have learned about the participants in order to maximize the chances for a successful transfer.

Some of the most successful transfers are unanticipated and very simple. For example, the motion of water and sediment in a bay was being debated. Because there was no agreement, a boatman was hired to spread dry coconuts about the water surface during a particular part of the tide. In a few minutes everyone learned (to their surprise) that the flow was highly complex and not as anticipated. Participants recalled that unplanned demonstration years afterwards, long after the expert had forgotten it.

For a person transferring technology (software), the most likely way to succeed is to assume that intended recipients will reject the new knowledge because it does not fit with their naive theories. The engineer-trainer can then begin by expanding the recipients' knowledge - theory, experience, or data base - thereby side-stepping the problem of imposing new technology from without. In this way, the knowledge becomes the recipients', and not the trainer's. When the expansion has progressed far enough, the specific new knowledge is presented with a high degree of certainty that it will be accepted.

If nothing more could be said about successful transfers (and there are never any guarantees), it would be: Assume nothing. Question everything, especially the naive theories of yourself or others.

References

Bulfin, R. L., (1977). "The role of technology in developing countries: an overview." In Developing Countries, Papers, Seminar Series, R.L. Bulfin and J.R. Greenwell, eds., University of Arizona, Office of Interdisciplinary Programs. Office of Arid Lands Studies, 1-8.

Resnick, L. B., (1983). "Mathematics and science learning: a new conception." Science, 220, 29 April, 477-478.

Scott-Stevens, S., (1987). *Foreign Consultants and Counterparts: Problems in Technology Transfer*. Westview Press, Boulder, CO., 229 pp.

Spicer, E.H., (1952). "Introduction." In *Human problems in technological change,* E.H. Spicer, ed., Russel Sage Foundation, New York, 301 pp.

A DROUGHT BIVARIATE EXTREME VALUE MODEL

Jose A. Raynal-Villasenor[1] , M. ASCE
Antonio Acosta[2]

Abstract

Drought frequency analysis has been carried out by using wide-known univariate probability distributions for the minima. A few examples of that are the Three-parameter Log-Normal, Pearson Type III and the ExtremeValue Type I (Gumbel) and Type III (Weibull) distributions. The methods for estimating the parameters of such distributions vary from the old method of moments to new options like the probability weighted moments and maximum entropy methods.Multivariate extreme value models for flood and drought frequency analyses have been explored towards their application in hydrological problems. That is the case of the bivariate extreme value model of the logistic type for the minima which is presented in this paper. The characteristics, properties and construction of the model as well are displayed in the paper towards its application in drought frequency analysis. The selected method of estimation of parameters is the method of maximum likelihood. An example of application is shown in the paper, which is illustrated with graphs to show the capabilities of modeling of the proposed model.

Introduction

During the last sixty seven years the use of the extreme value distributions types I (Gumbel), II (Frechet) and III (Weibull) in the field of frequency analyses for the maxima and minima has grown steadily. Such distributions, as three particular solutions to the Stability Postulate which any extreme must satisfy, they have been integrated in one form after Jenkinson(1955) found the general solution

[1]Department of Civil Engineering, Universidad de las Americas-Puebla, 72820 Cholula, Puebla, Mexico, [2]Water Management Under Direction, National Water Commission, Insurgentes Centro # 30-32, 06600 Mexico, D.F., Mexico

to that postulate giving birth to the so-called general extreme value (GEV) distribution.The GEV distribution has been used widely in flood frequency analysis and in other fields of the geophysical sciences, but such a distribution is applied to drought frequency analysis very rarely.

Recently, the multivariate models for extreme value distributions have been started to be explored for application in flood and drought frequency analyses (Raynal-Villasenor, 1985 and 1986, Raynal-Villasenor and Salas, 1987, Escalante-Sandoval, 1991, Acosta, 1993, and Escalante-Sandoval and Raynal-Villasenor, 1994). The results obtained so far shown the suitability of the use of these models for flood and drought frequency analyses. They have been explored up to the trivariate level for the case of the maxima and up to the bivariate level for the case of the minima.

Bivariate Extreme Value Distribution for the Minima

The general form of the Logistic model for bivariate extreme value distributions for the minima is , Gumbel (1962):

$$\Phi(x,y,m\ ) = \exp\left\{-\left[\left(-Ln\Phi(x\ )\ \right)^{m} + \left(-Ln\Phi(y\ )\ \right)^{m}\ \right]^{1/m}\ \right\} \tag{1}$$

where Φ(x,y,m) is the bivariate extreme value distribution, Φ(.) is the univariate extreme value distribution of (.), m is the association parameter for the Logistic model, and Ln is the natural logarithm. The particular form of eq.(1), when both marginals are GEV distributions for the minima is, Raynal(1986):

$$\Phi\left(x,y,\theta\ \right) = \exp\left\{-\left[\left(1-\left(\frac{\omega_1 - x}{\alpha_1}\right)\beta_1\right)^{m7\beta_1} + \left(1-\left(\frac{\omega_2 - y}{\alpha_2}\right)\beta_2\right)^{m/\beta_2}\ \right]\right\} \tag{2}$$

where ω, α and β are the location, scale and shape parameters of the marginal GEV distributions for the minima.

Maximum Likelihood Estimation of the Parameters

Using the following generalized log-likelihood function for bivariate distributions, when the sample sizes of the marginals are not equal, Raynal-Villasenor(1985):

$$L(x,y,\theta) = \left[\prod_{i=1}^{N_1} \phi(r_i,\theta_1)\right]^{I_1} \left[\prod_{i=1}^{N_1} \phi(x,y,\theta_2)\right]^{I_2} \left[\prod_{i=1}^{N_3} \phi(t_i,\theta_3)\right]^{I_3} \tag{3}$$

where I_1, I_2, I_3 are indicator numbers with value equal to one only if $N_i > 1$ and zero otherwise.

Computational Procedures for Maximum Likelihood Estimation of Parameters

Given the complexity of the mathematical expressions in eq.(3), and their partial derivatives with respect to the parameters, the constrained Rosenbrock method, Kuester and Mize (1973),was applied to obtain the maximum likelihood estimators for the parameters, by the direct maximization of the logarithmic version of eq.(3). The required initial values of the parameters to start the optimization of the logarithmic version of eq.(3) were provided by the univariate maximium likelihood estimators of the parameters for the case of the location, scale and shape parameters. The initial value of the association parameter was set equal to 2, following the procedure developed by Raynal-Villasenor(1985) for the case of the maxima.

Reliability of Estimated Parameters

The indicator selected to detect the reliability of estimated parameters when using the bivariate distribution as compared with to the univariate counterpart was the asymptotic relative information ratio. Table 1 shows a sample of relative information ratios obtained by using the following set of parameters:

$$\omega_1 = 10;\ \alpha_1 = 4;\ \beta_1 = 0.15;\ \omega_2 = 5;\ \alpha_2 = 2;\ \beta_2 = 0.10;\ m = 2$$

Table 1: Asymptotic relative information ratios of the parameters of the bivariate extreme value distribution for the minima

Parameter	N_1+N_2			
	25	50	75	100
ω_1	1.05	1.32	1.45	1.53
α_1	0.97	1.10	1.16	1.19
β_1	1.14	1.24	1.19	1.31
ω_2	1.05	1.02	1.01	1.01
α_2	0.97	0.96	0.96	0.95
β_2	1.16	1.07	1.04	1.02

Case Study

To apply the proposed methodology, five gauging stations were selected within the Fuerte River in the states of Sinaloa and Chihuahua in Northern Mexico. Table 2 show the results of the application of the BEV distribution for the minima to the data recorded in such gauging stations. The standard error of fit is that defined by Kite(1988).

Table 2. Univariate and Bivariate location, scale and shape parameters and standard errors of fit for stations considered in the case study

Station	Univariate Values				Bivariate Values			
	ω	α	β	EE	ω	α	β	EE
Huites	4.10	2.23	1.05	1.54	4.10	2.23	1.05	1.54
Choix	0.07	0.11	1.48	0.15	0.08	0.13	1.60	0.14
Palo Dulce	0.73	0.48	0.71	0.30	0.78	0.52	0.70	0.32
Chinipas	0.59	0.33	0.73	0.21	0.60	0.38	0.80	0.17
Urique	2.66	2.50	1.45	2.02	2.78	2.70	1.45	2.10

Conclusions

The Logistic model for bivariate general extreme value distribution for the minima has been presented for its application to drought frequency analysis. From the obtained values, both asymptotic and data based, the authors suggest the proposed model as a suitable option to be considered when performing drought frequency analysis.

Acknowledgements

The authors express their deepest gratitute to Comision Nacional del Agua, Universidad de las Americas -Puebla and Universidad Nacional Autonoma de Mexico for the support given for the publication of this paper. Founding for this study was provided by the Comision Nacional del Agua through the Agreement SARH-CNA-UNAM.

References

Acosta, A. (1993). Application of a Bivariate Extreme Value Distribution for Low-Flow Frequency Analysis, Ph. D. Dissertation, Engineering Graduate Studies Division, Universidad Nacional Autonoma de Mexico, 215p. (In Spanish).

Escalante-Sandoval, C. A. (1991). Trivariate Extreme value Distributions and its Applications to Flood Frequency Analysis, Ph. D. Dissertation, Engineering Graduate Studies Division, Universidad Nacional Autonoma de Mexico, 315p. (In Spanish).

Escalante-Sandoval, C. A. and Raynal-Villasenor, J. A., (1994). A Trivariate Extreme Value Distribution Applied to Flood Frequency Analysis, J. Res. Natl. Inst. Stand. Technol., Vol. 99, No. 4, pp 369-375.

Gumbel, E. J. (1958). Statistics of Extremes, Columbia University Press, 375p.

Gumbel, E. J. (1962). Statistical Theory of Extreme Values (Main Results), Chapter 6 in Contributions to Order Statistics, Sarhan, A.S. and Greenberg, B. G., editors, pp 59-63, John Wiley & Sons.

Jenkinson, A. F. (1955). The Frequency Distribution of the Annual Maximum (or Minimum) Values of Meteorological Elements. Quart. J. of the Roy. Met. Soc., Vol. 87, pp 158-171.

Kite, G. W. (1988). Frequency and Risk Analyses in Hydrology, Water Resources Publications, Fort Collins, Co., 257p.

Kuester, J. L. and Mize, J. H. (1973). Optimization with FORTRAN, McGraw-Hill Book Co., pp 386-398.

Raynal-Villasenor, J. A. (1985). Bivariate Extreme Value Distributions Applied to Flood Frequency Analysis. Ph. D. Dissertation, Civil Engineering Department, Colorado State University, 237 p.

Raynal, J. A. (1986). A Bivariate Extreme Value Model Applied to Drought Frequency Analysis. Multivariate Analysis of Hydrologic Processes, H. W. Shen, J. T. B. Obeysekera, V. Yevjevich and D. G. De Coursey, editors. Published by H. W. Shen, pp 717-731.

Raynal-Villasenor, J. A. and Salas, J. D.(1987). Multivariate Extreme Value Distributions in Hydrologic Analyses, Water for the Future, IAHS Publication No. 164, pp 111-119.

Matching Morphological Management and Funding Constraints, Red River Delta, Vietnam

by Eric J Lesleighter[1] F ASCE, and Tran Xuan Thai[2]

Abstract

The river works of the Red River Delta protect important areas of a major system of dykes. The budget allocated to them each year is significant for a limited economy such as that of Vietnam. The paper reveals how the scale of the river engineering works needed over the years cannot be matched even closely by the budget available. Accordingly, the national strategy has been to concentrate on critical areas, with the inevitable result that many needy areas miss out. The authors suggest that embarking on a course of large-scale river training would greatly compound the budgetary constraints, and would create a costly ongoing program of river 'control'. The paper points out the need to address the specific areas of concern, introduce economic river management works, and personnel training, in order to optimise the use of the budget.

Introduction

Flood and storm control, and natural disaster mitigation, are a central government responsibility for countries where there is intense monsoonal rainfall over a complex river system like that of the Red River in Vietnam. The realities of yearly flooding in the river system which networks throughout the Red River Delta have been a part of life for its 17 million inhabitants for centuries. The scale of the associated problems is reflected in the extensive system of major river levees (dykes), and the measures which are taken to safeguard the integrity of the system. Bank protection works and the dyke system form the last line of flood defence, and protection is directed at guarding the integrity of the dykes to avoid breaks and flooding of valuable lands. The total budget allocated to them each year is significant in the Vietnam context.

Figure 1 presents a general map of the Delta river system. The rivers are bordered by 3 000 km or so of river dykes, some up to 10 m in height. The major tributaries of the

---

[1] Chief Engineer Water Resources, SMEC International Pty Ltd, Cooma, Australia

[2] Deputy Head, River and Coastal Research Department, Vietnam Institute of Water Resources Research, Hanoi, Vietnam

River are the Da, Thao, and Lo Rivers. They have catchment areas (including those outside of Vietnam) of 52 900 km^2, 51 800 km^2, and 39 000 km^2, respectively. For the majority of the year, the Red River, due to its high sediment loads, is pinkish in colour, as are the River's tributaries and distributaries. The rivers have beds of silt and sand. The banks of the rivers are alluvial, 3 m to 4 m in height (but higher adjacent to the main dykes), and erodible. It is when the erosion occurs along banks where the dykes are close to the river's edge that measures are taken to stabilise the banks and avoid direct erosion of the river side of the dykes.

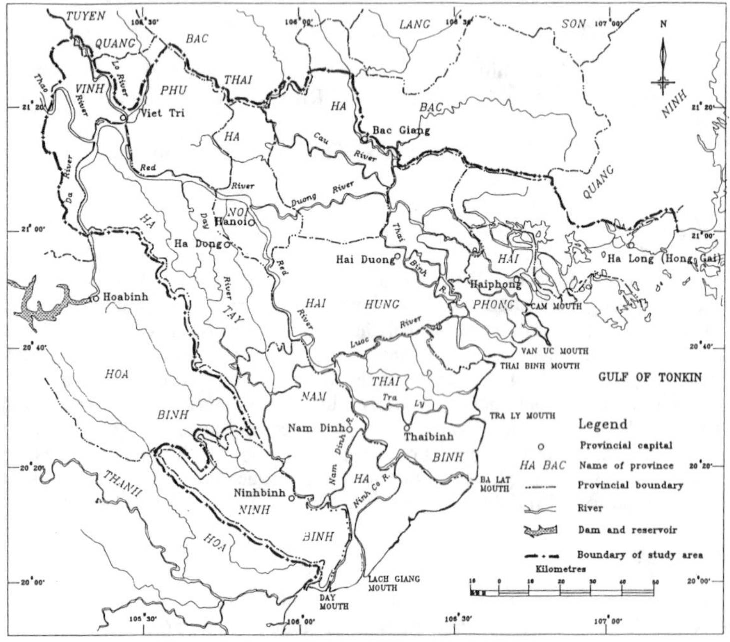

Figure 1. Area Map Red River Delta, Vietnam

Data gathering over many years has shown high levels of suspended sediment load in all the rivers, including large quantities of the fine grain sizes as wash load. Construction of the Hoa Binh Dam on the Da River some 50 km upstream of the Thao River confluence has cut off the sediment supply from that catchment, with the result that material has been 'taken up' from the Da river downstream of the Dam and the bed has degraded considerably over the last decade.

Management

Bank protection works are built and managed by the Ministry of Water Resources (MWR). The Vietnam Institute of Water Resources Research (VIWRR - within MWR) handles investigation, design and construction of river works. The budget was Vietnam Dong (VND) 50 billion (about US$5 million) in 1994, apart from provincial budgets. The

Department of Dyke Management and Flood Control (DDMFC - also within MWR) has a vital role in safeguarding the integrity of the dyke system. This is largely a 'balancing act' between the available budget and the actual needs.

Safeguarding Dykes for Flood Protection

The Red River has carried discharges in excess of 30 000 m^3/s. Many of the river reaches display depths of 20 m or more, and this makes the construction of river engineering works both difficult and expensive. In the whole dyke system of Ha Tay Province (Figure 1) there is a total length of 294 km of main and sub-dykes. Every year new and repair dyke work is necessary. Figure 2 shows the reach of the Red River from the entry of the Da and Thao Rivers to Hanoi. Bank protection works generally consist of revetment of riprap and short groynes spaced at about 150 m. There are many locations of bank erosion where the dykes are very close to the river's edge on the Da, Thao and Lo rivers. Numerous impermeable groynes have been constructed where the Da and the Thao Rivers meet. Between the groynes, severe bank caving is common, and revetment which had been there has been undermined and removed.

The 100 m high Hoa Binh Dam on the Da River is used for power and for flood mitigation. While the Hoa Binh reservoir provides flood mitigation in the Red River system, it is a trap for the large amount of sediment originating in the Da catchment and the classical degradation now prevails in the Da River. This has resulted in the lowering of river bed levels in the first few kilometres by around 10 m. More degradation of the river bed in the downstream direction will likely aggravate bank erosion.

Review of River Engineering Practice in the Red River

The main types of bank protection that have been used in the Red River system are revetments or groynes, or a combination of these. Groynes are almost invariably of an impermeable construction. For revetment work, the construction has been of riprap with toe protection in the form of long cylindrical rock-filled baskets called 'dragons', about 0.8 m in diameter. The annual budget for the various provinces has been allocated for construction of new bank protection works on severely eroded river bank reaches; repair of bank protection works which have been attacked and damaged; repair of dyke reaches which are damaged, and there is the danger of dyke failure by sliding or undermining, and maintenance of dykes and bank protection works. In addition to the limited number of groynes which can be funded on a particular reach, resulting in groynes which are spread too far apart, the complex nature of the river geomorphology means that it is difficult to foresee problem areas. In order to illustrate the ongoing challenge, the history of events in the Thao-Da area (Figure 2) is presented below.

The Thao-Da group of works, the most extensive in the Red and Thai Binh River system, includes bank protection on the right bank, in the Co Do village portion of Ha Tay Province (Figure 2), and works on the left bank at Le Tinh in Vinh Phu Province. The morphological processes in this area, being at the confluence of the Thao and Da rivers, are particularly complex. The thalweg and formation of mid channel bars are subject to major changes from year to year. Degradation of the bed of the Da River was particularly dramatic in the upstream reaches near the dam in that period, resulting in a surfeit of sediment around the Thao-Da confluence.

From 1960 to 1975, 13 groynes were constructed in the area to protect 6 km of dykes protecting Co Do village. Almost all of them were short (6 m to 8 m in length), with the spacing ranging from 180 m to as much as 400 m. Figure 3 shows the confluence of the Thao and Da Rivers, and the several kilometres downstream to the Lo River and beyond. It illustrates the characteristic of the moving thalweg within the wide waterway, and how the main stream can be directed at different sections of the river bank at different periods. For 1979 (Figure 3), a large point bar existed alongside groynes 1 to 8, where previously the flow pattern was such that protection works were necessary. By 1985 (Figure 3), the point bar had extended almost to groyne 13, the main stream directed towards the groyne 16 and 17 area, the mid-channel bar across from groynes 13 to 16 to had enlarged, and the main channel to be almost all directed across the river towards the mouth of the Lo River.

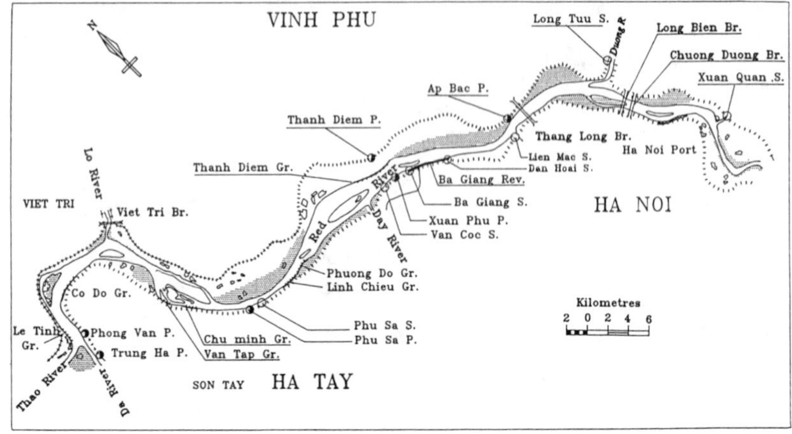

Figure 2. Red River from Thao-Da Confluence to Hanoi

Improved River Engineering Practice

The works which have been implemented, the unpredictable nature of where the next trouble spot will occur, and the limited budget, have imposed a 'reactionary' situation of responding to evident needs. There is the need for better technical solutions, with economies, to address the imbalance between a limited budget and the demand to stabilise rivers. Given the immense scale of the river system and the protective works required, the management of such a program points up a clear need for training of local engineers in river engineering and river management. Large-scale river training would compound the budgetary constraints and create a compelling ongoing program of river 'control' that would be 'never-ending'. The authors' view is that major, wide, fine-sediment rivers such as those of the Red River Delta should not be made the subject of large-scale efforts to control them.

Budget enhancement may be considered a priority management measure. However, in this context, the authors suggest that this means more efficient use of funds tied into more durable and effective methods, and organisational streamlining. This is a management objective, even though increased funding may be required to handle critical problem areas in the short term. The authors' experiences on the Red River and its tributaries suggest possible modifications of policy and planning (which can be applied with benefit in other major river systems), particularly when funds are limited. The following guidelines, based on the perceived deficiencies in the present practice, are advanced:

- there is no one correct answer to the river engineering problems
- embrace a policy of management of the rivers' problem areas
- recognise the economic disadvantages of completing insufficient works in specific trouble spots, ie, avoiding spreading of the works 'too thinly'
- maximise the budget and optimise the budgetary allocations by sound management of the costs of design, materials, and construction
- allocate a small proportion of the yearly budget to field measurements, survey, and research into better methods, and
- incorporate a program of training of key personnel.

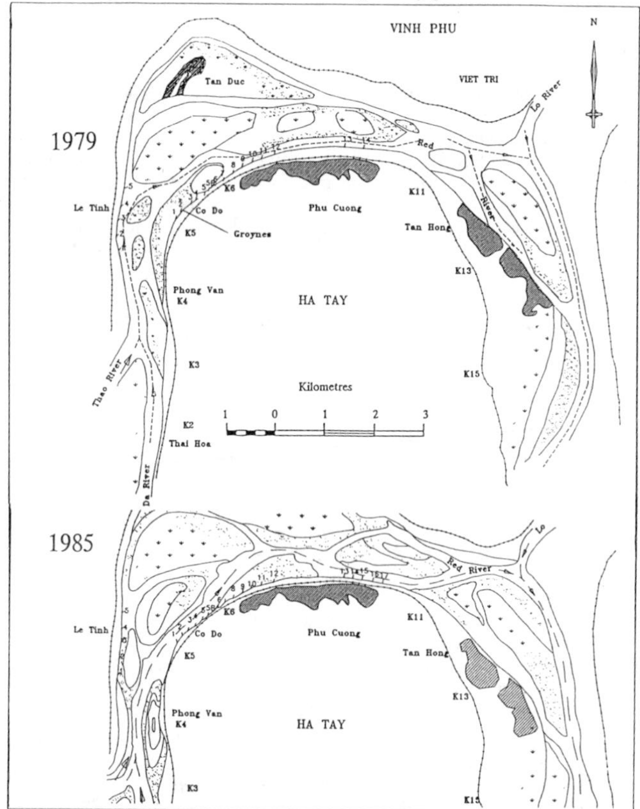

Figure 3. River Features Thao-Da Confluence 1979 and 1985

MINING: AN OPPORTUNITY FOR CONSTRUCTED WETLANDS

Danuta Leszczynska and Andrew A. Dzurik, Member ASCE [1]

Introduction

Process water and stormwater runoff from surface mining areas pose substantial problems for water quality in many parts of the country. The large areas involved and the high cost of treating water from mining operations make traditional water treatment schemes unattractive. Constructed wetland (CW) systems became a promising alternatives for some of the wastewater treatment facilities that are traditionally used for wastewater treatment. In the past decade, with over 400 wetland treatment systems having been built on mined lands, the biogeochemical treatment of acidic, iron-rich drainage from coal mines has proven to be a feasible alternative to the standardized chemical treatment (Frederick and Egan, 1994). The beneficial processes provided by plants and microorganisms in these constructed wetlands include iron removal from the water and neutralization of acids. In comparison, acid-neutralizing chemicals such as calcium carbonate, calcium hydroxide, sodium carbonate, or sodium hydroxide are typically used for acid mine drainage (Skousen 1988). Mine operators generally have found that constructed wetlands treatment reduces chemical costs by more than enough to repay the cost of the CW in less than a year. What is it that makes CWs such an attractive alternative for wastewater treatment at mines? This paper will discuss the needs for treating mine drainage and the opportunity for using CWs to satisfy those needs.

Wetland Functions

Properly designed wetland for metal mine drainage treatment should: (1) elevate pH by neutralization, and (2) remove soluble metals from water mainly by oxidation and biooxidation, adsorption, exchange, precipitation, and plant uptake . In balanced wetland ecosystem all processes should be in an equilibrium. Among important parameters which are responsible for maintaining this equilibrium are: pH, oxygen demand, temperature and hydraulic parameters.

pH : Neutralization of Acids

During the mining process, vast amounts of soil are disrupted together with the extracted coal, and naturally occurring iron sulfide becomes exposed to the atmosphere and is oxidized. Subsequently, rainwater percolates through this soil to form acid mine drainage which ultimately runs off into surface waters and possibly ground water. The neutralization of acids in wetlands occur mainly by chemical reaction of acid with ammonia and bicarbonates that are generated by bacteria.

Metals: Role of Water and Soil.

A number of earlier studies focused on total metal concentration that assess ecological impact of heavy metals. More recent studies have shown that the ecological influences of metals were more

[1] Department of Civil Engineering, FAMU/FSU College of Engineering, P.O.Box 2175, Tallahassee, FL. 32316,

determined by the specific forms rather than total metal concentration (Schalsoga at al, 1982; Calmano at al, 1986). Chemical fractionation of metals is the most important factor of ecological impact in the wetland ecosystem.

Water. In water, the heavy metals are present mostly in colloidal forms or as complexes with organic or mineral substances (Kabata-Pendias and Pendias, 1992). A variety of biogeochemical processes such as precipitation, redox reactions, sorption to suspended solids, and complexation to organic ligands in the water could determine the fate of metals entering a wetland. Dissolved organic carbon, presence of iron, calcium, and especially sulfides and pH values greatly affected metal immobilization rates (Hemond and Benoit, 1988; Best, 1994)). Precipitation of most transition elements and metals occurs in the presence of excess sulfides, because in general, metal sulfides are insoluble in water. In waters that do not contain appreciable sulfide ions, onset of anoxia in bottom sediments can also lead to release of reduced metal ions back into solution.

Soil. The solubility of trace metals in soils generally depends on the number of soil properties of which soil pH and redox potential are the most important parameters. Therefore, the solubility of metals is usually shown as a function of pH linked with other parameters, mainly with type organic matter. For example, the most mobile fractions of ions occur at lower range of pH and lower redox potential. Also, the microbial alkylation, as given examples for Hg, Se, Te, As, and Sn, that occurs mainly in sediments and on suspended particles in waters is a cause of ion's mobility (Jernelov, 1975). With increasing pH of the soil, the solubility of the most cations will decrease, but as reported mainly for Cu, Zn and Cr (Gemmell, 1975; Craze, 1977), the metals that most likely occur in soil as organic chelates may become soluble after raising pH by heavy liming. Pendias and Wiacek (1986) have shown that total concentration of Fe, Mn, Zn, Pb, Cu and Cd in very acidic soil was over 500 times higher than in the neutral range of pH.

Other soil factors playing important roles in mobility of trace elements are salt content, concentration of carbonates and Fe and Mn hydrous oxides, types of clay minerals, amount and quality of organic matter, temperature, types of plant species, micro and mezobiota activities.

The soil characteristic is one of the major criteria used for identification of the wetland ecosystem. Wetland soils should demonstrate hydric or waterlogged characteristics. These conditions have a great influence on soil chemistry and on the conditions for plant life. The typical hydric soil is characterized by oxygen-poor or anaerobic conditions because of high chemical and biochemical oxygen demands. In comparison, the dissolved oxygen in the water column remains high. Thus the oxygen diffusing through the water column is rapidly used at the soil surface. As a consequence, wetland soils have two different soil layers: (a) an oxidized or aerobic surface soil layer where oxygen is present, and (b) an underlaying reduced or anaerobic soil layer without available free oxygen. The intensity of soil anaerobiosis is expressed by redox potential, E_h. Depending on climate and season, the same type of soil can be characterized by different E_h values. Usually as soil wetness increases during wet season, the E_h decreases. In wetland soils, oxygen disappears at $E_h < 300$ mV. Depending on organic matter content and organic loadings of waterlogged soils, E_h values can reach values as low as -250 mV (Hemond and Benoit, 1988).

Metals: Role of Plants and Microorganisms.

Plants: Plants can accumulate metals in or on their tissues due to their great ability to adapt to variable chemical properties. Therefore , plants may be treated as a kind of passage for trace elements from soils, and partly from water and air, to animals and humans. Both phytoplankton and vascular water plants are known to selectively concentrate trace elements. As a result of this selectivity, concentration of some elements in waters may decrease in some seasons, while other elements may become soluble during the decay of plants. Concentration of trace elements in both bottom sediments and aquatic plants are known to be useful tools in biogeochemical exploration and environmental research.

Aquatic plants used in constructed wetlands vary widely, depending upon climate and soils, but the most common emergent plants are reeds, cattails, rushes, bulrushes and sedges. The emergent plants have the ability to absorb oxygen and other needed gases from the atmosphere through their leaves and stems above water, and conduct those gases to the roots. Thus the soil zone in immediate contact with the roots can be in aerobic and anaerobic environments.

A number of researchers reported results of laboratory experiments on uptake of some metals and nonmetals by water hyacinths *(eichhornia crassipes)*, a typical wetland plant. The results recounted by different scientists have shown a great ability to absorb and concentrate such metals as: cadmium, mercury, lead, nickel, zinc, and chromium (Wolverton, at all, 1977; Street, at al, 1977, and Delgado at al, 1993), and nonmetals such as: arsenic and selenium (Reed at al., 1988). Another plant, *sphagnum* is well known for its ability to accumulate iron to the concentration which petrifies plants (Kleinmann at al, 1991). Iron and manganese can be also accumulated by algae. However, despite a high accumulation of metals, the total algal biomass in a wetland is limited, therefore their contribution to metal removal is marginal (Reed at al., 1988). One of the most popular plants tolerant to mine water is cattails *(Typha)*. They can tolerate copper and nickel ions up to the concentration of 50 mg/L and 150 mg/L, respectively (Taylor and Crowder, 1983).

Although the plants can uptake nutrients and other constituents, perhaps the most important plant component in CWs are the submerged portions which serve as the substrate for attached microbial growth.

Microorganisms: Microorganisms in wetlands can help to reduce high levels of BOD, suspended solids, nitrogen, and significant levels of metals, trace organics and pathogens. The dissolved oxygen along with temperature and pH are major parameters that have an influence on type of microorganisms living in certain zones, and their activities. In aerobic layers, bacteria can catalyze oxidation of some of the metals and precipitate them as hydroxides. In contrast, the anaerobic zone is a home of sulfate-reducing bacteria. The hydrogen sulfide released during their metabolisms may react with dissolved metal cations and form insoluble metal sulfides. Also, another metabolism products, such as: ammonia and bicarbonates can neutralized acids which are the cause of low pH.

Mining Applications. Overview.

Pollutants from various types of mining activities have had a significant effect upon America's surface waters. Over 11,000 miles of streams have been contaminated by acid as a result of mining (Squillace, 1990). The most troublesome long-term threat is acid mine drainage from abandoned mine sites (over 550,000 in the West) and waste-rock tailings. Throughout the West, some 70 million tons of tailings have been dumped; these turn acidic with exposure to air and water, leaching heavy metals into ground and surface waters (U.S. News 1995). Coal mining has had a substantial impact in the states of Tennessee, Virginia, West Virginia, Pennsylvania, and Kentucky. Acid mine drainage from hundreds of sites in the coal regions of the eastern states is being treated with CWs. In some cases, CWs remove sufficient iron and acidity to meet discharge standards. Coal mining is perhaps the major application of CWs for mining operations. The Tennessee Valley Authority (TVA) has been active in developing and promoting the use of CWs for coal mine sites, especially in the above mentioned states. As of 1990, the TVA was operating seven CW systems for acid mine drainage at several of its reclaimed coal mines, and had an additional eight CWs in the planning stage. Treatment efficiencies at the TVA facilities ranged from 82% to 99% removal of total iron and 9% to 98% for manganese (Brodie et al. 1988). A review of current literature shows that applications to coal mining predominate. (See, for example, Hammer 1989 and Moshiri 1993). Non-coal mining applications are varied, but not as common as coal mining CWs. Among the applications are copper-zinc mines (Wildeman and Laudon 1989) and precious metals (Howard 1989). As a specific example, the Dunka Mine near Duluth, MN is a large open pit taconite mining operation of about 160 hectares. An overlying layer is removed to extract the taconite ore and is stockpiled along one side of the open pit. This stockpile covers about 120 ha and contains over 32 million tons of copper, nickel and iron sulfides. Nickel is the primary trace metal in the drainage at concentrations of 3 to 30 mg/l, with copper, cobalt and zinc also present, but less than 5 % of the nickel concentrations. Rather than building CWs, natural wetlands were modified to become controlled wetlands which were shown to reduce trace metal concentrations in stockpile drainage (Eger 1993).

In different application, a sulfate-reducing CW at a clay mining facility was used to test its feasibility as an alternative to chemical treatment of acid mine drainage. The results showed that a CW with locally available organics for use as a substrate were effective in treating acid runoff (Gross 1993).

Another CW demonstration project was designed and built near Duluth to test the ability of a peat/wetland treatment system to remove heavy metals from rock stockpile drainage . Unlike the majority of CW facilities associated with coal mining that have very low pH and high metal content, this system has a moderate pH (5 to 6) and low metal concentrations. Excellent results were shown for heavy metal removal (Frostman 1993).

Summary and Conclusions

We have shown that constructed wetlands are becoming increasingly important as a technology for improving water quality. Mining discharges are sole examples among of the long list of different applications. CWs can help mining facilities to meet more stringent water quality standards at a reasonable cost and can help to reduce some of the damage accumulated from years of mining throughout the country. Vast areas of mining continue to contaminate the nation's waters, and the application of CWs provides an excellent opportunity for the beneficial use of this comparatively low-cost technology. One of the unanswered questions with CW technology for mining applications is the ultimate fate of bioabsorbent materials. Continued research in this area should provide appropriate answers and alternatives for final disposal of removed contaminants.

References

Best R.G., 1994, “Wetlands Ecological Engineering: An Approach for Integrating Humanity and Nature through Wastewater Recycling Through Wetlands” Wetlands Ecological Engineering for Wastewater Treatment 1.

Brodie,G.A.et al.,1988."Constructed Wetlands for Acid Mine Drainage Control in the Tennessee Valley"

Calmano W., Forstner U., Kersten M., 1986, "Metal Associations in Anoxic Sediments and Changes Following Upland Disposal" Toxicol. and Environ. Chem. 12, 313.

Craze B., 1977, “Restoration of Captains Flat Mining Area”, J.Soil Conserv., N.S.W., 33, 98.

Delgado M., Bigeriego M., Guarddiola E., 1993, "Uptake of Zn, Cr and Cd by Water Hyacinths" Wat. Res., 27, 269.

Eger, P. et al., "The Use of Wetland Treatment to Remove Trace Metals from Acid Mine Drainage," in G.A. Moshiri (ed.),Constructed Wetlands for Water Quality Improvement, Lewis Pub. 1993.

Frederick, R.J. and M. Egan, 1994. "Environmentally Compatible Applications of Biotechnology," BioScience, 44(8), 529.

Frostman, T.M., "A Peat/Wetland Treatment Approach to Acidic Mine Drainage Abatement," in G.A.

Gemmell, R.P., 1975, “Novel Revegetation Techniques for Toxic Sites" International Conference on Heavy Metals in the Environment, Toronto, 579.

Gross, M.A. et al., 1993 "A Comparison of Local Waste Materials for Sulfate-Reducing Wetlands Substrate," in G.A. Moshiri (ed.), Constructed Wetlands for Water Quality Improvement, Lewis Pub.

Hammer, D. (ed.). 1989, Constructed Wetlands for Wastewater Treatment, Municipal, Industrial and Agricultural, Lewis Publishers, Chelsea, MI.

Hemond H.F., Benoit J., 1988, "Cumulative Impacts on Water Quality Functions of Wetlands" Environmental Management, 12, 5, 639.

Howard, E.A., J.C. Emerick and T.R. Wildeman., 1989, "Design and Construction of a Research Site for Passive Mine Drainage Treatment in Idaho Springs, Colorado" in D.A. Hammer (ed.), Constructed Wetlands for Wastewater Treatment, Lewis Pub.

Jernelov A., 1975, "Microbial Alkylation of Metals", International Conference on Heavy Metals in the Environment, Toronto, 845,.

Kabata-Pendias A., Wiacek K., 1986, "Effect of Sulphur Deposition on Trace Metal Solubility in Soils" Environ. Geochem. Health, 8,95.

Kabata-Pendias A., Pendias H., 1992, Trace Element in Soils and Plants, 2nd Edition, CRC Press.

Kleinmann, R.L.P, R.S.Hedin, and H.M. Edenborn, 1991, "Biological Treatment of Mine Water - an Overview", Proceedings second International Conference on the Abatement of Acidic Drainage, Montreal, 28.

Moshiri (ed.), Constructed Wetlands for Water Quality Improvement, Lewis Pub. 1993.

Reed, S.C, et al, 1988. Natural Systems for Waste Management and Treatment. McGraw Hill Book Co., New York, N.Y.

Satchell, M. 1995, "A new Battle Over Yellowstone Park," U.S. News and World Report, March 13, 1995, 34-42.

Schalscga E.B., Morales M., Vergara I., Chang A.C., 1982, "Chemical Fractionation of Heavy Metals in Wastewater Affected Soils" J.Water Pollution Control Federation, 54, 175.

Skousen, J. 1988. "Chemicals for Treating Acid Mine Drainage," Greenlands, 18(3):36-40.

Squillace, M. 1990. The Strip Mining Handbook, The Environmental Policy Institute and Friends of the Earth, Washingtron, D.C.

Street J., Lindsay J.W., Sabey B.R., 1977, "Solubility and Plant. Uptake of Cadmium in Soil Amended with Cadmium and Sewage Sludge" J.Envir.Qual., 6,72.

Taylor, G.J, and Crowder A. A.,1983, "Uptake of Accumulation of Heavy Metals by Typha Latifolia in Wetlands of the Sudbury, Ontario Region", Canadian Journal of Botany, 61, 63.

Wildeman,T.R and L.S.Laudon, 1989, "Use of Wetlands for Treatment of Environmental Problems in Mining: Non-Coal-Mining Applications," in D.A. Hammer (ed.)Constructed Wetlands for Wastewater treatment, Lewis Pub.

Wolverton B.C., McDonald R.C., Rebeca C., 1977, "Wastewater Treatment Utilizing Water Hyacinths" Proceedings National Conference on Treatment and Disposal of Industrial Wastewaters and Residue, 205.

Mathematical Theory and Numerical Methods for the Modeling of Wetland Hydraulics

Lisa C. Roig[1]

ABSTRACT

Current interest in wetland protection, restoration, and creation has created a demand for engineering analysis tools that can be used for wetland planning, design, and management. This paper introduces tools that have been developed by the USACE Waterways Experiment Station for the prediction of flow and sediment transport in wetlands.

Wetland surface flows differ from other open channel flows because of the following characteristics: 1) shallow flow spreading over the wetland surface; 2) networks of small drainage channels emanating from the marsh surface; 3) flow through emergent vegetation; 4) intermittent flooding and draining of the marsh surface; 5) flow control structures at wetland boundaries and within the wetland system. The unique characteristics of wetland flows require modification of the standard shallow water equations and sediment transport relations. In this paper the mathematical theory behind these modifications is described. Some unique features of the numerical methods that are being used for the solution of the resultant equations are also presented.

INTRODUCTION

Numerical hydrodynamic modeling of wetland surface flows can be used to plan, design, and manage wetland projects. When choosing a modeling tool it is important to understand the capabilities and limitations inherent in the model formulation to determine if the model is appropriate for the intended application. Some areas of concern for wetland modeling are listed below:

1. Dynamic variation of the wetting front over the wetland surface;
2. Frictional resistance formulation for submerged and emergent vegetation; and
3. Specification of control structures at wetland boundaries and in the wetland interior.

---

[1] Research Hydraulic Engineer , Hydraulics Laboratory, Waterways Experiment Station, 3909 Halls Ferry Road, Vicksburg, Mississippi 39180 (601) 634 - 2801.

In the past several years the results of research in each of these areas have coalesced into a comprehensive package for the modeling wetland surface flows. The package is based on the TABS-MD numerical modeling system that has been used extensively for the simulation of estuarine and riverine hydrodynamics. At the heart of the TABS system is the RMA2 two-dimensional finite element model for vertically integrated free surface flows, originally developed by Dr. Ian King and Mr. William Norton (Norton, *et al.*, 1973). A version of this model, RMA2-WES, has been modified and maintained at the USACE Waterways Experiment Station and it is this version that incorporates the most recent wetland simulation tools.

GOVERNING EQUATIONS

The two-dimensional, vertically integrated shallow water equations govern gradually varied flow in water courses where the water column is vertically well mixed. If one assumes that the gradients of density are small then the equations can be written as follows:

the continuity equation,

$$\frac{\partial h}{\partial t}+h\frac{\partial u}{\partial x}+h\frac{\partial v}{\partial y}+u\frac{\partial h}{\partial x}+v\frac{\partial h}{\partial y} = 0 \tag{1}$$

and the two momentum equations in the horizontal directions,

$$\frac{\partial u}{\partial t}+u\frac{\partial u}{\partial x}+v\frac{\partial u}{\partial y}+g\frac{\partial z_{ws}}{\partial x} - \frac{1}{\rho}\left[\varepsilon_{xx}\frac{\partial^2 u}{\partial x^2}+\varepsilon_{xy}\frac{\partial^2 u}{\partial y^2}\right]+\frac{1}{\rho h}\{\sum_i(-F_i)_x\} = 0 \tag{2}$$

$$\frac{\partial v}{\partial t}+u\frac{\partial v}{\partial x}+v\frac{\partial v}{\partial y}+g\frac{\partial z_{ws}}{\partial y} - \frac{1}{\rho}\left[\varepsilon_{yx}\frac{\partial^2 v}{\partial x^2}+\varepsilon_{yy}\frac{\partial^2 v}{\partial y^2}\right]+\frac{1}{\rho h}\{\sum_i(-F_i)_y\} = 0 \tag{3}$$

where

x = first horizontal Cartesian coordinate direction (L),
y = second horizontal Cartesian coordinate direction (L),
t = time (T),
z_{ws} = water surface elevation (L), where $z_{ws} = a_o + h$,
a_o = channel bed elevation (L),
h = local water depth (L),
u = time and depth averaged velocity component in x direction (L/T),
v = time and depth averaged velocity component in y direction (L/T),
ρ = density of water (M/L^3),
g = acceleration due to gravity (L/T^2),
$\varepsilon_{xx}, \varepsilon_{xy}, \varepsilon_{yx}, \varepsilon_{yy}$
= turbulent exchange coefficients in the x and y directions (M/LT), and
F_i = unit forces, other than gravity, that act on the fluid (M/LT^2).

The time scale for averaging of the velocity components is greater than the time scale of random turbulent velocity fluctuations, but less than the time scale of macroscopic current variations induced by tides, gravity waves and upstream flows. These equations apply everywhere within the wetted boundary of the flow domain.

WETTING FRONT

Let $\kappa\,(z_{ws})$ be a function describing the fractional wetted area associated with a node k as a function of water surface elevation. Let a_{abs} be an elevation that is lower than the minimum expected water surface elevation in the vicinity of node k. For example, a flat bottom could be described by the function $\kappa\,(z_{ws}) = 0.0$ for $a_{abs} \le z_{ws} < a_o$ and $\kappa\,(z_{ws}) = 1.0$ for $a_o \le z_{ws} < \infty$. Let R be a length describing the range of bathymetric variation in the vicinity of a node with mean elevation a_o, then a sloping bottom could be described by the function $\kappa\,(z_{ws}) = 0.0$ for $a_{abs} \le z_{ws} < a_o\text{-}R/2$; $\kappa\,(z_{ws}) = (1/R)z_{ws} + (1/2 - a_o/R)$ for $a_o\text{-}R/2 \le z_{ws} < a_o + R/2$; and $\kappa\,(z_{ws}) = 1.0$ for $a_o + R/2 \le z_{ws} < \infty$. For an irregular bed an empirical function could be developed for $\kappa\,(z_{ws})$. One can now define the effective water depth h_σ in the vicinity of this node as

$$h_\sigma(z_{ws}) = \int_{a_{abs}}^{z_{ws}} \kappa(z)dz \,. \tag{4}$$

By defining the bathymetry in terms of a_{abs} and $\kappa(z_{ws})$ one is assured that node k never experiences a water surface elevation that is lower than the lower limit for the vertical integration. To this point no new approximations have been introduced but the pressure variable has been recast in terms of a_{abs}, , κ and z_{ws}. One can now rewrite the governing equations either in terms of the absolute depth $\tilde{h} = z_{ws} - a_{abs}$, or the effective depth h_σ defined by equation 4. In this research the effective depth h_σ was retained as the primary pressure variable because it is exactly equivalent to the usual water depth h for submerged nodes.

To ensure that any node associated with a partially wet element remains in the computational flow domain until all of the nodes in an element are dry an approximation is introduced. One can define a minimum fractional wetted area κ_{min}. Thus for a node with a flat bed one could approximate the bathymetry as $\kappa\,(z) = \kappa_{min}$ for $a_{abs} \le z < a_o$ and $\kappa\,(z) = 1.0$ for $a_o \le z < \infty$, where a_{abs} is below the minimum water surface elevation. The parameter κ_{min} should be small to minimize approximation error and will be chosen according to numerical considerations. When all nodes associated with the element have reached the minimum wetted area the element can be considered to be dry.

It is useful to define a domain coefficient σ which dynamically describes the vertically integrated flow domain at node k:

$$\sigma(z_{ws}) = \frac{h_\sigma}{\tilde{h}} = \frac{h_\sigma(z_{ws})}{(z_{ws} - a_{abs})} \,. \tag{5}$$

Problems of numerical instability may arise if the functional form for $\sigma(z_{ws})$ given by equation 5 is not smooth. A numerical code which relies on an iteration scheme for the solution of a system of nonlinear equations requires the derivative $\partial\sigma/\partial\tilde{h}$ which is undefined when $\kappa\,(z_{ws})$ is discontinuous, as it is for the flat bottom case described above. To avoid this potential difficulty a second approximation is suggested. One can require that the flat bed be approximated as a gently sloping bed. The function $\kappa\,(z_{ws}) = \kappa_{min}$ for $a_{abs} \le z_{ws} < a_o\text{-}R/2$; $\kappa\,(z_{ws}) = (1/R)z_{ws} + (1/2 - a_o/R)$ for $a_o\text{-}R/2 \le z_{ws} < a_o + R/2$; and $\kappa\,(z_{ws}) = 1.0$ for $a_o + R/2 \le z_{ws} < \infty$ can then be used, assigning a small positive value to the parameter R. This approximation only arises when the bed in the vicinity of the drying node is flat.

In wetlands, a discretized description of the bathymetry will rarely capture all of the small scale bathymetric variations observed in the field. The function $\kappa(z)$ can be used to account for uneven topography that is not fully described by the numerical grid. Wetted area curves for wetland nodes are naturally smoothly varying and therefore do not suffer from the numerical instabilities described for flat beds. In the current research, the form $\kappa(z) = \kappa_{min}$ for $a_{abs} \leq z < a_o - R/2$; $\kappa(z) = (1/R)z + (1/2 - a_o/R)$ for $a_o - R/2 \leq z < a_o + R/2$; and $\kappa(z) = 1.0$ for $a_o + R/2 \leq z < \infty$ has been adopted as a generic wetted area curve. For wetlands the parameter R represents the length scale of microbathymetric variations in the vicinity of the node, and a_o is the mean bed elevation. Many other functional forms could be developed for $\kappa(z)$ that are more realistic, and this is an area of continuing research. The form suggested here was chosen for simplicity in specifying the parameters and because it is an easily integrated function of elevation.

The integration of the wetted area curve at node k over the depth effectively converts a depth of water associated with the partially wet nodal area to an equivalent depth of water over the fully wet nodal area. Thus mass continuity is preserved at the node. However the flow behaviors for these two cases are not equivalent and it is important to recognize that an approximation has been made. The effect of this approximation is minimized when the nodal area is small. Increasing the resolution of the grid over the marsh surface will yield a more accurate solution. Fortunately, most wetland flows have very slow moving wetting fronts and the approximation does not have a large affect on the overall solution.

Incorporating the transition element approximations into equations 1 through 3 results in the following modified forms of the equations for the conservation of mass and momentum:

$$\frac{\partial h_\sigma}{\partial t} + h_\sigma \frac{\partial u}{\partial x} + h_\sigma \frac{\partial v}{\partial y} + u \frac{\partial h_\sigma}{\partial x} + v \frac{\partial h_\sigma}{\partial y} = 0 \,, \quad (6)$$

$$\frac{\partial u}{\partial t} + u\frac{\partial u}{\partial x} + v\frac{\partial u}{\partial y} + g\frac{\partial z_{ws}}{\partial x} - \frac{1}{\rho}\left[\varepsilon_{xx}\frac{\partial^2 u}{\partial x^2} + \varepsilon_{xy}\frac{\partial^2 u}{\partial y^2}\right] + \frac{1}{\rho h_\sigma}\{\sum_i (-F_i)_x\} = 0 \,, \quad (7)$$

$$\frac{\partial v}{\partial t} + u\frac{\partial v}{\partial x} + v\frac{\partial v}{\partial y} + g\frac{\partial z_{ws}}{\partial y} - \frac{1}{\rho}\left[\varepsilon_{yx}\frac{\partial^2 v}{\partial x^2} + \varepsilon_{yy}\frac{\partial^2 v}{\partial y^2}\right] + \frac{1}{\rho h_\sigma}\{\sum_i (-F_i)_y\} = 0 \,. \quad (8)$$

THE EFFECTS OF EMERGENT VEGETATION

The drag force due to vegetation varies with the level of submergence, the plant form, and the spatial distribution of the vegetation. The spatial distribution of vegetation can be parameterized via measurable characteristics such as

1. the number of stems per unit area (stem density),
2. the distribution of stem diameters, and
3. the distribution of stem lengths.

The characteristics of plant form, which encompass features such as the leafiness and the stem surface roughness, are more difficult to parameterize in a consistent way. For the purpose of this study stiff, single stemmed plants were considered The flow resistance forces presented by other forms of vegetation are the subject of continuing research.

A laboratory experiment was conducted to determine the vegetation resistance force as a function fluid density, acceleration due to gravity, fluid viscosity, bulk flow velocity through the vegetation, depth of flow, stem length distribution, stem diameter distribution, and stem spacing (Roig, 1994). The surface roughness of the plants was assumed to be homogeneous and similar to the surface roughness of wooden dowels. For flows through emergent vegetation with uniform stem length and uniform stem spacing the resistance force per unit width was found to be

$$F_v = -0.0909\,\rho\,u^2\,(1-\eta)\,\mathcal{R}_\ell^{1/2}\left(\frac{h}{s}\right)^{1/2}$$

where

η	=	the void space of the control volume,
u	=	the spatially averaged bulk velocity through the vegetation,
h	=	the water depth,
s	=	the stem spacing,
$\mathcal{R}_\ell$	=	a Reynolds number based on the mean submerged stem length and the bulk velocity.

An extension of this relationship for non-uniform plant densities and plant forms has also been developed.

CONTROL STRUCTURES

Many wetlands are regulated by hydraulic control structures at inflow and outflow points, and frequently between storage areas within the wetland. A numerical hydrodynamic model for wetland planning and design must incorporate procedures to account for the effects of control structures. The RMA2 hydrodynamic model incorporates control structures in the form of a user specified stage discharge relationship that is applied at the location of the structure.

CONCLUSION

The results of research in the numerical simulation of flooding and drying fronts, of flow resistance due to emergent vegetation, and of hydraulic control structures have coalesced to produce a comprehensive package for the modeling wetland surface flows. The mathematical theory behind these developments has been summarized. The results of an example application of the numerical model are described by Roig and Richards (1995).

ACKNOWLEDGMENTS

The information presented herein, unless otherwise noted, was obtained from research conducted under the Wetlands Research Program of the United States Army Corps of Engineers by the Waterways Experiment Station. Permission was granted by the Chief of Engineers to publish this information.

REFERENCES

Norton, W. R., Ian P. King and G. T. Orlob (1973) *A Finite Element Model for Lower Granite Reservoir.* Prepared for U. S. Army Engineer District, Walla Walla, Washington by Water Resources Engineers, Walnut Creek, CA, 138 pp.

Roig, Lisa C. (1994) *Hydrodynamic Modeling of Flows in Tidal Wetlands*. Ph. D. Dissertation, University of California, Davis, California, 177 pp.

Roig, Lisa C. and Richards, D. R. (1995) Software for the Modeling of Flows and Sediment Transport in Wetlands. *Proceedings of the International Conference on Water Resources Engineering*, San Antonio, Texas, ASCE, (see table of contents).

PLANT SURFACE AREA AND WETLAND TREATMENT EFFICIENCY

E. Jinx Kuehn, Member[1]
Marshall J. English, Member[2]
Kenneth J. Williamson, Member[3]

Abstract

This study investigated the effect of plant surface area (plant density) on the efficiency of organic carbon removal in a bench-scale constructed wetland. Until oxygen is depleted, the rate of carbon removal and the efficiency of carbon removal increases as plant surface area increases. At higher plant densities, oxygen depletion occurs in as little as six hours. This study suggests that the land area requirements for constructed wetlands could be reduced by 75% if plant densities are increased and supplemental oxygen is provided.

Introduction

Constructed wetlands have proven to be effective for advanced treatment of wastewater (Reed 1992, Gearheart 1992). Present research supports the idea that constructed wetlands act as biofilm reactors in which most treatment occurs within attached microbial biofilms (Reed 1988). In such reactors, the treatment rate is proportional to the surface area and concentrations of electron donors and electron acceptors. For aerobic degradation of carbon, removal of carbon should be proportional to the biofilm surface area, providing oxygen is present in stoichiometric proportions (Williamson and McCarty 1976, Meunier and Williamson 1981).

- - - - - - - - - - - -

[1]Assoc. Engr., Oregon Dept of Transportation, 885 Airport Rd, Bldg 35, Salem OR 97310
[2]Professor, Dept. of Bioresource Engr., Oregon St. Univ., Corvallis, OR 97331
[3]Professor, Dept. of Civil Engr., Oregon St. Univ. Corvallis, OR 97331

The currently recommended design equation (WPCF 1990) for constructed wetlands is:

$$Ce/Co = F \times \exp(0.7 \times k_t \times Av^{1.75} \times n \times HRT)$$

where: Ce = concentration of BOD after a given HRT; Co = influent BOD conc.; F = unsettled fraction at inlet; k_t = rate (usually 0.005/day); Av = specific surface (m^2/m^3); n = porosity; HRT = hydraulic retention time (days).

The EPA (1991) determined plant surface area as a constant (about 5% plant density). The EPA found the equation insensitive to this parameter. However, the EPA only reviewed a very narrow range of densities (3.8-5.1%). Densities as high as 15% are reported and do affect treatment efficiency (Gearheart 1992).

In its analysis, the EPA (1991) also assumed that oxygen was not limiting because wetland plants transport oxygen through their tissues to the root zone. However, some researchers believe that most of the oxygen is used by the plants for survival, leaving little excess for wastewater treatment (Brix 1993). Researchers also cite lack of oxygen for nitrification as the limiting step for nitrogen removal in wetlands (Gearheart 1992, Reed 1992).

Objectives and Experimental Design

The main objective of this study was to examine the relationship of plant surface area to treatment rate and efficiency in constructed wetlands. Each of 14 tanks was filled with mature bulrush, young bulrush, or acrylic rods (neutral surface). Each tank held about 50 liters of influent. ENSUR (TM) was used as an organic carbon source. After a continuous flow acclimation period, batch tests were run at 6 different plant densities under four different conditions, using high or low organic loadings, and presence/absence of light. Dissolved oxygen and total organic carbon were monitored versus time.

Results

The batch rate data were expected to fit a first order degradation model. Most data did fit such a model, but tanks with dense bulrush became oxygen limited (O_2 < 1 mg/l) after only 5.5 hours (Figure 1).

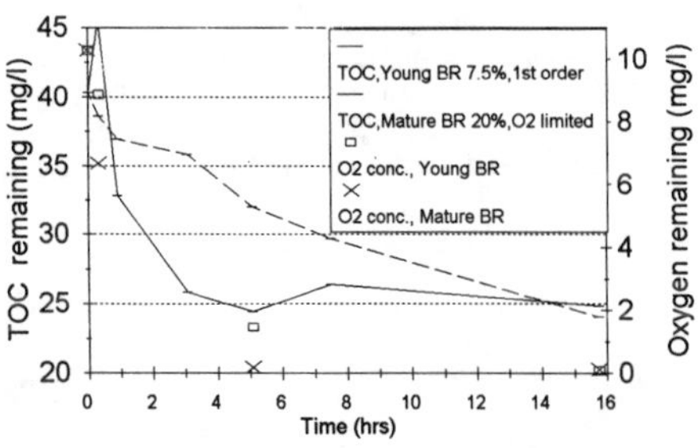

Figure 1: Tanks with and without oxygen limitation

Rates then were calculated for all tanks before oxygen limitation (assumed as oxygen concentration > 1 mg/l). Before oxygen became limiting, removal rates and removal efficiencies for the combined data from rods and mature bulrushes were linearly related to density. Significant interactions resulted because bulrushes and rods behaved differently for some parameters. As a result, each type was analyzed separately. Figures 2 and 3 show the raw means separated by type of stems for removal rates and efficiency.

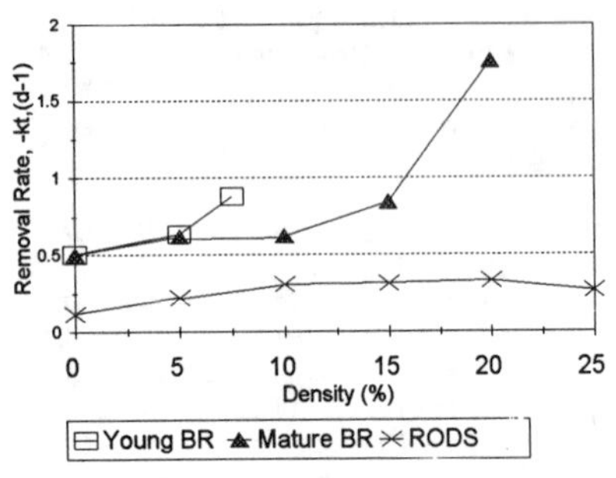

Figure 2: Rate of removal, density by type

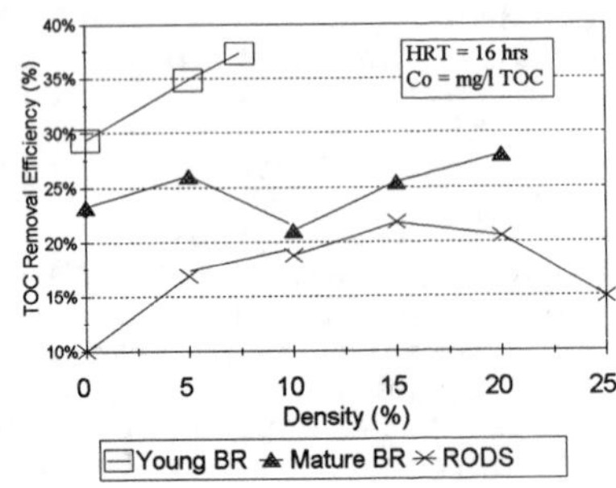

Figure 3: Efficiency of removal, density by type

Mature Bulrush

Mature bulrush showed a significant linear relationship ($r^2 = 0.72$) between increasing plant density (surface area) and higher treatment rates until oxygen became limited ($p = 0.0001$). However, efficiency of removal once oxygen was limiting was not related to plant density ($p < 0.85$). This supports the hypothesis that constructed wetlands are biofilm reactors able to treat wastewater at high rates, but that they are very susceptible to oxygen limitation.

These results showed higher removal rates with higher organic loadings ($p < 0.0004$). The presence of light was not significant for removal rates ($p < 0.24$) but was significant for removal efficiencies ($p < 0.009$).

Young Bulrush

Young bulrush reflected similar trends to the mature bulrush but showed less effects of oxygen limitation. The young bulrush may be less oxygen limited due to greater oxygen transport rates to their roots or due to lower sediment detrital oxygen demand, or some other factor.

Rods

Rods removed carbon at much slower rates and had lower overall efficiencies than the bulrush. In this study, the microbial communities on the rods consisted of many algae and some bacteria, while those on the bulrush were highly diverse including few algae but significant populations of fungi and rotifers as well as bacteria. At 25% density, removals were reduced largely due to two factors. First, natural changes in the microbial population occurred over time. Second, light, which boosts treatment rates in rods, could not penetrate at the highest densities.

Implications

This study supports the hypothesis that constructed wetlands act as biofilm reactors in which treatment rates and efficiencies are proportional to plant surface area (density), as long as oxygen does not become depleted. However, high density constructed wetlands contain such effective biofilm surface areas that influent oxygen is used up within 5.5 hours.

Figure 4 compares the data from this experiment with the behavior expected from the WPCF design equation. The solid lines represent the rates expected with the WPCF equation if the specific surface area varies with plant density (upper line) or is treated as a constant as in current designs (lower line). The dashed lines represent the data from this experiment. Young bulrush almost follow the expected upper line. Mature bulrush which were not oxygen limited also show increasing rates as surface area increases. Mature bulrush which were severely oxygen limited degraded carbon at constant low rates as is observed in existing wetlands designed using the WPCF equation.

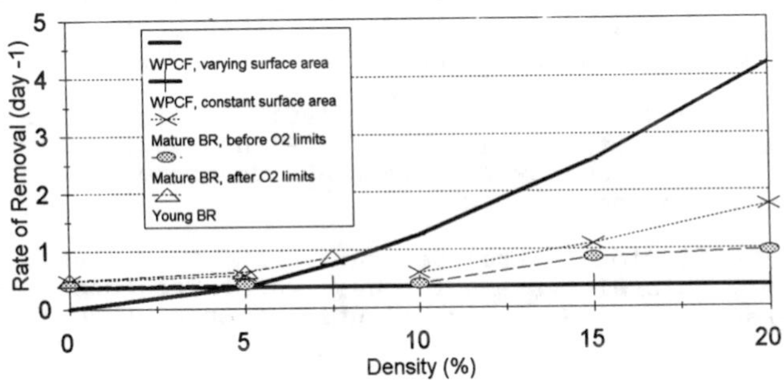

FIGURE 4: Comparison of WPCF Equation and Data

Currently, wetlands are designed with four to eight day retention times and achieve 41-65% BOD removal efficiency (Gearheart 1992), while in this experiment 20-60% of organic carbon was removed in 6-16 hours. This suggests that most of the degradation is occurring in the first day and the remaining retention time provides only surface reaeration.

A major drawback to the expanded use of constructed wetlands is that current designs require large land areas. This study suggests that land area for constructed wetlands could be reduced substantially (perhaps by 75%) if more oxygen were provided. Researchers have suggested using multiple cells with alternate loadings, large open water areas, or sand filters for reaeration. If optimal performance is to be achieved, future wetland design should include some mechanism for providing additional oxygen.

Appendix

Brix, H. (1993), "Macrophyte-Mediated Oxygen Transfer in Wetlands: Transport Mechanisms and Rates." *Constructed Wetlands for Water Quality Improvement*, ed. G. A. Moshiri, Lewis Publishers, CRC Press, London, England, 391-398.

Gearheart, R. A. (1992) "Use of constructed wetlands to treat domestic wastewater, City of Arcata, California." *Water Science and Technology,* 26, 1625-1637.

Meunier, A. D. and K. J. Williamson (1981) "Packed Bed Biofilm Reactors: Simplified Model." *Journal of Environmental Engineering* 107, 310-316.

Reed, S. C. and D. S. Brown (1992) "Constructed wetland design - the first generation." *Water Environment Research* 64(6), 776-781.

Reed, S. C., E. J. Middlebrooks, and R. W. Crites, eds. (1988), *Natural Systems for Waste Management and Treatment*. McGraw Hill Inc., New York, N.Y.

US EPA and R. W. Crites, eds. (1991), *Design Manual: Constructed Wetlands and Aquatic Plant Systems for Municipal Wastewater Treatment*. US Government Printing Office, Washington, D.C.

Water Pollution Control Federation and S. C. Reed, eds. (1990), *Natural Systems for Wastewater Treatment: Manual of Practice FD-16*. Water Pollution Control Federation, Alexandria, VA.

Williamson, K. and P. L. McCarty (1976) "A model of substrate utilization by bacterial films." *Journal of the Water Pollution Control Federation* 48(1)(Jan), 9-25.

Design Considerations for Ecosystem Restoration

Robert A. Laura, P. E., M. ASCE, Peter David, William Helfferich, Jennifer Barone[1]

Abstract

The Dupuis Reserve in South Florida historically represents the head of the Everglades Ecosystem. This area was drained and operated as a cattle and sheep ranch. The South Florida Water Management District is restoring the ecosystem by building a 12.9 km (eight miles) long levee to impound water and enhance hydroperiods. This paper discusses the hydrology and engineering design efforts involved in the restoration.

Introduction

This paper presents a case study of the hydrology and design considerations in restoring a wetland ecosystem on the Dupuis Reserve State Forest in South Florida. The Dupuis Reserve encompasses 8,746 ha (21,865 acres) in northwest Palm Beach County and southwest Martin County (see Figure 1). The property is located 64 km (40 miles) northwest of West Palm Beach. Dupuis Reserve is surrounded by intensive agriculture (e.g., citrus, sugar cane) except for a 12.9 km (eight miles) border with the J. W. Corbett Wildlife Management Area.

Historically, the property consisted of pine flatwoods, wet prairie, broadleaf marsh, and cypress that conveyed sheet flow southwest into an extensive broadleaf and sawgrass marsh. This marsh represented the northern extent of the vast Everglades ecosystem that originated at the eastern shore of Lake Okeechobee and included nearly 1,000 ha (2,500 acres) of the present Dupuis Reserve, referred to as the L-8 Marsh.

In the more recent past, the Dupuis Reserve operated as a cattle and sheep ranch. Wetlands were drained by swales and ditches in order to provide more pasture. The result was a loss of wildlife habitat, particularly bird rookeries and foraging areas, and encroachment of exotic vegetation into former wetlands.

---

[1] Sr. Civil Engineer, Sr. Environmental Scientist, Supervising Professional and Sr. Civil Engineer, respectively, South Florida Water Management District, P.O. Box 24680, West Palm Beach, Florida, 33416-4680

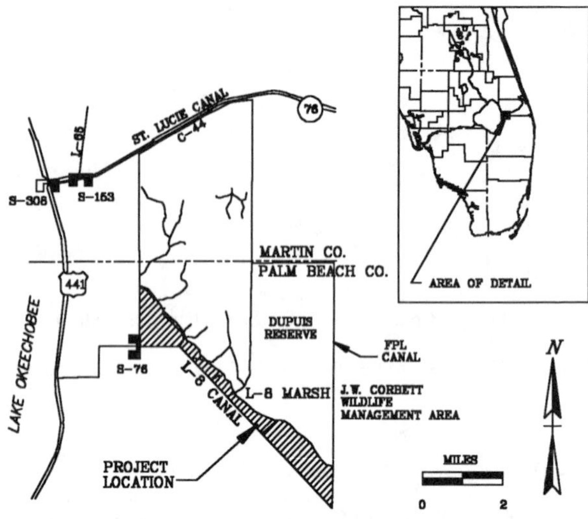

Figure 1. Location Map

The South Florida Water Management District (the District) purchased the Dupuis Reserve in 1986 through the Florida Save Our Rivers Land Acquisition Program. This program, enacted in 1981, allows the District to acquire land for water supply, water management, conservation and protection of water resources. The overall mission of the District is to manage water and related resources for environmental protection and enhancement, water supply, flood protection, and water quality protection. Save Our Rivers objectives set forth in Chapter 373.59 Florida Statutes mandate that these lands be managed with the intent to restore and protect the natural state and condition.

Restoration goals include recreating historic hydrologic conditions to promote the natural propagation of native wetland species, restoring Everglades marsh, preserving native wildlife habitat, impeding encroachment of exotic species, improving aquifer recharge, and maintaining existing flood protection for adjacent agricultural lands. A three-phase hydrologic restoration plan includes: (1) Plugging drainage ditches and swales to reflood interior wetlands, (2) reestablishing connections and sheet flow with adjacent wetland areas to the east, and (3) constructing a 12.9 km (eight miles) levee to reflood approximately 1,000 ha (2,500 acres) of drained Everglades Marsh. This third phase presented the greatest challenges to restoration and is the primary focus of this paper.

Hydrology

Hydrologic design criteria for the marsh restoration consist of establishing the water surface elevation to reflood the marsh, the discharge rates allowed to exit the marsh site, and the size of the control structures. Natural ground elevations vary from approximately 4.5 m (15 feet) above the National Geodetic Vertical Datum of 1929 (NGVD) at the L-8 Canal to over 7.6 m (25 feet) at inland areas. Water surface elevations required to recreate the historic L-8 Marsh are 5.8 m (19.0 feet) above NGVD. This elevation was obtained from reviewing historical records and aerial photographs. Currently the site drains directly into the L-8 Canal, which conveys the runoff away from the Dupuis Reserve.

Hydrologic design included assessing the 10-year and 100-year storm events. Rainfall amounts for these two storms are 19.6 cm (7.7 inches) and 28.4 cm (11.2 inches) respectively. Design of the primary outlet considered peak stages and runoff from the 10-year storm. Peak allowable discharge from the L-8 Marsh into the L-8 Canal during the 10-year storm is 2.54 cm per day (one inch per day) which corresponds to a flow rate of 26 cms (919 cubic feet per second). Runoff rates allowed to discharge from the site are based on environmental concerns for managing the marsh during periods of high water levels. Projected marsh water levels range from 4.6 m (15 feet) to 5.8 m (19 feet) above NGVD. Stages greater than 5.8 m (19 feet) for prolonged periods are undesirable for optimal environmental management of the marsh.

Design of the overflow structure used peak stages and runoff from the 100-year storm. All hydrologic calculations utilized a spreadsheet model based on Soil Conservation Service (SCS) unit hydrograph methods for computing runoff and stages in the study area.

Structural Features

In 1989 and 1990, 41 earthen ditch plugs were constructed in various drainage ditches within the Dupuis Reserve for the Phase I restoration in order to stop the overdrainage of isolated wetlands. These plugs greatly reduced the rapid flow of water off the property, and are changing the character of the former cattle and sheep ranch to its pre-drainage condition. Vegetation maps for 1940 indicate that this site was considerably wetter than the partially restored site is now, however, the plugs are responsible for reflooding approximately 1,600 ha (4,000 acres) of wet prairies, broadleaf marshes, and low pine flatwoods.

Phase II of the hydrologic restoration was to reestablish sheetflow conditions between the Corbett Wildlife Management Area and the Dupuis Reserve. In 1978, a canal was excavated along the Corbett-Dupuis border to provide fill material for a Florida Power and Light (FPL) access road. This road provides access to an electric transmission line for maintenance. Prior to the road and canal construction, runoff from the Corbett Area flowed into the Dupuis Reserve and the L-8 Marsh.

The elevated road now blocks this flow. The borrow canal drained directly into the L-8 Canal, lowered ground water levels, and shortened hydroperiods in this area.

In 1991, the District constructed three earthen plugs in the FPL borrow canal to stop the ground water drawdown and prepare the site for a future connection. Four swales were planned to cross the FPL road to equalize the flow of water between the two areas. Each swale is 30.5 m (100 feet) long and is stabilized by geoweb, a plastic material that provides a stable road bed in wet conditions. Geoweb is used by FPL and other utility companies under similar conditions. Two swales were constructed in 1993 upstream of the southernmost ditch plug. Two additional swales will be constructed south of the southernmost plug after the L-8 Marsh levee is constructed. Bottom elevations of the swales range from 5.6 m (18.5 feet) to 7.0 m (23.0 feet) above NGVD. The swales are designed to convey approximately 6.8 cms (240 cfs) each with a water depth of 0.4 m (1.3 feet).

Planning for the Phase III restoration looked at two structural alternatives for reflooding the L-8 marsh. One alternative considered a single water control structure across the L-8 Canal downstream of the L-8 Marsh. This structure would hold stages in the L-8 Canal high enough to overtop the north bank and reflood the marsh. The south bank of the L-8 Canal has an existing levee that separates the canal and marsh from the adjacent agricultural areas. Holding higher stages in this reach of the L-8 Canal would allow runoff from the Dupuis reserve to gravity flow into Lake Okeechobee for water supply purposes.

However, the L-8 Canal is a major water supply conveyance from Lake Okeechobee to West Palm Beach during low rainfall periods. No assurances can be given that the stages in the L-8 Canal and Marsh would not be lowered to provide water supply to coastal areas. Because of these conflicting operational requirements and schedules, separate systems are needed for reflooding the marsh and operating the canal.

The second alternative, and the selected option, is to construct a 12.9 km (eight miles) long levee on the northern bank of the L-8 Canal to hold back water and reflood the L-8 Marsh. The levee has three control structure sites to control the marsh levels and discharge stormwater runoff.

Control structures designed for this project consist of six 182.9 cm (72-inch) diameter culverts with 243.8 cm (96-inch) full can risers, and four 182.9 cm (72-inch) diameter culverts with 243.8 cm (96-inch) wide flashboard risers. These structures are split up at three locations. All risers are set at 5.8 m (19 feet) above NGVD and are designed to pass the 2.54 cm per day (one inch per day) runoff from the 10-year storm event together. The flashboards can be removed to pass the peak 100-year storm discharge while maintaining 0.76 m (2.5 feet) of freeboard at the top of the levee. The levee crest is at 7.0 m (23.0 feet) above NGVD. The levee has a top width of 3.0 m (10 feet) and side slopes of three to one (horizontal to vertical). Invert elevations of the culverts are at elevation 3.0 m (10 feet) NGVD.

Construction of the levee and control structures commenced in September 1994 and is scheduled for completion in September 1995. Reflooding the L-8 Marsh in Phase III will have more dramatic effects on the Dupuis Reserve than the first two phases.

Funding

The District solicited construction bids for Phase III and the low bid was approximately $1.5 million. This entire amount is currently budgeted from state taxes through the Florida Save Our Rivers Program. Additional funds to replace the tax funds are available from the District Regulation Department Surface Water Division as off-site mitigation from other development projects in Palm Beach County. Since this project has a single objective for marsh restoration, this site is considered an off-site mitigation area for other development projects. The District is collecting off-site mitigation money from developers in lieu of on-site wetland mitigation for certain projects. This type of funding may help to pay for all or part of the L-8 Marsh levee construction. Off-site mitigation at this scale was not tried by the District previously and may ultimately provide funding for other ecosystem restoration projects in the future to supplement or replace tax money.

Conclusions

This paper presented some of the engineering considerations in an ecosystem restoration project on the Dupuis Reserve in South Florida. This project is one of the largest ecosystem restoration projects in Florida, consisting of hydroperiod enhancement in 1,600 ha (4,000 acres) of isolated interior wetlands, reestablishing sheet flow conditions, and reflooding 1,000 ha (2,500 acres) of remnant Everglades Marsh. This project will have major positive impacts on one of South Florida's premier natural areas.

Restoration has three phases, plugging drainage ditches, reestablishing sheet flow connections with adjacent wetlands to the east and reflooding former Everglades Marsh. Major construction activities involved a 12.9 km (eight miles) levee with three control structure sites used to impound water. This impoundment required considerable time for planning, hydrologic modeling, and design activity. Hydrologic monitoring, as a habitat indicator for fish, waterfowl and wetland vegetation, will be undertaken after construction is complete to assess the effectiveness of this type of restoration.

Hydrologic Evaluation of Wetland Restoration Measures by Continuous Simulation

Ananta K. Nath[1], MASCE; Gail C. Abbott[2], AMASCE; and Rao K. Gadipudi[3]

Abstract

Water availability within the root zones of vegetation for optimum depth and duration is a key element for successful functioning of wetlands. Overdrainage due to urban and agricultural developments on the fringe of coastal wetlands adversely impact their sensitive ecosystem. The continuous process hydrologic simulation model HSPF was tested and applied to simulate long term runoff patterns and soil storage components of a large wetland system under existing conditions and with alternative structural measures to restore sheetflow patterns, wetland hydroperiods, and to reduce freshwater shock load to estuaries. With certain limitations HSPF can be used to evaluate the effectiveness of restoration measures in quantifying relative changes in runoff and soil moisture storages.

Introduction

The rapid urban and agricultural developments on the depressional watersheds and coastal wetlands in south Florida have been possible by introduction of large drainage canals. Most of these watersheds are cypress wetlands and low pinelands characterized by extremely flat topography, sandy soils and high water table with continual wet and dry cycles. The network of roads and canals alter the runoff pattern, cause rapid drawdown of the water table, and reduce wetland hydroperiods.

The study area (Figure 1), Southern Golden Gate Estates (SGGE) in southwest Florida, is part of a large real estate development undertaken in the early 1960's. Approximately 114 kilometers of canals were constructed to drain this predominantly wetland area. The long term drainage has lowered the water table, shortened hydroperiods of wetland vegetation and degraded the ecology of the coastal estuaries by fresh water shock loads from the canals.

---

1. Senior Civil Engineer, South Florida Water Management District, Naples, FL
2. Civil Engineer, South Florida Water Management District, Naples, FL
3. Associate Engineer, Montgomery Watson, Lakeworth, FL

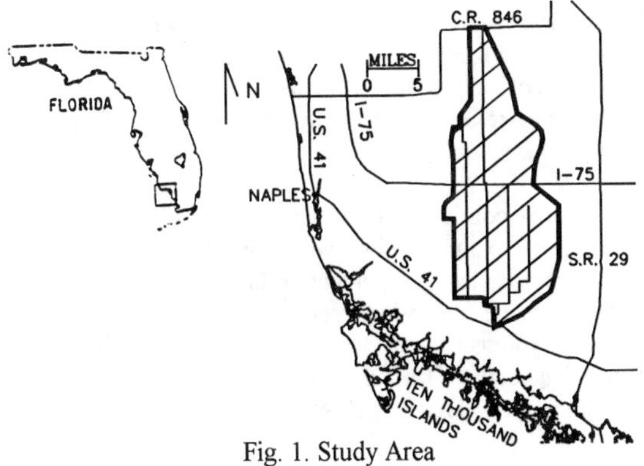

Fig. 1. Study Area

The SGGE area was identified as an important location for the region's ecosystem restoration plans, particularly for its key location as the headwaters of the Ten Thousand Islands National Estuarine Preserve, part of the western Everglades National Park. The major goals set for the project are: restoration of wetland hydroperiods; reduction of fresh water shock load discharges to estuaries; reduction of overdrainage; enhancement of aquifer recharge; and maintenance of existing levels of flood protection for areas north of the project.

A long term continuous process simulation model was developed using the U.S. Environmental Protection Agency's watershed modeling program Hydrological Simulation Program-Fortran (HSPF) to quantify the soil storage components and runoff patterns under existing conditions and to evaluate the effectiveness of three alternative structural measures in achieving the project goals.

Model Application

HSPF was chosen as the primary modeling tool for development of restoration measures for SGGE due to its ability to continuously simulate the hydrologic-hydraulic behavior of various soil storage zones. The PERLND and RCHRES modules of HSPF were used respectively to simulate the hydrologic processes on the land surface and hydraulics of the canal and water control structure network. Simulation was performed in 37 pervious land segments (PLS) and 36 reach-reservoir segments (RCHRES). The three restoration measures were represented in the RCHRES module and the NETWORK block of the program.

Calibration was performed at six stations within the watershed. The calibrated parameters showed relatively high infiltration rates (INFILT= 0.04-0.10), variable Lower Zone Nominal Soil Storage (LZSN = 2-8) and relatively low Active Ground Water Recession Coefficients (AGWRC) which result in a rapid release of surficial groundwater. The Lower Zone Evapotranspiration (LZETP) varied monthly. Calibration was done for periods ranging from 11 months to 15 years. Overall, calibration was fair. Undocumented operation of water control structures and changes to canal geometry due to weed growth affected the results. The model was verified using 15 years of flow data for the outlet of the basin.

The first alternative considered a partial restoration plan intended to reduce the point source discharge of voluminous freshwater into the estuaries. A flow diversion structure would divert approximately 50 percent of the base flow from the main canal to a spreader channel for eventual dissipation through wet prairies to the estuaries.. A user defined input time series represented the flows being diverted from the main channel into the spreader channel. The outflow from the spreader channel was then routed onto the land segment as a surface lateral inflow (SURLI).

The second alternative considered a full scale hydrologic restoration with filling up of all canals in SGGE and removal of the roads that blocked the natural sheetflow. Pump stations would be required to push the water from the northern canals into a large spreader channel to initiate sheetflow (Figure 2).This alternative is simulated by first

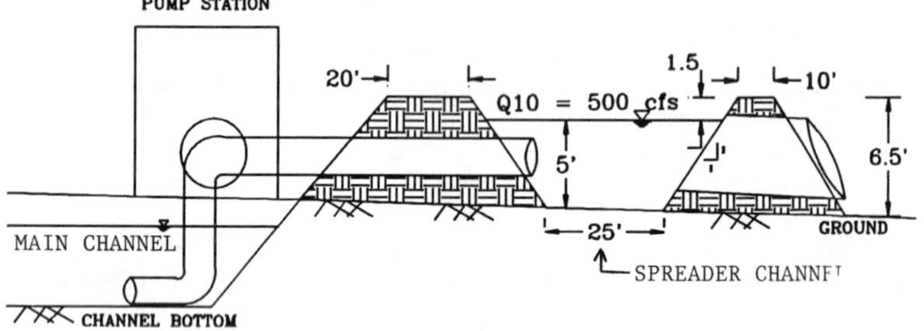

Fig. 2. Typical Configuration of Pump and Spreader Channel Arrangement

representing the spreader channel as a new reach which discharges onto the land segments. Runoff was routed from one PLS to another using surface, interflow and groundwater lateral inflows.

The third alternative (Figure 3) is a modification of alternative two with pumps, spreader channels, removal of only strategically selected roads and placement of canal plugs instead of filling an entire canal and drainage improvements on some roads. This is represented by two new reaches that collect water from the north canals and discharge onto the land surface. Some PLS were routed from one to another. However, new reaches with very wide shallow channels were simulated in the RCHRES module to represent broad movements of sheetflow along the historic flowways.

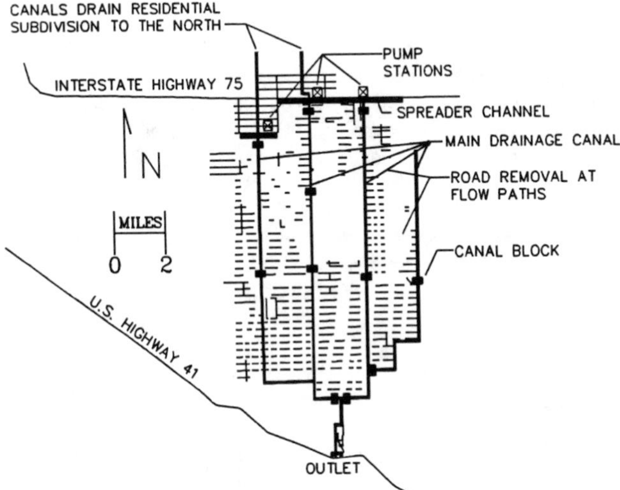

Fig. 3. Schematic Representation of Alternative Three

Model Results

The application of PERLND and RCHRES modules to represent alternative configurations of pumps, spreader channels, canal blocks, and road removal provided unique challenges different from the traditional applications of HSPF. The representation of alternative one by routing large volumes of channel outflow to a relatively smaller PLS resulted in unreasonable inflow depths and artificially high soil storage and runoff values. In the subsequent alternatives as the receiving PLS area became larger, the inflow depth was more reasonable. As such, the effect on soil storage and hydroperiods for alternative one are not comparable to the results of the other two alternatives.

In alternative two, the routing of channel outflow to PLS, and PLS outflow to subsequent PLS, resulted in marked increases in soil storage and runoff values for those PLS that received the lateral inflow. Basinwide, there was slightly more evaporation and less runoff due to the higher storage values. The average daily upper zone soil storage (UZS) increased by ten percent, lower zone soil storage (LZS) by six percent and active groundwater storage (AGWS) by 205 percent. The AGWS was extended for an average of one to two month period longer that the existing conditions.

In alternative three, the use of the RCHRES module to represent the historical flowways resulted in increased evaporation and less runoff. UZS increased by six percent, LZS increased by four percent, and AGWS increased by 62 percent. The duration of AGWS was extended for approximately one month longer than existing conditions.

Discussion

Modeling of existing conditions could be improved with a fully dynamic routing algorithm to accurately simulate the flow regimes of free surface, pressure flow and flow reversal conditions encountered in the complex canal network. Simulation of the hydrologic processes of the wetlands could also be improved by taking into account the ponding effects on the PLS in the PERLND module. Wetland conditions are unique due to the influence of a seasonally varying water table with large amplitudes. These cyclic groundwater levels affect such input parameters as LZSN and INFILT which cannot be seasonally varied in HSPF. HSPF also cannot model evaporation from overland flow and the transitional features of wetlands behaving like a PLS at one time, and as a RCHRES at other times.

Since any changes to the RCHRES representation do not affect computations of the PERLND module, the routing from a RCHRES to a PLS or from one PLS to another seemed to be the only feasible way to evaluate the effects of spreader channel on the soil moisture storages of the land segments. However, the introduction of the flow component as SURLI implied that water from the spreader channel is distributed evenly over the entire land segment, and not along a "line" as in the real physical world.

Routing from PLS to wide shallow RCHRES that resemble historical flowways is a more reasonable way to represent these alternatives. In this case, however, the reach begins to inundate the surrounding PLS as it spreads out over several miles. When this happens, precipitation will be counted twice. Therefore, precipitation calculations were not activated for the historical reaches that might overlap a PLS. Another suggested way to represent these alternatives would be to change the input parameters in both modules. The plugging of the canals results in an increase in groundwater levels which affect INFILT, LZSN and other PERLND parameters. These are calibrated parameters and would have to be re-calibrated for conditions reflecting these higher groundwater levels.

Conclusions

Within the purview of certain limitations, the PERLND and RCHRES modules of HSPF can be used to simulate the hydrologic-hydraulic processes of impacted wetlands. Modifications of the PERLND module to allow ponding and saturated soil conditions, and of the RCHRES module to include effects of continual wet and dry cycles will improve modeling of wetlands. Inclusion of a dynamic routing algorithm in the RCHRES module will enhance simulation of the multiple flow regimes encountered in a complex canal network.

Acknowledgments

The project was made possible by a subgrant from the U.S. Department of Commerce, National Oceanic and Atmospheric Administration, in cooperation with the Florida Department of Community Affairs.

Modeling Wetland Hydrodynamics Using SWMM-EXTRAN

Vassilios A. Tsihrintzis[1], David L. John[2], and Paul Tremblay[2]

Abstract

Application of SWMM-EXTRAN in modeling hydrodynamics of a wetland area in South Florida is presented. The project involves restoration of a wetland adjacent to an existing development, and required careful evaluation of drainage conditions. EXTRAN allowed sizing of hydraulic structures, and computation of maximum water surface elevations under various flood scenarios and under both existing and proposed conditions. The model was verified using measured field data, and was found to be a valuable tool for hydrodynamics of wetland restoration projects.

Introduction and Project Description

Due to the low success of freshwater wetland mitigation projects (less than 12% in Florida) a new approach to preservation is tested, the creation of "wetland mitigation banks." These are areas of restored or created wetlands that are undertaken expressly to provide off-site compensation credits for wetland losses caused by permitted projects.

The first privately operated wetland mitigation bank in Florida (and the second in the country) has received all the required permits, including the US Corps of Engineers dredge and fill permit, a Broward County, Department of Natural Resource Protection, Environmental Resource Mitigation Bank License, and permits from the South Florida Water Management District (SFWMD) and the South Broward Drainage District. It is developed by Florida Wetlandsbank on a 350-acre area owned by the City of Pembroke Pines, Broward County, Florida. The area is located in the C-11 canal basin, south of Sheridan Street, approximately 1 mile east

[1] Assistant Professor, Department of Civil & Environmental Engineering and Drinking Water Research Center, Florida International University, University Park, VH 169, Miami, FL 33199.

[2] Respectively, Senior Associate and Project Engineer, Robert H. Miller and Associates, Inc., 1800 N. Douglas Road, Suite 200, Pembroke Pines, FL 33024

of the Water Conservation Areas. This area is now a degraded, sterile, non-functioning site, infested with a monoculture of exotic plant species (melaleuca). It will be restored into a balanced, healthy, self-sustaining ecosystem with diverse habitats, representative of those normally found in the Everglades (cypress stands, emergent marshes, forested wetlands, sawgrass prairies, wading bird feeding areas, and tree island habitats). Parts of the area will be open to the public for recreation and education opportunities.

Creation of the proposed Wetlandsbank located adjacent to and downstream of the existing Chapel Trail development required careful evaluation of drainage conditions. Chapel Trail is approximately 1,300 acres in surface area, and contains a series of lakes, interconnected with culverts, which store the 100-year/3-day storm runoff, and deliver it through an open cut, approximately 146 feet long, to the existing downstream preserve which will be restored to the proposed Wetlandsbank. A nearly rectangular open channel, approximately 50 feet wide and 7,800 feet long, runs through the preserve conveying drainage from Chapel Trail development and adjacent areas. The channel ends to a structure containing 3 weirs, two of them 4 feet wide and the third 2.5 feet wide, which discharge into a reinforced concrete box that connects to 2x60-inch corrugated metal pipes (CMP), 228 feet long. These two pipes end into a 50-foot wide open channel for which the design water surface control elevation is set at 3.5 feet.

Under proposed project conditions, most of the Wetlandsbank area will be graded to elevations approximately 1 foot lower than existing conditions, significantly increasing its storage volume. The open channel currently crossing the Wetlandsbank will be backfilled, and the connection to Chapel Trail development will be relocated. A berm will be constructed around the Wetlandbank to isolate it from the surrounding sites. A diversion canal will be constructed along the east periphery of the Wetlandsbank connecting to the lakes of Chapel Trail. This canal will merge into the existing 50-foot wide channel in Wetlandsbank, upstream of the existing weir structure. A diversion weir set at an elevation of 5.5 ft, will be constructed at about mid-distance in this diversion canal, spilling excess water from this canal to the Wetlandsbank when water surface elevations in the canal and the Wetlandsbank permit this to happen by gravity. A small notch in this weir will be set at crest elevation 4.0 feet. The water surface in the Wetlandsbank then will be raised to a minimum of 4.0 feet.

Hydrodynamic Model Development and Application

To accurately compute the available flood detention storage and maximum water surface elevations, the Environmental Protection Agency's (EPA) SWMM, module EXTRAN, was used for flood routing through existing and proposed wetland channels and cells. Detailed description of the capabilities, limitations and application of this model is given by Roesner et al. (1989). The data required to run the model are geometric data to form the network of conduits connecting nodes. Chapel Trail and the wetlands were represented in the model as storage junctions. Schematic geometries used in the model for existing and proposed project are presented in Figure 1. Inflow hydrographs into each junction were specified.

Model predictions depend primarily on model geometry and Manning's roughness coefficients for culverts and open channels which were selected at 0.024 and 0.030, respectively. These values were adjusted based on procedures described by Tsihrintzis et al. (1995) to allow accounting of minor head losses (entrance and exit) which are not computed by EXTRAN. Model verification was done using field measurement under existing conditions. Field flow and water surface elevation measurements were taken once every 3 to 5 days during most part of 1993 (approximately 7 months) in the existing site. The measured hydrograph in 1993 was a result of both groundwater recharge into the lakes and canals, and direct rainfall. Flow was estimated utilizing the downstream weir and culvert structure rating curve based on the stage upstream of the structure. Stage measurements were taken in the canal within the existing preserve (proposed Wetlandsbank area) and in the lake of Chapel Trail that is located upstream of the 146-foot open cut. Figure 2 shows comparison between computed and measured flow hydrographs through the existing downstream weir structure and culverts. Figure 3 shows comparison between observed and computed water surface elevations in the existing preserve. Agreement is considered very good.

Figure 4 presents computer water surface elevations for existing and proposed condition for the 100-year/3-day storm for both Chapel Trail and Wetlandsbank as a function of time. Existing computed water surface elevations for Chapel Trail and the existing preserve coincide. The proposed project water surface elevations for both Chapel Trail and Wetlandsbank are below or at the existing water surface elevations. This is a result of the excess storage in the Wetlandsbank, compared to the existing preserve condition.

Conclusions

EXTRAN was proven to be a valuable tool in effectively designing the restoration of the Florida Wetlandsbank. Link-node model use in wetland design by practicing engineers seems to be a good alternative to more cumbersome two-dimensional models. The reasons are: input data preparation is easier and computer time is shorter; additional computation detail given by more advanced models may not be justifiable for studies similar to the one presented.

Acknowledgements

Funding was provided by the Florida Wetlandsbank (FWB) and Robert H. Miller and Associates, Inc. (RHM). Valuable comments by Lew Lautin (FWB), Kevin Hart (RHM), Leo Schwartzberg, Carlos de Rojas (SFWMD) and Eduardo Lopez (SFWMD) are appreciated.

Appendix. References

Roesner, L. A., Aldrich, J. A., and Dickinson, R. E. (1989). *Storm Water Management Model User's Manual Version 4: EXTRAN Addendum*, Environmental Research Laboratory, Office of Research and Development, U.S. Environmental Protection Agency, Athens, Georgia.

Tsihrintzis, V.A., Vasarhelyi, G.M., and Lipa, J. (1995). "Hydrodynamic and constituent transport modeling of coastal wetlands." *Journal of Marine Environmental Engineering*, (in press, to appear in Spring).

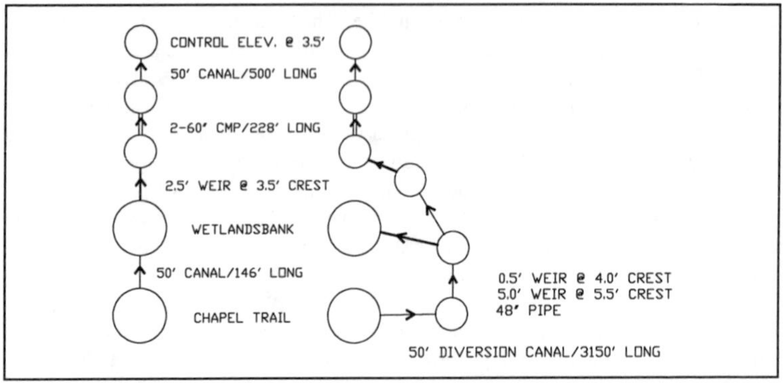

Figure 1. Existing and Proposed Schematic Model Diagrams

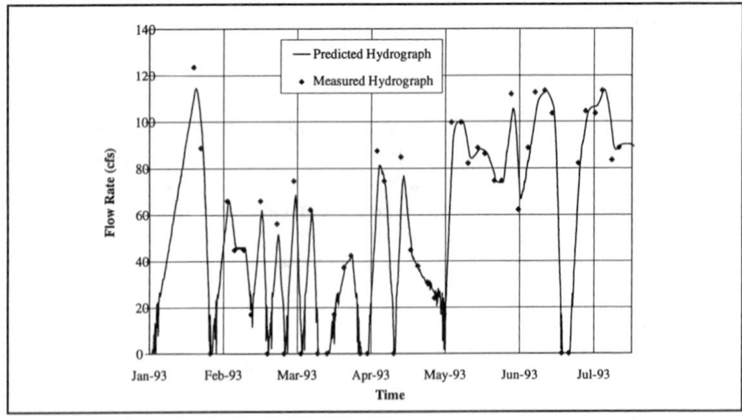

Figure 2. Measured and Predicted Flows through the Existing Culverts

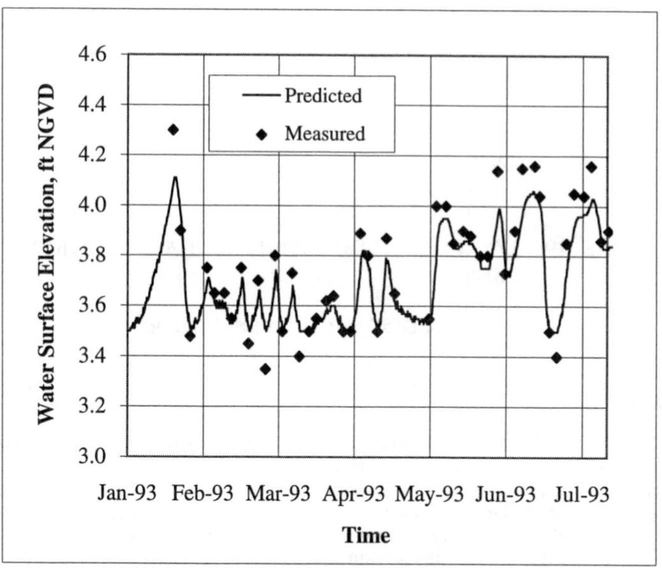

Figure 3. Measured and Predicted Water Levels in Existing Preserve

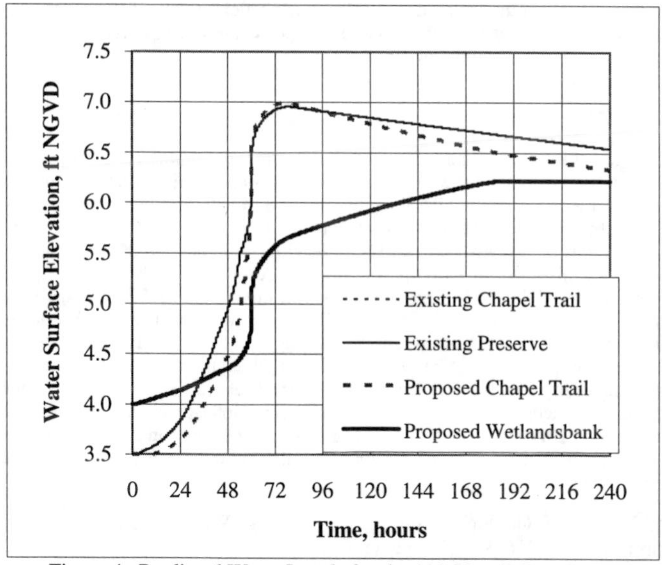

Figure 4. Predicted Water Levels for the 100-Year/3-Day Storm

PRECIPITATION DEPTH-DURATION CHARACTERISTICS, ANTELOPE VALLEY, CALIFORNIA

James C. Blodgett[1] and Iraj Nasseri[2]

Abstract

To document the changes in runoff characteristics of basins subject to urbanization, streamflow and precipitation data were collected at eight small basins in Antelope Valley, California, for the period 1990-93. The data collected at U.S. Geological Survey stations were supplemented by data collected at 35 long-term precipitation stations. These data will be used to calibrate and verify rainfall-runoff models for the eight basins and for estimating basin runoff characteristics throughout Antelope Valley.

Annual precipitation in Antelope Valley varies from more than 50 cm in the mountains to less than 10 cm on the valley floor. Most precipitation in the valley occurs during the winter months, December through March, but cyclonic storms in the fall and convectional storms in the summer sometimes occur.

Introduction

Changes in the peak magnitude and volume of runoff from newly urbanized drainage basins occur because of the increase in impervious areas and the modification of the drainage channels. An analysis of the hydrology of Antelope Valley (Figure 1), based on rainfall-runoff model studies, is the subject of a study by the U.S. Geological Survey in cooperation with the counties of Los Angeles and Kern, the cities of Lancaster and Palmdale, the Los Angeles City Department of Airports, and the U.S. Air Force at Edwards Air Force Base. The initial phase of the study began in 1990 with the selection of 8 drainage basins for collection of detailed precipitation and runoff data. In order to use the historic daily or hourly precipitation data for storm runoff modeling, a method of disaggregating the time interval of these data was necessary. Depth-duration ratios were calculated by disaggregating daily total precipitation data for intervals of 1, 2, 3, 4, 6, 12, and 18 hours for the storms that occurred during 1990-93. The hourly total precipitation data were then disaggregated at 5-minute intervals. Precipitation data for stations in the valley indicate that about 70 percent of the daily total occurs in the first 6 hours of a storm. Depth duration

[1] Hydrologist, U.S. Geological Survey, 2800 Cottage Way, Sacramento, California 95825, (916) 979-2615 extension 382.

[2] P.E., Los Angeles County Department of Public Works, Alhambra, California 91803-1331, (818) 458-6124.

relations for the USGS stations in the valley are about 19 percent greater than regional relations developed by NOAA in 1973.

Antelope Valley, California is about 80 km north of the City of Los Angeles in the northwestern Mojave Desert. The valley, with an average elevation of about 762 m, is a closed, inland drainage basin that covers about a 6,216-km^2 area. The Tehachapi Mountains, with elevations as high as 2,438 m, form the northern and western borders, and the San Gabriel Mountains, with elevations to 3,048 m, form the southern border (Figure 1). Areas in the western and southern part of Antelope Valley, particularly along the foothills and on the alluvial fans, are being urbanized. These new urbanized areas generally are located higher on these fans and on the foothills above the older subdivisions. Accordingly, storm runoff from the newly urbanized areas is causing flood problems in the older, downslope, urbanized areas.

Seasonal Variation in Precipitation

Precipitation in the Antelope Valley is the result of one of three types of storms. The first type of storm occurs when low-pressure frontal systems move eastward from the Gulf of Alaska or from near the Hawaiian Islands. The storms that are formed by these systems are affected by orographic uplift as they move from the Pacific Ocean across the Tehachapi and San Gabriel Mountains (Figure 1). The storms are most prevalent during the winter months

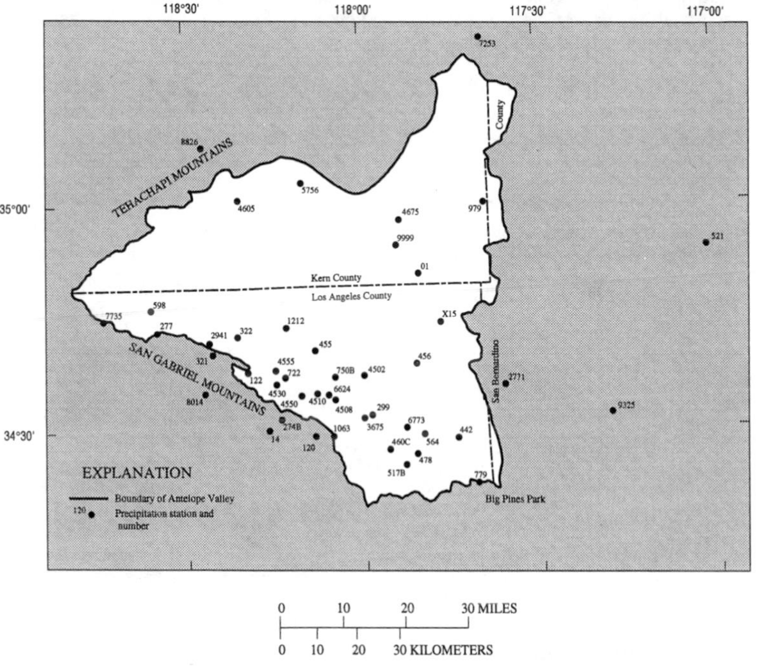

Figure 1. Locations of precipitation stations in Antelope Valley, California, and vicinity.

of December, January, February, and sometimes March and generally produce more than 2.54 cm of precipitation in a day. Frontal systems from the Gulf of Alaska and the tropics near Hawaii sometimes merge before moving inland and cause increased precipitation intensity, such as occurred during the storms of February and March 1983.

Tropical cyclones (hurricane type storms) that originate in the South Pacific also cause precipitation in the valley. This type of storm usually moves in a northeasterly direction from the South Pacific and Baja California areas. One such storm (Octave) occurred during September 28 to October 2, 1983.

Convectional storms are the result of convective uplift and cooling of air masses; such a storm occurred September 19, 1990, near Mojave. Precipitation during this type of storm generally covers a small area, and the intensity ranges from light showers to cloudbursts. These storms usually occur during the summer or early autumn in the foothill areas.

Disaggregation of Daily Total Precipitation Data

Rainfall-runoff models are used to calculate flood flows which are then used in the design of drainage facilities for areas subject to development. For smaller basins, the accuracy of model results depends, in part, on the availability of precipitation data that were collected at time intervals less than 24 hours. In Antelope Valley, historical hourly precipitation data are collected at only four stations. Therefore, the reliability of calculated flood flows on an areal basis is uncertain without some knowledge of precipitation depth-duration characteristics for intervals of less than 24 hours.

In order to use daily or hourly precipitation data from historic storms for rainfall-runoff modeling analyses, a method of disaggregating the time interval of these data was necessary. Depth-duration ratios were calculated by disaggregating daily total precipitation data for intervals of 1, 2, 3, 4, 6, 12, and 18 hours for the storms that occurred during 1990-93.

A precipitation depth-duration relation (Figure 2) for the NOAA hourly station at Palmdale was developed using data for 17 historic storms for the period of record, 1952-93. To compare the depth-duration intensities recorded at USGS stations for the various storms

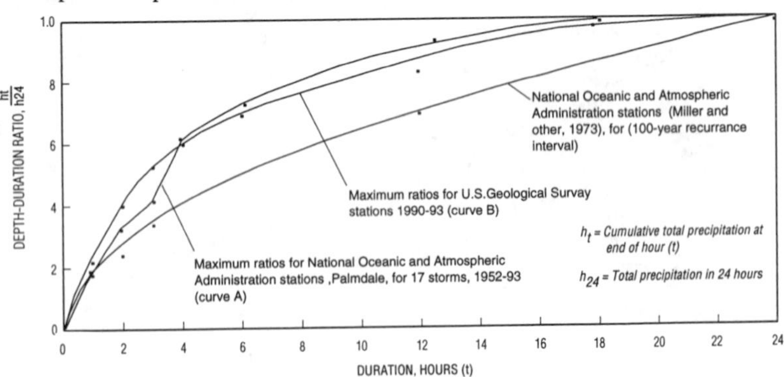

Figure 2. Depth-duration ratios for maximum recorder hourly precipitation depths in Antelope Valley, California.

during 1990-93 with those recorded at long-term NOAA stations, the maximum precipitation depths for each time interval were expressed as a ratio of the total precipitation for the 60-minute and 24-hour interval by the equation:

$$hn = \frac{ht}{hd}$$

where

ht is total maximum precipitation, in centimeters, recorded during a storm for selected time interval, and

hd is total precipitation depth, in centimeters, for the 60-minute or 24-hour duration that includes the interval *ht*.

The depth-duration curves for 24-hour periods (Figure 2) represent the maximum ratios of recorded precipitation for the indicated durations for the stations in the valley. These curves indicate that about 70 percent of the total daily precipitation occurs during the first 6 hours of the storm.

Depth-duration ratio curves based on 5-minute interval precipitation data collected at the 8 USGS stations for selected periods less than 60 minutes are shown on Figure 3. Curve A shows the average depth-duration relation for ratios based on data collected at USGS stations. Curve B shows the relation for the maximum ratios at the USGS stations that were calculated for storms that occurred during 1990-93. Hourly precipitation data were disaggregated to provide maximum as well as average ratios for preparation of Figure 3.

The maximum recorded ratio for any interval of time may occur at any gage location in the valley. The curve for maximum ratios calculated for USGS stations (Figure 3, curve B) is about 19 percent greater than the curve calculated for 200 NOAA stations (Figure 3, curve C) (Hersfield, 1961; Miller and others, 1973). Ferro (1993) presents a similar relation of depth-duration ratios for intervals of less than 60 minutes for the United States (Figure 3) that closely approximates the data by (Miller and others, 1973) for durations of less than 10 minutes.

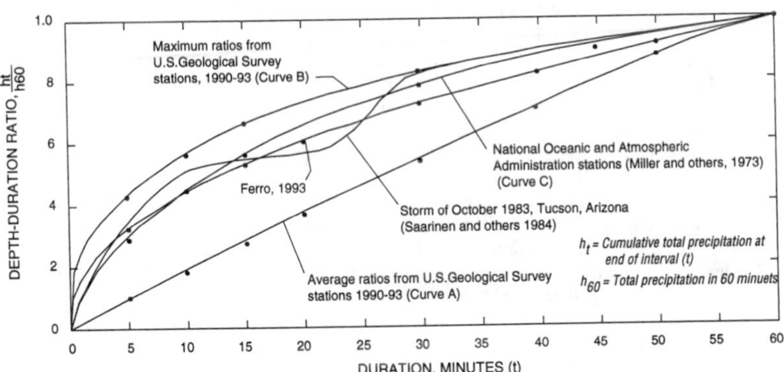

Figure 3. Maximum and average observed depth-duration ratios for intervals of less than 60 minutes in Antelope Valley, California.

Ferro (1993) presents a depth-duration relation

$$\frac{h_{t,T}}{h_{60,T}} = \left(\frac{t}{60}\right)^{s}$$

where

ht,T is precipitation occurring in a period of *t* hours and which has a return period of *T* years, and

s is a coefficient that has a different value for various geographical regions.

For the USA, Ferro (1993) indicates a value of s = 0.451. Ratios using maximum precipitation data for the U.S. Geological Survey stations, Figure 3, indicates a coefficient s = 0.272, which is similar to values of the coefficient s represented by Ferro (1993) for the regions of Molise-Abruzzo and India.

A precipitation depth-duration relation based on the ratios for the cyclonic storm "Octave" of October 1983 at Tucson, Arizona, is shown on Figure 3 (Saarinen and others, 1984) to indicate the relative significance of a cyclonic storm in desert-type regions with documented short-interval data. The precipitation depth-duration relation for this storm, which also occurred in Antelope Valley, can be used as an indicator of ratios that might occur in Antelope Valley during large storms for which no sub-hourly precipitation data are available.

References

Ferro, Vito, 1993, Rainfall intensity-duration-frequency formula for India: Journal of Hydraulic Engineering, v. 119, no. 8, p. 960-962.

Hersfield, D.M., 1961, Rainfall frequency atlas of the United States for durations from 30 minutes to 24 hours and return periods from 1 to 100 years: U.S. Department of Commerce Technical Paper no. 40, 61 p.

Jensen, R.M., Hoffman, E.B., Bowers, J.C., and Mullen, J.R., 1992, Water resources data--California, water year 1991. Vol. 1. Southern Great Basin from Mexican Border to Mono Lake Basin, and Pacific Slope Basins from Tijuana River to Santa Maria River: U.S. Geological Survey Water-Data Report CA-91-1, 312 p.

Miller, J.F., Frederick, R.H., and Tracey, R.J., 1973, Precipitation-frequency atlas of the western United States: National Oceanic and Atmospheric Administration Atlas 2, v. XI, California: 71 p.

National Oceanic and Atmospheric Administration, 1990, 1991a, 1991b, 1991c, Climatological data, monthly precipitation departure from individual station normals (1951-1980), California, September 1990: v. 94, no. 9.

Saarinen, T.F., Baker, V.R., Durrenberger, Robert, and Maddock, Thomas, Jr., 1984, The Tucson, Arizona, flood of October 1983: National Academy Press, 112 p.

REGIONAL FREQUENCY ANALYSIS OF FLOODS IN A MULTIVARIATE FRAMEWORK

by Gustavo E. Diaz [1], Jose D. Salas [1] and William R. Hansen [2]

Abstract

A regional flood frequency analysis for Zion National Park and surrounding areas is presented. Two regional models were developed using a multivariate lognormal distribution that explicitly accounts for the spatial correlation among the series of annual peak flows. This distinctive characteristic of the models is particularly important in regions like Zion, where the largest peak flows have been recorded at several gaging stations during the same water-year. The multivariate models incorporate regional assumptions derived from the observed series of peak flows employing a relatively small number of parameters and at the same time gaining in reliability by using readily available geographical and meteorological information. Parameters of the regional models are estimated using a maximum likelihood procedure derived from a set of stations that are known to preserve the assumed structure of the regional models.

Introduction

This paper presents a regional flood frequency analysis (RFFA) for Zion National Park (Zion) and surrounding areas. Specific objectives of the study were to develop methods to estimate peak flows quantiles at ungaged sites and to increase the reliability of estimates for sites with short periods of record. Zion is located in the southwestern Utah, within the Upper Virgin River Basin, see Figure 1. The study area contains the Virgin, Upper Sevier, Escalante, Paria, Kanab, Cedar-Parowan and Wahweep hydrologic units. Watersheds were selected within each hydrologic unit based on the availability of flow records and specific criteria of hydrological homogeneity. Zion is located in a region with numerous gaging stations with short periods of record. The problem of insufficient flow information or even the complete lack of data at a specific site has been tackled by using additional information from neighboring stations. RFFA has been found very effective in improving the reliability of peak flow frequency estimates as demonstrated by a solid body of publications in the field.

---

1 Assistant Professor and Professor, respectively. Department of Civil Engineering, Colorado State University; Fort Collins, Colorado 80523.

2 Hydrologist., WRD-National Park Service; 1201 Oak Ridge Dr., Ft.Collins, CO 80525.

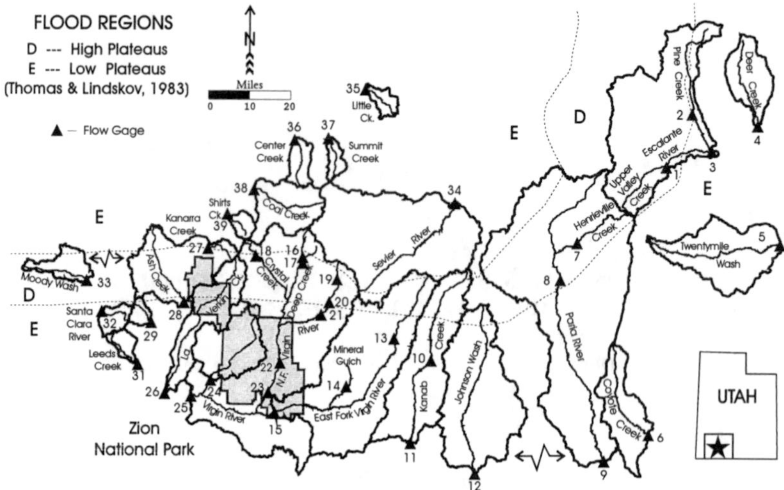

Fig. 1 Watersheds in the Study Area, Streamflow Gaging Stations and Flood Regions

Hydrological and Geographical Information

Figure 1 shows the location of 39 stations where maximum annual instantaneous peak flows were measured. Records vary in length from 2 to 77 years. Despite the large differences in record length, the study benefited from using all existing data rather than arbitrarily discarding stations with short records.

The study area is affected by two distinct types of storms: cyclonic storms generating precipitation that spread over a long period of time, generally during winter months; and thunderstorms, characterized by violent downpours of brief duration and high intensity (cloudbursts), occurring mostly during the warmest months (Wooley, 1946). Meteorological events that have a noticeable impact on hydrological series are regional in nature. For instance, the storm that caused the flood of December 1966 in southwestern Utah was a slow moving system traveling eastward across the state. Peak discharges at several neighboring sites including sites 23, 25, 29, 30, 32 and 33 were the largest on record as a result of that same storm. Under these conditions, the customary assumption of independence of flood events is not satisfied. This emphasizes the importance of utilizing a model that addresses the effect of interstation correlation between flow peaks.

All geographical and geomorphological information for the watersheds required for the analysis were obtained from a geographical information system (GIS) developed for the study area, including: watershed boundaries, drainage network, elevation, slope, location of flow gage stations and frequency and duration of precipitation.

Formulation of the Multivariate Models

The lognormal distribution in its univariate form was found suitable for representing the frequency of annual peaks at single sites within Zion. The methodology developed by Pons (1992) was used and is based on multivariate lognormal models with 2 and 3 parameters, MLN2 and MLN3, respectively. These models have a relatively simple form while still maintaining the most important characteristics needed for RFFA. The MLN3 model is expressed in Eq.(1), where X_j represents the peak flows at site j, $\mu_j(y)$ and $\sigma_j^2(y)$ are the corresponding mean and variance of the log-transformed peaks, and τ_j denotes the lower bound of each marginal distribution. The MLN2 model assumes τ_j's equal to zero.

$$Y_j = \ln(X_j - \tau_j) \sim N(\mu_j(y), \sigma_j^2(y)) \quad (1)$$

The MLN2 and MLN3 models assume that the mean of the annual peak series can be related to geographic and climatic characteristics of the catchments. The decision as to which characteristics to consider is based on the contribution of each added parameter to the reduction of error variance in the regression equations for the mean and the basin characteristics. The model adopted for the Zion region has the form shown in Eq.(2), where A_j is the drainage area, D_j is the elevation at the flow gaging site, P_j is the 2yr-24hr precipitation index and $\mu_j(x)$ is the mean of the X_j , all of them computed for site j. In addition, the model assumes that in regions where the flood producing mechanisms are unique, a relation between the cross-correlation coefficients of the logs of peaks ρ_{ij} and the intergage distances can be built as shown in Eq.(3), where b is a regional parameter and d_{ij} is the distance between stations i and j.

$$\mu_j(x) = c\, A_j^{\,a}\, D_j^{\,d}\, P_j^{\,p} \quad (2)$$

$$\rho_{ij}(y) = e^{-b\, d_{ij}} \quad (3)$$

Pons (1992) assumed that $\sigma_j^2(y)$ is constant for the region, which in turn implies a constant coefficient of variation Cv and a constant coefficient of skewness $\eta(y)$ for the region. The lower bounds of the MLN3 model can not be regionalized. However, under these underlying assumptions, the τ_j's can be computed analytically. In summary, the series of peak flows will be multivariate lognormal-3 with marginal distributions having the mean expressed as a function of basin characteristics, and a common coefficient of variation and a common coefficient of skewness throughout the region. The basic assumptions outlined above were confirmed for the Zion region. For instance, Cv was found to be independent of the hydrological characteristics and approximately constant within the region. Furthermore, skew coefficients were better described by a regional constant value rather than by at-site estimates.

Parameter Estimation

Since the equations derived from the maximum likelihood (ML) procedure cannot be solved explicitly, preliminary estimates of the model parameters are required prior to their final estimation using the ML procedure. Initial estimates of the parameters c, a, d, p, b, $\sigma^2(y)$ and $\eta(y)$ were obtained via multiple linear regressions and weighted average methods. The preliminary estimates were used to initialize a multivariate optimization algorithm that searches for the optimum set of parameters. The method consists of finding the parameter set that maximizes the log-likelihood function. The asymptotic properties of the ML estimators are preserved only for the MLN2 model, since a full ML estimation procedure for the MLN3 is not feasible.

Selecting an Homogenous Region Based on Preserving the Model Structure

Two criteria were used to select the stations for the RFFA. Thomas and Lindskov (1983) classification of flood regions for the State of Utah was used as the first criterion. This classification provided homogeneity among peak records based upon the dominant flood producing mechanism in the region. The second criterion consisted of analyzing the effect that the peak series have on the assumed regional structure. The objective was to find a group of stations that satisfy the level of homogeneity needed for the ML procedure to perform satisfactorily in estimating the parameters for the multivariate models. According to Pons (1992), three types of error can cause the ML procedure to yield improper values for the parameters, namely: (1) error in defining the mean of the log-peaks $\sigma^2(m)$; (2) error in assuming a constant variance of the log-peaks $\sigma^2(v)$; and (3) error in assuming a fixed relationship of cross-correlation versus distance $\sigma^2(r)$. The procedure consisted in minimizing the variance of the three error terms simultaneously. For that purpose, the model was run repeatedly, starting with the set of 39 stations, and dropping one station at a time by evaluating the associated reductions in the variance of the errors. This approach defined the most desirable subset of stations necessary to compose the regional models. Figure 2 shows the variance of the three error terms as a function of the number of stations remaining in the set. The curves display a smooth decrease as stations are progressively removed from the model. The model started rejecting the stations located the farthest from the Park, and yielded a subset of ten stations used for the estimation of the parameters based on the ML procedure. The selected stations cover a wide range of drainage areas and altitudes, including sites 11, 16, 17, 19, 21, 23, 24, 25, 28 and 32. The subregion encompassed by the selected stations can be considered homogeneous from the point of view of the structure of the models.

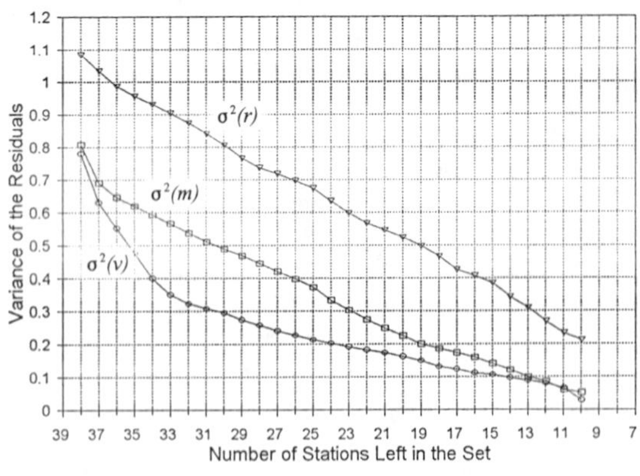

Fig.2 Variance of the Error Terms vs. Number of Stations

Model Application

The ML parameter estimation procedure was applied to the subset of ten stations and yielded the estimates shown in Table 1. Goodness of fit of the models was judged by graphically comparing the frequency curves obtained from the regional models versus the empirical frequency distributions at the stations. For example, Figure 3 shows the results of regional and univariate fittings at the North Fork of the Virgin River at Springdale (site 23). Both regional models provided very good and similar fits. Note that the regional models provide a better prediction for recurrence intervals greater than 10 years than the univariate model. For values smaller than 10 years, the univariate and multivariate models provide similar results.

Table 1. Models Parameters

Coeffi.	MLN2	MLN3
ln c	32.645	23.979
a	0.500	0.576
d	-3.512	-2.433
p	1.377	0.748
b	0.047	0.041
$\sigma^2(y)$	0.634	0.581
$\eta(y)$		0.917

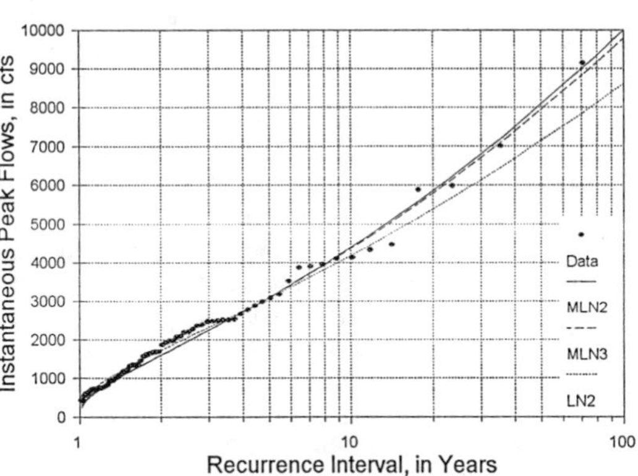

Fig.3 Regional and Univariate Fitting of Annual Peaks

Conclusions

Severe local effects of cloudbursts in Zion can yield erroneous results when flood frequency analysis is conducted at a single site, particularly if dealing with small watersheds and short periods of record. The regional models introduced here remove this effect from the estimation of flood frequencies and suggest that multivariate lognormal models are valuable tools in regional flood frequency analysis. The obtained regional models can be used to estimate flood frequency curves at ungaged sites within the same geographical region used to estimate the parameters of the models.

Acknowledgments

The authors would like to acknowledge the financial support to the project "Regional Flood Hazard Analysis" by NSF Grant #BCS-9101741; and from the National Park Service, Water Res. Division.

References

Pons, A. F., 1992. "Regional Flood Frequency Analysis Based on Multivariate Lognormal Models". PhD. Dissertation, Dept. of Civil Eng., Colorado State Univ., Ft Collins, CO.

Thomas, B. E. and K. L. Lindskov, 1983. "Methods for Estimating Peak Discharge and Flood Boundaries of Streams in Utah". USGS, Water Resources Investigations Report 83-4129.

Woolley, R. R., 1946. "Cloudburst Floods in Utah, 1850-1938". USGS, Water-Supply Paper 994.

Development of a Drought Warning System for Indiana

A.R. Rao[1] and T. Voeller[2], J.J. Hebenstreit[3]

Abstract

To manage water resources under scarcity conditions in Indiana, a set of drought indicators are proposed. Monthly precipitation, high temperature, low river levels, ground water levels and palmer hydrologic drought index (PHDI) series were analyzed to develop these indicators.

The data were standardized by subtracting the mean of each series from the observed value and by dividing the result by the standard deviation. The standardized data were averaged over different regions of the state. The standardized regional series were analyzed used to develop the drought indicators.

Three levels of drought severity indicators were established for all the series except temperature. These levels are called *drought watch, drought warning* and *drought emergency*. In general the 75% exceedence level is used for drought watch, 90% level for drought warning and 95% level for drought emergency. Because the exceedence level of temperature is inversely related to drought severity, a 25% exceedence level is used for a watch, 10% level for a warning and 5% level for an emergency. The drought warning system is being presently implemented in the state of Indiana by the Department of Natural Resources for managing instream use and for water quality maintenance.

Introduction

The drought severity is dependent on both water shortage and water use.

---

[1] School of Civil Engineering, Purdue University, W. Lafayette, IN 47907

[2] State of Montana, Helena, MT

[3] Division of Water, Indiana Department of Natural Resources, Indianapolis, IN 46204

It is obvious that if there is no need for water there is no question of shortage. To illustrate this relationship consider the effects of a three month rainfall deficit. Although stream flows may be seriously affected by this deficit, its duration or magnitude may not be sufficient to affect ground water levels. However, separate severity levels can be established to determine the level of drought severity for ground water levels.

Because the indication of drought using one time series does not necessarily indicate drought if another series is used, the following time series are used in this study: precipitation, high summer temperatures, low river flows, low ground water levels, low reservoir volumes and Palmer Hydrologic Drought Index (PHDI).

The National Oceanic and Atmospheric Administration has divided Indiana into 9 climatological regions, and PHDI values have been computed for these 9 regions for the period of record (1895-1988). Because Indiana is not climatologically homogeneous, the state is divided into three drought regions in consultation with the Indiana Department of Natural Resources.

The methods used to develop drought indicators in Kibler, et al. (1987) were used in the present study. The computation of drought indicators involves assembling the data bases well as regionalizing the monthly time series and establishing drought severity levels for each series. The procedures used to develop each of the time series which are applicable to a region are discussed below.

Assembling the Data Base

For most of the drought indicators, daily data are the basic data which are available. Daily values reflect short term and not the long term conditions needed for development of droughts. The daily time series are transformed into monthly values, and the drought indicators are based on monthly series.

Precipitation

Daily precipitation information was analyzed by examining the length of the historical records, the amount of missing records, and the gage location within the drought region. Data from seven stations in region 1, eight stations in region 2 and five stations in region 3 were used in the study. The daily precipitation time series is transformed into 3-, 6-, 9-, and 12-month duration series. Therefore, each precipitation time series is used to develop four separate drought indicators based on 3-, 6-, 9-, and 12-month rainfall series. An example of a 3-month duration rainfall for February, 1972 is the total precipitation for the months of December, 1971, January, 1972, and February, 1972.

Temperature

Daily high temperatures were available for the same stations that have daily precipitation data. Because monthly high temperature varies little within a region, data from only two stations are used in region 1 and from one gage each is used in regions 2 and 3. The daily high temperature time series are transformed into series of 1-, 2-, 3-, and 4-month durations. A two month temperature for July, 1986 is the average daily high temperature for June 1, 1986 through July 30, 1986. Similar to precipitation, the missing information is filled in by using temperature gages in the vicinity. The period of record used for all temperature data is 1950 -1988.

Stream Flow

Four criteria were used to select stream flow data used in the study: (1) the flows should not be very much affected by man, (2) the period of record should begin on or before 1958 so that 30 years of data are used in the statistical analysis, (3) the stream should represent a unique location and drainage basin within the region, and (4) data from intermittent streams were not used because they are obviously not useful for determining drought severity.

There are no rivers in the state of Indiana with basin areas which exceed 1500 square miles and which are not affected by man's activities. Consequently, although the data from these streams are affected by man's activities, a few of these large rivers are also included in the analysis. Large rivers with dams immediately upstream of the gaging site, however, are not included in the analysis. Data from four stations in region 1, 3 in region 2 and 1 in region 3 were used in the study. Apart from these, two stations, which were common to regions 2 and 3, were also used in the analysis.

Palmer Hydrologic Drought Index (PHDI)

The monthly PHDI data were used. The index is determined from precipitation, evapotranspiration, soil water recharge, runoff, and water loss from soil. Although PHDI records exist prior to 1931, the index was computed in a different manner prior to 1931 (Kibler, et al., 1987). Therefore, the period of record used for PHDI is 1931-1988.

Computation of Drought Indicators

Regionalization of Time Series

With the exception of PHDI, the time series established for each indicator is transformed into a dimensionless time series. PHDI data are not transformed because it is already a dimensionless index. Other series are transformed to eliminate the differences in means and standard deviations of these series. The differences are eliminated because only one averaged or regionalized times series is used for each indicator in a drought region.

Each of the time series, with the exception of PHDI, is transformed by using the monthly averages and standard deviations to normalize the data as shown in eq. 1

$$N_{i,k} = \frac{X_{i,k} - \bar{X}_k}{S_k} \tag{1}$$

where, i = Order in the times series, k = month associated with i, $N_{i,k}$ = normalized time series, $X_{i,k}$ = observed time series, $\bar{X}_k$ = average for month k, and S_k = standard deviation for month k.

Severity Levels of Droughts

Three levels of drought severity were established for each time series. These levels are called drought watch, drought warning, and drought emergency. The severity levels are established for each indicator using the regionalized time series. Exceedance levels using the regionalized times series were determined for monthly values. For all indicators with the exception of temperature, the 75% exceedance level is used for a drought watch, the 90% exceedance level is used for a drought warning, and the 95% exceedance level is used for a drought emergency. The drought watch exceedance level allows for a drought watch, warning, or emergency to occur 25% of the times, and a drought warning exceedance level allows for a drought warning or emergency to occur 10% of the time. Because temperature is inversely related to drought severity, a 25% exceedance level is used for a watch, a 10% exceedance level is used for a warning, and a 5% exceedance level is used for an emergency.

Exceedance curves are developed in a similar manner for all other months. After exceedance curves are developed, the drought thresholds for watches, warnings, and emergencies are computed for each of these months. An example of the variation in thresholds in shown in Figure 1 for river flow in region 3. The dashed line is the regionalized river flow boundary between a normal situation and a drought watch. Any regionalized river flow value that is below the solid boundary would produce a drought emergency response. The regionalized river flow of -0.427 which indicates a drought watch in August would not cause a drought response during any other month.

Summary

Ten indicators are used in each drought region. Drought indicators are developed for precipitation, temperature, river flow, and Palmer Hydrologic Drought Index (PHDI). Precipitation indicators are established for 3-, 6-, 9-, and 12-month durations and temperature indicators for 1-, 2-, 3-, and 4-month durations. After the data base are assembled, the times series are normalized. Then they are regionalized by averaging the time series. PHDI is not normalized before it is regionalized. Monthly exceedance levels are used in determining drought severity. These exceedance levels are the same as those used by Kibler et al. (1987). Other exceedance values are also presented for future adjustment to the severity levels. Further details of the study are available in Voeller et al. (1990).

References

Kibler, D.F., E.L. White, and G.L. Shaffer, "Investigation of the sensitivity, reliability and consistency of regional drought indicators in Pennsylvania", Tech. Rept., Dept. of Civil Engineering, Pennsylvania State University, University Park, PA, 132 pp. 1987.

Voeller, T.L., A.R. Rao, and J.W. Delleur, "Instream flow requirements and drought indicators for Indiana", Tech. Rept. No. CE-HSE-90-3, School of Civil Engineering, Purdue University, W. Lafayette, IN 47907, pp. 201, 1990.

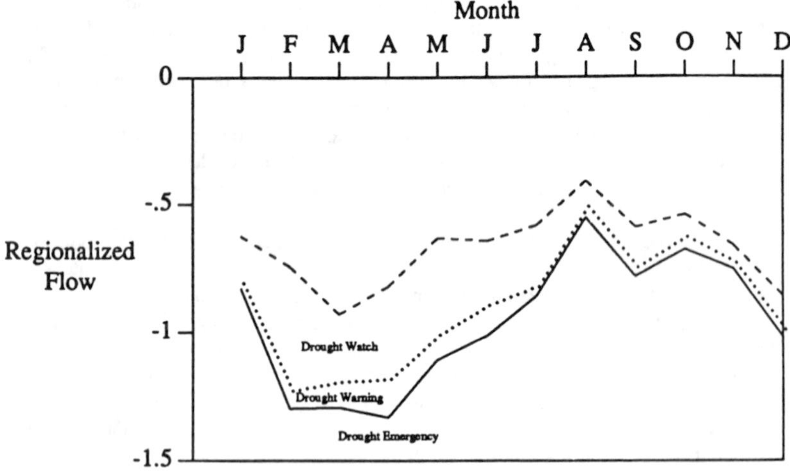

Figure 1. Drought indicator severity levels for river flow in region 3.

REAL TIME WARNING SYSTEMS FOR LOW WATER CROSSINGS

John L. Rutledge, P.E.[1]
Henry Bain, P.E.[2]
Curtis Beitel, E.I.T.[3]

ntroduction:

San Antonio is located at the transition of the Texas hill country and the Gulf Coast lains. As its name implies, the hill country consists of small watersheds with very steep slopes nd relatively sparse vegetative cover. In addition, the Balcones Escarpment located just north f San Antonio provides orographic lifting of moist gulf air masses, producing intense onvective thunderstorms. This combination of geography and climate produces runoff ydrographs which have very high peaks and short durations. Many roads which cross ntermittent streams are overtopped on a relatively frequent basis, threatening the safety of otorists in these areas. Effective and timely responses to the flash floods at these locations are ery difficult and costly. The City of San Antonio (City) has identified 286 low water crossings f various sizes throughout the city. There have been numerous rescues at these sites and ccasional fatalities. Since the cost of upgrading all of the roads to pass floodwaters is beyond he means of the City to perform, the City, funded through a 1987 bond issue, decided to install n site warning systems beginning with seventeen of the low water crossing sites.

In 1993, the City retained Bain Medina Bain of San Antonio and Freese and Nichols of ort Worth to design a real time warning system that could be adapted to the proposed seventeen ites as well as other future sites. Phase I of the project included identifying and analyzing the vailable alternatives, developing design criteria, researching existing systems, and ecommending a model warning system. Phase II included adapting the model warning system o fit each of the seventeen sites and producing final plans and specifications for construction of he warning systems. The final design was completed in 1994 and MICA Contractors was warded the contract.

---

[1] Principal, Freese and Nichols, Inc., 4055 International Plaza, Suite 200, Fort Worth, Texas 76109 (817) 735-7300.

[2] Principal, Bain Median Bain, Inc., 1026 Central Parkway South, San Antonio, Texas 78232 (210) 494-7223.

[3] Project Engineer, Freese and Nichols, Inc., 4055 International Plaza, Suite 200, Fort Worth, Texas (817) 735-7300.

Phase I - Development of a Model Warning System

The warning system at a low water crossing was planned to serve two major function to warn motorists when water is over the roadway, and to prevent those who do enter the hig water from being carried downstream. The focus of the system is on the warning functio because adequate warning should greatly reduce the number of motorists entering a floode crossing. The form of the warnings needs to be familiar to the motorists, in order to reduce t amount of recognition time required to understand the warning. This narrowed the range dow to existing traffic warning signs. The more significant warnings, such as "DO NOT ENTER or "ROAD CLOSED", would also need to be displayed only at an appropriate time, not all the time.

A variety of different warning systems were reviewed. The simplest form of warnir at the low water crossing is the flood gauge sign. A typical flood gauge is a scaled sign five fe in height, located at the crossing which allows motorists to directly read the depth of water ove the road. They need to be highly visible, even at night, and be clearly marked with units feet. Flood gauge signs have already been installed at each of the seventeen crossings in th study. Due to the large depths of flow at the crossings, sometimes more than 20 feet, multip flood gauge signs in five foot increments are to be installed to allow the motorists to read dept at least up to the 25 Year flood level.

Warning can also be provided with a yellow caution sign. These signs are the standar 36" x 36" yellow diamond sign which reads "WATCH FOR WATER ON ROAD", as specifie in the Manual on Uniform Traffic Control Devices. The signs utilize a high intensity sheetin surface treatment to make them highly visible at night. Additional warning is conveyed throug the use of flashing lights on the warning sign, triggered by a sensor in the stream channel.

The highest level of warning reviewed was the inclusion of automatic barricades acros the roadway, also triggered by high water levels. The barricades were planned to be withi sight of the water on the road, to reinforce the purpose of the barricade in the driver's min There are several types of automated gates available, including a swing gate on rollers, a slidin gate on a track, or a railroad crossing gate arm. The railroad gate arm is the least likely to b blocked by debris in the roadway and appeared to be the most dependable. The bases of th gate arms were located so that the mechanism would be flooded infrequently, and they wer sealed to be watertight for additional protection. The bases were located a minimum of four fe from the roadway for safety purposes. The length of the gate arms was sized to block a oncoming traffic lanes at each location, either one or two lanes.

There are several existing suppliers of warning devices in the San Antonio area, but non can supply an entire system specifically designed for low water crossings. Any system woul have to be a combination of components from several suppliers. Research of existing systems which the city used automated warning systems instead of manually placing barricades, was erformed. This allowed for the identification of the possible devices for each component of e Model Warning System, and the selection of the device which best satisfied the design riteria.

With any automated system, a method of control must be established. With a warning ystem that is automatically controlled at the site, there is no assurance to the City that the arning system is working properly during a storm. A purely remote controlled system would equire a person to monitor the system constantly to determine when to activate the warning evices. A combination of the two allows the system to be automatic at the site, with the ability monitor the system and override the on-site controls from the base station, if necessary. The east expensive connection for urban areas is the use of UHF/VHF radio telemetry, such as an ndustrial SCADA system.

Many telemetry systems are available which can monitor the sensors at the low water rossings, including industrial SCADA systems and ALERT systems. The ALERT telemetry ystem was selected for the Model Warning System because there is an existing ALERT network f rainfall gauges which covers the entire city. No additional frequency permits would be equired from the Federal Communications Commission for the low water crossing control ansmissions, which will shorten the construction time for the system. The ALERT software ncludes a wide range of warning functions, to activate alarms when set conditions in the atabase are exceeded. It uses advanced database management techniques to screen incoming ata for errors, and provides a graphic display of the database, along with report generation. he ALERT software also has other features such as remote dial-in access, Local Area etworks, paging warning systems, and control capabilities. The system was designed to be imple to operate, allowing the operator to monitor the numerous crossings quickly and ccurately. Graphic display of critical information helps the operator prioritize the crossings, nd quickly inspect the water level and status of the devices at each crossing.

For the telemetry system, a Remote Terminal Unit (RTU) is needed to send and receive he data at each crossing. The RTU contains a radio to transmit the sensor data, along with ircuitry to convert the input sensor signals to the radio transmission format. For a two way onnection, the RTU also contains a receiver to receive the incoming radio signals. Most RTUs ontain data loggers which store the data in case the transmission does not reach the base station.

The RTU selected for the Model Warning System is an intelligent RTU, which also ontains a microprocessor which can be programmed with control functions. The RTU is rogrammed to activate all needed warning devices when the water level sensors indicate that vater is over the road. It sends the sensor data to the base station, along with the status of the varning devices at the crossing. If the operator at the base station decides to override the on ite control commands, based on all of the information available including rain gauges in the rea, the RTU receives and executes the new control command. The RTU at the crossing is lesigned to have sufficient capacity to provide the control functions to operate all of the warning levices at that site.

There are two types of water level sensors which can be used to determine the level water at the crossing, binary and continuous sensors. Binary sensors indicate whether water present at the level at which they are mounted and provide an on/off signal. Continuous senso measure the water level at constant time intervals and produce a 4 to 20 mAmp signal which proportional to the relative height of the water. In order to reduce maintenance, it is desirab for both the continuous and binary water level sensors to be as free from clogging as possibl The binary sensors are mounted in a standpipe so that their trigger point is at the warning lev they measure. This makes their signal as reliable as possible, to prevent unwanted activation the warning functions. They also require minimal signal processing. The sensors chosen fo the system require minimal maintenance, such as cleaning, calibration, or replacement. Multip binary water level sensors were used to provide a low cost redundancy, since they ar considerably less expensive than the continuous sensors.

There are numerous binary sensors available, many of which are designed for use i storage tanks and controlled environments. The binary sensors selected for the Model Warnin System are tilt-type mercury switches, because they are fully enclosed to prevent clogging an require very little signal processing. The binary sensors are mounted in a standpipe to prote them from debris and vandalism. The continuous water level sensor selected was an ultrason sensor which is mounted above a pad. The City currently uses this sensor in a separat stormwater tunnel project to monitor open channel flow and requested its use for this project

The warning water level at which the system is to be activated was also determined. Fo complex systems, caution level devices are activated when water first overtops the road and th critical level devices are activated when water becomes 6 inches or deeper. For locations wit only one warning level, the devices are activated at the critical level. These levels were set i order to avoid frequent operation of the system, which may cause drivers to ignore the warnings

A railroad type gate arm was selected as the primary warning component at the crossin for the Model Warning System. This device is very familiar to drivers, and has been shown t provide reliable operation with very little maintenance. To provide additional warning, cautio signs with flashing lights were placed at the nearest intersection on either side of the crossing to allow drivers to conveniently select an alternate route. The control signals for these signs ca be provided through either a hard-wired or telemetry connection, whichever is less expensive

Power for the warning system is available at each of the crossings. In order to provid uninterrupted power to the system during storm events, the power supply selected for the systen is a bank of batteries. The batteries are kept fresh by a trickle charger connected to an AC meter loop. The size of the bank of batteries is designed to provide power for 24 hours in th event of a power outage.

A small paved space was provided in front of the gate arms to allow motorists to tur around. This space was located so that it required a minimum amount of additional right-of way. In addition, a barrier was located downstream of each crossing to prevent a car whic entered the flooded crossing from being swept downstream. The barrier was constructed of stee

ble between I-beams in order to minimize the amount of debris caught during a flood event.

Once components were selected for the Model Warning System, a preliminary cost timate was prepared based on 17 systems, one at each crossing.

daptation of the Model Warning System to Each Site

Each site was analyzed to determine the locations of the gate arms, warning signs, and ater level sensors to provide the Model Warning System at that site. The 10 Year, 25 Year, d 100 Year flood levels were determined from FEMA floodplain maps at each location. The se of the gate arms was located above the 25 Year flood level wherever possible. epresentatives of the City Department of Public Works were also consulted to help determine e appropriate configuration for each location.

The site adaptation involved placing more than two gate arms at several of the locations. ne location did not require gate arms at all, because it presented more danger to pedestrians an motorists. The number of RTUs at each site was also determined, with additional RTUs ed to transport the control signals to remote caution signs at intersections. Once the number warning devices at each location was known, the number of inputs and outputs of the RTU uld be specified.

Once the design was complete, detailed cost estimates were produced for each location. ecifications for each component of the system were prepared, and the plans and specifications ere reviewed by the City. The gate arm component of the warning system was removed before ds were opened. Once the City's comments were incorporated, bid documents were prepared d the project was put out for competitive bids. MICA Contractors was awarded the contract.

ummary

These warning systems should provide a specific and timely warning to motorists to not nter a flooded crossing, protecting both the public and the City's interests. Some motorists are oing to enter the flooded crossing no matter what warning they receive, but the flashing lights ould attract their attention. The original design included the automatically controlled gate ms, but these were removed from the final construction package.

The ALERT telemetry system will allow City personnel to check the status of the arnings at each of the crossings, and allow them to issue overriding control commands from e base station. The system is designed with a graphic user interface to allow quick and curate inspection of the water level and status of the devices at each crossing. The system is so expandable, allowing additional sites to be included at a later date.

INVESTIGATION OF POTENTIAL FLASH FLOODS IN THE SCIOTO RIVER BASIN

Tiao J. Chang[1], M. ASCE and Hong Y. Sun[2]

ABSTRACT

For a small rural community situated on a flood plain, expensive flood protective structures are not economically justifiable. A community warning system by the use of existing precipitation gaging stations could be an alternative. This study investigated the regional distribution of potential flash floods based on several existing precipitation gages in the studied area. Daily truncation levels of 99.9- and 99.99-percent were used to derive potential flash floods, where a 99.9-percent level means that 0.1% of historically recorded daily precipitation are larger than or equal to the truncated value. The higher the truncation level, the greater the magnitude of a potential flash flood. Daily records from twenty-one precipitation gaging stations in the studied basin were selected for the study. The kriging method based on the minimum unbiased estimation was used to estimate spatial interpolation for ungaged areas. These estimations form a regional distribution of potential flash floods, which are expressed by contour lines at varied levels. The result shows that the estimation errors associated with the regionalization were reasonably small.

INTRODUCTION

Fatalities and damage in the 1990 Shadyside flash flood in Ohio of the United States raised a serious

[1]Associate Professor, Civil Engineering Department, Ohio University, Athens, OH 45701-2979

[2]Engineer, R.D. Zande Associates Inc., 1237 Dublin Road, Columbus, Ohio 43215

question of flood protection for an ungaged watershed. The Shadyside area, lying on two tributaries of the Ohio River, Wegee and Pipe Creeks, has no rainfall gages. The only rainfall estimate in the watershed for this particular event came from a resident of upper Wegee Creek where daily accumulation in a child's wading pool was about 100 to 130 mm. The county sheriff had an unofficial report of 70 mm rainfall in a hour in the town 30 kilometers northwest of the Shadyside (NOAA, 1991). While neither of these estimates were reliable, 26 fatalities and enormous property damage in the region are certainly known.

Small rural communities that are not economically justified for expensive flood protective structures are extremely vulnerable to floods. An early warning system for such communities may reduce possible flood damage. BY on the method of truncation level, this study used time series of precipitation to derive potential flash floods at a gaging location. Then, the kriging method was applied to obtain regional distributions based on estimated potential flash floods at known locations.

The method of truncation level for deriving potential flash floods has been used for the construction of hydrologic floods and droughts (Chang, 1987, 1990; Chang and Kleopa, 1991). The kriging method for the estimation of a regional distribution was successfully performed by Delhomme (1978) and Karlinger and Skrivan (1980) in the study of spatial precipitation.

POTENTIAL FLASH FLOODS

Based on partial duration series, potential flash floods are derived from time series of precipitation. By applying a specified truncation level to a time series of daily precipitation, the sequence of consecutive days whose precipitations exceed the specified truncation level is defined as a potential flash flood. To obtain truncation levels, daily precipitation are sorted in a descending order. Then, a Y-percent truncation level is the value that corresponds to the ith rank of the sorted data with a total record length N, where the relationship between i^{th} rank and Y-percent truncation level can be expressed as:

$$i=\frac{(100-Y)}{100}N \qquad (1)$$

Consequently, at the Y-percent truncation level, Y% of the

historic data are less than this truncation level. A potential flood occurs whenever precipitation are greater than the specified truncation level.

For the purpose of comparison, potential flash floods were also defined based on annual duration series. Extreme storm events of 50-year and 100-year return periods were estimated by the Gumbel Method. Magnitudes of these storms served as truncation levels for deriving potential flash floods. The following equation was used to estimate such truncation levels:

$$X=X_m+(0.7797Y-0.45)\sigma_x, \tag{2}$$

where X_m is the mean of annual potential flood duration series and σ_x is its standard deviation; Y is the reduced variate, which is a function of return period by the following relation:

$$Y=-\ln[-\ln(1-P)], \tag{3}$$

where P is the probability of a given value being equaled or exceeded; P = 1/T, T is the return period; ln is the Napierian logarithm.

FLOOD REGIONALIZATION

To obtain regional distribution of potential flash floods, the kriging method that is based on the linear minimum variance unbiased estimation was applied to derived potential floods at gaging stations in the studied region (Kitanidis, 1983). Twenty-one precipitation gaging stations were selected from the Scioto River Basin, a tributary of the Ohio River in the United States, for this study. Truncation levels for deriving potential flash floods, i.e., values of 99.9- and 99.99-percent, and magnitudes of 50-year and 100-year returned periods, were estimated based on methods described in Equations 1 to 3. Table 1 lists magnitudes of these potential flash floods in mm along with their monthly means and record lengths. It can be seen that magnitudes of potential floods increase as their corresponding levels of truncation increase. This implies an increase of severity of potential flash flood. Based on values from twenty-one gaging stations at each severity level, the kriging method was applied to obtain a regional distribution to determine the regional threat of potential flash floods. Figures 1 and 2 show regional distributions of 99.99-percent and 100-year potential flash floods expressed by contour lines of daily storm in mm, respectively. The example of estimation errors associated with the regionalization of 99.99-percent potential flash floods in Figure 3 shows that they are less than 10%.

CONCLUSION

Based on both partial duration series and annual duration series, potential flash floods were derived from time series of daily precipitation by the method of truncation level. The use of magnitude of precipitation is to reflect the severity of potential flash flood since severe storms result in flash floods in the studied region. Table 1 shows that daily magnitudes of storms for the derivation of potential flash floods are mostly greater than their monthly mean values. The severity level of potential flash flood increases as its truncation level increases, i.e. an increase of potential threat of flash flood.

The severity level of potential flash flood obtained from historic records at a gaging station is a localized measure since the daily precipitation is observed only at a certain location. Based on magnitudes of estimated potential flash floods at 21 selected gaging stations, a regionalization was made by the use of kriging method. This results in a regional distribution of potential flash floods for ungaged areas. Errors associated with the kriging estimations are relatively small. This increases the confidence for using such a regional distribution to determine the potential threat of flash floods for an early warning system in ungaged areas.

REFERENCES

Chang, T. J., 1987. "Analysis and Simulation of Three-Component Floods in the Ohio River Basin," Hydrologic Frequency Modeling, edited by V.P. Singh, D. Reidel Publishing Co., pp.583-594.

Chang, T. J., 1990. "Effects of Drought on Streamflow Characteristics," Journal of Irrigation and Drainage Engineering, ASCE, Vol.116, No.3, pp.332-341.

Chang, T. J. and X. Kleopa, 1991. "A Proposed Method for Drought Monitoring, "Water Resources Bulletin, AWRA, Vol. 27, No. 2, pp.275-281.

Delhomme, J.P., 1978. "Kriging in Hydrosciences", Advances in Water Resources, Vol.1, No.5, pp.251-266.

Karlinger, M.R., and Skrivan J.A. 1980. "Kriging Analysis of Mean Annual Precipitation, Power River Basin, Montana and Wyoming," U.S. Geological Survey Report 80-50, Reston, 25p.

Kitanidis, P.K., 1983. "Statistical Estimation of Polynomial Generalized Covariance Functions and Hydrological Applications," Water Resources Research, Vol.19, No.4, pp.909-921.

NOAA, 1991. Natural Disaster Survey Report: Shadyside, Ohio, Flash Flood June 14, 1990, National Weather Service, National Oceanic and Atmospheric Administration, Department of Commerce, Silver Spring, Maryland, 125p.

Table 1. Gaging Stations for Flash Floods in the Scioto River Basin

No.	Index No.	Record Length	Monthly Mean (mm)	Potential Flash Floods (daily values in mm)			
				99.9%	99.99%	50-yr	100-yr
1	152809	1973-86	81	52	103	93	102
2	159205	1951-86	80	61	111	120	133
3	178305	1949-86	80	64	116	117	130
4	178605	1949-86	79	58	86	101	111
5	211901	1936-86	78	62	99	107	117
6	375808	1940-86	91	69	119	129	143
7	400409	1956-86	87	65	75	95	103
8	418904	1973-86	76	56	94	104	115
9	468105	1936-86	82	66	90	111	121
10	494205	1951-86	74	58	97	103	114
11	497905	1936-86	77	61	87	102	112
12	678109	1936-86	87	73	106	162	181
13	879405	1936-86	82	62	104	118	131
14	212405	1950-86	76	62	79	93	102
15	302102	1953-86	81	71	107	119	131
16	440905	1949-86	97	60	85	103	113
17	553505	1953-86	80	60	93	104	114
18	663009	1949-86	81	67	102	113	125
19	678609	1957-86	88	71	92	130	144
20	686105	1957-86	79	58	90	100	111
21	753805	1949-86	83	71	96	117	129

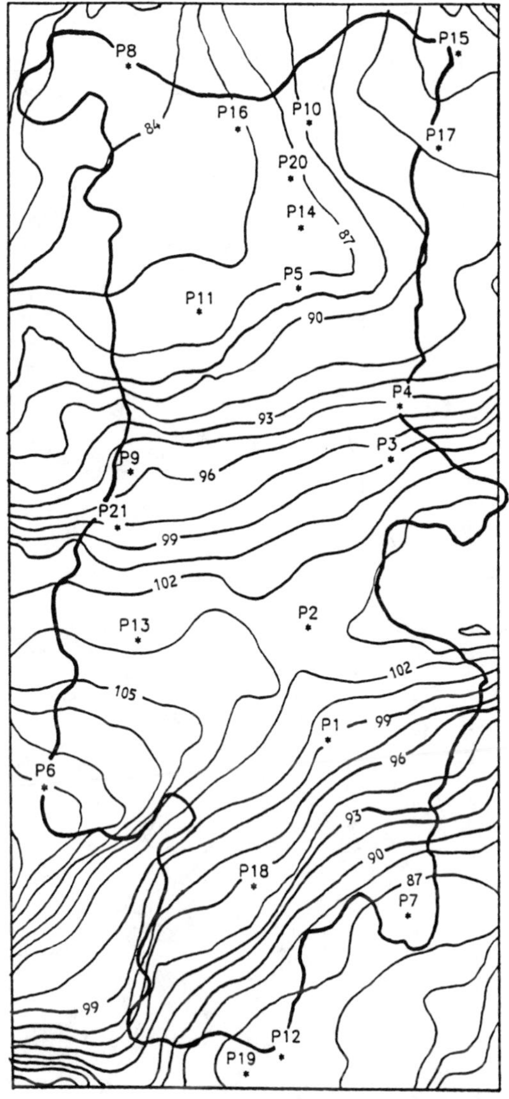

Figure 1. Regional distribution of potential flash floods derived by 99.99% truncation level in the Scioto River Basin (mm)

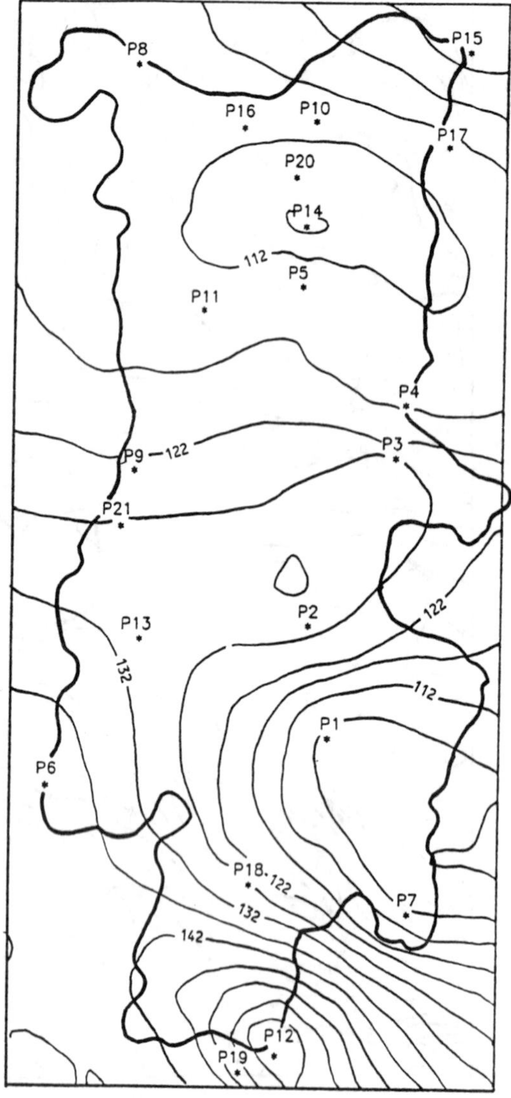

Figure 2. Regional distribution of potential flash floods derived by 100-year truncation level in the Scioto River Basin (mm)

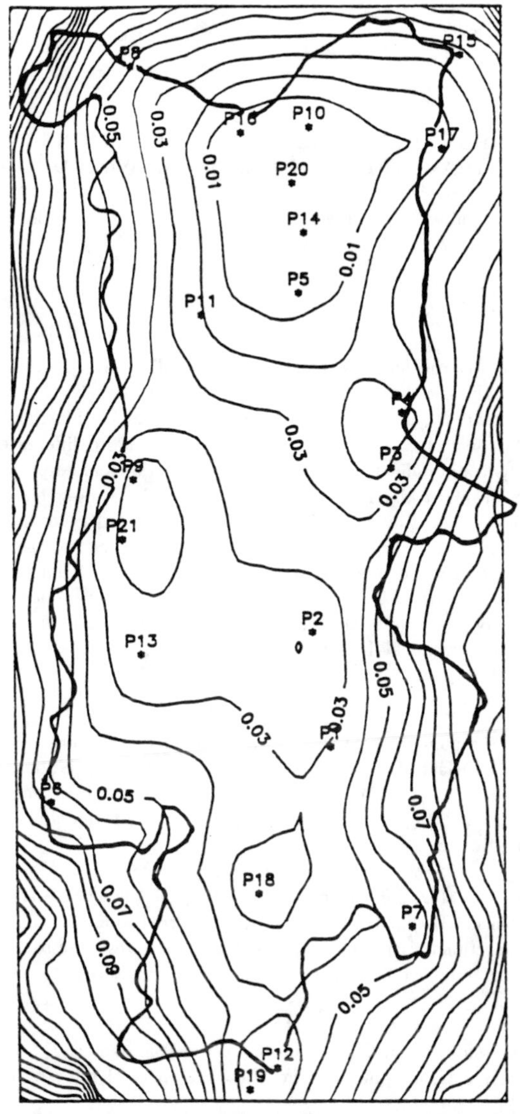

Figure 3. Kriging estimate errors associated with regional flash floods derived by 99.99% truncation level in the Scioto River Basin (mm)

REDUCING NUMERICAL OSCILLATIONS IN RAPIDLY–VARIED UNSTEADY FLOW MODELING

Chintu Lai,[1] M. ASCE, Chin-Lien Yen,[2] F. ASCE, and Wen-Cheng Lee[3]

Abstract

With a focus on the findings made in the preceding investigations on numerical modeling of rapidly-varied unsteady open-channel flows, namely, the findings that the nonprismatic factor, P_a, is most sensitive to numerical oscillations, and that, under a setup where the effect of P_a is highlighted, two terms, the finite-differenced depth gradient and velocity-head gradient, are most prone to inception and amplification of the oscillations. Introducing to these two terms a fully-forwarded time-weighted second-order difference term multiplied by a suitable smoothing factor, has each shown a noticeable improvement in reducing oscillations. Furthermore, it has been revealed that the model can be improved best if the above technique is applied to both the depth-gradient term and the velocity-head-gradient term combined; next best if it is applied to the first term alone; and third if it is done to the second term alone. This technique has been put to test using field data from the Hsintien Creek, Taiwan.

Introduction

In the previous numerical investigations on rapidly-varied unsteady open-channel flow modeling, a special attention is placed on numerical oscillations, also known as spurious oscillations or parasitic waves, i.e., to look into factors causing such oscillations, parameters with high sensitivity to oscillations, and terms prone to inception or growth of oscillation.

Through a series of numerical experiments using five key parameters (all dimensionless), it has been found that the parameter, P_a, referred to as the nonprismatic factor reflecting cross-sectional variation, has the most sensitive response to numerical oscillation. As a distant second comes a B. C. number pertaining to the downstream-boundary depth variation.

Having successfully identified the most sensitive parameter, P_a, another series of numerical experiments, designed for such that the effect of the parameter is empha-

[1]Visiting Professor, Dept. of Civil Engrg. and Hydraulic Research Laboratory, National Taiwan University, Taipei, Taiwan.

[2]Professor, Dept. of Civil Engrg. and Hydraulic Research Laboratory, National Taiwan University, Taipei, Taiwan.

[3]Graduate student and Research assistant, Dept. of Civil Engrg. and Hydraulic Research Laboratory, National Taiwan University, Taipei, Taiwan.

sized, have been conducted with a focus on the disturbance inception and amplification. These experiments have revealed that two finite difference forms in the IFDM model, corresponding to the longitudinal flow-depth gradient term, $\frac{\partial h}{\partial x}$, and the velocity-head gradient term, $\frac{\partial}{\partial x}(\frac{u^2}{2g})$, are most prone to numerical oscillation.

The main objective of the present study is to devise some numerical techniques to reduce the troublesome numerical oscillations, by focusing on the findings made in the preceding modeling investigations of rapidly-varied unsteady channel flows, which are just outlined in the preceding paragraphs.

Methods for Reducing Numerical Oscillations

Numerical Viscosity: One of the most classical methods is introduction of sufficient artificial flow-resistance or numerical viscosity to damp out rapidly-varying flows or sharply-changing profiles. The idea originates from von Neumann and has long been widely used among numerical modelers. [e.g., see Roache (1972)] However, the modeler is cautioned not to use the technique indiscreminately, as an unrealistic simulation may evolve.

Weighted Higher Order Derivatives: As stated in the Introduction, the nonprismatic factor, P_a is found to be most sensitive to numerical oscillations. A rectangular channel with variable constriction was designed for the next step of numerical experiments. With rectangular cross section, nonprismatic factor reduces to nonconstant channel width factor, $\Delta B_r = (\Delta B/\Delta x)L_o/B$, where B = top width and L_o = a characteristic length. The IFDM model was then selected to scrutinize what term or terms in the governing equations to be most responsible for inception or growth of numerical oscillations. The governing equations used in this model are:

$$\frac{\partial A}{\partial t} + \frac{\partial Q}{\partial x} = 0 \tag{1}$$

$$\frac{1}{gA}\frac{\partial Q}{\partial t} + \frac{1}{2gA^2}\frac{\partial Q^2}{\partial x} + \frac{1}{2g}\frac{\partial}{\partial x}(\frac{Q}{A})^2 + \frac{\partial h}{\partial x} = S_o - S_f \tag{2}$$

in which A = cross-sectional area of the flow; Q = discharge; h = flow depth; S_o = bed slope; and S_f = friction slope. Two terms $\frac{\partial h}{\partial x}$ and $\frac{\partial}{\partial x}(\frac{u^2}{2g})$, where $u = Q/A$, in their discretized forms were identified to be most apt to generate numerical perturbation, or most prone to numerical oscillation. This finding closely approximates what was mentioned in Henderson (1966).

From the foregoing findings, it is evident that in order to reduce numerical oscillations in the IFDM model, one should begin with improvement of the finite difference terms corresponding to $\frac{\partial h}{\partial x}$ and $\frac{\partial}{\partial x}\frac{(Q/A)^2}{2g}$ in the set (1) and (2). The method and procedures toward this end are described in the following paragraphs.

(a) Improvement on the flow-depth gradient $\frac{\partial h}{\partial x}$:

A weighted second-order finite difference term of $\frac{\partial h}{\partial x}$ was added to the original linear expression:

$$W_1 \left[h_{j+1}^{n+1} - 2h_j^{n+1} + h_{j-1}^{n+1} \right] \quad (3)$$

in which W_1 is a weighting factor, or more specifically, the depth-gradient-term smoothing factor. The second-order expression is to increase the accuracy of this term so as to more accurately respond to the rapidly changing depth-gradient, and the smoothing factor is to damp out its tendency to numerical oscillation.

Figure 2(a) and (b) show the relationships between the depth-gradient term smoothing factor, W_1, and the maximum depth-oscillation, $(O_s)_{max}$, for subcritical and supercritical flows, respectively. It can be readily observed from the plottings that in subcritical flow the $(O_s)_{max}$ decreases as W_1 increases, and in supercritical flow the $(O_s)_{max}$ decreases as W_1 decreases.

(b) Improvement on the velocity-head gradient $\frac{\partial}{\partial x}\frac{(Q/A)^2}{2g}$:

In a similar way as done in (a), a weighted second-order finite difference term of $\frac{\partial}{\partial x}\frac{(Q/A)^2}{2g}$ was added to the original linear expression:

$$W_2 \left[\left(\frac{Q^2}{A^2}\right)_{j+1}^{n+1} - 2\left(\frac{Q^2}{A^2}\right)_{j}^{n+1} + \left(\frac{Q^2}{A^2}\right)_{j-1}^{n+1} \right] \quad (4)$$

in which W_2 is a velocity-head-gradient smoothing factor.

As with the depth-gradient term, the same set of geometrical data (Fig. 1) was used to evaluate $W_2 - (O_s)_{max}$ relationships. Two sample plots are given in Figs. 3(a) and (b), one depicting a subcritical flow case and the other portraying a supercritical flow case. A trend opposite to the preceding case (i.e. the depth-gradient case) is observed from these plots; that is, in subcritical flow the $(O_s)_{max}$ decreases as W_2 decreases, and in supercritical flow the $(O_s)_{max}$ decreases as W_2 increases.

(c) Improvement on the combined effects:

The optimum value for the smoothing factor is such that it can supress the largest depth oscillation, $(O_s)_{max}$, without causing excessive numerical dispersion. Analysis of the above two sets of numerical experiments indicate that the depth-gradient smoothing factor W_1 has a wider range of smoothing effects than that of the velocity-head-gradient smoothing factor W_2. Hence, in carrying out the model improvement in terms of reducing numerical oscillations, first a best W_1 value is selected under various ΔB_r values, followed by the adjustment of the W_2 value. The proceduce is repeated until satisfactory results are obtained.

Summerizing the foregoing experimental results and findings, it can be concluded that in the numerical solution of unsteady-flow equation set, addition of second-order finite-difference expressions for the flow-depth gradient term and the velocity-head gradient term, each weighted by appropriate smoothing factors, W_1 and W_2, respectively, serves to reduce numerical oscillations or impede disturbance propagation. The model can be improved best by combined modification of both the depth-gradient term and velocity-head-gradient term; modification of the first term alone is the next best, followed by modification of the second term alone.

The above numerical technique has been put to test using the field data of the Hsintien Creek in Taiwan. The new model appears capable of simulating flows in steep channels often encountered in mountainuous zones.

Other General Stability Improvements: A model will be less prone to numerical oscillations when it is well behaved, and well under standard stability limits. Hence, any improvements in standard stability criteria are conducive to reducing chance of oscillation occurrence. However, it is to be noted that suprious oscillations or parasitic waves are different from numerical instability resulting from violating standard stability criteria. Therefore, a set of well designed numerical experiments under the concerned criterion should be conducted to find the best remedy for the oscillation, similar to the foregoing examples. These are out of the present study scope and will not be discussed further.

Summary and Conclusions

Focusing on the findings made in the preceding investigations on numerical modeling of rapidly-varied unsteady open-channel flows, some numerical studies to reduce numerical oscillations have been made. Introduction of a fully-forwarded time-weighted second-order difference term multiplied by a smoothing factor to the depth-gradient term and to the velocity-head-gradient term, each has shown a noticeable improvement in reducing oscillations. Furthermore, it has been revealed that the model can be improved best if the afore-mentioned technique is applied to both the depth-gradient term and the velocity-head-gradient term combined; the next best if applied to the first term alone; which is followed by application to the second term alone. This technique has been put to test using the actual flow data from the Hsintien Creek, Taiwan.

Acknowledgement: Support for this research by the National Science Council, Taiwan, ROC, under grant NSC 82-0414-P-002-015-B, is gratefully acknowledged. The computing facilities provided by the Hydraulic Research Laboratory for the this research are highly appreciated.

References

Henderson, D. A., (1966). "Open Channel Flow," Macmillam Co., N. Y. Collier Macmillam Limited, Londen.

Rouche, P. J., (1972). "Computational Fluid Dynamics," Hermosa Publishers, Albuquerque, New Maxico.

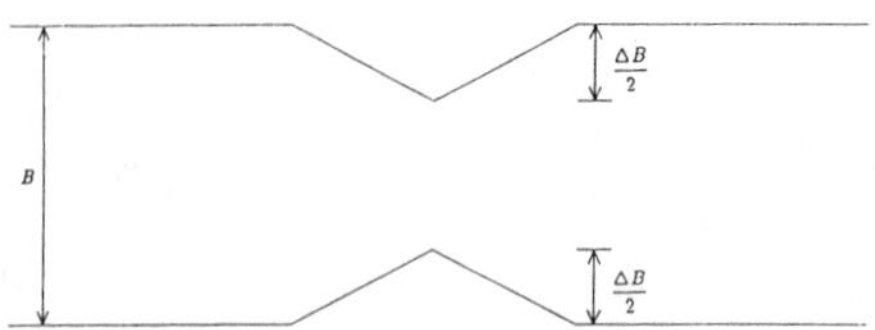

Figure 1. A Rectangular Channel with Variable Constriction, Used for Invesigation of Numerica Oscillations

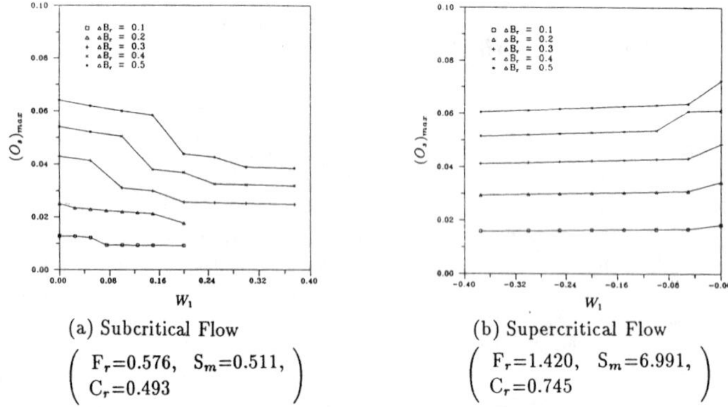

(a) Subcritical Flow
$\left(F_r=0.576, \; S_m=0.511, \; C_r=0.493 \right)$

(b) Supercritical Flow
$\left(F_r=1.420, \; S_m=6.991, \; C_r=0.745 \right)$

Figure 2. The Maximum Flow-depth Oscillation vs Depth-Gradient Smoothing Factor; Under Various Values of Channel Constriction (ΔB_r)

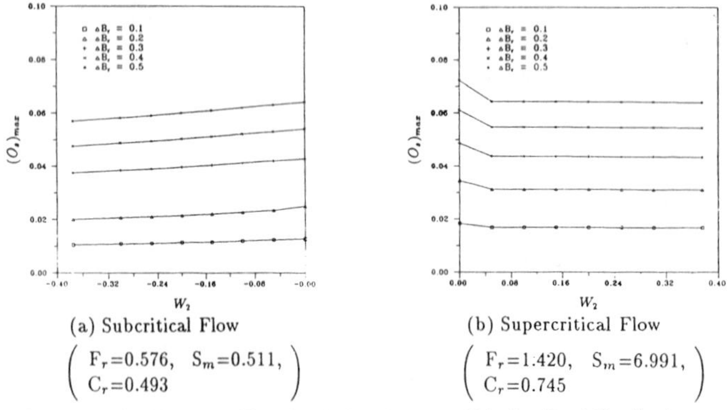

(a) Subcritical Flow
$\left(F_r=0.576, \; S_m=0.511, \; C_r=0.493 \right)$

(b) Supercritical Flow
$\left(F_r=1.420, \; S_m=6.991, \; C_r=0.745 \right)$

Figure 3. The Maximum Flow-depth Oscillation vs Velocity-Head-Gradient Smoothing Factor; Under Various Values of Channel Constriction (ΔB_r)

THEORETICAL AND NUMERICAL ASPECTS OF STEEP-CHANNEL FLOW MODELING

By Task Committee on Theory and Numerical Modeling of Flow in Steep Channels[1]
Chairman: Chintu Lai,[2] M. ASCE,

ABSTRACT: Numerical modeling of steep channel flows is an extremely difficult subject and a rather challenging research field. Not only numerical stability has to be satisfied, but hydrodynamic stability needs to be carefully examined as well. Even the validity of the governing flow equations is sometimes questioned. The pressing hydraulic, hydrologic and environmental problems continue to increase the importance of the riverine flow modeling, which would demand the subject area, largely neglected in the past, to be adequately investigated without further delay.

An ASCE task committee entitled Task Committee (TC) on Theory and Numerical Modeling of Flow in Steep Channels was organized with the objectives of conducting literature survey concerning hydrodynamic, numerical and general aspects of steep-channel flow; critical review of steep-channel flow equations, with the aid of several key parameters for their systematic and uniform analyses; study of hydrodynamic and numerical stability behavior of the equation or model system, with the final goal of preparing some guidelines for "steep-channel flow investigations through numerical modeling." The technical areas in which the TC conducts its investigation tasks should include: (a) steep-channel flow, (b) high-velocity flow, (c) rapidly-varied flow, and (d) flows in related areas. The present paper is prepared for brief statement of the TC objectives and functions, and as an "intermediate report," summarizing generally the first half of the TC activities and accomplishments.

Introduction

The past modeling activities for stream and riverine flows show a tendency for the activities to arise according to the population distribution rather than to the geographical distribution of the fluvial systems. Inspection of topographical maps reveals that mountainous zones occupy larger portions of the global land mass, which in the hydraulics context implies that hill-slope, high-gradient channel flows are very common "occurences" on the earth. However, the real reason for the numerical modeling of steep channel flows not gaining much interest in the past appears to be in the great

[1]Parent Committee: ASCE Hydraulic Div., Technical Committee on Computational Hydraulics

[2]Visiting Professor, Department of Civil Engineering, National Taiwan University, Taipei, Taiwan.

difficulty attached to the treatment of the subject flow modeling, thus making research into this field rather difficult. Unlike in many other open-channel flow modelings, where numerical stability is a chief concern of model developers, in steep-channel flow modeling, hydrodynamic stability has to be investigated as well. Further complicating the situation, even the validity of the governing flow equations is sometimes questioned. For many hydraulic engineers, the need to study both hydrodynamic and numerical aspects of steep-channel flow modeling is not altogether clear; worse yet, for a number of numerical modelers, the distinction between the two kinds of stability criteria – hydrodynamic and numerical – is often not explicitly made in their modeling efforts. (cf. Liggett 1975)

Because of the aforementioned various factors reflecting natural, human, social and technical attributes or backgrounds, the modeling records or publications of the steep-channel flow or related flows such as high-velocity or rapidly-varied unsteady flows are rather scanty, and for those that were published they could differ greatly in concepts, hypotheses, and approaches, to the extent that the same things are addressed as different and different things are confused as same. The pressing hydraulic, hydrologic and environmental problems continue to increase the importance of the riverine flow modeling, which would demand the subject area, largely neglected in the past, to be adequately explored. It is fair to say that the modeling study of steep-channel flows is a critical task whose time has come, or is overdue.

In view of the aforementioned background, this task committee called Task Committee (TC) on Theory and Numerical Modeling of Flow in Steep Channels was organized with the objectives of conducting literature survey concerning hydrodynamic, numerical and general aspects of steep-channel flows; critical review of steep-channel flow equations, with the aid of several key parameters or "numbers" for their systematic and uniform analyses, study of hydrodynamic and numerical stability behavior of the equation or model system, with the final goal of preparing some guidelines for "steep-channel flow investigations through numerical modeling." The technical areas in which the TC conducts its literature survey, formulates modeling concepts, develops investigation procedures, and identifies research needs, should include: (a) steep-channel flow, (b) high-velocity flow, (c) rapidly-varied flow, and (d) flows in related areas.

The work schedule of the TC was laid out, which consisted of several 3∼4 month work units, starting from the literature survey to the final TC report that would take a form of a journal article, and also include two TC meetings and an "intermediate TC report." The present paper is composed for this "intermediate report," summarizing generally the TC activities and accomplishments up to the end of 1994, in addition to the foregoing general statements of the TC. The paper is also planned for an oral presentation at the 1995 ASCE International Conference on Water Resources Engineering.

Literature Survey

As evident from the Introduction, references and publications can be classified into three categories — hydrodynamic, numerical, and general.

The hydrodynamic category concerns the theoretical aspect of steep channel flow, with hydrodynamic stability of the flow in particular. Validity of the governing flow equations should be another important and interesting subject to be discussed in this category.

The numerical category involves numerical modeling and simulation of steep channel flows, with all necessary or supporting knowledge of numerical mathematics, numerical analysis, numerical techniques, and other numerical aspects.

The general category includes articles that deal with both hydrodynamic and numerical aspects, that deal with neither, or that just concern steep-channel flows in general aspects. This category can have a broader or narrower scope; the latter pertains only to the theoretical and numerical aspects of steep-channel flows, which is essentially the scope of this TC, whereas the former, in addition to these areas, extends to include experimental, empirical, practical, and hydraulic engineering aspects of steep-channel flows. Because the TC on "Applied Hydraulic Analyses of Flow in Mountain Rivers," a sister task committee concurrently set up by the ASCE, complements much of the above extended areas, the broader scope then amounts to a comprehensive aspect of steep-channel flow, resulting from combining study areas of both TCs. Although for actual literature survey this TC more heavily centers on the narrower scope, it also considers subjects of the broader scope deemed appropriate for advancing the TC investigations.

By the summer of 1994, i.e., at the first TC meeting, approximately 150 literatures were reviewed, two thirds of which belonged to the hydrodynamic category. The remaining part was shared by the other two. More are expected to enter by the second TC meeting.

Governing Flow Equations

Because numerical flow modeling begins with selection, examination, or derivation of basic partial differential equations (PDEs) describing the flow of concern, it is a natural sequence after literature survey to conduct systematic review or research of steep-channel flow equations. Although the subject study is a branch of general unsteady open-channel flow analyses, the special attributes of these flows demand attemtion to the validity of the prospective governing equations. This work should begin with reexamination of the basic assumptions used or to be used for derivation of the governing PDEs. Assumptions needed for steep-channel flows, high-velocity flows, or rapidly-varied flows of interest may be different from those used for general unsteady open-channel flows (Chow 1959); consequently, the equations derived may differ in form, in number, in kind of terms involved, in function, and in application. And finally, for the modeling purposes, they should be of the form usable as governing equation sets of computer programs, permitting implementation into workable simulation models.

Dimensionless Parameters for Flow Model Analysis

There are several key numbers that serve as critical indices to the stability and other behavior of hydrodynamics and numerical modeling of steep-channel flows, such as the Vedernikov number, the Ursell number, the Courant number, the Froude number, and the Reynolds number. To review and investigate these numbers (Basco 1989; Chen 1994b; Lai 1993) is an important aspect of hydrodynamic and/or numerical stability dealing with the steep-channel flow study. Among these numbers, some are mainly used for hydrodynamic analyses (e.g., the Vedernikof number, the Ursell number), and some for numerical analyses (e.g., the Courant number, etc.). Their use is not limited to such analyses, but to a wider range of design and development of numerical models as well as of guides to successful model operation.

Hydrodynamic aspects – Hydrodynamic Stability

As noted in Introduction, the modeling study of steep-channel flows requires that not only numerical stability but hydrodynamic stability must also be given due attention and appropriate review and examination. Items requiring such attentions are: (i) The relation with the key "numbers" that are identified in the preceding paragraph (Chen 1994b); (ii) Application ranges (or limits) of unsteady-flow equation sets, such as the St. Venant equation, the Boussinesq equation, and other shallow-water equations (Basco 1989; Berger 1994); and (iii) Addition of more nonhomogeneous or nonlinear terms.

The importance of examining or testing the hydrodynamic behavior of steep-channel flow equations as to their suitability for modeling has been mentioned earlier. Other theoretical studies on hydrodynamic stability of steep-channel flows are of interest and significance (Chen 1994a; Kyotoh 1994).

Numerical Aspects – Numerical Stability

The ultimate objective of this study being numerical modeling of unsteady flow in steep channels, numerical stabilty still is a critical focal-point of the TC investigations. The investigation can proceed along two avenues — first, about numerical analysis per se (e.g., Hosoda and Yoneyama 1994; Yen and Hsu 1994; Zhang et al. 1994) and second, advancing numerical modeling techniques (e.g., Berger and Stockstill 1994; Chaudhry and Mays 1994; Lai et al. 1994; Li et al. 1994).

A mere increase of channel slope can result in numerical instability; conversely, a model can run successfully in an extremely steep slope for which the governing equations are clearly invalid hydrodynamically. Likewise, an increase of flow velocity or that of time rate of parameter variation can incur the similar effects. A combination of some parameters functioning normally at a mild slope may cause numerical troubles at a steep slope, or may upset the mass conservation ability of the model even the governing PDEs are well within the stability range. These are some of the challenging problems model developers may often encounter. Some papers of the 1994 ASCE National Hydraulics Conference are in fact attempting to address such problems. (e.g., Lai et al. 1994) Others discuss related numerical aspects of steep-channel flows.

National Hydraulics Conference, 1994

In the 1994 ASCE National Hydraulics Conference at Buffalo, NY, this TC organized three special sessions on the subject matters. They are grouped according to sessions in the References list at the end of this article. More from the 1995 Conference are expected to add to this list, and with the progress made in the all aforementioned areas, accompanied by the fruit of the second TC meeting, the TC hopes to be able to assemble a final report (in a form of journal article) in which some guidelines for steep-channel flow investigation through numerical modeling will be included.

Concluding Remarks

Because of the great difficulty attached to the treatment of numerical modeling of steep channel flows, the subject matter did not gain much interest in the past. In steep-channel flow modeling, not only numerical stability has to be properly taken care of, but hydrodynamic stability need to be scrutinized as well. To complicate the situation, even the validity of the governing flow equations is sometimes questioned. The pressing hydraulic, hydrologic and environmental problems continue to increase the importance of the riverine flow modeling, which would demand the subject area to be adequately investigated without further delay. In view of this, the Task Committee on " Theory and Numerical Modeling of Flow in Steep Channels" was organized with the objectives of conducting literature survey concerning hydrodynamic, numerical and general aspects of steep-channel flow; critical review of steep-channel flow equations, with the aid of several key parameters or "numbers" for their systematic and uniform analyses, study of hydrodynamic and numerical stability behavior of the equation or model system, with the final goal of preparing some guidelines for "steep-channel flow investigations through numerical modeling." The technical areas in which the TC conducts its investigation tasks include: (a) steep channel flow, (b) high-velocity flow, (c) rapidly-varied flow, and (d) flows in related areas.

A review at the end of 1994, which is a little over the mid-way of the TC function perod, has indicated that the TC work schedule is generally followed, and increasing understanding of, and interesting findings on, the subject matter are continually made. The present paper, serving as an "intermediate report," is herewith respectfully submitted to ASCE by the TC Control members:

Rutherford C. Berger, U.S. Army Corps of Engineers, Waterways Exp. Sta.
M. Hanif Chaudhry, Washington State University
Cheng-Lung Chen, U.S. Geological Survey, Menlo Park, CA
Chintu Lai, Chair, National Taiwan University

References

Basco, D.R. (1989). "Limitation of De Saint Venant equations in dam-break analysis." *J. Hydr. Engrg.*, ASCE, 115(7), 950-965.

"Computer Modeling of Free-Surface and Pressurized Flow." M. H. Chaudhry and L. W. Mays Ed., Kluwer Academic Publishers, Dordrecht, Netherlands, 1994.

Chow, V.T. (1959). "Open-channel hydraulics." McGraw-Hill Book Co., New York, NY

Lai, C. (1993). "The Courant number and unsteady flow computation." Proc., 1993 *Nat. Hydr. Engrg. Conf.*, ASCE, July 26-30, 1993, San Francisco, CA 2196-2201.

Liggett, J.A. (1975). "Stability." in *Unsteady Flow in Open Channels.* K. Mahmood and V. Yevjevich, eds., Vol. I, Chap. 6, Water Resources Publ., Fort Collins, CO.

The following articles are all from Proc., *Hydr. Engrg.*, vol.2, 1994 ASCE Hydr. Div. Nat. Conf. – Sessions, Theory and Numerical Modeling of Flow in Steep Channels

Berger, R.C. (1994). "Strengths and weaknesses of shallow water equations in steep open channel flow"

Chen, C.L. (1994a). "Stability principle of critical flow depth in a moving coordinate system."

Chen, C.L. (1994b). "Free-surface instability in flow down steep rivers."

Kyotoh, H. (1994). "Theoretical study of unstable waves in the rapids of the river."

———.

Hosoda, T., and Yoneyama, N. (1994). "Numerical analysis of high velocity flow with zero-depth line and oblique shock wave in steep open channel."

Yen, C.L., and Hsu, M.H. (1994). "Numerical experiments on instability in unsteady flow modeling."

Zhang, Y., Yang, Y., and Wu, C. (1994). "Investigation of an anisotropic k-e model."

———.

Berger, R.C., and Stockstill, R.L. (1994). "Consideration in 2D modeling of hydraulically steep flow."

Lai, C., Young, D.L., and Chang, J.T. (1994). "Fortification of MMOC models for steep-channel flows."

Li, K., Zhao, J., and Luo, L. (1994). "Numerical modeling and laboratory verification for the turbulent flow field in the branching channel."

Application of HIVEL2D to the Rio Hondo Flood Control Channel

Scott E. Stonestreet[1] M. ASCE

Abstract

This paper discusses the use of the two-dimensional, high velocity flow model HIVEL2D in a case study of a bridge constriction in a supercritical flood control channel. The model was used to develop alternatives which improve flow conditions at the bridge.

Introduction

The Rio Hondo Flood Control Channel, located within the Los Angeles County Drainage Area (LACDA), is part of a comprehensive flood control system which includes concrete-lined channels with supercritical and subcritical flow regimes. The US Army Engineer District, Los Angeles (USAEDLA), plans to increase the level of flood protection provided by the existing system. Increasing the capacity of the Rio Hondo Flood Control Channel is included in this plan and consists of constructing parapet walls on top of existing channel levees and modifying existing bridges which constrict the flow.

Numeric and physical model studies were conducted during the detailed design phase of the study. At one railroad bridge crossing, results from the physical model study indicated that flow conditions were not satisfactory nor easily correctable. The main problem appeared to be concentration of flow due to channel curvature upstream of the railroad bridge. The purpose of this study was to develop alternatives, using numerical analysis, which would otherwise require significant effort to construct for the physical model tests. After a satisfactory alternative was developed, then the final design could be tested in the physical model to verify its performance.

---

[1]Hydraulic Engineer, US Army Corps of Engineers, Los Angeles District, PO Box 2711, Los Angeles, CA 90053

The Model

The high velocity channel model, HIVEL2D, is a depth averaged, 2-D, finite-element flow model designed specifically for flow fields containing supercritical and subcritical regimes as well as the transitions between the regimes (Berger & Stockstill, 1993). The version of the model discussed herein was developed for open channel flow and did not compute losses associated with pressure flow or impact with the bridge superstructure. HIVEL2D was developed at the Waterways Experiment Station (WES) in Vicksburg, MS.

Study Approach

The prototype channel is trapezoidal with a concrete invert and grouted stone sideslopes. The study reach is shown in Figure 1 and includes a series of compound curves (circular with spiral transitions), base width transitions near the bridge, bridge piers, and access and bike ramps located on the sideslopes. The design Manning's n-value is 0.017 with an average Froude number of about 1.04 associated with normal depth for a deisgn flow rate of 1400 m^3/s. The vertical clearance through the railroad bridge is 4.9 m. Results from the physical model indicate that an unstable pressurization condition occurs in the left bay (i.e. left and right looking downstream) for the existing configuration.

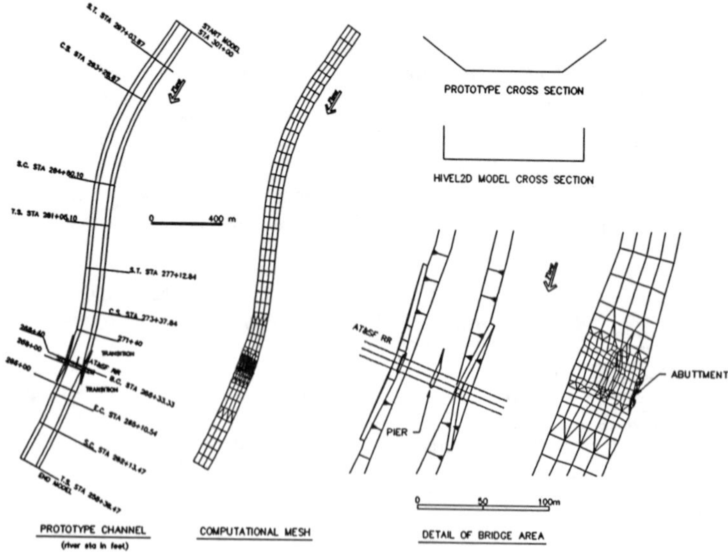

Figure 1. Channel geometry of prototype and HIVEL2D model

A rectangular cross section with an area approximately equal to the trapezoidal area was used for the 2-D model to eliminate the instabilities associated with element wetting and drying expected to occur on the sideslopes. An n-value of 0.0156 was used for the rectangular section to reproduce the Froude number of 1.04. As a matter of simplification, the ramps and one pier located on the sideslope were ommitted from the 2-D geometry. However, the effect of channel curvature, pier and abutment loss, slope, and roughness were included in the 2-D model.

For each simulation, a "cold-start" procedure was used to begin the numerical model. The boundary conditions included supercritical inflow and a subcritical tailwater condition. The initial velocities in the x- and y-directions were set to zero and a constant subcritical depth was set at each node. As the model "warmed up", the tailwater boundary was dropped and flow exited the model as supercritical flow.

Alternatives which would counteract the flow concentration and therefore reduce the flow depth through the left bay were investigated in this analysis. This paper discusses the HIVEL2D modeling of the existing configuration along with two alternatives.

Existing Conditions Analysis

The flow conditions, in terms of depth and velocity contours, computed by HIVEL2D for the existing configuration are shown in Figure 2. In general, the predicted flow conditions match the results from the physical model, especially with respect to wave location and orientation. For example, the model predicts the wave pile-up on the nose of the pier and a V-shaped "wake" wave from the tail. Additionally, superelevation from the upstream curve is noted by the depth contours. The model indicated a flow distribution at the bridge constriction of 56 percent through the right bay and 44 percent through the left bay.

Alternatives Analysis

Alternative 1 consisted of superelevating the invert through the curve upstream of the bridge. The USAEDLA has successfully used invert superelevation in many existing concrete-lined, supercritical channels in order to minimize transverse wave formation and to produce a more uniform velocity distribution through channel curves.

Depths computed from this analysis are shown in Figure 3 as well as depth differential contours which show the difference in depths between this alternative and the existing condition. The maximum depth under the bridge was about 4.3 m with an average decrease in the depth of 0.5 m which was a slight improvement. The model indicated a flow distribution very similar to the existing configuration though the bridge constriction. However, depths upstream of the superelevated invert were up to 1.5 m deeper with this

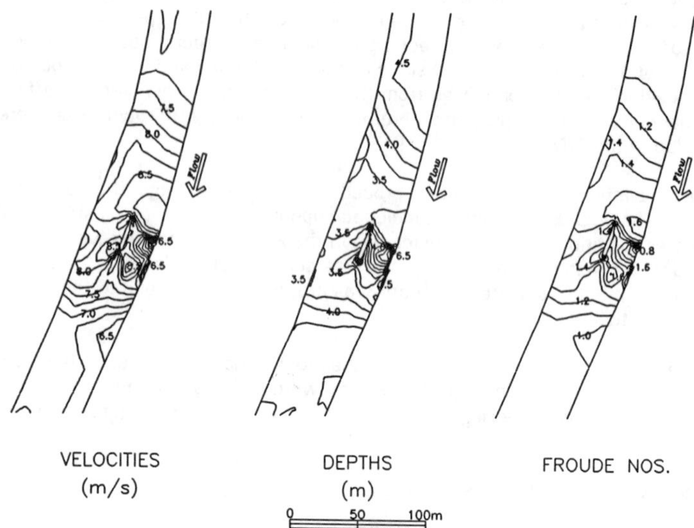

Figure 2. Existing flow conditions at vicinity of railroad bridge

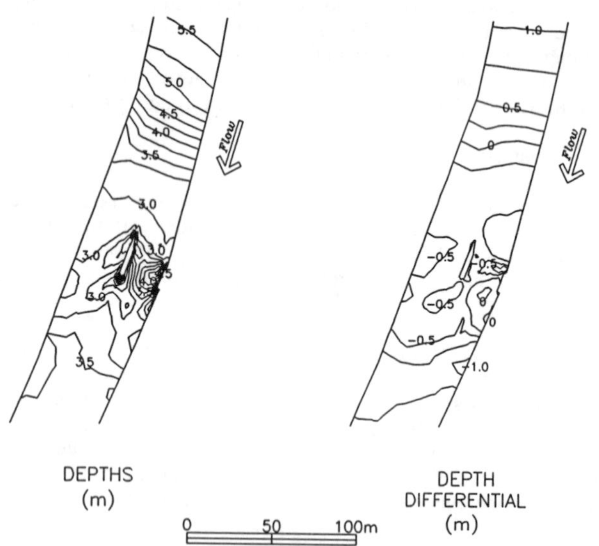

Figure 3. Alternative 1 depth & depth differential contours

alternative apparently due to the adverse slope created by rotation of the invert about the inside of the curve.

Alternative 2 consisted of simply extending the pier well upstream of the bridge in order to restrict flow through the left bay. Depths computed for this alternative are shown in Figure 4 as well as depth differential contours. The maximum depth under the bridge was about 3.4 m with an average decrease in depth of 1.0 m under the left bay coupled with an average increase in depth of 1.0 m under the right bay. The model indicated a flow distribution of 62 percent through the right bay and 38 percent through the left bay.

Conclusion

Numerical modeling of the Rio Hondo Flood Control Channel was performed with the HIVEL2D model. Alternative 2 appeared to produce better flow conditions than Alternative 1 at the railroad bridge without significnalty impacting the channel upstream or downstream of the bridge. Further testing of this alternative should be conducted in the physical model.

Reference

Berger, R.C. and R.L. Stockstill. 1993. "A 2-D Numerical Model for High Velocity Channels." ASCE Proceedings National Conference on Hydraulic Engineering, San Francisco CA, pp 1085-1090.

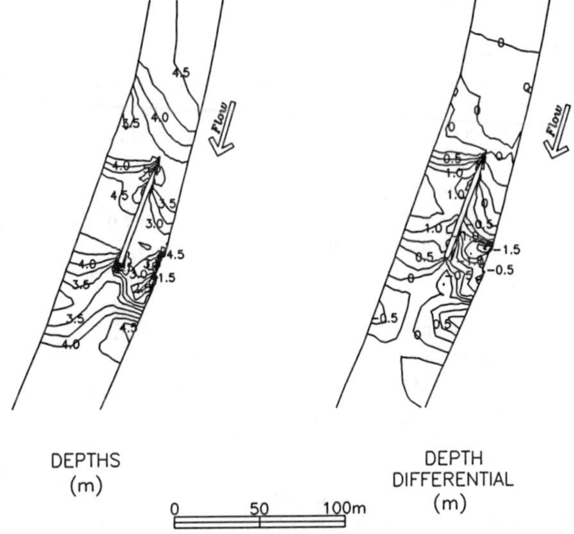

Figure 4. Alternative 2 depth & depth differential contours

Screening Criteria for Embankment Overtopping Protection

By Dave Campbell[1], M.ASCE and John Harrison[2], M.ASCE

Abstract: The use of overtopping protection for embankment dams as a means of increasing spillway capacity has become common practice, due to the recent development and application of precast block systems and roller compacted concrete technology. Each dam project is uniquely characterized by its site setting, construction materials, foundation characteristics and project facilities arrangement. The appropriateness of armoring techniques will depend upon these characteristics and the magnitude and duration of floods that need to be accommodated at the dam site.

Projects where armoring has been found to be both cost-effective and technically viable generally have common characteristics (moderate embankment height, significant need for additional flood passage capacity, lack of critical facilities immediately downstream of areas designed to be overtopped, presence of potential liability concerns related to increased downstream flood levels, etc.). Based on past applications of embankment overtopping protection techniques, preliminary screening criteria are presented to provide a means of early identification for dam projects where overtopping protection should be considered. A general framework for evaluating embankment armoring alternatives is also presented to assist in the selection of appropriate treatments.

Screening Criteria

When an embankment dam has been identified as requiring additional spillway capacity, the next logical step is to identify technically viable, cost effective solutions to remediate this deficiency. Additionally, in some cases, limitations on peak downstream discharges need to be considered to protect the owner and engineer from litigation due to downstream flood impacts. The elements presented below have been identified as important factors that can have considerable effect on the appropriateness of overtopping protection as a viable alternative.

o **Increased Spillway Capacity**-Where spillway capacity needs to be increased significantly, constructing additional conventional spillway capacity (overflow control

[1]Principal and [2]Project Engineer, Schnabel Engineering Associates, 882 S. Matlack St., West Chester, PA 19382

structures, chutes and channels, and energy dissipating works) can be very costly. The need for a significant increase in discharge capacity therefore tends to be favorable for provision of overtopping protection. For the purposes of this discussion, an increased discharge capacity of at least 25 percent or greater than 1000 cfs is considered significant. More moderate capacity increases can many times be more cost-effectively accommodated by moderate embankment raising (earth fill, crest parapet, etc.) or by increasing weir capacity (higher efficiency crest shape, collapsible crest control, etc.).

○**Downstream Impact Area Sensitivity**-Where flood impacts to developed areas downstream of the dam are sensitive to peak discharge, the provision of additional conventional spillway capacity can create a potential for litigation due to flood damages. Downstream property owners can be considered to have come to rely upon the flood attenuation provided by the dam for the range of floods that can be safely handled by the existing spillway. Where post-modification peak discharges significantly exceed pre-modification discharges for flood events up to the existing safe capacity, the potential for litigation against the owner and engineer should be recognized and considered (Campbell, 1991). Embankment armoring tends to minimize this concern, since increased discharges are generally limited to those circumstances where the unmodified dam would otherwise have been overtopped and likely have failed. If a portion of the embankment crest is to be significantly lowered to accommodate armoring, pre- and post-modification flood routings are recommended where nearby downstream development is present.

○**Embankment Height**-Armoring material quantities for crest protection are generally related to the crest width and maximum unit discharge. Slope protection material quantities are generally related to the embankment height and unit discharge (areal coverage and flow acceleration considerations). Quantities for toe protection are also generally related to embankment height and unit discharge (peak velocity and energy considerations). Since the required quantity of armoring materials (per unit width) is greatly influenced by the height of the embankment to be armored, the cost of armoring for a given discharge capacity is significantly less for low embankments (This is also many times true for conventional spillway additions). Reductions in unit price for larger quantities does provide a moderate economy of scale. Armoring has historically been a very common solution for embankments less than twenty-five feet high, reasonably common for embankments from twenty-five to fifty feet high, and less common for higher structures. Armoring of higher embankments has also been influenced by caution on the part of engineers in scaling up use of newer technologies in a gradual manner.

○**Embankment Slope Stability**-Embankment dams that do not meet slope stability requirements and require a significant increase in spillway capacity are excellent candidates for armoring with roller compacted concrete (RCC). In these cases, the RCC armoring can provide for the necessary increase in discharge capacity and improve embankment slope stability. In some case, the embankment slope must still be flattened, but the extent of this work can be significantly reduced.

○**Overflow Impacts to Site Facilities**-Where important site facilities are located at or in the vicinity of the toe of the embankment, the choice for designed overtopping must be weighed against the estimated frequency and magnitude of overtopping anticipated, and the impact on and criticality of these facilities. If overtopping flows cannot readily be diverted away from critical site facilities (water treatment plant, manufacturing plant, etc.), the risks to these facilities must be accommodated. If they would likely be affected by flood backwater levels, and overtopping flows would not likely increase damage to already inundated facilities, a lesser impact exists. The ease of providing flood flow diversion works also needs to incorporated in this rating.

○**Project Aesthetic Appearance**-Generally, both owner and designer share concerns about the finished appearance of a rehabilitated project; but where this is a significant issue, the owner's input is vital. Concrete blocks systems and exposed RCC can be considered attractive to some and an eyesore to others. Where considered unattractive, exposed RCC faces can be formed; or the armoring system covered with earth fill, topsoiled and seeded. While these treatments are, in general, not overly expensive, they can tip the scales where other viable alternatives exist. Acceptable finishes also depend upon the estimated magnitude and frequency of overtopping flows. Recognize that an owner's perception of aesthetics could entail far more than a seeded slope.

Armoring Alternatives

A brief overview of common armoring techniques is presented to assist in the assessment of armoring options, advantages and disadvantages for those projects conducive to overtopping protection.

Protection Method	Advantages	Disadvantages
Roller Compacted Concrete (RCC)	•Cost-effective in larger quantities; especially where on-site aggregate borrow is available. •Typically provides armoring thickness of 1.0 m (3 feet) or more normal to the slope. •Ease and speed of construction. •Some energy dissipation provided by steps. •Has successfully withstood significant overtopping events. •Curves and contouring reasonably easy to accommodate.	•Requires relatively large staging and production area for stockpiling, batching and delivery. •Appearance generally considered unappealing if left exposed. •Not usually economical in small quantities.
Cellular Concrete Mat Systems (CCM)	•Has performed well in flume studies for up to 1.2 m (4 feet). •Can provide some energy dissipation. •Lower mobilization and setup costs than RCC. •Cable anchoring keeps blocks aligned. •Design guidelines available for sizing components.	•Forming curves and irregular shapes can be difficult. •Limited working space between mats for cable splicing. •Requires close monitoring of cable connections. •Limited prototype exper-ience. •Attention to detail required to address turbulence at toe.

Armoring Methods (con'd)

Protection Method	Advantages	Disadvantages
Interlocking Wedge Shaped Blocks	•Reported to be successful for overtopping depths up to 10 m in Soviet installations. •Configuration noted to be hydrodynamically self stabilizing. •Relatively simple and quick construction. •Provides moderate energy dissipation. •Have been used for armoring partially completed embankments during winter shutdown; and can be relocated and reused.	•Requires attention to detail regarding erodibility of foundation material; and destabilizing effects of turbulence at the toe. •Little prototype experience outside of the former Soviet Union. •May be subject to vandalism; unless large and/or interlocking blocks are used.
Riprap	•Good for low embankments subjected to limited overtopping. •Cost-effective where inexpensive riprap source available.	•Paved crest protection generally required. •Typically requires geotextile separator with sand or gravel bedding to protect geotextiles from punctures.

Screening Factors

Following is a listing of the screening factors discussed, together with simplified project rating and weighting factors to allow a rapid assessment of the general merits of overtopping protection for a given dam site. However, since the setting and characteristics of each dam project are unique, other factors of importance to the facilities being examined may need to be examined and added to the assessment matrix, or weighting factors may need to be modified to reflect unique site conditions and considerations of importance. However, for most projects, the following tabulations can provide rapid and effective guidance with regard to the merits of further examination of overtopping protection to enhance spillway capacity.

Screening Factors (Points and weighting factors in **bold**)

1.Spillway Capacity Increase (**3**)

0 pts. < 25% and < 1000 cfs
1 pt. 25% to 100% or > 1000 cfs
2 pts. >100% or »1000 cfs

2.Downstream Impact Area Sensitivity (**1**)

0 pts. Low Sensitivity
1 pt. Moderate Sensitivity
2 pts. Highly Sensitive

3.Embankment Height (**3**)

0 pts. >50 feet
1 pt. 25 feet to 50 feet
2 pts. <25 feet

4.Embankment Slope Stability (**2**)

0 pts. Stable
1 pt. Marginally stable
2 pts. Unstable

5.Overtopping Impact to Site Facilities **(2)**
0 pts. Facilities significantly impacted
1 pt. Extreme flood impact only
2 pts. No significant facilities impacts

6.Project Aesthetic Appearance **(1)**
0 pts. Project appearance very sensitive
1 pt. Appearance to be considered
2 pts. Appearance not important issue

If Total Rating on the following table is from:

0-10 Embankment armoring is unlikely to be the most effective alternative.

11-17 Embankment armoring merits further consideration. If rating for either factor #1 or factor #5 is 0, carefully consider other alternatives.

18-24 Armoring is very likely to be appropriate for the site. If rating for factor #5 is 0, other alternatives may merit consideration.

SCREENING RATING TABLE

Screening Factor	Factor Rating	Factor Weighting	Factor Subtotal
1)Spillway Capacity Increase		**3**	
2)Downstream Impact Area Sensitivity		**1**	
3)Embankment Height		**3**	
4)Embankment Slope Stability		**2**	
5)Overtopping Impacts to Site Facilities		**2**	
6)Project Aesthetic Appearance		**1**	
TOTAL RATING			

References

Campbell, David B., ASDSO Newsletter, Technical Review, "Potential Changes in Dam Owner Liability Must Be Considered Before Modifications Are Made", January, 1991.

Campbell, David B., ASDSO Newsletter, Technical Review, “Key Factors for Assessing Embankment Armoring”, March, 1995.

Alternatives for Overtopping Protection of Dams, Task Committee of Overtopping Protection, ASCE, 1994.

RCC Overlays for Embankment Dams

Steven H. Snider, P.E.[1]
Member

Abstract

The use of roller compacted concrete (RCC) to armor low-height embankments against overtopping has become common in the United States (about 37 to-date). This technology is particularly cost effective for increasing emergency spillway capacity for large flood events required by modern regulatory criteria. Both water supply and hydropower dams can benefit from armoring of their existing embankments. Many hydropower sites consist of a large concrete gravity overflow section damming the river channel with comparatively long earthfill sections impounding the adjacent floodplain. This situation is ideal for an RCC overlay where a significant increase in project discharge capacity can be added, often without creating major backwater.

This paper will discuss the experience gained from the application of RCC to three water supply dams and one hydropower dam. The experience gained from these case studies will be used to discuss a variety of topics specific to the application of RCC for this type of remediation. Emphasis will be placed upon the development of designs to realize the economic benefit of RCC.

Hydraulics

The practicality of placing RCC in relatively thin lifts has been instrumental in the profession's rediscovery of step faces. Many masonry structures from the 19th

[1]O'Brien & Gere Engineers, Inc., 5000 Brittonfield Parkway, Syracuse, NY 13221

century in the northeastern US have stepped downstream faces on their spillway sections. Clearly, the builders recognized the energy dissipation accrued from the step configuration which was easily incorporated into the construction technique. The development of cast concrete and a better understanding of hydraulic principles appear to have "conspired" to sidetrack stepped spillways for nearly a century. Now, we can incorporate formed steps into RCC overlays to benefit from the energy dissipation. The US Bureau of Reclamation and others have demonstrated through model studies that steps can be effective. On small overlays overtopping heads are often small and the precision of step construction may not be particularly important. On high spillways, low overflow heads generate a cascade effect. As overflow depths approach design head, horizontal rollers develop on each step; maintenance of these rollers absorbs considerable energy. With low-height spillways, it is unlikely that sufficient verticality is available to develop the rollers. Secondly, the hydraulic advantage of stepping may be limited to impact dissipation if overtopping heads are small. Thus, forming a perfect 90 degree angle between step and riser may not be justified. At the Lighthouse Hill RCC overlay, compaction of fresh unformed RCC steps to a near-vertical riser using a hand-operated impact tamper was considered to be an economic alternative to forming.

Geometry

The plan layout of the RCC placement should be kept as simple as possible. Often, the embankment toe is not at a fixed elevation, changing grade with the valley slopes. Extreme care should be exercised in defining the limits of each RCC lift such that access and egress is possible by construction equipment. This credo cannot be emphasized strongly enough! Difficult access may force the contractor to either use very expensive means to deliver the RCC to meet specification restrictions on compaction time (obviating its economy), or RCC quality will suffer due to an unacceptable duration for placing, spreading and compaction. If vehicular access is not possible, then it is the designer's responsibility to change the geometry or budget a larger unit price than is quoted in the literature.

With a stepped downstream slope, any reasonable ratio of step tread to riser is possible. However, care should be taken such that the placement lane width is at least ten feet. Anything narrower causes extreme difficulty for the placement equipment, especially where a nonlinear longitudinal alignment is required. At Lighthouse Hill, the designer specified eight-foot lane widths even around a small-radius curve in the embankment. The placement and spreading machinery could not safely negotiate the turn such that considerable time was lost during spreading and compaction. The contractor was always racing to meet the specified interval between mixing and compaction at this location.

Finally, a tracked excavator equipped with a trench bucket for leveling and spreading the concrete was mobilized. This machine and more importantly, its operator, did a great job with these tasks such that time constraints were achieved routinely. The excavator worked so well on the narrow lanes that it was used for nearly the entire project.

RCC Aggregates

RCC mixes vary considerably depending upon the application. Even within overlays, the demands upon the mix can differ depending upon overflow hydraulics, climate, material availability, aesthetics and designer preference. Typically, RCC overlay volumes are relatively small. Taking economic advantage of roller compacted concrete often requires that the mix design incorporate local aggregates from commercial sources, even though this means hauling aggregates to the site. The cost of developing an on-site borrow or quarry source, and special on-site or off-site processing to meet a "nonstandard" aggregate gradation, will severely impact economy. The designer should seriously consider, and research, using standard aggregate gradations, or at least blends, from local suppliers. Often, quarries and borrow pits will produce and stockpile large quantities of aggregates meeting State DOT gradations. Although these materials may be higher quality than is necessary, their mass production means economy.

Incorporation of pozzolans may not be a necessity. Unless there is a strong need or desire for long term strength gain, why spend the capital for flyash importation, and maybe more importantly, the necessary quality control? Depending upon the distance to the fossil fuel site, flyash costs can approach that of cement. Given the small cross-section of typical overlays and the low concern for thermal constraints, it may simply be better to use more cement. Moreover, most portable continuous mix plants (pug mills) are equipped with one silo; adding flyash to the mix design will require additional silo mobilization, setup, weighing and monitoring costs.

Specifications

Specifications are probably the hardest thing to prepare for an RCC overlay since the only ones published are those from Federal agencies for new dams. These specifications are often too demanding for the needs of an overlay project of modest size. Remember, the overlay section is often oversized to meet the hauling and placing equipment dimensions. Thus, it will have adequate mass to resist potential sliding and overturning forces, especially if a drainage layer is provided beween the earth slope and the RCC.

The specifications writer should carefully consider what requirements he needs to meet the intent of the overlay. For example, it is expensive and unrealistic to require that a contractor mobilize a portable mix plant, produce RCC, then hold production for 28-day, or greater, strength results. Good correlation can be realized between careful design-phase mix studies and RCC production on-site, especially if the mix design uses commercial aggregates.

Mix criteria should be reasonable. The overlay functions primarily by mass with lift joint friction sufficing for all but the last few lifts near the crest where uplift exceeds vertical stress. It is often more economical to include a bedding layer to improve bond on the upper few lifts than it is to demand this function from the RCC mix itself.

Quality Control

As stated earlier, it is easy to demand that the RCC in an overlay meet the same high standards that are dictated by its use in a new dam. Usually, the overlay will be unchallenged except for short duration, statistically infrequent events. A relaxation in quality should be considered accordingly. For example, lift joint cleanliness, which received great attention during early RCC development, should be of little concern where sliding is resisted by friction only.

Aesthetics

Roller compacted concrete has never been considered to be a particularly attractive material by most critics. In fact, efforts to make it so have been mixed. Selection of a flat slope combined with low overtopping frequency offers the opportunity to cover the downstream slope with a thin layer of earth to create a more pleasing grassed or otherwise vegetated surface. Such an approach was used at one site where the dam was in a park-like setting. The downstream slope was covered with about two feet of soil and then seeded. The RCC was completely obscured such that the dam presented its pre-remediation appearance.

Conclusion

Probably the key word for implementing an RCC overlay is simplicity. The challenge of the designer is to achieve that simplicity, and its resulting economy and elegance, through careful consideration of constructability and a realistic assessment of predicted demands upon the structure.

RCC Mix Design With Aggregate Base Course

Y. Kit Choi, M., ASCE[1]

Abstract

An approach for laboratory mix design of roller compacted concrete (RCC) using standard aggregate base courses is introduced and discussed in this paper. This approach has been successfully used in several laboratory RCC mix designs for dam safety rehabilitation projects. Some of the advantages of this approach include: (1) ready availability of the aggregate; (2) optimization of cementitious materials; (3) optimization of strength and durability characteristics; (4) minimization of segregation problems during handling in construction; and (5) improvement of field compaction and field quality control. The ready availability of the aggregate can be a particularly significant advantage where small RCC quantities do not justify the expense of developing on site borrow. Laboratory data will be presented to illustrate the material properties of the RCC.

Introduction

The purpose of this paper is to introduce a method for using standardized highway aggregate base course in roller compacted concrete (RCC) construction, particularly in the applications for overtopping protection of embankment dams and as emergency spillways. The most common method of overtopping protection for embankment dams and spillways is armoring with RCC. Since 1980, there are over 40 RCC projects in the United States with this application. The most costly component in RCC is the aggregate. Using aggregate base course in RCC represents, in most cases, a cost-effective approach for these types of projects. Four RCC projects with mix designs using this approach are described in this paper to illustrate its technical basis. Laboratory test data for the RCC aggregate and RCC mixes are presented.

---

[1]Senior Project Engineer, Woodward-Clyde Consultants, Denver, Colorado

RCC Aggregate and Mix Design Data

The aggregates and mix designs for four RCC projects are presented in this section. The four projects include: (1) Harrison Lake Dam, Virginia; (2) New Elmer Thomas Dam, Oklahoma; (3) Umbarger Dam, Texas; and (4) Lake Ilo Dam, North Dakota.

RCC mix designs were performed during final designs for these projects. The aggregates used in the mix design were base course aggregates that were standardized by the highway department of the respective states. Table 1 contains the designations and gradation specifications for these materials.

The Virginia Type 21A Dense Grade Aggregate for Harrison Lake Dam, and the Oklahoma Type A Aggregate for New Elmer Thomas Dam were specified for construction in their respective standard gradations. For Umbarger Dam, the Texas Flex Base, Type B, Grade 2 material, which has a wide gradation band, was modified by narrowing the gradation bands and adding a requirement for percent fines (minus No. 200 sieve). For Lake Ilo Dam, the North Dakota Class 13 Surface Aggregate material, which does not have adequate gravel content, was modified by blending two parts of this material with one part of a 25-mm (1-inch) crushed rock; the resultant blended material is more well-graded.

The gradation curves for the four aggregate materials used in the mix designs are shown on Figure 1.

Note that all of these curves have the following features:

a. The maximum particle size is 50 mm (2 inches).
b. The materials are well-graded.
c. The percents fines (minus No. 200 sieve) are between 5 and 10.

Table 2 contains a summary of the index properties of the four aggregates. Note that the aggregate have the following index characteristics:

a. The specific gravity values all exceed 2.60, which indicates sound and dense particles.
b. The maximum dry density (ASTM D1557) values range from 2208 to 2256 kg/cu. m, which indicates that the materials can be compacted to a very dense matrix, which has a direct bearing on RCC strength characteristics.
c. The L.A. abrasion loss values are less than 50, which is the maximum allowed in ASTM C33 for concrete aggregate.
d. The sodium sulfate soundness loss values are typically less than 12 percent, which is the maximum allowed in ASTM C33 for concrete aggregate. For Lake Ilo aggregate, this value is slightly

higher because of the presence of some shale materials in the aggregate.

For each RCC aggregate, RCC batches and cylinders were prepared using various trial mixes with different cement contents, which are expressed in terms of percents of dry weight of aggregate. Wet-dry (ASTM D559) and freeze/thaw (ASTM D560) durability tests, and compressive strength tests (ASTM C39) were performed on the cylinders. The results of these tests are summarized in Table 3.

Several notable observations of the RCC cylinder properties are listed below:

a. Both the wet-dry and freeze-thaw percent losses are significantly less than the maximum allowable value of 14 percent (Portland Cement Association guidelines), even for cement content as low as 5 percent.
b. Typically, 28-day compressive strengths of 7000 to 14,000 kPa can be obtained, and in the case of Oklahoma Type A aggregate, a 28-day strength of 21,000 kPa was achieved.

Proposed Methodology

Based on limited data presented in this paper, the author proposes that for RCC mix design studies, using a suitable aggregate base course for RCC aggregate would likely produce a mix, when well compacted and properly cured, with adequate strength and durability for most applications in overtopping protection in embankment dams. Since typically each state's highway materials specifications contain several base course aggregates or surface aggregates, Figure 1 would provide a guide to select a suitable material that meets maximum particle size, uniformity, and percent fines criteria. If it is not possible to select a standardized material to meet those criteria, such as in the case of Umbarger Dam and Lake Ilo Dam, then the standardized gradation can be modified so that the modified gradation curve would be within the desirable limits shown on Figure 1.

The desirable gradation limits shown on Figure 1 was developed based on the following reasoning:

a. A maximum particle size of 50 mm will minimize the segregation problems during mixing, transportation, unloading, and spreading of the RCC during construction.
b. A well-graded material, which is inherent in most base course aggregates, will result in a densely compacted matrix, and hence, higher strength.

c. A range of fines between 5 and 10 percent provides optimal filler materials for a densely compacted matrix, and reduces the cement requirement in the paste.

After a base course aggregate is identified, then index tests should be performed to evaluate whether the material conforms to ASTM C33 for concrete aggregate. Although most base course aggregates are non-plastic, it is also useful to test the plasticity of the minus No. 40 sieve materials, since the presence of plastic fines will adversely affect the performance of the RCC. In the opinion of the author, a maximum plasticity index of 5 should be used.

In conclusion, the advantages of using the proposed approach in RCC mix design can be summarized as follows:

a. The methodology applies for every state in the United States, since every state has similar specifications for base course aggregates.
b. The material is readily available in most sand and gravel pits or quarries. The commercial availability of the aggregate should always be evaluated in a borrow investigation, especially in small overtopping protection projects, where small RCC quantities do not usually justify the expense of developing on-site borrow.
c. Selecting a suitable base course material as RCC aggregate would likely result in a mix with adequate strength and durability for an overtopping protection application, and also may contribute to some savings in cement cost.

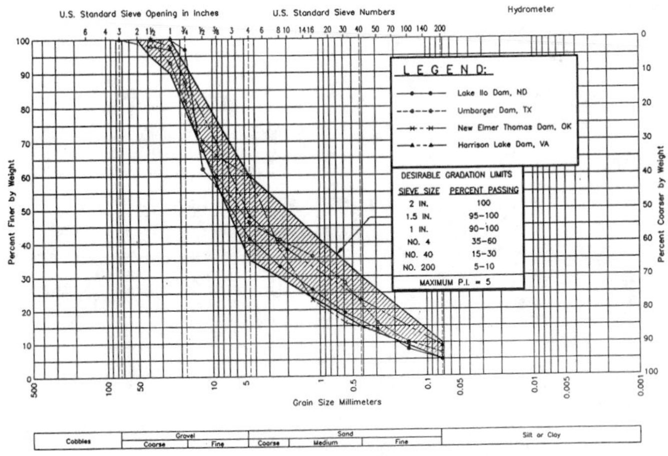

FIGURE 1 – GRADATION CURVES FOR RCC AGGREGATES

TABLE 1
SUMMARY OF STANDARD BASE COURSE GRADATIONS

U.S. Standard Sieve Sizes	Percent Passing By Weight			
	Virginia Type 21A	Oklahoma Type A	Texas Flex Base Type B, Grade 2	North Dakota Class 13
2 in.	100	--	--	--
2-1/2 in.	--	100	95-100	--
1 in.	94-100	--	--	100
3/4 in.	--	40-100	--	70-100
3/8 in.	63-72	30-75	--	--
No. 4	--	25-60	25-70	38-75
No. 8	--	--	--	22-62
No. 10	32-41	20-43	--	--
No. 30	--	--	--	12-45
No. 40	14-24	8-26	15-30	--
No. 100	--	--	--	--
No. 200	6-12	4-12	--	7-15

TABLE 2
INDEX PROPERTIES OF RCC AGGREGATES

Index Property	Harrison Lake Dam, VA	New Elmer Thomas Dam, OK	Umbarger Dam, TX	Lake Ilo Dam, ND
Specific Gravity	2.71	2.72	2.61	2.65
Maximum Dry Density (kg/m^3), ASTM D1557	2256	2240	2224	2208
L.A. Abrasion Loss (%)	--	39	35	29
Sodium Sulfate Loss (%)	--	0.2	7.9	15
Absorption, + No. 4 (%)	--	1.2	2.3	4.7

TABLE 3
TEST RESULTS OF RCC MIXES

Test	Harrison Lake Dam % Cement		New Elmer Thomas Dam % Cement			Umbarger Dam % Cement			Lake Ilo Dam % Cement		
	5	9	5	7	9	5	7.5	10	6	8	10
Wet-Dry Loss (%)	--	--	0.8	1.1	0.2	1.2	1.4	2.9	1.2	0.5	0.9
Freeze-Thaw Loss (%)	0	0	3.2	0.3	0.3	0.9	0.4	1.1	0.1	0.5	0.8
Compressive Strength (kPa)											
2-Day	5382	9246	--	--	--	1587	3174	6210	1725	4347	5589
7-Day	6486	12,489	5934	12,903	14,352	1656	8280	11,109	3036	7038	7728
28-Day	7452	15,249	8211	15,735	23,184	2898	12,558	14,835	4140	8970	12,903
56-Day	7797	15,870	7935	14,766	23,598	2553	10,833	16,284	5106	10,557	13,110

Alternative Materials for Overtopping Protection

Daniel M. Hill, M.ASCE[1]

Abstract

The City of St. Clairsville chose to armor their water supply dams to withstand overtopping as more cost-effective than providing additional spillway capacity. Both gabion and roller compacted concrete (RCC) armoring have been used. Cost comparisons and the limited experience of contractors with new construction materials are discussed.

Introduction

St. Clairsville is located in eastern Ohio near the Ohio River border with West Virginia. This is in the unglaciated area of Ohio, with rugged topography and narrow valleys. The City owns two dams, both found to have deficient spillway capacity in Ohio Department of Natural Resources' (ODNR) dam safety inspection reports. Preliminary cost estimates prepared by Burgess & Niple (B&N) in 1987 indicated armoring the embankments might cost only about 40 percent of the usual approach to add emergency spillways. The City thus embarked on an armoring program to bring their dams into compliance.

Provident Dam

Provident Dam was the first armoring project for the City and was the first armoring project approved by ODNR. The dam was originally built in the early 1920s, and is an earthen embankment 79 meters (260 feet) long, with a maximum height of 9.4 meters (31 feet) and a concrete core wall. It has gentle downstream slopes, 3.5H:1V in the upper portion and 9H:1V in the lower portion. The existing chute spillway has a 5.6 meter (18.5 feet) long weir. Normal pool storage volume is

----

[1]Director, Water Resources Group, Burgess & Niple, 5085 Reed Road, Columbus, OH 43220

76,000 cubic meters (62 acre-feet). The embankment crest is 1.4 meters (4.5 feet) above the spillway crest; storage volume at the embankment crest is 120,000 cubic meters (97 acre-feet). ODNR designated this a Class I dam, based on downstream hazards, which requires Probable Maximum Flood (PMF) spillway capacity. They determined overtopping depths to range from 0.4 meters (1.3 feet) for the 25 percent PMF to 1.4 meters (4.6 feet) for the PMF.

A site visit was made by ODNR, B&N and City staff to view gabion armoring on the Southwestern Pennsylvania Water Authority's Wisecarver Dam. The flat slopes of Provident Dam made it a good candidate for similar gabion armoring. Modifications requested by ODNR included: grout topping the entire armoring surface to prevent vandalism of the gabion baskets, armoring the hillside slopes at each end of the dam, and placing the 0.46 meters (18 inches) thick gabions on a filter fabric to preclude possible seepage piping problems.

Provident Dam's crest was excavated so the finished armoring surface would match original grade. This also allowed the core wall to protect the leading edge of the gabions against undercutting. An existing large headwall for the lake drain pipe provided undercutting protection at the downstream toe. There was concern the requested grout topping would create a long-term maintenance problem of crack repair and weed control, with a resulting unpleasant appearance. A seeded topsoil cover over a filter fabric was instead proposed, which was accepted by ODNR. Figure 1 shows the gabion armoring details.

The plans included installation of an inclined valve on the lake drain pipe inlet and placing a subsurface seepage collection pipe in the left downstream groin. Six bids were received at the opening in early October 1990; they ranged from \$155,000 to \$180,000. Construction started in early December 1990 and placement of all 2,320 square meters (2,770 square yards) of gabions was completed a month later. Remaining construction resumed in March 1991 and all work was completed in early April 1991. The final total cost was \$151,000, of which \$132,000 was for the gabions and soil cover. Total cost of the gabion armoring with soil cover was 57 $\$/m^2$ (48 $\$/yd^2$).

South Dam

South Dam was originally built in 1929. It is also earthen embankment dam with a concrete core wall, 87 meters (285 feet) long with a maximum height of 8.8

meters (29 feet). The downstream slope is steep, at 2H:1V. The existing side channel spillway has a 7.6 meter (25 feet) long weir. Normal pool volume is 127,000 cubic meters (103 acre-feet). The minimum embankment crest is 1.2 meters (4 feet) above the spillway crest; storage volume at the embankment crest is 226,000 cubic meters (183 acre-feet). ODNR also designated this a Class I dam based on downstream hazards. They determined its maximum overtopping depths to range from 0.12 meters (0.4 feet) for the 25 percent PMF to 1.6 meters (5.2 feet) for the PMF.

ODNR had accepted gabion armoring at Provident Dam with no increase of discharge-storage capacity. They required any armoring of South Dam, however, to include increasing discharge-storage capacity to the 50 percent PMF. South Dam impounds the main water supply reservoir for the City and is just upstream of the water treatment plant; protecting this vital public utility was important. An emergency spillway channel was designed to be excavated in the right hillside abutment on existing City property. The emergency spillway crest was set 0.61 meters (2 feet) above normal pool. Excavation would be into shale with side slopes set at 0.5H:1V; a gabion wall was included to retain the soils above the shale. A wire mesh would be rock bolted to the shale sides; this mesh and the gabions would receive a shotcrete coating, with drilled weep holes. Base width of the emergency spillway channel was 12 meters (40 feet), and the embankment crest would be raised a maximum of 1.2 meters (4 feet) to contain the 50 percent PMF. Residual overtopping depth for the PMF was 0.91 meters (3 feet).

Stepped armoring was necessary because of the steep embankment slope, and would also provide energy dissipation. Plans were prepared to bid both gabions and RCC armoring. Each would include a filter fabric against the stripped embankment surface. Gabion steps, or lifts, were 0.46 meters (18 inches) high and 1.8 meters (6 feet) wide, with all exposed surfaces to receive a shotcrete coating. RCC lifts were 0.23 meters (9 inches) high and 1.5 meters (5 feet) wide, with an additional 0.91 meters (3 feet) width of porous fill against the filter fabric. This configuration was selected to economize concrete use and to allow concrete/porous fill placement with a split spreader box. Both alternates had the same provision for lifts totaling 0.91 meters (3 feet) deep below existing grade to prevent frost heave or undercutting at the upstream edge of the crest and the downstream toe. The RCC armoring included a styrofoam/wood cushion on top of the existing core wall to minimize settlement cracking of the RCC. Figure 2 shows the RCC armoring details; the soil cover was an optional bid item requested by the City (also for the gabion alternate) to achieve a "natural" appearance.

The plans included repairs to the existing concrete principal spillway and concrete outlet tower plus replacement of the lake drain valve. Initial plans were to bid the project for construction in 1992, but lack of adequate funding postponed the bidding date to mid-May 1993. Although several contractors had purchased plans, no bids were received. A survey of the plan holders indicated they declined to bid because they already had an adequate amount of work, but also that they were not familiar with RCC (even through it was not necessary to bid both gabions and RCC). Additional efforts were undertaken to increase awareness of the project and the RCC alternate in particular. This included listing Mr. Ken Hansen of the Portland Cement Association as an information reference, and sending copies of the revised advertisement to the 70 some contractors originally notified plus additional, more remote, contractors.

Only two bids were received at the second opening in early February 1994. Their base bids (RCC without optional soil cover) were $693,000 and $758,000. The low bidder also bid the gabion alternate at $19,000 more; the high bidder did not bid the gabion alternate. The City deleted much of the principal spillway repair and all the outlet tower/drain valve work to arrive at a $546,000 package they could fund.

General construction started May 1, and included trial mixes to arrive at the following approved mix, units are kg/m^3 (lbs/yd^3): coarse aggregate = 1,320 (2,220); fine aggregate = 740 (1,250); cement = 150 (250); water = 145 (245). The first RCC test lifts were placed in early August and all the RCC work was completed by mid-September. The RCC operations included placement of extra earthfill as a ramp over hardened RCC lifts to deliver concrete to the higher lifts. This allowed the final earth cover slope to be flatter than 2H:1V, which will improve mowing operations. All of the construction was completed in mid-October 1994. The final total cost was $549,000, of which $407,000 was for the RCC and soil cover. An area of 1,870 square meters (2,240 square yards) was armored; total cost of the RCC armoring with soil cover was 218 $\$/m^2$ (182 $\$/yd^2$).

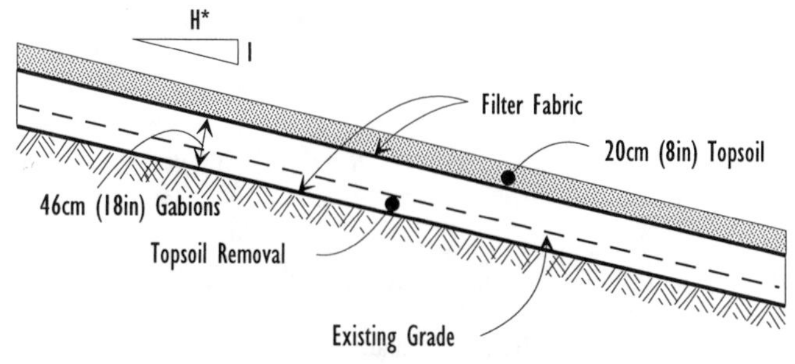

Figure 1
Provident Dam
Armoring Typical Section

*H varies, 3.5 to 9.0

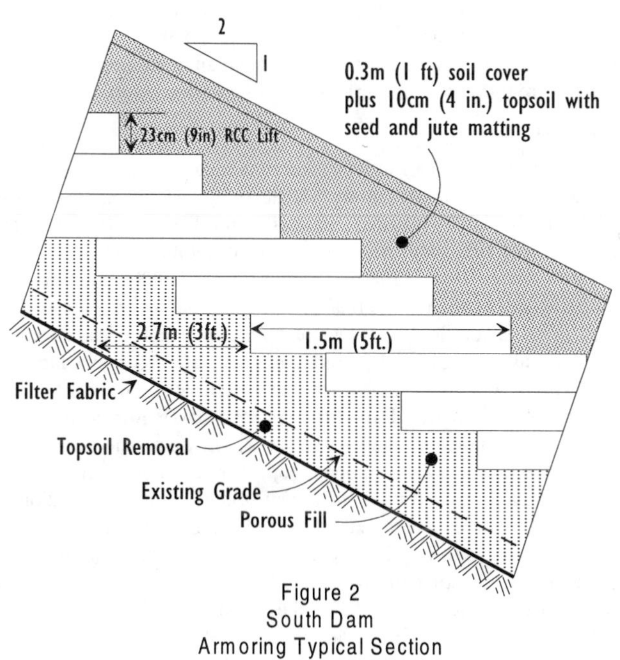

Figure 2
South Dam
Armoring Typical Section

« PILOTE »: optimal control of irrigation canals

By P.O. Malaterre[1]

ABSTRACT

This paper presents an application of optimal control theory to the automatic control of a 8-pool irrigation canal. The model used to design the controller is derived from Saint-Venant's equations discretized through the Preissmann implicit scheme. A Kalman Filter is used to reconstruct the state variables and the unknown perturbations from a reduced number of observed variables. Both perturbation rejection and tracking aspects are handled by the controller. Known offtake withdrawals and future targets are anticipated through an open-loop. The controller and Kalman filter are tested on a linear model and proved to be efficient.

INTRODUCTION

Significant research efforts in irrigation canal automation can be recognized in the literature and were summarized by Malaterre (1994). In particular, since it handles easily multivariable systems, optimal control was considered by the following authors. The first published application of optimal control on irrigation canals was from Corriga et al. (1980). Balogun et al. (1988) and Garcia et al. (1992) did not consider explicitly external perturbations acting on the system, such as unpredicted offtakes' outflows inherent in on-demand deliveries. Reddy et al. (1992) considered unknown external perturbations. A state observer which did not include perturbations was used. So, when the system was perturbed, the state reconstruction error was expected to differ from zero. However this was not commented on. Furthermore, variable targets (i.e. tracking) and anticipation on future known offtake withdrawals (i.e. open-loop) were never explicitly studied.

In this paper, a discrete time optimal control algorithm for irrigation canals under predicted and unknown external perturbations is presented. It is named « PILOTE » for « Preissmann Implicite scheme, Linear Optimal control, Tracking of variables and Estimation of perturbations ». The control algorithm is tested on canal 2 (Figure 1) of the test cases set by the ASCE Task Committee on Canal Automation Algorithms (ASCE 1995).

[1] Cemagref, BP 5095, 34033 Montpellier Cedex 1, France

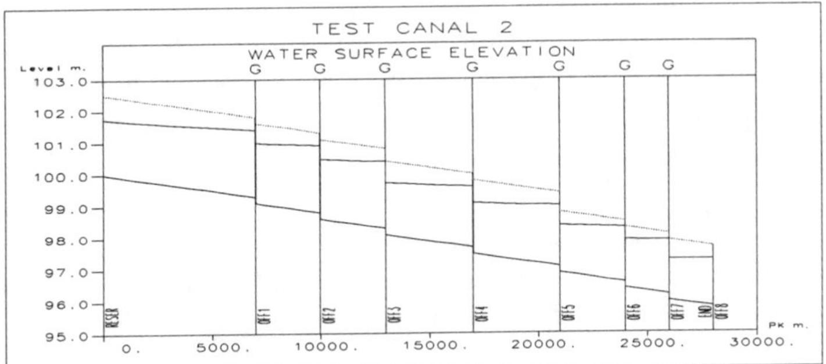

Figure 1. Longitudinal profile of the tested irrigation canal.

LINEAR MODEL

To apply the optimal control method, a linear model is required. The latter can be obtained analytically (Malaterre 1994) or from transfer function identification (Kosuth 1994). The first option is selected in this paper. Saint Venant equations are discretized with Preissmann's implicit scheme, replacing the partial derivatives by finite differences. The canal is divided into j cross sections and two variables are considered at each section: the water elevation δZ_i and the flow discharge δQ_i. Appropriate boundary conditions need to be specified. The control action variables are the upstream discharge δQ_1 and the cross regulator gate openings δw. The perturbation variables are the unknown offtake outflows δQ_p. All δ variables are relative to the reference steady state.

CONTROL ALGORITHM

Linear optimal control theory applied to a perturbed system

Discretized Saint-Venant equations can be written under the usual representation of a linear discrete dynamic system:

$$\begin{cases} x_{k+1} = Ax_k + Bu_k + B_p p_k \\ y_k = Cx_k \end{cases} \tag{1}$$

where $x_k \in \mathbb{R}^n$, $u_k \in \mathbb{R}^m$, $y_k \in \mathbb{R}^l$, $p_k \in \mathbb{R}^p$, are respectively state, control action, and controlled variables, and perturbation vectors at each time step k. A, B, B_p, and C are matrices of appropriate dimensions. The sequence of control vectors u_k can be calculated using the well known linear discrete-time optimal regulator theory. Performance index J is minimized:

$$J = \sum_{k=0}^{N} \left[\left(y_k - y_k^*\right)^T Q_y \left(y_k - y_k^*\right) + u_k^T R u_k \right] \tag{2}$$

where N is the optimization horizon, Q_y and R are respectively a non-negative and a positive definite symmetric weighting matrix. In the case of tracking, a controlled variable's reference trajectory y^*_k (k = 0 to N) different from zero is defined. The optimization constraints are imposed by the system linear dynamics (1). By using standard optimization procedures, the optimal control variable u_k is obtained as:

$$u_k = - K x_k + H_k \tag{3}$$

K is the steady feedback matrix gain obtained from the algebraic Riccati equation, and H_k is a feedforward control vector.

Discrete-Time Observer

The control law (3) assumes that the complete system state vector x_k can be measured accurately, which is often unrealistic. Most frequently, only certain linear combinations of states, denoted observed variables z_k, can be measured:

$$z_k = Dx_k \tag{4}$$

where $z_k \in \mathbb{R}^q$, and D is a (q, n) matrix. From variable z_k, the state vector x_k can be reconstructed. Then, the actual state x_k is replaced by the reconstructed state $\hat{x}_k$ in (3). Due to unknown perturbations acting on the system, a state Kalman filter including a perturbation observer is designed.

The state Kalman filter is defined as:

$$\hat{x}_{k+1} = A\hat{x}_k + Bu_k + B_p\hat{p}_k + L\,[z_k - \hat{z}_k] \tag{5}$$

where $\hat{p}_k$ is the perturbation vector estimation:

$$\hat{p}_{k+1} = \hat{p}_k + L_p\,[z_k - \hat{z}_k] \tag{6}$$

L and L_p matrices can be computed by pole placement (Luenberger observer), or through the minimization of the reconstruction error (Kalman filter). The second option is tested in this paper. In steady conditions the global observer (5) plus (6) guarantees the vanishing of the reconstruction error and reconstructs accurately the perturbation acting on the system.

SIMULATION RESULTS AND ANALYSIS

The process to be controlled is a 8-pool open-canal receiving water from a source located upstream. The control system aims to match the water level at the downstream end of each pool with a target value. It adjusts upstream inflow and opening of cross gates. The observed variables are water levels at the upstream and downstream ends of each pool.

The discrete-time linear model (1) is generated by the computer package SIC, developed by Cemagref (1992), with a sampling interval of 15 minutes. The controller design and simulations using a linear model are carried out with the commercial package MatLab & Simulink (1992).

Simulation results for scenario 1

After 2 hours, offtakes 5 and 6 increase their discharge by 1.5 and 1 m3/s, respectively, as scheduled. Then, at time 14 h, offtake 6 reduces its discharge by 2 m3/s without prediction. Figure 2 shows the evolution of the controlled variables and of the control actions as a function of time. The obtained performance indicators (ASCE 1995, Liu et al. 1995) are presented in table 1.

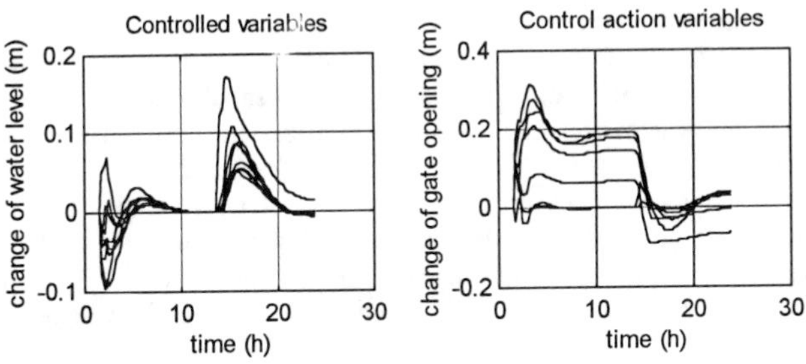

Figure 2. Results on scenario 1.

Simulation results for scenario 2

After 2 hours, offtakes 1, 2, 3, 6, 7 and 8 increase their discharge by 1.5, 1.5, 2.5, 0.5, 1 and 4 m3/s, respectively, as scheduled. Then, at time 14 h, the same offtakes reduce their discharge by the same value, without prediction. Figure 3 shows the evolution of the controlled variables and of the control actions as a function of time. The obtained performance indicators are presented in table 1.

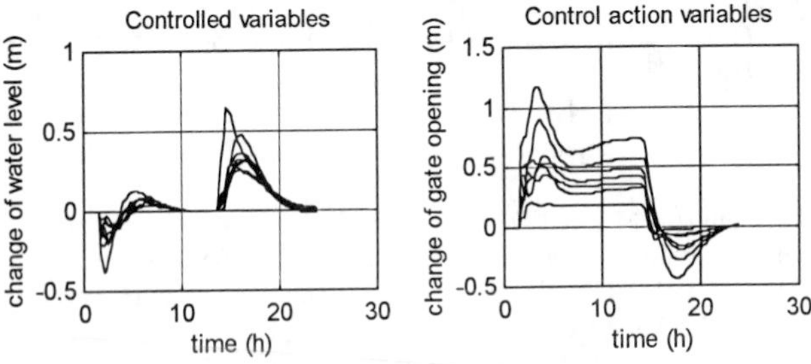

Figure 3. Results on scenario 2.

	MAE (%)		IAE (%)		StE(%)		IAW		IAQ	
Period	0-12	12-24	0-12	12-24	0-12	12-24	0-12	12-24	0-12	12-24
Scenario 1	5.1	10.1	0.3	0.4	0.1	1.0	0.3	0.2	7.5	3.5
Scenario 2	17.9	37.7	1.0	1.6	0.3	1.5	1.1	0.9	29.5	15.8

Table 1. Performance indicators.

SUMMARY AND CONCLUSION

This paper proposes a method for automatic control of irrigation canals. By adjusting the upstream discharge and the gate openings a target water level is maintained at the downstream end of each pool. Linear optimal control theory provides an elegant way to tackle this multivariable control problem. Under external perturbed conditions, such as on-demand deliveries, a perturbation observer is added to improve the state reconstruction. The controller and the Kalman filter are found efficient. The resulting control algorithm can be suited for real-time operations. Some features supporting this potential of applications are: (1) It is designed for water levels and discharge control, but only feasible hydraulic measurements (water elevations) are used for feedback; (2) Its performance is satisfactory in terms of a quick evolution to target water levels; (3) The algorithm is able to deal with unknown perturbations such as unpredicted water withdrawals. Results presented in this paper are obtained using a linear model. Simulation are currently being carried out on a non-linear model, and the results will be reported in the future.

REFERENCES

ASCE (1995). Test cases and procedures for algorithm testing and presentation. First International Conference on Water Resources Engineering, San Antonio, USA, 14-18 August 1995. 5 p.

Balogun O.S., Hubbard M., DeVries J.J. (1988). Automatic control of canal flow using linear quadratic regulator theory. J. of Irrigation and Drainage Eng., 114 (1), 75-101.

Cemagref (1992). SIC user's guide and theoretical concepts. Cemagref Publication. 191p.

Corriga G., Sanna S., Usai G. (1982). Sub-optimal level control of open-channels. Proceedings International AMSE conference Modelling & Simulation, Vol 2, p 67-72.

Garcia A., Hubbard M., and DeVries J. J. (1992). Open channel transient flow control by discrete time LQR methods. Automatica, 28, 255-264.

Kosuth P. (1994). Techniques de régulation automatique des systèmes complexes : application aux systèmes hydrauliques à surface libre. Thèse de Doctorat, Institut National Polytechnique de Toulouse - Cemagref - LAAS CNRS, 330 p.

Liu F., Malaterre P.O., Baume J.P., Kosuth P., Feyen J. (1995). Evaluation of a canal automation algorithm CLIS. First International Conference on Water Resources Engineering, San Antonio, USA, 14-18 August 1995. 5 p.

Malaterre P.O. (1994). Modélisation, Analyse et Commande Optimale LQR d'un Canal d'Irrigation. Ph.D. thesis of the Ecole Nationale du Génie Rural, des Eaux et des Forêts (ENGREF), Paris, France. Presented January 1994. 220 p.

MatLab & Simulink (1992). A program for simulating dynamic systems. MathWorks Inc.

Reddy, J.M. (1990). Local optimal control of irrigation canals. J. of Irrigation and Drainage Eng., 116 (5), 616-631.

Reddy J.M, Dia A., and Oussou A. (1992). Design of control algorithm for operation of irrigation canals. J. of Irrigation and Drainage Eng., 118 (6), 852-867.

Simulation and Optimal Operation of Canal Systems

Zihui Lin [1] David H. Manz[2]

Abstract

Open channel systems are frequently used to deliver water for irrigation, municipal and industrial use. Operating open channel systems optimally is a very important and complex area of water resources management. It is very difficult or impossible to operate large open channel systems at maximum efficiency using techniques which rely solely on operator judgement and experience. A nonlinear programming (NLP) Model has been developed and integrated into in a dynamic simulation model. The objective of the NLP Model is to minimize water losses and minimize the difference between water supply and demand, with consideration of operational and physical constraints. The Irrigation Conveyance System Simulation (ICSS) Model was selected to simulate the hydraulics, hydrology and operation of open channel conveyance system as required by managers, planners and operators. The NLP Model integrated into in the ICSS Model is capable of real time simulation and optimal operation of open channel systems in order to improve the performance of open channel systems and to achieve optimal water management. The computer model has been successfully demonstrated on large open channel systems operated using constant volume control.

Introduction

Constant volume control has been used to operate large open channel systems. Constant volume control is based on maintaining a relatively constant volume of water in each canal reach at all times using a simultaneous gate agjustment technique and has the ability to quickly change the flow conditions in the entire canal system (Rogers, 1988). Using only operator's judgement or experience, it is very difficult to maintain a constant volume in each canal reach, and also satisfy user's water demands. The purpose of this paper is to describe a nonlinear programming model for automatic operation of constant volume control, which is called **system constant volume control**.

The NLP Model For Constant Volume Control

A nonlinear programming (NLP) Model was used to assist in making

---

[1] Ph.D Graduate Student, Department of Civil Engineering, University of Calgary, Alberta, Canada.

[2] Associate Professor, Department of Civil Engineering, University of Calgary, Calgary, Alberta, Canada,

decisions necessary for the operation of a complex open channel system using constant volume control. The objective of the NLP Model was to minimize water losses from the system and to minimize the difference between water supply and demand with consideration of operational and physical constraints. The objective function minimizes the difference between initial canal storage and canal reference storage and the difference between water demand and water delivery for each time step, that is

$$Min\ Z = \sum_{r=1}^{N} \{ [Q_r - Q_{r+1} - q_r - \frac{K_r}{TC_r} \times (S_{ref,r} - S_{0r}]^2 + (q_r - q_{req,r})^2 \} \tag{1}$$

where Q_r is the gate inflow for reach number r, m^3/s; Q_{r+1} is the gate outflow for reach number r, m^3/s; q_r is the water delivery for reach number r, m^3/s; $q_{req,r}$ is the water demand for reach number r, m^3/s; K_r is the control parameter; TC_r is the wave travel time for reach number r; $S_{ref,r}$ is the reference canal storage for reach number r, m^3; and S_{or} is the initial canal storage for reach number r, m^3.

Both physical and operational constraints are used. Physical constraints include canal storage and gate inflow/outflow constraints. The canal storage for each reach can not be larger than the maximum allowable canal storage or less than the minimum allowable canal storage, that is

$$S_{r,min} \le S_r \le S_{r,max} \tag{2}$$

where $S_{r,min}$ is the minimum allowable canal storage for reach number r, m^3; and $S_{r,max}$ is the maximum allowable canal storage for reach number r, m^3. Gate inflows/outflows should be less than the capacity of gate discharge and greater than the minimum gate discharge, that is

$$Q_{r,min} \le Q_r \le Q_{r,max} \tag{3}$$

where $Q_{r,min}$ is the minimum allowable gate inflow/outflow for reach number r, m^3/s; and $Q_{r,max}$ is the capacity of gate inflow/outflow for reach number r, m^3/s. Gate operational constraints considered are

$$\begin{aligned} Q_r &\le Q_{or}\left(1 + \frac{OG(t)}{GOP}\Delta t\right) \\ Q_r &\ge Q_{or}\left(1 - \frac{OG(t)}{GOP}\Delta t\right) \end{aligned} \tag{4}$$

where Q_{or} is the initial gate inflow/outflow for reach r, m^3/s; OG(t) is the maximum gate movement speed, m/s; GOP is the initial gate opening, m; and Δt is time step, s.

Solution Algorithm

The NLP Model applied to constant volume control is a constrained nonlinear programming problem. A penalty method, which is an exterior-point method, is used to transform a constrained nonlinear programming problem into an unconstrained nonlinear programming. Nelder and Mead's method was selected to solve the unconstrained nonlinear programming problem. In the Nelder and Mead' method, a set of (n+1) mutually equidistant points in n-dimensional space is initially set up, which is called a general regular simplex. The idea of the method is to compare of the function values at the (n+1) vertices of a general simplex and adapt the simplex towards the optimal point during the iterative process and contract to the final minimum. The adaptation of the simplex in this method is achieved by the application of four basic operations: reflection, expansion, compression and contraction. The accuracy, convergency, and stability characteristics of the method of Nelder and Mead have been thoroughly investigated by Lin (1991). The solution of the above NLP Model for constant volume control appeared to be globally optimal.

Unsteady Flow Simulation

In order to test the system constant volume control method, it was necessary to simulate steady flow and unsteady flow in the canal. There are a variety of steady-flow and unsteady-flow models available in the world. For this study, we used the Irrigation Conveyance System Simulation (ICSS) Model, Manz and Schaalje (1994), which has successfully simulated many types of irrigation canal systems. No limitations are placed on the hydraulic and operational characteristics of the hydraulic control structures considered by this model. The NLP Model for constant volume control was easily integrated into the ICSS Model. This combination made possible real time simulation and optimal operation of open channel systems to improve the performance and achieve optimal water management.

Results and Discussion

To demonstrate the applicability of the technique discussed above, the test was carried out in a 42 km long canal system with the following characteristics (see Figure 1): reach number= 6; turnout number =5; gate number =7; length of each reach = 7000 m; bottom width =12.5 m; side slope (H: V) = 2.5 : 1; bottom slope = 0.0001; Manning's coefficient = 0.025; width of gate (rectangular gate) = 18.25 m; width of turnout = 5 m. The water delivery schedule is listed in Table 1.

Table 1 The Water Delivery Schedule (unit: m^3/s)

Schedule	Duration (Hours)	Turnout #1	Turnout #2	Turnout #3	Turnout #4	Turnout #5	Gate #7
Initial	1.0	8.0	6.0	6.0	6.0	6.0	15.0
#1	12.0	4.0	3.0	3.0	3.0	3.0	15.0
#2	12.0	15.0	12.0	9.0	9.0	9.0	15.0

The results of the simulation are presented in Figures 2 and 3. For schedule #1 which is 50% decrease water demand for all turnouts, the maximum change in

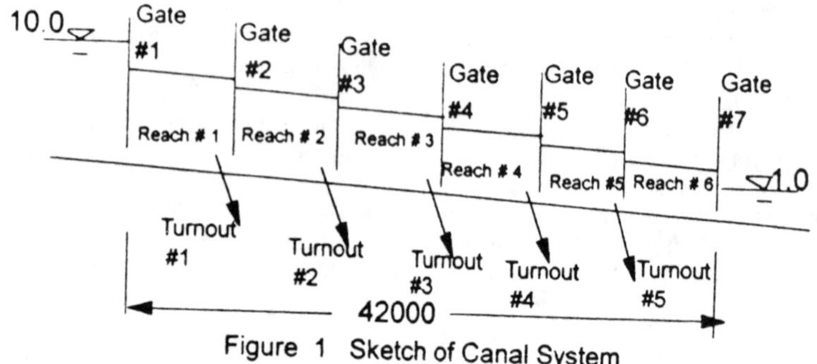

Figure 1 Sketch of Canal System

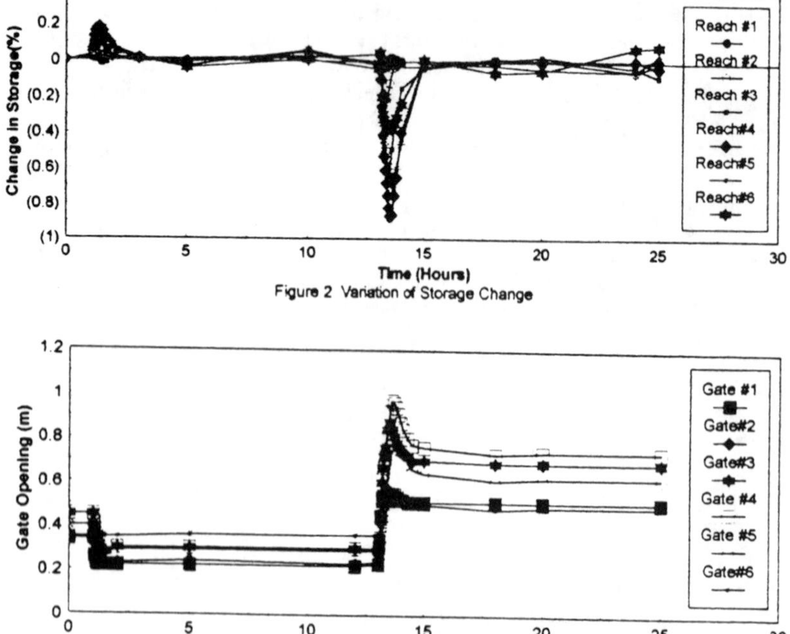

Figure 2 Variation of Storage Change

Figure 3 Variation of Gate Opening

reach storage is 0.18% of initial storage in reach #4, which is negligible. After one hour of operation, the changes in storage are within 0.1 % of initial storage for every reach. The maximum variation in gate opening is 0.17 m for gate #3 followed by a variation of 0.16 m for gate # 4. For schedule #2 which is a more than 200% increase in water demand for all turnouts, the maximum change in reach storage is 0.87% of initial storage in reach #4. The maximum variation in gate opening is 0.67 m for gate #4 followed by a variation of 0.57 m for gate #3. After three hours, the change in storage is less than 0.1% of initial storage for every reach and the gate openings are almost unchanged. Also, the water delivery through each turnout exactly satisfies water demand within 0.2 hours after the scheduling change. These results indicate that the performance of system constant volume control is excellent. Under field conditions, the water demand change for all turnouts may not be decreased/increased at the same time in which case the performance of the system would be better.

Summary and Conclusions

System constant volume control was successfully demonstrated on a large open channel system (42 km long) with 6 reaches. The maximum change of storage in reaches was less than 0.87 % when the total water demand change on the canal system was 38 m^3/s, which is 200 % more than the water demand in previous scheduling. The performance of this technique was found to be excellent.

System constant volume control has the ability to minimize the variation of canal storage and to satisfy water demand immediately while considering all of the canal physical and operational constraints.

References

Lin, Z. (1991), Optimum Operation of Irrigation Canal Systems, thesis as part of the requirements for Master of Science in Civil Engineering, University of Calgary, Alberta, Canada, 154pp.

Manz, D. H. and Schaalje, M., (1994), Modelling Irrigation Conveyance System Using the ICSS Model, Irrigation Water Delivery Models, Food and Agriculture Organization of the United Nations, Rome, Italy, pp. 225-240.

Rogers, D. C. (1988), Operation of Canal Systems, Water Conveyance Branch, Civil Engineering Division, Denver Office, Bureau of Reclamation, USA, 86pp.

An Improved Hydrodynamic Simulation Tool for Controlled Irrigation Channels

M.G.F. Werner[1], J. Schuurmans[2], R. Brouwer[3]

Abstract
The design and evaluation of control algorithms for use in hydraulic systems can be very complicated. Present hydrodynamic simulation tools have the disadvantage that the implementation of new or revised control algorithms is often laborious.

A model developed at the Delft University was combined with a powerful existing computational package, MATLAB. The hydrodynamic computational element of the original model, based on the discrete Saint Venant equations was maintained, while the calculation of control algorithms is done by MATLAB. The graphical output is also performed by MATLAB. The possibility to plot variables, 'real time' during computation has increased the ease with which the control methods can be evaluated.

This paper will discuss the structure of an improved model and demonstrate some of its capabilities through a simple case study

Introduction
In the design of water systems the desire to determine the effect of various alternatives before actual implementation has led to the development of a wide variety of programs to predict the movement of water. Most of the programs developed to meet this need are able to simulate non-steady flow in open channels, based on the Saint-Venant equations.

The effort to increase the efficiency of water systems has led to the desire to be able to evaluate the effectiveness of proposed operational strategies. This has put extra demand on simulation programs, as simple modelling of hydrodynamic flow is inadequate.

A special issue of the Journal of Irrigation and Drainage Engineering, ASCE was devoted to 'canal system hydraulic modelling'. The editorial task committee formulated

---

[1]Research assistant, [2]Ph.D Candidate, [3]Professor, Section of Land and Watermanagement, Delft University of Technology, Dept. of Civil Engrg., P.O. Box 5048, Stevinweg 1, Delft, The Netherlands.

comparison criteria to evaluate the various existing models. Among the criteria indicated were computational accuracy and robustness, user friendliness, ease of definition for a canal system in the model and the flexibility of graphical output (Clemmens, 1993)

It was found that although the existing hydrodynamic models usually simulate the unsteady flow of water through the water system accurately, the facilities to implement an operational algorithm are generally insufficient. To model an operational strategy the user is often required to program the strategy into the model, using a low level computer language. This can be very frustrating as hydrodynamic models are not often well documented, and are rarely user friendly (Burt, 1993). To alleviate these problems a link has been created between a hydrodynamic simulation package and an existing mathematical package.

A Description of the Hydrodynamic Model Used

A hydrodynamic model which gives the user the opportunity to change the parameters of for example structures during simulation was developed at the Delft University of Technology by W. Schuurmans (Schuurmans, 1992). This program, know as MODIS (Modelling Drainage and Irrigation Systems) is based on the RUBICON modelling package developed by HASKONING Engineering Consultants and Architects (Verwey and Haperen, 1988). The model uses the full Saint-Venant equations to calculate the behaviour of water in open channels.

This model has been used at the Delft University of Technology for a number of studies, mainly in the scope of research on projects involving the modelling of not only irrigation systems, but also river systems. These projects often involved the implementation of control algorithms to maintain the water system at a predefined desired state.

As an example of a control algorithm, the sill level of a gate may be manipulated by a certain amount, depending on the deviation of water level at the downstream end of the pool directly below the gate in question. The amount by which the gate is manipulated is calculated by a Proportional Integral algorithm (PI). El-Flow (Lowell F. Ploss, 1987) uses this principle.

It became apparent that although the actual hydrodynamic calculation was robust and accurate (Schuurmans, 1993), the implementation of algorithms in the model was laborious and took a disproportionate large amount of time. To simplify the implementation of algorithms a link was made between the hydrodynamic model MODIS and a mathematical package MATLAB.

The MATLAB package has been under development for a number of decades. There are a large a number of advanced control methods and mathematical functions available within the MATLAB Package and these can now readily be combined with the hydrodynamic simulation model. Additional routines are easy to develop. The advantage it has in the development of programming new procedures is that the language is non-compiled and script based. As a consequence the actual programming is interactive and user-friendly. Another advantage of the MATLAB package is that it has advanced graphical routines; with the combination of MODIS and MATLAB it is possible to give a graphical representation of the simulation results at each time step. The user can thus 'follow' the simulation while it is in progress, and in case of a modelling error the simulation can be aborted. This in contrast to running the simulation and analysing the results afterwards.

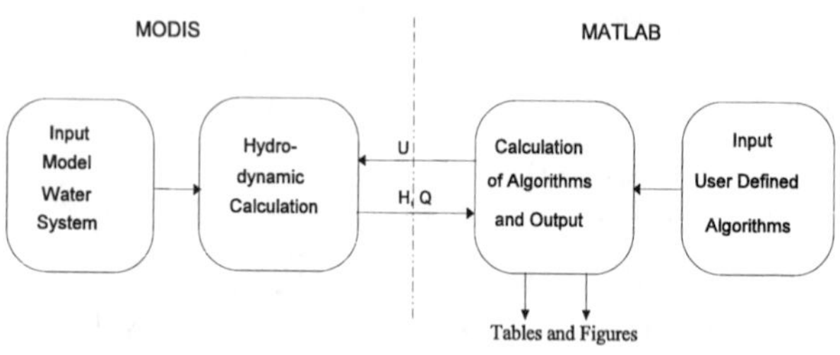

Figure 1 Structure of link MODIS and MATLAB (from Werner, 1995)

Communication between MODIS and MATLAB

The actual hydrodynamic calculation is performed by the MODIS side of the model. This is called at each time step as a sub-routine from MATLAB. MODIS then calculates the discharges and water levels at each grid-point in the system modelled and sends these back to MATLAB. If a control algorithm is in use then this is evaluated in MATLAB and the new parameter values (for example manipulated sill levels) are passed to MODIS. These are subsequently used for the next time step. The updating of plots showing water levels and discharges is done by MATLAB. See figure 1 for the structure of the link between MODIS and MATLAB. To increase user friendliness further a shell is under development through which the user can operate the model.

Creating a user-friendly shell

Because the MATLAB version used works in a WINDOWS operating environment there are possibilities of creating a shell to send controls to MATLAB. The development of a user-friendly interface increases the usability of the model as the user is not required to familiarise him/herself with complicated computer procedures. For the combination of MODIS and MATLAB a menu driven shell is under development through which the user can implement the model of the water system, as well as run the simulation. A graphical representation of the water system can also be generated so that the user can easily see if the input is correct. Figure 2 shows a part of the menu driven shell.

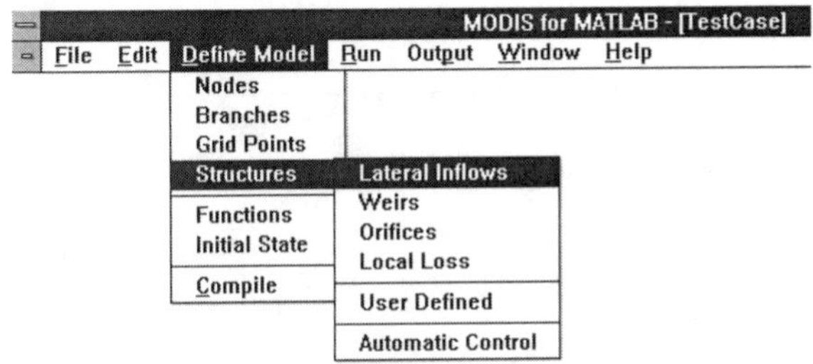

Figure 2 Part of the menu driven shell for the MODIS/MATLAB link

The user can select the desired operation and the shell then simply sends commands which are processed by MATLAB as if they were typed in at the MATLAB command line. Setting up the communication in this manner gives the user the choice between running the simulation from the shell program, or directly from MATLAB. Thus giving user-friendliness via the shell on the one hand, while not losing any flexibility if the user would like MATLAB to perform an option not included in the shell on the other.

A simple case study as an Example

To give an idea of the graphical output as produced by the simulation model a simple case is discussed. It considers an elementary channel layout of a main channel with a length of 6 km. There is a 1 km long side branch at the halfway mark. In the main channel there are four weirs set at regular intervals. Figure 3 illustrates the layout.

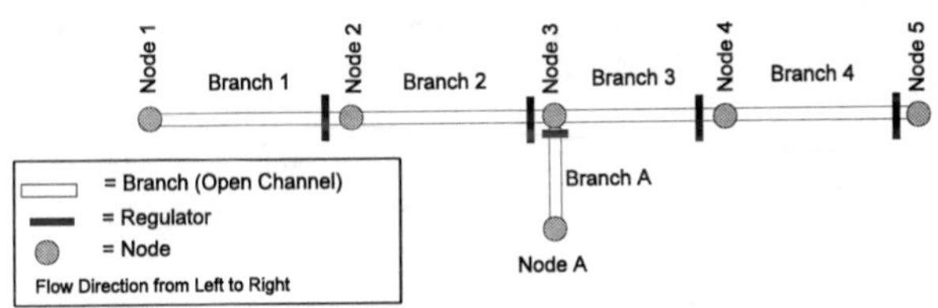

Figure 3 Layout of channels in this simple case (not to scale)

The discharge from the top of the channel is set at 1.5 m^3/s and remains constant. The orifice at the start of Branch A is set to a discharge of 0.25 m^3/s. At a time of 40 minutes this orifice is set to a discharge of zero. Figure 4 shows the longitudinal discharge line resulting from a simulation run (calculation time step is 60s, total simulation time is 240 minutes). Until time step t = 2400s there is a sharp drop in the discharge line. This is the 0.25 m^3/s going down Branch A. After the orifice is closed the discharge in the lower half of the main channel increases with time. The figure below was actually created during the simulation run. The user can follow the simulation in this manner. The water levels and discharge at any location in the system can also be plotted against time.

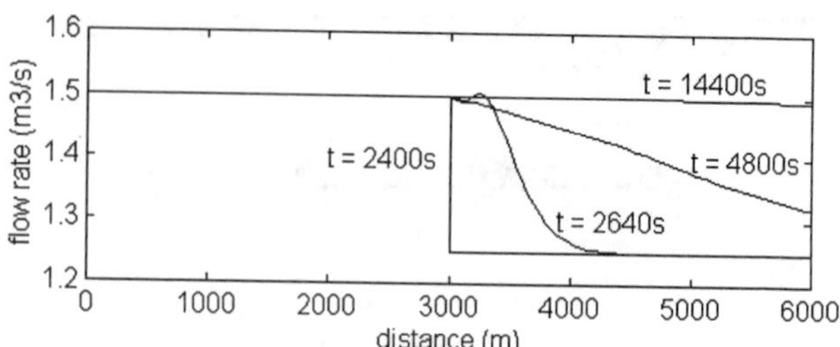

Figure 4 Longitudinal discharge lines for test case

Conclusions

With the combination of the hydrodynamic model MODIS and the MATLAB package improvements have been made to the model in three aspects. Firstly, the implementation of controller algorithms is greatly simplified, giving the user more time to design and investigate the algorithms, rather than spending time on laborious programming. Secondly, the user friendliness of the model has been improved. The input of a water system into the model is less complicated Thirdly, the versatility of the graphics produced by the model is the same of the MATLAB package. The user can define a wide variety of plots (even including 3D plots) and link these with the screen, the printer or files for use in word processing programs. With the model in its present form the simulation of water systems has become much more accessible.

References

Burt, C.M. (1993), "Irrigation-Canal-Simulation Model Usage" , *Journal of Irrigation and Drainage Engineering*, ASCE, 119(4), 631-636.

Lowell F. Ploss, M. (1987), "Canal Automation using the Electronic Filter Level Offset (El-Flo) Method" , Planning, Operation and Automation of Irrigation Water Delivery Systems, Portland, Oregon, USA.

Schuurmans, W.,. (1992), "A Model to Study the Hydraulic Performance of Controlled Irrigation Canals", Ph.D Thesis, Delft University of Technology, Delft, the Netherlands

Schuurmans, W. (1993), "Description and Evaluation of Program MODIS", *Journal of Irrigation and Drainage Engineering*, ASCE, 119(4), 735-742

Verwey, A., and van Haperen, M.J.M. (1988), "HD-System Rubicon", *J. Hydrosoft*

Werner, M.G.F, Schuurmans, J, Campfens, H.J., Brouwer, R., (1995), "Application of a hydrodynamic flow model in the operational design of water systems", ICID (to be published)

Linear Approximation Model of the WM Canal for Controller Design

J. Schuurmans, M. Ellerbeck[1]

Abstract
A linear approximation model is presented that describes open-channel flow with backwater effects; the approximation model allows the application of effective control synthesis methods. The accuracy of the approximation model for the WM canal is verified using field data.

Introduction

Automatic control can increase the efficiency of irrigation systems considerably. However, canal automation appears to be rather difficult. Currently, both hardware problems and theoretical problems exist.

The authors believe that the major theoretical problem is how to design a control system that (1) satisfies the requirements of the irrigation water management, (2) performs in the most optimal way and (3) can actually be implemented in the field.

For this problem an open-channel flow model is needed that is both sufficiently accurate for designing control systems and, at the same time, as simple as possible. Simplicity of the model is required to obtain a simple control system when applying automatic controller synthesis methods; the more simple the control system is, the higher the possibility that the control system can be implemented in the field.

Most feedback controller synthesis methods require the model to be described by linear ordinary differential or difference equations. However, the physical balances that describe open-channel flow, the St. Venant equations (Chow, 1959), can not be used in most synthesis methods, because they are nonlinear partial differential equations. To obtain a linear model that is in the desired form, the St. Venant equations could be linearized and discretisized (among others: Reddy 1990; Balogun 1988). However, these models are usually of high order, and as a result the control system is complex.

[1] PhD Candidate, Delft University of Technology, Department of Civil Engrg., P.O. Box 5048, Stevinweg 1, Delft, The Netherlands; Msc Candidate, Delft University of Technology.

Therefore, it is useful to consider simplified models. Corriga *et al* (1982; 1983; 1988) presented a model that might be considered as a reservoir model for each reach. Ermolin (1992) and Papageorgiou (1985) proposed an approximation for reaches with uniform flow. However, in all these approximations it is assumed that flow is uniform, whereas in most controlled channels flow is non-uniform due to the structures; in fact, the water upstream of the structures is usually backed up, causing the backwater effect (Chow, 1959).

This paper presents an approximation model for flow in an open-channel with backwater effects, of which the model parameters can be derived analytically. The accuracy of the model is verified using field data measurements of the WM canal in the Maricopa Stanfied Irrigation District.

Linear approximation model

The system under consideration is an open-channel with motor-operated structures. Flow is described by the St. Venant equations. It is assumed that the system is operated close to steady state conditions in which the water profile is a backwater profile. Figure 1 presents a schematic representation of a backwater profile. The part of the reach affected by backwater will be referred to as the backwater part, the unaffected part is denoted here as the part with uniform flow.

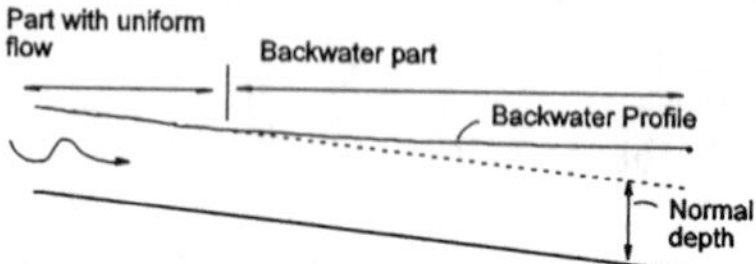

Fig. 1 Backwater profile

For the system under consideration an approximation model is proposed that describes the relation between flows through structures, such as gates and hydropower stations (these flows serve as the inputs of the model) and the water levels at the downstream ends of each channel reach (the outputs of the model).

For the backwater part the following approximation is proposed (Schuurmans et al, 1994):

$$\frac{dh(x,t)}{dt} = \frac{1}{A_e}\left(q_{in}(t) - q_{out}(t)\right) \tag{1}$$

where: A_e = area of the water surface in the backwater part (m^2); h = variation of water level with reference to its initial level (m); q_{in} = variation of inflow with reference to its initial value (m^3/s); q_{out} = variation of outflow with reference to its initial value (m^3/s).

In the part of the reach with uniform flow the dynamics is approximated by the kinematic wave model:

$$q(x,t) = q(0,t-t_d)$$
$$t_d = \frac{2x}{(1+a)V_0} \qquad (2)$$

in which the parameter a is given by:

$$a = 1 + \frac{4}{3}\frac{P_0}{T_0}\left(\frac{dR}{dY}\right)_0$$

A = cross sectional area (m^2); P = A/R wetted perimeter of cross section; R = hydraulic radius (m); t = time (s); T = width of channel at surface level (m); V = Q/A mean velocity of flow (m/s); x = distance in the direction of flow (m); Y = depth of flow (m);
The subscript zero is used to indicate initial values.

This model expresses a pure time delay between the inflow at the upstream head of the reach and the outflow at x. To derive an approximation model for a reach it must be known whether the backwater affects the whole reach or just part of it. If it affects the whole reach, the reservoir model, given by Equation (1), can be applied; otherwise, the inflow of the reservoir model is delayed according to Equation (2).

Application: linear model for the WM canal

An approximation model has been derived for the WM canal consisting of eight reaches.

The geometry of the canal was gathered from design drawings. Figure 2 shows a side view of the channel and a sketch of its initial water profile. The channel has relative steep slopes. Flow is mostly supercritical in the uniform parts, showing a hydraulic jump when entering the backwater parts.

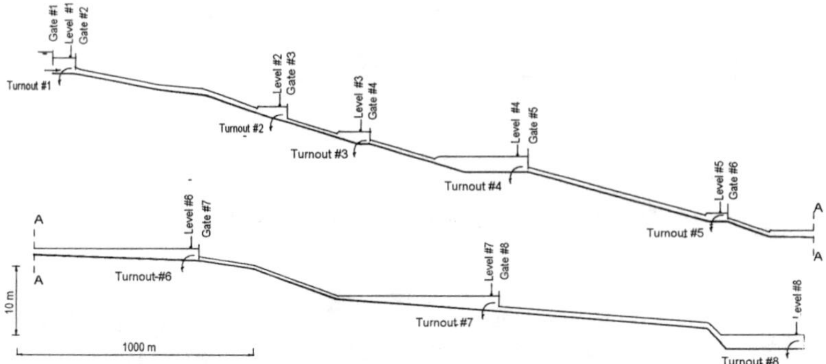

Fig. 2 Longitudal section of the WM canal (to scale).

For each individual part of the channel in which the slope differs from other parts, a delay time is computed using Eq. (2). The water profile in the backwater part is assumed horizontal and to run from the downstream end up to the location where it intersects the water profile in the part with uniform flow (which runs parallel to the bottom of the canal at normal depth); these assumptions made it possible to derive analytical formulae for the delay times and the surface areas of each backwater part.

Experimental verification

Part of the model was verified experimentally just before this paper was written. The results of the verification experiment are shown in Figure 3. The parameters of the linear model (delay times and surface areas of the backwater parts) were computed for the initial flow conditions as estimated in the canal. During the experiment, gates # two and three were manipulated. The response of the model is reasonably accurate.

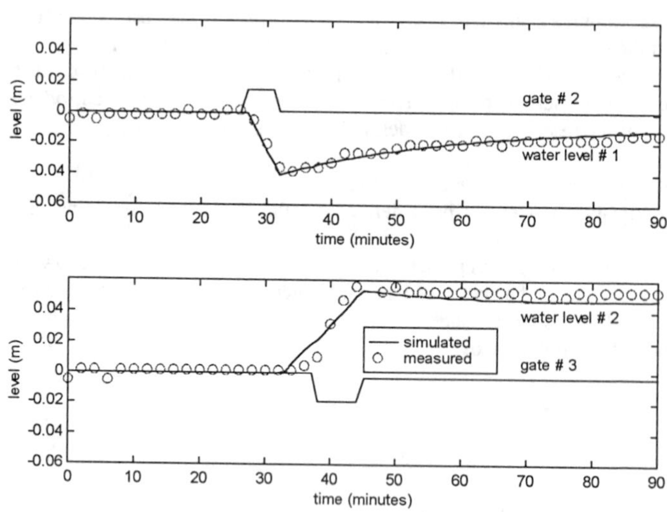

Fig. 3 Variations of gate openings and water levels with respect to their initial levels.

Conclusions

An approximation model of open-channel flow with backwater effects has been proposed. For the WM canal a linear model has been derived, of which part of it was verified experimentally. The results are promising.

References

Balogun O.S, Hubbard M, De Vries J.J. (1988). *Automatic control of canal flow using linear quadratuc regulator theory*. J. Hydr. Engrg. ASCE, 114(1), 75-102

Chow, V.T. (1959). *Open-channels hydraulics*. McGraw-Hill Book Co., Inc., New York

Corriga, G., Fanni, A., Sanna, S. and Usai, G. (1982). A constant-volume control method for open-channel operation. *Int. J. Modelling and Simulation* **2**, 108-112

Corriga, G., Sanna, S. and Usai, G. (1983). Sub-optimal constant-volume control for open- channel networks. *Appl. Math. Modelling* **7**, 262-267

Corriga, G., Salembeni, D., Sanna, S. and Usai, G. (1988). A control method for speeding up response of hydroelectric stations power canals. *Appl. Math. Modelling* **12**

Ermolin, Y.A. (1992). Study of open-channel dynamics as controlled process. *J. of Hydraulic Engineering* **Vol. 118**, 1, 59-71

Papageorgiou, M., Messmer, A. (1985). Continous-time and Discrete-time Design of Water Flow and Water Level Regulators. *Automatica* **Vol. 21**, 6, 649-661.

Reddy, J.M. (1990). *Local optimal control of irrigation canals*. J. Irrig. and Drain. Engrg., ASCE, 116(5), 616-631.

Schuurmans, J., Bosgra, O.H. and Brouwer, R. 1994. Open-channel flow model approximation for controller design. Draft copy, Delft, The Netherlands.

Neurocontrol of Irrigation Canals

J. Mohan Reddy[1] and N. Saratchandra Babu[2]

Abstract

A neural network based system (irrigation canal) identification model was obtained using a set of simulated input-output data from a canal. This model was then used to derive the control input (gate opening) to maintain a constant water level at the downstream-end of the pool. The results obtained are very promising in terms of obviating the need for collecting detailed information regarding the physical properties of an irrigation canal.

Introduction

Design of efficient control algorithms for irrigation canals operated under demand delivery has been a challenging task, particularly in the absence of accurate information on canal properties. In the past, tedious trial and error methods were used to derive canal control algorithms. Recently, optimal control theory has been applied to derive control algorithms for irrigation canals. Application of optimal control theory provides a direct solution for canal control algorithms. However, when lumped parameter models are used to derive control algorithms for irrigation canals, the number of state variables that must be used in the feedback loop becomes large. Consequently, it becomes expensive to implement feedback control algorithms when the number of state variables that must be measured is very large. To minimize the cost of implementing feedback control algorithms, the number of measurements that must be used in the feedback loop must be kept to an absolute minimum. Since two flow depths per pool are normally measured in practice, it is preferable (and possible) to derive control algorithms that require only two flow depth measurements per pool. One method of reducing the number of measurements required per pool is to use a state estimator or observer. The design of observers based upon the pole-placement technique (Reddy et al 1992) and the Kalman Filter (Reddy 1995) was presented earlier. An alternative method to minimize the cost of implementing the feedback control algorithms is to derive control algorithms that require only two flow depth measurements per pool. In addition, in many irrigation projects, accurate information on canal properties is not available. The

---

[1]Professor, Department of Civil Engineering, University of Wyoming, Laramie, WY 82071, USA

[2]Joint Director, Microprocessor Applications Engineering Program, Department of Electronics, New Delhi, India.

performance of the feedforward or feedback control algorithms derived based upon the inaccurate information may not be satisfactory. Since it is expensive and time consuming to obtain accurate data on canal geometry and roughness values, alternative methods for deriving canal control algorithms that use least amount of data on channel geometry and, at the sametime, require smaller number of measurements per pool are desirable. Here, a neural network based methodology for the identification and control of an irrigation canal is presented. This methodology does not require any information on canal properties, and it uses only two flow depth measurements per pool.

Regulation of Irrigation Canals Using an Optimal Controller

In the operation of irrigation canals, decisions regarding the gate opening in response to random changes in the water withdrawal rates into lateral or branch canals are required to maintain the flow rate into the laterals close to the desired value. This is accomplished by maintaining the depth of flow in a given pool at the target value. This problem is similar to the process control problem in which the state of the system is maintained close to the desired value (or the deviations from the target values of the state variables) by using real-time feedback control. Here, the problem is formulated in terms of the deviations in the values of the state variables. The Saint-Venant equations of open-channel were linearized using a finite-difference technique and the Taylor series. The discrete-time version of the system dynamic equation is given as:

$$\delta \mathbf{x}(k+1) = \Phi\ \delta \mathbf{x}(k) + \Theta\ \delta \mathbf{u}(k) + \Psi\ \delta \mathbf{q}(k) \tag{1}$$

in which Φ, Θ and ψ = discrete-time versions of the system feedback, control distribution, and disturbance distribution matrices, respectively; k = sampling instant; $\delta \mathbf{x}(k) = \ell \times 1$ state vector; $\delta \mathbf{u}(k) = m \times 1$ control vector; $\delta \mathbf{q}(k) = p \times 1$ matrix representing external disturbances (changes in water withdrawal rates) acting on the system; ℓ = number of dependent variables in the system; m = number of controls; and p = number of distributed disturbances acting on the system. Equation 1 can be used to simulate the evolution of the system as a function of time given the initial conditions ($\delta \mathbf{x}(0)=0$), the external disturbances acting on the system ($\delta \mathbf{q}$), and the change in gate opening ($\delta \mathbf{u}$). In Eq. 1, the lateral withdrawals were modeled as follows:

$$\delta q_{i,N}(k) = q_{i,N}(k) - \bar{q}_{i,N}(k) \tag{2}$$

in which $q_{i,N}$ (k) = historical or calculated withdrawal rate from node N of pool i at time instant k; $q_{i,N}$ (k) = historical or calculated average withdrawal rate from node N of pool i; and $\delta q_{i,N}(k)$ = deviation from the average flow rate at node N of pool i at time instant k, and is called the disturbance acting on the system. In canal operations, the gate opening is the unknown. Application of linear optimal control theory helps in the derivation of a direct expression for the control algorithm, which is given below:

$$\delta \mathbf{u}(k) = -\mathbf{K}(k)\ \delta \mathbf{x}(k) \tag{3}$$

where K(k) = $m \times \ell$ controller gain matrix. Once the elements of the K matrix are derived, measured deviations in the values for the state variables must be available to calculate the required change in the gate opening to maintain the depth of flow at the downstream end of the pool at the target value.

System Identification and Control Using a Neurocontroller

A neural network is a massively parallel distributed processor that is designed to model, in a primitive fashion, the functioning of a brain; hence the name artificial neural network. The network has a natural propensity for storing experiential knowledge and making it available for use. Neural networks basically consist of several neurons which are the basic information processing units. The three basic elements of a neuron are explained below.

Synapses or connecting links: These are characterized by the strength or weight of its own, and are used to weigh the input received from a sensor in producing the specified output from the system. The weight is positive if the associated input is excitary ; it is negative if the input is inhibitory. A signal x_j at the input of synapse j connected to neuron k is multiplied by the synaptic weight w_{kj}. The first subcript referes to the neuron in question, and the second subscript refers to the input end of the synapse to which the weight refers.

$$u_k = \sum_{j=1}^{p} w_{kj} x_j \tag{4}$$

Adder: The net effect of all the inputs on a given neuron (output) is obtained by summing the products of the synaptic weights and the associated input strengths. This is basically a linear operation.

Activation function: This function is used to limit the amplitude of the output of a neuron. Usually, an externally applied threshold (similar to a sensory input) is used to lower the net input to the activation function. Several types of activation functions such as piece-wise linear, sigmoidal, and hard-limiting type functions are used. The general activation function is defined as follows:

$$y_k = \varphi(u_k - \theta_k) \tag{5}$$

in which y_k is the output from neuron k, θ_k is the threshold used at neuron k, and φ is the user defined function. The sigmoid function is by far the most commonly used form of the activation function, and is given as

$$\varphi(v) = \frac{1}{1+\exp(-av)} \tag{6}$$

Artificial Neural Networks(ANN) can be placed into one of three classes based on their feedback link connection structure: recurrent structure (global feedback connections), local recurrent structure (local feedback connections, e.g., cellular neural networks), and nonrecurrent (no feedback connections). A special type of nonrecurrent ANN is the feedforwardneural network(FNN), which consists of layers of neurons with synaptic (weighted) links connecting the outputs of neurons in one layer to the inputs of neurons in the next layer. Once a structure is selected for the given problem, the neural network must be trained to achieve the desired task. This is done in several ways. One of the methods used to train (or learn) the network is to use a set of input-output pairs or patterns, and modify the synaptic weights of the network until the error between the predicted and the desired (target) output is minimized.

The procedure used to perform the learning process is called a learning algorithm or a training algorithm, the function of which is to modify the synaptic weights of the network in an orderly fashion so as to attain a desired design objective such as pattern recognition, and control of a chemical process, etc. Here, a feed forward neural network is used to develop an on-line identification model of an irrigation canal. The model structure selected is as follows:

$$\delta \hat{y}_{dn}(k+n1) = f(\delta y_{up}(k), \delta y_{dn}(k), \delta y_{dn}(k-1), \delta u_{up}(k), \delta u_{up}(k-n2), \ldots) \quad (7)$$

in which $\delta y_{dn}(k+n1)$ = variation in flow depth at the downstream end of the pool at time - instant k+n1; and $\delta u_{up}(k)$ = variation in gate opening at time instant k. A neural network structure of the form shown in Figure 1 was used for the model. Once the system model (which is in terms of gate opening and the flow depths in the given pool), the equation can be solved for the required variation in gate opening (control input) to maintain the desired water level at the downstream end of the pool. The Widrow-Hoff learning algorithm (Zurada 1992) was used to identify the canal model.

Results and Discussion

An example irrigation canal with a single pool was considered. First, Eq. 1 alongwith the linear optimal control theory was used to derive a constant-level control algorithm for an irrigation canal with a single pool. In liu of data from an actual canal, the linearized Saint-Venant Equations were used to generate a set of input-output data for use in the system identification process. Using the network structure shown in Figure 1, and the Widrow-Hoff training algorithm (Zurada 1992), the following predictive model was obtained between the input-output variables:

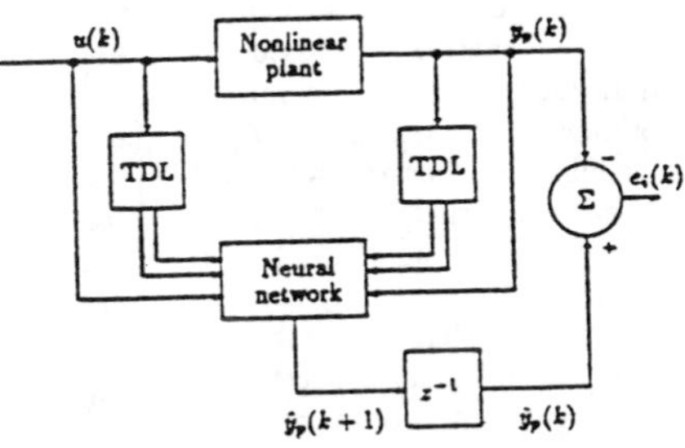

Figure 1. A Neural Network Based System Identification Model.

$$\delta y_{dn}(k+n1) = 0.0432\delta u_{up}(k-n2)+0.5642\delta u_{up}(k)-0.1961\delta y_{up}(k)-0.6970\delta y_{dn}(k-1)+0.9638\delta y_{dn}(k) \quad (8)$$

The above identification model was obtained after training the neural network (with 5 inputs and one output; no hidden layers) until the error between the actual and the predicted deviation in the depth at the downstream end of the pool was less than 0.0001. Equation 8 was re-arranged to obtain an expression for the control input, δu_{up} (k), as a function of the current condition of the canal (flow depths at the upstream and downstream ends of the pool), the time history of gate opening, and the future desired deviations in the water level at the downstream end of the pool.

The performance of the above controller (Eq. 8) in terms of maintaining a constant water level at the downstream end of the pool was simulated using a linearized model of the canal. The difference in the gate opening computed using the neural network based model and the optimal control theory based model is insignificant in the steady state (Figure 2). At the beginning of the simulation period, there was a large difference between the two models, but the difference in the predicted gate openings disappears very quickly. Similarly, the simulated water levels at the downstream end of the pool are presented in Figure 3. Once again, the difference between the neural network based algorithm and the optimal control based algorithm is insignificant in terms of maintaining a constant-level at the tail-end of the pool.

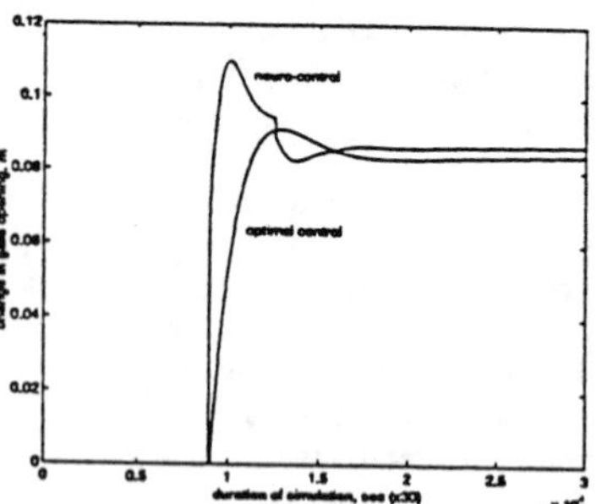

Figure 2. Predicted Gate Openings.

Summary and Conclusions

The performance of the neural networks based identification and control algorithms in terms of maintaining a constant water level at the downstream end of the pool was found to be acceptable. Since the system identification was based upon the set of input-output data collected from the given canal of interest, there is no need for detailed information on channel properties in order to design a control algorithm for irrigation canals. Research on the extension of the present technique for multiple pool canals is currently underway.

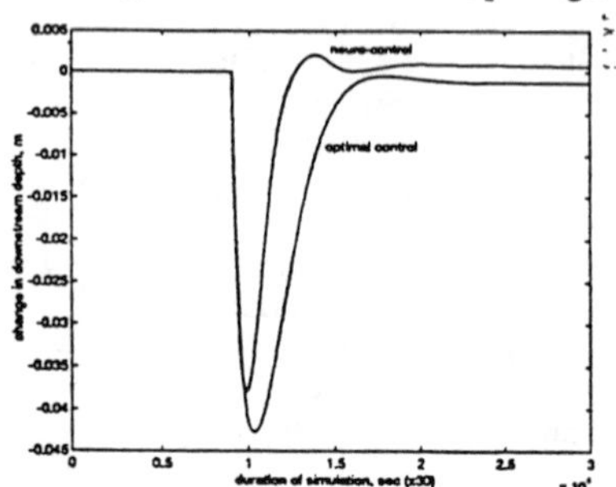

Figure 3. Simulated Water Levels at Downstream end of the Pool.

References

Kwakernaak, H., and Sivan, R. (1972). *Linear optimal control systems.* John Wiley, NY.

Narendra, K.S., and Parthsarathy, K. (1990). "Identification and control of dynamical systems using neural networks." *IEEE Transactions on Neural Networks*, 1(1):4-27.

Reddy, J.M., Dia, A., and Oussou, A. (1992). "Design of control algorithm for operation of irrigation canals." *J. Irri. and Drain. Engrg., ASCE*, 118(5),852-867.

Reddy, J.M. (1995). "Kalman filtering in the control of irrigation canals". *Applied Mathematical Modeling*, (in press)

Rudd, J.B. (1994). "Prediction and control of paper machine parameters using neural network models." Proc. of the *World Congress on Neural Networks*, June, San Diego, California.

Zurada, J.M. (1992). *Introduction to artificial neural systems.* West Publishing Company,NY.

Coping with EPA's Storm Water Permit for Hazardous Waste Facilities, Industrial Landfills, Wastewater Treatment Plants, and Recyclers

James P. Amick, P.E.
Member ASCE[1]

Abstract

The Environmental Protection Agency (EPA) has released NPDES storm water monitoring data from group applicants. This paper will analyze that data to determine which pollutants are present in storm water runoff from disposal, treatment, and recycling facilities.

EPA created 29 sectors of industry based on similar types of facilities. EPA assigned all members of all groups to one of the sectors. The monitoring data will be analyzed by sector to determine specifically which industry discharged which pollutants in storm water.

The paper's primary focus will be four sectors that included approximately 3,000 facilities. Several hundred facilities representing the hazardous waste treatment, storage, or disposal facilities; industrial landfills, land application sites, and open dumps; domestic wastewater treatment plants; and scrap and waste materials sectors submitted monitoring data. Each sector will be discussed and appropriate best management practices will be presented with the capability to reduce or eliminate the pollutants in storm water discharges.

Introduction

The Environmental Protection Agency (EPA) published National Pollutant Discharge Elimination System (NPDES) storm water regulations in the *Federal Register* on November 16, 1990 in response to the Clean Water Act amendments of 1987. These regulations required most industries and cities with a population of 100,000 or more to obtain NPDES permits for their storm water discharges. Three options were described for industrial dischargers to obtain permit coverage: (1) an individual application with considerable monitoring and other requirements, (2) file for coverage under a general permit that was not available until September 9, 1992, or (3) join a group of like facilities that had requirements similar to the individual application but with significantly less monitoring requirements for

---

[1]Project Manager, HNTB Corporation, 500 Throckmorton, Suite 1909, Fort Worth, Texas 76102 (817/429-9287)

most group members if the group contained a minimum number of facilities as specified in the regulations.

Group Monitoring Results

Approximately 45,000 industrial facilities chose to become members of group applications. Approximately 2,000 of these facilities submitted monitoring data to EPA as part of their group application. All group members who monitored their storm water discharges were required to analyze for a minimum of eight pollutants including: Oil and grease (O&G), pH, five day biological oxygen demand (BOD5), chemical oxygen demand (COD), total suspended solids (TSS), total phosphorus (Total P), total Kjeldahl nitrogen (TKN), and nitrate plus nitrite nitrogen (N+NN). Some facilities were also required to analyze for other pollutants.

The facilities in the four sectors analyzed for this paper were reported by EPA as follows: hazardous waste treatment including storage or disposal facilities (Haz-Waste) - two groups with 15 facilities; industrial landfills, land application sites, and open dumps (Landfill) - 36 groups with 1,358 facilities; scrap and waste materials (Recycler) - five groups with 297 facilities; and domestic wastewater treatment plants (Wastewater) - 44 groups with 1,337 facilities. Based on this analysis of the submitted monitoring data, the approximate number of facilities (varies depending on pollutant) that submitted data is: Haz-Waste - 9, Landfill - 58, Recycler - 137, and Wastewater - 105. All results are based on grab sample data.

This paper presents the number of samples, the mean values, the median values, the percentage of samples that are greater than the mean values published in the *Results of the National Urban Runoff Program* (NURP) in December 1983 by EPA, and the observed 95 percent value in the following two tables. Minimum values reported for Recyclers were non-detect or zero. At least one pollutant was detected at all facilities for the other three sectors. The results (except for COD and TSS since those numeric values are large compared to the other pollutants) of the mean values are also presented graphically following the tables.

This analysis used the following assumptions: (1) All reported values are observed values and have not been adjusted for standard statistical distributions, (2) Raw data contains some blank fields that were considered to be equal to 0.0, and (3) Values were compared to the NURP mean values for all pollutants except oil and grease, which were compared to the storm water effluent limit of 15.0 mg/l. The NURP mean values (all are in mg/l) used were: BOD5 - 10.2, COD - 77, N+NN - 0.84, TKN - 1.96, Total P 0.40, and TSS - 123. This paper does not present an analysis for the pH of the samples.

All sectors analyzed had at least two pollutants that exceeded the NURP mean value by more than 50 percent (the benchmark value in this paper). The reported values indicate that the Haz-Waste sector has the least polluted storm water discharges (only BOD5 and TSS were reported above the benchmark). The Landfill sector was the next best, exceeding the benchmark for COD, N+NN, TKN, Total P, and TSS. The Wastewater sector had the second worst discharges exceeding the benchmark for BOD5, COD, N+NN, TKN, and Total P. The Recycler sector has the most polluted discharges, exceeding the benchmark for BOD5, COD, N+NN, TKN, Total P, and TSS. Interestingly, only the Wastewater sector exceeded the effluent limit for O&G, and only by approximately 50 percent.

Pollutant	Sector	No. of Samples	Mean (mg/l)	Median (mg/l)
BOD5	Haz-Waste	9	16.3	10.0
BOD5	Landfill	59	14.3	7.0
BOD5	Recycler	137	22.9	8.0
BOD5	Wastewater	114	30.2	13.0
COD	Haz-Waste	8	104	41.0
COD	Landfill	59	119	31.0
COD	Recycler	137	252	120
COD	Wastewater	103	124	61.3
O&G	Haz-Waste	9	10.1	2.0
O&G	Landfill	55	4.0	2.0
O&G	Recycler	137	9.0	5.0
O&G	Wastewater	105	22.5	3.3
N+NN	Haz-Waste	9	0.46	0.47
N+NN	Landfill	58	1.44	0.49
N+NN	Recycler	144	1.73	0.61
N+NN	Wastewater	102	17.51	0.94
TKN	Haz-Waste	9	1.43	1.30
TKN	Landfill	58	3.28	1.10
TKN	Recycler	137	3.60	2.10
TKN	Wastewater	91	9.67	1.60
Total P	Haz-Waste	9	0.24	0.07
Total P	Landfill	58	0.92	0.50
Total P	Recycler	137	0.81	0.29
Total P	Wastewater	105	0.92	0.50
TSS	Haz-Waste	9	300	82
TSS	Landfill	59	2,922	646
TSS	Recycler	137	439	148
TSS	Wastewater	115	174	69

Pollutant	Sector	% of Samples >NURP Mean	Observed 95% Value (mg/l)
BOD5	Haz-Waste	44	43.0
BOD5	Landfill	25	46.0
BOD5	Recycler	41	86.0
BOD5	Wastewater	56	51.0
COD	Haz-Waste	25	129
COD	Landfill	34	617
COD	Recycler	66	1,100
COD	Wastewater	42	370
O&G	Haz-Waste	11	7.0
O&G	Landfill	4	10.0
O&G	Recycler	11	32.0
O&G	Wastewater	12	21.0
N+NN	Haz-Waste	0	0.65
N+NN	Landfill	40	3.75
N+NN	Recycler	40	3.30
N+NN	Wastewater	53	116.0
TKN	Haz-Waste	11	1.71
TKN	Landfill	28	12.0
TKN	Recycler	55	11.1
TKN	Wastewater	45	13.0
Total P	Haz-Waste	11	0.14
Total P	Landfill	53	3.35
Total P	Recycler	39	2.19
Total P	Wastewater	63	2.23
TSS	Haz-Waste	44	924
TSS	Landfill	71	11,040
TSS	Recycler	50	2,096
TSS	Wastewater	33	575

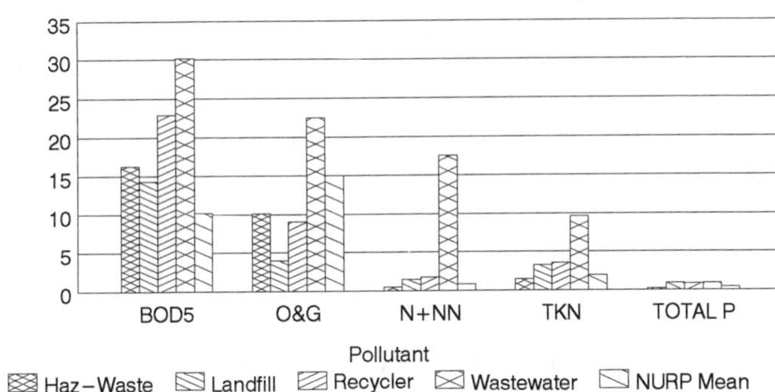

All four sectors have analytical wet weather monitoring requirements in the proposed Multi-Sector Permit published in the *Federal Register* on November 19, 1993. The only pollutants discussed below are the seven that were analyzed for this paper. All four sectors have monitoring requirements for additional pollutants. The Haz-Waste sector has monitoring requirements and cutoff concentrations (all in mg/l) for COD - 65 and TKN - 1.50. Based on the results of this analysis, less than half of those facilities will exceed the cutoff concentrations and be classified as "dirty". The Landfill sector has monitoring requirements and cutoff concentrations (all in mg/l) for Total P - 0.33 and TSS - 100. Based on the results of this analysis, more than half of those facilities will exceed the cutoff concentrations and be classified as "dirty". The Recycler sector has monitoring requirements and cutoff concentrations (all in mg/l) for COD - 65, N+NN - 0.68, and TKN - 1.50. Based on the results of this analysis, more than half of those facilities will exceed the cutoff concentrations for COD and TKN, but less than half for N+NN. However, a facility that exceeds any cutoff concentration will be classified as "dirty". The Wastewater sector has monitoring requirements and cutoff concentrations (all in mg/l) for N+NN - 0.68 and Total P - 0.33. Based on the results of this analysis, more than half of those facilities will exceed the cutoff concentrations and be classified as "dirty".

Conclusions and Recommendations

Many of the facilities that submitted monitoring data show significant concentrations of pollutants to be present in their storm water discharges. Several types of Best Management Practices (BMP) could be used to improve the quality of storm water discharges from these facilities. Suggested BMP for Haz-Waste and Wastewater facilities include vegetative swales and practices, reuse of collected storm water (such as for a process or as an irrigation source), inlet controls (such as oil/water separators), snow management activities, infiltration devices, and wet detention/retention devices. Suggested BMP for Landfills include seeding, mulching, geotextiles, silt fences, earth dikes, gradient terraces, drainage swales, sediment traps, check dams, pipe slope drains, level spreaders, storm drain inlet protection, rock outlet protection, reinforced soil retaining systems, gabions, and temporary or permanent sediment basins. Suggested BMP for Recyclers include diversion devices, covers, sand filters, dry absorbents, containment trenches, and many of the BMP listed above for Haz-Waste, Landfill, and Wastewater facilities.

AN OVERVIEW OF WATERSHED HYDROLOGY AND TRANSPORT MODELS NON-POINT SOURCE LOADINGS TO RECEIVING WATERS

Edward Z. Hosseinipour[1] and Conrad Heatwole[2]

Abstract

This paper provides an overview of hydrologic and Non-Point Source (NPS) water quality models with emphasis on the practitioners need for models that are functional and meet the needs of engineering projects often in lieu of regulatory needs. Protection of water and natural resources are mandated by national and state laws. The professional community in the analysis of stormwater management, waste load allocations, and treatment requirements of processed waters work on the basis that the results of their analysis must meet or surpass the regulatory guidelines. Therefore, often the main goal of the projects are to come up with solutions that are economically feasible and technically defensible for natural resource management, waste treatment, and implementation of Best Management Practices (BMPs) that meet regulatory criteria for receiving water quality. Application of models to better understand the system being analyzed and evaluate various alternatives and management scenarios are often an integral part of environmental projects. In an effort to facilitate the selection of modeling tools for the engineering/scientific and regulatory professionals, ASCE surface water hydrology committee has formed a Task Committee to review and evaluate the cadre of mainly public domain available watershed and NPS models.

Introduction

Watershed hydrology and NPS transport modeling may be performed for a variety of objectives. Some of the most important are:

a. Characterizing the runoff quantity/quality as to temporal and spatial extent (localized and total hydrographs), and the resulting water quality as it relates to concentration or load range for different contaminants (pollutographs, sedigraphs), etc.

---

[1]Water Resources Eng., Law Eng. & Environ., 114 Town Park Drive, Kennesaw, GA 30144.

[2]Associate Professor, Biological Systems Engineering Department, VPI, Blacksburg, VA 24061.

b. An extension of the above to provide information for the analysis of receiving water quality; that is, input to drive a surface water or groundwater quality model.

c. Using the information provided by the model to choose among the different management alternatives, that is, determining the effects, magnitude, extent, etc. of control and/or remediation options in response to environmental regulations.

d. Economic analysis of management options, i.e., cost-benefit studies.

In recent years as control of point source discharges has reached its practical threshold, assessment of loadings on a watershed scale has become a challenge for which there are not well developed methods. At the same time emphasis is being placed on the reduction of non-point sources of pollution. In fact, even in the regulation of effluent discharges to the receiving waters, more and more the holistic approach of watershed management and control of all sources of loadings are being pursued due to the newly changed focus from treatment to source control. In part, this is due to the fact that no matter how much treatment is regulated upon the dischargers, the receiving waters at times are at/or beyond their assimilative capacity limits to receive effluent of any kind. Regulatory agencies (EPA) are now stressing more integrated water quality management that consider all sources of pollution within a watershed. To make the case, basin wide hydrologic studies are recommended to show that storm water loadings and other component of NPS needs to be considered and managed if the water resources of the basin are to be protected and the economic base preserved. To this end, many models have been developed to address some of these concerns. Although these models are reported and discussed in technical literature to a great extent, there are no standard criteria for applied professionals for the selection of the most appropriate model for a specific project at hand. Part of the complexity involves the fact that the available models are similar enough to the point that could confuse the person looking for the right tool, and are very different with respect to spatial and temporal scales, type and methods of constituent evaluation and modeling, and treatment of different geographical settings, and water flow in the hydrologic cycle, etc. Although several conferences and symposia have been held in recent years on hydrologic and NPS modeling advances most of the papers deal with latest research findings rather than application of models. Recent literature dealing with reviews of models include: US EPA (1992), Imhoff (1993), and Hosseinipour (1993).

NPS Modeling Approach

In general, the objectives of a watershed NPS modeling project are to study pollution trends and consequent water quality impact, develop plans for pollution abatement/reduction, prioritized source controls and geopgraphic areas for remediation, and to enhance water quality and ecosystem. To this end, a well established strategy for developing a comprehensive watershed scale modeling system that enables evaluation of the full impact of multiple chemical releases to the aquatic environment is needed. In brief, the common steps to be followed in a hydrologic and diffuse source water quality modeling project are: (a) having a clear objective for the modeling effort; making sure that modeling is really necessary, (b) if it is determined that modeling is necessary then start with the simplest model that will satisfy the project objectives, (c) choose models consistent with the objectives of the project and availability of data; in other words no oversight with regard to the means to support the model and no overkill, (d) start with one or two of the most important quality parameters for a suitable time period and when satisfied with the results build upon it, try to keep the modeling work simple and manageable from the beginning, (e) once a calibrated and tested model of a site is in hand, perform a sensitivity analysis of the model parameters, and if need be to recalibrate and verify the model for the site, (f) use the functional model of the site for evaluation of desired scenarios and comparison of alternative management methods, (g) plan for a modeling data base and integration of Geographic Information System (GIS) with the model. Most of the recently developed models are more comprehensive than older models and require a lot more input data that are not often available for the site. Therefore, care must be taken in the early stages of the project to select a model which input requirements are consistent with the available data at the site. Development of most available models have been through public funding and therefore most functional models are in the public domain. In the following sections after classification of models, a few in each category are briefly discussed.

Classification of NPS Models

Nonpoint source loading models may be classified by spatial, temporal, purpose/scope, and component processes considered. They are categorized by spatial configuration into three groups of field, watershed, and basin scale. From temporal point of view, the models may be event based or continuous simulation, or have capability for both; event based models are not recommended for long term water quality assessment. The range of processes considered include physical, chemical, and biological. In terms of purpose and scope, the models may be either screening, management, or research models. Screening models rely on readily available inputs and represent only the most important processes, and range from index models to simulation models. This group is generally used in comparative studies for spatial delineation (i.e., identifying areas in a watershed with high erosion potential) often in association with GIS to compare possible outcomes from alternative BMPs at a specific site.

Management models incorporate more of the processes pertinent to the system being modeled and also represent those processes in greater detail in the model algorithm and input parameters. Management models can thus evaluate a much wider range of scenarios. Research models are most detailed in terms of processes used and parameters required. These models are used to study the response of the system and to evaluate different modeling techniques. Though research models are not readily applicable because of the difficulties encountered in determining appropriate parameter values, they may be used in cases where the best technology is needed. While models in the extremes (screening, research) may be restricted in their range of applicability, models in the middle in terms of complexity may be used in applications that range from screening to research. The focus of the rest of this paper will be on models that are generally characterized as management/functional models.

Field Scale Models

Field scale models are used to evaluate the effects of land use or management alternatives for "fields", with a field being an area small enough to be considered spatially uniform, although multiple layers in the soil profile can generally be considered. In fact, the size and spatial configuration of the site is a very minor factor since most calculations are performed on a unit area basis. These "unit-source-area" models typically use long-term (multi-year) simulation as a means of accounting for the impacts of meteorological variability on upland hydrology, and as a means of representation for long-term scenarios. Brief discussion of a few of the more popular field scale models are given here. **CREAMS**- The CREAMS was the first of the field-scale models developed by the USDA as a management tool to evaluate the effects of agricultural practices on receiving water quality due to loading from runoff, erosion, nutrient and pesticide transport. CREAMS modeling concepts were subsequently used in the development of other models. CREAMS has been widely applied and customized for site specific applications to the point that several years ago a CREAMS symposium was held in Athens, Georgia. GLEAMS in based on CREAMS hydrology and erosion but incorporates a new and more detailed pesticide transport and transformation component in the root zone. In its latest release (1993) the model includes a nutrient component that is one of the most comprehensive available. This release of GLEAMS in effect is a replacement for CREAMS and older GLEAMS. GLEAMS considers water balance and chemical dynamics through multi-layers in the soil profile using a daily time step for upto 50 years of simulations. **PRZM**- The Pesticide Root Zone Model was developed by EPA to support evaluation of fate and transport of pesticides for registration purposes. Enhanced versions of PRZM can simulate chemical transport below root zone, the maximum depth considered by GLEAMS.

Watershed Scale Models

At the watershed scale, defined here as an area having minimal base-flow in terms of its percent contributions to annual watershed runoff, a number of models have been developed. The focus here is on distributed parameter models which in effect link multiple "unit source areas" together, providing overland flow and channel routing to generate a hydrograph or total flow estimate at the outlet. These models have been largely event based models simulating runoff and sediment transport for individual storm events. The complexity of the algorithm and processes used in these models varies greatly as does spatial resolution and time scale. Among these models are: **ANSWERS**- The Areal Nonpoint Source Watershed Environment Response Simulation model is a watershed runoff and transport model for evaluation of the effects of land use and BMPs on the quantity and quality of water from various land uses. The distributed structure of the model allows for analysis of spatial and temporal variability of pollution loads and sources. ANSWERS has not been widely used in part due to file management requirements for large watersheds. **AGNPS**- The Agricultural Non-Point Source model provides an event-based prediction of runoff, erosion, nutrient, and COD in surface runoff using a square grid-cell representation of the watershed. The model is based on a simplified hydrology and nutrient release from CREAMS. A structured user interface facilitates data input, editing, and visualization of model results. The model has been widely used in part due to its user friendliness. With about 18 parameters defining the characteristics of cells, the range of alternatives that can be evaluated are more restricted than other models.

Basin Scale Models

Basin scale models are applied to study the runoff and water quality trends on large areas containing multiple watersheds and land uses. These models are continuous simulation packages that require time series of hydrologic and meteorologic data for watersheds within the basin. Among the models in this group are: **HSPF**- The Hydrologic Simulation Program-Fortran is a comprehensive modeling package for simulation of hydrology and water quality for a wide range of pollutants and land uses. It is one of a few models with capabilities for agricultural and urban modeling. The model performs continuous simulation of runoff, transport and transformation, and its hydrologic component includes groundwater component. The model also includes in-stream water quality and sediment transport processes. Data requirements of the model are extensive and it requires a highly skilled modeler to perform successful simulations. It is currently being modified to simulate wetland transport processes as well as hydrologic simulations in areas with high groundwater table. The model has been widely used especially in the north west and mid-west USA, and the Chesapeake Bay region.

SWRRB/SWRRBQ- The Simulator for Water Resources in the Rural Basins is an adaptation of the CREAMS model to simulate the complex rural watersheds. The model considers surface runoff, percolation, irrigation return flow, evapotranspiration, transmission losses, pond and reservoir storage, sedimentation, nutrient and pesticide movement, and in stream routing and water quality processes. Input requirements of the model are relatively high and experienced personnel are needed for successful simulations. The model is useful for estimation of the order of magnitude of pollutant loadings for relatively small basins or basins with fairly uniform properties. It uses a daily time step to evaluate the effects of management decisions on water, sediment yields, and pollutant loadings.

Conclusions

Preliminary observations and findings of the task committee in regard to the needs for the practitioners for an integrated watershed water quality and environmental assessment are as following:

a. Modeling systems are needed that are flexible and can be used in support of various regulations. This need will continue regardless of present and future regulations by EPA.

b. Large scale continuous simulation models are dynamic packages that require steady nurturing and technical support in order to remain useful in practice. These models need a sponsoring agency that will maintain them and provide technical assistance to the user community.

c. The models need to be enhanced if they are to be used as a comprehensive tool that can evaluate the full impact of multiple chemical releases from both point and non-point sources to the aquatic environment.

d. Appropriate assessment of the impact of pollution on the environment requires deterministic and continuous simulation models coupled with data bases, supporting software and graphical user interface.

e. Design of model codes should be flexible in order to permit integration/linking of different software to develop a unified system that for consistent user interface, data management, and integration of different software components.

g. Ability for multiple models to interact directly with each other and with a common data base.

h. Making software more universally usable through on-line technical support and use of new advances in computer programming such as object oriented coding.

REFERENCES

Hosseinipour, E.Z. (1993), "Irrigation, Agrichemicals, and the Environmental Impact; The Role of Hydrologic and Non-Point Source Models," Proceedings of the ASCE Engineering Hydrology Symposium, San Francisco, California, USA.

Imhoff, John C., Editor (1993), Developing Dynamic Watershed modeling Capabilities for Great Lakes Tributaries, Proceedings of a Workshop Sponsored by US EPA at Heidelberg College, Tiffin, Ohio.

US EPA (1992), Compendium of Watershed-Scale Models for TMDL Development, Office of Wetlands, Oceans and Watersheds, Office of Science and Technology.

Calibration and Verification of Watershed Quality Model SWMM in Sub-Tropical Urban Areas

Vassilios A. Tsihrintzis, Rizwan Hamid, and Hector R. Fuentes[1]

Abstract

Nonpoint source pollution modeling of urban runoff is studied using four watersheds in South Florida, representing high and low density residential, commercial and highway land uses. The US Environmental Protection Agency (EPA) Storm Water Management Model (SWMM) was used for each watershed to simulate flow hydrographs and the fate and transport of the following four quality constituents: 5-day Biochemical Oxygen Demand (BOD_5), Total Suspended Solids (TSS), Total Kjeldahl Nitrogen (TKN), and Lead (Pb). Calibration and verification were completed using various storm events, measured by the United States Geological Survey (USGS) at all four sites. Predicted hydrographs and constituent loadings are presented and compared with measured data.

Introduction

Urban runoff carries various pollutants into the receiving waters typically generated from urban activities. Urban activities add pollutants in amounts that are threatening to the integrity of the receiving waters. Urban constituents, such as suspended solids, are found in higher concentrations with increases in urban population and activities such as construction. Similar patterns can be ascribed to metal concentrations related to increased traffic volumes, and BOD concentrations related to increases in urban population.

Hamid et al. (1995) presented a methodology for modeling nonpoint source pollution from urban runoff, using the SCS hydrology method and simple empirical pollutant loading formulations. Measured hyetographs, hydrographs and pollutant

[1] Respectively, Assistant Professor, Graduate Assistant, and Associate Professor, Department of Civil & Environmental Engineering and Drinking Water Research Center, Florida International University, VH 160, Miami, FL 33199, USA.

loadings from various watersheds in Broward County were used in that study. This study concentrates in validating EPA's urban watershed model SWMM using field measured data at four South Florida watersheds. The validated model can be used, with a high degree of confidence, to applications in South Florida.

SWMM is a comprehensive water quantity and quality simulation model developed primarily for urban areas (Huber and Dickinson 1988). Single-event and continuous simulations can be performed for almost all components of the rainfall, runoff and quality cycles for a watershed. The RUNOFF Block of the model was used in this study, which can simulate both the quantity and quality of runoff from a drainage basin, and routing of flows and contaminants through the major storm drains. The program can accept any arbitrary rainfall hyetograph and accounts for infiltration losses in pervious areas, surface detention, overland flow, channel flow, and constituents washed to inlets, leading to the calculation of inlet hydrographs and pollutographs.

Methods and Materials

Four watersheds, representing high and low density residential, commercial and highway land uses, located in Broward County, Florida, are used to predict hydrographs and associated pollutant loadings. Required hyetographs, hydrographs and pollutographs were obtained from studies conducted by the USGS (Miller et al. 1979; Hardee et al. 1978; 1979; Mattraw et al. 1978). Ninety storm events, ranging in total precipitation from less than 0.5 inch to about 2.0 inches, were selected to be used in this study to validate SWMM. Input parameters, specific to each land use, are found during model calibration for both quantity and quality computations.

The high density residential watershed comprises an apartment complex covering an area of 14.7 acres, with 10.4 acres of total impervious area and 6.48 acres of hydraulically effective impervious area. The soil cover is lawn sod with some garden shrubbery and trees, and the hydrologic soil group is D. The stormwater network comprises circular and corrugated metal pipes. The streets have no curb and gutter, but are formed such that the centerline is the lowest point on the cross section. Stormwater is drained toward, and conveyed along the center of the street. Street material is bituminous concrete. The low density residential watershed comprises single family housing spread over an area of 40.8 acres, with 17.9 acres of total impervious area and 2.41 acres of hydraulically effective impervious area. The soil cover is lawn, shrubbery and trees, and the hydrologic soil group is A. The stormwater network comprises circular and rectangular concrete pipes. The streets have no curb and gutter, and water is drained through swales along them.

The commercial land use watershed is 20.4 acres of shopping center, with 20.0 acres of hydraulically effective impervious area. Stormwater is carried through a network of circular concrete pipes. The highway land use watershed is 58.3 acres of a highway, and adjacent business establishments, light industrial facilities and open lots; 21.1 acres are impervious area with 10.5 acres hydraulically effective impervious area. The soil cover is mostly unvegetated with some native shrubs, and the hydrologic soil group is A. The stormwater network is made of circular concrete pipes. The east-west portion of the highway, Sample Road, has six lanes with adjacent curbs and gutters. The north-south street, NE 3rd Avenue, has four lanes with swale and lawn drainage.

Each watershed was divided into various sub-catchments and physical characteristics of each catchment were used to create an input file defining the whole network. These characteristics include total catchment area, percent impervious areas, average catchment slopes, overland flow widths of each sub-catchment, pipe lengths and diameters, pipe invert slopes, and Manning's roughness coefficients for both pipe flow, and overland flow on pervious and impervious surfaces within each sub-catchment. As an example, Figure 1 shows the stormwater conveyance network of the commercial land use watershed. This watershed was divided into 25 sub-catchments, each draining into the respectively numbered inlet, ultimately collecting to point A.

Results

Preliminary results from SWMM model predictions are shown in comparison with the observed data in Figures 2 and 3. Figure 2 shows the predicted and observed hydrographs used for the verification run at the commercial land use watershed. The predicted peak shows a slight delay, however the trend and the total runoff volume show a very good agreement. Figure 3 shows a bar chart of four water quality constituents, from the storm event used in the verification run at the commercial site. Overall quality results from all the watersheds showed the highway land use as having the highest amounts of BOD_5 while the commercial and high density residential land use showed high amounts of TSS. Although Pb and TKN amounts are relatively low at all land uses, commercial land use seems to show higher values for these two constituents.

Since this study is still in the preliminary phase, the input parameters are not calibrated to show the best fit. As a result, quality constituents show over-predicted values. Input parameters will be improved for better fit, using all ninety rainfall-runoff events. Best parameters, specific to each land use, will then be developed.

Conclusions

Calibrated models, such a SWMM, can be applied in obtaining regional loading estimates by running the model in a continuous mode over an extended period of time. Such methods provide important pollutant loading information to plan for implementing best management practices for pollution prevention and reduction.

Appendix. References

Hamid. R., Tsihrintzis, V.A. and Fuentes, H.R. (1995). "Model Validation for Runoff Pollution from Urban Watersheds." *American Society of Civil Engineers, 22nd Annual Conference of the Water Resources Planning and Management Division*, Boston, MA, May 7-11.

Hardee, J., Miller, R.A. and Mattraw, H.C. (1978). "Stormwater-Runoff Data for a Highway Area, Broward County, Florida." *U.S. Geological Survey, Open-File Report 78-612.*

Hardee, J., Miller, R.A. and Mattraw, H.C. (1979). "Stormwater-Runoff Data for a Multifamily Residential Area, Dade County, Florida." *U.S. Geological Survey, Open-File Report 79-1295.*

Huber, W.C. and Dickinson, R.E. (1988). "Storm Water Management Model, Version 4: User's Manual." *Environmental Research Laboratory Office of Research and Development, U.S. Environmental Protection Agency, Athens, Georgia.*

Mattraw, H.C., Hardee, J. and Miller, R.A. (1978). "Urban Stormwater Runoff Data For a Residential Area, Pompano Beach, Florida." *U.S. Geological Survey, Open-File Report 78-314.*

Miller, R.A., Mattraw, H.C. and Hardee, J. (1979). "Stormwater-Runoff Data For a Commercial Area, Broward County, Florida." *U.S. Geological Survey, Open-File Report 79-982.*

Miller, R.A. (1979). "Characteristics of Four Urbanized Basins in South Florida." *U.S. Geological Survey, Open-File Report 79-694.*

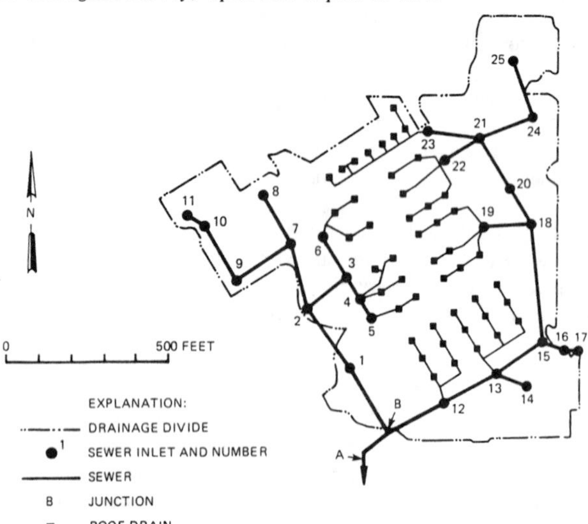

Figure 1. Stormwater Drainage Network at the Commercial Watershed.

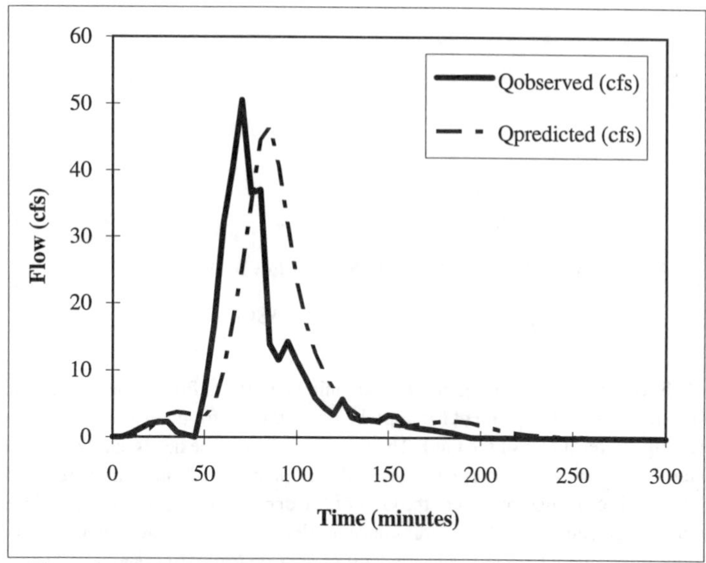

Figure 2. Observed and Predicted Hydrographs - Commercial Land Use

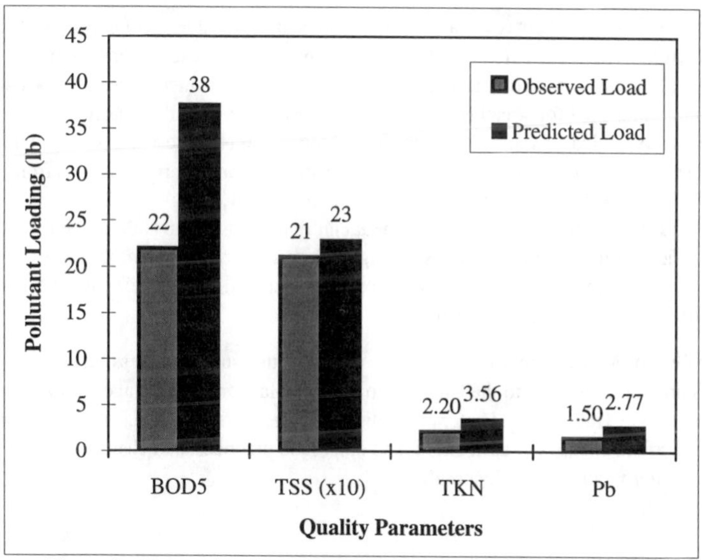

Figure 3. Observed and Predicted Loadings - Commercial Land Use

Assessment of Nutrient Loads in Streamflow to the Gulf of Mexico

David D. Dunn[1], A.M. ASCE

The U.S. Environmental Protection Agency initiated the Gulf of Mexico Program in 1991 to develop and implement a comprehensive strategy for managing and protecting the resources of the Gulf. The strategy balances the needs and demands of human activities and the preservation and enhancement of the living marine resources of the Gulf. Freshwater inflows to the Gulf of Mexico contain nutrients, defined here as total nitrogen and total phosphorus concentrations (dissolved and suspended). The Nutrient Enrichment Committee of the Gulf of Mexico Program is sponsoring an investigation of nutrient loads to the Gulf.

The U.S. Geological Survey is analyzing monthly and annual loads of nutrients to the Gulf from each of 37 major streams. The drainage area of these streams composes about 95 percent of the drainage area to the Gulf from the United States. At least 40 years of data exist for streamflow-gaging stations near the mouth of each of these streams. An unbiased multiple regression technique that uses the Minimum Variance Unbiased Estimator procedure is used to estimate long- and short-term loads from periodic nutrient analyses and daily values of streamflow. Long- and short-term temporal trends in total monthly and annual nutrient loads to the Gulf are being identified. Trends in nutrient loads are highly dependent on trends in streamflow. Accordingly, trends in streamflow and nutrient concentrations are being identified in addition to trends in nutrient loads.

The Gulf of Mexico Program can use the results of this study to assess factors affecting nutrient loads to the Gulf and to develop plans for the management and protection of Gulf waters. Factors that might be affecting nutrient loads include urbanization, changes in agricultural and other land practices, and changes in wastewater treatment and disposal.

---

[1] Hydraulic Engineer, U.S. Geological Survey, 8011 Cameron Rd., Austin, TX 78754

Hydrological Impacts of Climate Changes on a Semi-Arid Region of Brazil

Y.D.P. Medeiros[1]

Abstract

Global climate change may be expected to occur within the next few decades. This hypothesis is based on the increase in atmospheric CO_2 and other greenhouse gases. These increases will likely have different degrees of effects on the water resources in different places.

This paper focuses on the sensitivity of runoff and soil moisture to the climate change in the semi-arid region of North-East of Brazil (NEB). Climate scenarios, based on the results yielded by the General Circulation Models (GCMs), were used to translate the potential changes in temperature and precipitation for the NEB into runoff and soil moisture predictions.

Introduction

In recent years, there has been increasing scientific, public and political attention paid to the theory that the increasing concentrations of greenhouse gases, principally carbon dioxide, methane and chlorofluorocarcons (CFCs) will lead to global warming and consequently changes in climate.

The IPCC (Houghton *et al.*,1990) declared that the unequivocal distinction of the enhanced greenhouse effect from the natural climate variability is rather unlikely for a decade at least. There are large uncertainties not only in the modelling of the effect of the greenhouse gas enhancement, but also in the estimates of the rate at which these gases will be both emitted by human activities and absorbed by oceans and plants.

However, the IPCC presented several hypothetical emission scenarios with which the impact on the environment and global society could be assessed in quantitative terms through sensitivity analyses. In these scenarios, the cumulative effect of the

[1] Senior Lecturer, PhD, Universidade Federal da Bahia (UFBa) and Universidade Estadual de Feira de Santana (UFBa)

gas emissions was calculated using the concept of equivalent CO_2 concentration, the global growth rates were taken from World Bank projection, and population estimates were taken from United Nations studies.

General Circulation Models (GCMs)

The most highly developed tool to predict future climate is known as a general circulation model or GCM. These models are based on laws of physics and use descriptions in simplified physical terms of the smaller-scale processes such those due to clouds, deep mixing in the ocean, and land-surface processes. Although the GCM models so far are of relatively coarse resolution, the large-scale structures of the ocean and the atmosphere can be simulated with some skill. However, on a smaller scale there are significant errors in all models. For this reason, regional details of CO_2-induced hydrometeorological changes are virtually unknown. At the present time, the capability to produce specific regional forecast does not exist. What is currently possible, however, is the capability to assess water resources sensitivities, by linking GCM outputs with the regional (or local) hydrological models. Such assessments are especially necessary in environmentally fragile arid and semi-arid regions, where the conflicts associated with low water availability and increasing water demands are high.

Hydrological Models

Many climate change studies have used hydrological models of varying degrees of complexity to estimate the sensitivity water resources to a change in climatic inputs. These models permit the direct study of specific catchments and focus on basin-scale process. The effects of climate change expressed as scenarios (hypothetical and generated from GCM outputs) can be modelled.

The choice of models for use in the climate studies, in particular, depends primarily on the availability of regional data and models such as the NWSRFS model are data intensive. The lack of data is a severe constraint to the use of sophisticated hydrological models. To overcome the problems of insufficiency in the data base it is recommended that as simple a model as possible is used and the number of parameters kept at a minimum.

The Case Study Catchment

The Paraguaçu river basin is situated between the parallels 11° 11'S and 13° 42'S and meridians 38° 48'W and 42° 01'W, totally within the state of Bahia, in the north-east region of Brazil. In contrast to other regions in this latitude range, this region has a semi-arid climate. Although some eastern coastal areas annually receive 1600 mm or more of rain, some interior valley areas have an annual mean rainfall of less than 400 mm.

Impact Analysis of the Climate Change

Simple generalised procedures can give some insights into the effect of changing climate on average annual runoff, but more detailed analysis can be based on hydrological models. Such models allow the investigation of the effect of different seasonal distributions of changes and the importance of catchment characteristics. This section presents results from the application of the MODHAC model (Lanna and Schwarzbach, 1989), conceptual runoff model, modified by Medeiros (1994). Scenarios representing possible future climate conditions were represented by altering the model input data and the effects of the changes on model output were examined by comparison with the output for the calibrated model, representing current conditions.

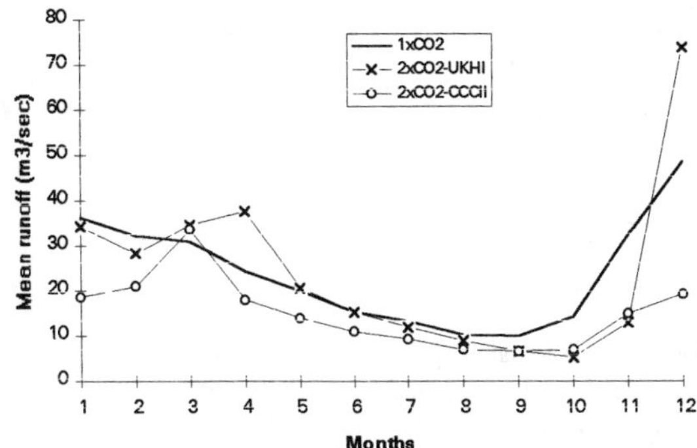

Figure 1. Monthly mean runoff for the UKHI and CCCII based scenarios respectively compared with the mean for the current CO_2 atmospheric concentration level.

Simulations denoted $1xCO_2$ and $2xCO_2$ were carried out to assess the impact of the change in climate predicted by UKHI and CCCII models.

Catchment Hydrological Response Analysis

To simplify the analysis of the results the following variables were selected to summarise the hydrological response of the catchment under the alternative climate scenarios: (1) monthly total runoff, (2) monthly evapotranspiration and (3) end of month soil moisture storage.

a) Runoff: Figure 1 shows the simulated changes in the seasonal distribution of runoff for the study catchment. The effect of the changes in rainfall pattern, augmented by the changes in the evapotranspiration, is immediately apparent on the shape of annual hydrographs; both the $2xCO_2$ scenarios show an abrupt rise in the autumn and a prolonged dry period until the middle of spring. Runoff decreases substantially in the two cases for the months of winter and spring. In the CCCII scenario, the decrease is observed through the year (except in March), which ranges from approximately 25% during the autumn to 50% during the summer. This leads to the significant reduction on total annual runoff of nearly 40%. In the UKHI scenario, on the other hand, the runoff increases during autumn to compensate decreases during the rest of the year.

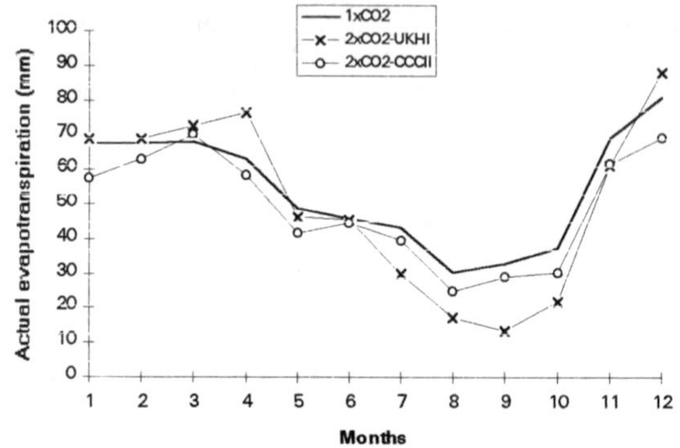

Figure 2. Monthly total actual evapotranspiration for the UKHI and CCCII scenarios respectively, compared with the total for the current CO_2 atmospheric concentration level.

For this reason, the total annual runoffs remain almost the same as for the $1xCO_2$ scenario. It should be noted that the change in seasonal distribution of runoff can be more critical for water resources management than the annual runoff change.

b) Evapotranspiration : Actual evapotranspiration (ET), as simulated with the soil moisture model, depends on soil moisture as well as on potential evapotranspiration (PET). In semi-arid catchments there exits an excess of atmospheric evaporative capacity (relative to the supply); as a consequence, the ET is extremely sensitive to changes in rainfall and the rate of ET declines as the soil dries. For this reason, although PET increases for all months for both the scenarios because of the increased temperature (Figure 3), the direction of changes in ET varied according to the changes in rainfall. In the UKHI scenario, ET increased during the wet December-April period and decreased during the dry April-November period in relation to current actual evapotranspiration. In the CCCII scenario, ET decreased through the year, except in March, as the rainfall did. The net result, despite the change in seasonal distribution, was relatively little change in annual total ET, in the CCCII scenario.

c) Soil Moisture Storage : Generally, soil responds to variations in rainfall and evapotranspiration, accepting or rejecting more or less moisture during the storm and interstorm periods. Figure 3 shows long-term simulated average soil moisture storage

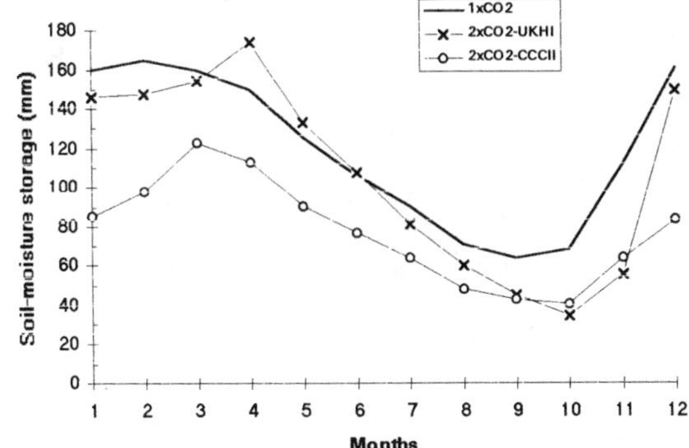

Figure 3. Monthly total soil-moisture storage for the UKHI and CCCII scenarios, respectively, compared with the total for the current CO_2 atmospheric concentration level.

at the end of month, for the prescribed scenarios. Compared with the current scenario, the warmer and generally drier climate in the CCCII scenario caused severe soil moisture shortages, ranging from 25% to 50% throughout the year. The UKHI scenario showed a slight increase in soil moisture available during the autumn period and significant decrease, varying from nearly 10% to 50% through the rest of year. Reductions in soil moisture storage may diminish the possibility for many species of plants to extract moisture from the soil, although some agronomists point to the improved conservation of water practised by plants in an atmosphere which contains double the present amount of carbon dioxide.

References

Houghton, J.T., Jenkins, G.J. and Ephraums, J.J (eds). (1990). Climate Change: The IPCC Assessment. The Policymakers' Summary of the Report of the IPCC Working Group I. Cambridge University Press. 26pp.

Lanna, A. E and Schwarzbach, M. (1989) MODHAC - Modelo Hidrologico Auto-Calibravel. Instituto de Pesquisas Hidraulicas, Universidade Federal do Rio Grande do Sul, Brazil, 55pp

Medeiros, Y.D.P (1994). Modelling the Hydrological Impacts of Climatic Change on a Semi-arid Region, PhD. Thesis University of Newcastle upon Tyne, U.K, 220pp.

Evaluation of Force-Restore Methods for the Prediction of Ground Surface Temperature

Zhenglin Hu and Shafiqul Islam[1]

Abstract

The computational efficiency and simplicity offered by the force-restore approximation of the diffusion equation have resulted in its wide-spread application in modeling ground surface temperature. Assumptions regarding the definition of ground surface temperature, however, have led to different versions of the force-restore method. Here, four existing versions of the force-restore method for ground surface temperature are compared and contrasted. An improved version of the force-restore method is developed by minimizing the error produced by the force-restore approximation of the heat diffusion equation.

Introduction

Representation of surface moisture and energy processes in atmospheric models is crucial for the partitioning of atmospheric forcing at the land surface. Usually, the ground surface temperature is determined from the solution of energy balance equation. A troublesome component, as noted by Deardorff [1978], however, is the specification of soil heat flux term, which require the time-dependent solution of balance equation. Such a solution with a high resolution multiple layer model demands substantial amount of computing resources to solve the land surface balance equation. Consequently, computational parsimony is sought by parameterizing (and sometimes ignoring) soil heat flux term.

Commonly used force-restore method essentially reduces the partial differential equation formulation into an ordinary differential equation for the ground surface temperature T_g of a soil slab with thickness δ (Bhumralkar 1975; Blackadar 1976). This is accomplished by using the solution of heat diffusion equation, in response to purely sinusoidal forcing. However, different assumptions regarding the definition of ground surface temperature have resulted in different versions of the force-restore method. A logical question one may ask is that: are there any significant differences among various versions of the force-restore method? Since the

----

1) Both at Department of Civil and Environmental Engineering, University of Cincinnati, P. O. Box 210071, Cincinnati, Ohio 45221-0071

force-restore method is essentially an approximation of heat diffusion equation, how well are these different versions compared with the analytical solution?

Existing Force-Restore Methods for Ground Surface Temperature

Different ways to approximate ground surface temperature have resulted in various versions of the force-restore method. Different versions of the force-restore method described in the literature (Bhumralkar 1975; Deardorff 1977; Deardorff 1978; Lin 1980) can be generalized as

$$\frac{dT_g}{dt} = C_1 G(0,t) - C_2(T_g - \overline{T}) \qquad (1)$$

The first term on right hand side of (1) is the forcing term and the second term is the restoring term. For comparison purposes, we list the corresponding coefficients for different versions of the force-restore method in Table 1.

Table 1 Coefficients of Different Versions of the Force-Restore Method

Reference	C_1	C_2
Bhumralkar [1975]	$\left(\frac{1}{1+2\delta/d_1}\right)\frac{2}{cd_1}$	$\left(\frac{1}{1+2\delta/d_1}\right)\omega_1$
Blackadar [1976]	$\left(\frac{1}{0.95}\right)\frac{2}{cd_1}$	$1.18\omega_1$
Deardorff [1977, 1978]	$\frac{2}{cd_1}$	ω_1
Lin [1980]	$\left(\frac{1}{1+\delta/d_1}\right)\frac{2}{cd_1}$	$\left(\frac{1}{1+\delta/d_1}\right)\omega_1$

Use of the force-restore method, instead of solving the explicit heat diffusion equation, to predict ground surface temperature greatly reduces the computation time. Consequently, different versions of the force-restore method were widely used in land surface modeling for the past two decades. In the derivation of the force-restore method, two assumptions were made. One is to parameterize the soil heat flux, and the other is the approximation of ground surface temperature. Errors resulting from these two assumptions will be evaluated next.

Figures 1 and 2 show the relative amplitude and phase shift of the ground surface temperature in the force-restore method compared to the solution of the explicit heat diffusion equation as a function of the scaled soil thickness. A relative amplitude of unity and a phase shift of zero would imply an exact match between the solution of force-restore method and the solution of the original heat diffusion equation. In Deardorff's and Blackadar's versions of the force-restore method, amplitudes and phase shifts are significantly distorted (Figures 1 and 2). Since

Deardorff's and Blackadar's versions are very similar, we will no longer discuss Blackadar's version. Lin's version of the force-restore method enlarges the amplitude, whereas Bhumralkar's version reduces the amplitude. The temperature phase shift in the Lin's version of the model is positive compared to the heat diffusion equation. In contrast, the phase shift in the Bhumralkar's model is negative for smaller scaled soil thickness.

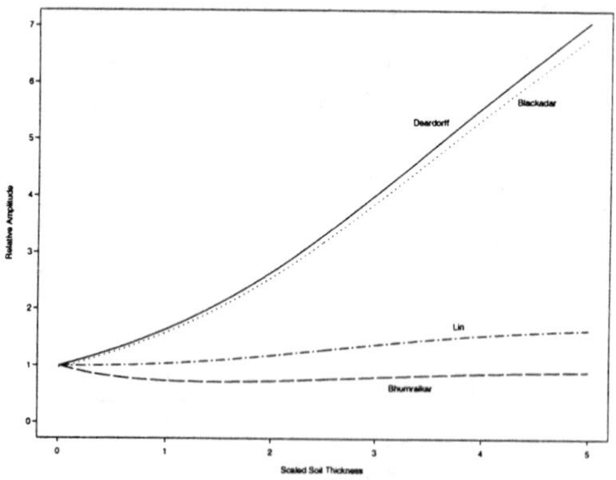

Figure 1 The relative amplitude of ground surface temperature for different versions of the force-restore method as a function of the scaled soil thickness

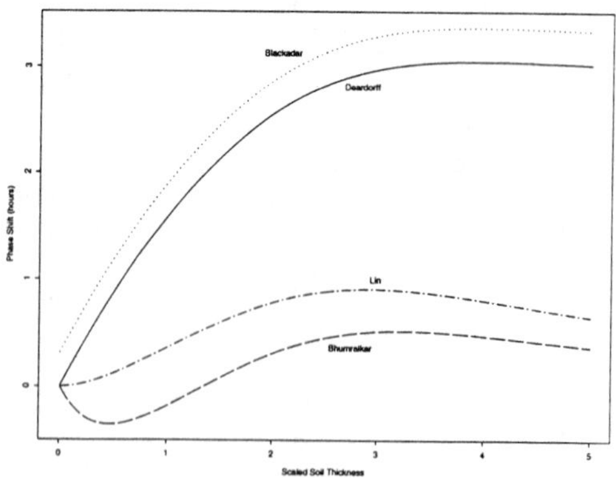

Figure 2 The phase shift in various versions of the force-restore method as a function of the scaled soil thickness

A Proposed Force-Restore Method for Ground Surface Temperature

Different versions of the force-restore method described above primarily differ in α. For example, the α value is unity for Deardorff's, $1+\delta/d_1$ for Lin's and $1+2\delta/d_1$ for Bhumralkar's respectively. Our proposed method estimates α by minimizing the error between the solutions of the force-restore method and those from the heat diffusion equation for the single frequency forcing case.

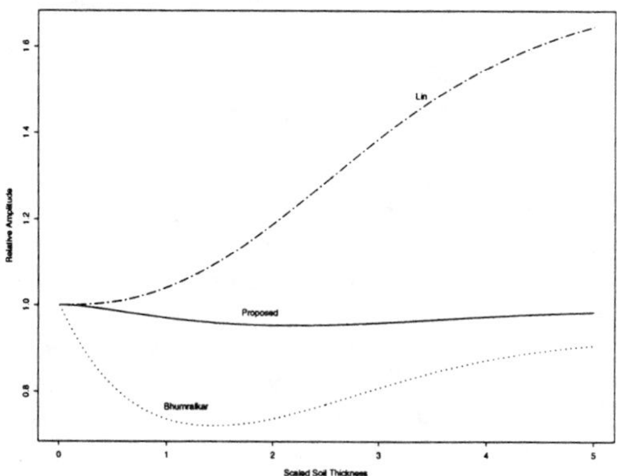

Figure 3 Relative amplitude of the ground surface temperature in the proposed version of force-restore method as contrasted with the Lin's and Bhumralkar's versions

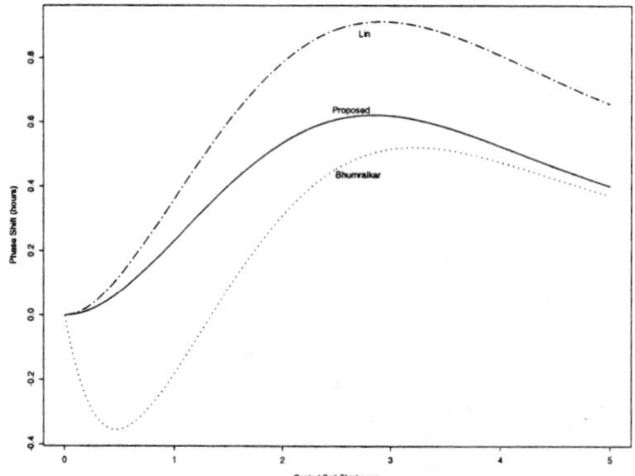

Figure 4 Phase shift of the ground surface temperature in the proposed version of the force-restore method as contrasted with the Lin's and Bhumralkar's versions

Figures 3 and 4 show the relative amplitude and phase shift for the proposed version of the force-restore method. The relative amplitude of the ground surface temperature response in the proposed version is quite close to unity and much better than the Lin's and Bhumralkar's versions (Figure 3). Lin's version of the force-restore method enlarges the amplitude while Bhumralkar's version reduces the amplitude. The phase is also better reproduced by the proposed method. Compared to the Lin's and Bhumralkar's versions, the range of phase shift is smaller in the proposed case (Figure 4).

Discussion and Concluding Remarks

The parsimony and simplicity offered by the force-restore approximation of the classical heat diffusion equation have motivated hydrologists and atmospheric scientists to apply this approximation for modeling ground surface temperature. The force-restore method essentially has only two free parameters whereas the solution of original heat diffusion equation requires at least 12 to 15 layers to obtain reasonable prediction of ground surface temperature. Four versions of the force-restore method are compared and contrasted. The result indicates that for a single frequency diurnal forcing, different versions produce reasonable prediction of temperature for the different upper soil thickness. Deardorff's and Blackadar's versions are adequate for the prediction of the surface skin soil temperature; Lin's version performs well when the soil thickness of the upper slab is less than the damping depth of the diurnal forcing; Bhumralkar's version is adequate when upper layer soil thickness is on the order of several damping depths.

In view of the above depth-dependence for the force-restore methods, a generalized version of the force-restore method is developed, which is applicable for a wide range of upper soil thickness. This is achieved by minimizing the errors between the analytical solutions from the force-restore method and from that of the diffusion equation under diurnal forcing. Bhumralkar's, Deardorff's and Lin's versions are the special cases of the proposed version of the force-restore method. The proposed version of the force-restore method performs well for the entire range of scaled soil thickness and reproduces the amplitude and phase that are quite close to the solution from the diffusion equation under a single periodic forcing.

Reference

Bhumralkar, C. M., Numerical experiments on the computation of ground surface temperature in an atmospheric circulation model, *J. Appl. Meteorol. 14*, 1246-1258, 1975.

Blackadar, A. K., Modeling the nocturnal boundary layer, in *Proceedings of the Third Symposium on Atmospheric Turbulence, Diffusion and Air Quality*, pp. 46-49, American Meteorological Society. Boston, Mass., 1976.

Deardorff, J. W., A parameterization of ground-surface moisture content for use in atmospheric prediction models, *J. of Appl. Meteorol., 16*, 1182-1185, 1977.

Deardorff, J. W., Efficient prediction of ground surface temperature and moisture with inclusion of a layer of vegetation, *J. Geophys. Res. 83*, 1889-1903, 1978.

Lin, J.D., On the force-restore method for prediction of ground surface temperature, *J. of Geophysical Research, Vol. 85*, No. C6, pp 3251-3254, 1980.

Climate Change: What the Water Engineer Should Know

Maurice Roos, Member ASCE[1]

Abstract

Long-range forecasts of future global warming have been made based on the increase in carbon dioxide and trace gases in the air from human activities. Potential changes which would especially affect water resources systems are changing runoff patterns, sea level rise, and (less sure) possibly larger floods. These changes, if they occur, would have a substantial effect on water supply and other aspects of a water engineer's vocation. Although the uncertainty of future climate change and impacts is high, water engineers should look at their systems to see how vulnerable they may be to climate change and what can be done to meet today's needs but still ensure a measure of protection from potential future problems.

Introduction

Two years ago, a task committee on climate change effects on water resources systems was established in the ASCE's Climate and Weather Change Committee. This paper is a summary report of the group.

Global warming is an important environmental issue for water engineers. Traditionally, engineers tend to think weather and climate patterns are stable, although their high short-term variability is one of the reasons for our profession. Climate change means that the

---

[1]Chief Hydrologist, Calif. Department of Water Resources, PO Box 942836, Sacramento, CA 94236-0001.

hydrologic past may not be the best guide for the future.

Scientific research and the popular media tell us to expect warmer temperatures and changes in precipitation in the next 50 to 100 years. Alterations in streamflows, season of snowmelt, soil moisture, rain intensity, evapotranspiration, ground water recharge, and other hydrologic items could have widespread impacts.

Causes of Climate Change

The cause of potential climate change is higher concentrations of carbon dioxide (CO_2) and other trace gases such as methane, nitrous oxides, and the chlorofluorocarbons (freons) from burning fossil fuel and other human activities. Chart 1 shows estimated worldwide CO_2 emissions, now about 6.2 billion metric tons per year. If all of this CO_2 stayed in the atmosphere, it would add about 3 parts per million per year; however, the actual increase during the past 10 years seems to have been about 1.5 ppm. The current CO_2 content of nearly 360 ppm is up from about 280 ppm some 200 years ago.

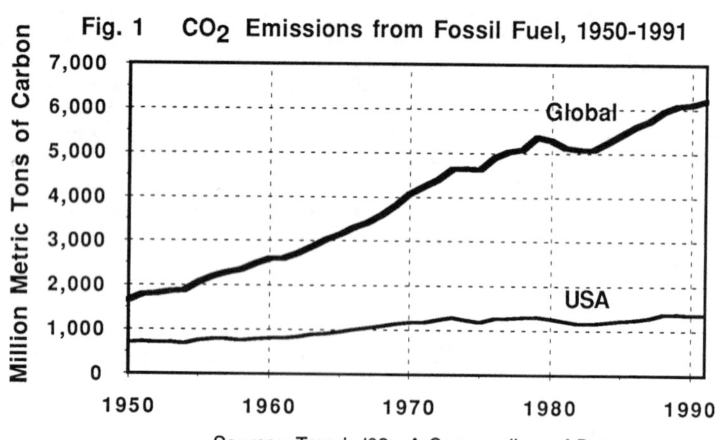

Source: Trends '93: A Compendium of Data on Global Change, CDIAC, Oak Ridge, TN

Weather records so far do not show an unambiguous temperature increase linked to CO_2 levels. The primary support for global warming comes from of atmospheric general circulation computer models (GCMs). The GCMs predict increases in temperature and changes in regional precipitation due to increased CO_2. Unfortunately, regional hydrology changes from model work are quite uncertain -- and this is the information most needed by the water engineer.

Potential Changes

The most robust results of the climate model studies are a rise of about 3 degrees Celsius over the next century, with a likely range of 1.5 to 5 degrees. (IPCC, 1990 and 1992). Regional changes are less sure. Sea level rise is more uncertain, but average projections indicate about 0.5 meter in the next century (OTA, 1991 and 1993) compared to the 0.1 to 0.2 meter rise estimated for the past 100 years.

The most likely impact of climate change on water resources in the temperature zone is a shift in runoff patterns, with less and earlier runoff from snowmelt and more winter season runoff. Predicted warmer temperatures mean higher snow levels, more direct cool season storm runoff and less in late spring and early summer (Roos, 1990). Since many reservoir storage systems in the western U.S. and elsewhere have been developed around the natural regulation of the mountain snowpack, the change will adversely affect water supply unless replacement storage is built. Hydroelectric power, too, could be affected.

Water demands for irrigated crops and landscaping are likely to increase slightly with warming. The temperature-induced increase in evapotranspiration could be partly offset by less transpiration (plant pores don't have to open as much) with higher CO_2 levels in the air. Some annual crops may just be planted a few weeks earlier with ET demands similar to today. The longer growing season may lead to more double cropping if water can be obtained.

Sea level rise could require relocation or protection of coastal and estuary facilities, including waste water and storm sewer works and developments on low lying ground. For years, however, the major problems would be of short duration during storm events at high tide. Hurricane damage would rise.

Fresh water supply in river delta and estuary areas could be affected by rising sea level and perhaps by the shorter duration of uncontrolled larger outflows, particularly where snowmelt is a significant fraction of outflow. Potential salinity changes and impacts on fisheries could create new water conflicts.

Some increase in the intensity of flood-producing storms could be expected with warmer temperatures and the related greater moisture-holding capacity of air. In mountain areas, higher snow levels during winter storms could produce more direct rain runoff than now. It would be wise to prepare for some increase in the large floods used for 100-year floodplain zoning and for flood control and dam spillway design.

Ways for Engineers to Cope With Change.

Because of the uncertainty on possible future climate change, major outlays to prevent what only might be a problem during the design life of facilities do not seem warranted now. Most conventional water supply/demand solutions would be useful in a warmer world. More storage would help.

Offstream storage facilities, whether surface or underground, also have to consider the time period during which surplus flows can be diverted. The number of days when water is available could be less.

Use of existing water supplies can be stretched by programs fostering water conservation. Evaluation of true water savings requires a look at the entire hydrologic system. Fresh water is lost when tail water, deep percolation, or sewage goes to saline underground or surface water bodies. In most interior areas tail water or deep percolation is not lost but returns to usable surface or ground water supplies.

There is a need to periodically reevaluate flood risk estimates. It is suggested that reevaluations be conducted every 20 to 30 years to update flood hydrology. The same is true for the rainfall depth-duration-frequency data widely used for designing storm water control and drainage facilities. In this way, global climate changes will be gradually incorporated into the record and the rainfall statistics.

In coastal areas, design of new facilities should consider existing local sea level trends. At any site, it is the net effect of tectonic movement and ocean changes which is important, not global averages. The amount of future global sea level rise is uncertain, but sometimes an initial modest change can add significant insurance against a moderate sea level change.

For agriculture, the response of common crops to higher carbon dioxide content of the air, especially on their water consumption, would be a good area of study. Such research could include development of strains for a warmer enriched carbon dioxide environment.

Monitoring research by others on possible impacts of changing climate on natural systems, including vegetation and fish (Univ. of Calif., 1993) can also help engineers adjust water project operations.

Finally, the prospect of long-term and widespread climatic change suggests that more water resource planning be on a regional scale. Thus, resources can be pooled and risks spread across more agencies. This may require interconnecting water systems.

Summary

The problem for the water engineer is that the possible changes are not sure and the range of forecasted changes is quite large. Institutional and social barriers may limit adjustments for climate change. Engineers will have to broaden their interdisciplinary contacts in developing solutions to water supply problems. Potential climate change is just one factor. An expanding world population, growth, environmental needs and a desire to restore past environmental losses, quality of water and land resources, and economics place new pressure on the water engineer to find additional supplies and to better manage existing water projects and uses to provide for human needs. Being aware of possible climate change should help us better prepare for an uncertain 21st century.

References

Intergovernment Panel on Climate Change (IPCC). WMO/UNEP (1990). *Climate Change, the IPCC Assessment.* Cambridge, UK.

IPCC (1992). *Climate Change 1992: Supplementary Report to the IPCC Scientific Assessment.* Cambridge, UK.

Roos, M. (1990). "Possible Climate Change and Its Impact on Snowmelt and Water Supply in California." *Proceedings of 58th Annual Western Snow Conference,* 130 - 136. Sacramento, CA.

U.S. Congress, Office of Technology Assessment (OTA) 1991. *Changing by Degrees, Steps to Reduce Greenhouse Gases.* Washington, D.C.

OTA (1993). *Preparing for an Uncertain Climate.* Washington, D.C.

U.S. Environmental Protection Agency (1989). *The Potential Effects of Global Climate Change on the U.S.* Report to Congress. Washington, D.C.

University of California (1993). *Integrated Modeling of Drought and Global Warming, Impacts on Selected California Resources,* N. Dowling, editor. National Institute for Global Environmental Change, Davis, CA.

Assessing Climate Change Impacts on Water Resources in Jamaica

Reginald A. Blake[1], Reza M. Khanbilvardi[1],
Cynthia Rosenzweig[2] and David Rind[2]

Abstract

Anthropogenic activities have caused and are causing an increase in the atmospheric concentrations of carbon dioxide and other "greenhouse gases" resulting in a shift of the atmosphere's radiative balance. Consequently, global and regional temperatures, rainfall patterns and other climatic variables will be altered. This paper sets forth the approach taken and some preliminary results obtained from an on-going study assessing climate change impacts on Jamaica's water resources. Preliminary results indicate that the island may already be undergoing climate change.

Introduction

There has been a relative paucity of research on the possible effects of global climate change on the Caribbean region. Gable (1987), Granger (1991), Wigley and Santer (1993) and Maul (1993) have done initial studies. No studies have been done to assess climate change impacts on the water resources for any individual island within the region. Given the poor condition of water resource management in the Caribbean, it becomes rather important that some insight be gained as to whether or not regional climate change may exacerbate or alleviate the water resources problems. An assessment for the individual island is imperative.

Before the middle of the next century, the mean temperature for the Caribbean region is expected to increase within the range of 1.0 - 3.0°C. The regional rainfall pattern is expected to be spatially varied, but no major shifts are anticipated. However, with warmer ocean temperatures come the concern of increased frequency and intensity of hurricane activity. Sea levels are also expected to rise by about 30 cm (Granger, 1991), resulting in the inundation of the islands'

[1] Graduate Student and Professor, respectively, Center for Water Resources and Environmental Research, Dept. of Civil Engineering, CCNY of CUNY.
[2] Research Scientists, Goddard Institute for Space Studies, New York

coastlines and the intrusion of salt water into coastal aquifers.

The island of Jamaica is the largest of the English-speaking Caribbean islands. It is located in the north-western part of the Caribbean sea in the Greater Antilles. Jamaica is elongated in an east-west direction. It is approximately three times as long as it is wide, and has a surface area of approximately 11,000 sq km. The principal mountain chain stretches from the WNW to the ESE portions of the island. The highest elevation on the island is 2,257 m and is located at the peak of the famous Blue Mountains of eastern Jamaica. Jamaica's climate is classified as maritime tropical with a mean daily temperature of 27°C. The rainy season lasts from May to November, while the dry season dominates the other months of the year. Mean island-wide rainfall is approximately 350 cm/yr. To facilitate this research, one of the island's major river basins, the Hope River basin, was chosen as a case study to assess the impacts of climate change on Jamaican water resources.

Basin Characteristics

The Hope River basin is located in the south-east end of Jamaica where most of the island's rainfall occurs. Its catchment area extends 41 sq km and is comprised of alluvium and limestone aquifers and basement aquicludes. Annual precipitation and runoff average about 193 cm and 65 cm respectively. This basin is of extreme importance since it is a major contributor to the domestic water supply of Jamaica's capital city Kingston which has well over 600,000 inhabitants.

Methodology

Trends in the observed data are important evidence in the study of climate change. General Circulation Models (GCMs) are the most sophisticated tools used to obtain possible scenarios of future climates. However, despite their high level of sophistication, they do exhibit limitations (see Mitchell et al., 1990). GCMs provide limited information on regional hydrology because their spatial resolution is too coarse and their parameterizations of hydrological processes are over-simplified. The use of GCM outputs alone to assess climate change on a regional scale does not produce acceptable results. Figure 1 is an example of how poorly a GCM can do in replicating observed regional rainfall data.

GCMs undergo perpetual refinement, and it will take years before their limitations are significantly improved. Until then and even then, an accepted way to study climate change impacts on a regional scale is to link GCMs to regional hydrologic models (see Gleick, 1989). The approach of using the output from GCMs to drive a regional hydrological model is used in this study. Hypothetical scenarios of climate change is also used to obtain plausible input data for the regional hydrological model. The output from the regional hydrological model is then used to drive a water management model. The broad outline of this methodology is shown in Figure 2. The Sacramento Watershed Model is currently being used to study the river basins of Jamaica, and it is, therefore, selected for this study.

Preliminary Results

At this time, only the trend analyses of the observed mean annual temperature at two opposite parts of the island are available. The climate change scenarios and the basin simulations will be presented in a later paper. Figures 3 and 4 clearly indicate that a gradual warming trend has been taking place at these two important sites on the island at a rate of about 0.15°C per decade. Montego Bay is located in the NW regions of the island while Kingston is located in the SE.

Concluding Remarks

Water resources managers and policy makers in the Caribbean need to know that the global climate is predicted to change and that the implications of global climate change for the region may be significant. It is expected that the regional temperature will increase and that rainfall and streamflow patterns will change. As is seen from these results, a warming trend is revealed in the Jamaica temperature data (~ 0.15°C per decade) . Contamination of freshwater resources and changes in coastal geomorphology are all likely to occur. The livelihood of the people in the region may be severely altered as the major economic sectors of tourism and agriculture undergo change. It, therefore, behooves managers and policy makers to start considering and implementing adaptation and mitigation strategies.

References

Gable, F. (1987). "Changing Climate and Caribbean Coastlines," *Oceanus*, 30 (4), 53 - 56.

Gleick, P.H., (1989). "Climate Change, Hydrology and Water Resources," *Rev. Geophys.,* 27 (3), 329 - 344.

Granger, O. (1991). "Climate Change Interactions in The Greater Caribbean," *The Environmental Professional,* vol 13, pp 43 - 58 .

Maul, G. A., (1993). "Implications of the Future Climate on the Ecosystems and Socio-Economic Structure in the Marine and Coastal Regions of the Intra-Americas Sea". In: Climate Change in the Intra-Americas Sea. N.Y. Edward Arnold.

Mitchell, J. F. B., Manabe, S., Meleshko, V. and Tokioka, T., (1990). "Equilibrium Climate Change - and its implications for the future ." In: Houghton, J. T., Jenkins, G.J. and Ephraums, J.J.(eds), *Climate Change: The IPCC Scientific Assessment,* Cambridge Univ. Press, pp. 131 - 164.

Wigley, T. M. L. and Santer, B. D., (1993). "Future Climate of the Gulf/Caribbean Basin from Atmospheric General Circulation Models. In: Climate Change in the Intra-Americas Sea. N.Y. Edward Arnold.

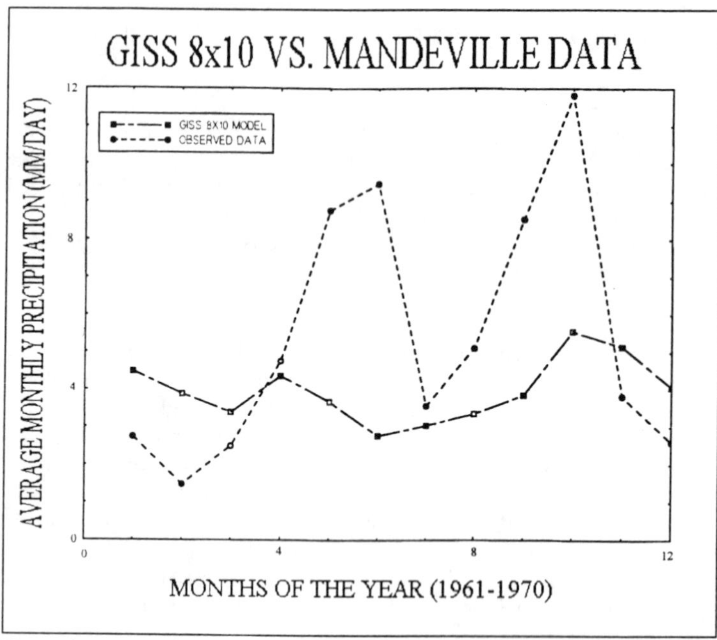

Figure 1. A comparison of observed rainfall data and the GCM output for ten years (Mandeville, Jamaica)

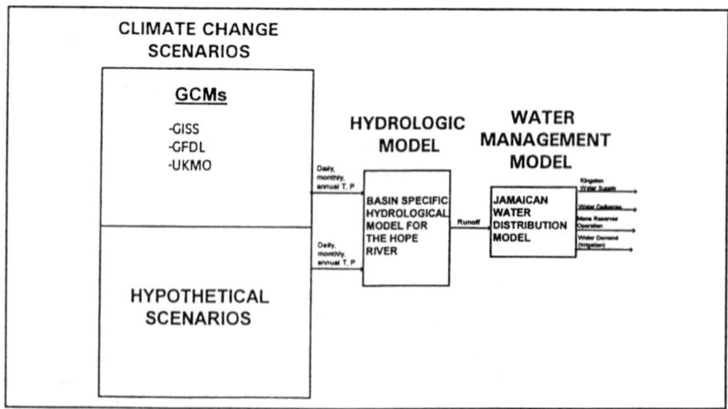

Figure 2. Linking of Various Models in Assessing Climate Change Impacts on Jamaican Water Resources

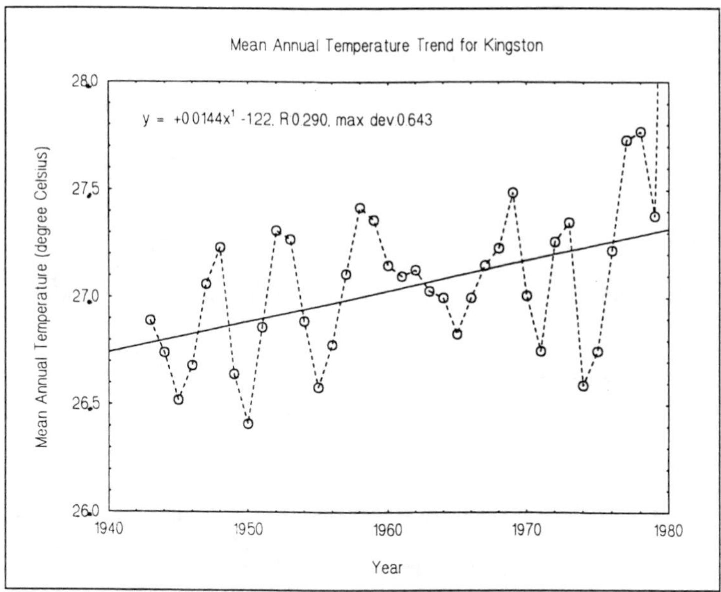

Figure 3. Kingston, Jamaica Temperature Data (1940-1980)

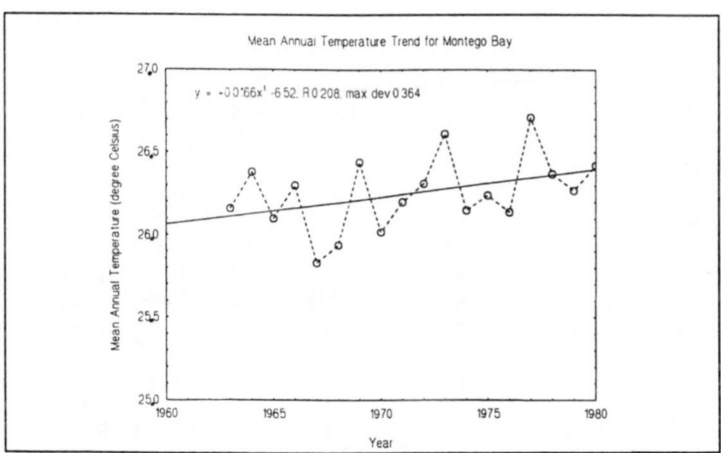

Figure 4. Montego Bay, Jamaica Temperature Data (1960-1980)

AN EFFICIENT APPROACH TO MODELING THREE-DIMENSIONAL HYDRODYNAMICS

Michael Amein, M. ASCE[1] and William Grosskopf, M. ASCE[2]

Abstract

Three-dimensional flows can be efficiently modeled with limited computer resources by linking a three-dimensional model to a sophisticated one-dimensional network model. Only limited number of three-dimensional cases were run, and the outputs of the three-dimensional runs and the field measurements were used to calibrate the one-dimensional model.

Introduction

The purpose of this paper is to demonstrate that three-dimensional flows can be efficiently modeled with limited computer resources by linking a three-dimensional model to a sophisticated one-dimensional network model. DYNLET (Amein and Kraus (1991)) is the one-dimensional hydrodynamic model used in this study. It is an implicit finite difference model. Although it is based on the one-dimensional flow equations, however, it is cast in a two-dimensional framework, so that the input data and the results of the computations are given over a two-dimensional grid. The flow domain is represented by a network of channels and junctions. Because the grid spacing in both directions can be varied in this model, the bathymetry can be represented at a high resolution in the region of interest.

Although any shallow water flow field can be simulated by a system of one-dimensional network of channels, with results ranging from very precise to approximate, however, the one-dimensional representation does not provide two types of information that could be important in particular applications. First, the one-dimensional representation assumes that the flow directions are known or can

---

1. Principal, Civil Analysis Group, Inc. Raleigh, NC 27607
2. Vice-President, Offshore & Coastal Technologies, Inc.-East Coast, Avondale, Penn 19311

be defined by the user. Second, the one-dimensional equations calculate the vertically integrated velocity at each grid point, and details of the velocity variation over the depth are not given by the on-dimensional model. Two- and three-dimensional models are needed to provide flow directions and the vertical structure of the velocity field.

In the efficient approach to modeling hydrodynamics, most of the computations are performed by DYNLET, and two- and three-dimensional models are used to delineate the DYNLET grid network and to obtain the vertical structure of the velocity field. The three-dimensional model used for the purpose of this study is the finite difference model SIM3D (Leendertse and Liu (1979), Nelson and Liu (1977)) developed at Rand Corporation for the Office of Water Technology. Using the two- and three-dimensional models for all computations is not practical with limited computer resources, and in some cases even with ample computer resources, because general-purpose multi-dimensional models are still under reearch.

Description of Models

1. DYNLET:

The shallow-water hydrodynamic equations for one -dimensional depth-averaged flow consist of the equations for the conservation of mass, momentum, and energy. For describing flow at tidal inlets and estuaries, the momentum and mass equations may be written as,

$$\frac{\partial Q}{\partial t} + \frac{\partial}{\partial y}\left(\frac{Q^2}{A}\right) = -gAS_f + gB\tau_s - gAS_e - gA\frac{\partial z}{\partial y} \qquad (1)$$

$$\frac{\partial Q}{\partial y} + \frac{\partial A}{\partial t} - q = 0 \qquad (2)$$

where Q = volume flow rate, t = time, y = horizontal distance (along a channel), A = cross-sectional area, g = acceleration due to gravity, S_f = friction slope, B = width of top of channel cross-section, τ_s = surface shear stress due to wind, S_e = transition loss rate with distance, z = water surface elevation, and q = lateral inflow or outflow per unit channel length per unit time. The equations must be solved numerically for arbitrary bathymetry and forcing conditions. The numerical scheme implemented in DYNLET extends and enhances the implicit finite-difference technique of Amein (1975) for solving the shallow-water wave equations.

2. Model SIM3D

SIM3D is a multi-layer finite difference model, with fixed grid spacing dx in the x and fixed spacing dy in the y- direction. The depth in the vertical is divided into layers. The thickness of each layer and the number of layers can be varied according to the physical characteristics of the study area. In this model, the vertical accelerations are neglected, and the momentum equations are vertically integrated over each layer.

In addition to the hydrodynamic equations, SIM3D also includes equations for subgrade turbulent energy, pollutant transport, temperature and salinity. Details of the numerical procedure and applications are given by Leendertse and Liu (1979) and Nelson and Liu (1977).

The hydrodynamic equations as used in the model are given as follows.

Conservation of mass:

$$\frac{\partial(hu)}{\partial x}+\frac{\partial(hv)}{\partial y}+\frac{h\partial w}{\partial z}=0 \tag{3}$$

Momentum Equations

$$\frac{\partial(hu)}{\partial t}+\frac{\partial(huu)}{\partial x}+\frac{\partial(huv)}{\partial y}+\frac{h\partial(uw)}{\partial z}-hfv=-\frac{h}{\rho}\frac{\partial p}{\partial x}+\frac{h}{\rho}\left|\frac{\partial\tau_{xx}}{\partial x}+\frac{\partial\tau_{yx}}{\partial y}+\frac{\partial\tau_{zx}}{\partial z}\right| \tag{4}$$

$$\frac{\partial(hv)}{\partial t}+\frac{\partial(hvu)}{\partial x}+\frac{\partial(hvv)}{\partial y}+\frac{h\partial(vw)}{\partial z}-hfu=-\frac{h}{\rho}\frac{\partial p}{\partial y}+\frac{h}{\rho}\left|\frac{\partial\tau_{xy}}{\partial x}+\frac{\partial\tau_{yy}}{\partial y}+\frac{\partial\tau_{zy}}{\partial z}\right| \tag{5}$$

where f is the Coriolis coefficient.

Applications and Conclusions

Efficient approach to hydrodynamic modeling has been used on several recent national and international projects to investigate the degree to which multi-dimensional modeling is required. The estuaries were modeled as multi-layer systems in the vertical. It was not practical to conduct these calculations with the three-dimensional model. Only limited number of three-dimensional cases were run, and the outputs of the three-dimensional runs and the field measurements were used to calibrate DYNLET. The vertically averaged velocities from DYNLET were almost identical to the vertically averaged velocities obtained from the three-dimensional model. Results from Indian River and Masonboro Inlet Systems, as shown on Figs. 1-5, are used to illustrate the main points.

References

Amein, Michael (1975), " Computation of flow through Masonboro Inlet, NC", Journal of the Waterways, Harbors and Coastal Engineering Division, Proc. ASCE, Vol 1, No. WW1, pp. 93-108

Amein, M. and Kraus, N. C (1991), "DYNLET1: Network Model for Tidal Inlet Dynamics", Estuarine and Coastal Modeling, Proceedings of the 2nd International Conference , pp. 644-656, ASCE

Civil Analysis Group Inc.(1995), DYNLET User's Manual, 7424 Chapel Hill Rd, Raleigh, NC 27607.

Leendertse, J. J. and Liu, S-K (1975), "A three-dimensional Model for Estuaries and Coastal Seas: Volume II, Aspects of Computation, Report No. R-1764-OWRT, The Rand Corporation.

Leendertse, J. J., Liu, S-K and Nelson, A. B.(1975), "A three-dimensional Model for Estuaries and Coastal Seas: Volume III, The Interim Program, Report No. R-1884-OWRT, The Rand Corporation.

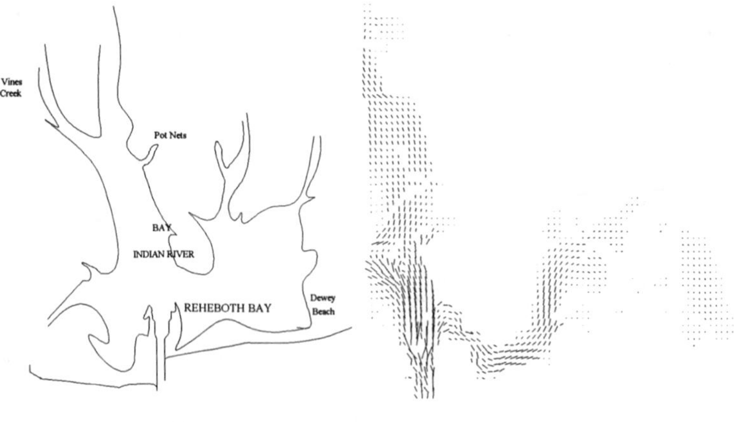

Figure 1. Outline of Indian River Inlet-Bay System

Figure 2. Currents from 3-dimensional model

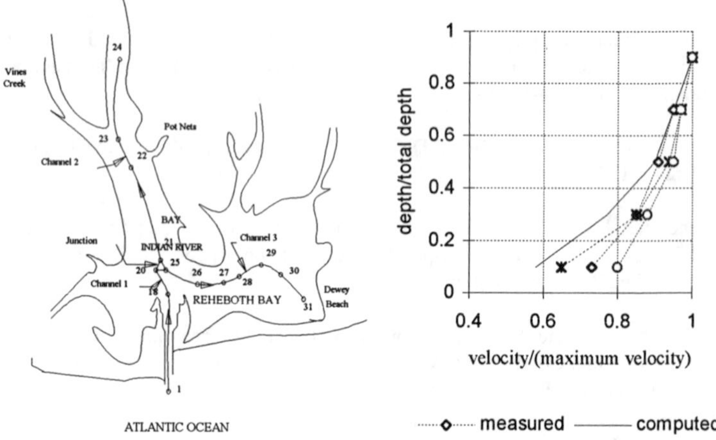

Figure 3. DYNLET grid for Indian River Inlet-Bay System

Figure 4. Comparison of computed and measured vertical velocity distribution

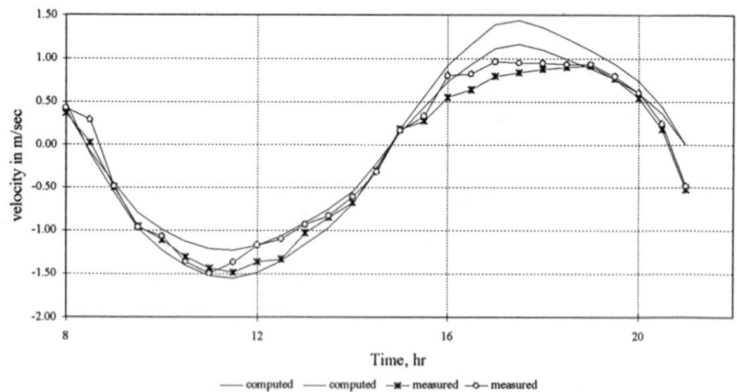

Figure 5. Comparison of vertically integrated measured velocities and velocity computed by DYNLET at inlet throat in Masonboro Inlet, NC

A COMPREHENSIVE GRAPHICAL ENVIRONMENT FOR SURFACE WATER FLOW MODELING

by Norman L. Jones[1], Alan K. Zundel[2], and Robert M. Wallace[3]

ABSTRACT

A comprehensive graphical environment for surface water flow modeling has been developed by the Engineering Computer Graphics Laboratory of Brigham Young University in cooperation with the U.S. Army Engineer Waterways Experiment Station and The Federal Highway Administration. The environment is called the Surface-water Modeling System (SMS). SMS is a graphical pre- and post-processor with a consistent interface for several finite element and finite difference surface water flow modeling codes. The system provides sophisticated pre- and post-processing capabilities for two-dimensional finite element and boundary-fitted finite difference grids. The system is designed to support a variety of surface water modeling codes including the RMA-2 and FESWMS models.

INTRODUCTION

In 1989, the Engineering Computer Graphics Laboratory (ECGL) began the development of a computer program which eventually became known as FastTABS (ECGL, 1994). FastTABS was developed as the result of a partnership between ECGL and WES. FastTABS is a graphical pre- and post-processor for the RMA-2 surface water modeling program. RMA-2 was originally developed by Norton, et al. (1973) of Resource Management Associates, Inc. of Davis, California. RMA-2 was eventually made part of the TABS-MD suite of surface water modeling software (Thomas & McAnally, 1990) supported by the U.S. Army Engineer Waterways Experiment Station (WES) in Vicksburg, Mississippi.

Prior to the development of FastTABS, modeling with RMA-2 was very expensive. Accurate modeling of the flow using RMA-2 requires the use of large two-dimensional finite element meshes. Most automatic mesh-generation tools are not well-suited for surface water flow modeling because the regions modeled are typically highly complex and irregular. As a result, the meshes were often constructed

[1] Assistant Professor, Dept. of Civil and Env. Eng., 368 CB, Brigham Young University, Provo, Utah, 84602.
[2] Research Associate, Dept. of Civil and Env. Eng., Brigham Young University.
[3] Graduate Research Assistant, Dept. of Civil and Env. Eng., Brigham Young University.

manually by coding the mesh in an ASCII file. For large, complex meshes, this could take several weeks to complete.

The development of FastTABS drastically reduced the cost involved in flow simulation studies by more than a factor of ten. Suddenly, it became possible to generate large meshes of several thousand elements in a relatively short period of time. A variety of sophisticated post-processing tools were eventually included in FastTABS to complement the mesh generation capability.

Surface Water Modeling System

As the FastTABS program matured and the use of FastTABS become more widespread, significant pressure developed to redesign FastTABS so that it would support a variety of surface water flow modeling codes rather than being customized to a single code, RMA-2. This interest has come primarily from WES and the Federal Highway Administration (FHWA). WES has developed several surface water flow models over the years. In many cases, the authors of the codes would either attempt to develop pre- and post-processing utilities themselves or contract out the development of such utilities. This approach resulted in a large number of graphical tools with a considerable range in quality and sophistication. Furthermore, few standards existed and, as a result, the interfaces and file formats associated with the programs differed substantially, hindering the transfer of information between programs and requiring retraining of users as they attempted to use a variety of flow models. A single, consistent system for surface water modeling that was developed for a large number of WES codes would clearly result in cost savings and increased efficiency.

The FHWA has also shown interest in an open system for two-dimensional flow modeling. The FHWA has invested considerable effort and expense in the development of a surface modeling code called Finite Element Surface Water Modeling System (FESWMS). FESWMS is a two-dimensional code that is similar to RMA-2 but it has been customized for application to highway design problems. Although there has been considerable interest in FESWMS in state departments of transportation and in private industry, utilization of FESWMS has been hindered due to the lack of a sophisticated pre- and post-processor. As a result, many hydraulics studies are ultimately completed with a one-dimensional model when a two-dimensional model would clearly be more appropriate.

In response to these needs, the ECGL has developed a new flow modeling system in cooperation with WES and FHWA. The new program is called the Surface Water Modeling System (SMS). SMS includes all of the features of FastTABS but also includes a large number of new utilities. The organization of the program and the underlying data structures have been redesigned so that custom interfaces to new models can be added to the system with relatively little effort. In order to properly support varying data input and mesh and grid generation requirements, SMS is divided into modules. Each module supports a particular type of data and the user can quickly switch from one module to another. The four modules in SMS are: the finite element module, the finite difference module, the scatter point module, and the map module.

Finite Element Module

The finite element module contains all of the tools for generating and editing two dimensional finite element meshes. Several automated mesh generation utilities are provided including triangulation, Coons' patches, and adaptive meshing to a polygonal boundary. Utilities are also provided for interactively editing meshes once they are constructed. A sample finite element mesh constructed with the tools provided in SMS is shown in Figure 1.

In order to make the finite element module applicable to a large number of analysis codes, the functions or commands have been partitioned according to scope of use. The functions which are applicable to any type of finite element analysis regardless of the code being used are grouped into two menus: one for mesh generation and one for mesh editing. A separate menu is then included for each of the analysis codes supported by SMS. These menus contain file i/o, material properties, boundary conditions, and other utilities which are unique to a particular code.

FINITE DIFFERENCE MODULE

The finite difference module in SMS is used for pre- and post-processing of boundary-fitted finite difference grids. A sample boundary-fitted grid is shown in Figure 2. Grids are typically constructed by subdividing the region to be modeled into quadrilateral subregions and filling each subregion with cells with a mapped sequence of cells. Tools are provided for repositioning cell boundaries and adding new cell boundaries to refine the grid where necessary. As with the finite element module, the commands for the interfaces to specific codes are separated from the general purpose gridding tools.

SCATTER POINT MODULE

With both the finite element and the finite difference approach, each node must be assigned a z coordinate corresponding to the bathymetry of the region being modeled. The scatter point module allows the user to enter a series of scatter points representing bathymetry measurements which are independent of the mesh or grid nodes. These points could be digitized from a bathymetry map or they could be direct measurements from a bathymetric survey. Several interpolation schemes, including kriging, are provided to interpolate from the measurements to the nodes or cells

MAP MODULE

The map module in SMS is used for importing and displaying DXF files and TIFF files containing scanned or photographic plots. It is often useful to display these types of plots in the background while constructing a mesh or grid to ensure that the nodes are properly located in the xy plane. The map module also contains tools for drawing circles, rectangles, arrows, text, etc. to add annotation to a plot prior to including the plot in a final report.

POST-PROCESSING

Several post-processing utilities are provided in SMS including vector plots, contour plots, time-history curves, and a variety of animation utilities. All of the post-processing functions are implemented in a generic fashion so that the steps involved in using the functions are identical regardless of the analysis code being used and regardless of whether the computational grid is a finite element mesh or a finite difference grid.

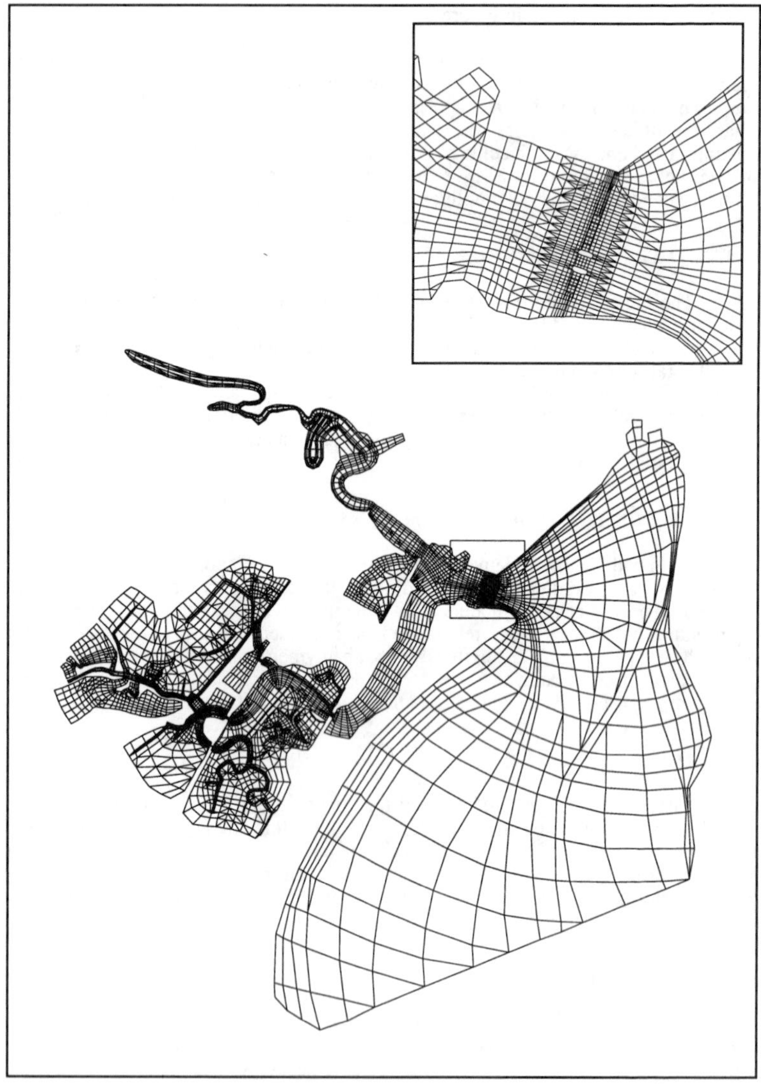

Figure 1. Finite Element Mesh of the Saugus Estuary Near Boston, MA.

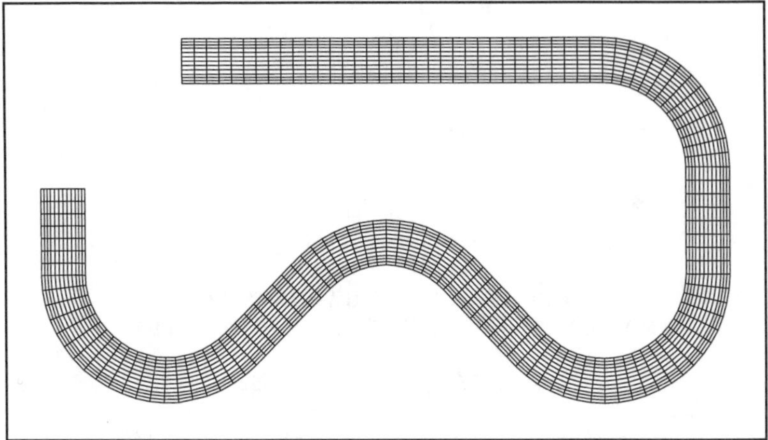

Figure 2. Sample Boundary-Fitted Finite Difference Grid.

CONCLUSIONS

The development of SMS has resulted in a sophisticated pre- and post-processing system which can be used for both finite element and boundary-fitted finite difference two-dimensional surface water flow modeling. SMS has been designed so that customized interfaces can be added to numerous analysis codes at a relatively low cost.

ACKNOWLEDGMENTS

The authors wish to thank WES and FHWA for their support and encouragement for the work described in this paper.

APPENDIX. REFERENCES

The Engineering Computer Graphics Laboratory, (1994), "FastTABS Reference Manual," Brigham Young University, Provo, Utah.

Norton, W. R., King, I. P., and Orlob, G. T. (1973). "A Finite Element Model for Lower Granite Reservoir", Water Resources Engineers, Inc., Walnut Creek, California.

Thomas, W. A., and McAnally, W. H., Jr. (1990). "User's Manual for the Generalized Computer Program System: Open-Channel Flow and Sedimentation, TABS-2." US Army Engineer Waterways Experiment Station, Vicksburg, Mississippi.

A MODIFIED NEWTON-RAPHSON SCHEME FOR UNSTEADY RIVER FLOW SIMULATION

Desheng Wang [1] and Hung Tao Shen [2]

Abstract:

Hydraulics of one dimensional river flow can be modeled by the Saint-Venant equations. Numerical methods for solving the complete St. Venant equations exist. One of the most popular schemes for solving the Saint-Venant equations is the four-point implicit-finite difference method. When this method is applied, a system of non-linear algebraic equations results from St. Venant equations. The Newton-Raphson iteration technique is then used to solve this non-linear system of equations (Fread 1985). This scheme is rather stable and fast for a river of large flow depth and relatively uniform channel geometry. However, when this scheme is applied to a river of shallow flow depth with severe changes in river cross sections along the river, the scheme may not give a stable solution. A modified Newton-Raphson scheme is therefore developed and implemented in an unsteady flow model. This refined unsteady flow model is applied to simulate winter flow conditions in the lower Yellow River. Simulation results show that the new scheme is more stable and gives good agreement with the field observations.

Introduction

Hydraulics of one dimensional river flow can be modeled by Saint-Venant equations (1871). Numerical methods for solving the complete St. Venant Equations exist. One of the most popular schemes for solving the Saint-Venant equations is four-point implicit finite difference method. When this scheme is applied, a system of non-linear algebraic equations results from St. Venant equations. Newton-Raphson iteration technique is then used to solve this non-linear system of equations (Fread 1985) This scheme is rather stable and fast for a river of large flow depth and relative uniform channel geometry. However, when this scheme is applied to a river of shallow flow depth with severe changes in river cross sections along the river, the scheme may not give a stable solution (Barkan 1991).

[1]Evironmental Engineer, Carr Research Laboratory, Inc., 5 Wethersfield Rd., Suite 5, Natick, MA 01760

[2]Professor, Dept. of Civil and Environ. Engrg., Clarkson University, Potsdam, N.Y. 13699-5710

The lower Yellow River is one of the most difficult rivers to simulate in the world. Due to relative small discharges in the winter, the flow depth can be very shallow. In some sections a depth of less than 0.4 m can occur. The river is a typical alluvial river. Its cross sectional geometry changes significantly year to year. Rapid changes in cross sections along the river also exist. The existing unsteady flow model (Lal and Shen 1991), which has been applied successfully to other large rivers fails to apply to the lower Yellow River. A modified Newton-Raphson scheme is therefore developed and implemented to resolve this problem. This refined unsteady flow model has been applied to the lower Yellow River with a success. This shows that the new scheme has a better stability.

Governing Equations

For a river with floating ice covers, the one-dimensional continuity equation is given as

$$\frac{\partial Q}{\partial x} + \frac{\partial A}{\partial t} = 0, \tag{1}$$

in which, Q = discharge,m^3/sec; A = net flow cross-sectional area, m^2; x and t = distance and time, respectively.

The momentum equation is given as

$$\rho\frac{\partial Q}{\partial t} + \rho(\frac{2Q}{A}\frac{\partial Q}{\partial x} - \frac{Q^2}{A^2}\frac{\partial A}{\partial x}) + \rho g A(\frac{\partial H}{\partial x} + S_e) + p_b\tau_b + p_i\tau_i + B_o\tau_a = 0, \tag{2}$$

in which ρ = density of water; H = water level, m; S_e = local loss of slope; p_b and τ_b are bed wetted perimeter and shear stress, respectively; p_i and τ_i = cover wetted perimeter and shear stress, respectively; B = width of the open water surface; τ_a = wind drag on the water surface.

Modified Newton-Raphson Scheme

In the four-point implicit-finite difference method, Eqs. 1 and 2 are written into finite difference form with water levels and discharges at nodal points which discretize the computation domain. Together with nodal and external boundary conditions, a system of nonlinear equations can be expressed as the following:

$$\mathbf{F}(\mathbf{x}) = \mathbf{0}, \tag{3}$$

in which, $\mathbf{x} = (x_1, x_2, \ldots x_n)^t$ is the vector containing the unknown variables, Q^p and H^p and $f_1, f_2, \cdots f_n$ are the coordinate functions of $\mathbf{F}$. Using Newton's method, the equation to find $\mathbf{x}$ at the k th iteration can be written as (Burden and Fairs 1985)

$$\mathbf{x}^{(k)} = \mathbf{x}^{(k-1)} - J(\mathbf{x}^{(k-1)})^{-1}\mathbf{F}(\mathbf{x})^{(k-1)}, \tag{4}$$

in which, $J(\mathbf{x})$ = Jacobian matrix. For computational efficiency, $\mathbf{x}$ is solved in a system of linear equations. The (i, j) element of the Jacobian matrix can be written as

$$J_{i,j}(\mathbf{x}) = \frac{\partial f_i(\mathbf{x})}{\partial x_j}. \tag{5}$$

At each computational time step, the above iterative procedure is followed to obtain the unknowns, $\mathbf{x}$, which are the H and Q values of the river reaches. To determine the initial values $\mathbf{x}^0$, known initial condition values should be the first choice. Otherwise, a steady state solution is commonly used as the initial condition.

At each iteration, the correction of each variable is computed by the iterative formula, Eq. 4. The corrections become smaller when the solution approaches the correct value. However, the solution for water level for each iteration can not be lower than the river bottom (Fread 1985, Lal and Shen 1989). This rarely happens in a deep river. When a shallow river is considered, especially when the geometry changes rapidly along the channel, this physical constrain can be easily violated in the computation. A simple method to avoid this is to raise the calculated water level to maintain a minimum flow depth, whenever it falls below the river bed at a cross section. However, this may create a disturbance in the whole scheme by creating a locally reversed water surface slope, which can make the scheme unstable. In this study, a modified iteration scheme is developed. Rather than raising the water level at a local point and creating a possible disturbance to the whole equation system when the physical constraint is violated, the new scheme proposes a method to smooth any local disturbances by attenuating changes in water level simultaneously over the whole domain. The revised values are then consistent at all computation points. The same principle can be applied to the flow discharge.

The physical constraints on water level and flow discharge are:

$$H_i^k = H_i^{(k-1)} + \Delta X_{H,i}^k > H_b + D_{min}, \tag{6}$$

and

$$Q_i^k = Q_i^{(k-1)} + \Delta X_{Q,i}^k > Q_{min}, \tag{7}$$

where, i = node number; k= iteration number; H_b = bed elevation; D_{min}, Q_{min} = specified minimum flow depth and discharge, respectively. $\Delta X_{H,i}^k$, $\Delta X_{Q,i}^k$ = water level and discharge corrections for the kth iteration, respectively. These corrections are calculated simultaneously for all cross sections using Eqs. 4.

$$\Delta X_i^k = -J(\mathbf{x}^{(k-1)})^{-1}\mathbf{F}(\mathbf{x})^{(k-1)}, \tag{8}$$

where, $X_i^k = [X_{H,i}^k, X_{Q,i}^k]^T$.

Whenever conditions in Eqs 6 or 7 are not satisfied at one or more locations during an iteration, the correction values at all cross sections are modified by a factor f_{min}. The revised corrections are:

$$\Delta X_i'^k = -J(\mathbf{x}^{(k-1)})^{-1}\mathbf{F}(\mathbf{x})^{(k-1)} f_{min} \tag{9}$$

and, Eq. 4 becomes

$$\mathbf{x}^{(k)} = \mathbf{x}^{(k-1)} + \Delta X'^{k}, \tag{10}$$

in which, f_{min} is the minimum value of the weighted correction factors for all cross sections, where

$$f_{min} = min(f_{H,i}, f_{Q,i}). \tag{11}$$

The weighted correction factors for water levels and discharges are:

$$f_{H,i} = \Delta X^k_{H,i,max}/\Delta X^k_{H,i}, \tag{12}$$

and,

$$f_{Q,i} = \Delta X^k_{Q,i,max}/\Delta X^k_{Q,i}. \tag{13}$$

The maximum corrections for water levels and discharges can be calculated from Eqs 6 and 7:

$$\Delta X^k_{H,i,max} = H_i^{(k-1)} - (H_b + D_{min}), \tag{14}$$

and,

$$\Delta X^k_{Q,i,max} = Q_i^{(k-1)} - Q_{min}. \tag{15}$$

Applications

This new scheme is applied to the lower Yellow River, which has a total length of 820.78 km. This is the most problematic and disastrous part of the second largest river in China. The river is contained by two long dikes, and its bottom is well above surrounding ground areas. The channel slope changes rapidly in the upstream part of the study reach, from 7.7×10^{-4} between Xialongdi and Tiexie to 2.1×10^{-4} between Tiexie and Jiahetan, within a distance of 26 km. The slope becomes milder further downstream. The slope is about 1.1×10^{-4} in the lower half of the reach between Jiahetan and Shibagongli. Cross sections vary severely along the river. The width of cross sections varies from a couple hundred meters to several km. The depth varies from less than half a meter to more than 5 meters.

Flow conditions of 9 typical winters (4 months each) are simulated. Typical simulated water levels and discharges are presented in Fig. 1. Solid lines represent observed values, dashed lines for simulated ones. Results show good agreement between simulations and observations.

Summary

This paper presents a refined Newton-Raphson scheme for solving one-dimensional unsteady river flows. The new scheme is proved to be more stable than the

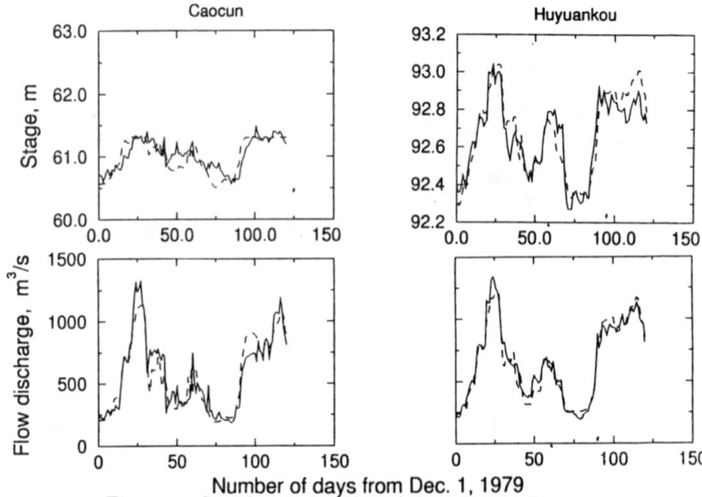

Figure 1: Simulation Results of the lower Yellow River

traditional Newton-Raphson Scheme. An application of the new technique to the lower Yellow River showed a good agreement between observations and simulations. More applications to other rivers will be encouraged to further verify and improve this method.

Acknowledgement: This study was supported by the World Bank Project CPR /91/621, through the UNDP Umbrella Funding for the Preparation of Xiaolangdi Project.

References

Barkau, R. L. (1991) "UNET: One-dimensional unsteady flow through a full network of open channels (user's manual)," Hydrologic Engineering Center, US Army Corps of Engineers.

Burden, R. L., and Faires, J. D., (1985) "Numerical Analysis," Prindle, Weber & Schmidt.

Fread, D. L. (1985) "Channel Routing," in *Hydrological Forecasting*, John Wiley and Sons Ltd.

Lal, A. M. W., and Shen, H. T., (1991) " Mathematical model for river ice processes," *J. of Hydraulic Eng.*, ASCE, Vol. 117, No. 7, 851-867.

Artificial Neural Network Simulation of Alluvial River Characteristics

David C. Hoffman[1] and Panagiotis D. Scarlatos[2]

Abstract

The stable geometry of an alluvial river is estimated using an artificial neural network. This approach is superior to the traditional curve-fitting regime formulas since it has the ability to re-adjust itself to new data inputs and to recognize complex patterns. The neural network used was trained using data from Colorado River. The model predicts width and depth of a channel given inputs of flow discharge, bed slope and median particle diameter. The number of network hidden layer nodes and training set size are varied to obtain the optimum network configuration. After the optimum network architecture is reached, the model predicts the regime depth and width with a less than 10% error.

Introduction

Design of new stable channels in alluvium or modification of natural streams and rivers have always been a challenge to hydraulic engineers. As a consequence, there have been many attempts to develop equations and methods that would allow prediction of a stable cross section geometry of an alluvium channel. One of the first methods for predicting stable channel geometry was developed by engineers designing irrigation canals in India and Pakistan (Chang, 1992). This method is called regime theory and is based on the premise that an alluvial channel at equilibrium maintains a stable, or regime geometry. Employment of the regime geometry during design guarantees stability of the channel. There have been many regime formulas proposed since the original theory was conceived, including those by Lacey (1926), Blench (1970) and Simons and Albertson (1960). One reason for the profusion of equations is that regime formulas are based on empirical data. Analysis of the regime data was traditionally accomplished by statistical curve-fitting type methods. As a result, the applicability of regime formulas is limited only to riverine systems that are similar to those for which the equations were developed.

A more general approach for analysis of regime canal data can be achieved

[1]Graduate Student, [2]Associate Professor, Department of Ocean/Civil Engineering, Florida Atlantic University, Boca Raton, Florida 33431

by means of an Artificial Neural Network (ANN). ANNs use digital computers to simulate the ability of a biological brain to learn and recognize patterns. ANNs have been applied to a wide variety of problems including scouring around a pier (Trent et al., 1993) and sediment transport in streams (Trent et al., 1993).

In this paper, an artificial neural network is utilized for prediction of the regime geometry of an alluvial channel. The ANN is trained to predict the stable channel width and depth for a wide variety of hydraulic conditions. The training set is taken from existing alluvial channel data.

Regime Theory

Regime channels are defined as those which adjust themselves to a certain stable configuration according to flow discharges, sediment inputs, and bank stability. The stable configuration is independent of the initial conditions. One of the first methods of stable canal design was proposed by Kennedy in 1895 (Vanoni, 1975). However, Kennedy's equation was inadequate for design of stable channels. Later, Lacey (1926) undertook the task to analyze data from existing canals and develop procedures for stable design. These procedures estimate the width, depth and slope which produces a stable channel cross section (Blench, 1969). Regime equations work reasonably well when used under conditions very similar to those for which they were developed. A compilation of regime formulas applicable under a wider variety of conditions was presented by Yalin (1992).

Generally, a regime channel can be defined by means of six characteristic quantities:

$$g,\ \rho,\ \nu,\ Q,\ \gamma_s,\ d$$

where g is the gravitational acceleration, ρ is the density of water, ν is the viscosity of water, Q is the flow discharge, γ_s is the specific weight of sediments, and d is a characteristic particle diameter e.g., d_{50}. Usually, regime formulas express the width B, depth D and river slope S as a function of the flow discharge Q,

$$B = m_1 Q^{n1}, \quad D = m_2 Q^{n2}, \quad S = m_3 Q^{-n3}$$

where m_i and n_i are dimensional experimental coefficients that incorporate the effects of the other five characteristic quantities besides discharge. Very often however, the range of particle diameters for which the regime formula is applicable is explicitly stated. Other regime formulas may correlate the geometric features to both the discharge and particle diameter (Yalin, 1992). Another feature that is considered in some regime channel studies is river meandering.

ANN Methodology

Artificial neural networks are comprised of a set of building blocks called nodes. The function of a node is to receive information, perform some operation (usually arithmetic) on this information and generate an output. The way in which the nodes are arranged in the ANN is called the network architecture. The ANN used in this study uses a three-layer feed-forward architecture. This means that the nodes are combined into three groups, or layers, and that information flows forward from one layer to the next but not back to a previous layer (Freeman and Skapura, 1991). The first layer of nodes is the input layer. The function of the input layer is to receive the network inputs and pass them to the next layer. The number of input layer nodes is set by the type of problem being solved. For this study, network inputs are: channel discharge, mean particle diameter and channel slope. Thus, the network has three input layer nodes.

The next network layer is the hidden layer. Each node of the hidden layer receives an output from each node in the input layer. These outputs are multiplied by weights, summed and become the input to the hidden layer node. The input is then passed through a filtering function and an output is generated. The number of hidden layer nodes is a network variable determined by trial and error. If there are insufficient hidden layer nodes, the network will be unable to learn the mapping to the required level of accuracy. If there are too many nodes, network training time will increase and accuracy may decrease as well (Freeman and Skapura, 1991).

The final network layer is the output layer. Each output layer node receives an input from each hidden layer node. These inputs are multiplied by weights, summed and become the input to the output layer node. This input is passed through a filtering function and an output is generated. This output is the network output. The number of output layer nodes is set by the problem. For this study, the network generates values for channel width and depth. Therefore the network has two nodes. Finally, the network output is compared to the real values and a network error is calculated. The error is propagated back through the network and the relative contribution of each node to the network error is calculated. Based on this contribution, the network weights are modified and the inputs are applied to the network again. This process continues until the error reaches an acceptably small value. This is called back-propagation learning. Another important factor of an ANN simulator is the training set. The training set must cover the entire range of values that the network is likely to encounter during normal operation. An ANN is good at generalization, but does not extrapolate well beyond its training space (Caudill and Butler, 1990). The size of the training set can also make a difference in the ability of the network to learn the mapping. If the training set is too small, the network will not generalize well and network accuracy will be reduced. On the other hand, if the training set is too large, the network has a tendency to "memorize" the data. This also will reduce the ability of the network to generalize. Finally, training data must be presented to the network in a random order. If the training data is presented in an ordered fashion, the network will learn one group and this learning will be abolished when it learns the next group. A set of verification data is used once training has been completed to measure the generalizing ability of the network. This gives an indication of how well the training has been accomplished. The procedure then, is to take a set of data and split it into different size combinations of training and verification data. For each size training set, the number of network hidden layer nodes is varied to determine the number that gives the most accurate results. The optimum combination of training set size and number of hidden layer nodes can then be found, producing the most accurate ANN for this problem.

Applications and Results

Regime equations are still, however, empirical relations and are most accurate for conditions that produced the equations. In this project, an artificial neural network is used to predict stable channel geometry. Using 105 data points from the Colorado River collected by the United States Bureau of Reclamation, the network is trained to predict river width and depth for inputs of discharge, slope and mean sediment diameter (Brownlie, 1981). Two parameters were varied to determine the optimum network configuration. The optimum network being defined as the one that produces the most accurate results. The first variable is the number of nodes in the network hidden layer. This number has a direct relation to the level of accuracy that the network will be able to achieve. The number of hidden layer nodes was varied from 4 to 11 to determine the optimum number. Along with the number of hidden layer nodes, the size of the network training set was also varied. The size of the training set also has a significant impact on network accuracy. If there are too few data points, the network will not have enough information during training therefore unabling it to generalize to the required degree of accuracy. If there are too many data points, the network tends to "memorize" rather than "learn" and its ability to generalize will be decreased. The river data was divided into a training set and verification set. The training set sizes used were 20, 40, and 60 data points. The remaining data points were used to verify that the network can generalize beyond the training set and to determine the network error. Figure 1 shows the percent error in width and depth for the various combinations of training set size and number of hidden layer nodes. The best combination appears to be 60 data points in the training set and 11 hidden layer nodes. This combination was able to achieve a width error of 9.9 percent and a depth error of 7.5 percent.

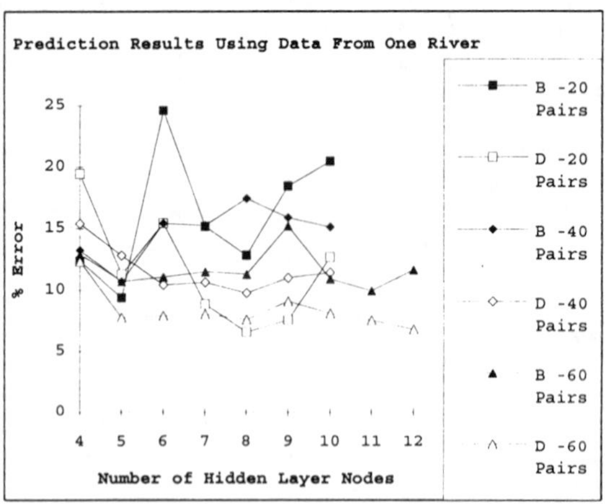

Figure 1. Prediction error of regime width (B) and depth (D).

Summary and Conclusions

An artificial neural network was applied for prediction of the regime geometry of alluvial channels. The ANN was trained using data from the Colorado River. After optimization of the network architecture, the stable riverine width and depth were predicted with an accuracy of approximately 90 percent. However, a more comprehensive test of the effectiveness of this trained network would require its application to riverine systems that were not included in the training set. The model is presently being extended to training sets for different rivers subject to a wide variety of flow discharges, bed slopes and median particle diameter. This would generalize the applicability of the model. In addition, the shape of the activation function is treated as a variable to be optimized. A possible improvement of the model's performance may require additional parameters in the network training set such as sediment concentration or type and amount of bank vegetation.

Literature Cited

Blench, T., 1969. Coordination in Mobile Bed Hydraulics, J. Hydraulics Div., ASCE, 95(HY6), pp. 1871-1898.

Blench, T., 1970. Regime Theory Design of Canals With Sand Beds, J. Irrigation and Drainage Div., ASCE, 96(IR2), pp. 205-213.

Brownlie, W.R., 1981. Compilation of Alluvial Channel Data: Laboratory and Field, Report No. KH-R-43B, W.M. Keck Lab., California Institute of Technology, Pasadena, California.

Caudill, M. and Butler, C., 1990. Naturally Intelligent Systems, The MIT Press, Cambridge, Massachusetts.

Chang, H.H., 1992. Fluvial Processes in River Engineering, Wiley Interscience, New York, New York.

Freeman, J.A. and Skapura, D.M., 1991. Neural Networks Algorithms Applications, and Programming Techniques, Addison-Wesley Publishing Company, Reading, Massachusetts.

Lacey, G., 1926. Stable Channels in Alluvium, Proc. Inst. of Civil Engineers, 229(I), London, England, pp. 259-292.

Simons, D.B. and Albertson, M.L., 1960. Uniform Water Conveyance Channels in Alluvial Material, J. Hydraulics Div., ASCE, 86(HY5), pp. 33-71.

Trent, R., Gagarin, N. and Rhodes, J., 1993. Estimating Pier Scour with Artificial Neural Networks, in: Hydraulic Engineering, Vol. 1, H.W. Shen, S.T. Su and F. Wen (Eds), ASCE, San Francisco, California, pp. 1043-1048.

Trent, R. Molinas, A. and Gagarin, N., 1993. An Artificial Neural Network for Computing Sediment Transport, in: Hydraulic Engineering, Vol. 1, H.W. Shen, S.T. Su and F. Wen (Eds), ASCE, San Francisco, California, pp. 1049-1054.

Vanoni, V.A., Ed., 1975. Sedimentation Engineering, ASCE Manual 54, New York, New York.

Yalin, M.S., 1992. River Mechanics, Pergamon Press, Oxford, England.

Application of Wetlands Model to Cache River, Arkansas

R. Walton[1], R.S. Chapman[2], and J.E. Davis[3], Members ASCE

Abstract

As part of the Corps of Engineers' Wetland Research Program, a wetlands dynamic water budget model was developed. The model simulates surface water, vertical, and groundwater flow processes. The surface water module was applied to the Cache River, Arkansas, and used to develop long-term information for use by other wetlands researchers. The results demonstrated that this wetland is generally inundated by the backwater from downstream constrictions during high river flows.

Introduction

Water availability and the processes by which it moves through wetlands are critical to the quantity and quality of functions provided. A quantitative understanding of wetlands hydraulics and hydrology (H&H) can be obtained by analyzing field data, and by applying numerical models.

The need to study wetlands in more detail led to the formation of the Corps of Engineers' Wetland Research Program. During this program, the Black Swamp wetlands along the Cache River, Arkansas, were monitored, and the Wetlands Dynamic Water Budget Model developed and applied to identify major system processes and their relative importance. The model simulated four years of gage record to extend the observational database for a variety of ecological analyses, and to investigate how various 'hypothetical' alterations would affect H&H responses.

---

[1] Senior Water Resources Engineer, WEST Consultants, Inc., Seattle, WA 98121

[2] President, Ray Chapman & Associates, Vicksburg, MS 39180

[3] Research Hydraulic Engineer, CERC, Waterways Experiment Station, Vicksburg, MS 39180

Model Development

The Wetlands Dynamic Water Budget Model was developed (Walton et al., 1994) to simulate the long-term response of various wetland types to hydrologic forcing. This suggested emphasizing efficiency, and reducing grid resolution and dimensionality. The model is based on the link-node method, which has been previously been used to simulate tidal wetlands, and has three modules.

The surface water module includes tidal forcing, river and basin inflows, channel and overland flows, and hydraulic structures such as weirs, gates, and culverts. The vertical processes module includes canopy interception, drainage, infiltration, and evapo-transpiration. The groundwater flow module simulates variably-saturated flow, and included wells and fixed-head boundaries. The model is explicit, and can use a variety of solution techniques and boundary conditions.

Application to the Cache River

The Black Swamp wetlands are located on the Cache River, between Patterson and Cotton Plant in eastern Arkansas (Figure 1). The river is an underfit stream, flowing in an old channel of the present-day Black and St. Francis Rivers. Much of the area has recently become a U.S. Fish and Wildlife refuge, and has been designated a site of critical biological importance (Kleiss 1993). The wetlands contain about 60 square kilometers of bottomland hardwood forests, typical of wooded wetland systems in the lower Mississippi River Valley (Kleiss 1993).

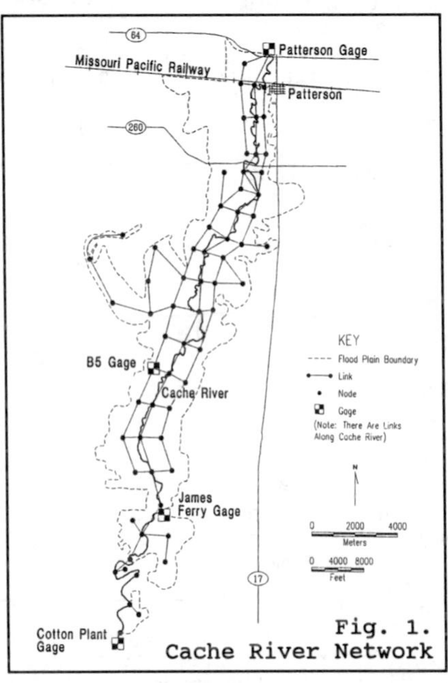

Fig. 1. Cache River Network

There is a long-term USGS streamflow gage at Patterson, and gages with shorter records at James Ferry and Cotton

Plant (Figure 1). As part of the Wetlands Research Program, the Black Swamp wetlands were extensively monitored for a variety of physical, chemical and biological variables, including surface water levels.

A grid was developed with 66 nodes and 115 links (or channels), some of which are pairs along the main channel to represent both the deeper channel and the immediate flood plain. The surface water module of the dynamic water budget model was run using river flows at Patterson and stages at Cotton Plant, and was calibrated to the interior stages measured at James Ferry and the B5 station (Figure 1). The results at James Ferry (Figure 2) compare well to observations.

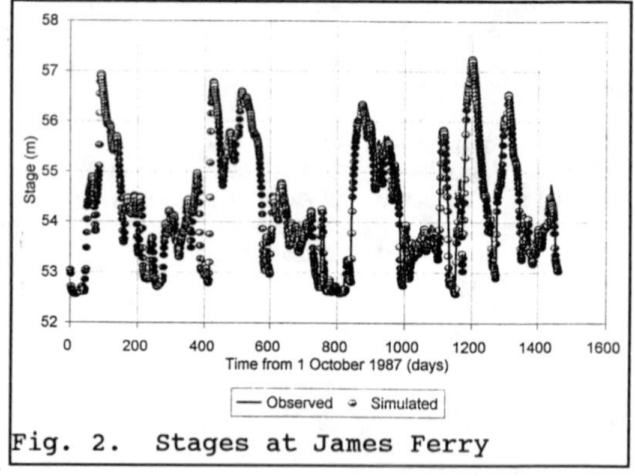

Fig. 2. Stages at James Ferry

One of the interesting aspects of this system, is that at high river flows, the stages at Cotton Plant, James Ferry and B5 become similar (Figure 3). Plotting stages at Cotton Plant against surface water gradients between James Ferry and Cotton Plant (Figure 4) clearly shows that the gradient becomes significantly smaller at higher flows. This process was simulated by the model. At James Ferry and at Cotton Plant, the flood plain narrows as the banks become relatively high. The effect is that, at higher

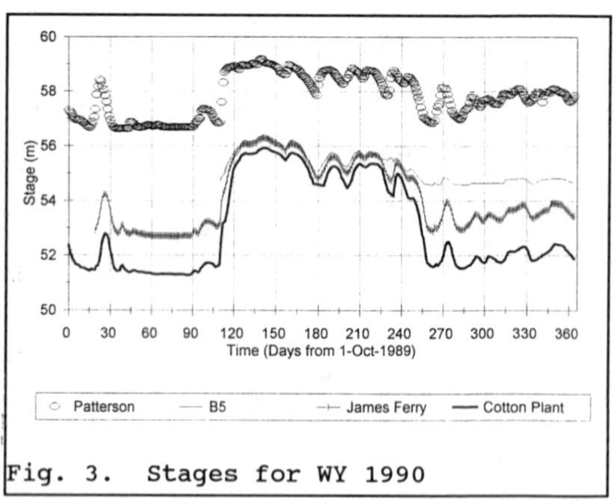

Fig. 3. Stages for WY 1990

flows, these constrictions cause a backwater (and the tendency for the water surface to become flatter), which floods the wetlands upstream.

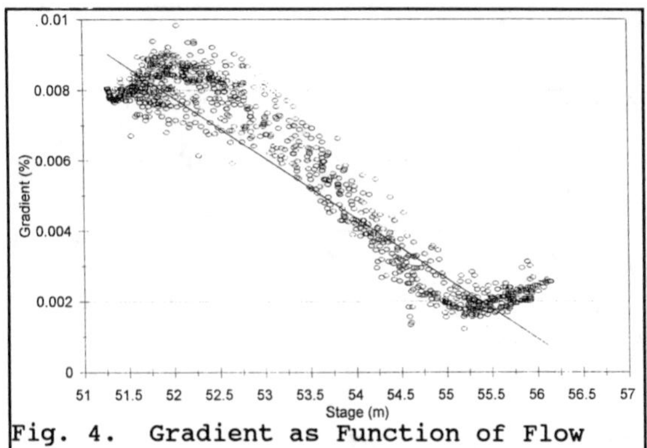

Fig. 4. Gradient as Function of Flow

In one scenario we simulated a new 'road' over the flood plain, with various types of openings such as a bridge over the main channel, culverts and road overtopping. Upstream (Figure 5), the stages depend on the head required to pass the flow. Downstream (Figure 6), however, there was virtually no change in stage. This is because the downstream channel constrictions at James Ferry and Cotton Plant will still create a backwater that floods the lower system once the system above the road has adjusted its elevation to pass the flood flow.

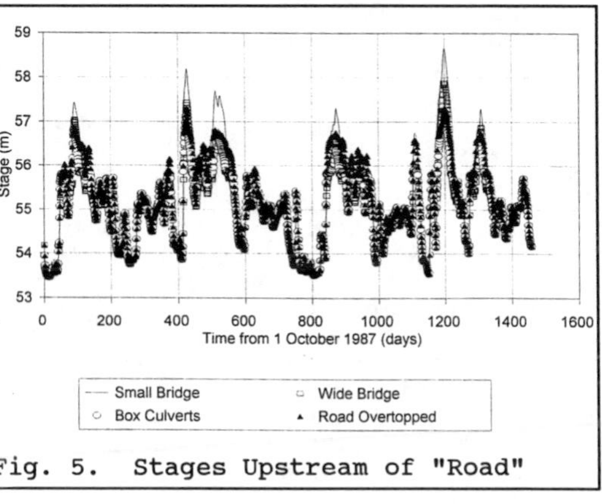

Fig. 5. Stages Upstream of "Road"

Discussion

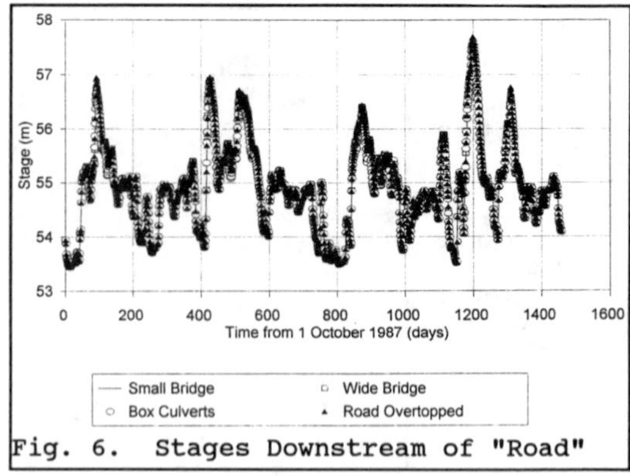

Fig. 6. Stages Downstream of "Road"

A database containing nodal volumes and stages, and link velocities and flows, was developed from the simulation to provide additional information to support data collection efforts and other wetland studies. The information was used, for example, to examine the frequency of inundation at the B5 gage, and to develop a method of determining hydroperiods anywhere in the system from stages at the Patterson gage. The model can also be used to interpret the system between discrete sampling periods and during periods of instrument failure, such as during January 1991, when the B5 gage was overtopped.

The model confirmed that flooding within the Black Swamp is generally produced by backwater from the downstream constrictions, rather than from inundation due to the downstream propagation of the flood wave. Flood events on the Cache River typically affect the system for a period of a few days to a few weeks. This time scale is much longer than the time for floodwaters to move laterally over the floodplain, and therefore studies requiring only water level information could perhaps use a one-dimensional model with complex channel and overbank geometries and conveyances. However, a two-dimensional application of the model is very useful in providing lateral variations in hydraulic parameters for other processes, such as water quality and sediment transport.

References

Kleiss, B., (1993), "Cache River, Arkansas: Studying a Bottomland Hardwood (BLH) Wetland Ecosystem", The Wetlands Research Program Bulletin, Vol. 3, No. 1., USAE Waterways Experiment Station, Vicksburg, MS.

Walton, R., T.H. Martin, and R.S. Chapman, (1994), "Investigation of Wetland Hydraulic and Hydrologic Processes, Model Development and Application", CERC Report, USAE Waterways Experiment Station, Vicksburg, MS.

EROSION CHARACTERISTICS OF COHESIVE SEDIMENTS

Mohammed A. Samad[1], Drew C. Baird[2], Tracy B. Vermeyen[3], and Brent W. Mefford[4]

1.0 Introduction:

A study has been undertaken to help solve sediment management and water delivery problems to the Elephant Butte reservoir associated with sediment deposition. Modeling transport of silts and clays is a critical element of the study. As part of the sediment modeling study laboratory flume tests and rotating cylinder tests were performed on silts and clay samples collected from the upper end of the reservoir. Critical shear stress for erosion, particle and mass erosion rates were determined in the flume tests. Rotating cylinder tests were performed for determining critical shear stress for erosion and particle erosion rate, and the two types of test results are compared.

2.0 Flume Tests

The flume has a channel length of 12 ft (3.7 m), a depth of 1.5 ft (0.5 m), and a width of 1 ft (0.3 m). The flume walls are constructed of clear acrylic. The flume floor is made of a high density, closed-cell urethane and has a surface roughness similar to fine sand. Water is recirculated using a 15-hp centrifugal pump, with a capacity of 5 ft^3/s(0.14 m^3/s) under 10-ft (3.3 m) of pressure head.

The results of the tests on three samples from the river channel are presented in figures 1 through 3. The critical shear stress for initiation of erosion for these three samples varied from 0.10 lb/ft^2 to 0.15 lb/ft^2. The particle erosion and the mass erosion rates, as shown in the figures, varied widely for these samples. For the River Site No. 2 sample (figure 2), the particle and the mass erosion phase could not be distinguished.

The results of the tests on the two overbank samples and the four samples from the reservoir area are shown in figures 4 and 5, and figures 6 through 9, respectively. The critical shear stress for the sample from Overbank Site No. 1 was 0.10 lb/ft^2 (figure 4), and the critical shear stress for the sample from Overbank Site No. 2 was 0.20 lb/ft^2 (figure 5). The erosion rates for these two samples did not show any consistent trend.

[1, 3, 4] Hydraulic Engineer, Bureau of Reclamation, P.O. Box 25007, Denver CO 80225.

[2] Chief, River Analysis Branch, Bureau of Reclamation, 505 Marquette N. W., Suite 1313, Albuquerque NM 87102.

The test results obtained from the reservoir samples were scattered as shown in figures 6 through 9. The critical shear stress could be estimated anywhere from almost zero for the surface sample from Reservoir Site No. 2 (figure 7) to 0.30 lb/ft^2 for the consolidated clay sample from Reservoir Site No. 2 (figure 8). No definite trend for particle erosion or mass erosion could be identified from the test results.

3.0 Rotating Cylinder Tests

A total of four samples were sent to Dr. Ray B. Krone at the University of California at Davis (UC-Davis) for erosion testing by using the rotating cylinder apparatus. Out of the four samples that were sent for the test, two fell apart when setting them up in the test apparatus. The two remaining samples were tested, one was 5 feet below the grade (River Site No. 1), and the other 2 feet below the grade (River Site No. 2). The test results for the two samples are shown in figure 10 and 11, respectively.

In addition to erosion tests for samples from River Site No. 1 and No. 2, the cation exchange capacity (CEC) and the particle size distribution for all the four samples were also determined. The results of the CEC test along with the results of the erosion tests, the Unified Soil Classification, and the plasticity index of each sample are presented in table 1.

TABLE 1
Rotating Cylinder and CEC Test Results

Sample Identification	Critical Shear Stress (lb/ft^2)	CEC (meq./100 gm of Soil)	Plasticity Index	Unified Soil Classification
River Site #1	0.13	34.8	54	CH
River Site #2	0.12	36.4	53	CH
Reservoir Site #1	-	38.6	72	CH
Reservoir Site #2	-	38.4	Non Plastic	ML

4.0 Comparison of Results

The results from the flume tests and the rotating cylinder tests are compared in table 2. As shown in the table the critical shear stress obtained from the flume tests and the rotating cylinder tests for samples taken from the same location are comparable. Also shown in the table is the shear stress obtained from the relation

$$\tau_c = 0.0034\ (PI)^{0.84} \tag{1}$$

where:

τ_c = critical shear stress in pounds per square feet, and
PI = the plasticity index.

Equation (1) was obtained by Smerdon and Beasley (1959) by using results from a number of flume tests on cohesive sediments.

The Bingham shear strength for the clay samples tested at UC-Davis was calculated using the relation presented by Krone

$$S_B = .0021\ [3.92 + 0.8447(CEC)] \tag{2}$$

where:

S_B = The Bingham shear strength in pounds per square feet.

CEC = The cation exchange capacity of the clay material in milliequivalents per 100 grams.

The calculated Bingham shear strength for the four samples are also presented on table 2.

As shown in table 2, critical shear stress obtained from rotating cylinder test and the flume test are nearly the same. However, there were only two samples to compare. Also, the shear stress computed by equation 1 is about the same as the shear stress obtained by rotating cylinder and flume test.

The Bingham shear stress values are somewhat lower than the critical shear stress values obtained in rotating cylinder test and the flume test.

TABLE 2
Comparison of Critical Shear Stress Obtained by Different Approaches

	Critical Shear Stress (lb/ft^2)			
Sample Identification	Rotating Cylinder	Flume Test	Equation1	Equation2
River Site #1	0.13	0.10	0.10	0.070
River Site #2	0.12	0.15	0.10	0.073
Reservoir Site #1	-	0.25	0.12	0.077
Reservoir Site #2	-	0.30	-	0.076

Conclusion:

The test results show that clay from the upper end of Elephant Butte Reservoir has high critical shear stress. The critical shear stress for erosion varied from 0.10 lb/ft^2 to 0.30 lb/ft^2 with the exception of one sample, which has critical shear stress of nearly zero. It was also found that Smerdon and Beasley's (1959) equation (Eq. 1) is a good predictor of critical shear stress if the plasticity index of the sample is known.

The erosion rates of the samples were not consistent. They varied widely. With the exception of two samples, the shear stress at which particle erosion changes to mass erosion could not be clearly defined.

Good agreement was found between the critical shear stress obtained by rotating cylinder test and the flume test for samples collected from the same location.

References

Krone, R. B., "Cohesive Sediment Properties and Transport Processes.

Smerdon, E. T., and Beasley, R. P., Oct. 1959, "Tractive Force Theory Applied to Stability of Open Channels in Cohesive Soils," Research Bulletin No. 715, Agricultural Experiment Station, University of Missouri, Columbia, Missouri.

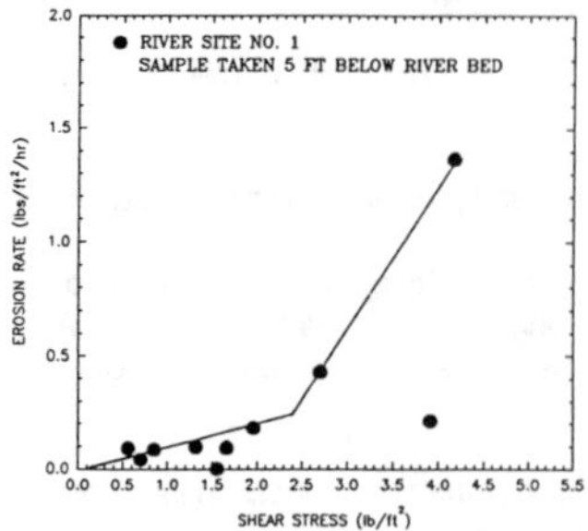

Figure 1. Erosion characteristics for clay sample collected at River Site No. 1. This sample was collected 5 ft below the river bottom.

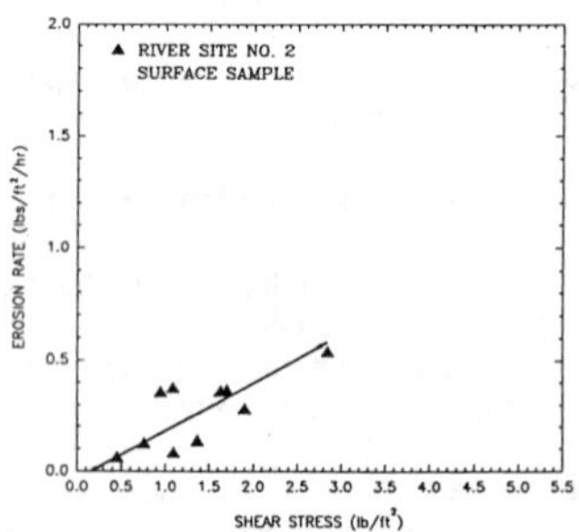

Figure 2. Erosion characteristics for clay sample collected at River Site No. 2. This sample was collected 1 ft below the river bottom.

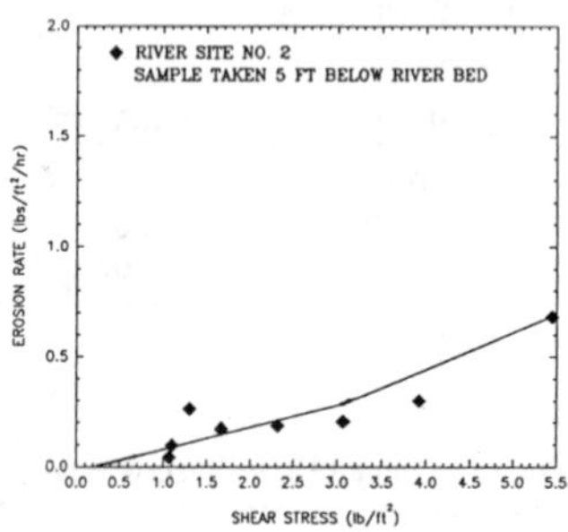

Figure 3. Erosion characteristics for clay sample collected at River Site No. 2. This sample was collected 5 ft below the river bottom.

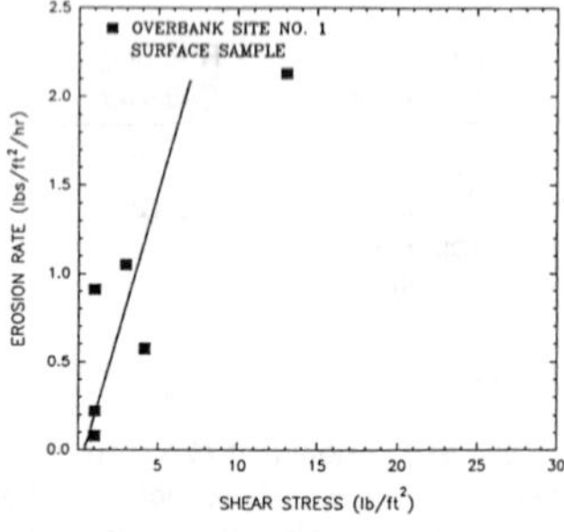

Figure 4. Erosion characteristics for clay sample collected at Overbank Site No. 1. This sample was collected 1 ft below grade.

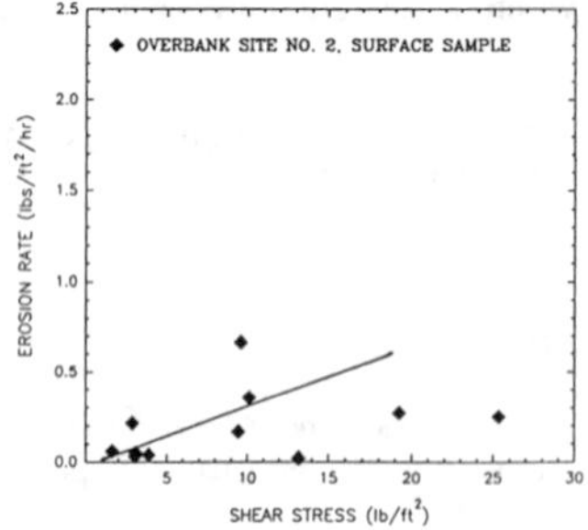

Figure 5. Erosion characteristics for clay sample collected at Overbank Site No. 2. This sample was collected 1 ft below grade.

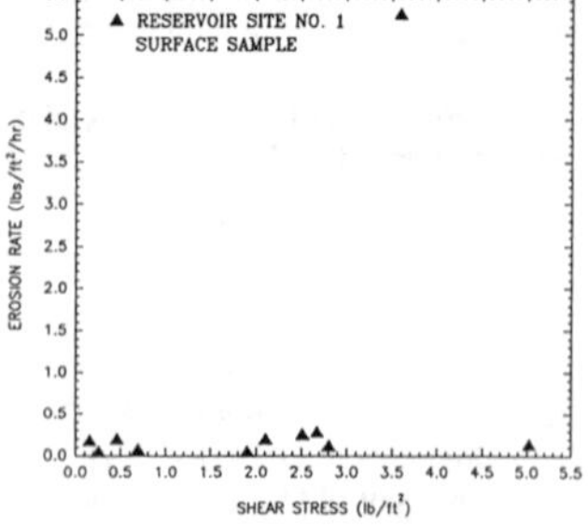

Figure 6. Erosion characteristics for clay sample collected at Reservoir Site No. 1. This unconsolidated sample was collected 2 ft below the reservoir bottom.

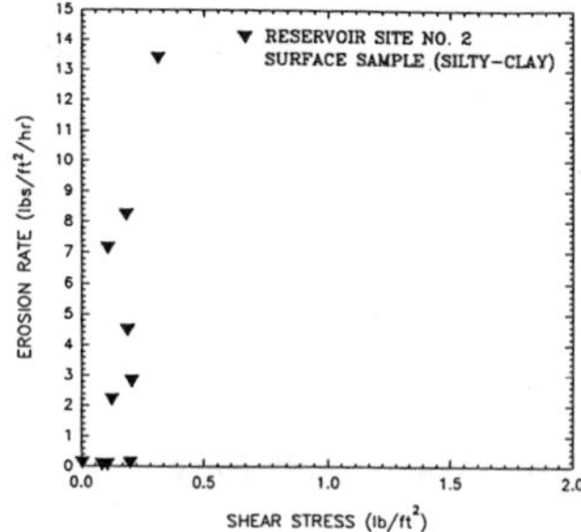

Figure 7. Erosion characteristics for silty-clay sample collected at Reservoir Site No. 2. This sample was collected 1 ft below the reservoir bottom.

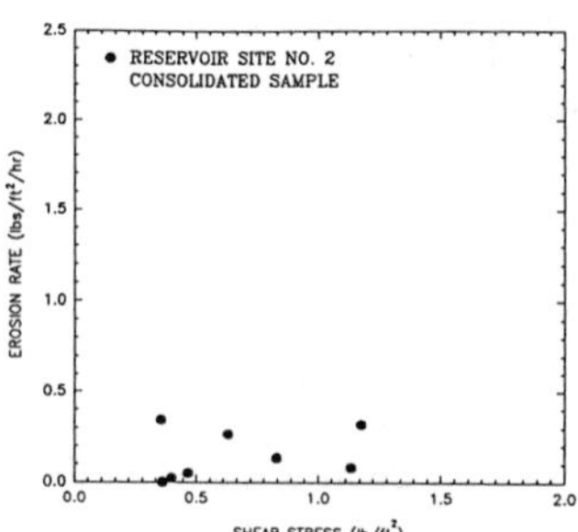

Figure 8. Erosion characteristics for clay sample collected at Reservoir Site No. 2. This consolidated sample was collected about 8 ft below the reservoir bottom.

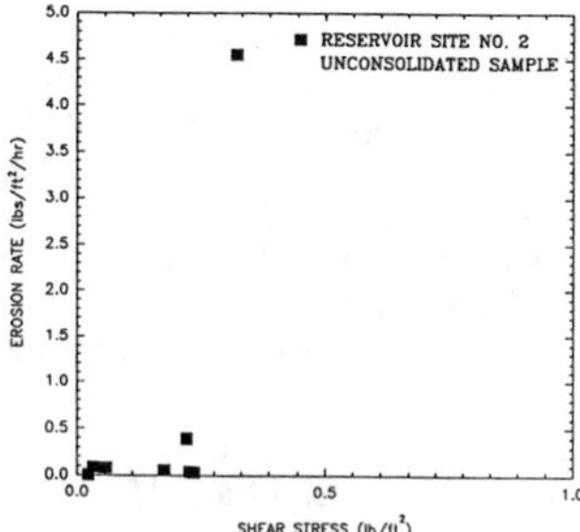

Figure 9. Erosion characteristics for clay sample collected at Reservoir Site No. 2. This disturbed, unconsolidated sample was collected 1 ft below the reservoir bottom.

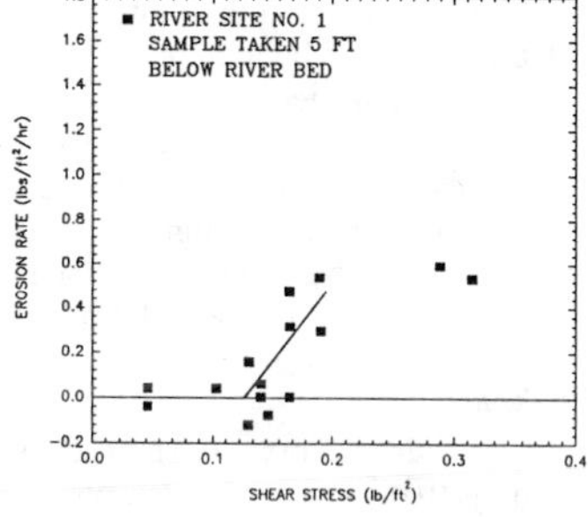

Figure 10. Erosion characteristics for rotating cylinder test No. 1. This sample was collected 5 ft below the bed at River Site No. 1.

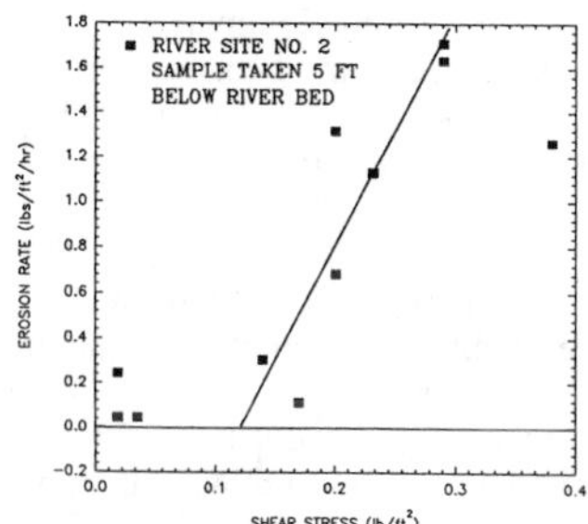

Figure 11. Erosion characteristics for rotating cylinder test No. 2. This sample was collected 5 ft below the bed at River Site No. 2.

Contraction Scour at a Bridge over Wolf Creek, Iowa

Edward E. Fischer[1], Member, ASCE

Abstract

Contraction scour at the State Highway 14 bridge over Wolf Creek in south-central Iowa was caused by a large flood on September 14 and 15, 1992. The bridge is a 30.5-m, single-span steel structure supported by vertical-wall concrete abutments with wingwalls. Approximately 6 meters of scour resulted from the flood. The peak discharge was estimated by water-surface profile analysis to be 2,200 cubic meters per second. A crude stage hydrograph is depicted by a line of gravel deposits created by water overflowing the downstream face of the highway embankment. The fall of the water surface at the bridge was 3.4 m when water began to flow over the highway.

Introduction

Intense thunderstorms in south-central Iowa on September 14 and 15, 1992, produced floods that caused considerable damage in the region. More than 75 mm of rain fell over a large area (fig. 1). Floodwaters closed many highways and local roads, including an Interstate Highway, and caused scour that damaged many bridges throughout the area. For example, 3 m of piling were exposed at the left abutment at the State Highway 2 bridge over the Weldon River in Decatur County (fig 1) (Fischer, 1993). The same storms produced floodwaters that caused 6 m of contraction scour at the State Highway 14 bridge over Wolf Creek in southern Lucas County. This paper describes the flooding and resulting scour at the Wolf Creek site.

Storm Description

The storms that caused the flooding in south-central Iowa are described by Fischer (1993) and are summarized here. Because of antecedent conditions, the ground in south-central Iowa was nearly saturated on September 14, 1992 when the first rainfall that caused the flooding was recorded early in the day. By evening, meteorological conditions caused thunderstorms to develop and move eastward over the same area most of the night. The largest official rainfall total for the storms was 292 mm at Derby in Lucas County (fig. 2). A total of 236 mm of rain was recorded at that site between 23:00 hrs September 14 and 05:30 hrs the next morning. The average rainfall

[1] Hydrologist, U.S. Geological Survey, P.O. Box 1230, Iowa City, Iowa 52244

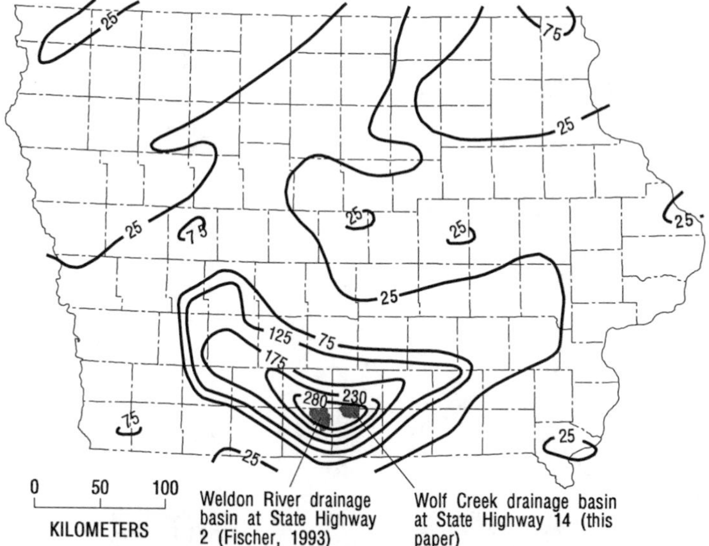

Figure 1. Areal distribution of rainfall from intense thunderstorms in Iowa, September 14-15, 1992. (Lines of equal total rainfall in mm, variable intervals, from Hillaker, 1992, p. 3.)

intensity for this 6.5-hour period was 36.3 mm/hr and the maximum rainfall intensity was 30.5 mm in 15 minutes.

Basin Characteristics and Site Description

Nearly the entire drainage basin of Wolf Creek at State Highway 14 was within the area of greatest total rainfall determined for the storms (fig. 1). The drainage area is 138 km^2 (Larimer, 1957) and the drainage-basin perimeter is about 52 km. A major tributary, Brush Creek, drains into Wolf Creek about 10 m upstream of the bridge (fig. 2). The topography of the drainage basin is rolling hills with a total relief of approximately 46 m as determined from U.S. Geological Survey 1:24,000-scale topographic maps. The width of the valley at the study site is approximately 400 m. Land use in the basin is primarily agricultural.

The State Highway 14 bridge over Wolf Creek is a 30.5-m by 7.9-m single-span deck plate girder bridge supported by vertical-wall concrete abutments with 45-degree wingwalls (Iowa Department of Transportation (IDOT), 1946). The abutments and wingwalls were built in 1925 for a single-span truss bridge and are supported by piling. The tops of the abutments were raised 2.00 m when the current structure was built in 1947; however, the wingwalls were not raised. The roadway is on an embankment that is approximately 5 m above the valley floor and is approximately perpendicular to the principal axis of the stream and floodplain. The stream flows through dense woods on both the upstream and downstream sides of the highway in the vicinity of the bridge. Upstream of the bridge, the right floodplain is wooded near the stream and

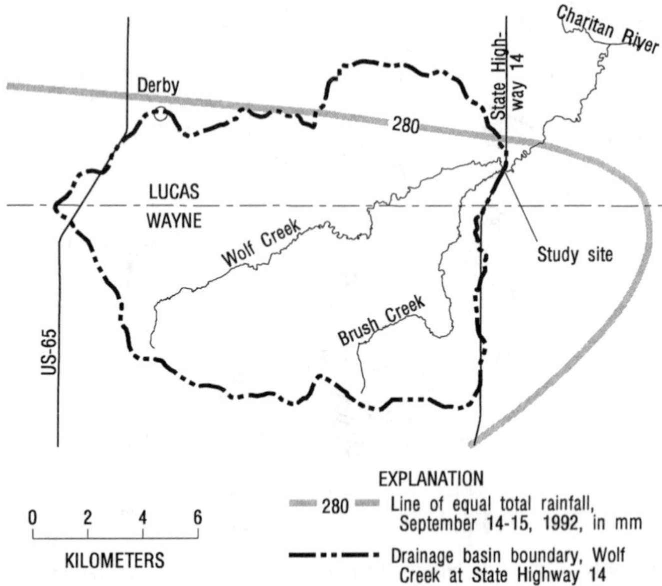

Figure 2. Location of study site

is a pasture away from the stream. No design-flood information for the bridge is available. The bridge plans show an extreme high-water elevation of 296.69 m that occurred in June 1946.

Flood Description

The peak discharge of the September 14-15 flood in Wolf Creek at State Highway 14 was 2,200 m^3/s, which was determined on the basis of a water-surface profile analysis using the computer model WSPRO (Shearman, 1990). The upstream high-water elevation used in the peak discharge computation was 299.66 m, which was the extreme high-water elevation for the flood (Ronald Chapman, Resident Construction Engineer, IDOT, Chariton, Iowa, written commun., September 1993). The downstream high-water elevation used was 297.72 m, which was the highest elevation of a line of gravel deposits created by road overflow on the downstream face of the right highway embankment. These marks indicate a fall of the water surface of 1.94 m. According to information provided by IDOT, the eventual high-water elevation at the downstream side of the bridge was 299.4 m, which was caused by backwater created by channel conditions farther downstream.

The form of the gravel deposits depict a crude stage hydrograph of the flood as floodwaters rose. When water began to flow over the road on the right embankment, road-shoulder gravel near the lowest elevation of the crown of the highway was washed down the side of the embankment until it became submerged under the downstream water surface. The slope of the embankment is about 1:3.5. As the upstream

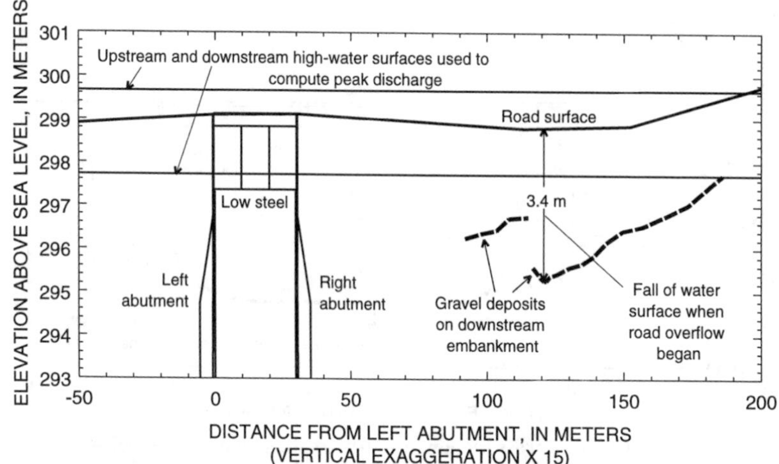

Figure 3. Partial view of cross section at downstream side of State Highway 14 bridge over Wolf Creek during flood of September 14-15, 1992 showing line of gravel deposits and high-water surfaces used to compute peak discharge

water surface continued to rise, the edge of water of the road overflow moved along the highway away from the bridge. The road overflow washed more road-shoulder gravel down to submergence under the downstream water surface, which also was rising. Based on this interpretation of the form of the gravel deposits, the fall of the water surface at the bridge section as floodwaters rose was at least 3.4 m, which occurred when water began to flow over the road (fig. 3). The hydraulic analyses suggest that the flow may have been supercritical through the bridge opening at this time during the flood. The elevation of the streambed at this time, however, is unknown.

Scour Description

Using the streambed profile shown in the bridge plans (Iowa Highway Commission, 1946) as a reference, it is concluded that contraction scour lowered the streambed in the bridge opening about 6 m (fig. 4). The scour profile shows that piling at the right abutment may have been exposed, although this was not verified. A large area downstream of the bridge was scoured by the high-velocity discharge exiting the bridge opening, and a pond about 75 m long and 50 m wide remained after floodwaters receded. Upstream of the bridge, the stream channel and vegetation appeared unaffected. It is surmised that the water surface rose very quickly during the flood, creating a high head at the bridge. The resulting high-velocity discharge through the bridge opening scoured the streambed and adjacent area downstream of the bridge. The water surface downstream of the bridge also rose so that eventually the bridge section became totally submerged.

The theoretical contraction scour for a flood of this magnitude is 9.1 m, as determined using the live-bed sediment transport scour equation in the manual "Evaluating

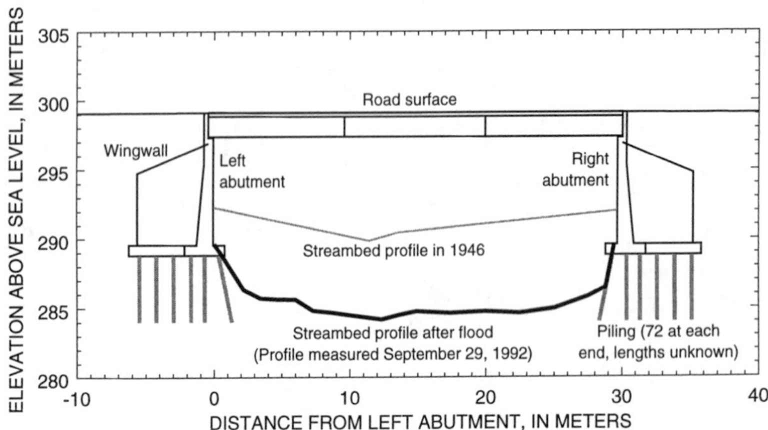

Figure 4. Channel cross section at downstream side of State Highway 14 bridge over Wolf Creek after flood of September 14-15, 1992. View is downstream. (Bridge dimensions and 1946 streambed profile from Iowa Highway Commission, 1946)

scour at bridges, second edition" (Richardson, et. al., 1993). Soil borings were not available, so it is unknown whether the limit of scour at this site was caused by resistant substrata. It is also possible that the downstream water surface rose at a rate sufficient to reduce the velocity of the discharge through the bridge opening, limiting the scour that otherwise might have occurred.

References

Fischer, E.E. (1993). "Scour at a bridge over the Weldon River, Iowa." *Conference Proceedings, Hydraulic Engineering '93, Volume 2*, American Society of Civil Engineers, New York, New York, pp. 1854-9.

Hillaker, H. (1992). "Rainfall totals for 60 hours ending 7 a.m. CDT September 16, 1992." Iowa Climate Review, Iowa Department of Agriculture and Land Stewardship, Des Moines, Iowa, 6(9), 3.

Iowa Highway Commission (1946). "Design for 100-foot by 26-foot deck plate girder bridge, Lucas County." *File No. 13123, Design No. 2346.* Iowa Department of Transportation, Ames, Iowa.

Larimer, O.J. (1957). "Drainage areas of Iowa streams." *Bulletin No. 7*, Iowa Highway Research Board, Ames, Iowa.

Richardson, E.V., Harrison, L.J., Richardson, J.R., and Davis, S.R. (1993). "Evaluating scour at bridges, second edition." *Hydraulic Engineering Circular No. 18: FHWA-IP-90-017.* Federal Highway Administration, McLean, Virginia.

Shearman, J.O. (1990). "User's manual for WSPRO--a Computer Model for Water Surface Profile Computations." *FHWA-IP-89-027.* Federal Highway Administration, McLean, Virginia.

MANAGEMENT OF SEDIMENTATION AND DEGRADATION PROBLEMS OF THE NILE RIVER IN EGYPT

BY

Dr. M. EL-KORANY GOUDA
Deputy Director, Nile Research Institute
National Water Research Center
Ministry of Public Works and Water Resources
Cairo, Egypt

1. Introduction

The Nile River is one of the longest rivers in the world. The modern Nile measures about 6825 km from the sources of Luvironza River in Tanzania to the Mediterranean Sea, of this length, about 1530 km lie within the boundaries of Egypt (Fig. 1). Since the complete control on the Nile River flows by the construction of Aswan High Dam (A.H.D.) in 1967, changes in the characteristics of the river have taken place.

The Nile River reaches can be summarized as follows :

1. The part upstream A.H.D., this region is the sedimentation region.

2. The part downstream A.H.D. of length about 984 km. This region is sub-divided into four reaches separated by control structures, Esna barrage, Naga-Hammadi barrage, Assiut barrage and Delta barrages.

3. The two main branches, the Rosetta and Damietta, each of length about 240 km.

After A.H.D., the hydraulic conditions of the Nile is changed and instead of passing about 1100 million m^3/day in flood period (July to October), only a maximum of 240 million m^3/day passes the Nile River channel in these monthes. Accordingly, morphological changes have been happened to the Nile channel as formation of some islands and migitation of others, scour and or degradation in some locations specially downstream (D.S.) barrages and sedimentation in some locations upstream (U.S.) barrages and erosion of the embankments of the Nile River specially in bends. All these problems which can be sedimentation and or degradation problems have to be monitored, managed and solved or it may cause very severe disaster to the Nile course and its control structures.

In this paper, problems of Nile River degradation, sedimentation and banks erosion were discussed briefly. By the end of the paper, some conclusions and recommendations were presented.

2. Nile River Problems

The Nile River in Egypt is an alluvial stream and gets about 80% of its water from the Ethiopian plateau and the other 20% from the Equatorial plateau. As Egypt is the last country on the Nile River, it gets a limited share of water which is 54.5 milliard m^3/year according to the last agreement between Egypt and Sudan. The classical problem of waters is that the supply is not sufficient to satisfy the demands. The Nile waters have to be used rationally in agriculture (@ 6-7 million feddan), municipal waters (@ 60 million capita), navigation (@ 250 tourist boats and @ 250 cargoes), electricity from High Dam and Esna barrage and industrial (@ 75 factories). After A.H.D. operation in year 1967, the regime of the Nile changes and the clear water passes through the Nile River barrages causes degradation and sedimentation problems in the reach, scour holes directly D.S. barrages, erosion problems on the embankments of the Nile. On the contrary, sedimentation problems occurs U.S. the A.H.D. in Lake Nasser.

3. Sedimentation Studies in Lake Nasser

Since the partial operation of the AHD in 1964, monitoring and data collection regarding sediment transportation and degradation are taking place annually by an expert team from NRI and A.H.D. Authority by taking cross-sections along Lake Nasser at fixed locations, taking sediment samples and water samples, and making hydrographic survey using falcon 4. The results of the study show that the total sediment deposited up to the year 1993 amounts to 2.4 milliard m^3, (Fig. 2) shows the longitudinal cross-section of the bed of the lake.

4. Degradation Downstream Barrages on the Nile River

After the A.H.D. construction clear water becomes hungry and try to substitute the sediment loss by degrading the stream channel. It is noticed that degradation occurs D.S. barrages and according to the sediment balance, the degraded sediment accumulates in some places specially in the upstream boundary. Before A.H.D. construction, researchers try theoretically to estimate the degradation that will happen D.S. barrages using some theoretical approaches or emperical approaches. Unfortunately, they estimated the degradation that will happen in the D.S. of the barrage to be about 8 to 16 ms, which is very dangerous to the barrages.

Recent studies after A.H.D. using mathematical modelling technique shows that degradation D.S. Gaafra gauging station in the first reach of the Nile at 34 km D.S. old Aswan Dam will be in the range of 1.8 ms. after a period of 150 years and 2.00 ms. D.S. Esna barrage, 3.00 ms. D.S. Naga-Hammadi barrage and 3.5 ms. D.S. Assiut barrage. Two mathematical models were used in this study, i.e. Modified Hec 6 and HSRI model, (Korany 1990).

The observations after 20 years from the operation of the A.H.D. shows that degradation is in the range of 60 and 80 cms respectively at Gaafra and D.S. Esna barrage (Shalash, 1985) and the studies at NRI also shows that drop in water surface at gauging discharge stations D.S. barrages decreased with time from 2 cm/year after AHD operation to 0.2 cm/year in year 1990 in the average. Also the researchers of Nile Research Institute (NRI) find out from their researches that bed armouring is going on in the four reaches of the Nile but with different rates and this causes the decrease in the drop in water level with time. More studies is going on at NRI as investigation of navigation bottlenecks using one-dimensional model. Also two-dimensional model is being used in studying some reaches of the Nile River which suffers from navigation bottlenecks, and according to the results of this model which estimates the velocity at different locations of the cross-section after doing some dredging - the alignment of the navigable channel is selected.

5. Bank Erosion Problem

Since the operation of A.H.D., clear water passes in the Nile course causing degradation and erosion problems. The bank erosion problem is one of the problems that have a great concern to study and solve. Follow-up of the eroded banks have taken place before and after the construction of A.H.D., but in year 1988, the Nile Research Institute (NRI), conducted a surveillence and find out that 240 km. of the banks of the Nile have been eroded by different rates.

To solve the bank erosion problem before A.H.D., groins (spurs) were constructed in some locations to direct the current far from the eroded banks, and in other locations traditional revetment protection was used. It was found out that the spurs cause problems in most of the areas they used in, as they creates a reversable spiral current which causes scour holes between spurs and consecutively bank failure. For this NRI through the Nile River Protection and Development Project developed a new method for bank protection using revetments with filter (Fig. 3). The alignment of the revetment toe is very important in the new method as smooth alignment should be done to get smooth flow. The method is tested in three pilot reaches with different lengths in Luxor, Souhag and Beni-Mazar. In Beni-Mazar area, piezometric wells is constructed to measure the variation of the ground water levels near the protected area and to test the filter performance. The work in these pilot projects is monitored and it shows good performance. Sustainable bank protection project is going on since year 1992 to protect 100 km. from the eroded banks and 30 km. of water front. The management steps taken to finish this work is as follows :

5.1. Surveillance.

5.2. Collecting data and preliminary survey.

5.3. Design

Getting the field data, it is processed using computer facilities and the base line is drawn. From the base line and the bank levels, a smooth alignment of the toe is done, then the cross-section is drawn and the design is put on the C.S.. The slope of the revetment is used to be 3 : 2 which was found to be stable slope after checking some areas. The design is done and the stone and back fill quantities is calculated, and then the tender document and specifications and construction requirements is explained. When all these documents are ready for bids, it have to be sent to the Nile Directorates responsible for bidding and give orders to the contractors to begin the work. Before this step, the contracts were selected according to a criteria done to choose the best one from every governorate.

5.4 Execution

The contractor begin the work with the supervision from the Nile Directorate staff. Consultants and managers of NRI visit the field work from time to time to check the quality of the work and give any technical advise to the supervision staff. After the work is finished, Nile Directorate staff have to make the final survey and record and calculate the actual work of the contractor.

5.5. Follow-up

The final survey is conducted by the Nile Directorate team, cross-sections is taken and marked every 50 ms. or less according to the location requirements. Also visual test is done and recorded. The contractor is responsible for one year to do any remedy to the work. Every 6 months the work have to be monitored visually and by taking cross-sections in the same locations as the final survey. The status of the work have to be recorded and conclusions and remarks should be drawn to be taken in consideration in the new designs.

6. Conclusions

By the end of this paper, the most important conclusions are:

1. Nile Research Institute takes care of the Nile problems researches, studies, monitoring and solutions.

2. Nile problems can be summarized in degradation, sedimentation, scour holes, bank erosion problems and its consequence as barrage stability, navigation bottle necks, reservoir sedimentation ... etc.

3. Nile problems are monitored and solved by an expert team from NRI.

4. Bank erosion protection project is managed by NRI through a very good system, beginning from surveillance to design then execution and finally follow-up.

7. Recommendations

The Nile River is one of the longest alluvial rivers in the world. After A.H.D., change in the regime of the Nile takes place. In this brief study the effort that NRI is doing to solve Nile River problems is shown. It is recommended that :

1. Monitoring Nile River status is very important and should be continued.

2. Studies and researches should be continued and two and three-dimensional modelling should be used in a big scale.

8. References

1. International Symposium on River Water Front Development, September 15-17, 1994, Cairo, Egypt, NRI.

2. M. El-Korany Douda, "Sedimentation Study of High Aswan Dam, Researvoir", Dec. 1993, NRI Report.

3. International Conference on Protection and Development of the Nile and other Major Rivers, Cairo, Egypt, February 3-5 1992.

4. "Physical Responses of the River Nile to Interventions", National Seminar, Cairo, Egypt, 12-13 November 1990, NRI.

5. M. El-Korany Gouda, "Computer Flow Sediment Models for Nile River Degradation", Ph.D. Thesis, Cairo University, 1985.

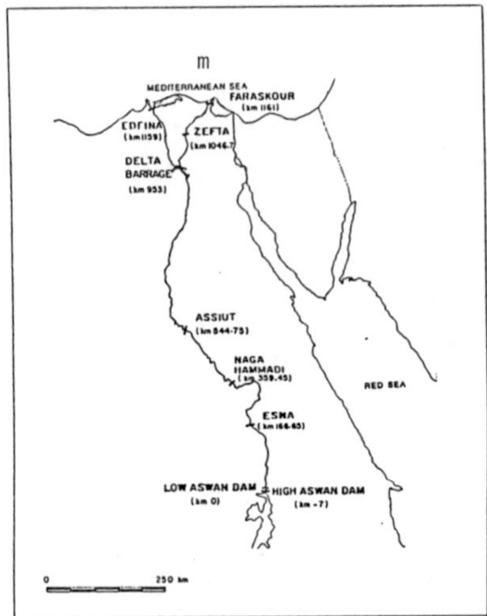

Figure 1. The River Nile in Egypt.

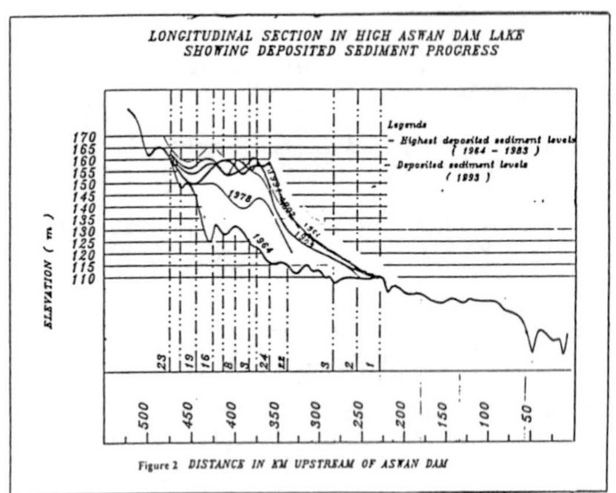

Figure 2 DISTANCE IN KM UPSTREAM OF ASWAN DAM

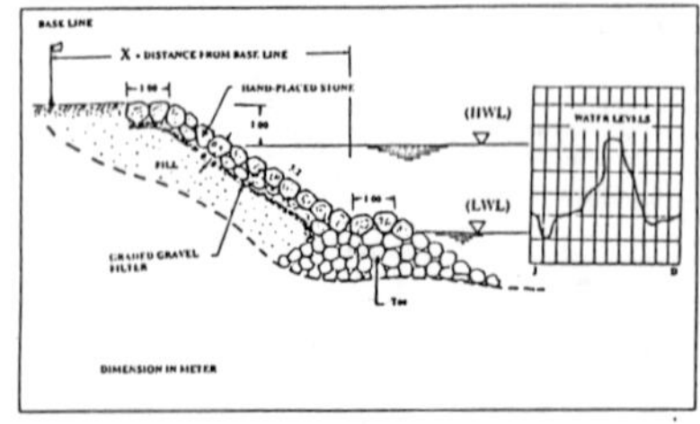

Figure 3. A typical new design for a revetment slope

Modeling Scour of Contaminated Sediments in the Ashtabula River

R. E. Heath[1], T. L. Fagerburg[1], and T. M. Parchure[1]

Abstract

The federal navigation project in the lower Ashtabula River at Ashtabula, OH, contains a break-water protected harbor in Lake Erie and a navigable, commercial waterway extending about 3.2 kilometers (2 miles) upstream. Dredging in the upper 2.4 kilometers (1.5 miles) was suspended in the 1970's, and the accumulated bed sediments are contaminated with heavy metals, chlorinated hydrocarbons (including, in some locations, toxic levels of polychlorinated byphenyls, PCB's), and polynuclear aromatic hydrocarbons. Limited dredging operations in this upper reach were conducted in 1993 to permit continued use of the waterway by recreational traffic. The most heavily polluted sediments are buried under relatively clean sediments. As part of a broader effort to evaluate the future of the waterway, the Waterways Experiment Station (WES) is conducting both field and numerical model investigations to determine the risk of scour through the relatively clean surficial sediments thus exposing and dispersing the underlying contaminants. This paper will discuss the methods used to determine this risk and significant findings from the ongoing study.

Introduction

The federal navigation project in the lower Ashtabula River at Ashtabula, OH, contains a break-water protected harbor in Lake Erie and a navigable, commercial waterway extending about 3.2 kilometers (2 miles) upstream (figure 1). Dredging operations required to

[1]M. ASCE, Research Hydraulic Engineers, US Army Engineer Waterways Experiment Station, 3909 Halls Ferry Road, Vicksburg, Mississippi 39180-6199.

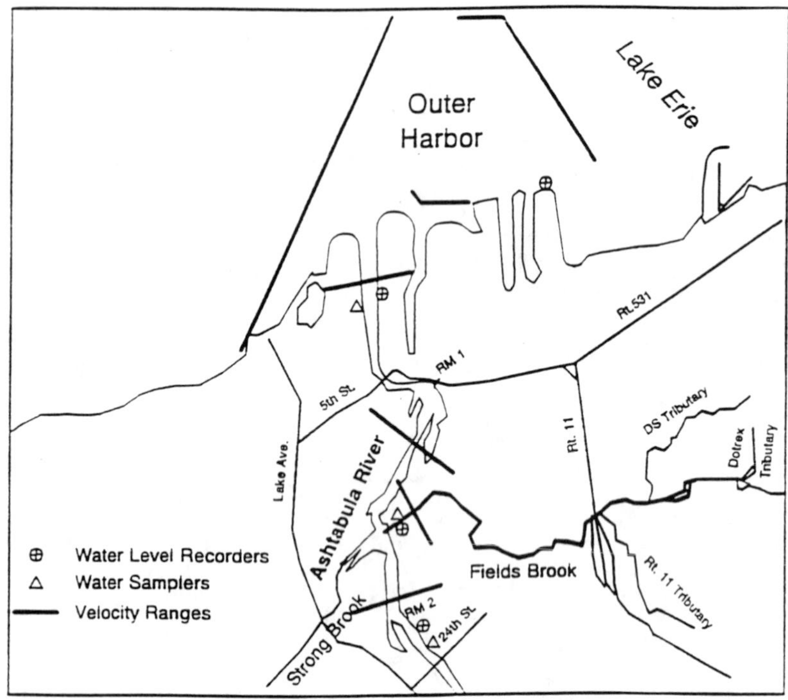

Figure 1. Project location and field data monitoring locations

permit commercial navigation were suspended in the 1970's in the upper 2.4 kilometers (1.5 miles) of the waterway, and the accumulated bed sediments are contaminated with heavy metals, chlorinated hydrocarbons (including, in some locations, toxic levels of polychlorinated byphenyls, PCB's), and polynuclear aromatic hydrocarbons. The most heavily polluted sediments are buried under relatively clean sediments. Contaminant concentrations are generally greatest in the upper turning basin located between Fields and Strong Brooks, gradually decrease as one proceeds downstream through the lower turning basin near river mile 1 (RM 1), and rapidly decrease upstream of the turning basin. Limited dredging operations in this upper reach were conducted in 1993 to permit continued use of the waterway by recreational traffic.

The objective of the on-going study of sediment

transport in the Ashtabula River is to determine the potential magnitude and extent of scour that may occur during a flood event or in response to rapid changes in Lake Erie stages potentially causing exposure and dispersal of contaminants buried in the channel bed sediments. This is to be accomplished by a combination of field data collection and analysis and numerical model studies.

Approach

The purpose of field investigation was to identify and characterize significant hydraulic and sediment transport processes in the river and determine the physical properties of the bed sediments. The ongoing field investigation includes both long-term continuous monitoring of water levels and suspended sediment concentrations with automated data collection equipment and short-term efforts to collect bed material samples along with velocity and suspended sediment concentration profiles.

Laboratory analysis of the bed material consists of testing to determine physical and chemical properties, i.e., gradation, pH, cation exchange capacity, etc., and to estimate the critical shear stress(es) for erosion and erosion rate. The field and laboratory investigations provide critical data needed to properly direct the overall effort toward resolution of the problem under study and to estimate coefficients used in numerical model simulations of hydraulic and sediment transport in the river.

The numerical model study is being conducted using the TABS-MD modeling system, a family of numerical models which provide multi-dimensional solutions to open-channel flow and sediment transport problems (Thomas and McAnally 1985). RMA-2V, a two-dimensional, depth-averaged hydrodynamic numerical model is used to generate water levels and current patterns. RMA-2V employs finite element techniques to solve the Reynolds Form of the Naiver-Stokes equations for turbulent flows. Input data requirements for RMA-2V include a finite element mesh describing system geometry, Manning's roughness coefficients, turbulent exchange coefficients, and boundary conditions. STUDH, a two-dimensional sediment transport model which solves the convection-diffusion equation with bed source and sink terms may be used in combination with the hydraulic forces computed by RMA-2V and input describing bed sediment characteristics to simulate the erosion, transport, and deposition of sediment.

In addition to these investigations being conducted by the WES, the U.S. Army Cold Regions Research and Engineering Laboratory has conducted a field investigation to determine if ice processes in the river have a significant impact on channel scour (Wuebben and Gagnon), and the U.S. Army Corps of Engineers, Buffalo District, is conducting hydrologic studies to develop flood hydrographs for the Ashtabula River.

Preliminary Results

A preliminary investigation was conducted to determine whether the channel scour is highly probable, warranting a detailed study of sediment transport, or whether the erosion is so improbable as to eliminate or reduce the requirement of a thorough investigation of this phenomena. A report describing the results of the preliminary investigation is in preparation (Heath, et. al.).

The results of laboratory erosion tests showed that the critical shear stress for commencement of surface erosion was as low as 0.2 to 0.3 Pascal (0.004 to 0.006 psf). Continued erosion over a 30 minute duration was observed at higher shear stresses on the order of 0.5 to 0.6 Pa (0.010 to 0.013 psf). Estimated bed shear stress with a 100 year flood discharge may vary between 1 and 24 Pa (0.02 to 0.5 psf) over the region of buried contaminated sediment. If the high flood duration extends over a long period of time, there is a very high probability that the relatively clean surface layer of bed sediments covering the contaminated sediments may be completely eroded, thus exposing and eroding the contaminated sediment.

Based on the preliminary results, the potential exists for substantial erosion of channel sediments during a large flood event. This potential increases for floods coincident with relatively low lake levels. The magnitude of scour will be dependent on a number of as yet unquantified factors, including the shape of the flood hydrograph, the erodibility of subsurface bed material layers, and the sediment yield from the watershed.

Laboratory analysis of the surficial bed material indicated that compaction under the weight of subsequent deposits may occur. Thus, subsurface material may be more resistant to erosion. Resolution of this issue will require obtaining deeper sediment samples for analysis.

Acknowledgments

The tests described and the resulting data presented herein were obtained from research conducted under an agreement between the US Army Corps of Engineers and the US Environmental Protection Agency. Permission was granted by the Chief of Engineers to publish this information.

Appendix I. References

Heath, Ronald E., Fagerburg, Timothy L., Parchure, Trimbak, Teeter, Allen M., Boyt, Bill (in preparation). "Ashtabula River, Ohio, Sedimentation Study, Interium Results of Field and Numerical Model Investigations of Channel Scour." U.S. Army Engineer Waterways Experiment Station, Vicksburg, MS.

Thomas, William A. and McAnally, William H., Jr. 1985. "User's Manual for the Generalized Computer Program System: Open-Channel Flow and Sedimentation, TABS-2," Instruction Report HL-85-1, U.S. Army Engineer Waterways Experiment Station, Vicksburg, MS.

Wuebben, James L. and Gagnon, John J. (in preparation). "Ice Regime of the Ashtabula River, Ashtabula, Ohio." U.S. Army Cold Regions Research and Engineering Laboratory Hanover, NH.

REPRESENTING POULTRY LITTER MANAGEMENT WITH GLEAMS

M. C. Smith,[1] W. G. Knisel,[2] Member, ASCE , D. L. Thomas,[3] Member, ASCE, and S. R. Wilkinson[4]

Abstract

The GLEAMS model was applied with data from a 7-yr field study to compare simulation results with observed runoff, nitrogen losses, and nitrogen uptake by coastal bermudagrass. Six treatments included an unfertilized control, two inorganic fertilizer and two broiler litter application rates, and a duplicate broiler litter with interseeded winter rye. Simulated and observed values were compared for surface runoff volume and runoff NO_3-N, NO_3-N mass and maximum monthly-weighted concentration leached below the 122-cm depth, harvested forage yield, and nitrogen content of the yield. All components agreed well for the unfertilized control plot. Runoff volumes and NO_3-N loss compared well for all treatments. Results varied considerably for NO_3-N leached, and for forage and nitrogen yield for the fertilizer and broiler litter treatments.

Introduction

Poultry production has increased tremendously in the U.S. in the past decade, and continues to increase rapidly especially in the southern and southeastern regions. In Georgia alone, where the industry has been predominately in the Southern Piedmont physiographic area, 960 million broilers were produced in 1993 (GASS, 1994). Poultry litter generation was estimated at about 1 million tons per year.

The GLEAMS model (Leonard *et al.*, 1987) was developed to assess the impacts of management alternatives on edge-of-field and bottom-of-root-zone loadings of non-

[1]Associate Professor, Bio. and Agr. Engr. Dept., Univ. of Georgia, Athens, Georgia 30602.

[2]Senior Research Sci., Bio. and Agr. Engr. Dept., Univ. of Georgia, Coastal Plain Expt. Sta., Tifton, Georgia 31793.

[3]Associate Professor, Bio. and Agr. Engr. Dept., Univ. of Georgia, Coastal Plain Expt. Sta., Tifton, Georgia 31793.

[4]Soil Scientist, U.S. Dept. Agr., Agr. Research Service, Southern Piedmont Conserv. Research Center, Watkinsville, Georgia 30677.

point source pollutants. Knisel (1993) developed a comprehensive plant nutrient component to consider runoff and leaching losses of nitrogen and phosphorus. The model additions included such comparisons as inorganic fertilizer and animal waste applications for crop production.

Research studies were conducted at Watkinsville, Georgia in the 1970's on fertilizer and broiler litter application on coastal bermudagrass (*Cynodon dactylon* L pers.) plots (Dudzinsky *et al.*, 1983). Nitrogen losses in surface runoff and percolate water were measured for different application rates and durations. These data were used to validate the GLEAMS model and to determine the capability of the model to represent the wide range of practices. This paper presents a summary of the results of comparing observed and model-simulated values for the study.

Available Data and Methods of Analyses

A 7-year study was conducted at the USDA-ARS Southern Piedmont Conservation Research Center at Watkinsville, Georgia, to measure the effects of broiler litter and inorganic fertilizer application on nitrogen losses from different practices on coastal bermudagrass plots (Dudzinsky *et al.*, 1983). The study was conducted on a Cecil sandy loam soil (Typic Hapludult, clayey, kaolinitic, thermic). Coastal bermudagrass was sprigged on all plots in 1970, and all plots were fertilized with ammonium-nitrate fertilizer three times each year in 1970 and 1971 (168 kg N/ha/yr) to insure a vigorous stand when the study began in 1972.

Surface runoff and percolate below the 122-cm soil depth were monitored for treatments which included different broiler litter and inorganic fertilizer applications four times a year in May, June, July, and August, and an unfertilized control as shown in Table 1 (Dudzinsky *et al.*, 1983). One treatment included fall interseeding of rye (*Secale cereale* L.) for winter uptake of nitrogen from spring and summer broiler litter application (Wilkinson *et al.*, 1985). Total nitrogen in the broiler litter was monitored at the time of each application. Bermudagrass forage production and nitrogen content were measured in multiple cuttings of hay each year.

Soil characteristic data for the study site were used to develop model parameters. Nitrogen content in the soil profile was not measured at the beginning of the study, and the initial values for the various conceptualized model nutrient pools were largely estimated. Nitrogen species data were not available for the broiler litter, and input parameter estimates were made from literature (Knisel, 1993) without fine tuning.

Results and Discussion

Model simulation results and observed data are summarized, and 7-yr totals are given in Table 2 for each treatment. Comparisons are made for 6 components: runoff volume and nitrogen in runoff, NO_3-N mass and maximum monthly-weighted NO_3-N concentrations leached below the 122-cm depth, and forage yield and nitrogen content of bermudagrass hay. Suction lysimeters were used to sample soil-water percolate when rainfall indicated probable availability of leachate, and NO_3-N concentrations were used with a daily water accounting procedure.

Table 1. Description of Watkinsville, Georgia, fertilizer and broiler litter treatments.

Code	Description	Duration (years)	Nitrogen (kg/ha)
UC	Unfertilizer Control	7	0
FC	Fertilizer Control No. 1 56 kg N/ha/Application 112 kg N/ha/Application	 2 5	 448 2,240
F2	Fertilizer Control No. 1 336 kg N/ha/Application Residual	 2 5	 2,688 0
B1	Broiler Litter No. 1 22.4 t Broiler Litter/ha/Year Residual	 4 3	 2,793 0
B2	Broiler Litter No. 2 44.8 t Broiler Litter/ha/Year Residual	 4 3	 5,586 0
B2R	Same Broiler Litter as B2, Interseeded to Rye in Oct. 44.8 t Broiler Litter/ha/Year Continued Rye, Residual N	 4 3	 5,586 0

Simulated runoff volumes and runoff nitrogen agreed well with observed data for all treatments. Slight adjustments in field capacity were made for some treatments. The winter rye interseeded with the bermudagrass (treatment B2R) resulted in less surface runoff volume. Adjustments had to be made to account for the observed reduction.

One suction lysimeter sample was taken at the 122-cm depth in each plot, and NO_3-N concentrations in these point (time and space) samples were used with percolation volumes estimated from water balance calculations to determine the observed NO_3-N mass percolate. Monthly flow-weighted concentrations were calculated from these data.

Maximum monthly weighted concentrations in the 7-yr study are given in Table 2 for comparison with model simulated values. Concentration and mass values agreed very well for the UC treatment. Results for the other treatments were mixed: relatively good agreement for mass leached for treatments F2, B2, and B2R, and relatively good agreement for maximum monthly-weighted concentrations for treatments FC and B1. The winter rye in treatment B2R obviously removed considerable NO_3-N that resulted in low mass and concentrations leached compared with simulated values.

Table 2. Observed and GLEAMS simulated data, Watkinsville poultry litter study.

Treatment	Component	Units	Observed	Simulated
UC	Runoff	mm	147	156
	Runoff NO_3-N	kg/ha	0.3	1.2
	Leach. NO_3-N	kg/ha	10.1	16.6
	Max. Concen.	mg/L	5	7
	Forage Yield	t/ha	11.0	12.5
	Yield N	kg/ha	187	133
FC	Runoff	mm	130	133
	Runoff NO_3-N	kg/ha	0.8	1.2
	Leach. NO_3-N	kg/ha	170	584
	Max. Concen.	mg/L	37	42
	Forage Yield	t/ha	75.2	81.6
	Yield N	kg/ha	2,015	1,027
F2	Runoff	mm	155	151
	Runoff NO_3-N	kg/ha	0.5	1.4
	Leach. NO_3-N	kg/ha	685	696
	Max. Concen.	mg/L	215	92
	Forage Yield	t/ha	61.3	61.8
	Yield N	kg/ha	1,771	1,408
B1	Runoff	mm	201	202
	Runoff NO_3-N	kg/ha	2.4	2.6
	Leach. NO_3-N	kg/ha	136	419
	Max. Concen.	mg/L	68	72
	Forage Yield	t/ha	82.2	81.0
	Yield N	kg/ha	2,175	882
B2	Runoff	mm	135	137
	Runoff NO_3-N	kg/ha	2.9	2.7
	Leach. NO_3-N	kg/ha	1,247	1,125
	Max. Concen.	mg/L	273	131
	Forage Yield	t/ha	97.1	114.9
	Yield N	kg/ha	2,850	1,296
B2R	Runoff	mm	46	45
	Runoff NO_3-N	kg/ha	0.7	0.7
	Leach. NO_3-N	kg/ha	582	700
	Max. Concen.	mg/L	38	118
	Forage Yield	t/ha	112.7	113.2
	Yield N	kg/ha	3,356	1439

The largest differences between simulated and observed data occurred in the nitrogen content of forage removed. Total forage yield in the multiple grass cuttings agreed well between simulated and observed except for the large over-simulation for treatment FC. However, total nitrogen in the yield was significantly under-estimated for all treatments except UC. The worst discrepancies occurred for the broiler litter treatments B1, B2, and B2R. Knisel (1993) stated that GLEAMS does not automatically simulate "flush" nitrogen uptake by crops for high fertilization rates, but that the coefficient in the exponential relation of nitrogen content to total dry matter can be user-defined to represent it. This was attempted in the present study, and it was found that an increased coefficient resulted in a high estimated N demand that oftentimes resulted in reduced yield due to nitrogen stress. Thus potential yield and the coefficient values are interdependent in GLEAMS, and are relatively sensitive parameters. GLEAMS simulated significant quantities of denitrification which reduces availability for N uptake. In the field study, denitrification was thought to be very small, and was not considered in the nitrogen balance calculations (Dudzinsky *et al.*, 1983; Wilkinson *et al.*, 1985).

Conclusions

The GLEAMS model can be used satisfactorily to compare management practices for land application of poultry litter. Also, the model gives a good comparison of forage yield with multiple cuttings of hay each year. Flush uptake of nitrogen can be represented for heavy fertilization, but simulated nitrogen stress may reduce yields.

GLEAMS should not be used for absolute predictions, but can be used effectively for relative comparisons of management alternatives. This study has shown that site-specific observed data are desirable for model parameterization and fine tuning.

Appendix.--References

Dudzinsky, M.L., S.R. Wilkinson, R.N. Dawson, and A.P. Barnett. 1983. Fate of nitrogen from NH_4NO_3 and broiler litter applied to coastal bermudagrass. In: R.R. Lowrance, R.L. Todd, L.E. Asmussen, and R.A. Leonard (Eds.), Nutrient Cycling in Agricultural Ecosystems. Univ. of Ga., College of Agr., Athens, Spec. Publ. No. 23, pp. 373-388.

Georgia Agricultural Statistics Service (GASS). 1994. Georgia Farm Report. Athens, Georgia. vol. 94, no. 5, May 5, 1994.

Knisel, W.G. (Ed.) 1993. GLEAMS: Groundwater Loading Effects of Agricultural Management Systems, Version 2.10. Univ. of Ga., Coastal Plain Expt. Sta., Biol. and Agr. Engr. Dept., Publ. BAED 5. 260 pp.

Leonard, R.A., W.G. Knisel, and D.A. Still. 1987. GLEAMS: Groundwater Loading Effects of Agricultural Management Systems. Trans. of the Amer. Soc. of Agr. Engrs. 30(5):1403-1418.

Wilkinson, S.R., M.L Dudzinsky, and R.N. Dawson. 1985. The effect of interseeding rye (*Secale cereale* L.) in coastal bermudagrass (*Cynodon dactylon* L. pers.) on nitrogen recovery from broiler litter applications. Proc., XV Intnl. Grassland Cong. The Sci. Council of Japan and the Japanese Soc. Grasslands Sci., Nishi-Nasuno, Tochigi-ken, Japan. pp. 537-538.

Modeling Riverine Transport of a Pesticide Plume

Camilla M. Saviz[1], Student Member, ASCE
John F. DeGeorge[2], Gerald T. Orlob[3], Fellow, ASCE,
Ian P. King[4], Member, ASCE

Abstract

A pesticide spill into the Upper Sacramento River was simulated using coupled finite element hydrodynamic and water quality models. The water quality model was extended to represent the physical and chemical processes determining the fate of the pesticide metam and its conversion to the toxic compound, MITC. Model results include estimates of MITC concentration in the water column and mass emission rates of MITC to the atmosphere as functions of time and distance along the river.

Introduction

At 9:50 p.m. on July 14, 1991 an accidental spill of about 27,000 kg of the pesticide metam sodium (sodium N-methyl dithiocarbamate) occurred by derailment of a tank car at the Cantara Loop near Dunsmuir, California. During transit in the Upper Sacramento River downstream toward Shasta Lake, decomposition by hydrolysis and photolysis produced the major toxic breakdown product, methyl-isothiocyanate (MITC), which was then diminished in aqueous concentration by volatilization. Exposure to the pesticide resulted in mortality of virtually all aquatic organisms in the river including fish, crustaceans, algae and plant species which were entrained in the plume and transported downstream. The greatest risk to people in the area resulted from exposure to MITC volatilized to the atmosphere from the river surface. Assessment of human and ecosystem health risks required quantification of the mass of MITC volatilized as a function of time and distance downstream from the spill. Model simulations of

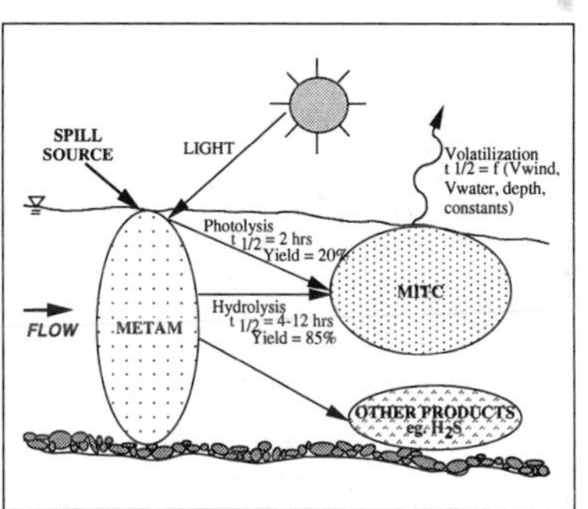

Figure 1. Fate of Metam and MITC in the river

---

[1] Ph.D. Candidate, Department of Civil and Environmental Engineering, University of California, Davis, CA 95616.

[2] Resource Management Associates, 4171 Suisun Valley Road, Suite C, Suisun, CA 94585.

[3] Professor Emeritus, Department of Civil and Environmental Engineering, University of California, Davis, CA 95616.

[4] Professor, Department of Civil and Environmental Engineering, University of California, Davis, CA 95616.

the contaminant plume's fate and riverine transport provided estimates of MITC production and volatilization rates for air quality modeling and risk assessment.

Metam Sodium and MITC

Two constituents were simulated in this study, metam and MITC. Metam sodium, a pesticide used in California principally as a preplanting soil fumigant, is stable in solution, but upon dilution with water and exposure to sediments, suspended matter, and light, rapidly decomposes to its major breakdown product, MITC with production of other by-products, as shown in Figure 1 (CEPA, 1992; DFG 1993). Upon exposure to light, it decomposes by photolysis with a half-life varying from 30 minutes to 2.5 hours, yielding 15% to 60% of the volatile compound, MITC (CEPA, 1992). Certification tests and published literature on metam have generally concentrated on photodecomposition and decay in soil, although recent tests indicate that hydrolysis is also an important reaction pathway of metam to MITC, resulting in yields of 60% to 90% with hydrolysis half-lives of 4 to 12 hours. Higher MITC yields and a longer half-life are expected in the presence of sediments (Miller, 1994).

Methodology

Flow behavior in the Upper Sacramento River is typical of mountain streams with non-uniform geometry and gradient and a rough stream bed consisting largely of cobbles and boulders. It is characterized by run-riffle-pool sequences where varying bed slopes create rapidly-flowing riffles interspersed with slow-flowing pools. To simulate the hydraulics of the river, a finite element model was adapted to the 70.6 km (43.9 mile) reach from Lake Siskiyou to Shasta Lake (Figure 2). A preliminary finite element grid was generated from 1:24000 USGS topographic maps, bed slopes were estimated by interpolation between topographic contours, and cross sections were estimated from a 1:3600 GIS riparian habitat map. The network was refined in the vicinity of the spill and near tributary junctions, resulting in a mesh of 546 elements ranging in length from 30m to 250m.

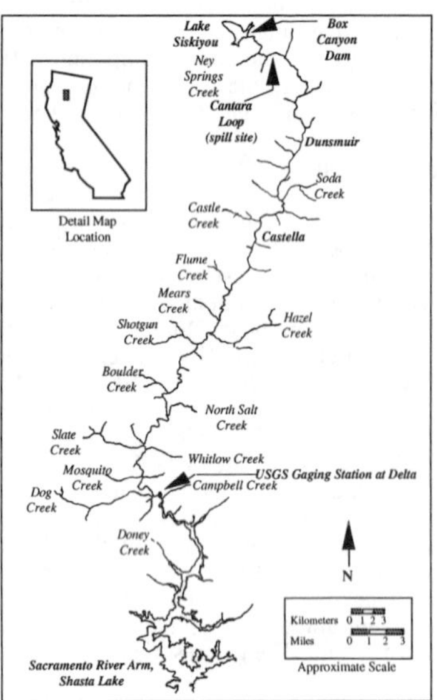

Figure 2. Upper Sacramento River and Tributaries

The network geometry was input to the finite element hydrodynamic model, RMA2, (Norton, 1973; King 1993) which solves the depth and cross sectionally averaged shallow water equations for continuity and momentum to describe velocities and depths within the water body. RMA2 was modified for this project to simulate run-riffle-pool sequences of a mountain stream by applying a scaling factor to change the effective bed slope within elements, increasing or decreasing the effective head loss to represent pools or riffles.

During the period of the spill, Box Canyon Dam discharged water at 1.17 $\pm$0.06 m^3/s (41.3 $\pm$2 cfs). Steady flows from fourteen tributary streams (Figure 2) were incorporated into the model as element side flows. To simplify calculations, ungaged groundwater seepage and flows from minor tributaries were neglected. Because releases from Box Canyon Dam, flows from the tributaries and measured flows at the USGS gaging station downstream at Delta (Figure 2) were nearly constant over the 72 hour period following the spill, steady-state flow was assumed. Simulated flow rates were checked against observations at Delta and Castella and then used in the water quality model, RMA4Q.

RMA4Q (King and DeGeorge, 1994), used to simulate transport of the plume over the 72 hour period following the spill, is a combined one- and two-dimensional finite element model capable of describing the transport and fate of multiple linked constituents. Decay rates of metam and MITC were approximated in the model using first order kinetics:

$$\frac{dC_{MITC}}{dt} = -k_{MITC}C_{MITC} + S_{Metam}$$

where $k_{MITC}C_{MITC}$ represents volatilization, considered to be the only cause of MITC loss, and S_{Metam} represents MITC yield from metam, considered to be the only source. Volatilization, generally occurring more rapidly than hydrolysis, is dependent on concentration, Henry's law constant, and gas and liquid phase exchange, the coefficients for which have been described empirically in terms of wind velocity, stream velocity and water depth (CEPA, 1992; DeGeorge and Saviz, 1994). In the simulated system, even very low wind velocities (<1 m/s) were found to cause rapid volatilization of MITC.

MITC data from field observations following the spill were used to calibrate and verify the travel time and plume concentration calculated by the hydrodynamic and water quality models. The most complete data set was comprised of hourly measurements of MITC in the river as the plume passed the sampling location at Doney Creek (Figure 2) from midnight to 10 a.m. on July 17, 1991. Only limited sampling results were available at other locations. The plume peak required 55.5 hours to travel from the spill site to Doney Creek (CVRWQCB, 1991).

The effects of MITC half-life and wind were examined and compared to observed MITC concentrations. Alternative spill scenarios were developed and simulated for extreme or 'worst case' scenarios, based on the assumption that 27,000 kg of active material spilled into the river over a 2 hour period. Scenarios were constructed assuming metam hydrolysis half-lives of 4 to 12 hours, considering first only hydrolysis, and then combined effects of hydrolysis and photolysis. The water quality model simulated photodecomposition of metam only during daylight hours, while decomposition by hydrolysis occurred continuously over the full period of transport. Initial model runs minimized the volatilization rate by setting the wind speed to zero. Because simulated concentrations at Doney Creek were found to exceed observed values, scenarios were repeated using successively higher volatilization rates with wind speeds up to 1 m/s. Wind velocity was used as a calibration parameter to examine the sensitivity of MITC concentration, as shown in Figure 3 for a 12 hour metam hydrolysis half-life. A wind velocity of 0.45 m/s produced closest agreement between the model result and the observed MITC peak concentration at Doney Creek.

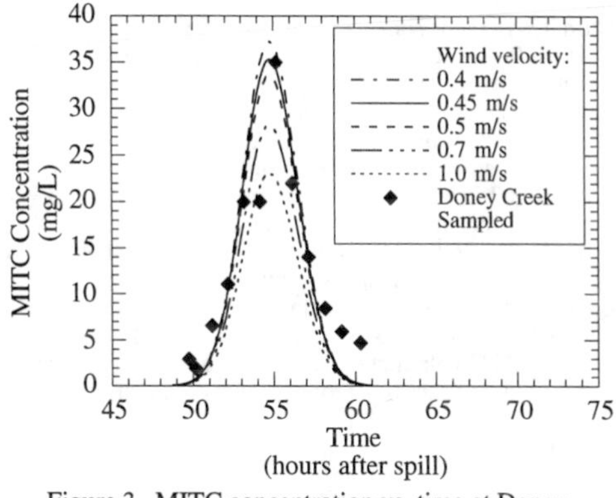

Figure 3. MITC concentration vs. time at Doney Creek, comparison of wind effects

Simulation results shown in Figure 4 indicate reasonable estimates of plume arrival time at Doney Creek, but some features of the pollutant pattern, like the long trailing edge, were not well-represented by the model. The discrepancy may be due to an actual spill configuration different from that assumed. For example, rather than a steady 2-hour discharge at a high concentration, the initially high rate may have been followed by a gradually decreasing mass flux. In

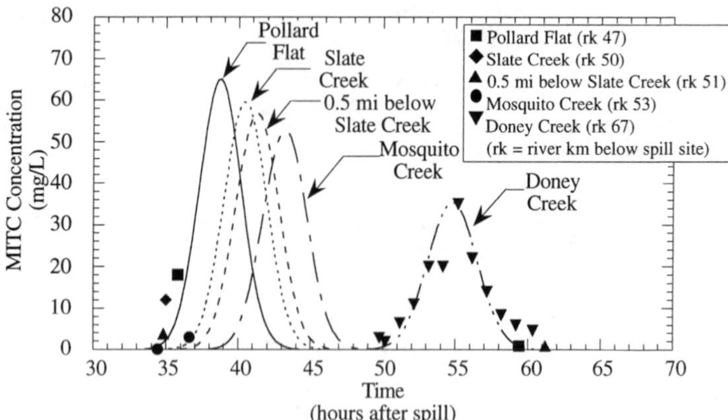

Figure 4. MITC concentration vs. time; 12 hr hydrolysis half-life, $V_{wind} = 0.45$ m/s

addition, some of the contaminant may actually have been stored temporarily in pools, behind rocks or along the river bank and released to the main channel after the bulk of the plume had passed. Also shown in Figure 4 are model results and field observations at four locations upstream of Doney Creek. Discrepancies between the simulated and measured MITC concentration may have been due to samples being collected at times when the plume appeared to arrive at a given location, but not necessarily at the actual time of peak concentration. The apparent difference in travel time may have been due in part to the assumed distribution of tributary flows.

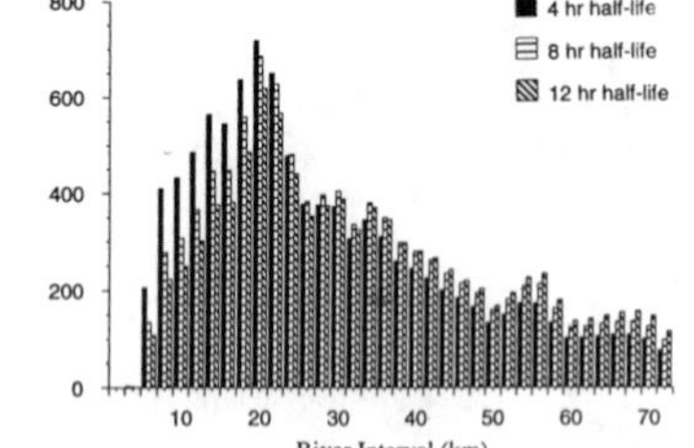

Figure 5. Mass of MITC (kg) volatilized: Degradation by hydrolysis

Simulation results were used to compute the mass of MITC volatilized over the length of the river during the 72 hours following the spill. The rates of volatilization are integrated over space (546 elements) and time (72 hours) to compute the total mass of MITC volatilized. Estimates ranged from approximately 5,000 kg to 10,000 kg, depending on the assumed decomposition mechanisms and ambient conditions, representing from 33% to 65% of the original mass of metam released to the river. The estimated mass of MITC volatilized over successive 2 km reaches along the river is plotted in Figure 5. When hydrolysis is the only decomposition mechanism, the effect of shortening the half-life is to generate more MITC earlier in the simulation, rather than strongly affecting the mass of metam that is converted to MITC. When photolysis is included, shortening the hydrolysis half-life has a greater effect because photolysis has a relatively short half-life and a much lower yield; more metam decomposes to MITC before the sun rises, making more MITC available for volatilization. The mass of MITC volatilized increases to a maximum at approximately 20 km below Box Canyon Dam, and then decreases rapidly downstream primarily due to dilution by tributary inflows.

Conclusions

Coupled hydrodynamic and water quality models provided estimates of the mass of MITC volatilized from the pesticide metam as the contaminant plume traveled downstream following the Cantara Spill. The models can be used to assess the fate and transport of MITC and to evaluate associated health risks. However, considering the uncertainty of boundary conditions and model parameters, model results should be regarded as preliminary estimates of probable concentrations following the actual spill. The following conclusions are derived from this study:

- The rate of metam hydrolysis strongly influences the mass of MITC volatilized, especially in the upstream reaches of the system;
- Because headwater flows are relatively cold and low temperatures decrease the rate of hydrolysis, a longer metam half-life is realistic for the Cantara spill episode;
- When photolysis is an important decomposition mechanism for metam, the total mass of MITC volatilized can be expected to be lower due to the relatively low yield by this process compared with hydrolysis;
- Between 33% and 65% of the metam released to the river during the Cantara Spill volatilized as MITC, with 10% to 26% entering the atmosphere within the first 20 km downstream of the spill site.

Experience with the Cantara Spill indicates the need for comprehensive monitoring of a contaminant substance in both water and air, and identification of those environmental conditions that influence its movement and concentration. Early deployment of monitoring teams at strategic locations, continuous sampling of pertinent contaminant states over critical periods, and knowledge of rate characteristics of important processes are needed to model and simulate such episodes. In cases where a high potential exists for accidental spills, advanced planning should include *a priori* modeling of aquatic systems, identification of strategic monitoring sites and organization of data bases of useful information, e.g. hydrologic, meteorologic and chemical data. The unique experience of modeling the Cantara Spill provides a foundation for improved assessment of future catastrophic episodes should they occur.

Acknowledgments

Modeling of the Cantara Spill episode was supported by the California Department of Fish and Game. Data, information, and assistance were generously provided by staff of the Central Valley Regional Water Quality Control Board, the California Department of Fish and Game, Dr. Glenn Miller of the University of Nevada, Reno and Dr. Robert Yamartino of Sigma Research Corporation.

References

California Department of Fish and Game (DFG), "Natural Resource Damage Assessment Plan, Sacramento River: Cantara Spill, Shasta and Siskiyou Counties," October 1993.

California Environmental Protection Agency (CEPA), Office of Environmental and Health Hazard Assessment, "Evaluation of the Health Risks Associated with the Metam Spill in the Upper Sacramento River", External Review Draft, September 21, 1992.

Central Valley Regional Water Quality Control Board (CVRWQCB). Final Water Sampling Report, Southern Pacific - Cantara Spill, November 1991.

DeGeorge, J. F., C. M. Saviz, G. T. Orlob, and I. P. King. "Modeling the Fate of Metam Sodium and MITC in the Upper Sacramento River, the Cantara-Southern Pacific Spill" In Publication, Center for Environmental and Water Resources Engineering, Report No. 95-2, University of California, Davis, 1994.

King, I. P., RMA2 - A Two Dimensional Finite Element Model for Flow in Estuaries and Streams, Version 4.4, 1993.

King, I. P. and J. F. DeGeorge, RMA4Q - A Two Dimensional Finite Element Water Quality Model, Version 1.0, 1994.

Miller, G., 1994 Personal Communication (May 19, 1994)

Norton, W. R., I. P. King and G. T. Orlob, "A Finite Element Model for Lower Granite Reservoir," prepared for the Walla Walla District, U.S. Army Corps of Engineers, Walla Walla, WA, 1973.

Sodium Hypochlorite at Seattle Water's Open Reservoirs

Jill M. Marilley[1]
Associate Member

Abstract
Seattle Water currently owns and operates nine open distribution reservoirs within the City of Seattle service area. Gas chlorination has been used by Seattle Water since 1950 at all reservoir sites. However, due to increased safety concerns regarding the handling and transport of gaseous chlorine and recent regulations, Seattle Water has chosen to change to sodium hypochlorite disinfection. The use of sodium hypochlorite will provide increased levels of safety since all of the facilities are located in densely populated areas. As public interest and involvement increases in the future, many utilities are being pressured to undertake similar projects.

Introduction

Seattle Water, in partnership with CH_2M Hill, Northwest, is in the process of designing and constructing sodium hypochlorination facilities for eight reservoirs. A ninth reservoir was already converted to sodium hypochlorite disinfection treatment in 1987 as a full scale, operational prototype. This site has assisted Seattle Water in developing an understanding of sodium hypochlorite disinfection and its operation and maintenance effects.

In 1982, Seattle Water realized the potential safety problem that chlorine gas can present in densely populated areas. As part of the planning process, a study was undertaken and completed in 1985 by James M. Montgomery Engineers (now Montgomery-Watson), which identified sodium hypochlorite as a technically feasible alternative to chlorine gas. Conclusions drawn from this report indicated that the short term installation of chlorine gas scrubbers and associated building construction at the reservoir sites was not economically advantageous. While scrubbers removed harmful gasses from the air, they did not answer the

---

[1] Jill M. Marilley, Associate Civil Engineer, Seattle Water, 710 Second Avenue, Room 900, Seattle, WA 98104

long term safety problem of the presence of chlorine gas. The sodium hypochlorite disinfection alternative would eliminate the need to store, handle and transport chlorine gas.

In 1991 the Seattle Fire Department (SFD) adopted the Uniform Fire Code with its revised Article 80, which addresses itself to the prevention, control and mitigation of dangerous conditions related to hazardous materials. Despite an excellent history without a single chlorine leak within the city limits, SFD informed Seattle Water that the use of gaseous chlorine would no longer be allowed. Permits for chlorine within the city would no longer be issued and Seattle Water was instructed to bring the facilities into compliance immediately. SWD worked out a compliance agreement and schedule to bring the facilities up to code as soon as was technically realistic. SFD continues to be actively involved in monitoring the progress of this project.

The basic components of the sodium hypochlorite disinfection system include: chemical storage, chemical dilution, mixing and transfer, chemical feed and delivery and instrumentation and control. In all of these areas, Seattle Water's primary objectives have been those of simplicity, uniformity and reliability throughout the system and among all the facilities. Each part of the design could provide for extensive discussion and review. However, this paper will address two areas: chemical storage and the minimization of the chlorate ion and the chemical feed and delivery system choices.

<u>Chemical Storage</u>

Seattle Water hired Montgomery- Watson in 1992 to assess the potential impact of chlorate ion formation associated with the use of sodium hypochlorite as a disinfectant for a potable water supply. The chlorate ion is considered carcinogenic and is listed as one of the compounds for regulation under the Disinfection By-Products rule under development by the US EPA. Therefore, concern must be taken when establishing controls on sodium hypochlorite in any system.

Tests have indicated that as the amount of available chlorine present in the sodium hypochlorite solution decreases, the amount of chlorine ion increases. Therefore, the impact on the drinking water system is that as the strength of the sodium hypochlorite solution decreases an increased amount of solution must be fed into the system. This, in turn, increases the amount of chlorate ion being put into the system. Any system to be developed to use sodium hypochlorite must develop it to reduce the maximum rate the system feeds the chlorate ion. This can be done through an appropriate balance of initial storage strength, temperature and chemical storage period

Analysis by CH2M Hill of degradation of sodium hypochlorite versus initial strength has given Seattle Water two definitive principles. First, the higher the initial strength of the solution, the more rapid the decay. Secondly, the higher the stored temperature of the solution, the more rapid the rate of decay. Using information gathered by CH2M Hill it was determined that an optimum storage strength of the sodium hypochlorite was between 6% and 10%. However, the demands of the system would dictate a delivery schedule of once every 2-4 weeks with the maximum tank size available to install in the buildings. It was therefore determined that delivery of a solution of a trade strength of 16% and immediate on-site dilution to 8% was the optimum operating method. The immediate dilution still gave the degradation characteristics of the final diluted strength.

Secondly, controlling the temperature of the building was required to control the chlorate ion production. It was highly unrealistic to install a temperature controlled system in Seattle's moderate climate. However, measures could be taken in the choices for ventilation and building materials to keep the temperature in the building materials to keep the temperature in the building to a mild to cool level.

Data was developed to illustrate the relationship between time and storage strength and the degradation of sodium hypochlorite. For example, for an 8% solution stored at 24 Deg C (75 Deg F), after sixty days had degraded to an approximately 6.85% solution. However, if that same 8% solution had been kept 9 Deg C cooler at 15 Deg C (59 Deg F) the degradation was only to an approximate 7.3% solution. This process continues but with more incremental differences. Therefore, it is Seattle Water's operational goal to keep these buildings at or below an internal temperature of 7 - 9 Deg C (55 - 59 Deg F).

Finally, since all calculations indicated that maximum degradation occurs by day 60 of the any storage period, a design decision was made to design the storage tanks around a 45 day cycle. Since the residential locations of all the reservoirs requires Seattle Water to limit the number of deliveries it accepts, a 45 day cycle would limit deliveries to once every six weeks to two months.

Chemical Feed and Delivery Systems

There are two options available for chemical feed systems. Variable speed, positive displacement metering pumps can be used to control the chemical feed rate. Multiple metering pumps of overlapping capacities would be required to cover the necessary flow ranges of Seattle Water's system. Alternatively, flow control valves in a pressurized chemical feed loop could be used to control the chemical feed rate and, again, multiple control valves of overlapping capacities would be required. The feed loop would be pressurized by a constant speed centrifugal pump.

These two options have been evaluated and compared based in the following three areas: technical feasibility, cost and operator familiarity with the equipment. In addition, Seattle Water's current experience with metering pumps at the full-scale prototype facility and its less than favorable results was taken into consideration.

It was determined that both options would meet the requirements outlined for the system by Seattle Water for CH2M Hill . Both metering pumps and control valves are commonly used in the chemical process industry. However, metering pumps are the most commonly used with sodium hypochlorite. Seattle Water has had experience with both types of systems and found that the operators were comfortable with both types.

In comparing the costs of both systems it was found that the installation costs of the two systems were very similar with the metering pump option being less than ten percent greater than the cost for the control valve option. The final decision by Seattle Water was based in large part on the expectation that a control valve system will have substantially lower operation and maintenance requirements than a conventional metering pump installation. Since control valves are commonly used in the chemical process industry for regulating chemical feed rates and they are available in sodium hypochlorite compatible materials, it was determined that they should provide effective chemical feed control.

Operation and Maintenance

It has already been discussed that the general operation and maintenance costs for a control valve system is exceptionally lower than for a metering pump systems. However, it must be noted there is a significant difference in the operation and maintenance requirements for any type of hypochlorination facility in comparison to gas chlorination facilities. It is expected that daily maintenance will increase from 20 minutes to 1 - 2 hours. Interval maintenance may increase as much 10 - 20 times the current requirements. Therefore, during design it was critical that the operations personnel became involved in the design to minimize operation and maintenance impacts wherever possible. Seattle Water personnel have been actively involved and included throughout the project as participants in regular meetings and reviews. As design has progressed, this has proven to be an extremely effective practice in creating a quality project and a system that personnel can successfully and effectively operate and maintain.

Conclusion

This project gives Seattle Water three areas of success. By removing the potential hazard that the presence of chlorine gas creates, and by replacing it with an equally effective disinfection method, Seattle Water ensures continued delivery of their high quality product while protecting the public's safety. With CH_2M Hill's design of the new hypochlorination facilities utilizing control valve operation, Seattle Water will be able to operate a disinfection system that has both low operation requirements and a high level of reliability. Finally, the project and its current schedule will enable Seattle Water to comply with the Fire Department Agreement and bring their facilities into compliance.

References

Seattle Water Department. *Sodium Hypochlorite Disinfection Treatment Facilities Design Criteria and Guidance.* November, 1992.

Seattle Water Department. *1985 COMPLAN, Seattle Comprehensive Regional Water Plan.* December, 1985.

James M. Montgomery Engineers, Inc. *Control of Chlorate Ion Formation at Hypochlorination Facilities.* November, 1992.

CH_2M Hill Northwest. *Sodium Hypochlorite Disinfection Treatment Plants, Basis of Design Report.* April, 1993.

CH_2M Hill Northwest. *Sodium Hypochlorite Disinfection Treatment Plants, General Predesign Report.* July, 1993.

TOXICANT IMPACTS ON PLANKTON: AN APPROACH TO MODELING

Yuri M. Plis[1]

ABSTRACT

An approach to simulating temporal-spatial variation of a plankton population in a toxicant polluted lake is presented. The mathematical model applies equations describing the transport of population numerical density and biomass characteristics along space and age independent variables and equations of plankton individual life-history. The model uses a curvilinear mesh to improve the computational efficiency of the numerical algorithm, which is based on the method of component-by-component operator splitting.

INTRODUCTION

The impact of toxicants on aquatic ecosystems is a serious problem for environmental protection. Ecological risk assessment was developed to evaluate the direct and indirect toxicant effects on individual organisms and populations. Mathematical models are effective tools for these evaluations. At the present time, the majority of these models do not take individual properties of organisms into account and aggregate them on the populational level. In ecotoxicology, this approach has not proved to be successful because chemical impact occurs at the level of the individual, not at the population level. Intra-population resolution is needed if one is to estimate the ability of a population, consisting of individuals of differing ages and physiological characteristics, to withstand severe environmental stress from toxicant impact.

The effectiveness of applications of existing individual-based plankton population models (Hallam et al.

----

[1]Senior Research Associate, National Research Council, National Academy of Science, c/o US EPA, 960 College Station Road, Athens, GA 30605-2700

1990) are decreased by the lack of consideration of mechanisms describing transport of organisms.

The spatially structured individual-based model employed here allows us to evaluate tendencies of distributions of physiological characteristics of zooplankton population in water bodies under the influence of spatial heterogeneity of food resources and toxicity.

BASIC EQUATION OF PLANKTON POPULATION MODEL

Independent variables of a plankton population model consist of time, t, spatial coordinates, $\mathbf{x}$, physiological properties of an individual plankton organism, $\mathbf{m}$, and individual's age, a. Introducing a distribution function $P(t,\mathbf{x},\mathbf{m},a)$, such that $Pd\mathbf{x}d\mathbf{m}da$ is the number of individuals of the plankton population in the volume of a linear space of our independent variables between $\mathbf{x}$, $\mathbf{m}$, a and $\mathbf{x}+d\mathbf{x}$, $\mathbf{m}+d\mathbf{m}$, $a+da$, we can describe the transport P by the classical advection-dispersion equation combined with the equation of a structured population dynamics

$$\frac{\partial P}{\partial t} + LP + \bar{L}P = G - M + Q,$$
$$LP=\nabla_\alpha (v^\alpha P - D^{\alpha\beta}\nabla_\beta P)\,, \quad \bar{L}P=\frac{\partial P}{\partial a}+\sum_{j=1}^{n_m}\frac{\partial}{\partial m_j}\left(\frac{dm_j}{dt}P\right), \tag{1}$$

where G, M, and Q are, respectively, birth rate, mortality rate, and rate of external source of plankton organisms; ∇_α is the covariant derivative; v^α is the contravariant component of a velocity of currents; $D^{\alpha\beta}$ is the tensor of coefficients of the turbulent diffusion; m_j are elements of the vector $\mathbf{m}$.

The term LP describes the transport along independent spatial variables $\mathbf{x}$. The term $\bar{L}P$ describes the transport along n_m independent physiological variables and age a.

ZOOPLANKTON INDIVIDUAL MODEL

Individual growth. For this research, we applied the mathematical model describing the life history and bioenergetics of an individual daphnid (Hallam et al., 1990). An individual daphnid is represented as consisting of two body components: structure m_s and lipid m_L. Its dynamics are depicted by the rates of change of these two components, as controlled by energy supply and demand. The dynamics of an individual lipid and structure are described by equations:

$$\partial m_L/\partial t = f_L - R_L\delta(t-t^*), \quad \partial m_S/\partial t = f_S - (R_S+C_S)\delta(t-t^*), \quad (2)$$

where f_L and f_S are rate functions, describing continuous input and output flows of lipid and structure, respectively; R_L, R_S and C_S are, respectively, losses of lipid and structure associated with reproduction and carapace formation, which occur at the discrete time intervals t^*, when the organism reaches appropriate physiological conditions; $\delta(t-t^*)$ is the Dirac delta function.

Toxicant bioaccumulation. The equation of toxicant balance in individual organisms of *Daphnia M.* (Hallam et al., 1990) was modified here to consider losses of toxicant via reproductive processes:

$$dm_T/dt = SK_w(C_w-C_a)+F-E-R_T\delta(t-t^*), \quad (3)$$

where K_w is the coefficient of conductivity across the organism's body surface; S is the effective exchange surface area of the organism's body; C_w and C_a are, respectively, the concentrations of toxicant in the ambient water and in the aqueous phase of the organism; F is the rate of toxicant inflow with feeding food; E is the rate of toxicant outflow with defecated material; R_T is the toxicant mass lost for reproduction.

Toxicant effects on an individual. Toxicant mortality was computed based on the amount of chemical concentration in the aqueous phase (blood) of the organism. Blood concentration was compared with the lethal blood concentration, C_L, obtained from the relationship between the octanol-water partition coefficient (K_{ow}) and the lethal blood concentration for an individual. We assumed that death occurs in all individuals for which

$$\log(C_a) \geq -0.8 - \log(K_{ow}). \quad (4)$$

Sublethal effect of toxicant on individuals was described following Lassiter (1990) by the dose-response model describing a fractional reduction r_j of the growth rate of jth physiological function (equations 2) relative to its optimum value f_j.

ADVECTION-DIFFUSION TRANSPORT

To improve the efficiency of the model numerical realization in regard to computational memory (Plis, 1992) the transport term LP of equation (1) was written using curvilinear orthogonal in the horizontal plane coordinates ξ, η

$$LP = \frac{1}{Jh}\left[\frac{\partial}{\partial\xi}(JhS^{\xi}P) + \frac{\partial}{\partial\eta}(JhS^{\eta}P) - \frac{\partial}{\partial\xi}(hD_L\frac{\partial P}{\partial\xi}) - \frac{\partial}{\partial\eta}(hD_L\frac{\partial P}{\partial\eta})\right], \quad (5)$$

where $J=\partial(x,y)/\partial(\xi,\eta)$ is the Jacobian of the transformation from Cartesian coordinates (x, y) to (ξ, η); h is a depth; D_L is the horizontal coefficient of turbulent diffusion; S^{ξ} and S^{η} are components of the full stream vector.

ALGORITHM OF SOLUTION

To reduce a very large number of dependent variables of physiologically structured population model's two major assumptions about aggregating an organism's properties were made. The first is an assumption about newborn organisms. The group of youngest organisms consist of offspring from individuals of all reproductive ages. Parents are distinguished, in addition to age, by lipid, structural body composition, and assimilated toxicant. Hence, they produce offspring that also vary somewhat. We average these properties to derive their values for the entire youngest age group of the physiological cluster. The second assumption is that the properties of the same-age organisms at a spatial site in a water body are the weighted averages of the properties of the same-age recruits transported from neighboring spatial sites and the remaining residents.

As a result of the foregoing assumptions, our model keeps homogeneous mass characteristics distributions among organisms of the same age at any spatial site in a water body (term $\tilde{L}P=0$ in equation 1).

Numerical algorithm. A component-by-component operator splitting method (Marchuk 1984) has been applied as an approximate technique to integrate equation (1) with the advection-diffusion transport term (5) and $\tilde{L}P=0$ over an arbitrary time interval Δt. This integration can be written as

$$P_{t+\Delta t}-P_t=-\int_t^{t+\Delta t} L_{\xi}P dt-\int_t^{t+\Delta t} L_{\eta}P dt-\int_t^{t+\Delta t}\frac{\partial P}{\partial a}dt+\int_t^{t+\Delta t}(G-M+Q)\,dt, \quad (6)$$

The method uses the decomposition of the positive semidefinite differential operator L into a sum of positive semidefinite operators L_{ξ} and L_{η}.

The integration (6) is split into four stages. Each of the stages uses solutions of the previous stage as initial conditions. The same procedure should be applied to integration of equations describing dynamics and spatial distribution of a population's lipid $C_1=Pm_L$, structure $C_2=Pm_S$, and accumulated toxicant $C_3=Pm_T$.

An attractive feature of the operator splitting procedure is that each stage can be solved using a different numerical technique that is specially suited to achieve high accuracy for each integral in (6).

APPLICATION AND CONCLUSIONS

The model was applied to zooplankton population dynamics simulations in the Twelve Mile Creek area of Lake Hartwell (South Carolina). This area was polluted by polychlorinated biphenyl (PCB) from a capacitor manufacturing plant located on the creek 39 km upstream from Lake Hartwell. PCB were detected at all levels of the food chain in both the Twelve Mile Creek watershed and Lake Hartwell. There is a clear dietary pathway from the allocthonous detritus and seston to the macroinvertebrates and plankton.

The results of simulations allowed us to obtain information about tendencies of temporal-spatial distributions of zooplankton population and individual physiological characteristics, including total biomass, ratio of lipid to total biomass, mass of the accumulated toxicant, number of organisms, and age-mass distributions in polluted and unpolluted areas of water body.

The adequacy of this model depends in large measure on the corresponding adequacy of the individual's life history and physiology models incorporated into the population model, but this level of assessment is still in its infancy.

Application of the approach provides, we believe, both diagnostic and prognostic information beyond that available from traditional approaches to the evaluation of ecological effects.

ACKNOWLEDGMENTS

This work was conducted under the Ecorisk Research Program of the Environmental Research Laboratory, US Environmental Protection Agency, Athens, Georgia.

APPENDIX

Hallam, T.G., R.R. Lassiter, J. Li, and W. McKinney. (1990) "Toxicant-induced mortality in models of Daphnia populations." *Environ. Toxicol. Chem.*, 9,597-621.

Lassiter, R.R. (1990) *A theoretical basis for predicting sublethal effects of toxic chemicals* U.S. Environmental Protection Agency, Athens GA.(Unpublished report).

Marchuk, G.I. (1982) *Methods of numerical mathematics* Springer-Verlag New-York Inc.,New York, N.Y.

Plis, Y.M. (1992) "An approach to calculating wind-driven currents and transport of substances in unstratified water bodies using curvilinear coordinates." *Water Resour. Res.*, 28(1), 83-88.

BALANCING BETWEEN CHANNEL MAINTAINABILITY AND NAVIGATION ON THE RED RIVER

Thomas J. Pokrefke, Jr.,[1] Member, ASCE

Abstract

This paper addresses studies conducted on John H. Overton Lock and Dam on the Red River. Those studies included physical and numerical movable-bed models and physical fixed-bed navigation and structural models. The solutions developed on the various models and tested on other models will be presented, including results and conclusions reached. As plans were refined, compromise solutions resulted, and those solutions were fully tested to develop a final plan. The results of the final plan tested will be compared to performance of the prototype.

Introduction

The Red River flows easterly from the northwest portion of Texas along the border between Texas and Oklahoma into southwestern Arkansas where it turns southwesterly to flow through the northwestern portion of Louisiana to Shreveport and then easterly to join the Old River and form the Atchafalaya River. From that point the Atchafalaya River flows through the southeastern portion of Louisiana to the Gulf of Mexico.

The Red River Waterway provides a navigation route from the Mississippi River at its junction with Old River via the Old and Red Rivers to Shreveport, LA. The project is comprised of a 380-km (236 miles) long, 2.7-m (9-ft) deep, and 61.0-m (200-ft) wide channel which includes five locks and dams to control water levels. During the construction and development of the Waterway the existing river channel was realigned to develop an efficient channel. Bank stabilization and river training structures were used to hold the newly developed channel in position. The project also provides flood control, recreation, fish and wildlife, and water quality control.

----

[1]Supervisory Hydraulic Engineer, Hydraulics Laboratory, U.S. Army Engineer Waterways Experiment Station, 3909 Halls Ferry Road, Vicksburg, MS 39180

The Purpose and Plan of Study

Development of the Red River required comprehensive modeling efforts on specific portions of the system due to the complexity and inordinate amount of fine sediments in the system. Early modeling efforts at the U.S. Army Engineer Waterways Experiment Station (WES) focused on the use of physical undistorted fixed-bed models to address navigation issues and physical distorted movable-bed models to address bed material sedimentation issues. After Lock and Dam No. 1 was opened in the fall of 1984, the depositional problem with the fine sediments in low velocity areas became very evident. Therefore, modeling efforts on John H. Overton Lock and Dam, formerly Lock and Dam No. 2, included an entire forte of models. This included a physical navigation model constructed to an undistorted scale of 1:100; a physical movable-bed model constructed to a horizontal scale of 1:120 and a vertical scale of 1:80; a structural model constructed to an undistorted scale of 1:50; and a numerical model to address fine-grained sedimentation issues.

Site Description and Limits of Models

John H. Overton Lock and Dam is located approximately 141 km upstream of the confluence of the Red and Old Rivers (Figure 1). On the Red River this is referenced as river mile 87.4. The lock and dam were constructed in the "dry" and then the channel was realigned through a cutoff to the lock and dam. The project consists of a single lock on the left descending bank line. The lock has a useable length of 208.8 m (685 ft) and is 25.9 m (85 ft) wide. The dam has five, 18.3-m (60-ft) wide gates and a 76.2-m (250-ft) long overflow weir on the right bank.

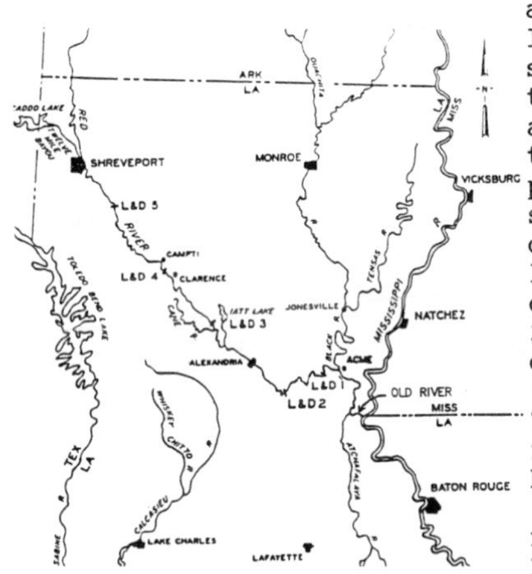

Figure 1. Location Map

As stated earlier, the modeling effort on the John H. Overton project included four types of models. The navigation model reproduced all necessary details of the channel, banks, overbank areas, and structural components from approximately 2.8 km (1.8 miles) upstream to 3.1 km (1.9 miles) downstream of the lock and dam. The physical movable-bed model reproduced channel details including the existing and cutoff channels, banks, overbank areas, and structures from about 4.5 km (2.8 miles) upstream to 3.9 km (2.4 miles) downstream of the lock and dam. The 1:50-scale structural model reproduced the project from 427 m (1,400 ft) upstream to 823 m (2,700 ft) downstream of the lock and dam. The TABS-2 numerical model reproduced about 3.2 km (2.0 miles) upstream and 1.1 km (0.7 mile) downstream of the lock and dam. For all of these models, it was felt that the area reproduced was sufficient to address the appropriate questions to be answered by the studies.

Model Results and Prototype Performance

As with many WES model studies, tests conducted for the John H. Overton project covered many aspects and concerns; thus, tests were conducted on numerous plans, alternatives, or modifications. This project was additionally complex since the final plan development was coordinated with all of the studies to ensure that navigation, bed-load sedimentation, suspended fine material sedimentation, and hydraulic structure considerations were satisfactory. The final plan layout developed through this series of model studies is presented in Figure 2; however, as with many designs, the final plan developed was not constructed exactly as tested. This was due to changes in conditions in the prototype which precluded construction of some final plan details. Therefore, this paper will present certain aspects of the results of those model studies and as it compares to prototype response. Due to limitations it will not be a complete comparison of model and prototype.

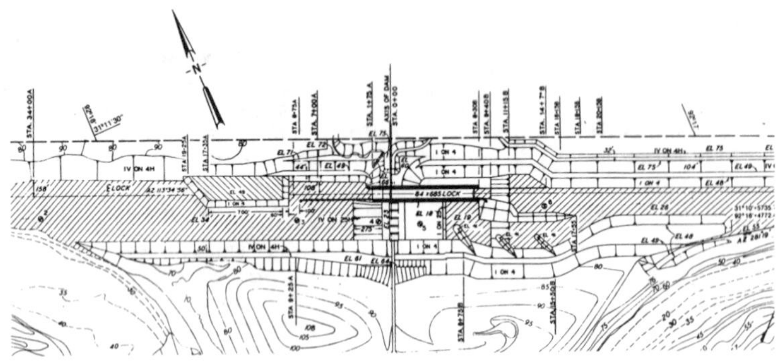

Figure 2. Final Model Plan

Sedimentation Results. The tests conducted on the physical movable-bed model with the final plan indicated that the navigation channel in the upper pool would meander from the right to the left bank immediately upstream of the lock. In the model, local scour occurred to about elevation 6.9 m (20 ft) NGVD[2] along the right bank about 1.9 km (1.2 miles) upstream of the dam. From that point the navigation channel crossed to the left bank with a crossing elevation of about 12.8 m (42 ft). The channel elevation along the left bank just upstream of the berm, about 0.9 km (0.6 mile) upstream of the dam, was about el 11.6 m (38 ft). The tests conducted on the numerical sediment model indicated that after a high flow condition, about 2.7 m (9 ft) of shoaling would occur just to the left of the lock along the left bank. Data presented on a September 1994 prototype survey of the upper pool indicated that local scour along the right bank 1.6 km (1.0 mile) upstream of the dam was at el 4.6 m (15 ft). The crossing of the navigation channel to the left bank was at about el 11.0 m (36 ft). Along the left bank just upstream of the berm, about 0.8 km (0.5 mile) upstream of the dam, the channel was about el 10.4 m (34 ft). Although the prototype data were limited, the September survey indicated that shoaling just to the left of the lock along the left bank was about 3.4 m (11 ft) above the constructed elevation.

These results indicate a good correlation between the model and prototype sedimentation results, and there are other areas within the model limits where the model predicted prototype response to a high degree of accuracy. One general model trend that to this point has not developed in the prototype, is general shoaling in the upper pool. The physical movable-bed model had predicted significant long-term shoaling in the upper pool based on average or typical discharge hydrographs. It is unclear if that development has not occurred because of insufficient time, the fact that the entire Red River Waterway is still responding to major changes in alignment and establishment of the pools, or that the model was over conservative in prediction of such long-term depositional patterns. As time passes, this particular trend can be reevaluated.

Navigation Results. The navigation tests conducted on the final plan (Figure 2) indicated that navigation conditions in the upstream lock approach were satisfactory. It should be noted that the guard wall in the upper approach was ported to aid navigation in approaching the lock and to reduce the potential for an outdraft condition at the upstream end of the wall. When the

---

[2] All elevations (el) cited herein are referred to the National Geodetic Vertical Datum (NGVD) of 1929.

Lock and Dam were put into operation in November 1987, the initial upper pool was held at el 17.7 m (58 ft), and the final pool elevation of 19.5 m (64 ft) was established about February 1989. While the interim pool conditions were being maintained, pilots reported an adverse flow condition on the ported guard wall which required excessive maneuvering to clear the wall and proceed upstream. To address this problem the navigation and structural models were reactivated. The structural model was originally used to assist in the development of plans for the stilling basin design, riprap protection, checking spillway discharge characteristics, preliminary powerhouse tests, and to develop a stable riprap plan for the downstream right bank dikes. Due to the model scale, 1:50, it was felt that the details of flow through the guard wall would be better reproduced in this larger-scale model than the 1:100-scale navigation model. Thus, the structural model was used to obtain detail flow conditions around and through the guard wall ports and for measuring forces required to hold a model tow off of the guard wall for various flow conditions. At the same time tests on the 1:100-scale navigation model optimized the location and height of two submerged sills upstream of the berm. This evaluation was based on providing satisfactory navigation conditions approaching the guard wall for a variety of flow conditions. Then, by coordinating the results from these two models, a satisfactory solution was developed and installed in the prototype. Performance of this modification in the prototype has been successful, and significantly reduced the required maneuvering along the guard wall.

Conclusions

Relative to sedimentation in the upper pool, the prototype response has the same trends and tendencies as predicted by the physical and numerical sediment models. Relative to navigation in the upper pool, the modifications to the upper lock approach and guard wall developed on the navigation and structural models have performed as predicted in the models.

Acknowledgements

The presented research was performed at WES and was sponsored by U.S. Army Engineer Districts, New Orleans and Vicksburg. The Chief of Engineers granted permission to publish this paper. The author thanks the various WES researchers who supplied model data and Vicksburg District personnel Phil Combs and Rick Robertson who provided prototype data.

Environmental Preservation of Red River Oxbows

Charles D. Little, Jr., P.E.*, Phil D. Dye, P.E.*, Charles F. Pinkard, Jr., P.E.*

Abstract

Development of navigation on the Red River from the Mississippi River to near Shreveport, Louisiana required the construction of channel realignments for navigation alignment improvements. The abandoned river bends that remain after realignment form oxbow lakes ranging from 1.6 kilometers to 11 kilometers in length. Utilization of these oxbow lakes for environmental and recreational purposes is a goal of the Corps of Engineers Vicksburg District and the project sponsor (Red River Waterway Commission). Management of fine grain sediments is essential in meeting these project goals. Two-dimensional finite element model studies are being conducted to establish deposition trends and estimate long-term quantities. Study results will be used to assess structural and non-structural sediment control alternatives.

Introduction

The lower Red River from the Mississippi River to near Shreveport, Louisiana has recently been opened to commercial navigation. Even though the primary purpose of the Red River Waterway Project is navigation, recreation demand and environmental considerations are also important aspects of the project design. During the past decade, environmental consciousness has dictated that preservation, and even enhancement, of the environment be included in the implementation of water resource projects. The Vicksburg District and the project sponsor are planning, designing and constructing various recreational facilities such as boat ramps, camping areas, picnic grounds and other public access

* Hydraulic Engineer, U.S. Army Corps of Engineers, Vicksburg District, Vicksburg, MS 39180

points. These areas have allowed recreational use of the river that was previously limited to begin to flourish. Ideal locations for many of these public sites are on the oxbow lakes formed by channel realignments constructed for navigation purposes. However, these slack water oxbows are subject to bothersome sediment deposition due to the fine grain suspended sediment load of the Red River. Determination of deposition trends and quantities, and formulation of viable alternatives to maintain oxbow-to-live river connectivity provides the challenge in preserving oxbows to meet environmental and recreational objectives.

Development of Red River Oxbows

The development of navigation on the river involved the construction of channel realignments (cutoffs) for navigation alignment improvement. The oxbow lake formed by the construction of John H. Overton Lock and Dam is shown as an example in Figure 1. Project design criteria requires that oxbows 1.6 kilometers or greater in length be preserved for environmental and recreational usage. Within the project limits there are 29 oxbows to be preserved. The lengths of the oxbows range from 1.6 kilometers to 11 kilometers. According to the project Environmental Impact Statement (EIS) approximately 3200 hectares of present river channel will be converted to oxbow lakes that meet requirements for preservation.

Figure 1. Oxbow lake at John H. Overton Lock and Dam

The typical deposition pattern for naturally-formed oxbow lakes is rapid filling of the upstream end with coarse sediments, followed by less rapid filling of the downstream end with fine sands, silts and clays. To prevent this from occurring in project-formed oxbows, non-overtopping closure dams are constructed across the upstream end of the oxbow. This design increases the longevity of the oxbow, and enhances utilization potential for recreation and environmental purposes. However, deposition continues to occur in the downstream end of the oxbow. According to the project EIS, approximately 30 percent of the created oxbow waters will gradually silt up over the project life. Shallow, narrow low water outlet channels typically develop through the deposition in the lower end of the oxbow. The channels may not provide adequate year-round access to the oxbows from the live river. Maintenance of a clear, reliable connection between the live river and the oxbow is paramount in ensuring long-term access for natural fish restocking, fresh water ingress, and recreational boaters.

Sediment Deposition Modeling

Studies have been initiated to analyze deposition trends and estimate quantities of fine grain sediment deposition in the oxbow outlets. Study results will be used to determine long-term maintenance requirements and/or develop structural features to make the outlets more self-maintaining.

Description of the 2-dimensional flow field in the oxbow outlet vicinity is necessary to address deposition trends of the fine grain Red River sediments. To accomplish this, the *TABS* modeling system is being used. *TABS* is a system of 2-dimensional finite element hydrodynamic and sediment transport codes maintained by the USAE Waterways Experiment Station (Thomas and McAnally, 1985). A detailed discussion of the *TABS* system is not attempted within the scope of this paper. Numerous applications of *TABS* models to Red River sediment problems were performed during design of the navigation project. This fact provided a basis for using the *TABS* system and also utilizing the model parameters developed during those studies.

A typical annual hydrograph was developed to use as the input hydrology for the model studies. The hydrograph was partitioned into discrete steady state flow steps, and a sequence of hydrodynamic and sediment transport solutions were performed to simulate the hydrograph. Updated model geometry after each sediment transport cycle was used to develop hydrodynamics for the next simulation period, and so forth throughout the entire hydrograph. Long-term simulations with the 2-dimensional model become extremely time and resource consuming, so the annual hydrograph results are used to determine trends and to estimate deposition quantities over the project life.

An important factor in the development of the outlet channel is the flow in and out of the oxbow resulting from rises and falls on the live river. The volume of water stored in the oxbow, the contraction of inflow and outflow currents, and the abruptness of the rise and fall all influence the development of the outlet. To account for this in the numerical model, flows in and out of the oxbow corresponding to changes in the typical annual hydrograph were calculated. A storage volume curve was developed for the oxbow, and the change in volume associated with the rise or fall from one simulation period to the next was determined based on the change in the live river level. The average flow rate required to produce this change in volume over the given time period was then calculated. The level of the oxbow was assumed to closely follow that of the live river.

The fine grain suspended sediment of the Red River was of primary interest because the main deposition problems are in slack water zones, and also because the *TABS* system currently considers one grain size only. Based on numerous samples of deposited material from slack water areas of the Red River, a grain size with a d50 of 0.07 mm was used in the models. Sediment concentrations were determined from suspended sediment measurements on the Red River. The percent of the total concentration to associate with the given grain size was based on how representative the grain size is of the total size distribution curve. Diffusion coefficients for sediment transport were similar to those used in previous model studies on the Red River.

Development of Alternative Solutions

At the writing of this paper all oxbow model investigations have not been completed, hence detailed solutions have not been formulated. Several potential alternatives have been addressed which may be utilized dependent on the severity of the deposition trends. The solutions will also be unique to each oxbow based on where within the navigation pool the oxbow is located and whether or not tributaries flow into the oxbow.

The first alternative is simply maintenance dredging. This non-structural option may be used throughout the navigation pools, but may be limited to areas where deposition quantities are minimal and the location of anticipated dredging is manageable.

Maintenance dredging may be more likely used at oxbows located in the lower ends of the navigation pools where water depths are greater and pool level fluctuations less prominent. Determination of relatively accurate long-term deposition quantities is essential in scheduling resources for maintenance dredging requirements.

Structural measures which may be used include contraction dikes in the oxbow outlet channel and raising the realignment revetment between the live river and the oxbow outlet. Contraction dikes will be spaced and sized to define a permanent outlet channel in the lower end of the oxbow. The dikes will confine flow in and out of the oxbow and generate sufficient velocities to keep the outlet free of sediment, thus creating a self-maintaining situation. This type of solution has the greatest potential for use at oxbows in the upper ends of the navigation pools where water depths are less and pool level fluctuations are more pronounced.

Raising the realignment revetment between the live river and the oxbow outlet will prevent frequent over-topping of the revetment during rises on the river. This will prevent eddies from developing in the outlet vicinity and retard the transport of fine sediments into the outlet area. This alternative will be used primarily in areas of severe deposition and where public access areas, such as boat launch ramps, are located immediately adjacent to the oxbow outlet. This method has already been successfully used in the navigation project.

Summary

Preservation of Red River oxbows formed by navigation improvements is an important goal of the Red River Waterway Project from an environmental and recreational standpoint. Management of the fine grain sediment deposition in the oxbow outlet vicinity is a necessity in realization of this goal. Two-dimensional finite element model studies are being performed to predict deposition trends and estimate long-term sediment deposition quantities. Both structural and non-structural alternative solutions will be evaluated with the model results. Solutions will be incorporated to ensure environmental and recreational utilization of project-formed oxbow lakes is successfully integrated as part of the Red River Waterway navigation project.

References

Thomas, William A. and McAnally, William H. Jr., “Open Channel Flow and Sedimentation, TABS-2”, Instruction Report HL-85-1, USAE Waterways Experiment Station, July 1985

Local Downstream Control Method for Irrigation Canals

Fubo Liu[1], *Jan Feyen*[1] *and J. Mohan Reddy*[2]

Abstract

A local downstream control method is presented for the operation of irrigation canal systems. A target water level is maintained at the downstream end of each pool. In contrast to a centralized control method which uses the status of the flow in the entire system, the local control method requires only the status of the flow in the pool immediately downstream of a given gate.

Introduction

A control technique can be referred to as centralized or local control method according to the data required for calculating the adjustment of each control gate. In a system operated with centralized control the adjustment of a gate depends on the information gathered from the entire or a large part of the system. Conversely, in a system operated with local control the adjustment of a gate only depends on the information from its neighbouring region. A number of local downstream control methods can be found in the literature (Burt 1983; Buyalski et al. 1979; Chevereau et al. 1987; Reddy et al. 1992; Zimbelman et al. 1983). However, most of them are still in their theoretical and experimental stage and it remains a challenging task to develop new control methods that are more effective and easier to implement.

Development of the concept

A local control method can be developed based on the hydraulics of local flow phenomena, or derived from a centralized control method assuming that only the status of the flow in the neighbouring region is known. A downstream control

---

[1]Postdoctoral Researcher and Professor, Institute for Land and Water Management, Katholieke Universiteit Leuven, Vital Decosterstraat 102, 3000 Leuven, Belgium
[2]Professor, Dept. of Civil Engineering, University of Wyoming, Laramie, WY 82071, USA

method was proposed recently by Liu et al. (1994, 1995) based on an inverse solution procedure of the St. Venant equations. The approach is centralized control since the adjustment of a gate depends on the water levels and flow rates of all the information nodes downstream of the gate. However, the data from different information nodes do not have the same importance. The farther the information node is from the gate, the less influential its data becomes, and the status of the flow in the immediate downstream pool is most influential. So, it is interesting to study the performance of the control method when the adjustment of a gate is only based on the information from its immediate downstream pool.

Following the inverse solution procedure presented in the previous paper, in order to compute the adjustment of the gate at the upstream end of a pool, the desired variation of water level and flow rate at the downstream end of the pool have to be specified. If the water level at the downstream end of pool *i* differs from its target value, the required change of water level at this section is the difference between the target and the current water level. With the centralized control logic, the required change of flow rate at the downstream end section of a pool is calculated according to the status of the flow at all the information nodes downstream of this section. With the local control method, the change of flow rate at the downstream end section is assumed as zero. This assumption implies that no direct effort is given to change the opening of the gate at the upstream end of a canal pool for adjusting the flow rate delivered to the next downstream pool. However, it is obvious that if water level at the downstream end of a pool can be maintained at its target value at the new steady state, the flow rate delivered to the next pool will be implicitly guaranteed. The time needed for the variations of discharge and water level is called recovery time and is noted as *R*. It is estimated based on the time needed for a wave to travel from the downstream to the upstream end of the pool.

With the specified variations of water level $\Delta Z_{i,p}$ and flow rate $\Delta Q_{i,p}$ at the downstream end of pool *i*, one may calculate the required variations of flow rate and water level at the upstream end of the pool, $\Delta Q_{i,1}$ and $\Delta Z_{i,1}$, using the same calculation procedure as in the centralized control algorithm. Then the adjustment of gate *i* can be calculated as:

$$\Delta u_i = \frac{\partial u}{\partial Q}\left[\frac{t_g}{R_i}\Delta Q_{i,1}\right] + \frac{\partial u}{\partial Z_d}\left[\frac{t_g}{R_i}\Delta Z_{i,1}\right] + \frac{\partial u}{\partial Z_u}\left[\frac{t_g}{R_{i-1}}\Delta Z_{(i-1),p}\right] + K_I\int_{t-t_g}^{t}(T_i - Z_{i,p})dt$$

where $\partial u/\partial Q$, $\partial u/\partial Z_d$ and $\partial u/\partial Z_u$ are derived from the gate flow equation; t_g = time interval for the adjustment of the gate; K_I = integral gain factor. The purpose of having the integral term in the above equation is to eliminate any possible offset of a maintained water level form its target value.

Illustrative example

In order to compare the local control method discussed here and the centralized control method presented in the previous paper, the same canal system and test events are used to test the performance of the local control method. The canal system has 6 identical pools separated by 7 identical gates and bounded by a constant level reservoir at both the upstream and downstream ends of the system. There is a lateral offtake at the downstream end of each pool which has a peak flow rate of 5 m^3/s. Physical parameters of the system are specified as: pool length = 3000 m, bottom width = 5.0 m, side slope (H:V) = 1.5:1, bottom slope = 0.0002, Manning's coefficient = 0.015, width of gate (rectangular) = 5.0 m, discharge coefficient of gate = 0.80. Water levels in the reservoirs, and target water level at the downstream end of each pool are the same as specified in the previous paper.

The simulation model CANSIM used here was developed in the PhD study of Liu (1995). It solves the complete St. Venant equations using the implicit Preissmann scheme. The model has been compared with the simulation model SIC (CEMAGREF 1992) developed in Montpellier, France. In a number of tests conducted, identical results were given by both models.

The same disturbances used to evaluate the performance of the centralized control method in the previous paper were used to test the performance of the local control method. For all the simulations, the distance interval was taken as 300 m, while the time interval was taken as 5 minutes. The time interval used for the adjustments of the control gates was taken as 5 minutes which was equal to the time interval of the simulation. Due to the limitation on the length of the paper, only the results from one of the tests are presented. The complete tests are described in the PhD thesis of Liu (1995). At initial steady state, the flow rate discharged to the downstream reservoir was 5 m^3/s while flow rates of all lateral outlets were 0. Within 5 minutes, the flow rate at lateral outlet in pool 3 increased from 0 to its peak value 5 m^3/s and the rest of the lateral outlets remained closed. For such a disturbance, the response of the control gates, the variation of water level at the downstream end of each pool, and the flow rate through each gate are plotted in Fig. 1a. As can be seen, a sudden increase of flow rate through outlet 3 caused a rapid drop of water level at the downstream end of pool 3. This resulted in a rapid increase in the opening of gate 3 at the next time step, with a simultaneous increase in the flow rate through gate 3 and a drop of water level upstream of it (at the downstream end of pool 2). The drop of water level at the downstream end of pool 2 in turn caused an increase of the opening of gate 2. In a multireach canal system, such a chain reaction continues from one pool to another until the most upstream intake. At the downstream side of the disturbance, the opening of a gate will not change until the propagation arrives. The variation of the flow at the downstream part of the system was rather small. About 4 hours later, the controlled water levels all gradually approached their target values. Flow rates through gates 1, 2 and 3 all increased 5 m^3/s as a result of the increase of the flow rate at outlet 3, and flow rates through gates 4, 5, 6 and 7 were the same as before. Compared with the same test for the centralized control method (Fig. 1b), when the system is operated with the local control method, fluctuation of the flow in the system is stronger. Similar conclusion was drawn by Merkley et al. (1991) for an upstream control method.

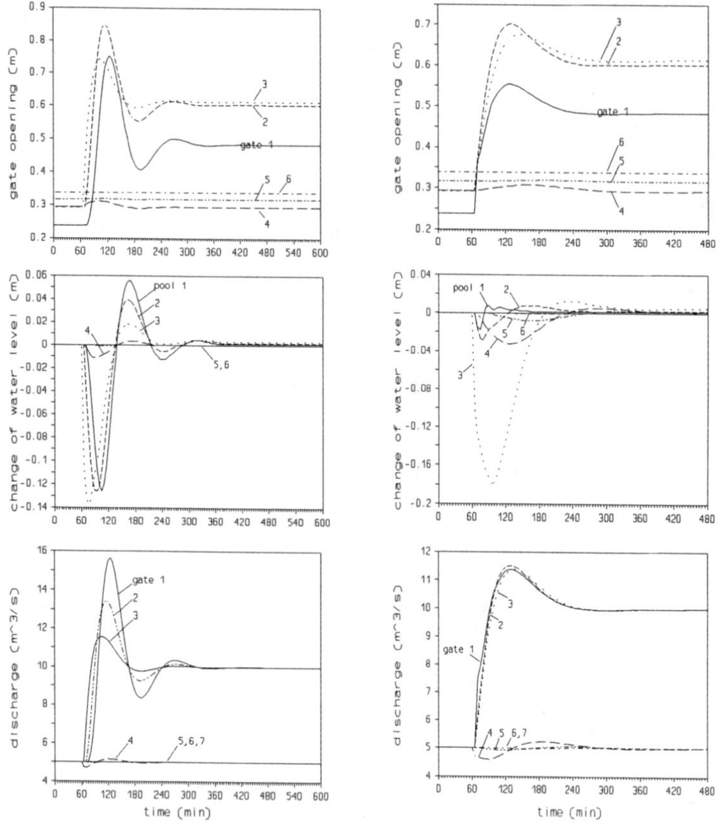

Fig. 1a Local control

Fig. 1b Centralized control

Summary and conclusions

Generally speaking, when the system is operated with the local control method fluctuations of the flow are more pronounced and it takes a longer time to reach new steady state. Nevertheless, the use of a local control method can be justified when flow in only part of a canal system can be monitored. Furthermore, a local control method is relatively simpler and cheaper to install. With the proposed local downstream control method, adjustment of each gate is based only on the status of the flow in the immediate downstream pool. If there is any disturbance in one pool, the gate at the upstream end of the pool and the gate immediately downstream of the disturbance will respond simultaneously, the rest of the upstream gates will respond one after the other with a time lag which is equal to the interval of the adjustment of the control gates, and the rest of the downstream gates will not respond until the propagation arrives. The control method can cope with unknown disturbances in the system. In the test presented, disturbances (changes of flow rates through lateral outlets) in the system were only used for the purpose of simulation, and this information is not needed for calculating the adjustments of the control gates.

References

Buyalski, C. P. and Serfozo, E. A. (1979). "Electronic filter level offset (EL-FLO) plus reset equipment for automatic control of canals". Engineering Research Center, U.S. Bureau of Reclamation, Denver, Color., 86 p.

Burt, C. M. (1983). "Regulation of slopping canals by automatic downstream control". PhD dissertation, Utah State Univ., Logan, Utah, USA, 171 p.

CEMAGREF (1992). "SIC: theoretical concepts & user's guide".

Chevereau, G., and Schwartz-Benezeth, S. (1987). "BIVAL system for downstream control". Planning, operation, rehabilitation and automation of irrigation water delivery systems, ASCE Symp. Proc., Oregon, USA, 155-163.

Liu, F. (1995). "Analysis of control algorithms for water delivery in irrigation canals based on hydrodynamic simulation". PhD dissertation, Katholieke Universiteit Leuven, Belgium, 166 p.

Liu, F., Feyen, J., and Berlamont, J. (1994). "Downstream control algorithm for irrigation canals". J. Irrig. and Drain. Engrg., ASCE, 120(3), 468-483.

Liu, F., Feyen, J., and Berlamont, J. (1995). "Downstream control of multireach canal systems". To be published in April, J. Irrig. and Drain. Engrg., ASCE.

Merkley, G. P., and Walker, W. R. (1991). "Centralized scheduling logic for canal operation". J. Irrig. and Drain. Engrg., ASCE, 117(3), 337-393.

Reddy, J. M., Dia, A., and Oussou, A. (1992). "Design of control algorithm for operation of irrigation canals". J. Irrig. and Drain. Engrg, ASCE, 118(6), 852-867.

Zimbelman, D. D., and Bedworth, D. D. (1983). "Computer control for irrigation canal system". J. Irrig. and Drain. Engrg., ASCE, 109(1), 43-59.

Controller Design for the WM Canal

J. Schuurmans, G. Liem[1]

Abstract
A controller is designed for the WM canal. The design method consists of a search over controllers that optimize a quadratic performance criterium while being constrained to a given structure. By forcing the control structure into a decentralised one, controllers are obtained that can be implemented in a decentralised way. Using this method, local PI controllers are tuned automaticly for the WM canal in the Maricopa Stanfield Irrigation District, Arizona. Initial experiments seem promising.

Introduction
A control system for an irrigation canal should be able to maintain the water levels within an acceptable range. A control system is judged better if it is able to perform this task using a minimum amount of information on future offtake changes. Such a control system provides a the highest degree of flexibility towards the water users. Thus, the control system should use as little anticipating feedforward as possible and consequently, the feedback part of the control system should be as good as possible.

However, the control system should also be implementable in the field without excessive hardware requirements. This often requires a decentralised control system, for the amount of water levels that should be controlled is usually large and the distances between the water levels are usually long. Furthermore, the problem how to handle constraints on the manipulated variables are much easier to solve for decentralised control systems than for centralised control systems.

In this article a decentralised control system is designed for theWM canal using an automatic synthesis method.

The control system consists of a water level controller that computes the necessary

---

[1] PhD Candidate, Delft University of Technology, Department of Civil Engrg., P.O. Box 5048, Stevinweg 1, Delft, The Netherlands; Msc Candidate, Delft University of Technology.

flow rate changes at each time level and passes that to a flow controller; the latter controls the flow rate changes through a gate by manipulating the gates.

This hierarchical structure of the control system has the following advantages:
- the interaction between the water levels in the reaches are removed, allowing better control performance
- the relation between flow rates and water levels is less nonlinear than the relation between gate openings and water levels.

In fact, this 'trick' of using flow controllers has been applied to a lot of canals.

Decentralised controller synthesis
In order to apply an automatic synthesis method, a performance measure is needed, which is clearly defined in a mathematical way. An open-channel flow control system is judged better if it reduces the maximum water level deviation, because such a control system can handle more disturbances. Furthermore, the control manipulations should be divided proportionally over the structures, taking into account each structure's capacity (allowable size of the control action). From these considerations a performance measure (J) can be defined that penalizes the water level deviations from setpoint (h) relative to their allowable fluctuations, and the relative sizes of control manipulations represented by the flow rate variations (q) close to initial conditions.

$$J = \sum_{i-1}^{N} w_i \|h_i(t)\| + r_i \|q_i(t)\| \tag{1}$$

Here, N is the amount of reaches, $\|.\|$ denotes a norm and w_i and r_i are weighting factors. The weights w_i can be chosen as follows:

$$w_i = \left(\frac{|h_{\max,\ 1}|}{|h_{\max,\ i}|} \right)^p \qquad i = 2,\ 3,\ 4,\ \dots$$

where p is a value that depends on the norm and $h_{maxc,\ i}$ is the allowable size of h_i. With this weighting the water level deviations contribute relatively equally to their maximum allowable deviations to the criterium.
The weighting factors r_i can be selected so as to weight control manipulations relative to the maximum possible control manipulations.

$$r_i = \rho \left(\frac{|q_{\max,1}|}{|q_{\max,\ i}|} \right)^p$$

in which ρ is used to weight the relative importance between water level deviations and control manipulations this factor must be determined by trial and error.

For automatic decentralised controller synthesis the 2-norm is selected here, as most literature on decentralised controller synthesis concerns this norm. Let K be the feedback gain matrix, defined by: $q(t) - Kx(t)$

The discrete decentralised optimal control problem can then be defined as follows:

$$\min_{K} \quad J = \sum_{t=0}^{\infty} \left(\boldsymbol{h}(t)^T W \boldsymbol{h}(t) + 0.2\,\boldsymbol{q}(t)^T R\, \boldsymbol{q}(t)\right)$$

with $h(0) = h_0 \neq 0$, and the following constraints:
$K \in K_C$ (the family of all constrained feedback matrices), h(t) and q(t) are related by a linear difference model:

$$x(t) = A\,x(t-1) + B\,q(t-1)$$
$$h(t) = C\,x(t)$$

Here, $\mathbf{h} = [h_1\ h_2\ h_3\ \ldots]^T$ the water level deviations from setpoint in reach 1, 2, 3, ..
$\mathbf{q} = [q_1\ q_2\ q_3\ \ldots]^T$ flow rate variations through gates 1, 2, 3, ..., x = state vector
A, B, C are the state space matrices, t= discrete time level (-),
$W=\mathrm{diag}(w_1^2, w_2^2\ w_3^2, \ldots)$ weight matrix, $R=\mathrm{diag}(r_1^2, r_2^2\ r_3^2, \ldots)$ weight matrix

There are several numerical search techniques to solve this problem, although none of these methods is guaranteed to find the global optimal solution. There are both hierarchical computation methods and decentralised computation methods that were especially developed for solving this problem when it concerns a large scale system (Trave et al 1989).

A problem with this design method is that the feedback controller may not be robust stable. The stability robustness can often be improved by choosing a higher weight for R (i.e. by choosing a higher value for ρ); another method is to change the performance criterium into:

$$\min_{K} \quad J = \sum_{t=0}^{\infty} \alpha^{2t} \left(\boldsymbol{h}(t)^T W \boldsymbol{h}(t) + \boldsymbol{q}(t)^T R\, \boldsymbol{q}(t)\right)$$

This forces the closed loop response to have a settling time of $100^{1/\alpha}$ samples.

Controller design for the WM canal

The control systems for the WM canal were designed using an approximating linear model of open channel flow, which is described in more detail in a c-paper. The performance criterium was optimized using a sequential programming search algorithm from the optimisation toolbox of Matlab.

A control system, consisting of local PI controllers was optimized using this technique. However, a memory problem occurred when optimizing a control system for the whole canal with 8 reaches. This problem was solved by applying a decentralised design method: first, a control system is designed for the first (upstream) four reaches; then, a control system was designed for reaches 2 to 5 while the control parameters for the PI controller that controlled reach 2 were constrained to the values obtained from the first optimisation cycle. This procedure was repeated until all reaches had been covered. Such a procedure is not ideal, i.e. it is never going to reach the global minimum, but at least it is tried to minimize the performance criterium it and it is able to design a decentralised control system for indefinitely many reaches.

The flow controllers that were developed for the WM canal could not use gate position measurements, as these were not available in real-time. Therefore, a flow controller was used that did not need gate positions. Linearizing the gate discharge equation gives:

$$q = C_u u + C_h h$$

where:

$$C_u = \left(\frac{\partial Q}{\partial U}\right)_0, \quad C_h = \left(\frac{\partial Q}{\partial H}\right)_0$$

with H = head of water (m), U = gate opening (m)
Thus, the requested flow changes by the water level controller can be translated into a gate opening change using:

$$u = \frac{1}{C_u} Q - \frac{C_h}{C_u} h$$

This feedforward controller is only a function of the water level change and the requested flow rate change. The disadvantage is that the coefficients are constant, whereas they should change considerably depending on flow rate and water level. Therefore, the coefficients were determined for each test: the flow rate through each gate was estimated using data from offtake flows; then assuming that the water level remains at setpoint (which it is reasonable if the controller functions fine), the coefficients can be determined.

The first control system that was tested on the WM canal were local PI controllers, which had been designed using the automatic synthesis method with a weighting value of $\rho = 3$.

At the time of writing this paper, not all controllers had been implemented yet (they were implemented sequentially, allowing better troubleshooting if something went wrong). However, the first five pools were controlled successfully. Figure 1 shows the water levels in the first five reaches while being controlled by local PI controllers.

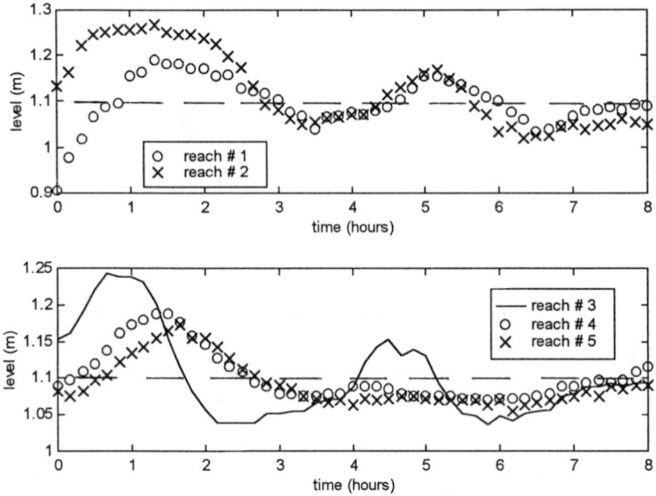

Figure 1 Water levels in first five reaches controlled by PI controllers (setpoint is 1.1 m).

Initially, the water levels were a little off from setpoint, but due to the controllers they approach setpoint more closely, although oscillations remain present. This is due to a minimum amount of gate movement that could be send out to the gates; in order to approach setpoint closer a smaller gate-adjustment would be needed.

Conclusions

A method to design decentralised feedback controllers for canal operations has been presented. The method avoids trial and error procedures. The first results seem promising.

Acknowledgements

The authors would like to thank A.J. Clemmens, R.S Strand of the US Water Conservation Lab, and G. Sloan of the Maricopa Stanfield Irrigation District for making it possible to perform experiments on the WM canal.

Reference

Travé L., A. Titli, A. Tarras Large scale systems: decentralization, Structure constraints and fixed modes, Springer-Verlag, Berlin, Heidelberg, etc, 1989

Evaluation of a Canal Automation Algorithm CLIS

Fubo Liu[1], Pierre-Olivier Malaterre[2], Jean-Pierre Baume[2], Pascal Kosuth[2] and Jan Feyen[1]

Abstract

The control algorithm CLIS has been evaluated using test cases set by the ASCE Task Committee on Canal Automation Algorithms (ASCE 1994). The control algorithm is designed for the operation of demand oriented systems. It calculates the adjustments of control gates using only two water levels measured in each pool, one of which at its upstream end and the other at its downstream end. No flow rates need to be measured. A target water level is maintained at the downstream end of each pool. The control algorithm has been tested under ideal, tuned and untuned conditions, respectively. It is able to take into account scheduled offtake changes, and provide an effective control of unknown disturbances in the system.

General description of the control algorithm

The control algorithm is called CLIS since it is a closed-loop control method and it is based on the inverse solution of the St. Venant equations. It was developed within the framework of the PhD research of Liu (1995) under the supervision of Dr. Feyen at the University of Leuven, Belgium. The work presented here were carried out in cooperation with the research team led by Dr. Kosuth at CEMAGREF, Montpellier, France. At this stage, the control method has not yet been applied in the field. However, from numerous computer simulations, it appears rather effective under severe conditions (Liu 1995; Liu et al. 1994, 1995).

The formulation of CLIS can be found in the references cited above. By solving the St. Venant equations inversely, for a required change of flow rate $\Delta Q_{i,p}$ and water

---

[1]Postdoctoral Researcher and Professor, Institute for Land and Water Management, Katholieke Universiteit Leuven, Vital Decosterstraat 102, 3000 Leuven, Belgium
[2]Hydraulics Engineers, Irrigation Division, CEMAGREF, 361 rue J.F. Breton, BP 5095, 34033 Montpellier Cedex 1, France

level $\Delta Z_{i,p}$ at the downstream end of a pool, the required change of flow rate $\Delta Q_{i,1}$, and water level $\Delta Z_{i,1}$ at the upstream end of the pool can be calculated. With which the required change of upstream gate opening is computed as

$$\Delta u_i = \frac{\partial u}{\partial Q}\left[\frac{t_g}{R_i}\Delta Q_{i,1}\right] + \frac{\partial u}{\partial Z_d}\left[\frac{t_g}{R_i}\Delta Z_{i,1}\right] + \frac{\partial u}{\partial Z_u}\left[\frac{t_g}{R_{i-1}}\Delta Z_{(i-1),p}\right] + K_I\int_{t-t_p}^{t}(T_i - Z_{i,p})dt$$

where $\partial u/\partial Q$, $\partial u/\partial Z_d$ and $\partial u/\partial Z_u$ are derived from the gate equation; t_g = time interval for the adjustment of the gate; R = expected recovery time; i = pool index; T, Z = target and current water levels, respectively; K_I = integral gain factor; t_p = time of previous gate adjustment.

The input variables of CLIS are water levels and flow rates at several information nodes along the canal, and the output variables are adjustments of gate openings. In practice, since flow rates are normally difficult to measure directly, they are often estimated by measured water levels. In this study, water levels and flow rates in the system are provided by an observer which is based on unsteady flow simulation. With this observer, only two water levels need to be measured in each pool, one of which at its upstream end and the other at its downstream end.

Three control parameters need to be specified for each pool in order to apply the CLIS control method. One is the anticipation time A which is related to a scheduled offtake change. The other control parameter is the expected recovery time R which is a compromise between effectiveness and stability. The third control parameter is the integral gain factor K_I which is similar to that of a PID controller. The control parameters are tuned by using an unsteady flow simulation model.

Specifications of the tests

The tests were carried out using the unsteady flow simulation model CASIM (CAnal SImulation Model) which was developed in the PhD study of Liu (1995). It solves the complete St. Venant equations discretized by the implicit Preissmann scheme. The model has been compared with the SIC simulation model developed by CEMAGREF (1992). In several tests conducted, identical results were given by both models.

A steep and a flat canal, with two test scenarios for each canal, were specified by the ASCE Task Committee on Canal Automation Algorithm (ASCE 1994). In addition to the specifications given by the ASCE, the following details were specified in the simulation. For each example canal, the system was fed by a constant water level reservoir upstream of the control gate at its head. In all simulations, the time interval was taken as 5 minutes. For canal 1 the distance interval was taken as 50 m considering its steep slope, and for canal 2 the distance interval was 200 m. The regulation interval used was 5 minutes for canal 1 and 15 min for canal 2. In the case of canal 2, the adjustment of a gate was completed in 5 min, then the position of the gates remained constant for 10 min until the next regulation step. Changes of the gate openings of lateral offtakes were completed in 5 min. Cross gates of canal 1 were in free flow condition due to the large drop at the end of each pool. Cross gates of canal 2 and offtakes of both canals were in submerged flow condition. It was assumed that embankments and gates were high enough and there was no overtopping.

Test results

For the two example canals, two test scenarios for each canal, three types of test have been carried out (12 tests in total). (1) Ideal condition: minimum gate movement is zero, all physical parameters of the canal system are correctly known and both water levels and flow rates in the main canal are accurately measured; (2) Tuned condition: minimum gate movement is 0.005 m and 0.01 m for canals 1 and 2, respectively (about 5% of the average hight of the control gates in each canal), water levels and flow rates in the system are estimated based on the water levels measured at the upstream and downstream ends of each pool; (3) Untuned condition: control parameters tuned from condition 1 and 2 are applied to canal systems which have different Manning coefficient n (changed from 0.014 to 0.018 and 0.02 to 0.026 for canals 1 and 2, respectively) and different gate discharge coefficient (10% less than under the tuned case). In addition to the constrains imposed in condition 2, the real scheduled offtake discharge changes are 5% higher than as scheduled (e.g. at offtake 1 of canal 2, the real change at 2:00 hrs is 1.575 m^3/s instead of 1.5 m^3/s as scheduled, and at 14:00 hrs, the unscheduled change is -1.575 m^3/s which is the same magnitude as the scheduled change. The same rule is applied to canal 1).

Due to the limitation on the length of the text, only results from canal 1 (steep canal), in scenario 1 (small change in offtake discharge), under untuned condition, and canal 2 (flat canal), in scenario 2 (big change in offtake discharge), under untuned condition are plotted in the figures. Fig. 1a and Fig. 1b represent gate openings and changes of water levels, respectively, in canal 1. As can be seen, an offset occurred at the new steady state. However, the offset was caused by the limitation of minimum gate movement, and not by the wrong estimation of the physical parameters of the system. Under the same untuned condition, when the minimum gate movement was set to 0, the offset was eliminated (Fig. 1c). Without the limitation of minimum gate movement, the response of the system under untuned condition is less good than, but similar to that under ideal condition. This is a positive indication of the robustness of the control algorithm. Fig. 2 shows gate openings and changes of water levels, respectively, in canal 2. As can be seen, large changes of water levels were experienced after both scheduled and unscheduled changes of offtake discharge. Actually, the gates started to open at the very beginning of the simulation which means that there was not sufficient anticipation time. The drops of water levels after the scheduled change can be reduced if the simulation starts some time earlier. Water levels increased rapidly after the unscheduled change despite that the gates were completely closed. This is not surprising since there was a total rejection of 11.55 m³/s of discharge (5% higher then under ideal condition) out of 14.25 m³/s in the system, and at the downstream end there was a rejection of 4.2 m³/s (2.1 m³/s by offtake 8 plus 2.1 m³/s by the pumped outlet) out of 4.9 m³/s.

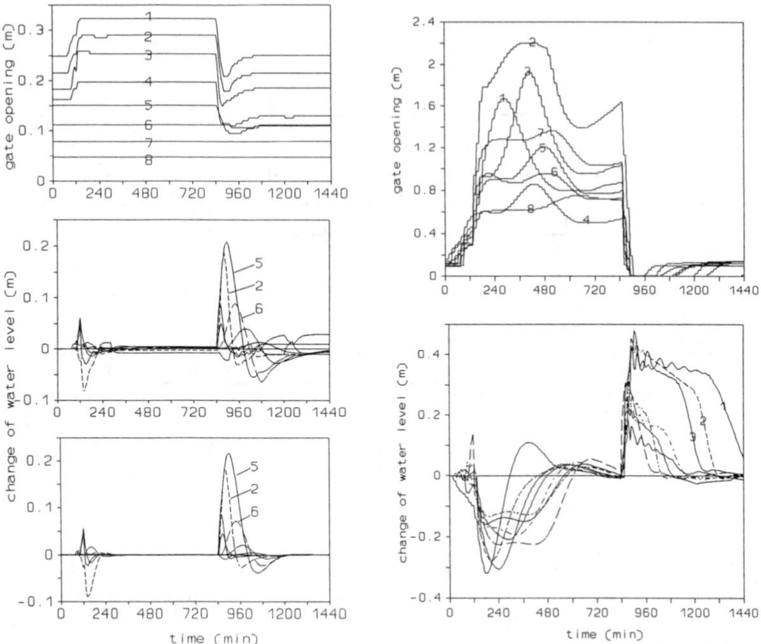

Fig. 1 Canal 1, scenario 1, untuned condition

Fig. 2 Canal 2, scenario 2, untuned condition

The performance indicators MAE (maximum absolute error), IAE (integral of absolute magnitude of error), StE (steady state error), IAW (integrated average absolute gate movement) and IAQ (integrated average discharge change) are defined by the ASCE Task Committee (1994). However, the term Δt/T is removed from IAW and IAQ of the original definitions in order to make them independent of the time parameters.

$$IAW = \sum_{t=t_1}^{t_2} |w_t - w_{t-1}| - |w_{t_1} - w_{t_2}| \qquad IAQ = \sum_{t=t_1}^{t_2} |Q_t - Q_{t-1}| - |Q_{t_1} - Q_{t_2}|$$

The maximum value (of values from each pool) of each indicator is given in table 1. The indicators are computed for each 12 hour period (0-12, 12-24) in order to compare the response of the system under scheduled and unscheduled disturbances. A relatively large offset still exist at the end of some simulations, but can be reduced (see Fig. 2b) or eliminated if a longer simulation period is used.

Table 1: Summary of test results

Indicators		MAE(%)		IAE(%)		StE(%)		IAW		IAQ	
Tests		0-12	12-24	0-12	12-24	10-12	22-24	0-12	12-24	0-12	12-24
canal 1	scenario 1 ideal	6.9	21.1	0.5	1.5	0.0	0.0	0.031	0.026	0.068	0.211
	scenario 1 tuned	7.7	19.4	1.0	1.7	1.0	0.8	0.044	0.060	0.158	0.203
	scenario 1 untuned	9.1	23.1	0.9	2.2	0.9	3.2	0.030	0.083	0.104	0.249
	scenario 2 ideal	17.4	36.1	2.3	1.6	0.0	0.0	0.238	0.073	0.493	0.192
	scenario 2 tuned	16.2	35.1	3.4	2.6	1.8	3.6	0.252	0.050	0.552	0.197
	scenario 2 untuned	17.1	39.5	3.6	2.6	1.4	3.1	0.315	0.089	0.565	0.217
canal 2	scenario 1 ideal	2.5	6.6	0.7	0.6	0.1	0.2	0.218	0.297	3.500	5.721
	scenario 1 tuned	2.4	7.1	0.9	0.8	0.8	0.9	0.193	0.331	2.518	5.704
	scenario 1 untuned	4.9	7.5	2.1	1.2	1.0	1.2	1.006	0.845	7.380	8.801
	scenario 2 ideal	10.7	19.6	2.3	3.8	0.3	0.5	0.641	0.310	4.191	7.886
	scenario 2 tuned	10.8	21.5	2.4	4.0	0.7	1.6	0.607	0.378	3.854	9.619
	scenario 2 untuned	15.2	22.9	7.5	6.1	2.2	8.1	1.917	0.707	8.435	9.243

Summary and conclusion

The CLIS control algorithm is able to provide a rather effective control using only two water levels measured in each pool. It is applicable to both steep and flat canals, although flat canals are easier to control than steep ones. The robustness of the control algorithm is rather appreciated. Relatively, it is more sensitive to the roughness coefficient of the canal than to the discharge coefficient of the gate.

References

- ASCE Task Committee on Canal Automation Algorithms (1995). "Test cases and procedures for algorithm testing and presentation". First International Conference on Water Resources Engineering, San Antonio, USA, 14-18 August 1995. 5p.
- CEMAGREF (1992). "SIC: theoretical concepts & user's guide".
- Liu, F. (1995). "Analysis of control algorithms for water delivery in irrigation canals based on hydrodynamic simulation". PhD dissertation, Katholieke Universiteit Leuven, Belgium.
- Liu, F., Feyen, J., and Berlamont, J. (1994). "Downstream control algorithm for irrigation canals". J. Irrig. and Drain. Engrg., ASCE, 120(3), 468-483.
- Liu, F., Feyen, J., and Berlamont, J. (1995). "Downstream control of multireach canal systems". Accepted for publication by J. Irrig. and Drain. Engrg., ASCE.

Application of Automatic Control on MSIDD WM Lateral Canal

A.J. Clemmens, M. ASCE, J. Schuurmans, R.J. Strand[1]

Abstract
An irrigation district in central Arizona was constructed with motorized gates for all gates on the main canals and all lateral check gates. These gates can be controlled remotely through radio communication. The canal system has many canals with a wide range of characteristics. This provides an ideal setting for testing canal automation. One lateral canal, WM, was chosen for initial testing. This paper is a progress report and describes initial experiences with the application of automatic controls on this canal.

Introduction
The Maricopa Stanfield Irrigation and Drainage District (MSIDD) is located in central Arizona about 50 km south of Phoenix. The irrigation district was formed to receive Colorado River water from the Central Arizona Project. The irrigated area of about 35,000 ha had previously received water from groundwater wells. The irrigation district took over the wells during 1989 and delivers a mix of groundwater and surface water. Construction of the canal system was completed in 1989. The system was designed so that all canal check structures (cross-regulators), including laterals and sublaterals, could be controlled by motorized gates remotely through radio communication. Farm offtake are operated manually and each includes a single-path ultrasonic flow meter, that provides both flow rate and accumulated volume readings. Water is relatively expensive in the district and users expect a high level of service. The system is operated with very little spill (< 1%).

Engineers designed the entire system to be operated by supervisory control (remote, manual control), but also provided an option for automatic downstream feedback control (remote, computer control)(Kishel, 1986). The district began delivering water in 1987 through manual operation, prior to installation of canal gate remote control equipment.

---

[1] Research Hydraulic Engineer, U.S. Water Conservation Laboratory, USDA/ARS, 4331 E. Broadway, Phoenix, Arizona 85040; PhD Candidate, Delft University of Technology, Delft, The Netherlands; Engineering Technician, U.S. Water Conservation Laboratory.

During 1990, the engineering consultants and district staff attempted to implement automatic downstream control. Initial tests were made on the WM lateral canal, which resulted in extremely unstable performance, even when starting from near steady flow. Similar tests were run on the Santa Rosa main canal, which receives water from the Central Arizona Project and distributes water to the lateral canals. Tests on this canal were also aborted due to unstable behavior. After several months of efforts by district staff and the design engineering firm, automatic control remained nonfunctional. Only one algorithm (Zimbelman's) for automatic downstream control was tried (see Kishel 1986), and it may not have been the best available downstream control algorithm for this canal. Also, significant hardware problems existed (e.g., gear lash or hysteresis, radio interference) which may have kept the control algorithm from providing the proper control. The district significantly altered the gate control hardware and software to make remote control possible. Supervisory control on the Santa Rosa canal was implemented during 1991 and 1992. A few lateral canal gates are also operated remotely, where difficulties with local, manual control exist. (See Clemmens et. al. 1994 for additional details).

During 1992, the authors collected data on the WM canal, from district drawings and from field observations. Unsteady flow models were used to simulate the performance of various existing control methods on the WM canal, as reported by Clemmens et al (1994). Further research has been done by the authors and others on the development of improved canal control algorithms. The WM canal was chosen to test some of these improved control algorithms because the district had a workable supervisory control system and they were willing to allow a series of automatic control tests. The purpose of this paper is to describe the real-time testing that was performed on the WM canal, problems encountered and to report progress on the implementation of canal automation within MSIDD.

MSIDD Supervisory Control System

MSIDD operates their main canal remotely with micro-computer-based supervisory-control software. The system uses radio communication to query remote sites for water levels upstream from each (radial) check gate and sends control actions to each check gate. Under the current system, the computer poles the remote sites on a periodic basis or when requested by the operator. Control actions are initiated by the operator. The supervisory control system is manned about 18 hrs per day. All lateral canals are operated manually by operators on site (locally). Canal operators are on duty primarily from 7 am to 3 pm. Changes outside this shift are handled by a reduced staff and are discouraged.

The system was originally operated manually throughout. Conversion to supervisory control was resisted because operators believed that they could more precisely control the canal than the remote system. Two major problems caused this perception. First, an 8-bit controller was used to detect gate position. Thus the gate travel could only be divided into 255 parts. For the large radial gates, they felt that this was not fine enough resolution. Second, there was significant gear lash and it was difficult from the central computer to know when and how much the gate moved. Third, communication problems sometimes resulted in the gates moving to the full open or closed positions.

The district modified their gate controllers by adding a local controller that used timers to open the gates a fixed amount. These were called strokers. Two timers were used to move a large amount (large bump) or a small amount (small bump). the timers were calibrated so that the moved the gates an amount approximately equivalent to 10 cfs (283 l/s) for a large bump and 1 cfs (28 l/s) for a small bump. The supervisory control software also keeps track of changes in pool volume (as computed from one water level measurement) to suggest changes in inflow to that pool. Changes in volume are converted to a flow rate and the operators request changes in flow rate at each gate, converted to large and small bumps. Further redundancy and safety is built into the communication with numerous verification signals for each action.

A new computer program was written by Ken Taylor to allow both supervisory control and automatic control to occur. This could allow supervisory control of the main canal simultaneously with automatic control of the WM canal. This software allows the operator to move about the program even when radio commands are being sent and awaiting verification. The operators can also route flow changes on the WM lateral while the feedback control system is still on-line. It also provides an interface to the automatic control algorithms, which can be modified independently from the rest of the program. This allowed for convenient testing.

The WM Canal

Lateral canal operators also operate in terms of flow rate. Gates are marked to indicate the number of handwheel turns or change in gate position per 1 cfs (28 l/s) change in flow. Operators usually make changes in flow by starting and the head of the canal and routing the change in the downstream direction. For a very steep canal with little storage, like the WM, this is entirely reasonable. For larger canals with more storage and milder slopes, changes can be made downstream and worked back to the head, but this is not done on the WM canal.

Automation of the WM canal required building new controller boards for the strokers. The timers were set so that a large bump was 3 cfs (85 l/s) and a small bump was 0.5 cfs (14 l/s). These gates have significant gear lash, so that changes in the direction of gate movement had to be accounted for. The command for a flow rate change had to include an additional amount to correct for the gear lash if a change in direction occurred. These turn-around bumps were typically equivalent to 2 to 4 cfs (56 to 113 l/s).

Simulation of Control Algorithms

Simulations were performed on several different unsteady flow programs. Studies with various controllers indicated several key points. First, control of flow rates provided better control results than control of gate position, even when flow rates are inexactly known. In doing this, some knowledge of gate hydraulics is incorporated in the control logic. Second, on this steep canal, accumulating requested control actions in the upstream direction (summing requested flow actions from downstream) provided better overall control. Third including knowledge of previous flow changes with prediction of their effects (i.e., modeling canal response delay) also greatly improved control. In the initial simulations, no water level deadband was specified and no restriction was placed on the minimum gate movement.

Field Tests

Field tests on the canal began in March 1995. The initial test were used to obtain data on open-loop conditions (i.e. as the operators routed flows with their normal operating procedures). This data was used to verify the linear model (linear, ordinary differential equations, see Schuurmans et al, 1995), which was used to develop controller constants. The linear model appeared to fit this canal fairly well. Testing of feedback control started on the first four pools.

The first few tests were used to debug the control software and to identify problems with control logic. Simple Proportional Integral (PI) controllers were used so that the control logic could be easily followed and the interactive effects between pools would be minimized. In these initial tests, the request for gate flow changes did not consider the effect of an error in water level upstream from the check gate on the gate discharge. Tests were run to bring the canal water levels to the set point from their initial conditions. In several tests, a groundwater well ($\approx$ 3 cfs or 85 l/s) discharging into the canal was turned off (i.e., whose discharge was to be replaced with water from the main canal).

Several problems were encountered in getting the gates to move properly, as requested by the control algorithm. The new software used in the automation tests had not been fully debugged and was not yet being used for the districts supervisory control. For these tests, two different computers were used to operate the main canal in supervisory mode and the WM canal in supervisory or automatic mode. The two computers shared the same radio transmitter, so testing had to be coordinated with district operator needs so that they would not interfere. During some tests, alarms were given, indicating that a command was not properly received at the field site. Since the new program was not yet complete, the automatic control logic did not know what was received and how much the gate really moved.

Calibration of the timers for gate movement was done fairly precisely, however for first few tests, the gate movement was half or a third of what it was supposed to be because of gearing differences between the handwheel (as used by operators) and the motor. Several tests were also run to determine the amount of gearlash at each gate. This proved to be more difficult than expected.

Once these problems were resolved, a reasonably successful test was run with a PIR (PI Retard) controller which included prediction of delay between upstream gate movement and downstream water level response. A test was started very far from set point. The controller responded with a very large increase in flow -- to increase the volume of water in the pools-- which were subsequently reversed, after the pools were filled. After some oscillations near or around the set point, the pools gradually approached the set points. The canal inflow started at 44 cfs (1.25 m^3/s), was increased by 17 cfs (0.48 m^3/s), and was subsequently reduced to around 46 cfs (1.30 m^3/s) (an increase in flow was requested by the operator prior to the test). However, the control returned very slowly toward setpoint, indicating that several problems remained unresolved.

Future Plans

Several details are currently unresolved, for example: Do remainders in control commands need to be accumulated (i.e. below minimum gate movement)? Does precise flow control need to be implemented? How large can the minimum gate movement be without affecting control performance? In tuning, how much should the controllers be damped (e.g., penalty for numerous gate movements)?

The computer program which is being used to test these automatic controllers will also be upgraded to provide proper feedback on what control actions were successful and to add additional logic for safety; e.g., filtering of water level measurement, limitations on flow changes, limitations on maximum flow, etc.-

The testing will be extended to the entire canal -- all 8 canal pools. Several different feedback controllers will be tested on the entire canal, for comparison. Further testing will also be done to improve the performance of the controllers on this canal, and to select one for eventual use. Procedures will be developed to account for a zero flow condition in downstream pool, which would cause considerable control error for some control algorithms if the levels there are not at set point. Finally, the controller must function properly over the full range of expected conditions, rather than for one condition -- typically used to design controller. With the variety of canals within this district and the willingness of the district staff to assist in testing, this provides a good site for testing of these various control algorithms.

Acknowledgements

The author would like to acknowledge the assistance of the following individuals, who assisted in making this real-time testing possible; G. Sloan, G. Wall and F. Horton, MSIDD; Ken Taylor, Central Arizona Irrigation and Drainage District; and G. Liem and M. Ellerbeck, Delft University of Technology.

References

Clemmens, A.J., Sloan, G. and Schuurmans, J. 1994. Canal-control needs: example. *J. Irrig. & Drain. Eng.*, ASCE, 120(6):1067-1085.

Kishel, J. 1986. Automated control for Central Arizona Project distribution system. *Water Forum '86: World Water Issues in Evolution.* ASCE Conf. Proc.., 2017-2024.

Schuurmans, J., Bosgra, O.H. and Brouwer, R. 1995. Open-channel flow model approximation for controller design. Draft copy, Delft, The Netherlands.

Fall Velocity of Sea Shells as Coastal Sediment

by

Khaled A. Kheiredlin[(1)]

INTRODUCTION

Fall velocity, which is defined as the rate at which a particle falls in a fluid, is considered one of the most important factors affect the understanding of the fundamental processes involving sediment. Fall velocity controls the time that a sediment particle will remain in suspension, the distance that the particle can be transported without lying on the bed. Also, it controls the bed roughness of an alluvial channel and the location of deposition in reservoirs.

Current information regarding the sediment transport processes in the coastal zones primarily depends on the characteristics of gravel, sand, and fine sediment as clay and silt. However, shells of various types are commonly found in the coastal regions includes the surface layer, tidal entrances, and estuarine water.

The purpose of this paper is to determine the fall velocity for thirty two sea shells samples representing eight different shells families. The selected families are Cyprea, Bankivia, Catharus, Architectonica, Noetia, Bathybembix, Polinices, and Tellina. Also, the relationship between the drag coefficient of the falling shell, Reynolds number, and the shape parameter is tested for the selected samples.

REVIEW OF THE PREVIOUS STUDIES

The problem of fall velocity for the sediment particles had been studied by several researcher in the last decays. For example, Rubey (1933) developed a formulae for calculating the fall velocity for sand and gravel particles. Wadell (1935) studied the relationship between the drag coefficient on a falling particle and Reynolds number. Corey (1949) studied the influence of the particle shape on the fall velocity. He was the first to us the photographic technique for the determination of fall velocity. Corey found that the fall velocity and the particle shape was correlated. Alger (1964) studied the effect particle surface area on its falling velocity. Stringham (1967) investigated the behavior of spheres, disks, oblate spheroids, cylinders, and prolate spheroids falling in quiescent viscous fluid. Alger and Simons (1968) studied the Reynolds number -Drag Coefficient relation for a series of irregular shaped particles up to Reynolds number of 400,000. Moreover, they defined a new shape parameter for the irregular shape particle.

Mehta et. al. (1981) determined the fall velocity of sea shells through an intensive experimental study. In these experiments three species of bivalve halves were used and the fall velocity was measured in a settling column. From this study, the fall velocity and the relationship between the drag coefficient and Reynolds number was presented for the Coquina, Chione, and Ponderous shells. Also, they defined a certain shape factor for the sea shells to include the effect of the shell base area,surface area, volume, and dimensions.

SETTLING BEHAVIOR OF SEA SHELLS

When a particle falls freely in a fluid, there are different factors affect its motion pattern. These factors are gravity acceleration, flow separation, vortex formation and shedding (in the downstream side of the particle), and circulation (along the particle surface). Gravity force is responsible for the particle to be felt vertically downward. Flow separation and vortex formation

---

(1)- Ph. D., Senior Researcher at the Strategic Research Unit, Water Research Center, El-Qanatir, Kalubia, Egypt, 13621/5.

nd shedding affect the shear and pressure distribution on the particle surface. Iowever, circulation along the particle wetted surface will provide additional side forces. The omposite of these forces on the falling particle will cause the particle to exhibit three types of notion, namely sliding, tipping and rotation.

Visually, two different falling pattern are observed: Stable and unstable. In the stable node the external forces on the particle are almost balanced to the submerged weight of the article, Thus, the particle is falling vertically downward along the fall centerline. In the nstable mode, the particle is falling in curved arcs in a motion pattern which is exhibited like a eaf or a paper falls in the air.

HAPE FACTOR

Alger and Simons (1968) defined a shape factor for the irregular particles in the form of:

$$Shape\ Factor = \frac{c}{\sqrt{a\ b}}\ \frac{d_A}{d_n}$$

vhere a, b, and c are the largest, intermediate, and smallest dimensions of the particle along three nutually perpendicular axes. d_A = diameter of a sphere has the same surface area as that of the article. d_n = diameter of a sphere has the same volume as the particle.

Mehta et.al. (1980) defined a shape factor β for the Bivalve sea shells as :

$$\beta = 0.39\ \pi^{1/3}\ \alpha_4^{\ 1/3}\ \frac{c}{V^{1/3}}$$

vhere, $\alpha_4 = \frac{2\ S_A}{\pi\ a\ b}$; S_A = particle surface area.

REYNOLDS NUMBER

Alger and Simons (1968) defined the Reynolds number (R) for a falling particle as:

$$R = \frac{W\ d_A}{\nu}$$

vhere W = particle fall velocity, and ν = kinematic viscosity of the fluid.

DRAG COEFFICIENT

Mehta et. al. (1980) defined the drag coefficient (C_D) for an the sea shells as:

$$C_D = \frac{8\ g\ (\frac{\rho_s}{\rho_f}-1)\ V}{4\ B_A\ W^2}$$

vhere g is the gravity acceleration.; B_A= shell base area; ρ is the mass density for the fluid; ρ_s = mass density for the particles.

SELECTED SEA SHELLS

Thirty two samples from eight different families are selected for this study. A picture for the selected samples is presented in Figure (1). The shells have been classified according to Eisenberg J. (1984).

EXPERIMENTS: APPARATUS AND TESTS

The fall velocity for the sea shells are determined in a clear PVC column with a 15 cm inner diameter and 130 cm height. The column was filled nearly to the top with tap water and the temperature is maintained to be in the range of 12-14 ° C. Prior to the test the shells were left under water for 24 hours thus all the pores were saturated. During the test, each shell was held under the water surface in the convex down attitude, (the natural way for falling), by using a twizzer. This is because if the shell is left to fall from any other initial attitude, it will correct itself to the convex down attitude. Then, the shell is allowed to be released and settled in the water column. The falling time is determined by using a sport stopwatch with an accuracy of 0.01 second. The test is repeated for 8 times for each shell and the average falling time is calculated.

EXPERIMENTAL ERROR ANALYSIS

a- Confining Effect

The confining effect of the size of the settling column on the fall velocity is evaluated (according to Mehta (1980) procedure): i- relates the ratio of d/D with W_d/W_t. In which d= the diameter of a falling sphere; D= diameter of a settling column; W_d= falling velocity in a settling column; W_t=fall velocity in a finite extent. ii- for the range of Reynolds of 2500 to 6500 (experiments range), and for d/D=0.19, the ratio of W_d/W_t is found to be 0.95. This means a margin of ±5% error is caused due to the use of a settling column.

b- Path Velocity

The real path for an unstable falling particle is a three dimensional path. Stringham (1967)determined the 3-dimensional falling path by using a complex expensive photographic setup. Due to the complexity in the determination of the real 3-D path, in this research the vertical path was used to estimate the value of the falling velocity. the path is determined by measuring the vertical distance through which a particle falls.

c- Time accuracy

The personal error is found to be in the range of 0.15 second. According to this, the error in the measured time for the different experiments is in the order of ±2% - ±4% related to the shortest and longest falling time.

RESULTS AND ANALYSIS

Table (1), Figures (2, 3, and 4) present the experimental results for the present study. In what follows a brief discussion of the experimental results will be introduced:

a- Fall Velocity

As presented in table (1) the value of the fall velocity for the Cyprea class is varied between 27.47 cm/sec and 38.08 cm/sec depending on the shell dimensions and weight. From Figure (2) it can be concluded that the heavier the shell weight ,(for the same shell family), the higher the fall velocity. The fall velocity for the other shells is varied between 25.82 cm/sec and 35.26 cm/sec except for the Tellina since the fall velocity is 18.96 cm/sec. however, in all the experiments the settling mode was unstable for all the different shells.

b- C_D-R_e Relation

Drag coefficient is calculated by using equation (4) and the relation between R_e and C_D shown in figure (3) for the different shells. Also, the obtained values of C_D for the presented shells is compared with those from Mehta et. al (1980) as shown in the figure. The graph shows that the higher the value of the shape factor β, the lower the value of C_D on the particle. For example,for β in the range of 0.3-0.5, the average value of C_D= 1.12; while for β in the range of 0.8-0.9, the average value of C_D=0.86. Also, a good agreement between the results of the present study and those from Mehta et. al. (1980), since for β=0.50, C_D=1.10 as shown in figure (3).

In Figure (6) C_D is plotted against β based on the data in table (1) for all the shells used in this experiment. It observes that C_D decreases while β increases. These observations are in agreement with the results of Stringham (1967), Alger and Simons (1968), and Mehta et. al. (1980).

SUMMARY AND CONCLUSION

The fall velocity for 32 samples representing 8 different sea shell families is determined. The fall velocity is varied between the same shell family according to the shell dimensions and weight. The fall velocity is varied between 18.0 cm/sec to 38.0 cm/sec. The mean falling velocity for the Cyprea family is 32.81 cm/sec.

The drag coefficient C_D is calculated for each sample and the relationships between C_D and R_e and β are studied. It is observed that C_D is decreased while β is increased, i.e. increasing the particle sphericity. However, C_D is independent of R_e since the range of R_e was 2500 to 6500.

REFERENCES

Alger, G. (1964); "Terminal Fall Velocity of Particles of Irregular Shapes as Affected by Surface Area", Ph. D. Dissertation, Colorado State University, Ft. Collins, Colorado.

Alger, G. and Simons, D. (1968); "Fall Velocity of Irregular Shape Particles"., ASCE, J. of Hyd. Division, Vol. 94.

Corey, A. (1949); "Influence of Shape on the Fall Velocity of Sand Grains", M.Sc. Thesis presented to Colorado Agricultural and Mechanical Collage, Ft. Collins, Co.

Esienbergy, J. (1984); "A Collector Guide for Sea Shells in the World".

Mehta A., Lee J., and Christemsen B. (1980); "Fall Velocity of Shells as Coastal Sediment", J. of Hyd. Division, Vol. 106., No. HY11.

Rubey, W. (1933); "Settling Velocities of Gravel, Sand, and Silt Particles", American J. of Science.

Stringham, G. (1967); "Behavior of Geometric Particles Falling in Quiescent Viscous Fluid", Ph. D. dissertation, Colorado State University, Ft. Collins, Colorado.

Wadll, H. (1935); "Volume, Shape, and Roundness of Quartz Particles", J. of Geology.

Sample Number	Class	Subclass	Dimension a (cm)	Dimension b (cm)	Dimension c (cm)	Dimension t (cm)	Base Area (cm^2)	Surface area (cm^2)	Volume (cm^3)	Weight (gm)	Density (gm/cm^3)	dA (cm)	dn (cm)	Corey S.F.	S.F.	Avg. Fall Velocity (cm/sec)	Reynold number	Drag Coef.
1	Cyprea	Erosa	2.26	1.57	1.18	0.24	2.38	7.17	1.10	2.69	2.44	1.51	1.28	0.63	0.74	35.95	5124	1.87
2	Cyprea	Erosa	2.07	1.30	1.07	0.14	1.71	5.97	0.50	1.09	2.18	1.38	0.98	0.65	0.92	27.47	3573	2.01
3	Cyprea	Erosa	2.51	1.74	1.24	0.24	2.99	8.45	1.90	3.35	1.76	1.64	1.54	0.60	0.64	33.67	5210	1.35
4	Cyprea	Erosa	1.56	1.18	0.76	0.21	1.18	3.13	0.50	1.30	2.59	1.00	0.98	0.56	0.57	37.47	3528	1.46
5	Cyprea	Erosa	1.94	1.33	1.06	0.16	1.96	5.82	0.80	1.51	1.88	1.36	1.15	0.66	0.78	32.42	4163	1.27
6	Cyprea	Erosa	1.88	1.28	1.07	0.16	1.81	5.72	0.60	1.24	2.06	1.35	1.05	0.69	0.89	31.81	4049	1.44
7	Cyprea	Erosa	1.80	1.26	0.93	0.17	1.48	4.48	0.60	1.20	2.01	1.19	1.05	0.62	0.71	32.71	3685	1.29
8	Cyprea	Erosa	2.47	1.82	1.27	0.26	3.27	8.63	1.60	3.47	2.17	1.66	1.45	0.60	0.68	36.71	5739	1.65
9	Cyprea	Erosa	2.15	1.56	1.17	0.24	2.21	6.73	1.10	2.30	2.10	1.46	1.28	0.64	0.73	36.77	5078	1.36
10	Cyprea	Erosa	1.79	1.28	0.98	0.15	1.73	4.93	0.50	1.32	2.64	1.25	0.98	0.65	0.82	28.41	3358	2.62
11	Cyprea	Erosa	2.06	1.30	1.05	0.17	1.96	6.05	0.70	1.47	2.11	1.39	1.10	0.64	0.81	30.48	3991	1.72
12	Cyprea	Erosa	2.42	1.73	1.24	0.21	3.08	8.31	2.00	3.41	1.70	1.63	1.56	0.61	0.63	39.83	6111	0.91
13	Cyprea	Erosa	2.13	1.52	1.06	0.16	2.32	6.21	0.80	2.12	2.65	1.41	1.15	0.59	0.72	34.18	4534	2.13
14	Cyprea	Erosa	2.06	1.34	1.00	0.18	2.11	5.80	0.70	1.33	1.90	1.36	1.10	0.60	0.74	26.62	3413	1.84
15	Cyprea	Erosa	1.77	1.30	1.24	0.18	1.68	6.44	0.70	1.41	2.02	1.43	1.10	0.82	1.06	33.73	4556	1.29
16	Cyprea	Erosa	1.72	1.10	0.86	0.11	1.40	4.15	0.40	0.74	1.85	1.15	0.91	0.62	0.79	25.74	2791	1.54
17	Cyprea	Erosa	2.05	1.47	1.02	0.20	2.14	5.71	1.00	2.08	2.08	1.35	1.24	0.59	0.64	34.44	4380	1.48
18	Cyprea	Erosa	2.20	1.51	1.15	0.22	2.26	6.83	0.90	2.20	2.44	1.47	1.20	0.63	0.77	35.08	4879	1.83
19	Cyprea	Erosa	1.98	1.34	1.06	0.18	1.75	5.70	0.80	1.78	2.23	1.35	1.15	0.65	0.76	38.09	4840	1.28
20	Cyprea	Erosa	2.17	1.37	1.07	0.17	1.95	6.25	0.80	1.57	1.97	1.41	1.15	0.62	0.76	28.78	3829	1.76
21	Cyprea	Erosa	1.70	1.12	0.84	0.14	1.57	4.15	0.50	0.91	1.82	1.15	0.98	0.61	0.71	30.77	3336	1.12
22	Bankivia	Fasianus	2.64	1.12	1.09	0.11	1.96	2.55	1.40	2.36	1.69	0.90	1.39	0.63	0.41	35.27	2997	1.01
23	Catharus	Tinctus	1.69	1.03	0.89	0.14	0.85	4.04	0.60	0.77	1.28	1.13	1.05	0.67	0.73	33.22	3554	0.34
24	Catharus	Tinctus	2.01	1.19	1.01	0.16	1.29	4.90	0.80	1.23	1.54	1.25	1.15	0.66	0.71	34.33	4045	0.69
25	Catharus	Tinctus	1.58	1.19	0.87	0.15	0.97	4.12	0.40	0.75	1.88	1.15	0.91	0.64	0.80	32.93	3558	0.97
26	Architecto	Perdix	1.81	1.82	1.08	0.16	2.56	8.18	0.75	1.48	1.97	1.61	1.13	0.59	0.85	29.19	4443	1.68
27	Architecto	Perdix	1.77	1.64	1.03	0.10	2.22	8.76	0.75	1.42	1.89	1.67	1.13	0.60	0.89	30.86	4862	1.38
28	Noetia	Ponderosa	2.59	2.34	0.72	0.23	3.99	10.36	1.20	2.62	2.18	1.82	1.32	0.29	0.40	24.82	4252	3.31
29	Noetia	Ponderosa	2.88	2.54	0.79	0.23	5.12	13.26	1.60	3.21	2.00	2.05	1.45	0.29	0.41	25.82	5005	2.86
30	Bathybemb	CrumpII	2.63	2.00	1.46	0.11	3.19	8.50	1.60	3.30	2.06	1.64	1.45	0.64	0.72	33.72	5232	1.78
31	Polinices	Helicoides	2.68	2.41	1.43	0.08	4.25	18.48	1.50	3.20	2.14	2.43	1.42	0.56	0.96	26.30	6017	3.05
32	Tellina	Pulcherrim	2.79	1.78	0.56	0.14	3.55	7.58	0.80	1.28	1.61	1.55	1.15	0.25	0.34	18.96	2779	2.54

Table (1): Experimental Results.

Figure (1): Picture for the Selected Samples from Left to Right : Cyprea, Architectonica, Cantharus, Bathybembix, Polinices, Bankivia, and Tellina

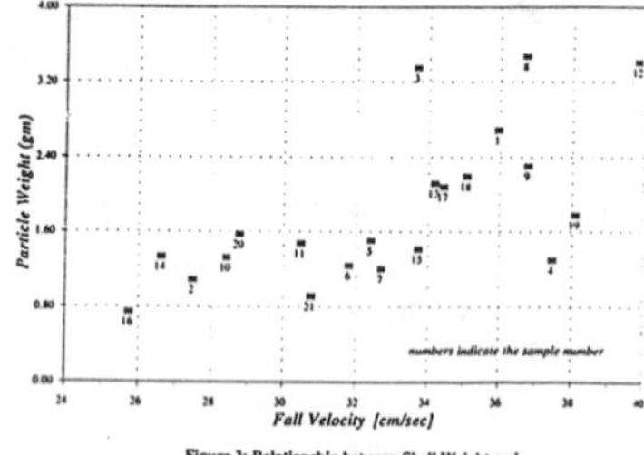

Figure 3: Relationship between Shell Weight and Fall Velocity for Cyprea Family.

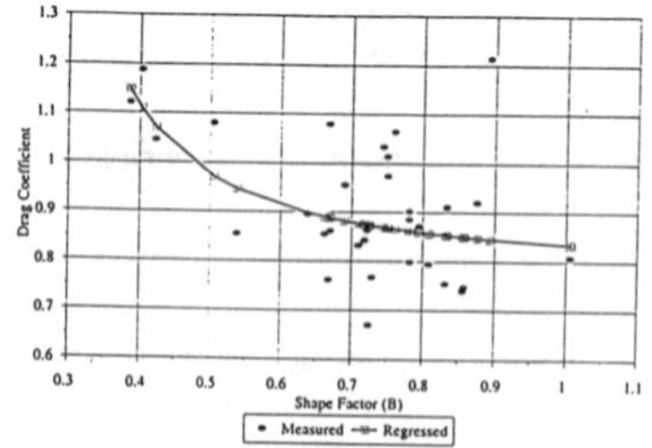

Figure (4): Relationship between Drag Coefficient and Shape Factor.

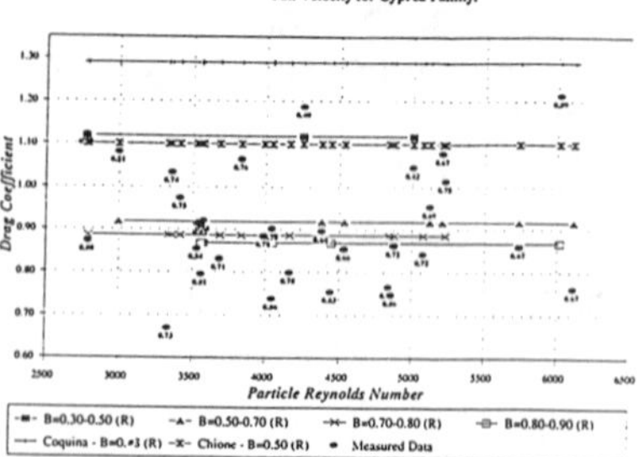

Figure 4: Relationship between the Drag Coefficient and the Particle Reynolds number for the different shell families classified according to the shape parameter.

Predicting Roughness in Concrete Channels with Bed Load

Ronald R. Copeland,[1] M. ASCE

Abstract. A sedimentation study was conducted to determine if deposition in the concrete-lined North Diversion and Embudo Channels, located in Albuquerque, New Mexico, would cause overtopping during the 1 percent chance exceedance flood (Copeland, in preparation). The flood-control project was originally designed and constructed in 1965-67 by the U.S. Army Corps of Engineers for the Standard Project Flood, but the effects of sediment deposition in the channel had been ignored. The primary source of sediment is unlined channels upstream from the concrete-lined channels. A recent flood deposited significant quantities of sediment in the Embudo Channel, raising concerns about the channel's ability to carry larger flood discharges.

The HEC-6 numerical sedimentation model was used to predict deposition in the concrete-lined channels. The effect of sediment deposits on boundary roughness was determined using analytical techniques. Calculated roughnesses were incorporated into the numerical model, and an iterative procedure was used to determine the effect of deposited sediment on conveyance. The numerical model was circumstantiated using data from an historical flood where significant deposition occurred in the Embudo Channel. Sediment inflow to the numerical model was determined using results of sediment yield and trap efficiency studies. Considerable uncertainty existed relative to the quantity of sediment delivered by the 1 percent chance exceedance flood. Therefore, sensitivity studies were conducted to assess the impact of different sediment loadings.

Calculated Deposition. Sediment deposits reduce channel capacity in two ways: by reducing the cross-sectional flow area, and by increasing the boundary roughness. At the peak flow, calculated sediment deposition was less than 0.2 m in the North Diversion Channel between the Embudo confluence and a location about 7.7 km

[1]Research Hydraulic Engineer, U.S. Army Engineer, Waterways Experiment Station, 3909 Halls Ferry Road, Vicksburg, MS 39180-6199

downstream. Because the deposition depths were relatively insignificant compared to the total depth of the channel, there was only a minor change in available cross-sectional area for this reach of the channel, but deposition was sufficient to cause an increase in roughness coefficient. Calculated deposition depths in the Embudo Channel were significantly greater than in the North Diversion Channel. Most of the deposition occurred during the recession of the flood hydrograph. Deposition depths calculated using the HEC-6 numerical model at the peak of the 1 percent chance exceedance flood varied between 0.2 and 1.1 m, with deposition depths decreasing in a downstream direction. Sediment deposition increased during the recession of the flood hydrograph. Calculated deposition depths at the end of the flood varied between 2.2 and 6.2 m. The total calculated volume of deposition in the North Diversion and Embudo Channels for the 1 percent chance exceedance flood was 37,500 m^3. The variation of stage and bed elevation during the 1 percent chance exceedance flood in Embudo Channel at a location 760 m upstream from the confluence with the North Diversion Channel is shown in Figure 1.

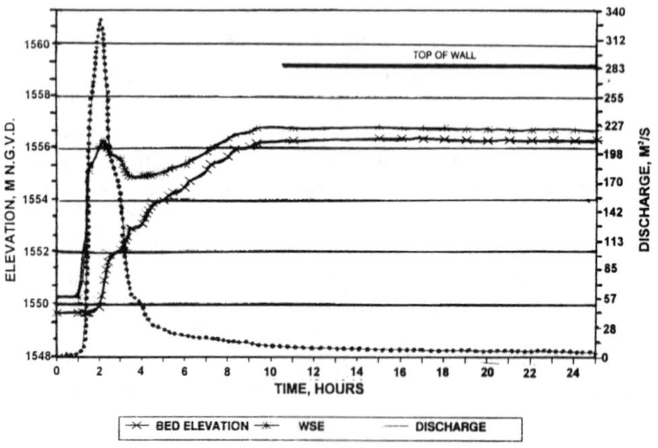

Figure 1. Variation of stage and bed elevation during 1 percent chance exceedance flood, Embudo Channel, 760 m upstream from North Diversion Channel

Design Roughness Coefficients. Manning's roughness coefficients in the numerical model were determined iteratively by considering the increasing effect of boundary roughness with increasing deposition in the concrete-lined channel during the course of the 1 percent chance exceedance flood. The model was run several times, with roughness coefficients adjusted as sediment deposition increased. This technique makes the roughness coefficients in the model unique to the particular hydrograph and sediment loading used in the study.

The Manning's roughness coefficient was calculated external to the HEC-6 numerical model by compositing the roughness of the concrete side slopes with the bottom deposits of sand. The Einstein-Horton compositing equation was used:

$$\bar{n} = \frac{\left(\sum_{i=1}^{3} P_i \, n_i\right)^{\frac{2}{3}}}{\left(\sum_{i=1}^{3} P_i\right)^{\frac{2}{3}}} \tag{1}$$

where $\bar{n}$ is the composite roughness coefficient for the cross section, P is the wetted perimeter, subscripts 1 and 3 associate variables with the side slopes, and subscript 2 refers to the channel bottom.

Roughness for concrete was calculated using resistance equations based on the Keulegan and Colebrook-White equations, as recommended in Engineer Manual (EM) 1110-2-1601 (HQUSACE, 1991). The equations have been modified here to calculate Manning's roughness coefficient directly:

$$n = \frac{-R^{\frac{1}{6}}}{32.6 \, \log_{10} \, (A + B)}$$

$$A = \frac{1.278 \, R^{\frac{1}{6}}}{n \, \mathbb{R} \, 10^{0.0960 A_s}} \tag{2}$$

$$B = \frac{k_s}{R \, 10^{0.0960 A_r}}$$

$$\mathbb{R} = \frac{4RV}{\nu}$$

where R is the hydraulic radius in meters, A_s and A_r are Iwagaki's coefficients for smooth and rough flow, respectively, k_s is the roughness height in meters, V is the average channel velocity in m/sec, and ν is the kinematic viscosity in m/sec. Iwagaki's coefficients vary with Froude Number (Chow 1959). A roughness height of 0.00213 m was assigned to the concrete boundary, as recommended in EM 1110-2-1601.

When sediment deposits were present, bed roughness was calculated using the Brownlie (1983) resistance equations:

for lower regime flow:

$$n = \left[0.07021\left(\frac{R}{d_{50}}\right)^{0.1374} S^{0.1112}\ \sigma_g^{\ 0.1605}\right] d_{50}^{0.167} \tag{3}$$

and for upper regime flow:

$$n = \left[0.04233\left(\frac{R}{d_{50}}\right)^{0.0662} S^{0.0395}\ \sigma_g^{\ 0.1282}\right] d_{50}^{0.167} \tag{4}$$

where d_{50} is the median grain size in meters, S is the energy slope, and σ_g is the geometric standard deviation of the bed-sediment. These equations, which account for both grain and form roughnesses, were developed for alluvial channels.

The sediment bed gradation used in the Brownlie equation was calculated with the HEC-6 model. The coarseness of the calculated bed-material gradation decreased longitudinally down the channel, and the calculated gradations at a given point became coarser with the progression of the hydrograph. Thus, bed gradations were finer in the North Diversion Channel, resulting in lower calculated roughness coefficients.

The effect of deposition on channel roughness was most significant in the Embudo Channel due to the greater quantity of deposition and because the deposited material was coarser than the material deposited in the North Diversion Channel. The bottom width of the channel increased with deposition in the Embudo Channel. This resulted in an additional increase in roughness due to the increased fraction of the wetted perimeter composed of sediment. The calculated variations of Manning's roughness coefficient with discharge for Embudo Channel are shown in Figure 2. As the hydrograph rose, roughness initially increased with discharge as bed forms grew. However, further increases in discharge eventually resulted in a decrease in roughness because as depth increases with discharge, a lesser percentage of the channel wetted perimeter is covered by sediment in the trapezoidal channel and bed forms begin to diminish in size. In order to demonstrate the influence of deposited sediment on the roughness coefficient, a roughness coefficient for a sediment-free channel is also shown in Figure 2.

Recommended design roughness coefficients for calculating maximum water-surface elevations for the 1 percent chance exceedance **peak** discharge varied between 0.017 for the North Diversion Channel and 0.019 for Embudo Channel. The roughness coefficient in Embudo Channel will increase with increased deposition and with falling discharge.

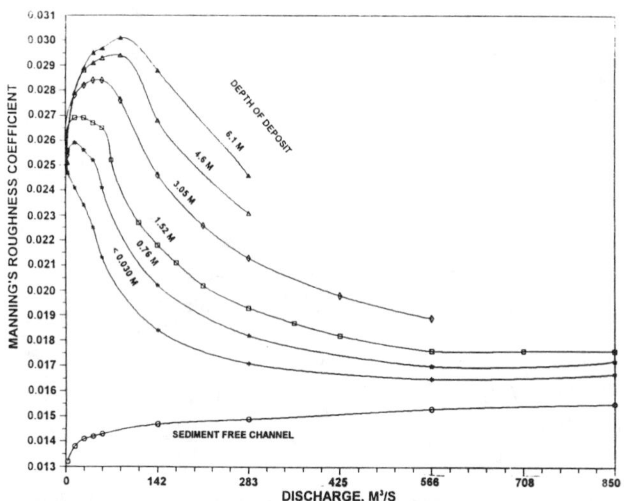

Figure 2. Variation of Manning's n with depth of sediment deposit, Embudo Channel

Conclusions. The study concluded that while sediment deposition caused an increase in channel roughness and a decrease in cross-sectional area, in certain reaches of the concrete-lined channel, the effects of the deposition were insufficient to cause over-topping of the existing channel during the 1 percent chance exceedance flood.

Acknowledgments. The study described and the resulting data presented herein were obtained from research conducted by the U.S. Army Engineer Waterways Experiment Station for the U.S. Army Engineer District, Albuquerque. Permission was granted by the Chief of Engineers to publish this data.

References.

Brownlie, William R. 1983. "Flow Depth in Sand-Bed Channels," *Journal of Hydraulic Engineering,* ASCE, Vol. 109, No. 7, pp 959-990.

Chow, Ven T. 1959. *Open Channel Hydraulics*, McGraw-Hill, New York.

Copeland, Ronald R. 1995. *Albuquerque Arroyos Sedimentation Study;* Numerical Model Investigation, (in preparation), U.S. Army Engineer Waterways Experiment Station, Vicksburg, MS.

Headquarters, U.S. Army Corps of Engineers (HQUSACE) 1991. *Hydraulic Design of Flood Control Channels*, Engineer Manual 1110-2-1601, Washington D.C.

The Effect of Approaching Flow Angles on the Local Scour at Semi-Circular Piers

G. W. Choi[1], C. J. Ahn[2], K. H. Kim[3], S. J. Ahn[4]

Abstract

In this paper, the effect of the approaching flow angles on the local scour at semi-circular piers, which are the typical piers in the rivers and large streams in Korea, was analyzed based upon the maximum scour depth. The pier models were obtained from San Gye bridge in the Bocheong stream which is a tributary in IHP experimental watershed in Korea. Four parameters, which are the flow depth, channel bed slope, pier skewness and opening ratio, were selected as the major parameters to investigate the effect of the approaching flow angles. Finally, for the maximum scour depth, three equations indicating multiplying factors for approaching flow angles are suggested based upon the Froude numbers. The suggested multiplying factors were compared with those suggested by Laursen and Toch, and Mostafa *et al.* In the low Froude numbers, the multiplying factors suggested in this paper are well agreed with the previous works. However, in the high Froude numbers, the multiplying factors suggested in this paper are larger than the previous works and bring the deeper scour depths compared to the previous works.

Introduction

Formation of vortexes is the basic mechanism causing local scour at bridge piers. The formation of these vortexes results from the pile up of water on the upstream face and subsequent acceleration of the flow around the nose of the pier. The action of the vortex removes bed materials away from the base region. If the transport rate of sediment away from the local region is greater the transport rate into the region, a scour hole develops. As the depth of scour is increased, the

1 Assistant Professor, Dept. of Civil Engineering, University of Inchon, Inchon, Korea.

2 Principal Researcher, Water Resources Research Institute, Korea Water Resources Corporation, Dae Jeon, Korea.

3. Professor, Dept. of Civil Engineering, Inha University, Inchon, Korea.

4. Professor, Dept. of Civil Engineering, Chungbuk National University, Chong Ju, Korea.

strength of the vortexes are reduced, thus reducing the transport rate. As equilibrium is reestablished scouring ceases and the scour hole will not enlarge further. For piers there is also an additional vertical vortex downstream of the pier, which is denoted as the wake vortex. Both vortexes remove material from around the pier. In many cases the material which is removed by these vortexes is redeposited immediately downstream of the pier.

In the natural stream or river, to minimize the local scour depth induced by constructing piers it is necessary to install the piers parallel to the flow direction. It is because the increment of the approaching angles decreases the opening ratio of the cross sectional area and increase the local flow velocity which cause the increment of the local scour depth at piers. However, there are several cases, such as connection to the road with angles, change of flow direction by time in the alluvial channel and so on, not to be parallel to the flow direction.

The research related to the maximum scour depth change by approaching angles was conducted by Laursen and Toch (1956). They suggested the multiplying factor K_α for the maximum local scour depth increment. The factor indicates the increased local scour depth with increasing the angle of attack in degrees. Also, Richardson *et al.* (1975, 1988) suggested K_α for angle of attack of the flow depending upon the ratio between the width and the length of the pier. As the ratio is increased, the suggested multiplying factor are also increased. Wang and Triweko (1986) suggested the local scour equation which included the projected pier width which is affected by the angle of attack. Recently, Mostafa *et al.* (1993) suggested the different type of K_α equations for the rectangular and oblong piers.

Experiments

The experiment was conducted in a glass-walled tilting sediment flume. The working section of the flume was 12m long, 40cm wide, and 40cm deep with a variable speed pump. The channel bed material in the Bo Chung stream in the Keum River basin in Korea was utilized for experiments. The average diameter was 0.8 mm and the geometric standard deviation was 1.95 and specific gravity was 2.58. The opening ratios, which are used in the experiments, were 90.0%, 92.5%, 93.8%, 95.0% and 97.5%. The channel bed slopes of 0.01%, 0.03%, 0.05%, 0.1%, 0.2%, and 0.4% were adopted and the approaching angles of 0 to 90 degrees with 15 degree interval were utilized in the experiments. Three types of experiments, which are the maximum local scour depth variation by changing channel bed slopes in the same approaching angle, approaching angles in the same channel bed slope and approaching angles in the same flow depth, were conducted in the same opening ratio. The experimental results were utilized to find out the multiplying factors depending upon the angles of attack and Froude numbers.

Experimental Results

The maximum local scour depth was compared depending upon channel bed slopes in the same approaching angles and opening ratios. In the first, the maximum scour depths in the different approaching angle were compared. The maximum scour depth is increased by increasing the channel bed slopes and water flow depth y. In the higher flow depth, the increasing rate is lower than in the lower flow depth.

In the second, the maximum scour depths depending upon the approaching angles in the same channel bed slope were compared. The maximum scour depths in 90 degree are much higher compared to the lower approaching angles.

The multiplying factors are shown in Figure 1 and Figure 2 in the ratio of the water flow depth over pier width (y/b). The ratios in the figure 1 and 2 were 2.5 and 3.0, respectively.

Based on above three types of experiments, three equations indicating multiplying factors for the angles of attack are defined as follows:

In Fr < 0.55;

$$K_\alpha = \left\{ \left(\frac{L}{b} - 1 \right) \sin \alpha + 1 \right\}^{1.24}$$

In 0.55 < Fr < 0.7;

$$K_\alpha = \left\{ \left(\frac{L}{b} - 1 \right) \sin \alpha + 1 \right\}^{0.97}$$

In the Fr > 0.7;

$$K_\alpha = \left\{ \left(\frac{L}{b} - 1 \right) \sin \alpha + 1 \right\}^{0.6}$$

Comparison with Existing Multiplying Factors

The generally accepted equation of multiplying factors, which is suggested by Laursen and Toch can be written as following equation with the limitation of 2 < L/b < 16 in the semi-curcular piers.

$$K_\alpha = 0.9 \left\{ \frac{L}{b} \sin \alpha + \cos \alpha \right\}^{0.68}$$

Also, Mostafa *et al.* suggested the following equation indicating multiplying factors.

$$K_\alpha = \left\{ \left(\frac{L}{b} - 1 \right) \sin \alpha + 1 \right\}^{0.9}$$

As indicated in the paper written by Mostafa *et al.*, the maximum scour depths calculated using Laursen and Toch's multiplying factors can be less compared to the field data. The multiplying factors suggested in this paper are similar with

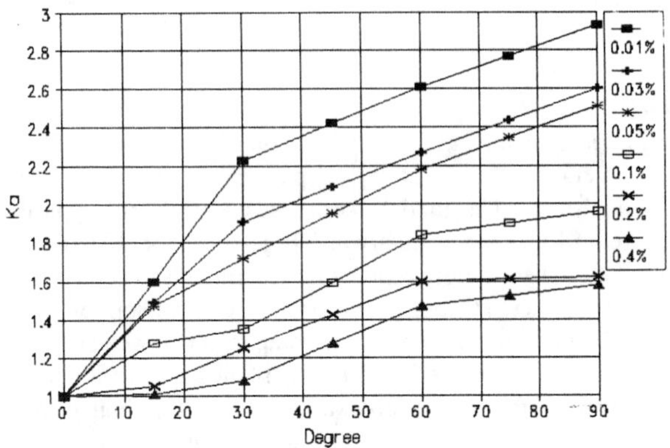

Figure 1. The multiplying factor K_α in dimensionless water flow depth y/b=2.5

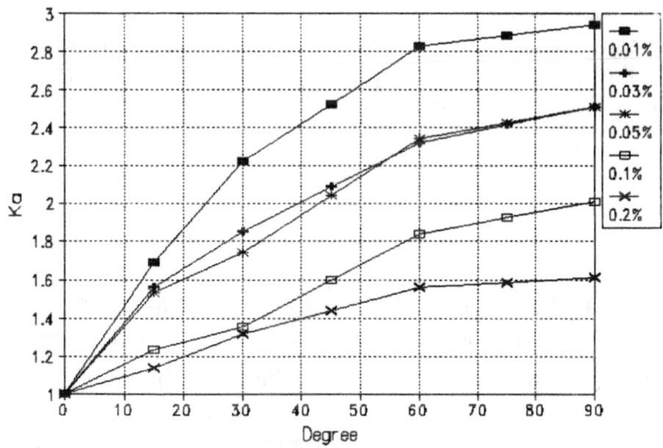

Figure 2. The multiplying factor K_α in dimensionless water flow depth y/b=3.0

Laursen and Toch's in the range of less than 0.5 in Froude number. Also, it is similar to Mostafa *et al.*'s in the range of 0.55 to 0.7 in Froude number. However, the multiplying factors suggested in this paper are larger than the previous researchers' and bring larger local scour depth in the range of larger than 0.7 in Froude number.

Conclusions

In this paper, the effect of the approaching flow angles on the local scour at semi-circular piers, which are the typical piers in the rivers and large streams in Korea, was analyzed. The models were selected from the typical piers in the Bocheong stream which is a tributary in IHP (International Hydrologic Program) experimental watershed. Before conducting experiments, several model verification processes were conducted through the roughness comparisons between the natural stream and the experimental channel, the comparison between experimental scour data and field measuring scour data pursuing scour depth variation with time, and so on. Four parameters, which are water flow depth, channel bed slope, pier skewness and opening ratio, were selected as the major parameters to investigate the effect of the approaching flow angles. Finally, for the maximum scour depth decision three equations indicating multiplying factors for approaching flow angles were suggested based upon Froude numbers. The suggested multiplying factors were compared with the multiplying factors suggested by Laursen and Toch, and Mostafa *et al.* In the high Froude numbers, the multiplying factors suggested in this research are well agreed with the previous works. However, in the lower Froude numbers the multiplying factors suggested in this research are larger than the previous works and bring the deeper scour depths compared to the previous works. The scour width extensions are severely changed depending upon the approaching flow angles. The suggested multiplying factors can be used for predicting the maximum scour depth in the design stage for bridge piers and in the safety investigation stage for the existing piers.

References

Laursen, E. M. and Toch, A., 1956, "Scour around Bridge Piers and Abutments", Iowa Highway Research Board, Bulletin No. 4.

Melville, B. W. and Sutherland, A. J., 1988, "Design Method for Local Scour at Bridge Piers", J. of Hydraulic Engineering, ASCE, Vol. 114, No. 10, pp. 1210-1226.

Richardson, E. V, et al., 1975, "Highways in the River Environment", First ed., U. S. Dept. of Transportation, FHWA, Ft. Collins, Co.

Richardson, et al., 1988, "Scour at Bridges", FHWA, U. S. Dept. of Transportation.

Wang, T. M. and Triweco R. W., 1986, "Maximum Depth of Scour around Bridge Piers", Proc. of 5th Congress, ADP-IAHR.

A FE Model of Moisture Flow Within a Landfill

Jeffrey J. Piotrowski, Ph.D.[1]

Abstract

Leachate generation at municipal solid waste facilities has become an increasing public concern, especially in light of numerous documented groundwater and surface water contamination incidents which were directly attributable to nearby landfill sites. Unfortunately, our understanding of the processes involved in the production of contaminated leachate is limited. Landfill excavation studies have indicated a strong correlation exists between refuse moisture content and the amount of degraded refuse, suggesting the amount and distribution of moisture within the landfill are key determinants of leachate formation. Current models of landfill systems are used to estimate total leachate production, however they are limited in their ability to characterize the patterns of moisture distribution within the landfill, partly due to the assumption of homogeneous landfill material properties. Accordingly, a two dimensional finite element (FE) computer model, which incorporates the hydraulic properties of various landfill materials, has been developed to examine the effects of anisotropic material properties on moisture distribution patterns within a landfill. The model is based on mathematical expressions of unsaturated flow in porous media and was used to calculate total leachate production and moisture distribution within a one-acre test landfill cell.

Introduction

Landfills have been a major component of waste disposal management for centuries (Senior 1990; Bagchi 1990) and, to date, remain the most cost effective means to dispose of solid waste (Senior and Balba 1987). Recent research on landfills has focused on reducing leachate generation and on leachate collection and treatment systems to minimize the impact of the landfill on the surrounding communities. Computer modeling has supported these efforts by estimating the potential

---

1. Manager, Science Group, ENSCO, Inc., 445 Pineda Court, Melbourne, FL 32940

amount of leachate production. However, few models address the issue of leachate quality, which depends on the types and rates of reactions occurring within the landfill. Bagchi (1990) identifies four important reasons for predicting leachate quality when designing landfills: (1) to determine if the waste is hazardous, (2) to choose an appropriate landfill design for the type of waste entering the landfill, (3) to design or gain access to a suitable leachate treatment facility, and (4) to develop a list of chemicals for the groundwater monitoring program.

The task of predicting leachate quality has proven challenging due to the enormous complexity of biological, chemical, and physical interactions and reactions within the landfill ecosystem (Senior 1990) and the lack of information concerning the physical properties and rates of refuse degradation. Studies of leachate formation have been made through laboratory testing (Raveh and Avnimelech 1979), field studies using lysimeters (Ham and Bookter 1982; Qasim and Burchinal 1970), and, more recently, through computer modeling (Demetracopolous et al. 1986; Straub and Lynch 1982; Ahmed et al. 1992). Leachate generation is dependent on moisture flow characteristics within the landfill and mechanisms of refuse degradation which release contaminants to the liquid phase. Thus, an important first step towards modeling the mechanisms of leachate generation involves the understanding and modeling of landfill moisture flow and distribution patterns..

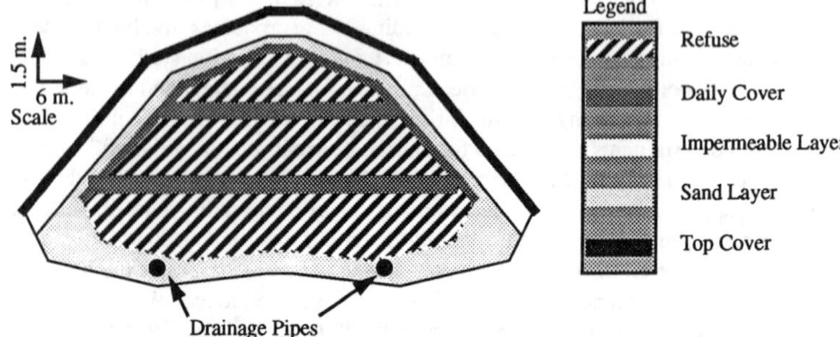

Figure 1. Schematic depicting landfill material properties.

Current models of landfill systems assume homogeneous material properties throughout the landfill, thus uniform moisture penetration into the landfill is depicted. However, landfills are composed of a number of different materials, e.g., refuse types, clay soil caps, sand drainage layers, etc., and each material has unique hydraulic properties (Figure 1). This study determined the effect non-homogeneous material properties had on modeling moisture distribution patterns within a landfill system and was supported by a comprehensive data set compiled by the Delaware Solid Waste Authority (DSWA). Their two one-acre test cells, located at the DSWA site in Sandtown, DE, have been in operation since 1989 and are typical of the larger on site landfills in their design, composition, and operation.

Methods

The mathematical expressions used to describe moisture flow through the landfill are based on the theory of unsaturated flow in porous media. These expressions are obtained from the conservation of mass equation and a motion equation that is rooted in Darcy's law. The two dimensional governing equation of unsaturated moisture flow is expressed

$$\frac{\partial \theta}{\partial t} = \frac{\partial}{\partial x}\left[D(\theta)\frac{\partial \theta}{\partial x}\right] + \frac{\partial}{\partial z}\left[D(\theta)\frac{\partial \theta}{\partial z}\right] + \frac{\partial}{\partial z}K(\theta) \tag{1}$$

where θ is the moisture content of the landfill (dimensionless), $D(\theta)$ is a diffusivity term ($m^2\ s^{-1}$), $K(\theta)$ is hydraulic conductivity ($m\ s^{-1}$), t is time (s), and x and z represent the horizontal and vertical directions (m) respectively. The diffusivity term is related to hydraulic conductivity as follows,

$$D(\theta) = K(\theta)\frac{d\psi}{d\theta} \tag{2}$$

where ψ is the pressure head (L). Equation (1) is non-linear due to the dependence of $K(\theta)$ and $D(\theta)$ on moisture content. Clapp and Hornberger (1978) developed the following power relationships to describe the hydraulic conductivity and pressure head terms in unsaturated porous media,

$$K(\theta) = K_s\left(\frac{\theta}{\theta_s}\right)^B \tag{3}$$

$$\frac{d\psi}{d\theta} = (-\psi_s)\, b\theta^{(-b-1)}\,\theta_s^{\ b} \tag{4}$$

where K_s, θ_s, and ψ_s represent the saturated hydraulic conductivity, moisture content and pressure head terms respectively, and B and b are positive empirical constants related to properties of the media (dimensionless).

Previous models of moisture flow through landfill systems employed finite difference techniques to solve Equation (1), and assumed homogeneous material properties throughout the landfill. A finite element solution method was chosen for this study to handle the more complex problem involving internal boundaries that develop due to incorporating varying hydraulic conductivities. Under anisotropic conditions, the hydraulic conductivity of the medium varies with direction ($K_x \neq K_y \neq K_z$), i.e., there is a directional quality to the overall structure of the medium, and K is a second tensor of hydraulic conductivity. However, it is assumed for the finite element method that isotropic conditions exist within each element reducing K to a scalar, and anisotropic conditions are simulated by varying hydraulic conductivities from element to element.

Results and Discussion

The FE model was configured to simulate the physical design of the DSWA test landfill cell (as seen in Figure 1). The model incorporated the hydraulic conductivity values for the different materials of construction, however values for the refuse are not known and model runs were made using assumed values. In general, lower refuse hydraulic conductivity values produced sharper effluent leachate peaks and shorter lag times between input precipitation and effluent leachate, a result of more moisture being directed around the refuse, through the sand drainage layers.

Assumed refuse conductivity values were used to make a direct comparison between leachate generation predicted by the model and DSWA measurements of leachate generation at the test landfill cells over a three year period. Although similarities in overall trends were observed, the measured values were consistently greater (Figure 2). However, the model did predict non-uniform moisture distribution patterns (Figure 3), consistent with what might be expected given antecedent rainfall amounts. While further investigation is warranted, these results highlight the importance of including landfill material properties, i.e., hydraulic conductivities, when modeling moisture flow and distribution within landfill systems.

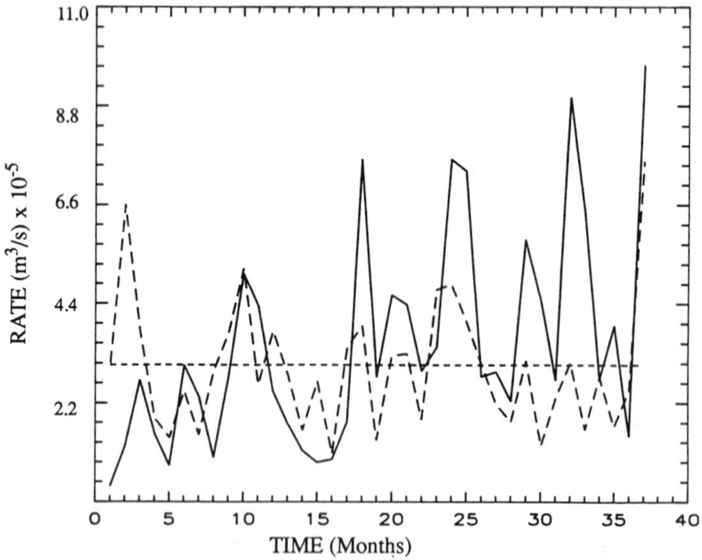

Figure 2. Comparison of leachate generation. Model calculated(long dashed line) vs. measured leachate (solid line), and average precipitation throughout the period (short dashed line).

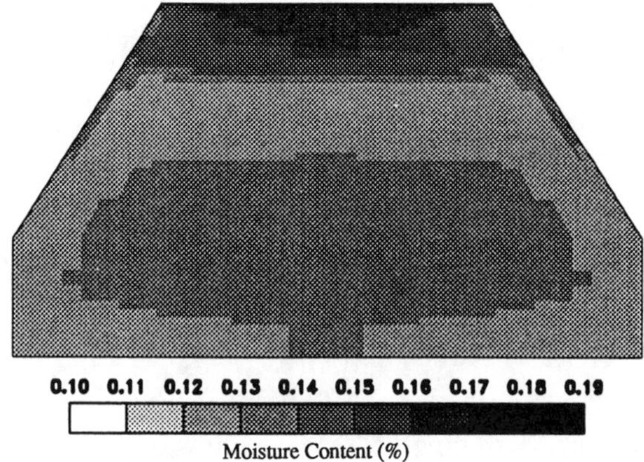

Figure 3. Moisture distribution patterns within the model landfill.

References

Ahmed, S., R.M. Khanbilvardi, J. Fillos, and P.J. Gleason, "Two-dimensional leachate estimation through landfills," *J. Hydraulic Engineering*, 118(2), 306, 1992.

Bagchi, A., "Design, construction and monitoring of sanitary landfills," John Wiley & Sons, Inc., New York New York, 1990.

Clapp, R.B. and G.M. Hornberger, "Empirical equations for some soil hydraulic properties," *Water Resources Research*, 14(4), 601, 1978.

Demetracopoulos, A.C., G.P. Korfiatis, E.L. Bourodimos, and E.G. Nawy, "Unsaturated flow through solid waste landfills model and sensitivity analysis," *Water Resour. Bull.*, 22(4), 601, 1986.

Ham, R.K. and T.J. Bookter, "Decomposition of solid waste in test lysimeters," *J. Environ. Eng.*, 108(EE6), 1147, 1982.

Qasim, S.R. and J.C. Burchinal, "Leaching from simulated landfills," *J. WPCF*, 42(3), 371, 1970.

Raveh, A. and Y. Avnimelech, "Leaching of pollutants from sanitary landfill models," *J. WPCF*, 51(11), 2705, 1979.

Senior, E. and M.T.M. Balba, "Landfill biotechnology, In Bioenvironmental systems, Volume II," D.L. Wise (eds.), CRC Press, Inc., Boca Raton, Florida, 1987.

Senior, E., "Microbiology of landfill sites," CRC Press, Inc., Boca Raton, Florida, 1990.

Straub, W.A. and D.R. Lynch, "Models of landfill leaching: moisture flow and inorganic strength," *J. Environ. Eng.*, 108(EE2), 231, 1982.

STREAM DISSOLVED OXYGEN MODELING AND WASTELOAD ALLOCATION

E. Zia Hosseinipour and Larry A. Neal[1]

Abstract

Effluent discharges to the receiving waters are regulated by National and State Environmental Protection Agencies. To discharge treated effluent, industrial facilities must obtain National Pollutant Discharge Elimination System (NPDES) permit that specifies discharge limits for specific chemical constituents on an individual basis. These limits depend on constituents type, receiving water flow regime and assimilative capacity, tolerance of aquatic species within the water body, background water quality, etc. This paper discusses water quality modeling in support of the NPDES permit application process. Water quality management alternatives were analyzed with the aid of a water quality modeling package to project discharge limits based on flow regime of the receiving waters, ambient water temperature, and the 5 day Biochemical Oxygen Demand (BOD5) of the discharged effluent. The goal was to determine the appropriate waste load allocations to meet the EPA guidelines for Best Practicable control Technology currently available (BPT) and protect the oxygen resources of the stream during critical conditions, while remaining cost effective.

INTRODUCTION

Along the stretch of the river that was the subject of this study, two industrial facilities that are 30 miles apart discharge treated effluent. The early operating NPDES permit limits for the mills were originally derived from EPA technology guidelines and were not based upon receiving stream water quality considerations. However, through river monitoring and water quality modeling it has been established that the segment of the river which is receiving discharge effluent is water-quality limited during certain low flow/high temperature conditions. **By definition**, on days that the river can not safely assimilate full technology based BOD load, the river is said to be water quality limited by the Dissolved Oxygen (DO) content. Currently both facilities provide treatment that produces significantly better effluent quality than that established by the EPA technology based requirements. Preliminary DO analyses suggested that the two plants would need to reduce their final effluent loads to about 1/3 of

[1]Law Engineering & Environmental Services, 112 Town Park Drive, Kennesaw, Georgia 30144.

the EPA technology guidelines in order to maintain a minimum instream DO of 5.0 mg/l during the 7 day 10 year low flow (7Q10). Because of the exceedingly high cost and fundamental questions about technical feasibility for meeting such stringent treatment requirements, the two mills informally asked the state department of environmental management to consider the use of permit limits that vary according to available stream flow and associated assimilative capacity. That is, the mills proposed to discharge in accordance with EPA technology guidelines unless the combination of low flow and high temperature in the river created a water-quality limited situation. When the river is water-quality limited for DO, the mills would further restrict their effluent loads and/or inject oxygen into the river in accordance with permit requirements designed to maintain instream DO criteria. The modeling study showed that by using flow-variable permitting, the mills can adequately protect receiving water quality without spending limited resources on unnecessary treatment or excessive oxygen injection.

Water Quality Management Alternatives

To meet the water quality criteria established by the regulatory authorities during critical conditions, several management alternatives were analyzed to develop an optimum management which is economically feasible and at the same time protective of water quality. Oxygen injection, storage and treatment of effluent, and artificial wetlands were the alternatives investigated. Oxygen injection is considered a "non-treatment" technique by the EPA and must therefore be justified in accordance with the requirement of non-treatment alternatives. Storage and further treatment of effluent was considered as an additional treatment. This option required that the mills discharge at the modeled allowable limit, and store any remaining excess BOD5, for subsequent discharge during periods that the river has more assimilative capacity. Artificial wetlands were considered as the third alternative. These systems have been demonstrated to be applicable for secondary and advanced treatment (Gersburg, 1989). The BOD5 removal capacity of municipal constructed wetlands is reported to range from 51% to 96% (Watson, 1989). Results of another study where constructed wetlands were used to further treat the secondary treated bleached kraft mill effluent, suggest that a lower BOD5 removal of 27% to 49% is attainable (Thut, 1989). In summary the costs for alternatives considered were: Storage $39.6 million; additional treatment $18 million; Oxygen injection $9.5 million; and artificial wetlands $34 million. Oxygen injection was selected because of economic feasibility. Two oxygen injection systems were proposed at locations where DO sag would occur 30 miles apart downstream of the mill discharge points. The two system would be operated jointly by the mills. The capacity of the systems were 12,000 and 40,000 lbs/day respectively. Oxygen would be injected only during the water quality limited periods, and would be firstly injected to the waste flow from the upstream system to offset excess BOD5. If the excess BOD5 is greater than the injection capacity of the upstream system, then the second system would come on line to offset the difference.

WATER QUALITY MONITORING FOR COMPLIANCE

As a part of their NPDES permit requirements both facilities monitor their effluent discharges to the river and perform a weekly river survey that include measuring water quality parameters such as DO, PH, and temperature within the river at specified locations downstream from the discharge point. A sample is collected for chemical analysis every day from the effluent prior to discharge. Each sample is analyzed for BOD5, Suspended Solids (SS), PH, and temperature. These data and the effluent discharge rate are collected and submitted to the State Environmental Protection Agency to fulfill the NPDES permit requirement. These reports are referred to as, the Discharge Monitoring Report (DMR) and the River Survey Report (RSR). River DO of 4 mg/l is required in the first 75 mile reach of the segment under study, because this stretch is classified as fish and wildlife, a DO of 5 mg/l is required in the following 21 miles, because it is classified as a drinking water segment. The water quality model of the stream is built using these data and historical flow data from United States Geological Survey (USGS).

FLOW RECESSION MODELING

Water quality modeling and waste load allocation analysis were done on a steady state basis. That is, time variant processes of river flow regime and kinetic reactions are not fully accounted for. To account for the time dependent characteristics of the river flow regime to some extent, a flow recession method was developed that recesses the flow for dilution by one day and for kinetics by N-1 day, where N is the flow recession period to be specified. Therefore, since the future flows are not always lower than the past flows, the waste load allocation results are inherently conservative and protective of the river water quality. The model used for the generation of recessed flows is called FLOWREC. This model uses the actual flow records compiled by the USGS for the whole period of record during the prescribed critical period, that is, May 1 through November 30 of every year. The available flow records cover the period 1928 to 1993. The output from the model is a table of flow ratio values for a number of chosen flow classes (i.e., 30), an specified exceedence probability, and a series of recession periods. The flow recession concept accounts for flow and velocity variations with time to some extent which the steady state model neglects. In practice, river flows do not always drop with time and therefore a margin of safety is maintained as far as river assimilative capacity is concerned. The N-day recession ratios are calculated for each day of the flow records as:

$$R_{i+N} = Q_{i+N}/Q_i$$

where: R_{i+N} = N-day recession ratio
Q = Average daily flow
i = The day being examined

The recession ratios (R) are then classified into 30 flow classes according to the corresponding flows. The ratios in each class are then sorted in ascending order, ranked and assigned an exceedence probability according to the following relationship:

$$P_E = 1 - [m/(n+1)]$$

where: P_E = The ratio exceedence probability
m = Rank; and n = The number of values in the class

When the model is run with the exceedence probability of 80%, it means that for the calculated recessed flow there is an 80% chance that the actual flow will be greater than the calculated flow. For predictive DO modeling at 7Q10 low flow conditions, water temperature data were obtained from a least square sinusoidal fit of the historical temperature data collected at the USGS gage. The water temperature value used corresponds with the 90th percentile upper bound of the sinusoidal best fit of the data.

DISSOLVED OXYGEN MODELING

Water quality modeling was performed with the aid of a conventional steady state model applied in a pseudo-dynamic fashion. The modeling scenarios were designed to provide information to be used for river water quality management during the periods that BPT guidelines would not maintain dissolved oxygen criteria within the stretch of interest. At this particular site, EPA's BPT guidelines would not maintain instream DO criteria during certain critical conditions associated with low flow and high temperature. The resulting water quality limited condition required water quality management approaches that not only would maintain the water quality standards for allowable effluent discharges based on BPT during non-critical conditions, but also maintain DO criteria during critical situations.

The DO model developed for the river assumes steady state conditions and does not account for flow variations with time directly. This limitation is a significant concern in a variable flow permit concept because the calculated hydraulic retention time from one mill, to a downstream location near the other mill, can be several weeks during low flow conditions. With such long retention times in the segment of interest, the direct use of the steady state model to predict day-to-day variations in assimilative capacity is not appropriate. The basic problem is that BOD discharged today at a given stream flow may not move through the river system before lower stream flows occur with a correspondingly lower assimilative capacity. What is needed is either a time variable (dynamic) water quality model, or some way to modify and use a steady state model in a conservative fashion.

The water quality analysis and waste load allocation model uses the modified Streeter-Phelps DO equation described in detail by Hosseinipour et al. (1995). The basic equation is modified to include the effects of oxidation of nitrogenous materials, sediment oxygen demand (SOD), and photosynthetic oxygen production and utilization. A model developed by the Georgia Environmental Protection Division called GADOSAG was modified to incorporate the enhanced equation in this study. The DO deficit equation in the new model (DOSAGIT) is solved sequentially for a series of ten river segments. Segmentation is selected considering the locations of major tributaries, changes in river morphometry, effluent inputs and water intakes.

The DOSAGIT model solves the DO deficit equation iteratively using the table of recessed flows produced by FLOWREC and the BOD loadings from both mills for all segments. It produces tables of waste load allocation for 30 flow classes, with temperature ranging from 15 to 32 C° at increments of 0.5 C°, and prevailing flows ranging from 900 to 50,000 cfs. Simulations were performed assuming that the "other" mill is always discharging at its allowable waste load, and that a minimum DO criteria is maintained throughout the river. Allowable waste loads are computed using DOSAGIT with the latest flow and water quality data on the river and the permitted effluent rates.

CONCLUSIONS

The water quality management alternatives enumerated above, were analyzed for cost effectiveness toward the selection of the most feasible method. Analysis of alternative management procedures has shown that oxygen injection is the preferred alternative which meet the NPDES requirements and improve/maintain the assimilative capacity of the river system. Modeling results using the water quality modeling package showed that the flow-variable discharge, in association with oxygen injection on occasions when instream DO levels require addition of oxygen, is practical as well as protective of the stream well being. Finally, in considering the flow-variable concept, it should be noted that there are several factors and assumptions, that collectively provide a margin of safety and protection for the DO resources of the river. Some of the factors are:

- The retention times used are longer than necessary; they are based on the retention time from the upstream boundary location above the mills rather than the retention time from each mill to the sag point. This assumption further reduces the recessed flow used to establish allowable BOD loads.

- For each mill, sufficient oxygen will be added to fully offset each pound of excess ultimate BOD. This provides more oxygen than necessary because no credit is taken for natural reaeration, and not all of the discharged BOD5 is exerted upstream from the sag point.

- The concept of using flow recession automatically assumes that river flow is always strongly decreasing. In fact, there are many days when river flow is either increasing or decreasing at a lesser rate than that computed from the flow recession data. This alone, tends to provide a significant margin of safety in computing the daily allowable BOD5 discharge.

- As a matter of course, the mills will tend to operate conservatively since they will not know today's BOD5 until five days from today. This means that the mills will actually operate at some fraction of the allowable BOD load and add more oxygen than necessary to avoid permit violations.

REFERENCES

Gersberg, R. M., et al; Integrated Wastewater Treatment Using Artificial Wetlands: A Gravel Marsh Case Study, in "Constructed Wetlands for Wastewater Treatment: Municipal, Industrial, and Agricultural", Hammer, D. A., editor, Lewis Publishers, Inc. Chelsea, Michigan, 1989.

Hosseinipour, E. Z., Heefner, S.C., Neal, L. N., and Olson, R. W., (1995) Water Quality Modeling in Support of NPDES Permit Applications in Southeastern USA, Georgia Water Pollution Control Association, Proceedings of Georgia Industrial Pollution Control Conference, Atlanta, Georgia.

Thut, R. N., Utilization of Artificial Marshes for Treatment of Pulp Mill Effluent, in "Constructed Wetlands for Wastewater Treatment: Municipal, Industrial, and Agricultural", Hammer, D. A., editor, Lewis Publishers, Inc. Chelsea, Michigan, 1989.

Watson, J. T., et al; Performance Expectations and Loading Rates for Constructed Wetlands, in "Constructed Wetlands for Wastewater Treatment: Municipal, Industrial, and Agricultural", Hammer, D. A., editor, Lewis Publishers, Inc. Chelsea, Michigan, 1989.

Use of Temperature Control Curtains to Modify Reservoir Release Temperatures

Tracy B. Vermeyen[1], A.M. ASCE

Abstract

Reclamation (U.S. Bureau of Reclamation) has constructed four temperature control curtains to reduce release water temperature at structures in the Sacramento and Trinity River drainages in northern California. Curtains can provide selective withdrawal at intake structures, control topography induced mixing, and control interfacial shear mixing associated with plunging density currents entering reservoirs. Comprehensive field monitoring has been conducted to measure curtain performance characteristics. Monitoring included extensive temperature profiling, and velocity profiling using an ADCP (acoustic Doppler current profiler). This paper presents and summarizes performance data collected near curtains in Lewiston and Whiskeytown Reservoirs.

Background

Increased water temperatures, especially during drought years, are one of many threats to endangered salmon species in the Sacramento River. As a result, much effort has been expended to lower river temperatures. A value engineering study (Reclamation, 1990) identified temperature control curtains as a potential alternative for reducing temperature of water released from Trinity Dam and routed through Lewiston and Whiskeytown Reservoirs into the Sacramento River. Fisheries biologists expect cooler releases will enhance late summer and fall salmon spawning and rearing conditions. As a result, Reclamation engineers developed lightweight and flexible curtain structures fabricated from nylon-reinforced rubber for installation in Lewiston and Whiskeytown Reservoirs.

Central Valley Project - Shasta and Trinity Division

Water from the Trinity River Basin is diverted to the Sacramento River Basin through two tunnels and three reservoirs. Trinity River water is diverted from Lewiston Reservoir through Clear Creek Tunnel to the Carr Powerplant and into Whiskeytown

---

[1]Hydraulic Engineer, Bureau of Reclamation, PO Box 25007, Denver, CO 80225.

Reservoir. From there, water flows through the reservoir and into the Spring Creek Tunnel and through Spring Creek Powerplant. Spring Creek Powerplant releases water into Keswick Reservoir where it combines with water released from Shasta Dam. Water released from Keswick Dam enters into the Sacramento River. Over the course of this diversion and prior to curtain installation it was not uncommon for water temperatures to rise 5-7 °C.

Applications

Four temperature control curtains, two in Lewiston Reservoir and two in Whiskeytown Reservoir, were installed in 1992 and 1993, respectively. At Lewiston, a 250-meter-long, 11-meter-deep curtain is suspended from flotation tanks and is secured in place by a cable and anchor system. This curtain was designed to block warm, surface water from entering the Clear Creek Tunnel intake. As a result, cold water from the bottom of the reservoir is diverted into Whiskeytown Reservoir. In addition to the reservoir curtain, a second curtain funded by the California Department of Fish and Game was installed surrounding the Lewiston Fish Hatchery intake structure. The fish hatchery desired both warmer and cooler water depending on the season and fish rearing requirements. Therefore, a curtain (90-meter-long and 14-meter-deep) was designed which could skim warmer water or underdraw cooler water depending on whether the curtain was in a sunken or floating position.

Ideally, cold water diverted from Lewiston is to be routed through the Whiskeytown's hypolimnion (deep, cold water layer) to the Spring Creek Conduit intake. To optimize this diversion, two curtains were installed in Whiskeytown Reservoir. The tailrace curtain (180-meter-long and 12-meter-deep) was installed to force cold water from Carr Powerplant into Whiskeytown's hypolimnion while limiting mixing with the epilimnion (warm surface water). This curtain restrains the epilimnion from moving upstream toward Carr powerplant. With the tailrace curtain in-place mixing is reduced where the cold density current plunges into the hypolimnion. The second curtain, a 730-meter-long, 30-meter-deep, surface-suspended curtain surrounds the Spring Creek Tunnel intake. This curtain, like the Lewiston curtain, was designed to retain warm surface water while allowing only cold water withdrawal. A more detailed description of these four curtains is presented by Johnson, et al., 1993.

Temperature Monitoring - An extensive temperature monitoring program has been conducted for several years. Monitoring was required to assure compliance with maximum allowable Sacramento River temperatures as specified by several regulatory agencies. Additional temperature profiling stations were installed near the curtains to gather data necessary to evaluate their performance.

Velocity Profiles - Velocity profiles were measured upstream and downstream from the curtains to evaluate the selective withdrawal and hydrodynamic performance. In the future, velocity profile data will be used to incorporate curtains into reservoir operation models. Hourly velocity profiles were collected using two upward looking, narrow-band ADCP's placed on the reservoir bottom. ADCP data were collected upstream of the Lewiston curtain and downstream of the Carr tailrace curtain. ADCP data were collected in a cooperative effort with USGS's Sacramento District Office.

Lewiston Reservoir Curtain Performance - Figure 1 presents hourly operations and outflow temperature data collected in the Clear Creek Tunnel intake. These data demonstrate the curtain's effectiveness in reducing the water temperature entering the intake. For similar operational conditions the average temperatures released through Carr Powerplant were reduced by about 1.4 °C after the curtain installation. 1.4 °C may appear to be a small improvement, but given the weak temperature stratification in Lewiston Reservoir 1.4 °C is a significant improvement.

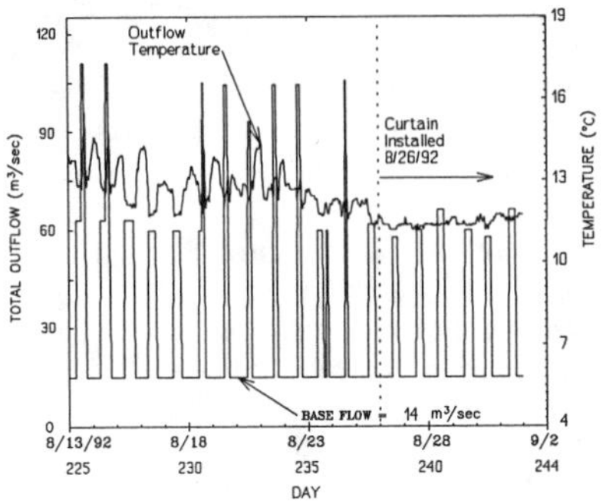

Figure 1. - Lewiston Lake outflows and hourly temperature data collected in the Clear Creek Tunnel intake structure during August 1992. Peaks in outflow represent Carr Powerplant operation.

Figure 2 illustrates the curtain's performance for three types of power operations at Trinity Powerplant in August 1994: 1) during days 220 through 227 the flows through the reservoir are baseload power operations at 90 m³/s; 2) days 228 through 242 had peaking power operations with one turbine on continuously at 50 m³/s and peaking was performed with the second turbine operating for 10 to 12 hours for a total flow of 100 m³/s; 3) days 243 through 260 had peaking operations between no-flow and peaking with one and occasionally two turbines operating for 10 to 12 hours. A comparison of reservoir inflow and outflow temperatures indicates that there is consistent 1.9 °C temperature gain through the reservoir for days 220 through 242 regardless of the two types of operations. However, when operations were changed on day 242 a steady increase in outflow temperature was observed. After day 252 the temperature gain through Lewiston Reservoir had stabilized at 3.6 °C. This additional 1.7 °C temperature gain occurs because warm water accumulates upstream and downstream from the curtain during no-

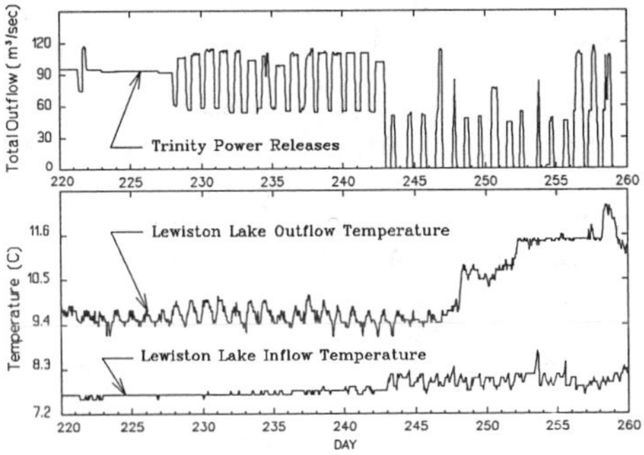

Figure 2 - Lewiston Reservoir inflows along with inflow and outflow temperatures illustrate the temperature gain of water diverted through the reservoir during mid-August 1994.

flow periods, and warm water is released during peaking operations. In addition, about 0.3 °C temperature gain can be attributed to increasing inflow temperatures. As a result of this significant temperature gain, we concluded that strictly peaking operation (both turbines on or off) should be avoided during periods when release temperature restrictions are in effect. Lastly, based on observations of a similar 1992 data set (fig. 1) it is reasonable to conclude that an additional 1.4 °C increase in release temperatures would occur if the curtain was removed.

Whiskeytown Reservoir Curtain Performance - Figure 3 illustrates how the tailrace curtain modifies the reservoir stratification. The temperature profiles collected upstream from the curtain (fig. 3a) indicate a weak thermocline, a relatively shallow epilimnion, fluctuations of underflow water temperatures, and diurnal fluctuations in epilimnion and thermocline temperatures. The isotherm plots of downstream temperatures (fig. 3b) indicate a stronger thermocline, a thicker epilimnion, reduced diurnal fluctuations, and slight warming of water passing under the curtain. The warming is probably generated by interfacial shear mixing. A review of pre-curtain temperature data identified 3.6 °C of warming in this same reach. Likewise, an analysis of temperatures measured between Carr Powerplant and the tailrace curtain indicated the inflow water warms 2.5 °C because of mixing between the powerplant and the tailrace curtain. Water leaking through a boat passage in the curtain has been identified as a large source of warm water. Alternative methods of boat passage which minimize leakage are currently being studied.

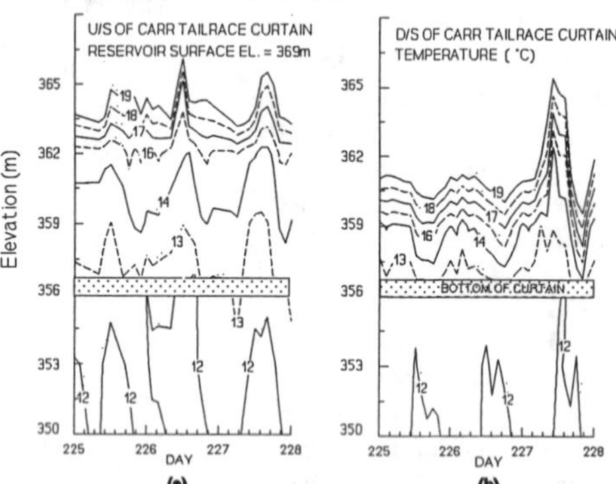

Figure 3 - Isotherm plots for locations (a) upstream and (b) downstream of the Carr tailrace curtain. These plots illustrate the difference in reservoir stratification across the curtain.

Figure 4 presents some typical reservoir operations (fig. 4a) and ADCP data (fig. 4b). Figure 4 illustrates how the velocity profiles downstream from the tailrace curtain change with peaking power operations. It is interesting how the velocities decrease to nearly zero at the curtain bottom when flows are reduced to 50 m^3/s. This indicates that the curtain no longer controls the underflow at this discharge, and flow passes the curtain as a density current. However, at flows of 90 m^3/s, curtain control is established and underflow velocities as high as 100 mm/sec extend into the thermocline (El. 359) where limited mixing with warmer water occurs. These flow conditions were confirmed by field observations that the curtain was heavily loaded during midday (high flows) and was slack during the early morning (low flows).

A comparison of pre- and post-curtain temperature profile data for Whiskeytown Reservoir for similar August power operations in 1988 and 1994 indicates that with the two temperature control curtains installed the epilimnion is warmer and the hypolimnion is slightly cooler. In addition, the average temperature gain of water routed through Whiskeytown Reservoir in August 1988 was 3.3 °C, while in August 1994 gains were 2.1 °C. These results are as expected because less warm water is being released from the reservoir.

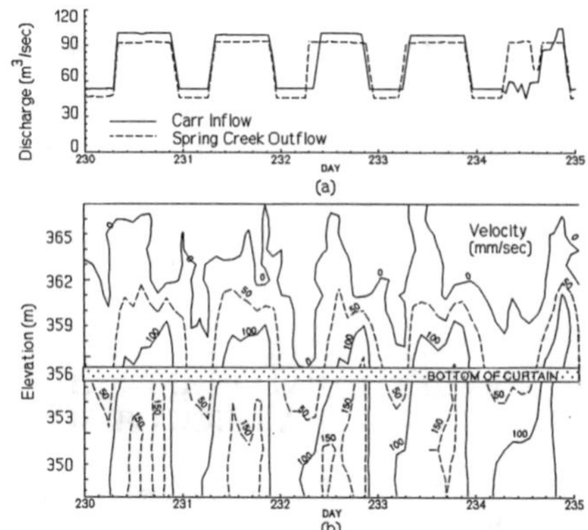

Figure 4 - (a) Whiskeytown Reservoir operations - baseload for one turbine and peaking the second turbine. (b) ADCP isovel data collected below the Carr tailrace curtain. Both plots are for the same 5 days in August 1994.

Conclusions

- Temperature control curtains allow project operators to manage power generation releases while controlling release water temperatures. Curtains have substantial economic advantages when compared to traditional selective withdrawal structures.

- In Lewiston Reservoir, peaking power operations result in a 1.6 °C temperature gain in water routed through the reservoir. As a result, two-unit peaking operations should be avoided for Carr Powerplant during periods when release temperature restrictions are in effect.

- For similar power operations, average temperature gain of water routed through Whiskeytown Reservoir in August 1988 (pre-curtain) was 1.2 °C higher than the curtain-controlled temperature gains measured in August 1994.

- ADCP data were useful in determining how powerplant operations affect temperature control curtain performance.

References

Johnson, P., T. Vermeyen, and G. O'Haver, *Use of Flexible Curtains to Control Reservoir Release Temperatures: Physical Model and Field Prototype Observations*, Proceedings of the 1993 USCOLD Annual Meeting, Chattanooga, Tennessee, May 1993.

U.S. Bureau of Reclamation, 1990. *Whiskeytown Temperature Control Value Engineering Study*. Memorandum to Regional Director, Mid-Pacific Region.

WAVE ATTENUATION AND NATURAL DISPERSION IN A THICK OIL SLICK

R.G. Jessup[1] and S. Venkatesh[1] (Member, ASCE)

ABSTRACT

In this paper, the process of wave attenuation by a thick layer of oil on a water surface is examined from a theoretical perspective. An expression is obtained for the viscous dissipation power associated with the wave induced deformation of the oil slick. Although it is not assumed that the oil layer is thin, the thickness is assumed to be small in relation to the wave length of the existing waves. The resulting formulation for the dissipation power is then used to set up a differential equation for the wave amplitude as a function of propagation distance and other relevant parameters such as oil viscosity, film thickness, water depth, wave number and initial amplitude. Despite the fact that the equation is highly nonlinear, it is nevertheless possible to obtain a useful first order solution by making only a few realistic assumptions. A discussion is given of the reduction in wave breaking as a result of wave damping and a consequent reduction in the dispersion of the surface oil into the water column.

1. INTRODUCTION

It is an observed fact that surface gravity waves are attenuated when the water surface through which they pass is covered by a layer of oil. When oil is spilt on a water surface, it eventually spreads to a thin film. Hence almost all of the previous work in understanding the wave attenuation by the oil have dealt with the presence of a very thin oil layer (e.g., see Lamb (1932), Barger et al. (1970)). The additional viscous damping, called the "Marangoni effect", introduced by the strong velocity gradient between the oil and the adjacent liquid layer has been described by Hühnerfuss et al. (1987) and others. Akin to damping in oil, viscous damping of waves propagating through an ice pack has been discussed by Liu and Mollo-Christensen (1988). The damping characteristics of oil on water, particularly the damping of the high frequency capillary waves, have been utilized to detect oil spills

[1] Environment Canada 4905 Dufferin Street, Downsview, Ontario M3H 5T4, Canada

through remote sensing (see Singh et al. (1986), Bern et al. (1993)). Depending on the volume of oil spilled, the time taken for the oil to spread into a thin film can vary from a few minutes to several hours. this time is further extended if the spilling of the oil itself takes place over an extended period of time. Thus there is a time window when the oil is relatively thick. If the reduction in wave amplitude produced by an oil layer of a given thickness can be predicted, then it becomes possible to estimate the time window available for surface slick containment/countermeasures before wave breaking and hence emulsification becomes significant.

2. THE VISCOUS DISSIPATION POWER

When an oil slick is forced to deform by the underlying water waves, its internal energy will increase due to the various dissipative processes associated with its viscosity. Because of these irreversible processes there will be a corresponding decline in the total wave energy, which in turn will be manifested by a gradual attenuation in the amplitude. In this section the focus will be on the viscous dissipation power, D , associated with the wave induced deformation of the oil cover, and on getting an analytical expression for it.

If u denotes the internal energy of the slick per unit mass and if r denotes its areal mass density, then the first law of thermodynamics for the slick may be written in the form

$$\rho\frac{du}{dt} = E^{ij}D_{ij} - \nabla\cdot \mathrm{q} + \rho r \tag{2.1}$$

where E^{ij} and D_{ij} are the 2-dimensional stress and rate of deformation tensors respectively, and where $\mathbf{q}$ and r describe respectively the heat flux and the internal heat sources. The quantity $E^{ij}D_{ij}$ is called the stress power and gives the rate at which the internal energy increases per unit area due to deformation of the 2-dimensional material element.

Assuming the oil to be a Newtonian fluid, the stress tensor for the slick may be written in the form

$$E^{ij} = -p g^{ij} + \tau^{ij} \tag{2.2}$$

where p is the pressure in the slick, g^{ij} is the metric of the curved 2-dimensional oil continuum, and τ^{ij} is the viscous stress tensor. This last tensor, τ^{ij}, in turn, may be expressed as

$$\tau^{ij} = \lambda^{*}g^{ij}\phi + 2\mu^{*}D^{ij} \tag{2.3}$$

where λ^{*} and μ^{*} are the viscosity coefficients of the film, and where ϕ is the contraction of the rate of deformation tensor, i.e.

$$\phi = D_k^k = g^{pk} D_{pk} \tag{2.4}$$

From equations (2.2)-(2.4), it can be shown that $E^{ij}D_{ij}$ has the form

$$E^{ij} D_{ij} = D - p\phi \tag{2.5}$$

where

$$D = \lambda^* \phi^2 + 2\mu^* D^{ij} D_{ij} \tag{2.6}$$

Equation (2.5) represents the decomposition of the stress power into its irreversible and reversible components. The quantity *D*, being always positive, is the part that describes the irreversible increase in the material element's internal energy, and is therefore referred to as the viscous dissipative power. The last term of (2.5), on the other hand, may be positive or negative, and therefore corresponds to the reversible component of $E^{ij}D_{ij}$.

In order to derive an expression for *D* it is necessary to make certain assumptions concerning the relative motion of the oil and water at the interface. In this paper it shall be assumed that there is no slippage between the slick and the underlying water surface, i.e., the two media move together with zero velocity differential. If this assumption is made, then the 3-dimensional velocity vector, **w**, of the oil slick will be identical to that of the water at the interface.

In Jessup (1995) it is shown that

$$D_{pq} = \frac{1}{2}\frac{d}{dt} g_{pq} - u K_{pq} \tag{2.7}$$

where K_{pq} is the extrinsic curvature tensor of the 2-dimensional oil slick continuum. It follows, therefore, that

$$\phi = \frac{1}{2} g^{pq} \frac{d}{dt} g_{pq} - u K_{pq} g^{pq} \tag{2.8}$$

When this last equation, along with (2.7) is substituted into (2.6) an expression is obtained for the viscous dissipation power *D*. This rather complicated equation, however, can be greatly simplified by taking note of the following fact:

$$D_{pq} D^{pq} = \phi^2 - 2\Theta_2 \tag{2.9}$$

where Θ_2 is the second invariant of D_{pq}. This invariant will be zero if the wave motion is assumed to produce no deformation along the line perpendicular to the propagation vector. This assumption will be valid for simple sinusoidal waves and gives the following simplified expression for *D*:

$$D = (\lambda^* + 2\mu^*)\phi^2 \tag{2.10}$$

It is this latter form of the viscous dissipation power that will be used in computing the attenuation in wave amplitude.

3. FORMULATION FOR SIMPLE SINUSOIDAL WAVES

In order to proceed further with the computations, it is necessary to have a mathematical transformation which defines the surface geometry of the water waves. It shall be assumed in this paper that the transformation is a simple sinusoidal vector function of the horizontal cartesian coordinates, X_p (p=1,2), the wave numbers, k_p (p=1,2), and the amplitude A. If $\mathbf{e}_p$ (p=1,2,3) denote the orthonormal Cartesian basis vectors, where $\mathbf{e}_3$ is in the vertical, then this vector function has the form

$$\mathbf{R}(X,t) = \mathbf{X} + A\sin(\mathrm{k}\cdot\mathrm{X} - \omega t)\mathbf{e}_3 \tag{3.1}$$

where

$$\mathbf{X} = X_1\mathbf{e}_1 + X_2\mathbf{e}_2 \tag{3.2}$$

and

$$\mathbf{k} = k_1\mathbf{e}_1 + k_2\mathbf{e}_2 \tag{3.3}$$

and where ω denotes the angular frequency. Note that in this formulation, X_p also plays the role of a surface material coordinate. It should also be noted that the horizontal damping distance is assumed to be large with respect to the wavelength.

In Jessup (1995) it is shown that the wave geometry defined above leads to the following expression for ϕ

$$\phi = F^2[1 + F^2]A^2\omega k^2\cos(f)\sin(f) \tag{3.4}$$

where

$$f = \mathbf{k}\cdot\mathbf{X} - \omega t$$

and where

$$F = \frac{1}{\sqrt{1 + A^2k^2\cos^2(f)}} \tag{3.5}$$

Substitution of (3.4) into (2.10) finally gives the expression for the viscous dissipation power. It should be noted, however, that this expression will give the rate of energy dissipation per unit of area tangent to the surface of the slick. Conversion of D to horizontal areal units is easily accomplished, however, and as shown in Jessup (1995),

$$D^* = (\lambda^* + 2\mu^*)F^3\{(1 + F^2)A^2k^2\omega\cos(f)\sin(f)\}^2 \tag{3.6}$$

where D^* denotes the rate of energy dissipation per unit horizontal area. It is this last form of the viscous dissipation power that will be used in the next section to derive the attenuation factor for the wave amplitude.

4. WAVE AMPLITUDE ATTENUATION

Using the final result of the preceeding section, it is now possible to derive a differential equation describing the manner in which an oil slick dampens the surface waves impinging upon it.

Consider a simple sinusoidal surface wave of amplitude A(x) propagating in the direction **p**, and let x and y be rectangular horizontal coordinates measured parallel and perpendicular, respectively, to **p**. Also, consider a water column of rectangular cross section dxdy aligned so that two of its vertical faces are perpendicular to **p** and horizontally located at x**p** and (x+dx)**p**. Let P(x,t)dy denote the total wave energy per unit time entering the column through the first of these faces and let P'(x) denote the time average of P(x,t). Then, if $D^{*\prime}$(A) denotes the time average of D^*, it follows that

$$P'(x+dx)dy = P'(x)dy - D^{*\prime}(A)dxdy \tag{4.1}$$

Dividing this last equation by dxdy yields the result

$$\frac{dP'(x)}{dx} = -D^{*\prime}(A) \tag{4.2}$$

However, according to Kandekar(1989)

$$P'(x) = \frac{1}{2}\rho g v A^2(x) \tag{4.3}$$

where ρ is the mass density of the water, v is the wave velocity, and g is the acceleration of gravity. Differentiating this last equation with respect to x and substituting the result into (4.2) gives the following differential equation for A

$$A\frac{dA}{dx} + \frac{1}{\rho g v}D^{*\prime}(A) = 0 \tag{4.4}$$

The solution of this ordinary differential equation will give the wave amplitude, A, as a function of the independent variable, x, and the initial amplitude A_0. Before the equation can be solved, however it is necessary to obtain the expression for $D^{*\prime}$. The method that shall be employed here involves expanding D^* in a trignometric power series and then computing the time average term by term. The expansion is straightforward and gives the result

$$D^* = (\lambda^* + 2\mu^*)\omega^2 \sum_{n=0}^{\infty} C_n [A^2 k^2]^{n+2} \cos^{2n+2}(f)\sin^2(f) \tag{4.5}$$

where C_n and B_n^M are quantities depending only on the values of their scripts. The time average of D^* is now obtained by integrating (4.5) over one wave period, T, and dividing the result by T. This gives

$$D^{*\prime} = (\lambda^* + 2\mu^*)\omega^2 \sum_{n=0}^{\infty} C_n S_n [A^2 k^2]^{n+2} \tag{4.6}$$

where

$$S_n = \frac{1}{2^{n+2}} \frac{(2n+1)!!}{(n+2)!} \tag{4.7}$$

These last two equations provide the required expansion for the mean viscous dissipation power. It should be noted that the terms in the series alternate in sign, and so the error produced by truncation may be readily obtained from the first term to be omitted. It can also be shown that the series converges for all situations in which the product of the amplitude and the angular wave number is less than unity.

For sufficiently small values of Ak, the series expansion of $D^{*\prime}$ may be approximated by its lowest order term. This gives

$$D^{*\prime} \cong (\lambda^* + 2\mu^*)\omega^2 C_0 S_0 A^4 k^4 \tag{4.8}$$

When this is inserted into (4.4), the following equation results

$$A\frac{dA}{dx} + \beta A^4 = 0 \tag{4.9}$$

where

$$\beta = \frac{1}{\rho g v}(\lambda^* + 2\mu^*)\omega^2 C_0 S_0 k^4 \tag{4.10}$$

However, the 2-dimensional viscosity coefficients λ^* and μ^* are related to the corresponding 3-dimensional quantities as follows:

$$\mu^* = \mu h\,, \qquad \lambda^* = \lambda h \tag{4.11}$$

where h is the thickness of the oil slick. Also,

$$\omega = kv = kQ(k) \tag{4.12}$$

where Q(k) defines the velocity as a function of the angular wave number, k. It follows therefore that

$$\beta = \frac{C_0 S_0}{\rho g}(\lambda + 2\mu) h k^6 Q(k) \tag{4.13}$$

Integration of (4.9) is straightforward and gives

$$A(x) = \sqrt{\frac{A_0^2}{2A_0^2\beta x + 1}} \tag{4.14}$$

where A_0 denotes the wave amplitude at the edge of the slick, where x is assumed to be zero. Substitution of (4.13) into (4.14) gives

$$A(x) = A_0\sqrt{\frac{\rho g}{2C_0S_0A_0^2(\lambda + 2\mu)hk^6Q(k)x + \rho g}} \tag{4.15}$$

From this last equation, it is immediately apparent that the attenuation will become greater if either the viscosity of the oil or the thickness of the slick is increased, or if the propagation distance, x, into the oil covered region becomes larger. These points are all in accordance with expectation. The dependence of the damping factor on k and A_0, on the other hand, has reasons which are less obvious and deserves some comment. The explanation for this dependence can be seen from equation (4.8) which shows that the mean viscous dissipation power depends on both the amplitude and the wave number. This in turn follows from the fact that the fluctuation in the area of the slick will be strongly dependent on both of these parameters.

It should also be noted that if the angular wave number and attenuation can be directly measured or obtained by remote sensing, then the above equation can be inverted to give the thickness of the oil slick as follows;

$$h = \frac{\rho g}{2C_0S_0(\lambda + 2\mu)k^6Q(k)A_0^2x}\left\{\frac{1 - [A(x)/A_0]^2}{[A(x)/A_0]^2}\right\} \tag{4.16}$$

Except in those situations where the wave height is large, it is to be expected that the remote measurement of the attenuation factor will be difficult or impossible, and so it may be that the above equation has only limited application.

5. DISCUSSION

In this paper a series expansion was derived for the mean viscous dissipation power associated with the wave induced deformation of the oil cover. The lowest order term in this expansion was then used in an energy balance equation to obtain a theoretical expression for the amplitude attenuation as a function of propagation distance and other relevant parameters.

When a gravity wave overcomes the damping effect of a thin layer of oil on the water surface and breaks on top of it, the oil is dispersed into the water column. The oil is thus moved from the surface to some level below it. On the other hand, if the oil layer were to be sufficiently thick, it will not only cause significantly higher damping of the gravity waves but also resist the tearing forces of a breaking wave, thus reducing dispersion of the oil into the water column. Higher the viscosity of the oil, lower the thickness required to resist dispersion of the oil into the water column by breaking waves (Delvigne and Hulsen, 1994). The thicker the oil layer on the surface, the greater the fraction of oil that can be succesfully removed by mechanical means (e.g, using skimmers) from the surface thus minimizing the pollution. Since the primary process responsible for dispersion and emulsification of the oil depends ultimately on wave development, it follows that an understanding of the damping process has significant practical application.

REFERENCES

Barger, W.R., W.D. Garrett, E.L. Mollo-Christensen and K.W. Ruggles. 1970. Effects of an artificial sea slick upon the atmosphere and the ocean. J. App. Meteor., Vol. 9, pp396-400.

Bern, T-I, T.Wahl, T. Anderson and R. Olsen. 1993. Oil spill detection using satellite based SAR: Experience from a field experiment. Photogram. Eng. & Remote Sens., Vol. 59, pp423-428.

Delvigne, G.A.L. and L.J.M. Hulsen. 1994. Simplified laboratory measurement of oil dispersion coefficient - Application in computations of natural oil dispersion. Proc. Seventeenth Arctic and marine Oil Spill Program (AMOP) Technical Seminar. Vancouver, Canada, June 8-10, 1994. pp173-187.

Hühnerfuss, H., W. Walter, P.A. Lange, and W. Alpers. 1987. Attenuation of wind waves by monomolecular sea slicks and the Marangoni effect. J. Geophys. Res., Vol. 92, C4, pp3961-3963.

Jessup, R.G. 1995. Attenuation of surface gravity waves by an oil slick. Internal report, Data Assimilation and Satellite Meteorology Div., Meteorological Research Branch, Environment Canada, Downsview, Canada, M3H 5T4.

Khandekar, M.L. 1989. Operational analysis and prediction of ocean wind waves. Coastal and Estuarine Studies, No.33, Springer Verlag.

Lamb, H. 1932. Hydrodynamics. Cambridge University Press.

Liu, A.K. and E. Mollo-Christensen. 1988. Wave propagation in a solid ice pack. J. Phys. Ocean., Vol. 18, pp1702-1712.

Singh, K.P., A.L. Gray, R.K. Hawkins and R.A. O'Neil. 1986. The influence of surface oil on C- and Ku-band ocean backscatter. IEEE Trans. Geosci. and Remote Sens., Vol. GE-24, pp738-743.

Evaluation of Gatewell Flows for Fish Bypass at a Large Hydroelectric Plant

M Allen[1], RA Elder[2], D Hay,[3] AJ Odgaard[1], LJ Weber[1] and D Weitkamp[4]

Abstract

During the evaluation of a juvenile salmon bypass system at a large hydropower dam it became apparent that very little information was available on the flow characteristics within the gatewell slot. The present paper will describe the general features of the bypass system and the final perforated porosities necessary to achieve a balanced discharge through the vertical barrier. Also, implication to successful fish passage will be made.

Introduction

Several large hydroelectric installations rely on intake screens to divert downstream migrating juvenile salmon away from the turbine runners. Once the fish have been diverted away from the turbine by the intake screen they are usually directed into a gatewell (see figure 1), which they must exit to continue their downstream migration. The gatewell slot is provided with a vertical barrier screen which extends from the intake screen to the top of the gatewell. The barrier screen retains fish in the upstream portion of the gatewell and allows water to pass through to the downstream part and back to the intake. It has a 51% perforated plate or bar screen on its upstream face and perforated plates with varying porosity on its downstream face to provide a uniform through-flow of water. The fish exit from the gatewell is usually a submerged orifice or an overflow weir. Currently, only limited available hydraulic data on the flow within the gatewell exists. These data are usually one-dimensional velocities at the gatewell entrance and are not sufficient to describe the gatewell flow pattern throughout the entire gatewell, or in particular,

[1] Iowa Institute of Hydraulic Research and Department of Civil and Environmental Engineering, University of Iowa, Iowa City, IA 52242-1585.

[2] Consulting Hydraulic Engineer, Tiburon, CA 94920.

[3] Hay and Company, Vancover, B.C. Canada V5Y 1L5.

[4] Parametrix, Inc., Kirkland, WA 98033.

near the exit. Also to minimize injury to fish resulting from contact with the vertical barrier screen it is important that the flow through the screen is uniform over the gatewell height and width.

To assist in the evaluation of the overall gatewell flow patterns and vertical barrier screen normal velocities, a scale model has been built of the Rock Island dam gatewell. The model is built of plexiglass with a structural steel framework. This allows measurement of all three velocity components throughout the entire gatewell using laser doppler anemometery. The purpose of this paper is to present velocity measurements that will be used to balance the vertical barrier screen such that the screen will have uniform through-flow over its height and width. Also, qualitative description of the gatewell flow patterns will be presented with and without the balanced vertical barrier screen.

Model Description

The model was constructed at an undistorted geometric scale of 1:4.43 at the Iowa Institute of Hydraulic Research's Model Annex. A section of an intake at the Rock Island Dam is shown in figure 1, with the model boundaries represented by the dashed lines. To achieve the correct flow pattern at the gatewell entrance the model included the upper portion of the diversion screen. During initial testing of the entrance flow in the gatewell it was determined that the diversion screen could be fully blocked, and provide the correct entrance flow conditions, while increasing the range of discharges available for model operation.

An overall elevation of the model is shown in figure 2. Flow to the model is supplied from an underground sump pit. Water is drawn from the sump using a variable speed pump which supplies the flow through a 14" diameter pipe. A transition is used to uniformly expand the flow to the rectangular inlet dimensions of the model. At the downstream end of the expansion a 4" thick honeycomb has been installed to further develop uniform flow. The water is then directed upward into the gatewell slot, travels through the vertical barrier screen, down the gatewell slot and then exits the model. The flow leaving the model is returned to the sump via a 14" diameter pipe.

The walls of the model were made out of clear acrylic plastic. This was done to allow for velocity measurements using LDV and visualization of flow patterns using dye. The vertical barrier screen (VBS) was constructed using Hendricks B-9 bar screen on upstream face and stainless steel perforated plate on the downstream face separated by a structural steel framework. The VBS is constructed such that the porosity of the perforated plates can be changed between horizontal frame members, resulting in 5 panels that may be used to develop balanced flow. Figure 3 provides details of the vertical barrier screen panels.

Experimental Results and Discussion

A balanced vertical barrier screen is one that has a uniform normal velocity everywhere on the screen face. Balanced flow through the vertical barrier screen is important for fish survival because if the VBS is not balanced 'hot-spots' may occur on the screen that result in the fish contacting, or worse, becoming impinged on the screen.

The initial unbalanced vertical barrier screen began with the flow entering the gatewell uniformly across the 577.86 elevation. However, as the flow progressed up the gatewell, a dominant flow pattern developed with higher velocity flow near the outside walls and near the vertical barrier screen face. Upon reaching the water surface, this flow turned downward near the upstream guidewall toward the center of the gatewell slot and exited the gatewell through VBS panel 4. This was seen with both dye injected at the gatewell entrance and also in the initial set of velocity data.

To balance the vertical barrier screen, two components of velocity (vertical and normal to the screen) were measured at elevations 577.86', 584.50', 590.90', 597.67' and 604.43'. At each elevation the velocities were measured at six locations across the 15.33' length and ten locations across the 2.75' width of the gatewell. Thus, a complete set of velocity data consisted of 300 velocity measurements for a given set of perforated plate porosities. It was felt that such an extensive data set was necessary to fully describe the flow patterns associated with the complex decelerating flows present in the gatewell slot.

The final data set is summarized in table 1. The discharge through a screen panel is simply the difference in discharge between the two elevations immediately above and below the panel, whereas, the screen normal velocity is the panel discharge divided by the panel area. As can be seen from table 1 the normal velocity is between 0.50 and 0.57 fps for all panels. The screen velocities have been normalized with respect to the overall screen average and the variation is less than 10%, as shown in table 1. Also, upon achieving a balanced screen based on normal screen velocities the overall flow pattern in the gatewell was much more uniform at each elevation, eliminated the high velocities at the walls and the downward flow near the upstream guidewall at the water surface.

Table 1. Final Velocity Data

Panel Number	Measurement Elevation (ft)	Average Vertical Velocity (ft/s)	Average Vertical Discharge (cfs)	Panel Porosity (% open)	Discharge Through Panel (cfs)	Screen Normal Velocity (ft/s)	Normalized Screen Velocity (ft/s)
	577.86	6.81	287				
1				7.5	53	0.52	0.98
	584.50	5.54	234				
2				19.6	56	0.57	1.07
	590.90	4.22	178				
3				29.3	52	0.50	0.94
	597.67	3.00	126				
4				32.6	54	0.52	0.98
	604.43	1.70	72				
5				26.1	72	0.55	1.03
	613.00*	0.00	0				

*Water Surface Elevation

Conclusions

An extensive set of velocity data has been used to balance the flow through the vertical barrier screen in the Rock Island Dam gatewell. This balanced screen improves the overall flow patterns in the gatewell. Also, the balanced design reduces the likelihood of fish impingement resulting from areas with high normal components of velocity through the vertical barrier screen. The balanced screen should provide an improved fish passage evironment.

Acknowledgments

The authors would like to thank Mr. Bill Christman Public Utility District No. 1 of Chelan County for providing financial support for this project. Also, the model construction was completed by the shop staff of the Iowa Institute of Hydraulic Research, Jim Goss, shop manager.

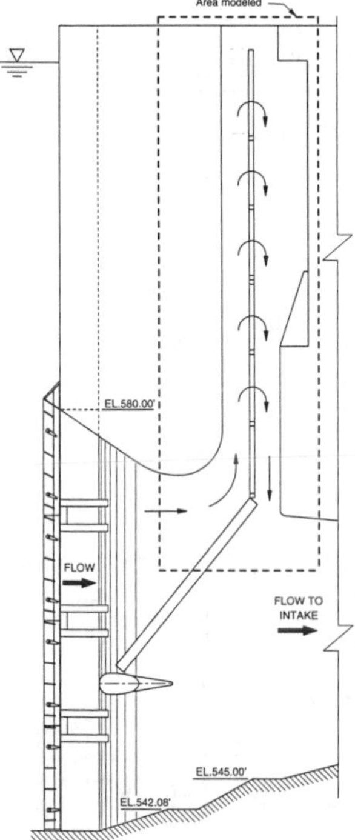

Figure 1. Section Through Rock Island Intake

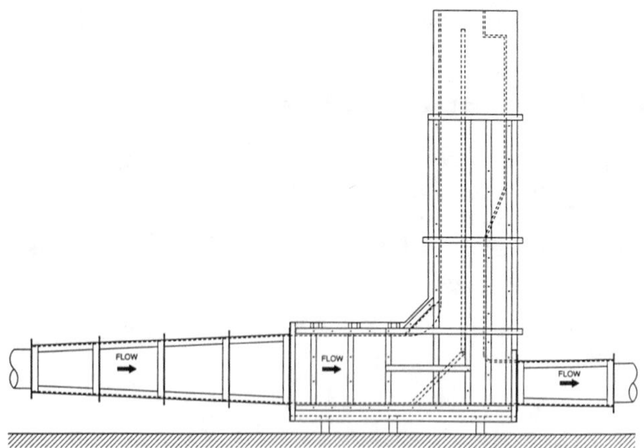

Figure 2. Elevation of Rock Island Gatewell Model

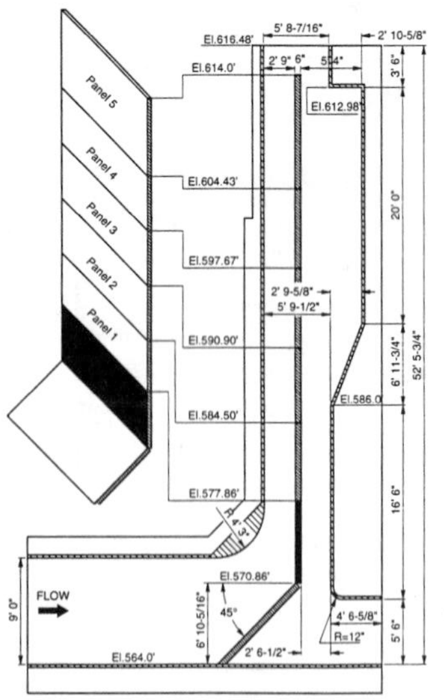

Figure 3. Vertical Barrier Screen Located in Gatewell Model

New Concepts for Bypassing Fish at Water Intakes

E. P. Taft[1], F. C. Winchell[2], T. C. Cook[3], S. V. Amaral[4], and R. A. Marks[5]

The passage of fish through hydraulic turbines, and the potential for resultant injury and mortality, has long been a concern in the hydroelectric industry. Since the 1970's, extensive research and development efforts have been ongoing to develop methods for preventing fish passage through turbines. This paper presents an overview of several new and emerging technologies that have the potential for widespread, cost-effective application: the high-velocity Eicher and Modular Inclined Screens; sonic fish deterrent systems; and strobe lights.

Eicher Screen. The Eicher Screen is a patented passive pressure screen designed for installation in penstocks. The first extensive evaluation of the screen was in a 2.7-m diameter penstock at the Elwha Hydroelectric Project near Port Angeles, Washington. This evaluation was funded by the Electric Power Research Institute (EPRI). All species and sizes were successfully diverted to a bypass pipe by the screen over the 1.2 to 2.4 msec^{-1} penstock velocity range. Results showed the following latent survival values for each species/lifestage: coho salmon smolts - 99.4 and 98.6% in 1990 and 1991, respectively; coho pre-smolts - 99.2%; steelhead smolts - 99.4%; chinook fingerling smolts - 98.6%; chinook pre-smolts - 98.7%; steelhead fry - 96.9%; coho fry - 91.0% (EPRI 1992a). The first full-scale Eicher screen installation (two screens in two, 10-ft diameter penstocks; total flow of 1,000 cfs) at B. C. Hydro's Puntledge Project showed similar results with chinook and coho salmon smolts: survival levels following diversion exceeded 99% at penstock velocities up to 6 ft/sec (Smith 1993).

---

[1] Vice President, Alden Research Laboratory, Inc., 30 Shrewsbury St., Holden, MA 01520

[2] Senior Fisheries Biologist, Alden Research Laboratory, Inc.

[3] Project Engineer, Alden Research Laboratory, Inc.

[4] Fisheries Biologist, Alden Research Laboratory, Inc.

[5] Project Engineer, Alden Research Laboratory, Inc.

The Eicher screen can be considered an available and effective technology for diverting salmonids in a penstock. However, based on hydraulic measurements made at Elwha, it was believed that further refinement of the hydraulic flow conditions along the screen would enhance the potential for general application at many sites. Therefore, EPRI conducted additional hydraulic model studies in 1992 at Alden Research Laboratory, Inc. (ARL) on the screen. These studies led to a revised screen and bypass geometry that creates improved hydraulic conditions. A complete description of the modified screen design and its hydraulic characteristics is presented in an EPRI report (1994a). These design changes can be considered for future applications.

Modular Inclined Screen. In 1995, EPRI was granted a patent on a new fish diversion concept known as the Modular Inclined Screen (MIS; U. S. Patent No. 5385428). The modular design is intended to provide flexibility for application at any type of water intake. Installation of multiple units at a specific site can provide fish protection at any flow rate (EPRI 1994b). The module consists of an entrance with trash racks, dewatering stop log slots, an inclined wedge-wire screen set at a shallow angle (15 degree) to the flow, and a bypass for directing fish to a transport pipe. The screen is mounted on a pivot shaft so that it can be cleaned by backflushing. The module is completely enclosed and is designed to operate at water velocities ranging from 0.61 to 3.05 ms^{-1}, depending on the species and life stages to be protected.

Hydraulic studies conducted in a 1:6.6 scale model at ARL demonstrated that the MIS design produces a uniform flow distribution over the entire screen surface without the need for graduating the porosity of the screen or baffling (Cook et al. 1993; EPRI 1994b). Biological evaluations were conducted in a test flume at ARL using rainbow trout, bluegill, walleye, channel catfish, mixed alosids (blueback herring and American shad), Atlantic salmon smolts and juvenile coho salmon, chinook salmon, brown trout and golden shiners. The average length of the fish groups tested ranged from 47 to 169 mm. Passage survival (diversion efficiency adjusted for 72-hour survival) for most test groups exceeded 99% at velocities of up to 2.44 ms^{-1}; for several species this survival rate was maintained at velocities of up to 3.05 ms^{-1}. Few injuries were noted, and delayed mortality after three days was comparable for test and control fish (Taft et al. 1993). Based on these excellent results, an MIS demonstration project is under construction at Niagara Mohawk's Green Island Hydro Project on the Hudson River for evaluation with blueback herring in the fall of 1995 (Sullivan et al. 1995).

Sonic Deterrent Systems. Various sound devices have been evaluated for their ability to repel fish at water intakes (EPRI 1994c). Although sound has been demonstrated to effectively and consistently elicit avoidance behavior from several fish species at a number of test sites, resource agencies have expressed a reluctance to accept this "experimental" technology pending further research and proof-of-concept. Two basic approaches are being used to identify sound signals that will repel fish.

One approach involves tuning a sound system to sound signals that fish produce.

American Electric Power (AEP) developed a patented sound "tuning" system used to identify sounds effective in repelling fish (Loeffelman et al. 1991a). The tuning method is based on the concept that fish use sounds for communication and assumes that a given species will be most sensitive to the types of sounds that it produces. Currently, the signal sound system is being evaluated to determine its ability to prevent outmigrating chinook salmon from entering the Georgiana Slough which diverts water from the Sacramento River for irrigation purposes. The deployment of the sound system consists of an 245-m long linear array of acoustic transducers suspended from buoys that are located beginning about 305 m upstream of the slough entrance. In 1993, barrier guidance efficiency was estimated to exceed 50% (Hanson Environmental, 1993).

Sound systems based on fish morphology and sounds which are expected to be biologically meaningful also have been successful in eliciting fish responses, particularly among species of the Genus *Alosa* (herrings). A high-frequency system has been evaluated as a means for reducing blueback herring entrainment during pumpback operation at the Richard B. Russell Pumped Storage Project located on the Savannah River between South Carolina and Georgia (Nestler et al. 1992). Blueback herring responded to high-frequency sound between 110 and 140 kHz at source levels greater than 190 dB//µPa. Similar studies were used to evaluate the use of a sound projection system to deter fish from the intake of the J. A. FitzPatrick Nuclear Power Plant (JAF) located on Lake Ontario (NYPA et al. 1991; Dunning et al. 1992). Test sounds were generated by an acoustic sound deterrent known as the FishStartle™ system developed and patented by Sonalysts, Inc. (U. S. Patent No. 4,922,468). The system ensonified the entire JAF intake at a minimum sound pressure level of 190 dB in a frequency band from 122 to 128 kHz. Results indicated that the sound system was 80 to 96% effective (Dennis Dunning, New York Power Authority, pers. comm.).

Alden Research Laboratory is presently conducting a large-scale sound deterrent study for Public Service Electric and Gas Company at the Salem Generating Station on Delaware Bay. The study involves cage tests followed by installation and evaluation at one of the plant's cooling water intakes. In cage tests conducted to date, weakfish, Atlantic croaker, bay anchovy and the three Alosid species (American shad, blueback herring and alewife) demonstrated strong avoidance responses to at least one of the 21, 1/2-octave frequency bands tested (between 100 Hz to 145 kHz), whereas spot, striped bass and white perch exhibited weaker responses. Avoidance responses were the most consistent for the clupeid species and Atlantic croaker.

Recently, there has been growing interest in the possible use of very low frequency infrasound to repel fish. This trend is based on the principle that, in the near field, fish response to sound is more related to particle motion than acoustic pressure and that such motion may elicit a response similar to that observed when a fish is confronted by an attacking predator. Several researchers have shown that particle motion at very low frequencies can effectively repel fish. Knudsen *et al.* (In press; 1992) demonstrated that "infrasound" at 10 Hz repelled Atlantic salmon smolts in Norway. The researchers constructed a simple piston and cylinder source to produce particle acceleration. In laboratory and field experiments, the smolts showed a repeatable and consistent avoidance of the source within the hydrodynamic near-field without habituation. Studies by NYPA *et al.* (1991) demonstrated a strong response by white perch to a 25 Hz single tone at 180 dB//μ Pa produced by a hydraulically-driven transducer.

The results of recent sonic fish deterrent system research are encouraging. Combined with ongoing efforts to develop a better understanding of the biological (evolutionary, morphological, behavioral) principles governing fish response to hydrodynamic and acoustic energy, researchers are in the process of developing innovative systems for creating low-frequency energy generators in a cost-effective manner.

Strobe Light Systems. Strobe lights have been shown to effectively repel selected fish species in laboratory and field experiments. The strongest avoidance response observed under field conditions at a hydroelectric facility was among juvenile American shad outmigrants at the York Haven Hydro Project on the Susquehanna River in Pennsylvania (EPRI 1992b). Studies from 1988 to 1991 have demonstrated that the shad avoided the underwater strobe lights strongly and repeatedly, and showed no acclimation to the light over long periods of time (hours). Further, at this site it was possible to periodically (once per hour) pulse the fish through an ice/trash sluiceway located adjacent to the downstream-most hydro unit and thereby prevent turbine passage. Netting results indicated that about 94 percent of the fish in the area of influence of the strobe lights passed through the sluicegate. Avoidance has also been demonstrated among Atlantic, chinook and coho salmon and steelhead trout in laboratory experiments at the University of Washington (EPRI 1994c).

Literature Cited

Cook, T. C., E. P. Taft, G. E. Hecker, and C. W. Sullivan. 1993. Hydraulics of a New Modular Fish Diversion Screen. Waterpower '93. p. 318-327.

Dunning , D. J., Q. E. Ross, P. G. Geoghegan, J. J. Reichle, J. K. Menezes, and J. K. Watson. 1992. Alewives avoid high-frequency sound. North American Journal of Fisheries Management 12:407-416.

Electric Power Research Institute. 1992a. Evaluation of the Eicher Screen at Elwha Dam: 1990 and 1991 Test Results. EPRI TR-101704, September 1992.

Electric Power Research Institute. 1992b. Evaluation of Strobe Lights for Fish Diversion at the York Haven Hydroelectric Project. EPRI TR-101703, Project 2694-1, November 1992.

Electric Power Research Institute. 1994a. Fish Protection/Passage Technologies Evaluated by EPRI and Guidelines for Their Application. EPRI TR-104120, May

1994.

Electric Power Research Institute. 1994b. Biological Evaluation of a Modular Inclined Screen for Diverting Fish at Water Intakes. EPRI TR-104121, May 1994.

Electric Power Research Institute. 1994c. Research Update on Fish Protection Technologies for Water Intakes. EPRI Report Number TR-104122, May 1994.

Hanson Environmental, Inc. 1993. Demonstration Project to Evaluate the Effectiveness of an Acoustic (Underwater Sound) Behavioral Barrier in Guiding Juvenile Chinook Salmon at Georgiana Slough: Results of 1993 Phase 1 Field Tests. December 1993.

Knudsen, F. R., P. S. Enger, and O. Sand. In press. Avoidance Responses To Low Frequency Sound In Downstream Migrating Atlantic Salmon Smolt, *Salmo salar* L. Submitted to the Journal of Fish Biology.

Knudsen, F. R., P. S. Enger, and O. Sand. 1992. Awareness Reactions and Avoidance Responses to Sound in Juvenile Atlantic Salmon, *Salmo salar* L. Journal of Fish Biology 40:523-534.

Loeffelman, P. H., D. A. Klinect, and J. H. Van Hassel. 1991a. Fish protection at water intakes using a new signal development process and sound system. WaterPower '91, pp. 355-365.

Nestler, J. M., G. R. Ploskey, J. Pickens, J. Menezes, and C. Schilt. 1992. Responses of blueback herring to high-frequency sound and implications for reducing entrainment at hydropower dams. North American Journal of Fisheries Management 12:667-683.

New York Power Authority, Normandeau Associates, Inc., and Sonolysts, Inc. 1991. Acoustic Fish Deterrents. Prepared for the Empire State Electric Energy Research Corporation. Research Report EP 89-30. April 1991.

Smith, H. 1993. Puntledge Hydro Fish Screens: Eicher Screen. Presented at the 1993 American Fisheries Society Symposium on Fish Passage Responsibility and Technology, September 1-2, 1993, Porland Oregon.

Sullivan, C. W. , E. M. Paolini, A. W. Plizga, and T. C. Cook. 1995. Modular Inclined Screen Test Facility Engineering and Design. Proc. ASCE 1995 International Conf. Wat. Res. Eng. San Antonio, TX, Aug. 14-18, 1995.

Taft, E. P., S. A. Amaral, F. C. Winchell and C. W. Sullivan. 1993. Biological Evaluation of a New Modular Fish Diversion Screen. AFS Bioengineering Symposium, September 1993, Portland Oregon.

MODULAR INCLINED SCREEN TEST FACILITY ENGINEERING & DESIGN

Charles W. Sullivan[1], Edward M. Paolini[2], Anthony W. Plizga[3], and Thomas C. Cook[4]

ABSTRACT

To prevent fish mortality at water intakes, the Electric Power Research Institute (EPRI) has developed and is constructing a fish diversion screen known as the Modular Inclined Screen (MIS). Hydraulic model testing and successful biological evaluation of the MIS has been completed in the laboratory. Following discussions with various federal and state agencies, the Green Island Hydroelectric Project, owned and operated by Niagara Mohawk Power Corporation (NMPC), was selected as the field test location. This project is located on the Hudson River, just north of Albany, in Green Island, New York. The MIS test facility is scheduled to be operational and ready for testing in the fall of 1995.

This paper describes the engineering and design considerations addressed during the recently completed design phase of the MIS Project. The important parameters included the size of the facility to minimize cost and allow for over-the-road shipment, removal after testing, and a level and stable foundation with minimal leakage. Developing a prototype facility representative of a fish diversion system to a bypass at a hydroelectric project presented many engineering challenges overcome by the project team during the engineering and design phase.

---

[1] Program Manager, Electric Power Research Institute, 3412 Hillview Avenue, Palo Alto, CA 94303 - USA, (415) 855-8948.

[2] Project Manager, Niagara Mohawk Power Corporation, 300 Erie Boulevard West, Syracuse, NY 13202 - USA, (315) 428-6000.

[3] Project Engineer, Stone & Webster Engineering Corporation, 245 Summer Street, Boston, MA 02210 - USA, (617) 589-8624.

[4] Project Engineer, Alden Research Laboratory, Inc., 30 Shrewsbury Street, Holden, MA 01520 - USA, (508) 829-6000.

Following biological evaluation of the MIS test facility at the Green Island facility, the project team will prepare and publish a report summarizing the results of the field test program and recommendations for improvements to this fish diversion concept.

INTRODUCTION

In 1991, the Electric Power Research Institute (EPRI) developed a new fish diversion concept known as the Modular Inclined Screen (MIS) (U.S. Patent No. 5385428). The modular design is intended to prevent fish mortality while providing flexibility for application at any type of water intake. Installation of multiple units at a specific site can provide fish protection at any flow rate (Reference 1).

To determine the viability of the MIS as a fish protection system, a laboratory testing program was conducted to evaluate its hydraulic performance and biological effectiveness. Eleven fish species, both resident freshwater and anadromous, were chosen for the assessment. Based on its effectiveness in diverting fish in the laboratory, EPRI and Niagara Mohawk Power Corporation (NMPC) are collaborating on a field evaluation of the MIS at NMPC's Green Island Hydroelectric Project located on the Hudson River. A strobe light deterrent system will be evaluated in conjunction with the MIS tests. Biological field evaluations are scheduled for the fall of 1995.

DESCRIPTION OF MIS

The Modular Inclined Screen consists of a streamlined entrance with a trash rack, upstream and downstream isolation gates, a wedge wire screen set at a shallow angle to the flow, and a bypass for diverting fish to a transport pipe or holding facility. The screen is set on a shaft so that it may be rotated and cleaned via backflushing. The module is completely enclosed and is designed to operate at water velocities from 2 to 8 ft/sec (0.6 to 2.4 m/sec). Plan and section views of the MIS are shown on Figures 1 and 2.

The intake of the module is rectangular in shape, but includes curved transition areas at the entrance to minimize flow separation. The distance from the entrance to the screen was kept short to reduce construction costs. The screen is a plane of commercially available 50 percent porosity wedge wire set at an angle of 15 degrees. The screen support structure under the leading (upstream) edge of the screen is recessed below the floor to provide a smooth transition from the entrance to the screen. While field testing is in progress, the downstream end of the screen is held in place using a hoist to secure the screen against the bottom of the bypass structure. For cleaning, the downstream end of the screen can be lowered to the floor, allowing water to flow through the underside of the screen.

In addition to the MIS, eight barrier wall panels are located between the MIS and existing hydro structures to form the tailwater area and collection area required to conduct MIS hydraulic and biological testing activities.

ENGINEERING AND DESIGN CONSIDERATIONS

Following successful hydraulic model testing and biological evaluation of the MIS in the laboratory, the next step was design and construction of a prototype test facility at a hydroelectric project. The primary engineering considerations in selecting a site were: 1) to select a typical hydroelectric project with features that allowed the MIS test facility to be representative of a fish diversion bypass system and 2) to select a project configuration which allowed construction and operation of a MIS test facility with minimal impact on the hydroelectric plant operations. Sizing of the test facility for shipment of the components to the site, design of the components for removal after testing, placement and stabilization of the MIS foundation, and construction techniques which provided minimal environmental impacts on the site were major considerations addressed during the final design of the MIS test facility.

The Green Island Hydroelectric Project was selected as the MIS test site because the project has: 1) a bulkhead structure with an adjacent ice sluice gate located upstream of the plant forebay; 2) a fixed crest spillway located between the ice sluice gate and the main spillway; and, 3) an inflatable rubber dam on the main spillway. The MIS is located upstream of the ice sluice gate which controls flows through the MIS. During the months of September, October and November when the primary test species (blueback herring) typically migrate downstream, Hudson River flows average about 9,500 cfs (269 m^3/sec). The hydroelectric plant has a hydraulic capacity of 6,000 cfs (170 m^3/sec). The main spillway also has a capacity of 6,000 cfs (170 m^3/sec) with the rubber dam fully deflated and the water level at the elevation of the fixed crest spillway. Since the ice sluice gate and the MIS have a maximum flow of about 190 cfs (5.4 m^3/sec), the test facility will have negligible impact on the hydroelectric plant operation. Furthermore, the location of the MIS structure adjacent to the bulkhead structure will not impact flow patterns to the plant intake and will only reduce the length of the fixed crest spillway by approximately five percent.

The MIS was designed to maximize shop fabrication, permit transportation by truck over public highways, and facilitate removal from the site. The MIS and barrier wall panels were fabricated in sections for field assembly. The sub-assemblies were shipped to the Installation Contractor's facilities for assembly on a barge and delivery to the site. The test facility has steel superstructure framing located above the operating deck which was entirely erected at the site after the MIS was positioned and anchored. In addition to the screen, other equipment and components which would be incorporated in a permanent MIS facility were shop installed on the MIS. These components included the trash racks, isolation gates, and bypass sluice bottom drop gate.

In order to pass the ice sluice gate flow through the MIS screen, barrier wall panels were installed between the MIS structure and the fixed crest spillway, and between the MIS and the pier between the ice sluice gate and the bulkhead structure. Placement, leveling, stabilizing, and sealing of the MIS structure and the barrier wall panels was essential for satisfactory operation of the test facility. The foundations and vertical and horizontal joints between MIS and barrier wall panels had to be watertight since a four foot differential water level was expected between the headpond and the fish collection area at the end of the bypass for the maximum velocity test condition. The design was further complicated by the U.S. Army Corps of Engineers' requirement not to transfer any loads to the existing dam structures. These concerns were addressed using a foundation system initially designed with leveling plates and pins, grout bags, and underwater grouting. However, the final design was revised to reflect the construction technique proposed by the selected Installation Contractor which consisted of compacted crushed stone layers covered with an impermeable membrane similar to liners used in landfills to control seepage. A steel frame was used to level the top layer of crushed stone.

Dewatering of the MIS was necessary to inspect and maintain the screen during the test period. A total of 40 rock anchors were installed 20 to 30 ft (6.1 to 9.1 m) into bedrock to stabilize the MIS and barrier wall panels against sliding, overturning, and flotation. The anchors were drilled within a casing between the rock and the MIS deck to contain drilling material.

An important design consideration to be evaluated at the site is the clogging of the 2 mm (0.08 in) wide bars with 2 mm (0.08 in) spacing. The screen is designed to be lowered with an electric hoist for backflushing and cleaning. The screen shaft is located at one-third of the screen length from the upstream end to assure that the screen weight is adequate to completely lower the screen such that water can pass through the screen in the opposite direction. When in the closed position, the hoist holds the screen in place.

The primary objective of the test facility at Green Island is to obtain hydraulic and biological data on the screen operation in a prototype facility representing a diversion system to a bypass. Gated ports for insertion of a pitometer to measure velocities approaching the screen and along the screen centerline have been installed in the MIS structure. The flow passages are equipped with cameras and lights to observe the screen face and evaluate the screen effectiveness. Pressure transducers have been installed to monitor water levels in the head pond, in the fish collection area, and downstream of the MIS, to monitor the screen differential pressures, and to monitor the bypass flow rate.

GREEN ISLAND TESTING PROTOCOL

The test program planned for the Green Island site represents the first opportunity to evaluate the effectiveness of the MIS in diverting naturally-migrating fish under field conditions. The MIS test facility is scheduled to be operational during August 1995. The primary test species will be blueback herring which typically migrate during September, October or November. The effectiveness of strobe lights for directing blueback herring to the MIS will also be evaluated in conjunction with this testing program.

After the test facility installation has been completed, pre-operational tests will be conducted to verify that the facility has been constructed in accordance with the design requirements. Operational and hydraulic testing will be conducted to verify that all mechanical and electrical equipment functions properly over the entire range of parameters required during the biological evaluation effort.

Following the pre-operational testing, approximately three weeks of field testing is planned, including ten days for shakedown testing. Depending on the size and duration of the 1995 blueback herring outmigration, the planned three week test duration should allow five replicate tests to be performed at four pre-determined test velocity conditions. Since it is not possible to predict the duration and magnitude of the blueback herring outmigration, every effort will be made to maximize the number of tests performed when the fish are present in numbers suitable for testing. The strobe light array will be used to increase the numbers of fish entering the MIS for evaluation.

The blueback herring tests will quantify impingement, injury, and immediate and delayed mortality. Testing will focus on determining these parameters for fish which have entered the MIS of their own volition since the level of handling required to mark and test control groups and introduce the fish into the MIS would contribute to mortality for this relatively fragile herring.

FUTURE WORK

Following biological evaluation of the MIS test facility at the Green Island Hydroelectric Project, the project team will prepare and publish a report summarizing the results of the field test program. Although the primary intent of this field test is to evaluate the ability of the MIS to allow fish to migrate downstream, the report will also address other aspects of the project, such as design concepts, installation methods, and strobe light effectiveness. In addition to summarizing the results of this field test, the project team will provide recommendations for further improvements to this unique fish diversion concept. Copies of the Modular Inclined Screen Field Test Report are expected to be available from EPRI in early 1996.

ACKNOWLEDGEMENTS

The project team would like to acknowledge the contributions of Steel-Fab, Inc. of Fitchburg, Massachusetts, for their assistance with fabrication and equipment procurement and the efforts of Steel Style, Inc. of Newburgh, New York for their alternative foundation scheme and installation support.

REFERENCES

1. Electric Power Research Institute (EPRI). Biological Evaluation of a Modular Inclined Screen for Protecting Fish at Water Intakes. EPRI Report No. TR-104121. May 1994.

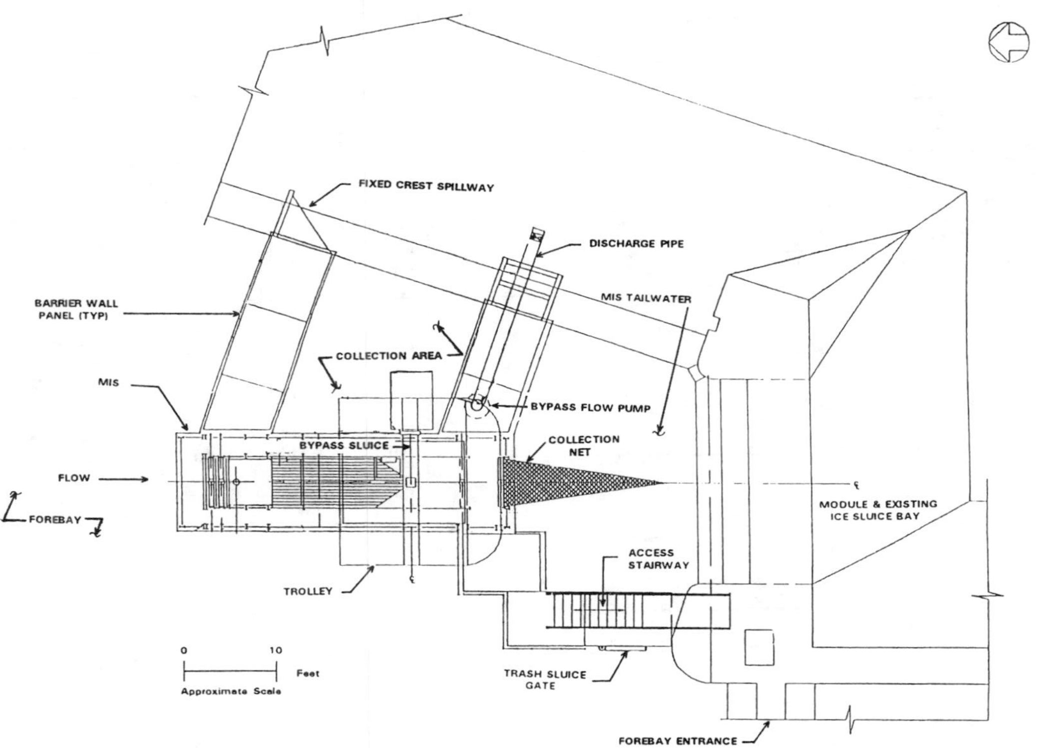

FIGURE 1 - MODULAR INCLINED SCREEN PLAN VIEW

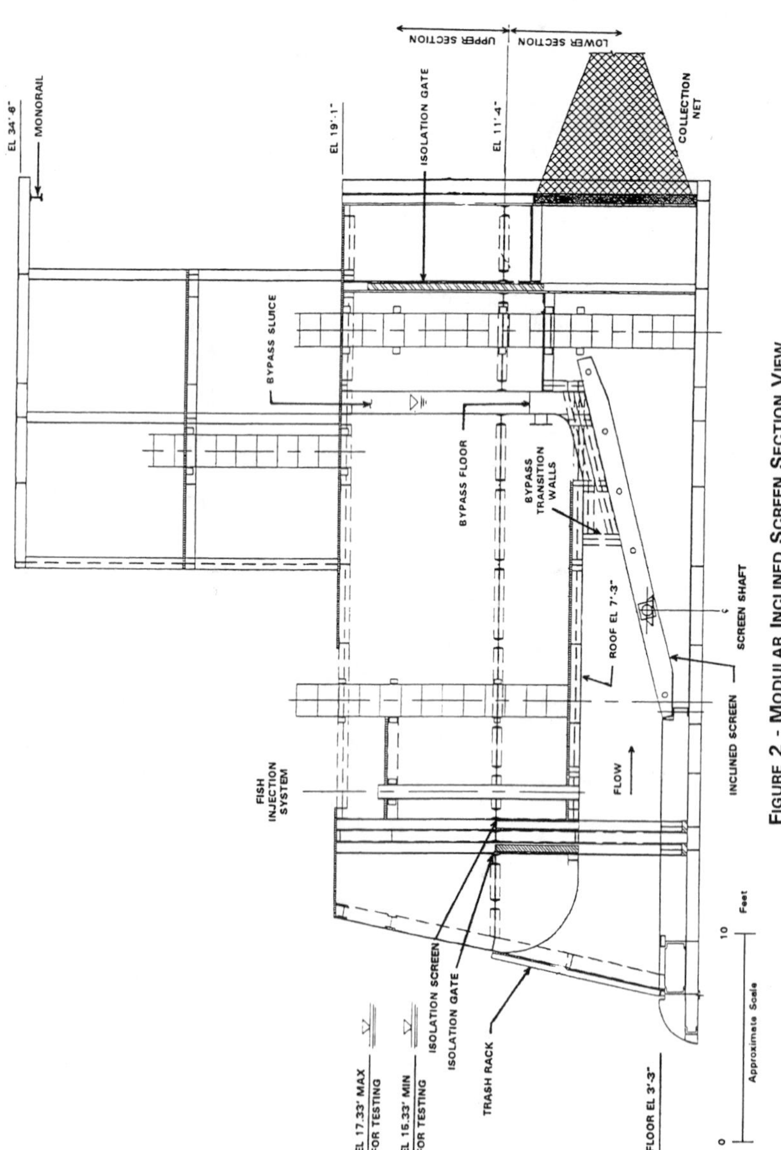

FIGURE 2 - MODULAR INCLINED SCREEN SECTION VIEW

DESIGN DEVELOPMENT OF VERTICAL BARRIER SCREENS

Cindy Philbrook, Affiliate[1] and Scott Ross

Abstract.

Fish diversion screens are an important element of the juvenile fish bypass systems used at the U.S. Corps of Engineers' dams on the lower Snake and Columbia River. This paper outlines a brief history and description of the design development of Vertical Barrier Screens scheduled for installation at McNary Dam on the Columbia River.

General.

Numerous locks and dams constructed within the Columbia River basin provide flood control, power generation, navigation, irrigation, and recreation benefits to the Pacific Northwest. However, these multiple purpose projects also form partial or complete barriers to migrations of highly valued anadromous salmonid species.[2] These species spawn in fresh water rivers. Juvenile fish travel (some over 1400 kilometers (900 miles)) past several dams on their migration to the ocean. One to four years later, adult fish return from the ocean to these same rivers. In order to reduce adverse impacts to migrants, most mainstem projects provide both juvenile and adult fish passage facilities. Diversion screens, located in turbine intakes, have been used extensively to improve survival of juvenile downstream migrants. This paper focuses on the design development of one of two turbine intake diversion screens of the US Corps of Engineers' McNary dam on the lower Columbia River.

Fish Diversion Systems.

Fish diversion systems located in the turbine intakes, are a critical part of the juvenile fish bypass system at McNary Dam. Each intake bay has a bulkhead slot and an

---

[1] Hydraulic Engineer and Mechanical Engineer, US Corps of Engineers, 201 N. Third, Walla Walla, WA 99362-1876

[2] Turner, *et al.*, (1993)

intake gate slot. These slots are essentially one large slot divided into two, with the bulkhead slot upstream of the intake gate slot and a large open cavity between them. See Figure 1.

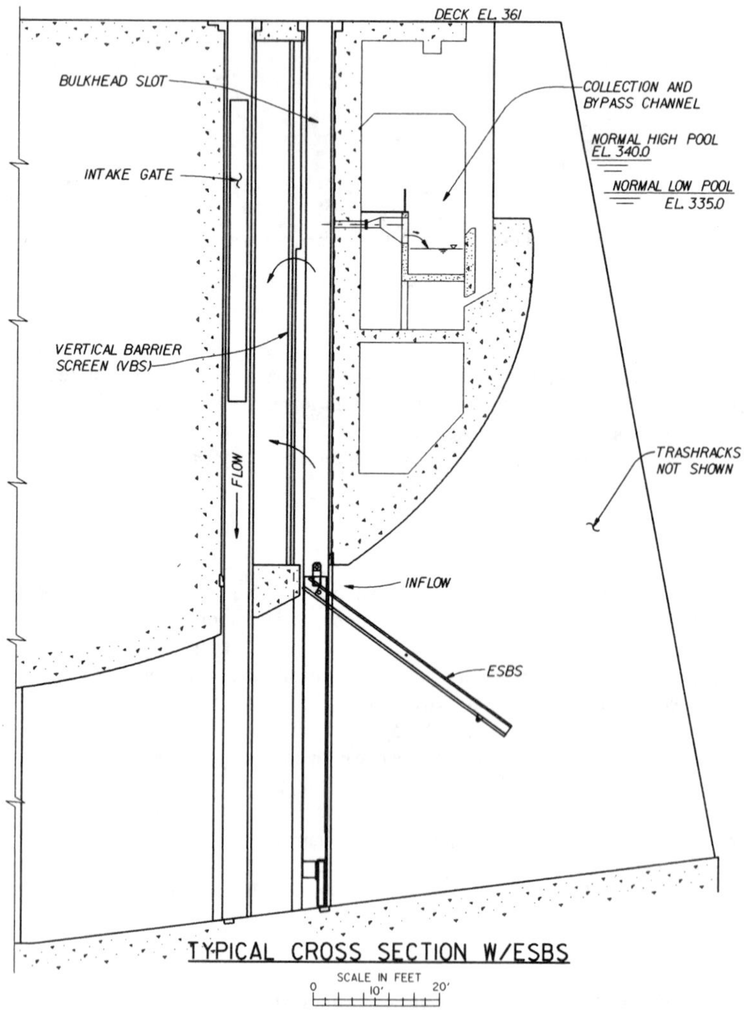

Figure 1.

The vertical barrier screen is located between these slots. It acts as a physical barrier and guide to fish while allowing water to flow from the bulkhead slot to the intake gate slot. Each intake bay inlet is partially screened using special fish diversion screens (for example, the 12.2 meter (40 ft) Extended-length Submerged Bar Screens, ESBS's). Diversion screens are inserted into the bulkhead slots and pivoted out into position in the turbine intakes. Water and fish from the turbine intakes are diverted up into the bulkhead slots. Fish pass out of the slots through submerged orifices near the water surface and exit into a collection and bypass channel.

Only a small percentage of the water diverted into each bulkhead slot passes through the fish orifices. For a turbine flow of 368 cubic meters per second (cms) (13,000 cfs), an average of 123 cms (4,333 cfs) enters each bay (turbine flow is split between three intake bays per turbine unit). Of the flow diverted into the bulkhead slots (approximately 10% of the intake bay flow), about 0.4 cms (14 cfs) passes through the orifices with the majority (around 11.3 cms (400 cfs)) passing through the VBS into the intake gate slot and flowing back to the turbine.

Diverting additional flow into the bulkhead slots is a key element in improving fish guidance. In order to divert additional flow, ESBS's were developed to replace the original 6.1 meter (20 ft) long diversion screens. Biological test results indicated an improvement in fish guidance, but also an increase in fish descaling. The additional flow being diverted by the extended-length screens resulted in severe turbulence in the bulkhead slot/VBS area. This turbulence was a likely cause of the increased descaling. New VBS systems were developed to improve flow characteristics throughout the bulkhead slot without adversely impacting fish guidance effectiveness.

VBS Design Development.

Information related to fish diversion systems was gathered from biologists and engineers within the Corps, outside agencies, and Public Utility Divisions. Due to the dynamic nature of the flow in the turbine intake and bulkhead slot, flow conditions could not be readily calculated or predicted. Two methods were used to observe existing flow conditions. One, video cameras were mounted to VBS's and ESBS's to record flow and fish during several turbine operating conditions. Two, scale models (1:25 and 1:12) of the turbine intake area were constructed out of plexiglas at Waterways Experiment Station (WES). Scale models of screening systems were constructed and tested in the plexiglas models. Dye was released into the model flow to determine flow patterns in the turbine intake and bulkhead slots. Laser meters were used to record flow velocities and directions. This non-intrusive method of flow measurement allowed accurate, detailed examination of hydraulic conditions in the models.

Video imaging and model test results indicated an unstable, expanding, vertical jet with high velocities along the VBS. This resulted in a large area of flow separation at the inlet to the bulkhead slot and turbulence throughout the slot. In some areas, fish were observed passing by the camera in one direction only to be forced back in the opposite direction as the flow shifted.

The ultimate goals when designing new VBS's included maintaining or improving fish guidance while minimizing damage to fish. Scale models provided an understanding of the complex hydraulics in the system. Biological judgment was then used to relate hydraulic behavior to fish behavior. Prototype tests were designed to evaluate the screen systems and verify biological assumptions.

Model Testing.

Initial model tests were done in the 1:25 scale model. Model testing focused on increasing bulkhead slot flows, reducing turbulence, and evenly distributing flow through the VBS. As previously mentioned, increased bulkhead slot flows resulted in increased fish guidance. However, the bulkhead slot was never meant to act as in inlet. A large area of flow separation was formed at the inlet to the slot as the increased flows turned sharply to enter the slot. Improving the inlet would, hopefully, improve slot conditions.

Several methods aimed at improving the inlet were tested with two recommended for prototype testing: an inlet flow vane and an expansion shape. The inlet flow vane helped direct flows into the bulkhead slot, increased slot flows, reduced turbulence at the inlet, and helped distribute the flow more evenly across the slot. However, it had the potential of harming fish through impact with the vane. The expansion shape provided a more rounded inlet which helped to reduce slot turbulence and also provided a shelter that appeared to calm the area in the upper part of the bulkhead slot.

Improvements to the screen itself were also investigated. A base was added to the screen to help flows transition into the bulkhead slot. This not only prevented a "hot spot" of high velocities through the lower part of the screen, but helped to turn the slot flow parallel to the VBS before screening water from the bulkhead slot. Interior vanes were added to one VBS design (VBS2) in order to help turn flows as they passed through the screen into the gate slot. An outlet flow control device was added to the downstream base of the screen. Its purpose was two fold. It helped to guide flow out of the intake gate slot (which reduced flow disturbances at the lower end of the VBS), and could be raised to reduce total slot flow if this was needed to reduce fish descaling (in the past, this was accomplished by lowering the intake gate).

In order to maximize inlet improvements with the inlet flow vane, it was necessary to lower the ESBS. This allowed the vane to intercept flow in the turbine intake, and provided more room for the flow to turn into the bulkhead slot. With the ESBS lowered, a beam extension was necessary to keep the same relationship between the

ESBS and the beam. If this was not done, the flow through the gap between the ESBS and the beam increased considerably. Video imaging in this area showed fish passing over the top of the diversion screen and through this gap. Reducing the flow through this gap was expected to further improve fish guidance. An added benefit provided by the inlet flow vane was a substantial reduction of flow through the gap.

From the 1:25 scale model tests, two options were chosen for prototype testing. The VBS1 option consisted of a bar screen and perforated plate VBS, inlet flow vane, beam extension, outlet flow control, and angled base. The VBS2 option consisted of a bar screen and perforated plate VBS (with vanes located between these materials), and an expansion shape. These designs were constructed for testing in the 1:12 scale model. Here the designs were refined and perforated plate arrangements selected to help spread the slot flow evenly across the VBS surface.

Impingement velocities for the smaller juvenile fish are very low (the maximum recommended velocity through the VBS is 0.15 meters per second (0.5 fps)). Selecting perforated plate arrangements to spread the flow evenly across the VBS reduces the chance of impinging fish on "hot spots" along the screen. This proved to be one of the more challenging aspects of the VBS design. Flow distribution was very sensitive to small changes in porosity. Several tests were required to finalize the perforate plate arrangements. Biologists from the Corps and NMFS (National Marine Fisheries Service) were brought to WES to observe the slot conditions. Their input helped optimize the designs to improve fish passage.

1994 Prototype Test Results.

Results of biological testing indicate both VBS alternatives reduce fish descaling and improve overall numbers of fish diverted. The success of these new VBS designs have improved the viability of the extended-length diversion screen program. This improvement is likely to have a substantial impact on a sensitive regional fish passage issue. Success of these designs has led directly to similar design efforts on other Snake and Columbia River projects. The final McNary VBS design is currently being prepared. Installation of new VBS systems is scheduled to begin in the fall of 1995.

References.

Turner, A. Rudder., Jr., Ferguson, John. W., Barila, Theresa. Y., and Lindgren, Mark. F., "Development and Refinement of Turbine Intake Screen Technology on the Columbia River", *Proceedings, Bioengineering Section Symposium. 123rd Annual Meeting, American Fisheries Society,* Portland, OR Aug. 29 - Sept. 2, 1993.

COMPUTER AIDED PIPELINE DESIGN: THE PROCESS IN DETAIL

Jorge A. Garcia, Ph.D., P.E.[1], Assoc. Member, ASCE

Abstract

State-of-the-art pipeline design is an integrated process involving field survey data and computer aided design software (CAD). The process begins with the development of terrain surfaces using field data. Horizontal alignment and vertical profiles are designed on the terrain surface utilizing a real world coordinate reference. The complete design is then transferred to a paper environment where views of plan and profiles are linked to produce construction plans at a desired scale and media size.

Introduction

Pipeline design involves a number of processes from hydraulic analysis to the development of final construction plans, all within a computer environment. The process of taking field survey data into CAD to generate final construction drawings is the subject of this paper. This process requires the manipulation of data to generate a working terrain model on which to lay the pipeline. A horizontal alignment which includes pipe centerline, stationing, and construction and permanent easements is created, and is then used in conjunction with the terrain model to generate vertical pipeline profiles at desired stationing intervals and vertical scale. Vertical layout of the pipeline follows, usually maintaining a uniform depth. Location of combination air valves, gate or butterfly valves and other fittings are designed on the final profile.

This paper will specifically describe the CAD design process utilizing AutoCad R.12 and AdCADD Design Modules. The commands and terminology

---

[1]Design Engineer, Utilities Engineering Dept., City of Las Cruces, P.O. Drawer CLC, Las Cruces, NM 88004.

employed are software specific. The intent of the paper, however, is to outline a design methodology and expose potential problems and limitations that may very well apply to other design environments.

Field Survey Data and CAD

The basis for field survey information are three dimensional space points with axes North, East and elevation. Points are collected along the proposed pipeline alignment utilizing modern surveying instruments and are electronically transmitted or downloaded to a computer for analysis. Surveyed points have a State Plane coordinate position and an elevation. A collection of points constitutes the basis for the development of a terrain model on which the pipeline will be designed. These points cover a wide enough area to provide terrain information for construction and permanent easements, existing utilities and other structures.

Utilizing a selected format, data points are imported into a coordinate geometry (COGO) point file where it can be fully utilized by the CAD system. Point data on the outside boundaries of the surveyed area are joined with a three dimensional (3-D) polyline to define fault data. Point data together with fault data enable the creation of a 3-D triangulated irregular network (TIN) surface where individual points are linked together by a series of contiguous triangles. The sides of these triangles are subdivided into elevation increments, thus creating an interpolation mechanism for the development of a contour surface. The TIN is developed using a digital terrain model (DTM) and can be edited as necessary to adequately fit project boundaries. Once a final TIN has been developed, existing ground contours are created at selected contour intervals.

Horizontal Alignment and Vertical Profiles

The pipe centerline is designed on the terrain model surface developed with point and fault data. Construction and permanent easements are also designed, sometimes simply using the OFFSET command on both sides of the pipe centerline. Using appropriate surveying convention, the pipe centerline is stationed, points of intersection (PI) are labeled and bearings are given for each straight section to create the horizontal alignment.

Design of the horizontal alignment complete with other related objects such as easements, is followed by the development of vertical profiles. The terrain must now be viewed in profile to address cut and fill areas and to design the finish grade. This process involves setting up a vertical scale, horizontal scale and reference frames for the development of plan and profile sheets.

The development of vertical profiles involves four phases. The first phase is to create the existing ground profile using the terrain model and the horizontal alignment. The second phase is to design vertical tangents and vertical points of intersection (PVI). The third phase involves the design of vertical curves using the tangents and a desired length of curve. Phases two and three may involve an iterative process where consideration is given to cut/fill volumes, haul distances, etc. The fourth and last phase is to lay the pipeline at the desired depth. If a uniform depth is utilized throughout the alignment, the OFFSET command can be used to accelerate the process. The design of combination air valves is dependent on the pipeline profile and is shown on corresponding profiles and on the horizontal alignment. Also, isolation gate or butterfly valves are shown on profiles and horizontal alignment.

Development of Plan and Profile Sheets

The design of the horizontal alignment and vertical profiles is performed in real world space, with no consideration given to output media size other than the reference frames utilized for setting up profile stationing. Linking the horizontal alignment to corresponding vertical profile to generate a plan and profile (P&P) sheet, however, requires consideration of the paper environment in which such sheets will be generated. This paper environment is called Paper Space in AutoCad R.12.

The paper environment is entered by switching the CAD system from Tilemode 1 to Tilemode 0, where the coordinate reference system will no longer be State Plane, but rather paper reference with distances in inches. Within Tilemode 0, a series of paper space frames within the appropriate layer are created with dimensions consistent with the desired output media size. For example, a D-size sheet will require a series of frames with dimensions 864x559 mm (34x22 in). Once the first frame is created, it can be copied as many times as required by the number of sheets in the construction plans. Each paper frame is then labeled with the VIEW command for later retrieval. When using the VIEW command, it is recommended to use the ASSIST/INTERSECTION feature so that exact dimensions of the frame are captured.

After the frames have been named, a VIEWPORT layer is created and made the current layer. Taking the first frame, two horizontal viewports are created. Remaining within Tilemode 0, the CAD system is switched to Model Space opening two windows into the real world environment. Making the bottom window active, the first vertical profile is retrieved using the command VIEW/RESTORE followed by the first profile name. The scale is then adjusted using the ZOOM/XP command. If the profile requires more space than the default one-half sheet, switch back to paper space and use the STRETCH command to expand the size of the viewport.

Next, making the top window active, the horizontal alignment is retrieved using again the VIEW/RESTORE command. In this case, the complete alignment as it exists in the real world appears in the window. Using the ZOOM command, the approximate portion of the plan view that will match the profile stationing is windowed. The plan view is then aligned horizontally with the paper axes using the DVIEW/TWIST command. The proper scale is set using the ZOOM/XP command as before. It is important to note that the MOVE and ROTATE commands should not be used in this case because this would move the horizontal alignment from its State Plane coordinate position in the real world.

The process just described is repeated for each P&P sheet. Figure 1 shows a completed sheet for stations 96+00 to 112+00 of a 457-mm (18-inch) water transmission pipeline in the City of Las Cruces, NM. The top portion of the sheet shows the plan view including pipe centerline, easements, contours, stationing and bearings. Note how the DVIEW/TWIST command was utilized to align the view with the horizontal axes of the paper, leaving the North arrow pointing to the right.

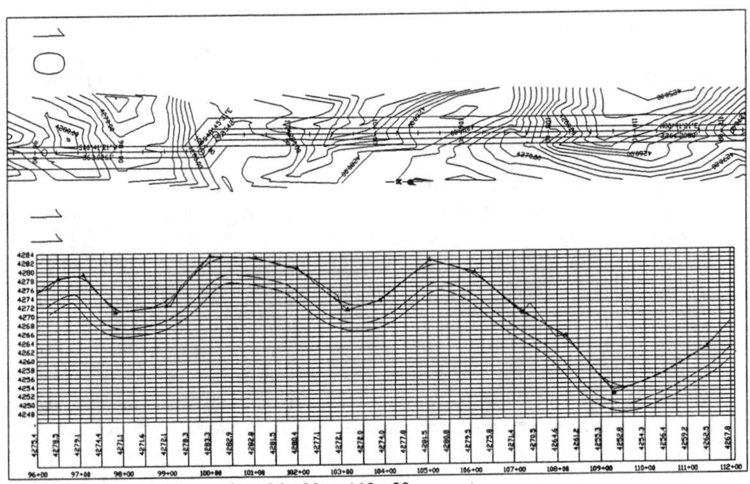

Figure 1. P&P sheet for stations 96+00 to 112+00

Some CAD Tips, Problems and Limitations

Some useful design tips, problems and limitations of the CAD environment are listed below:

1. Perform the complete design in real world reference coordinate system. Never MOVE or ROTATE any object or feature from its original survey position. Perform a complete project design on the same file if hardware is not a limitation.

2. Utilize the VIEW command to name the horizontal alignment, individual profiles and final P&P sheets in paper space. This allows easy retrieval when generating or plotting P&P sheets.

3. Always use the ASSIST feature in CAD to define views. Exact dimensions or intersections are obtained in this manner.

4. Use DVIEW/TWIST commands in model space within Tilemode 0 to develop plan views for individual P&P sheets. This process rotates the view rather than the object or feature.

5. Set proper scale in model space within Tilemode 0 using the ZOOM/XP command. This will insure exact dimensions when a drawing is generated on paper.

6. When field data has been collected in a band along a proposed centerline, changes in alignment will cause the TIN surface to inaccurately extrapolate the data. This limitation can be overcome by carefully editing the TIN outside fault line data, so that the terrain model contours accurately represent field topography.

7. Final P&P sheets must include construction notes that clearly indicate the stations where combination air valves, isolation valves and fittings are to be installed. Also, appropriate notes must be included pertaining to allowable joint deflections. Stations for horizontal points of intersection (PI) and points of vertical intersection (PVI) must be properly labeled.

Summary

Computer aided pipeline design is an interactive process that involves several steps. First, the engineer creates a representation of field topography utilizing field survey data along a proposed pipeline alignment. Point data and fault line data are used to create a TIN surface which is edited to reflect surveyed areas. Next, a terrain model is created by generating a contour surface at desired intervals.

Once a terrain model is created, a pipeline centerline is designed on this surface. Appropriate construction and permanent easements are also designed. The centerline is stationed and bearings are labeled using standard conventions.

Taking the terrain model and the horizontal alignment, vertical profiles are generated at selected station increments. Next, vertical tangents and vertical curves are designed for the finish grade. The pipeline is then designed at a given depth below finish grade, air and isolation valves are located and properly labeled.

After the design has been performed in a real world reference system, a paper environment is used to create the construction drawings. Frames are created in this paper environment taking into consideration media size and scale. Viewports are opened in these frames to accommodate plan and profile drawings for each sheet. Utilizing previously defined views, individual sheets are plotted.

Appendix

References

1. "AdCADD Civil Survey Modules", Softdesk Inc., 7 Liberty Hill Road, Henniker NH 03242.

2. "AutoCad Release 12 Reference Manual", Autodesk, Inc., publication 100752-01, August 6, 1992.

3. South Zone-1 18" Water Transmission Pipeline CAD design file, City of Las Cruces Utilities Engineering Department, Las Cruces, New Mexico.

NUMERICAL SIMULATION OF SUBCRITICAL AND SUPERCRITICAL FLOW IN A CONVERGING CHANNEL

Thomas Molls[1], R.C. Berger[2], Jean Castillo[3], Sean Cornell[3]

Abstract

Subcritical and supercritical flow in a contraction is simulated by numerically solving the 2D depth-averaged equations using a finite difference model DASH (Molls 1992; Molls and Chaudhry 1995) and a finite element model HIVEL2D (Stockstill and Berger 1994). In the finite difference model, time differencing is accomplished using a second-order accurate Beam and Warming approximation and spatial derivatives are approximated by second-order accurate central differencing. The equations are solved using an alternating-direction-implicit (ADI) scheme. HIVEL2D uses linear basis functions for depth and unit discharge and incorporates a SUPG type test function weighted along characteristics.

The numerical models are compared with experimental data reported by Ippen and Dawson (1951) for flow in a straight-walled contraction. The models demonstrate the effectiveness of the numerical schemes in simulating the 2D depth-averaged equations, but also reveal the weakness of these equations under severe conditions.

Introduction

Hydraulic engineers are commonly confronted with problems involving channel transitions. For example, the design of supercritical channel contractions is an important and complex problem. Due to the contracting sidewalls, standing waves appear in (and downstream of) the transition. Thus, the velocity and water depth vary considerably across the channel. On the other hand, a subcritical channel contraction

---

[1]Asst. Prof., Dept. of Civil Engr. and Mech., So. Illinois Univ., Carbondale, IL, 62901.

[2]Res. Hydraulic Engr., USAE - Waterways Experiment Station, Vicksburg, MS, 39180.

[3]Undergraduate Student, Southern Illinois Univ., Carbondale, IL, 62901.

does not exhibit standing waves. Consequently, the velocity and water depth are relatively constant across the channel and only vary longitudinally.

When designing a channel transition, it is important to estimate the velocity and water depth throughout the transition. A computer model can quickly assess the effect of varying design parameters (i.e. geometry, flow, channel material, etc.) on the velocity and water depth in the contraction. However, to apply the computer model with confidence, it must be demonstrated that the numerical results are sufficiently accurate.

Experimental Verification Data

The data used to verify the numerical models was obtained by Coles and Shintaku (1943) and reported by Ippen and Dawson (1951) for flow in a rectangular straight-walled contraction. The upstream and downstream channel widths were 0.610 m and 0.305 m, respectively. The transition section was 1.45 m long with walls angled in at 6°. Neither the channel slope or material were reported. The water depth, in and downstream of the contraction, was recorded for various flow conditions. To verify the computer models, a subcritical and supercritical data set were chosen. In both cases, the flow rate was $0.0411\,m^3/s$. For the subcritical case, the Froude number and water depth 0.305 m upstream of the contraction were 0.315 and 0.168 m, respectively. For the supercritical case, the Froude number and water depth 0.305 m upstream of the contraction were 4.0 and 0.0305 m, respectively.

Model Parameters

The numerical models were run using no-slip sidewall boundary conditions. Since the channel bottom slope and channel roughness (i.e. friction) information was not reported, the models assumed a horizontal channel and frictionless flow. The subcritical computations were performed on a 30x21 grid with the inflow boundary 0.305 m upstream of the contraction entrance and the outflow boundary 0.305 m downstream from the end of the contraction. The 72x21 grid used for the supercritical case was similar to the subcritical grid except the outflow boundary was extended to 3.66 m downstream from the end of the contraction. In both cases, Δx was 0.0726 m and Δy was variable.

Results

The computed water depths are compared with the experimental data in Figures 1-3. It should be noted that the experimental data was obtained from very small contour plots, which inevitably resulted in some error.

For the subcritical case, the depth was relatively constant across the channel and decreased in the longitudinal direction. From Fig. 1, it is evident that the computed solutions closely resemble the experimental data, with only a slight discrepancy at the downstream boundary. Interestingly, the flow became supercritical ($F_r \approx 1.1$) a few nodes upstream of the outlet boundary. The DASH model was more accurate than HIVEL2D near the outlet boundary. The computed solutions were not particularly sensitive to variations in the grid spacing (Δx), channel bottom slope, or channel roughness.

Figures 2 and 3 present results for the supercritical case. This flow is very complex, due to the formation of cross waves in the contraction, and discrepancies exist between the computed results and experimental data. From the upstream boundary to the middle of the contraction agreement between the computed results and experimental data is favorable. However, near the end of the contraction, the experimental and computed results differ significantly. HIVEL2D computed the water depth magnitude more accurately than DASH; but, neither model accurately predicted the location of the cross waves. Both models were sensitive to changes in channel bottom slope and channel roughness. In addition, reducing Δx caused both models to predict increased water depths.

Summary and Conclusions

It is obvious that both models performed better for the subcritical case. In this case, the models accurately predicted the water depth throughout the contraction. This suggests that the depth-averaged equations can be used with confidence to design subcritical channel contractions.

On the other hand, the supercritical computed results and experimental data differed significantly. In general, the models more closely predicted the water depth magnitude, but the predicted cross wave location did not coincide with the experimental data. This was probably due to the presence of substantial vertical accelerations resulting in a non-hydrostatic pressure distribution in the experimental flume. The depth-averaged equations become suspect under such conditions, since they assume negligible vertical accelerations and hydrostatic pressure. Although the models did not precisely match the supercritical experimental data, the authors believe they can still be useful design tools. For example, it is usually desirable to minimize the size and occurrence of cross waves in a channel contraction. The models can be used to identify a poor design (i.e. one with many cross waves), even though the model results will not be extremely accurate under these conditions. As the design is refined, the vertical accelerations and cross waves will diminish, and the model results will become more accurate.

Finally, the channel bottom slope and channel roughness were not reported. These are important parameters and the authors found that varying the bottom slope and roughness significantly affected the computed results for the supercritical flow. However, to avoid accusations of model "tuning", both models were run assuming frictionless flow and a horizontal channel. This inevitably contributed to the discrepancy between the computed and experimental results. In addition, no attempt was made to quantify errors in the experimental data. Even though the flow is assumed steady, the cross waves are not stationary in a supercritical contraction and a statistical estimate of the uncertainty in the experimental data would be very useful. In view of these consideration, a more complete experimental analysis of supercritical flow in a channel transition would be valuable.

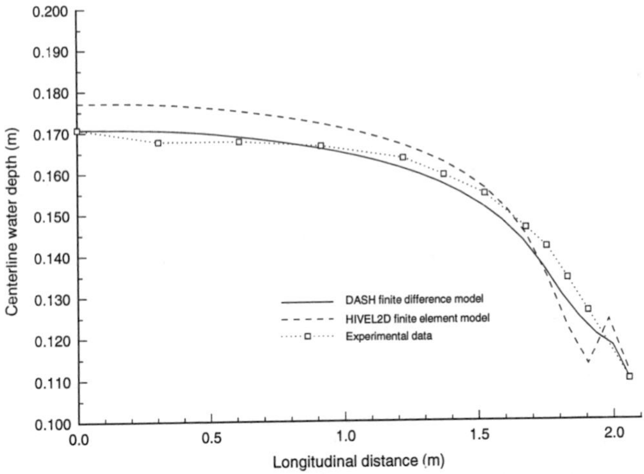

Figure 1 - Centerline flow depth in a contraction (subcritical flow).

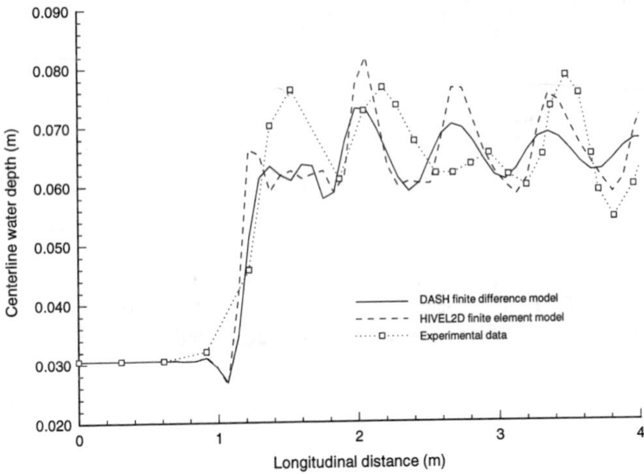

Figure 2 - Centerline flow depth in a contraction (supercritical flow).

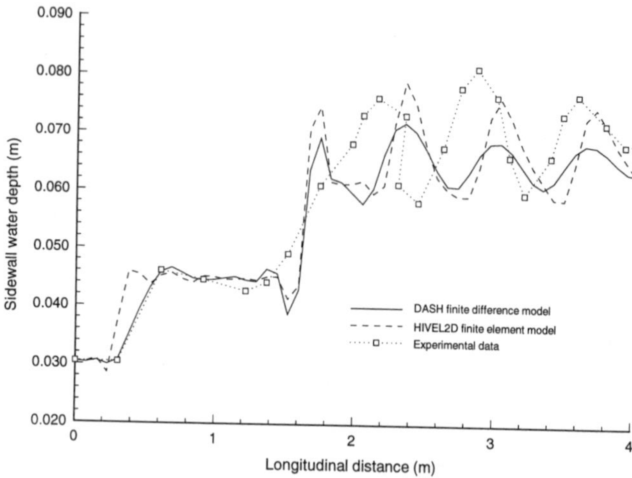

Figure 3 - Sidewall flow depth in a contraction (supercritical flow).

References

Coles, D., and Shintaku, T. (1943). "Experimental Relation Between Sudden Wall Angle Changes and Standing Waves in Supercritical Flow.", thesis presented to Lehigh Univ., Bethlehem, PA.

Ippen, A.T., and Dawson, J.H. (1951). "Design of Channel Contractions.", *Symp. on High-Velocity Flow in Open Channels*, Trans. ASCE, vol. 116, 326-346.

Molls, T.R. (1992). "A General 2D Free-Surface Flow Model for Solving the Depth-Averaged Equations using an Implicit ADI Scheme.", Ph.D. thesis, Washington State Univ., Pullman, WA.

Molls, T.R., and Chaudhry, M.H. (1995). "A Depth-Averaged Open-Channel Flow Model.", *J. Hydr. Engr.*, ASCE, 121(6), in press.

Stockstill, R.L., and Berger, R.C. (1994). "HIVEL2D: A Two-Dimensional Flow Model for High-Velocity Channels.", *Technical Report REMR-HY-12*, US Army Corps of Engineers, Waterways Experiment Station, Vicksburg, MS.

Software for the Modeling of Flows and Sediment in Wetlands

Lisa C. Roig[1] and David R. Richards[2]

ABSTRACT

Current interest in wetland protection, restoration, and creation has created a demand for engineering analysis tools that can be used for wetland planning, design, and management. In this poster session, numerical modeling software that has been developed by the USACE Waterways Experiment Station for the prediction of flow and sediment transport in wetlands is demonstrated. An example application of the software package to the Cache River watershed is described herein.

INTRODUCTION

The long term success of restored and created wetlands will depend upon our ability to manage these systems within the confines and restrictions of their managed surroundings. Trial and error management measures can have unexpected and undesirable long-term effects on the behavior of the wetland ecosystem. This research addresses the need for tools to allow managers to predict the impacts of alternative management plans on the wetland project. Numerical models have been developed at the USACE Waterways Experiment Station (WES) to predict the behavior of water and sediment in wetland regions under alternative management proposals.

The technology for simulating wetland surface flows is based on the solution of the equations for shallow water flow with some important modifications that allow one to properly simulate intermittent flooding and flow through emergent vegetation (Roig, 1995). The models are based on the RMA2 finite element model for the solution of two-dimensional, vertically integrated free surface flow problems (King, 1990). The code has been modified and enhanced to incorporate special features for the simulation of wetland surface flows (Roig, 1994, and Roig and Evans, 1994). The software package includes an advanced graphical pre- and post-processor called

1 Research Hydraulic Engineer , Hydraulics Laboratory, Waterways Experiment Station, 3909 Halls Ferry Road, Vicksburg, Mississippi 39180 (601) 634 - 2801.

2 Research Hydraulic Engineer , Hydraulics Laboratory, Waterways Experiment Station, 3909 Halls Ferry Road, Vicksburg, Mississippi 39180 (601) 634 - 2126.

FastTABS that permits the user to quickly develop the finite element grid and view the model results (Richards, 1993).

EXAMPLE APPLICATION

The Cache River watershed near Cotton Plant, Arkansas includes a large, bottomland hardwood wetland that has been protected from encroachment by agriculture. The wetland extends nearly 18 kilometers from north to south and 4 kilometers from east to west. The topography, and the distributions of vegetation and land use were obtained from a GIS database of the watershed. In dry seasons, the Cache River flow is contained within the banks of the central river channel. During the rainy season frequent floods overtop the channel banks, flooding the adjacent bottomland hardwoods. This wetland provides habitat for migratory waterfowl, songbirds, deer, small mammals, fish, amphibians and reptiles. The FastTABS numerical model for wetland surface flows was used to study flooding of the Cache River wetland.

In the following figures, two snapshots of simulated surface flows in the Cache River wetland are illustrated. These results are part of a continuous one-year simulation that was performed, extending from January 1, 1992 through December 31, 1992. The time step of simulation varied from 3 hours to fifteen minutes, depending upon the speed of the wetting front. The results of the numerical model provided quantitative information about the movement of floods through the wetland. The model could be used to test the effects of changing land use practices on the extent and duration of flooding in the watershed.

CONCLUSION

The results of research in the numerical simulation of flooding and drying fronts, of flow resistance due to emergent vegetation, and of hydraulic control structures have coalesced to produce a comprehensive package for the modeling wetland surface flows. In this poster session the software package FastTABS and the underlying finite element model for the simulation of wetland surface flows are demonstrated.

ACKNOWLEDGMENTS

The information presented herein, unless otherwise noted, was obtained from research conducted under the Wetlands Research Program of the United States Army Corps of Engineers by the Waterways Experiment Station. Permission was granted by the Chief of Engineers to publish this information.

REFERENCES

King, Ian P. (1990) *Program Documentation: RMA2 - A two dimensional finite element model for flow in estuaries and streams*, Version 4.3, November 1990, Resource Management Associates, Lafayette, California, 52 pp.

Richards, David R. (1993) *Wetlands engineering: FastTABS software for evaluation of wetland hydrodynamics.* Technical Note HL-EV-5.1, U.S. Army Corps of Engineers, Waterways Experiment Station, Vicksburg, MS.

Roig, Lisa C. (1994) *Hydrodynamic Modeling of Flows in Tidal Wetlands.* Ph. D. Dissertation, University of California, Davis, California, 177 pp.

Roig, Lisa C. (1995) Mathematical Theory and Numerical Methods for the Modeling of Wetland Hydraulics. *Proceedings of the International Conference on Water Resources Engineering*, San Antonio, Texas, ASCE, (see table of contents).

Roig, Lisa C. and Evans, Robert A. (1994) Environmental modeling of coastal wetlands. *Estuarine and Coastal Modeling*, Proceedings of the 3rd International Conference, Oak Brook, Illinois, ASCE, p. 522 - 535.

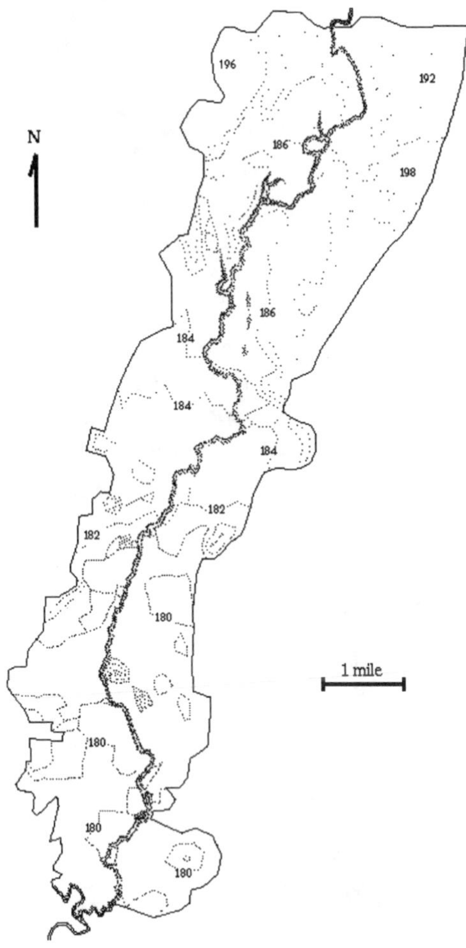

Figure 1. Topography of the Cache River wetland. Elevations are in feet.

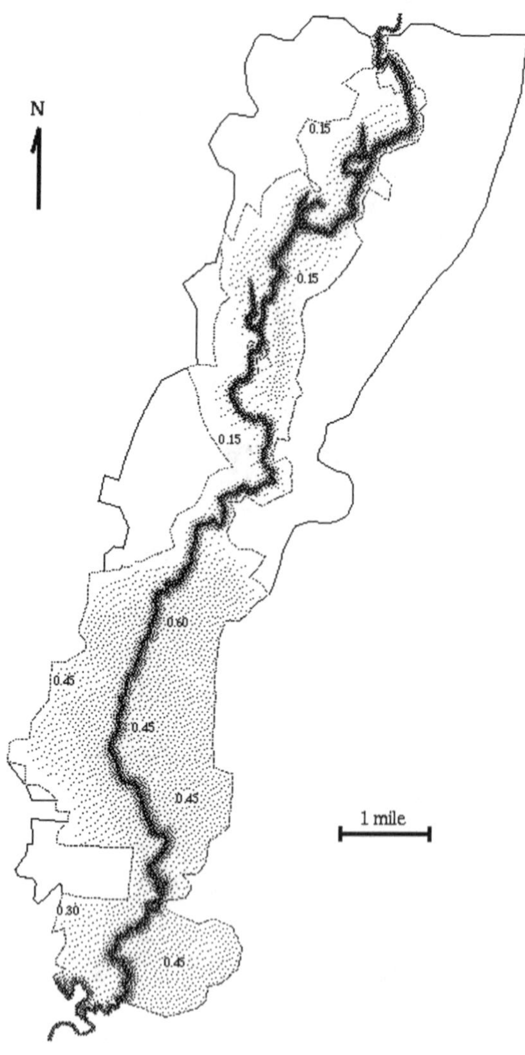

Figure 2. Simulated flood depths in the Cache River wetland on February 19, 1992. Depth contours are in feet.

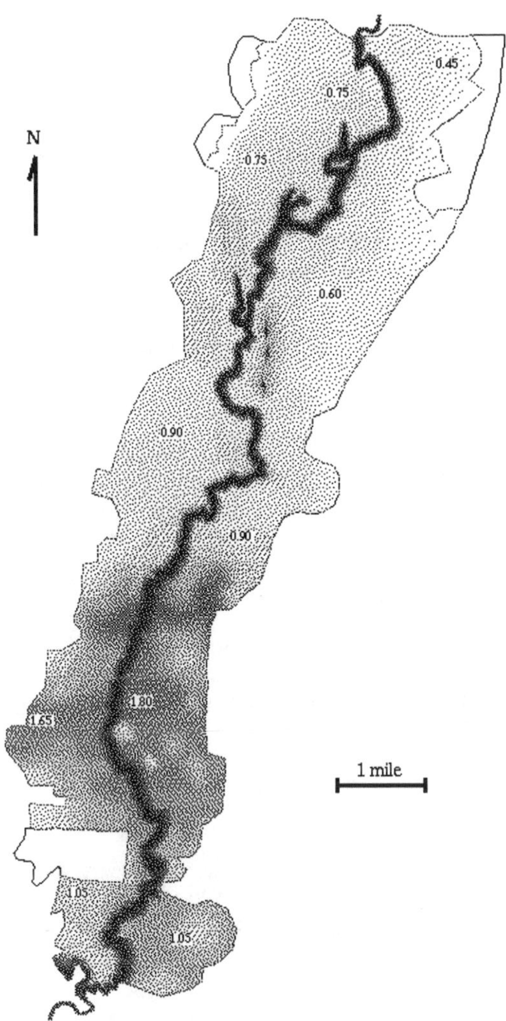

Figure 3. Simulated flood depths in the Cache River wetland on March 15, 1992. Depth contours are in feet.

EGYPT's WATER RESOURCES STRATEGIC RESEARCH PROGRAM[1]

by

Mona El-Kady & C. E. Israelsen[2]

ABSTRACT

The Nile River irrigation system is one of the world's oldest, and also one of its most productive. The construction of the High Aswan Dam (HAD) enabled man for the first time ever to completely control the river and to eliminate its annual fertile, but often destructive, flooding. Although for centuries the river has provided ample water for Egypt's inhabitants, the per capita amount is now decreasing rapidly because of the population increase.

A strategic research program (SRP) has recently been established within the Ministry of Public Works and Water Resources (MPWWR) for the purpose of fostering long-range research within the several institutes of the National Water Research Center (NWRC) to improve the utilization of Nile River water. One of its initial objectives is to locate excess water within the present system and make it available for bringing new lands under irrigation and cultivation. The five particular areas in which SRP is currently doing research are:

1) Integrated irrigation water reuse and efficiency; 2) Water conservation operations; 3) Desalinization; 4) Economic and environmental impacts; and 5) Integrated data systems. The paper presents details of specific activities currently being conducted by the SRP in each of these areas.

---

[1] A joint effort, with USAID and Government of Egypt funding, between the Environmental and Natural Resources Policy and Training Project (EPAT), Winrock International Environmental Alliance, and the National Water Research Center, launched the Strategic Research Program in April, 1994, in Cairo, Egypt.

[2] Director and Program Leader, respectively, of the Strategic Research Program Fum Ismailiya Canal, P.O. Box 74, Shoubra El-Kheima 13411 - Egypt.

Originating deep in the heartland of the African continent, the Nile river flows through 9 countries before it finally reaches Egypt. For countless centuries this mighty river responded only to the promptings of mother nature, and each year during the flood season provided another layer of rich, fertile silt to the farmlands of Egypt. The size and duration of the flood determined the amount of time Egyptian farmers could devote to raising crops, but there seemed to be plenty of food for everyone.

With the construction of the High Aswan Dam (HAD), man finally had the ability to control this mighty river. There would be no more floods, and for the first time ever, farmers could grow three crops per year. Although the new layer of soil no longer comes each year, fertilizer and good management have continued to increase production. But even with all of this, food production cannot keep up with the ever-increasing population, and large quantities must be imported. The irrigation efficiency in Egypt is one of the highest in the world, but there still is room for improvement.

Since its inception in 1975, the NWRC has been charged with providing the Ministry of Public Works and Water Resources (MPWWR) with the research capacity to support its role as Egypt's primary water management agency. MPWWR's mission is to ensure both adequate quantity and quality of water for agricultural, municipal, and industrial uses. Although provision of adequate water supplies has been the dominant focus of MPWWR's activities, water quality has recently emerged as an issue of increasing concern. This has evolved because of the effect of quality on the usable supply and the role of poor water quality in degrading Egyptian public health and physical environment.

The NWRC's mission mirrors that of the MPWWR. The NWRC's role is to provide planning-relevant data and analysis to the MPWWR to ensure long-term sustainable water management under the highly constrained supply conditions prevailing in the Nile river system. Toward this end the NWRC assesses long-term policies for managing water resources, solves technical and applied problems associated with the general policy for irrigation and drainage, and conducts research connected with agricultural land and water resource assessment.

The NWRC chairman and staff have identified the need of inter-institute collaboration that addresses strategic policy issues pertinent to water management in Egypt. "Strategic Research" is particularly focused on long-range options for resource management which integrate knowledge from various disciplines, in contrast to the specific technology or geographical focus of current institute research programs.

The development of a strategic research program within the NWRC will provide for such integration, and help to focus and sharpen the institute's activities to reduce redundancies, maximize complementarities, and economize on the limited resources NWRC has to implement its program. This in turn will help NWRC achieve its objective of more relevant policy orientation.

"Strategic research" as far as this project is concerned has to do with long-range (5, 10, 15 years or more) options of resources management which integrate non-conventional and innovative solutions along with others. The overall objective of the strategic research endeavor is the institutional strengthening of the Water Research Center to enable MPWWR decision makers and planners to better manage Egypt's water resources in a more sustainable way. Initially the research will focus on water supply, distribution, and utilization, but very soon should include all other factors related to the system including environmental and human-health impacts throughout the irrigation system.

The MPWWR faces a host of long-term policy and planning challenges that raise issues to which the Strategic Research Program through the NWRC may make some valuable contributions. Discussions between the NWRC Chairman, various NWRC institutes, the Soil and Water Research Center of the Ministry of Agriculture, and the MPWWR planning Sector indicated that achievement of the overall objective listed above might be accomplished by undertaking work in the following five research areas:

1. Improving the integrated water reuse and efficiency throughout the system, focusing on the possibility of effectively increasing the usable irrigation water supply in the Nile river system.

2. Water conservation operations for providing decision support to enhance MPWWR's ability to meet system-wide water demands. The focus will be on striving to put more flexibility and reliability into the system while reducing operational losses.

3. Desalinization studies for examining the cost and feasibility of its use as an option in Egypt's long-term water supply, utilization and management schemes.

4. Economic, human health, and environmental impacts on the irrigation and societal systems of activities identified in 1,2 and 3 above.

5. Integrated data systems that would provide reliable, usable, updatable and easily retrievable data to enhance MPWWR's policy and planning activities.

An advisory committee has been established for the SRP to review quarterly work plans, and individual research program priorities and progress.

Work on the project to date has focused on the following issues:

1. The agricultural output per farmer and per unit of land in Egypt is one of the highest in the world. Everyone operates as though he has plenty of water. In fact, many farmers probably use too much water but others at the ends of canals may not receive enough. In some areas too much is applied to the land, and water flows to the sea. The SRP is searching for extra water within the system that can be used for irrigating new land. To assist with this endeavor, present irrigation efficiencies must be known, and reliable salt and water balances for the entire system must be developed. Instead of utilizing the classical irrigation efficiency which is expressed in terms of the total amount of water handled by the system, the SRP proposes to use what is termed the "effective efficiency". This method takes into account the decreased productive capacity of water as its quality is reduced because of salinity. Using this approach it is believed that it will be possible to locate places within the system where water may be saved and made available for use on new land.

2. Gaining a better understanding of how the Nile irrigation system is currently operated. Water is released from the High Aswan Dam, and flows into each of the major and minor canals along the full length of the river according to a schedule which is determined by a prior assessment of irrigation needs. Each of the operators is experienced in operating his own gates and in maintaining the appropriate water levels for the designated flows. However , a lot of the "fine tuning" of the system is accomplished because of experience of the operators, and there is apparently no written set of instructions outlining how to do what they do. The SRP's initial efforts in this area of research will be to provide a written document is as much detail as possible of how the system is currently operated. Only then will it be possible to consider ways of conserving water by modifying or enhancing some of the operating procedures, and decreasing or eliminating operational spills.

3. By international agreement Egypt's supply of water from the Nile is constant, but it may decrease in the future as upstream developing nations demand larger shares. Even assuming a constant supply, the per capita amount will steadily decrease as Egypt's population continues to grow. Underground water is also limited and even under the most favorable conditions weather modification could not begin to make up the deficit. One of the viable long-term solutions for augmenting Egypt's water supply is the desalinization of sea water. Even though it is a very expensive process it may be profitably considered for use along sea coasts and water fronts to supply culinary needs to cities. When the cost of saving water by other means begins to approach that of desalting sea water, then

desalinization can be considered as a viable option for obtaining additional water. An abundance of literature is available on the subject, including the various methods currently being used and their power requirements, operating costs, etc. The SRP's objective on this subject is to attempt to identify specific key locations for initial installations, and the appropriate time to begin.

4. Studies and evaluations completed under components 1 and 2 will suggest various scenarios that, if implemented, will result in a saving of water in the system. Each of these will exert economic, environmental, and perhaps human-health impacts on the system. The nature, extent, and severity of each of these will need to be determined and evaluated. Several of these scenarios are currently being reviewed by the SRP, and others are being contemplated. Some of the more significant topics being considered include crop substitution, continuous instead of rotational irrigation, 24-hr versus daytime-only irrigation, and long-term effects of pollution on the system caused by salinity as well as by other constituents. It should be emphasized that each scenario will have a particular economic impact on the system as well as an environmental one, and both should be considered. Some will also have an effect on human health, either physical, societal, or emotional, and will be evaluated accordingly.

5. Each institute within the NWRC, and each research entity within the Ministry of Agriculture, has developed its own data base to meet the needs of its particular mandated responsibilities and type of research. However, as pressures continue to increase on the utilization of Egypt's limited supply of water, they, of necessity, will need to work together more closely than they have done in the past, and collaborate on strategic research problems. In that scenario, it would be advantageous to have a common data base that had been jointly developed and easily accessed by all of the collaborating entities. One of the SRP's immediate priorities is to develop, in collaboration with the institutes and other research groups, the framework for such a joint data base. It will require the complete trust and cooperation of each of the participating entities, but its potential benefits to all would be significant, and well worth the efforts required to get it started.

Water exists on the earth in all of its 3 phases; gas, liquid, and solid. The earth's supply of water is constant, and there will never be a shortage. But there will be a shortage of money. Water can be made available in whatever quantity and quality desired to any location on earth, provided there is enough money to do so. There will often be shortages of money, but never a shortage of water. Therefore, in strategically planning for Egypt's water supply into the next century, the economic evaluation of each supply scenario becomes extremely important. It has been said in the past that "the solution to pollution is dilution". Egypt can no longer afford such a luxury. There is not enough clean water remaining to be able to use it for dilution purposes, and other solutions are being found.

SOME RESEARCH ASPECTS OF XIAOLANGDI MULTIPURPOSE PROJECT

BY

Daniel Gunaratnam[1]

General

1. Since the early 1980's the World Bank has been providing financial and technical support for irrigation and water resources projects in the Peoples Republic of China. One of the most important of these is the Xiaolangdi Project on the Yellow River. The project is designed to manage the sediment, floods, water supply and peak power in the lower reaches of the Yellow River. The overall planning of this project took about 30 years and hundreds of Chinese experts with preparation starting from 1958 and was finalized in 1988. However the Bank using foreign consultants further reviewed the studies between 1989 to 1993 and the project was appraised in April/May 1993. The project however had a large number of complexities which are not observed in most dam projects. Two areas were considered rather important and needed further study. Firstly the sediment transport and deposition in the 800 km river channel needed to be verified using more rigorous mathematical models. Secondly the same stretch of river freezes in winter and as thawing starts in spring ice-jams occur and water backs up behind the jams and sometimes breaks the dikes. United Nations Development Program supported research efforts executed by the World Bank to develop and test mathematical models for sediment transport and river icing and river ice jam flooding, both of which are important in refining the operating rules for the reservoir. This paper describes some of the problems and issues involved in this research.

---

[1]Principal Engineer, China/Mongolia Department, The World Bank, 1818 H Street, Washington DC. 20433.

Background

2. The Yellow River rises on the northern slopes of the Bayankela mountains and falls 4,450 m over a length of 5,465 km, draining an area of 795,000 km^2. Because of intense soil erosion from of the 430,000 km^2 middle area, the Yellow River far exceeds any of the world's large rivers in terms of annual sediment transport. For example, it carries three times the sediment load of the Brahmaputra-Ganges with only 8 percent of the annual flow. In its lower reach, over a length of some 800 km, the river becomes a broad meandering channel contained by flood embankments and here the river drops part of its sediment load and the bed rises above the surrounding land at a rate of one meter every ten years.

3. **Floods in the Lower Yellow River Reach**. Reports of Yellow River floods and efforts to control them go back to the year 2297 BC. Historical records and archaeological investigations provide evidence of many major floods with some exceptional events in the years 223 AD, 1482, 1761 and 1843. Of these, the 1843 flood, with an estimated peak flow at the Xiaolangdi dam site of 35,000 m^3/s and a return period of 1,000 years, is regarded as the flood of record for the Yellow River.

4. Accurate reports of flood damage are available since the turn of this century. In the flood of 1933, the main dikes broke in 54 locations. The total flooded area was 11,000 km^2; 3.6 million people were affected and 18,000 died. Two years later in 1935, during a relatively small flood, the dikes were broken again in Shandong Province, and 12,000 km^2 were flooded, affecting a population of 3.4 million people.

5. Since 1935 there was a deliberate breach of the dike near Kaifeng in 1938 in an attempt to stop the advancing warring armies. As a result of the breaches the water swept repeatedly through 44 counties during the following nine years, submerging 1.3 million ha of cropland and leaving 12.5 million people homeless. Some 900,000 people were either drowned or died of hunger or diseases.

6. **River Icing and Spring Floods**. Apart from the summer floods, there are also possibilities of floods from ice jams. In the lower reach, the main stem of the Yellow River takes a turn to the northeast. There is a gain of 3° in latitude when the river reaches the Bohai Sea. According to statistics of the past 35 years, the probability of freeze-up in the lower reach is 86 percent. The differences in latitude and width, as well as the variation of discharge, contribute to the complexity of ice regimes and flooding on the lower reaches of the Yellow River.

7. According to historical records between 1883 and 1936, there were 21 years when dikes have breached during the ice-flood period. In 1951 and 1955, ice jams caused dike breaches in Lijin County (Shandong Province). Much land and many villages at Zhanhua, Lijin and Kenli were flooded during this disaster.

8. **Sediment**. The annual sediment load at Huayuankou, in the start of lower reach of the river, ranges from 0.4 to 4.0 billion tons and averages about 1.5 billion tons. Most of the sediment load occurs during extreme floods produced by storms in the middle region which lead

to sediment concentrations of more than 400 kg/m^3. Of the average annual sediment load of 1.5 billion tons, about 300-400 million tons is deposited in the lower reach of the river and in irrigation systems, and the remainder flows out to the sea.

9. **Xiaolangdi Dam Project**. The dam will have a reservoir with an initial live storage of 12.6 billion m^3 which would reduce to 5.1 billion m^3 within 20 years from first impounding. This volume of storage would be preserved indefinitely and the reservoir would thereafter mitigate catastrophic floods so that they are contained within the dikes. The reservoir would also be operated to flush fine sediments from the downstream channel into the sea and this, together with the interception of a large volume of coarse sediment, would make it possible to defer for 20 years the raising of the dikes in the lower flood plain. However, the deposition of sediment in the lower reach will eventually resume unless further steps are taken to reduce the river's sediment load or to intercept the sediment upstream. Several dam sites have been identified upstream of Sanmenxia to provide further long-term sediment storage. The timing of such dams will depend on the effectiveness of the ongoing soil conservation programs in the Loess Plateau.

Research Undertaken under the Xiaolangdi Project.

10. Sediment Transport Models. Considerable research was undertaken by the Yellow River Conservancy Commission (YRCC) during the development of the designs for reservoir sediment filling and flushing the sediment through the reservoir so as to maintain a stable reservoir volume. Most of these models were generally fairly well tested using physical models from three different laboratories. The density currents measured in the upstream Sanmenxia reservoir yielded data and empirical relationships to give enough confidence to accept the results of the analysis for the Xiaolangdi Reservoir sediment storage.

11. In addition there were several empirical mathematical models that were built by the YRCC Sediment Research Institute for transporting sediment through the downstream river channel. The sediment transport models were empirical and did not match any of the normal transport equations. Sediment transport equations as given below were used:

$$Q_s = A.Q^a.\rho^b.e^{c\Sigma\Delta ws}$$

A,a,b,c are constants
Q_s is the total sediment outflow in tons/s
Q is the water flow m^3/s
ρ is the sediment concentration in kg/m^3
Δws is the incremental sediment load inflow in the time period

12. The use of these empirical models were quite unorthodox. There were also different empirical equations with different terms for different sections of the river and for different seasons. The empirical models were not replicable on other rivers since most of them were derived for the given stretch of river and for a specific season. Most of the international sediment experts could not disagree with these models since the prototype data fitted so well with the model data. It was however felt that some more rigorous mathematical models with

a more uniform representation of all the various sections of the river should be used to verify the empirical equations. The GSTARS model developed by Chih Ted Yang and Albert Molinas at Colorado State University was selected as the basis for comparing and confirming the results with the empirical models and to then use the models for operational runs to refine the reservoir release operations.

13. The YRCC in the mean time applied for a UNDP grant of $115,000 to undertake the study. Two Chinese researchers were to undertake the studies at Colorado State University (CSU). The YRCC was to provide all the data and the study was to be undertaken under the guidance of Professor Molinas and Mr. Ted Yang which consisted of: (a) Review the Yellow River general geomorphology and sediment transport data for the proposed test sections; (b) Test and evaluate the sediment transport equations; (c) Apply minimum energy dissipation theory; (d) Test GSTARS model on the 1960-1964 data; (e) Modify GSTARS for the Yellow River applications to include the effects of hyperconcentration flows; and (f) Use the GSTARS model for sediment transport for the 1974-1984 data for the Lower Yellow River Channel between Gaocun to Lijin at the mouth of the river.

14. Two Chinese researchers, one modeler and another technical specialist worked with Colorado State University for approximately 18 months to undertake the various tasks given above. Despite the fact that the Chinese researchers had extensive experience in the data and empirical modelling, they were not exposed to the many models that were available in the US market. The computer power and the availability of software for processing data simplified their work greatly so that they could concentrate on the actual modelling. The Chinese could therefore concentrate on the work of adapting the models to the Chinese data and making runs to ensure good fits. The work was started in early 1993.

15. Four reports were produced, the first report verified the YRCC data, the second and third reports showed the derivation of a modified transport equation of Mr. Yang's 1984 transport equation for high sediment concentration which required modification of the fall velocity for high sediment concentration which was now called Yang, Molinas and Wu transport equation. Further modifications were also made for average bed load transport capacity using d_{50} particle size and the corresponding fall velocity, w_{50}, to obtain the total transport capacity. A similar transport function was also derived using the Wuhan Institute formula which was developed by Ruijin Zhang of China in 1958. The forth report indicated the computational results of using the modified transport equations in an non-equilibrium sediment transport equation using a steady state, non-uniform and concentration dependent diffusion equation for the two modified transport equations.

16. The GSTARS model was then applied to a 30 km stretch of the Yellow River above Sunkou where the river is about 6-7km wide between the dikes with a river channel varying between 200-500m and where the river channel could shift 4500m. The results of the computations showed very close correlation with observed data of cross sections showing that GSTARS gave a lot of promise for extensive use in the Yellow River. GSTARS model virtually integrated various functions to give a universal relationship that could be used to predict the sediment transport and deposition rates in the river channel.

17. The net result of the research was that the YRCC experts in collaboration with developed country researchers such as Molinas and Yang resulted in considerable transfer of technology to the Chinese in a short period. There was also very useful field data that was made available to Colorado State University which resulted in modification of the Yang's transport equations for hyperconcentration type flows in rivers.

18. River Ice Formation and Breakup Research . The Xiaolangdi Reservoir was designed to give the addition storage of 2.2 bcm above the 1.8 bcm given by the upstream Sanmenxia Reservoir to ensure that ice-runs are prevented during the spring season. However the early Xiaolangdi operation studies used for prediction of river ice formation and breakup were using very obsolete methodology using nomograms and charts. These charts could not provide the accuracy or reliability for forecasting the ice formation or breakup.

19. In 1991 a Chinese-Finish team undertook the development of a study called "Development of Ice Forecast Model and Ice Flood Prevention measures in the Lower Yellow River". The study was executed by YRCC and the consultants from Finland during the period from February 1991 to September 1992. A preliminary river ice model was developed and the YRCC Hydrology team who worked on this study began to understand the basic modeling techniques for ice modelling. However the model had to be improved and made more flexible.

20. The YRCC applied for UNDP Umbrella funds of $140,000 to undertake this study. The main focus was to develop a numerical unsteady state model for river flow, ice formation and breakup. Prof. Hung Tao Shen of Clarkson University in New York State who had wide experience, with the Corps of Engineers and other groups, in developing river ice formation and breakup models was appointed to provide services for the development of the models and to train two YRCC staff over the year and two others for different periods. Since the modelling involved a great deal of mathematical background, the model building was carried by Dr. Deshang Wang a research associate under the supervision of Hung Tao Shen. The YRCC staff from the Hydrology Bureau participated in various phases of the model development.

21. The program of research undertaken consisted of: (a) collection of hydrologic, meteorologic, river geometry, and ice regime data and initial field visits by Prof. Shen to the field to ensure that all data and researchers be adequately informed about the work to be undertaken; (b) short training on river ice modelling by taking some formal and informal courses; (c) model development; (d) simulation analysis and trial and final runs; and (e) final report preparation.

22. Initial model development required that the unsteady river flow model had to be modified to take into consideration of flow conditions with total, partial etc. ice cover. The model data had to be schematized for specific channels and nodal configuration and calibrated for specific flow conditions of the Lower Yellow River. Second, a model was developed for simulating water temperature using the surface, surrounding and intrinsic heat exchange parameters. Third, a model was developed for anchor ice simulation, and fourth, models were developed for ice cover formation, progression, under ice-cover transport and thermal growth and decay of ice cover. Since most of these models had been developed for the North American Rivers for the Corps of Engineers, called RICEN models, the model development

really required code changes which essentially provided for special features which were over and above those that were needed for several North American Rivers. Since the technicalities of the models were rather complex, and the lack of familiarity with the RICEN code, the Chinese researchers could not get involved in detail into the code modifications.

23. After model development it took almost 9 months to get the river flow model to be calibrated because of adjustments to flows which did not show the losses due siphoning of water from the river at different points. The calibration of the ice condition were then performed for winters of 1969-70, 73-74 and all subsequent winters till 1979-80. Special scenario runs were then performed for the Yellow River. The Chinese were involved in all aspects of the calibration and scenario runs for the Yellow River so that in the end they were very familiar with the models and handled the future modelling themselves. The Chinese were thoroughly trained within a short period and brought up to skilled operators on a fairly sophisticated model of a process such as river icing and breakup.

Conclusions

24. The inclusion of such research components in the project design has had some real returns. First, it exposes excellent research staff from countries such as China to the developed world and the technology available outside their countries. Second, a great deal of empirical work was tested with more rigorous modelling techniques and can now be made more widely available for other projects. Third, US researchers had access to data on a large scale to engineering problems which were then used to test some of the models on conditions not commonly observed in the US. For example the hyperconcentration flows of the Yellow River required changes to the transport equation and this was not generally observed in US river flows. The ice modelling occurred under shallow flow under a very wide elevated river which required modifications of the flow model. Finally the research components were extremely inexpensive. For about $255,000 two fairly sophisticated problems had mathematical models developed, tested and used for refining operations of a major $2 billion dam. What was undertaken here was remarkable since many institutions in North America have been trying to develop modelling techniques such as these but have had to spend much more time and money to come to get useful results. A focused approach to modelling such as in this case yield great returns to both the US and Chinese side. However one must admit there have been other cases where the Bank has not always been as successful as these two cases due to problems with data or lack of correct personal on especially the Chinese side.

25. Financial assistance to economically developing nations is an important concern to the Bank, but it is the writer's experience that strengthening the scientific, and institutional bases for water resources developments often is an equally important component of successful programs.

The Experience of a Water Resources Research Center in Colombia

Ricardo A. Smith Q.[1]

Abstract

The Water Resources Graduate Program was created at the National University of Colombia, Medellin Campus, in 1984. Since its creation about 50 master students has graduated, a doctoral program started in 1991 and research has been done for institutions all over the country. Research has been done in two directions: applied research done for institutions to analyze and make recommendations about specific practical problems and scientific research sponsor by COLCIENCIAS (the Colombian equivalent of the National Science Foundation). In the first case results are directly applied by the public institution. Examples of research done in this case are in the areas of hydrology evaluation and modeling, water resources systems management, groundwater evaluation and modeling, hydropower evaluation, and others. In the case of scientific research it has been difficult that the final results to be applied by public institutions. For the approval of a research project by COLCIENCIAS a support letter of a public institution is needed where it is stated that the institution is willing to apply the results of the project. Nevertheless it has been difficult that the support institutions use the results of the scientific research projects. Detail presentation of this experience, including description of the different projects and its final success in practice, is presented in this paper. The cost and the economical impact of the projects are also presented (if there is data) or discussed. Based on this experience some conclusions and recommendations are presented.

Introduction

A water resources graduate program was created in the National University of Colombia, Medellín Campus, in 1984. The program started with a master degree

---

[1]Water Resources Graduate Program, Universidad Nacional de Colombia, Facultad de Minas, Apartado Aéreo 1027, Medellín, Colombia.

program in 1985 and a doctoral program was initiated in 1991. Since its creation about 50 master students has graduated. The program has actually about 25 master students and 11 doctoral students. It is the first engineering doctoral program in Colombia (Water Resources Graduate Program, 1994). The graduate program has consolidated as the most important water resources research center in Colombia. In 1985 the program had 3 research projects for local institutions with a total cost of 10 millions Colombian pesos, and in 1994 the program ended up with 15 research projects for institutions all over the country and with a total cost of 300 million Colombian pesos. The staff of the graduate program includes 8 engineers with a master degree and 6 engineers with doctoral degrees (from universities in United States and Europe). The program has also the support of an hydraulic laboratory, a computer center and a specialized library. For the development of the research projects there is an intensive participation of our graduate students and in some cases the research project ends up as a master thesis.

Research has been done in two directions: applied and scientific research (Smith et al, 1992). Applied research has been done for local, regional and national institutions to analyze and to make recommendations about specific practical problems. Scientific researches has been sponsored mostly by COLCIENCIAS (the Colombian equivalent to the National Science Foundation) and the National University of Colombia. In both cases there are possibilities of transferring research results to practice. The experience of the water resources graduate program in both cases is now presented.

Applied Research

Applied research to solve specific problems or needs has been done for institutions all over the country. In this case the institutions totally finance the project and the results are directly applied by the institutions. research of this type done by the graduate program include:

- Regionalization studies
- Water balance studies
- Streamflow prediction models
- Water availability studies
- Synthetic streamflow generation models
- Aqueduct and sewer systems design
- Groundwater evaluation and modeling
- Hydrodynamic simulation models
- Hydropower evaluations
- Physical reduced models of hydraulic structures of different types

The institutions for which these applied research was done includes regional development corporations, municipalities, State governments, national ministries, decentralized institutions and electric generation companies. Because the institutions has to sponsor the projects in most cases the project results are directly applied to solve specific problems or to make recommendations about specific courses of action.

Scientific Research

Scientific research has a complete different approach. In the case of applied research most of the time the institution has a specific problem an looks out for the institution to be contracted to solve that problem. In the case of scientific research the researcher make a proposal that submits for founding to a given institution. The most known Colombian institutions to support scientific research are COLCIENCIAS (the Colombian equivalent to the National Science Foundation) and CINDEC (the National University Research Center). The graduate program has done researches with grants from these two institutions in areas such as:

- Regional energy planning (COLCIENCIAS)
- Expansion of the national electric generation system (COLCIENCIAS)
- Expansion of rural electric network (COLCIENCIAS)
- Multiobjective watershed planning and management (COLCIENCIAS)
- hydrodynamic and diffusion transport (salinity) simulation in coastal lagoons (COLCIENCIAS)
- River equilibrium (COLCIENCIAS)
- Hydrologic regionalization (CINDEC)
- Streamflow prediction modelling, the effect of El Niño phenomena (Regional Public Utility)

In order to approve the project most of the sponsor institutions ask for an intention letter from a public or private institution in which they state that the research to be done is of their interest and that they are willing to transfer the research results to practice. Nevertheless this is only an intention letter that do not represent a formal compromise for the institution. In most scientific researches cases the research results are not transferred to practice. In order to solve this problem and to guarantee that research results are transferred to practice, the sponsor institutions are asking for a real compromise to the result receiving institutions. Instead of sponsoring the whole project the sponsor institution is now co-sponsoring the projects. The result receiving institution has to sponsor part of the project.

Research Perspective

Research opportunities in Colombia are enormous. Most of the Colombian institutions are willing to transfer research results to practice, even research results that in some development countries stayed at a theoretical level. Many of the operational research developments for water resources planning and management in the United States was never transferred to practice. The transfer to practice of new developments in the

United States could take years because of the public awareness situation that has forced a conservative attitude in the public utility managers. In Colombia because of the low public participation utility managers are more liberals and willing to transfer research results to practice. In this sense it could be found that some methodologies developed in the United States are first applied in the developing countries.

Colombia is making efforts to increase its research capability as one of the strategic areas to develop the country. In 1994 Colombian government created a committee (Aldana et al, 1994) formed by ten (10) prominent Colombian citizens with the objective to recommend strategic actions for Colombian development. The committee recognized that Colombia has a very low research capability and recommended to increase government investment to rapidly improve that capability. In this respect Colombian government is sponsoring scientific research, doctoral programs, master and doctoral scholarships, and doing other actions.

Another aspects that recently has increase the demand for research in Colombia are some central government administrative changes. Colombian energy sector has a new structural organization that has generated many research needs. Colombia is changing from a centralized decision making and public owned energy sector to a energy free market scheme in which the energy generators (public or private) has to compete to sell the generated energy. The energy companies are sponsoring research in many different areas to understand the new energy environment such as: market regulations, open energy market, effects of El Niño phenomena, integrated energy planning, effects of global climate changes, optimal dispatch models, and others. The recent creation of the Environmental ministry has also generated many researches needs.

Conclusions

- When the result receiving institutions sponsoring or co-sponsoring the research projects, normally the research results are transferred to practice. Colombian scientific research sponsor institutions are changing its policies and are only founding projects as a cosponsor institutions in order to guarantee that research results are transferred to practice.

- When there is no real institutional support, with money and/or participation, the transfer of research results to practice is very difficult and most of the time is not done.

- There are many research opportunities in Colombia. Colombian government is willing to sponsor research projects as one of the actions to increase the country research capability. Other recent changes in the Colombian central administrative government has generated many needs for research in the water resources area.

- Applied research is done when an interested institution needs to solve a specific problem. In this case the institution looks for the research group that has the capability to solve that specific problem. The research results are directly applied by the institution and in this case almost all research results are transferred to practice. In scientific research a researcher present a proposal to a science supporting institution looking for a grant that will allow the researcher to do the proposed research. If this research does not have an interested institution that is willing to cosponsor the project and to transfer the research results to practice, most of the time the research results will not be applied.

References

- Aldana E., Chaparro L.F., Garcia G., Gutiérrez R., Llinás R., Palacios M., Patarroyo M.E., Posada E., Restrepo A., Vasco C.E., 1994. Colombia: At the edge of opportunity (in Spanish). COLCIENCIAS, Bogota, Colombia.

- Water Resource Graduate Program, 1994. Doctoral, masters and specialization courses in water resources (in Spanish). Universidad Nacional de Colombia, Facultad de Minas, Medellín, Colombia.

- Smith R.A., Valencia D. and Mesa O.J., 1992. A water Resources Program in Colombia: Experience and Needs. Water Resources and Environment: Education, Training and Research, Fort Collins, Colorado, July 13 to 17, 1992.

FLODRO: A USER FRIENDLY PERSONAL COMPUTER PACKAGE

Jose A. Raynal-Villasenor[1], M. ASCE
Carlos A. Escalante-Sandoval[2]

Abstract

The use of personal computers have changed dramatically the way to perform several studies related with hydrological problems.The increased capabilities of the current generation of personal computers allow the hydrological engineer to explore options that a few years ago were unthinkable without a mainframe computer. Within this stream of development, user-friendly personal computer package FLODRO was developed. FLODRO has to independent programs: FLOOD and DROUGHT. Program FLOOD performs flood frequency analysis by using eight typical distribution functions: Normal, Two and Three Parameter Log-Normal, Two and Three Parameter Gamma, Log-Pearson Type III, Extreme Value Type I (Gumbel) and General Extreme Value. Program DROUGHT which perform drought frequency analysis by using five typical distribution functions: Three Parameter Log-Normal, Pearson Type III, Extreme Value Types I (Gumbel) and III (Weibull) and General Extreme Value. The characteristics, properties and construction of the programs FLOOD and DROUGHT of computer package FLODRO are displayed in the paper towards its application in flood and drought frequency analysis. The selected methods of estimation of parameters are those of moments, maximum likelihood, probability weighted moments and sextiles for flood and drought frequency analyses. An example of application is shown in the paper.

[1]Department of Civil Engineering, Universidad de las Americas-Puebla, 72820 Cholula, Puebla, Mexico, [2]Engineering Graduate Studies Division, Universidad Nacional Autonoma de Mexico, 04510 Mexico, D.F., Mexico

Introduction

A subject of paramount interest in planning and design of water works is that related with the analysis of flood and drought frequencies. Due to the characteristic that design values have they are linked to a return period or to a non-exceedance or exceedance probability, and the use of mathematical models known probability distribution functions it is a must. Among the most widely used probability distribution functions for hydrological analyses are the following, (Kite(1988), Matalas(1976) and Salas and Smith(1980)):
For flood frequency analysis: Normal, Two and Three Parameter Log-Normal, Two and Three Parameter Gamma, Log-Pearson Type III, Extreme Value Type I (Gumbel) and General Extreme Value.
For drought frequency analysis: Log-Normal with 3 parameters, Pearson Type III, Extreme Value types I (Gumbel) and III (Weibull), and General Extreme Value distributions.

In the light of the personal computer applications in education and training in all the fields of science, a personal computer program was designed to take care of the processes of flood and drought frequency analyses, providing a wide number of options in the models to be used as in the analyses that can be done with such a tool as well. The resulting code has been named FLODRO as it will be refered herein. The paper contains the key features of FLODRO and one example for drought frequency analysis to show the main results that FLODRO can supply to the user.

Framework of FLODRO

FLODRO is written in GWBASIC, (GWBASIC is a registered trademark of Microsoft Corporation), a BASIC compiler compatible with IBM , (IBM is a registered trademark of International Business Machines) personal computers. The interactive mode in which FLODRO is written makes it to have a high user-friendly component. In any step, the user has control on the processes that the program executes, from data input to printing of results of the analysis. The personal computer package FLODRO has the structure shown in figure 1.

Both programs can perform the required computations to obtain, as shown in figure 2:
a) Estimation of parameters
b) Computation of probability distribution function for sample values or for any other values provided by the user
c) Computation of probability density function for sample values or for any other values provided by the user
d) Inverse of the probability distribution function for a fixed number of values or for any other values provided by the user

e) Confidence limits for design events
f) Goodness of fit tests based in the standard error of fit, Kite(1988), and based in a graphical comparison between the empirical and theoretical probability distribution and density functions

	INFORMATION
	NORMAL
	LOG-NORMAL 2 AND 3 PARAMETERS
FLOOD	GAMMA 2 AND 3 PARAMETERS
	LOG-PEARSON TYPE III
	EXTREME VALUE TYPE I
	GENERAL EXTREME VALUE

	INFORMATION
	LOG-NORMAL 3 PARAMETERS
	PEARSON TYPE III
DROUGHT	EXTREME VALUE TYPE I (GUMBEL)
	EXTREME VALUE TYPE III (WEIBULL)
	GENERAL EXTREME VALUE

Figure 1. Framework of computer package FLODRO

Personal computer program FLODRO has been designed to use minimum of memory and computer peripherals. Each computer program FLOOD or DROUGHT have less than 360K so there is no need to have a hard disk to run any of such programs. The graphs provided by FLOOD or DROUGHT are printed in a common printer there is no need to use costly plotters to get in paper these graphs. These features makes FLODRO very suitable in programs of hydrology education and training particularly in developing countries, and in continuing education as well.

	INFORMATION
	ESTIMATION OF PARAMETERS
FLOOD	COMPUTATION OF THE DENSITY FUNCT.
OR	COMPUTATION OF CUM. DIST. FUNCTION
DROUGHT	INVERSE OF THE CUM. DIST. FUNCTION
	COMPUTATION OF CONFIDENCE LIMITS
	GOODNESS OF FIT TESTS

Figure 2. Options of analysis in computer package FLODRO

Numerical Example

Gauging station Villalba, Chih., located in Northern Mexico, has been selected to analyze its annual one-day low-flows, in the period of record 1939-1981 using the General Extreme Value probability distribution function with its parameters estimated through the method of sextiles. The parameters obtained through the use of FLODRO are:

Location parameter = 0.3769; Scale parameter = 0.1622; Shape parameter=0.5030

Mean = 0.3355; Standard Deviation = 0.1534; Skewness = 0.6423

One of the graphs provided by FLODRO is contained in figure 3.

Conclusions

A personal computer program has been presented for flood and drought frequency analyses education and training. The computer code FLODRO has been applied successfully to train students coming from Latin American, Asiatic and African countries, making obvious the user-friendly component of such computer code given that most of the students had not any previous computer experience. Due to the minimum requirements of central memory and computer peripherals that the personal computer program FLODRO has, as it has been shown in the paper, makes it a versatile tool to train students or technical personnel in the field or with a personal computer without a hard disk nor a plotter. Due to the minimum requirements of central memory and computer peripherals that the personal computer program FLODRO has, as it has been shown in the paper, makes it a versatile tool to train students or technical personnel in the field or with a personal computer without a hard disk nor a plotter.

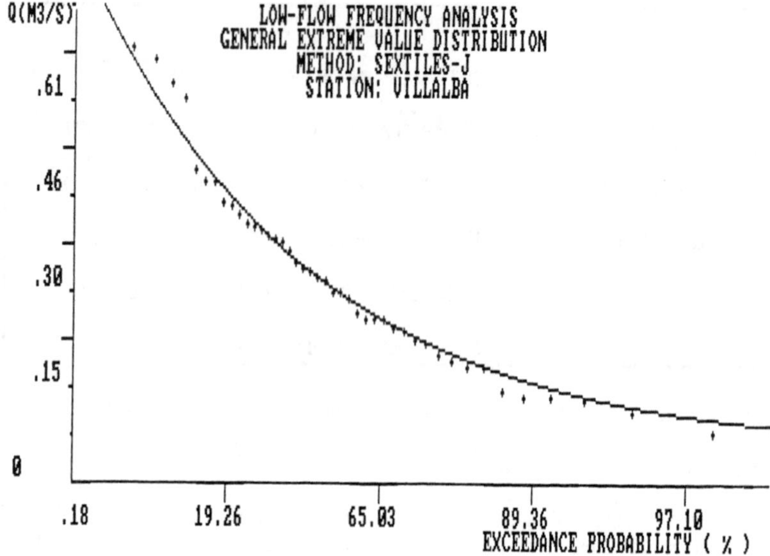

Figure 3. Empirical and Theoretical Frequency Curves for One-Day Low-Flows for Station Villalba, Chih., Mexico (1939-1981)

Acknowledgements

The authors wish to express their deepest gratitude to the Universidad de las Americas-Puebla and Universidad Nacional Autonoma de Mexico, respectively, for their support given to produce this paper.

References

Kite, G.W. (1988).Flood and Risk Analyses in Hydrology, Water Resources Publications, Littleton, Colorado.

Matalas, N. C.(1963). Probability Distribution of Low Flows, Statistical Studies in Hydrology, Geological Survey Professional Paper 434-4, pp A1-A27.

Raynal, J.A. and Escalante, C.A.(1989). FLODRO: USER'S MANUAL.

Salas, J. D. and Smith, R.(1980). Computer Programs of Distribution Functions in Hydrology. Colorado State University, Fort Collins, Colorado.

Inverse Computational Methods for Open-Channel Flow Control

E. Bautista, A. J. Clemmens, M. ASCE, and T. S. Strelkoff, M. ASCE[1]

Abstract

A study was conducted to compare three methods for solving the inverse problem of unsteady open-channel flow. Similar solutions resulted from the Bureau of Reclamation's GSM model and an implicit finite difference model, except under extreme transients in which case GSM performed better. However, GSM was unable to compute a solution for some examples. An explicit finite difference model was also examined and was found to be inherently unstable.

Introduction

A classical problem in the control of open-channel water delivery systems is determining how to operate the canal's control structures to satisfy predetermined demand schedules. Wylie (1969) was the first to propose a computational procedure to address this problem based on the unsteady, gradually-varied open channel flow equations. The method, named "gate stroking", uses the method-of characteristics to solve for the upstream depth and discharge, as well as the corresponding control structure motions, for given initial conditions and desired discharge and depth schedules at the canal's downstream boundary. Hence, this problem has been referred to as the inverse computation of open-channel flow. An operational gate stroking model (GSM) is currently available (Falvey and Luning, 1979) but it has not been widely used as a management tool because, among other reasons, exact demands are often not known far enough in advance, the computed inflow schedules are often impractical, the model is complex and requires some level of understanding by the user, etc..

Two computationally simpler alternatives for solving the inverse problem have been recently proposed. Chevereau (1991) developed a linearized implicit finite

[1] Respectively, Agricultural Engineer, Supervisory Research Hydraulic Engineer, and Research Hydraulic Engineer (and Research Professor, Univ. of Arizona), USDA-ARS U.S. Water Conservation Laboratory. 4331 E. Broadway Rd. Phoenix AZ 85040

difference solution based on Preissman's discretization scheme (Liggett and Cunge, 1975). Chevereau's strategy rotates the boundary value problem 90° in the x-t plane so that "initial conditions" are given by the required depth and discharge at the physical downstream boundary, while "boundary conditions" are given by discharge and depth at, respectively, the initial and final times.

Also using Preissman's scheme, Liu *et al* (1992) proposed an explicit finite difference solution in a rectangular region of the x-t plane. For the first cell, they used the steady-state discharge given at the last time of interest and the corresponding depth computed from backwater calculations to solve for two of the four unknowns. The problem is reduced then to solving, cell-by-cell, a system of two equations and two unknowns, moving first backward in space and then backward in time.

The purpose of this study is to compare the method-of-characteristics, implicit finite difference, and explicit finite difference inverse computation methods and test their applicability to single-pool canal systems differing widely in physical characteristics.

Methodology

Test cases were developed based on examples originally presented by the authors of each method (Table 1). Two of the original examples had to be modified to accommodate input data constraints imposed by the GSM model which allows the user to prescribe only discrete changes in flowrate and only at hourly intervals. For all examples, initial time was taken early enough to insure including in the calculations the characteristics relating the initial steady-state flow with conditions specified at the downstream boundary of the canal. Results were evaluated by comparing the upstream hydrographs computed with each method against each other and by testing whether the computed inflows, when fed to a simulation model, reproduced the desired outflow. A method-of-characteristics model was used to execute the forward simulations with the prescribed depth variations as the downstream boundary condition.

Table 1. Example canal data.

	Example 1	Example 2	Example 3
Reference	Wylie (1969)	Liu *et al* (1992)	Chevereau (1991)
Canal Length (m)	1,219	2,500	10,000
Slope	0.0002	0.001	0.00010
Bottom Width (m)	3.048	5.0	3.0
Side Slope	1.0	1.5	0.0
Manning n	0.015	0.025	0.0167
Initial checked-up depth/normal depth	1.00	1.91	N.A.

Results and Discussion

Figures 1a,b depict the solutions obtained for example 1. For this relatively short canal with small slope, GSM and the implicit model produced similar and relatively smooth solutions, but the explicit method exhibited oscillatory behavior. More importantly, the prescribed discharge hydrograph was less accurately reproduced with the explicit solution (Figure 1c). Greater numerical instability was observed with the other examples tested, causing the explicit solution to fail. The probable reason for the instability is that the explicit method ignores the backward characteristics emanating from the canal's downstream boundary. These errors appear to be less significant in cases where the computed transient wave travels down the canal without significant deformation, as in the example originally presented by the authors (Liu *et al*, 1992). Also, these authors used large time increments and a time weighting factor equal to unity, strategies that dampen the solution.

Some computational difficulties were experienced also with the implicit model of Chevereau. For the second example, a steep canal with a pronounced backwater curve, the computed upstream inflow at the final time did not match the given upstream inflow at the initial time -- the first "boundary condition" -- while the computed depths before and after the transient differed from the given final depth -- the other "boundary condition" (Figures 2a,b). These values, in fact, should agree since the initial and final (steady-state) demand flowrates and depths are the same. A loss of numerical accuracy with the linearized equations was identified as the source of the problem. New calculations were performed with smaller space and time steps but round-off errors destroyed the solution before adequate results were obtained. Because of these limitations, a fully non-linear implicit solution was developed. An additional modification was made to the implicit model: discharge was specified at both the initial and final time as "boundary conditions". This formulation has the advantage that it does not require the user or the program to compute a steady-state backwater curve, a step that can add errors to the calculations. With the non-linear method, correct values for the upstream discharges and depths were computed and the desired outflow schedule was well reproduced (Figure 2a).

Figures 1 and 2 show that the inverse solutions, when fed to a forward simulation model, produce outflow hydrographs that are smeared relative to the desired outflow functions. This smearing can be far more pronounced with the implicit model than with GSM due to the diffusive properties of the Preissman scheme. Such is the case of the transient of Figure 3. For this example, a constant depth of 3 m was imposed at the canal's downstream boundary. Results also show that the demand outflow hydrograph was less accurately reproduced by the implicit solution than by GSM. Note, however, that GSM also failed to restore the desired base-flow conditions. Inaccuracies in the computed inflows resulted in oscillations that decayed slowly with time due to the null-flow conditions. With nonzero flows, as in examples 1 and 2, frictional losses quickly eliminated the effect of these errors.

Tests with other nonzero flow examples showed generally good agreement

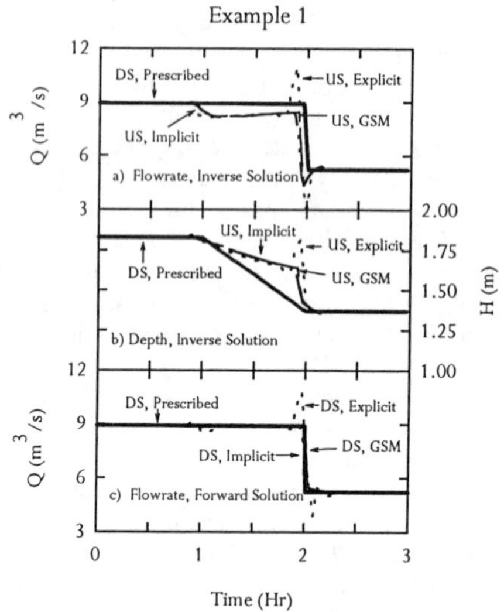

Figure 1. GSM, implicit and explicit finite difference inverse solutions for example 1. (DS = downstream; US = upstream).

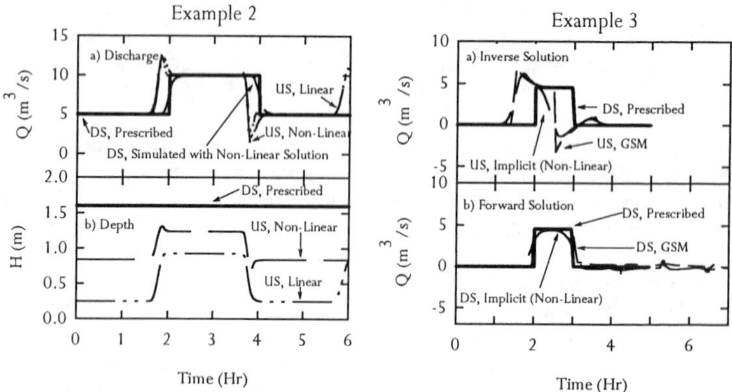

Figure 2. Linear and nonlinear implicit inverse solutions for example 2.

Figure 3. GSM and non-linear implicit finite difference inverse solutions for example 3.

between the implicit and GSM solutions. Also, solutions computed with the implicit model reproduced the demand outflow hydrographs with great accuracy, when the prescribed changes were gradual. This implies that both models can produce nearly equivalent and equally satisfactory results under an extensive range of canal conditions. Better results can be expected with GSM with extreme transients, but in such cases neither model is entirely accurate and the solution may be altogether impractical, such as in example 3, with negative inflows required.

Some examples were found in which a solution could be computed with the implicit model but not with GSM. Such is the case of the example of Figure 2. The fact that the implicit solution, when fed to characteristic simulation model provided a relatively accurate reproduction of the prescribed outflow, suggests that the problem lies in GSM's program logic and not with bore formation, a situation that cannot be handled by the characteristic scheme used. Further work is needed to identify clearly the source of the difficulty.

Conclusions

- The proposed explicit model is mathematically incorrect, leading to numerical instability and inaccurate results.
- The proposed linear implicit model can be inaccurate under certain flow conditions and, thus, a non-linear solution procedure is recommended with discharge specified as a "boundary condition" at both the initial and final time.
- Similar results can be obtained with GSM and the implicit model, except in cases where rapid and large changes in flow are demanded, case in which GSM is more accurate. However, GSM appears to be less robust as it failed to produce a solution for certain examples.

References

Chevereau, G. (1991). Contribution a L'Etude de la Regulation dans les Systemes Hydrauliques a Surface Libre. Doctoral Thesis. Institut National Polytechnique de Grenoble. Grenoble, France.

Falvey, H.T. and Luning, P.C. (1979). Gate Stroking. Report REC-ERC-79-7. USDI-U.S. Bureau of Reclamation. Washington, D.C.

Liggett, J.A., and Cunge, J.A. (1975). Numerical methods of solution of the unsteady flow equations. in: Unsteady Flow in Open Channels, Vol. I. K. Mahmood and V. Yevjevich, eds. Water Resources Publications Fort Collins, CO.

Liu, F., Feyen, J. and Berlamont, J. (1992). Computational method for regulating unsteady flow in open channels. J. of Irr. and Drain. Eng., ASCE 118(10): 674-689.

Wylie, E.B. (1969). Control of transient free-surface flow. J.Hydr.Div., ASCE. 95(1): 347-361.

Modernization of the Begoña Irrigation District Operation

S. Córdoba[1], M. Iñiguez[1], E. Mejia[1], A. Ramirez[1] B. de León M.[2], V. M. Ruiz C.[2T]

Abstract

This paper presents some aspects of the operation modernization project of the Begoña irrigation district. The Begoña district is divided in 6 operation sections. To improve the manual downstream water-depth pool operation method currently used in the Begoña district, the hydraulic infrastructure, flow rate control and turnout structures in each section, are to be changed for more efficient and uncomplicated devices. AMIL and overshot gates, long crest weirs and constant flow rate modules are used. A distributed downstream control system is used to regulate the flow rate between the sections. The downstream regulator, used in each section, is composed of two proportional-integral control algorithms and two feedforward actions. The control algorithm regulates the downstream water level in each section. The feedforward actions are based on the upstream propagation of the downstream flow rate changes at the regulation structures and a black box model to determine the flow rate evolution at the control structures needed to satisfy the future water demands. The simulation results obtained show the benefits of the operation modernization project.

Introduction

Since 1990, the "Colegio de Posgraduados" and the IMTA have been modernizing the operation of the Begoña irrigation district, in a pilot project, where different technologies are used to improve operations. The project is composed of five components: watershed management, main distribution system operation, on-farm irrigation management, real-time scheduling and environmental impact.

----

[1]P.de Hidrociencias, I.R.N., C. de Posgraduados,Montecillo, Edo. México, Mexico
[2]Riego y Drenaje, IMTA, Paseo Cuauhnahuac 8532, Jiutepec, Mor. 62550, México
[T]Author to whom correspondence should be addressed

The operation modernization project consists of the substitution of the control structures and outlets along the canal, the use of a distributed downstream control system and the rectification of the canal slope and bank irregularities.

The Begoña District

The Begoña district, built in 1968, is a 10,200 ha irrigation district located in central Mexico. The water is transported from the main dam (I.A.) by the Laja river and two main canals, the left bank (LBMC) and the right bank main canals (RBMC) (Figure 1). Two diversion dams channel the water from the river to the main canals. Both are trapezoidal concrete-lined canals. The RBMC is 20 km long with 28 reaches and 70 turnouts. The LBMC is 7.7 km long with 7 reaches and 34 turnouts. The main canals have slope and bank irregularities that make limit the flow below the maximum design value of 10.15 m^3/s for the RBMC and 3.0m^3/s for the LBMC.

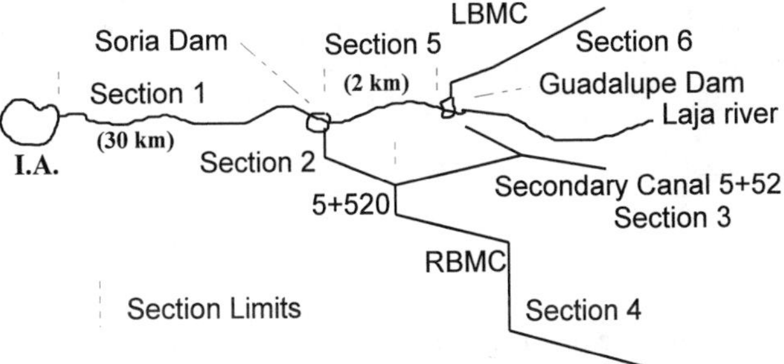

Figure 1. Main water distribution system of the Begoña district and the operational sections

The turnouts are secondary canals and farm outlets with a capacity of 0.1m^3/s to 4.7 m^3/s. The turnout structures are located at the downstream end of the reaches. The structures are slide or constant head gates for the secondary canals and Miller gates for the farm outlets. The flow rate through the structures is described by orifice flow equations. The turnout characteristics require the use of a constant downstream water-depth pool operation method (Buyalski, 1991). A manual upstream control method is used.

Modernization Project for the Begoña District Main Canals

The low capacity of the main canals, the great number of control structures and turnouts, the lack of field experience in modernization projects, and the funding for operation improvement motivated the used of a mixed solution for the operation modernization. The existing infrastructure was substituted by more efficient uncomplicated devices, and a distributed downstream control system was adopted.

To complete the project, the canal banks and slope are returned to the original design characteristics through maintenance. In the modernization project, the Begoña district was divided into six operational sections (Figure 1).

The control structures along the main canals, radial and slide gates, will be changed for AMIL and overshot gates, and duckbilled weirs. The physical and hydraulic characteristics of the canal determined which structure would be used in each case. The overshot gate was considered when high turnouts are operated intermittently, or the water-depth upstream of the structure must be kept constant over the complete flow rates range or both. For points where the flow rates are above $5m^3/s$ and the head greater than the minimum required for operation, the AMIL gate was selected. In the other cases, duckbilled weirs were selected. Constant flow rate modules will replace all the turnout structures on the main canals.

In the sections, an upstream control method will be used; between the sections, a downstream control method will be introduced. The modernization project calls for the decoupling of the canal operation variables flow rate and water depth. The water depth is to be regulated by the infrastructure and the flow rate, by the canal operation.

Distributed Downstream Control System

A distributed control system was chosen to solve the control problem. In each section, proportional and proportional-integral (P-PI) regulation algorithms were used. The section cross coupling effects were reduced by means of a feedforward action and the use of the flow rate as a control variable (Figure 2). The feedforward (I) transmits flow rate variations introduced by the local regulator to the local upstream regulators, improving the response time. The use of flow rate as control variable reduces downstream propagation of disturbances. From the desired flow rate at the control structure, the opening is determined using the gate model (Figure 2).

The goal of the P-PI regulator is to hold the downstream water level constant at the end of the section and to use the upstream water level (canals) to increase the stability margins, gain and phase (España and Ruiz, 1987). The regulation algorithm P-PI, in a discrete and recursive form, is implemented as (Isermann, 1982):

$$\Delta Q_i^r(t) = q_{0a}\, e_a(t) + q_{1a}\, e_a(t-1) + q_{0b}\, e_b(t) + q_{1b}\, e_b(t-1) \qquad (1)$$

where ΔQ_i^r is the upstream flow rate increment in the section; $e_a(t) = Y_{aref}(t) - Y_a(t)$; $e_b(t) = Y_{bref}(t) - Y_b(t)$; $q_{0x} = K_{px} + K_{ix}/2$; $q_{ix} = -K_{px} + K_{ix}/2$; Y_x and Y_{xref} are the water level and its reference profile at x respectively; K_{px} is the proportional constant; K_{ix} is the integral constant; 'x' = 'a' or 'b' specifies the upstream and downstream end of the section, respectively.

The most important variation in flow rate during operation takes place when farmers change their water requirements. Since these requirements can be determined in advance, it is possible to use a feedforward (II) to reduce the water level variation. The feedforward is based on a black box dynamic model of the flow rate evolution in a section where the water level, downstream of the reach, is kept constant. The models used are a pure delay system (Papageorgiou and Messmer, 1985) and a discrete first-order system with a zero (Rodeller et al., 1993). The inverse of the model for the different sections is used to determine the evolution of the upstream flow rate in each section needed to satisfy the downstream demand without producing variations in the water level downstream of the sections.

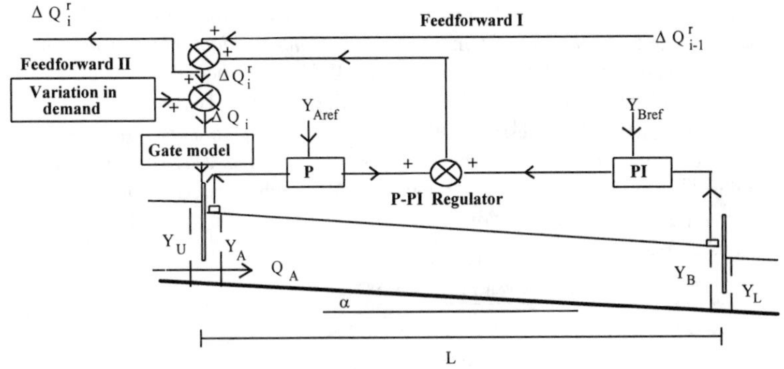

Figure 2. Regulation algorithm

Results

The regulator parameters used were: for the first section (P-PI) $K_{pa} = 0.0$, $K_{pb} = 0.5$, $K_{ib} = 0.01$; for the second section $K_{pa} = 4.00$, $K_{pb} = 1.5$, K_{ib}=0.07 and for the fourth section (P-PI) $K_{pa} = 3.5$, $K_{pb} = 1.0$, $K_{ib} = 0.05$. The sample time of the P-PI regulator was 12 min.

In Figure 3, the performance of the control system with the feedforward component I (IC) and without the feedforward component (NC) are shown for a variation of 100 l/s (t=5.0 h) at the turnout 6+860 of the RBMC (Section 4). The feedforward component reduces the perturbation effects over the canal and eliminates it faster. Without the feedforward action, the system is almost unstable.

Conclusions and Comments

The infrastructure modification and the distributed downstream control system proposed improved canal operation, by decoupling water depth and flow rate operation variables. The water depth was mainly regulated by the infrastructure (upstream hydromechanical control) and the flow rate by the operation. The use of the feedforward component I and the flow rate as a control variable reduced the cross coupling effects between the canal sections and improved the stability robustness of the control algorithm.

Other mono- and multi-variable control algorithms are being studied

References

Buyalski, C. P., D. G. Ehler, H. T. Falvey, D. C. Rogers and E. A. Serfozo (1991). *Canal Systems Automation Manual*. Bureau of Reclamation. Denver, USA.

España, M. and V. M. Ruiz-Carmona, (1987). *Control adaptable con modelo de regulación de un canal de riego*. II-UNAM, II-5142, Mexico.

Isermann R. (1989). Digital control systems, Springer Verlag, Berlin, Germany.

Papageorgiou, M. and A. Messmer (1985). *Continuous-time and discrete-time design of water flow and water level regulators*. Automatica, Vol. 21, No. 6: 649-661.

Rodellar, J., M. Gómez and L. Bonet, (1993). *Control method for on-demand operation of open-channel flow*. J. of Irrig. and Drain. Engrg, ASCE, Vol. 119, No. 2:225-241.

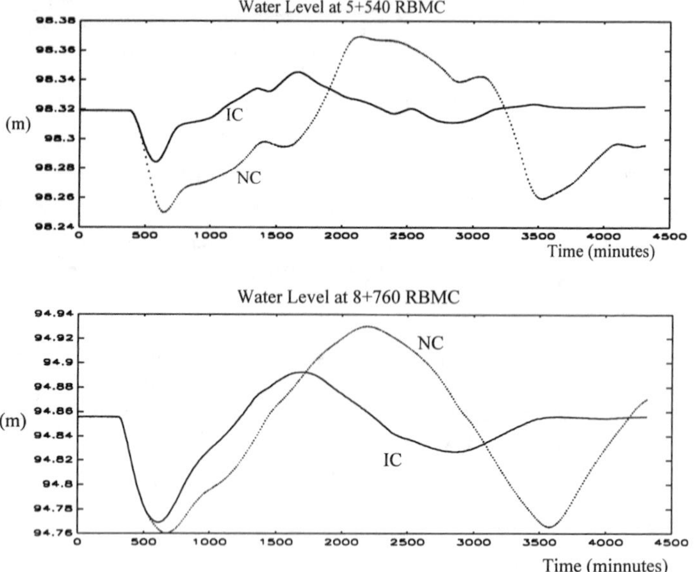

Figure 3. Demand variation of 100 l/s at turnout 6+860, Section 4.
IC with feedforward I; NC without feedforward I

FIELD DATA ON PARAMETERS OF STOCHASTIC OPEN-CHANNEL FLOW

By Timothy K. Gates[1], Member, ASCE and Muhammad Al-Zahrani[2]

ABSTRACT

The de Saint Venant model is presented as a set of stochastic partial differential equations whose parameters are spatiotemporal random fields. Extensive field data were analyzed to describe the statistical characteristics that must be captured in implementing an appropriate solution methodology. Results provide evidence that model parameters have high relative variability, are statistically nonhomogeneous, and commonly have non-normal residuals with strong lag-dependent correlation structure. Preliminary results from an application to a reach of the Columbia river are presented.

UNCERTAINTY IN DE SAINT VENANT OPEN-CHANNEL FLOW

To employ the well-known de Saint Venant model of one-dimensional open-channel flow requires specification of the values of the parameters representing physical properties, boundary conditions, initial conditions and sink/source terms. The values assumed by these model parameters are always, to a varying extent, ambiguous. The spatial and temporal variability that is inherent to natural phenomena, as well as that introduced by human intervention, present a wide variety of possibilities. Furthermore, attempts to quantify model parameters at space-time points in a system are always impaired by measurement error and limited samples. Parametric uncertainty is important because it generates through the governing equations an uncertainty in the predicted flow behavior.

Stochastic analysis provides a framework for formal consideration of the possibilities related to flow processes in an open-channel by assigning to each possibility an associated probability of occurrence. This paper describes de Saint Venant flow as a physically-based spatiotemporal stochastic process. Such a process incorporates the spatial and temporal structure of the random system parameters and relates them to the random dependent variables through deterministic mathematical models of physical behavior. This work was motivated by our desire to explore how parameter uncertainty affects analysis of unsteady flow for generalized and representative stream systems. Prerequisite to the development of a robust solution methodology, however, is a consideration not only of the structure of the governing equations but also of the actual statistical characteristics of the model parameters. In this paper, we analyze an extensive set of field data that reveals a statistical structure in the parameter random fields that is far more complex than has been commonly assumed.

[1]Assoc. Prof., Civil Engrg. Dept., Colorado State Univ., Fort Collins, Colo. 80523
[2] Graduate Student, Civil Engrg. Dept., Colorado State Univ., Fort Collins, Colo.

RANDOM FIELDS AND STOCHASTIC DIFFERENTIAL EQUATIONS

A spatiotemporal random field (STRF) $B(\mathbf{x}, t; \omega)$ is defined as a family of random variables indexed by their positions $\mathbf{x}$ in space and their positions t in time and dependent upon events ω in a Hilbert (mean square convergent) probability space. This notion can be extended to a set of spatiotemporal random fields which are stochastically correlated to one another. Such a set is defined as a vector STRF.

In de Saint Venant open-channel flow, each of the parameters are subject to uncertainty and can be modeled as a STRF, together forming a vector STRF. The de Saint Venant equations then may be posed in a stochastic setting as the following nonlinear stochastic partial differential equation (SPDE), where $\mathbf{x} = x$ for one-dimensional flow:

$$\partial h/\partial t + [A(x, t; \omega)/T_w(x, t; \omega)]\partial u/\partial x + u\partial h/\partial x = q_s(x, t; \omega)/T_w(x, t; \omega) \quad (1)$$

$$S_f(x, t; \omega) = S_o(x, t; \omega) - (1/g)\,\partial u/\partial t - (u/g)\,\partial u/\partial x - \partial h/\partial x + [q_s(x, t; \omega)/gA(x, t; \omega)]\,[u - u_s(x, t; \omega)] \quad (2)$$

where the channel friction slope (m/m) is evaluated with the Manning equation as

$$S_f(x, t; \omega) = u^2 n^2(x, t; \omega)/R(x, t; \omega)^{4/3} \quad (3)$$

and where $h = h(x, t; \omega)$ is the flow depth (m), $u = u(x, t; \omega)$ is the cross section-averaged x-direction flow velocity (m/s), $A(x, t; \omega) = A[\Gamma(x, t; \omega)$ is the area of the channel flow cross section (m^2), $T_w(x, t; \omega) = T_w[\Gamma(x, t; \omega), h]$ is the top width of the cross section measured at the water surface (m), $R(x, t; \omega) = R[\Gamma(x, t; \omega), h]$ is the hydraulic radius of the channel cross section (m), $\Gamma(x, t; \omega)$ is a vector of parameters used to describe channel cross-section geometry, $q_s(x, t; \omega)$ is the lateral seepage rate per unit length along the channel (a sink/source term) [(m^3/s)/m], $S_o(x, t; \omega)$ is the channel bed slope (m/m), and $u_s(x, t; \omega)$ is the seepage velocity (m/s).

For problems of practical interest, closed-form solutions to equations (1) - (3) are not available. They can be approximated by a set of algebraic difference equations applied to a discretization of the flow domain in space and in time. In deterministic problems, the difference equations are applied and solved in the continuous (x, t) domain at a discrete number of points in space and time. In stochastic problems, however, solutions on the probability space also must be obtained. That is, probability distributions or statistical moments of the dependent variables at the designated discrete set of space-time points in the flow domain must be described. The principal methodologies available for solving nonlinear SPDE of this type are finite-order analysis and Monte Carlo simulation. The methodology that is adopted must be able to solve a set of nonlinear SPDE whose parameter STRFs are statistically nonhomogeneous, have relatively large variance and complex covariance structure. The fact that the parameters of de Saint Venant flow often reflect these features is demonstrated in the next section using field data.

STOCHASTIC CHARACTERIZATION OF MODEL PARAMETERS

The results presented here are based upon a review and analysis of several representative data sets on hydraulic parameters derived from a variety of field conditions. While not exhaustive, they are indicative of the statistical character of the de Saint Venant parameters with regard to consideration of relative variability, statistical homogeneity and covariance structure. More details are presented in Al-Zahrani (1995).

Relative Variability of Parameter Data

Manning Hydraulic Resistance

It has been primarily in recent years that data sets have appeared that allow statistical descriptions of field variability in Manning's n. Data from eleven different field studies were analyzed. Each data set was found to best fit (Chi-square test) a lognormal probability distribution, with the exception of one set which was normally distributed. Coefficient of variation (ratio of standard deviation to mean),*CV*, values for the sample data sets range from 0.06 to 0.51 and in four cases exceed the maximum of 0.25 suggested in the literature for application of first-order analysis.

Bed Slope and Cross-Section Geometry

Compilation from a variety of available technical reports and agency files has resulted in data sets representing longitudinal and cross-section geometry in 26 distinct reaches, varying in length from 1.5 to 585 km, taken from 12 different rivers in the United States. Results of a statistical analysis of all 26 reaches are briefly summarized.

Values of S_o showed significant variability along the considered channel reaches. Channel reaches were partitioned into segments of approximately equal length, each containing about 30 or more surveyed cross-sections. *CV* values for S_o ranged from a low of 1.56 in a segment of the Colorado River to a high of 260.09 in a segment of the Missouri River, revealing a very high relative variability.

Cross section geometry was modeled by determining top width, flow area, and hydraulic radius from surveyed elevations in each cross section and fitting them to the following power functions of flow depth: $T_w = t_1 h^{t_2}$, $A = a_1 h^{a_2}$, $R = r_1 h^{r_2}$, where the parameters t_1, a_1, r_1 , t_2, a_2, and r_2 were determined by least-squares nonlinear regression with resulting coefficients of determination (r^2) typically exceeding 0.95. Hence, in this model $\Gamma = [t_1, t_2, r_1, r_2, a_1, a_2]$.

Results showed that the *CV* values of the parameters of Γ for the considered segments of the river reaches ranged from 0.25 to 1.95 for t_1, from 0.10 to 1.27 for t_2, from 0.30 to 3.04 for a_1, from 0.09 to 0.63 for a_2, from 0.11 to 1.56 for r_1, and from 0.09 to 0.62 for r_2. The guideline of 0.25 for maximum *CV* for first-order analysis was exceeded in the parameters of Γ, often by a large margin, in 78% of the river segments considered. In 45% of the cases the *CV* exceeded 0.50 .

Seepage

Data on seepage, q_s, to or from rivers is very sparse. However, Haskell (1994) statistically analyzed seepage outflow measured by seepage meters along 10 different earthen canals. The *CV* values for all of the considered segments indicate high relative variability in q_s, ranging from a low of 0.23 to a high of 2.36. Data sets of seepage from 20 earthen canals measured by ponding tests along reaches showed *CV* values ranging from 0.09 to 0.69.

Statistical Homogeneity of Parameter Data

Manning Hydraulic Resistance

None of the data sets for n were extensive enough to allow determination of statistical trends. Physical considerations, however, would suggest that trends in both the mean and variance of hydraulic resistance in open-channels should be anticipated. Systematic changes in geomorphological features, soil types, and vegetation are commonly observed in natural channels and canals. The well-established dependence of hydraulic resistance on these characteristics suggests corresponding variation in the statistical structure of Manning's n.

Bed Slope and Cross-Section Geometry

Channel bed slope commonly displays a random behavior showing no significant systematic variation over the channel reach. Tests indicated (significance level of 0.05) trend in the variance of S_o in only 3 out of 26 cases and significant trend in the mean in only 4 cases. A log transformation was used to remove the variance trend in each case. Cubic regression equations were used to model the trend in the mean of the transformed data. The probability distribution of the residuals were determined non-normal for only 12% of the cases, but could not be determined for 58%.

Similar analysis of the parameters of Γ for the considered river reaches revealed trends in the variance of t_1 for 35% of the cases, of t_2 for 27% of the cases, of a_1 for 35% of the cases, of a_2 for 27% of the cases, of r_1 for 27% of the cases, and of r_2 for 23% of the cases. In most instances, the trend was removed by a log transformation of the data; however, in 8 cases a power transformation was used. Significant trends in the mean of transformed (if required) values of the parameters of Γ were detected in the majority of cases: in 92% of the cases considered for t_1, in 88% of the cases for t_2, in 85% of the cases for a_1, in 88% of the cases for a_2, in 85% of the cases for r_1, and in 92% of the cases for r_2. Nonlinear regression analysis of the data yielded trend equations that were usually cubic but in a few cases quadratic. No periodic components have been detected in the data for S_o or Γ. The probability distributions of the residuals (transformed and detrended data) were found to be non-normal in 73% of the cases for t_1, in 65% of the cases for t_2, in 75% of the cases for a_1, in 65% of the cases for a_2, in 69% of the cases for r_1, and in 62% of the cases for r_2. Of the total non-normal residuals, about 46% were found to be distributed Weibull, 12% were Gumbel, 7% were gamma, and 2% were lognormal. Parametric probability distributions could not be determined for 11% of the cases.

Seepage

Variance trends were detected by Haskell (1994) in the seepage from two earthen canals. Power and log transformations were used to transform the data. Cubic trends in the mean were detected in the data for two canals. Quadratic and linear trends were found in data for two others, respectively. The data were not extensive enough to allow any other statistical trends to be detected. Distributions of the detrended data were lognormal in three cases, normal in one case, and gamma in one case.

Covariance Structure of Parameter Data

Manning Hydraulic Resistance

Vegetation density in earthen canals has been found to be significantly correlated in space and n to be significantly correlated with vegetation density. Available data sets, however, are too limited to allow direct inference of space-time correlation.

Bed Slope and Cross-Section Geometry

Autocorrelation and crosscorrelation coefficients were computed for different lag distances (separation distances between cross sections) for the residuals of S_o and of the parameters of Γ. Significant correlation structure was found in 9 river reaches where data were sufficient (50 or more cross-sections) to permit inference. The general trend of functions of correlation versus lag distance conforms to a typical exponential decay. The data indicate significant spatial correlation structure with dependence among parameter values extending over distances as great as 1.1 km.

Seepage

Detrended seepage data for one canal indicated a spatial autocorrelation length of 0.05 km. Remaining data sets were too small to allow inference of significant spatial autocorrelation structure.

EXAMPLE APPLICATION

Data along 10 km of the Columbia River allowed hydraulic parameters to be modeled as multivariate random fields located at 50 cross sections. The stochastic de Saint Venant equations were solved using Monte Carlo simulation (700 realizations) to consider flow uncertainty associated with installation of a cross regulator downstream. Flows were simulated using a stochastic upstream hourly peak flow hydrograph. Figure 1 shows the computed stochastic flow velocity over the reach at time 7 hours.

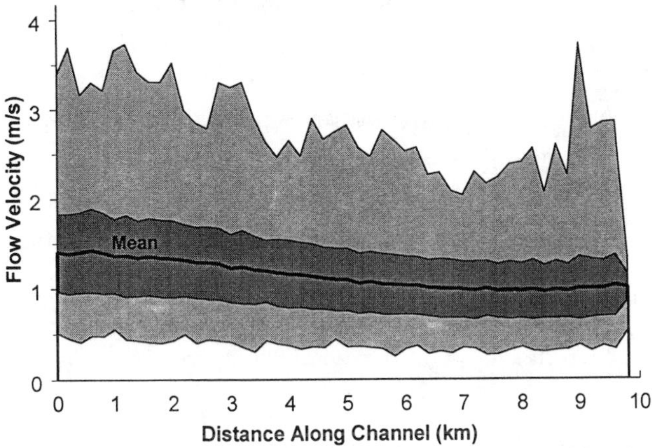

Figure 1. Computed Stochastic Flow Velocity for Peak Hydrograph, t = 7 hrs,Reach of Columbia River (Dark Shading: Values within ± One St. Dev. from Mean, Light Shading: Values within Max. and Min.).

SUMMARY AND COMMENTS

Field data were analyzed to describe the statistical characteristics that must be captured in implementing a solution to the stochastic de Saint Venant equations. Results showed that the *CV* typically exceeds by wide margins the limit of 0.25-0.50 required for first-order analysis; significant nonlinear trends in the variance and mean of the parameters are common, rendering the assumption of statistical nonhomogeneity invalid for the original data; probability distributions of the residuals are commonly non-normal; and lag-dependent autocorrelation and crosscorrelation in the residuals are strong. The stage is now set for exploring stochastic computational solutions to hydraulic engineering problems, as indicated by preliminary results from an application to a reach of the Columbia River using Monte Carlo simulation.

REFERENCES

Al-Zahrani, M. (1995). "Stochastic modeling of unsteady open-channel flow". Ph.D. Thesis, Civil Engineering Dept., Colorado State University, Fort Collins, Colo.

Haskell, W. C. (1994). "Statistical characterization of seepage losses from open channels." M. S. Thesis, Civil Engineering Dept., Colorado State University, Fort Collins, Colo.

Observations of initial sediment motion in a turbulent flow generated in a square tank by a vertically oscillating grid

D. A. Lyn[1]

Abstract

The turbulent flow generated in a tank of square cross-section by a vertically oscillating grid may provide a simple and convenient framework within which fundamental problems in sediment transport can be fruitfully studied. The present work describes such a laboratory flow configuration, and discusses its advantages and difficulties. Measurements in a sediment-free flow using laser-Doppler velocimetry are reported, and some preliminary observations, based on flow visualization, of initiation of motion are offered. Results on limited data suggest that a modification of the traditional Shields criterion might be applicable.

1. Introduction

Studies in sediment transport and river engineering have often been motivated by specific engineering problems, so that flow in a straight flume has been the standard flow configuration. A different approach has been followed in fundamental studies of turbulent flows, where simple idealized theoretical and physical models such as isotropic and/or homogeneous grid-generated flows have contributed to our understanding of turbulent flows. In the present study, a flow generated in a tank of square cross-section by a vertically oscillating grid (Fig. 1) is proposed for the study of sediment transport phenomena. A similar system was used as early as 1938 by Rouse and more recently by E and Hopfinger (1990). A portable device based on an oscillating-grid flow hasalso been proposed by Tsai and Lick (1986) for field estimates of sediment resuspension in lakes.

The simplicity of the flow stems in the ideal case from its zero mean velocity and mean shear, and its homogeneity in all except the vertical direction beyond a certain distance from the grid. Even in channel flows, local regions of zero or low bed shear stress may arise, e.g., in the vicinity of the reattachment point

[1] School of Civil Engineering, Purdue Univ., W. Lafayette, IN47907

following separation from the tops of bed forms, or in the flow in a scour hole. The oscillating-grid flow is however not primarily intended as a realistic model of river or even lake flows; rather it aims to abstract certain aspects of the suspension phenomenon and study them in isolation. A variety of problems could possibly be studied with the system, but the present paper is concerned with characterizing the basic (sediment-free) flow and with preliminary observations regarding initiation of sediment motion. The latter problem is traditionally treated using the Shields diagram (Vanoni, 1975), which focusses on a critical bed shear stress or mean velocity. In spite of its zero mean shear stress and zero mean velocity, the oscillating-grid flow can entrainment and support a suspension. Sediment entrainment and initiation of motion can therefore be posed within a wider context than the traditional. Alternatively, it may be asked whether the traditional approach can be easily modified to a quite different flow. Some first experiments with fine sands were performed to examine these questions.

2. Experimental considerations

2.1 The tank and flow conditions

In addition to its theoretical simplicity, the oscillating-grid flow has several practical advantages. It is inexpensive to implement, and its small size does not require large amounts of material. The tank was constructed of glass on all sides (except the top), permitting optical access for laser-Doppler measurements as well as flow visualization. Its square cross-section was 25 cm on a side, and it was 40 cm deep. The vertically oscillating grid was made from square plexiglass rods of side $\approx$ 0.95 cm and length $\approx$ 24 cm, with a mesh size of $\approx$ 4.8 cm. The resulting solidity was $\approx$ 36%, similar to that used in previous studies. In some respects, a greater control can be exercised over the flow characteristics than in the case of a laboratory flume. The stroke can be changed in several discrete steps from 5 cm to 11 cm, while the frequency of oscillation can be varied continuously from 1 Hz to 6 Hz, thereby enabling a study of the effect of changes in length and time scales of the generated turbulence. The grid can be placed either near the bottom depending on the problem of interest. In the present work, the stroke, S, was kept fixed at 5 cm and the midplane of the grid travel was located at an elevation $\approx$ 15 cm from the bottom of the tank. Flows at two oscillation frequencies, f, were examined: $f = 3.1$ Hz and $f = 5.9$ Hz. The depth of water in the tank was $\approx$ 20 cm, and the grid was always submerged throughout its travel.

2.2 Experimental techniques

A laser-Doppler velocimetry (LDV) system was used to characterize the turbulent flow field in the absence of bed particles in terms of time-averaged turbulence statistics. The system is a fiber-optic-based four-beam two-color TSI system with illumination from a 300 mW air-cooled Ar-ion laser and signal processing using digital burst correlation (TSI Model IFA–650 processor).

Sufficiently far from the bottom such that all four beams could be used, two components of the instantaneous velocity were measured. Near the bottom, blockage of one of the beams implied that only the horizontal component of velocity could be obtained.

3. Results

3.1 The sediment-free flow

Previous studies have noted that the ideal oscillating-grid flow (zero mean velocity and mean shear and statistical homogeneous in the horizontal plane) is only roughly approximated for z (the distance from the midplane of the grid travel) sufficiently large, with inhomogeneities still being observed even for distances two mesh sizes away from the grid (McDougall, 1986). Flow statistics at a given elevation were determined from averaging measurements at several points on the horizontal plane. The root mean squares of the velocity fluctuations in the horizontal and the vertical directions, u' and w', normalized by a velocity scale, fS, are shown in Fig. 2 for the two frequencies as a function of non-dimensional distance, z/S, from the grid mid-plane. Substantial scatter is evident. For $z/S < 2$, approximate decay laws, $u' \sim z^{-1.2}$ and $w' \sim z^{-1}$ may be estimated, in rough agreement with previous work. No attempt was made to determine a virtual origin of the decay.

The effect of the solid boundary can be found in the change in decay and isotropy of the flow statistics. For $z/S > 2$, the decay behavior deviates markedly from the decay law, particularly at the higher frequency. u' and w' become approximately constant in this region, attributed to a 'reflection' of turbulent kinetic energy from the solid boundary. The generated turbulence is also not isotropic. For $z/S < 2$, the ratio, $w'/u' \approx 1.2$ or less, in agreement with previous work; for $z/S > 2$, however, $w'/u' \approx 1.5$ or more. The bottom exerts a a stronger influence on the vertical component. The effect of a solid boundary has not received much experimental study, with only Hannoun et al. (1988) observing a tendency very close to the solid boundary for w' to decrease and u' to increase. Such tendencies are not especially evident in the present results, but measurements very close to the solid boundary were not taken.

Energy spectra for both components of fluctuating velocity were also estimated. For $z/S > 1$, the oscillating frequency of the grid could not be detected, suggesting that direct effects of the grid frequency is unimportant in this region.

3.2 Preliminary observations with bed particles

The first experiments with sediment have been conducted with spherical glass particles with a specific gravity, $s \approx 2.46$, in two size range, 0.149 mm–0.177 mm and 0.063 mm–0.074 mm. It was attempted to place a thin and uniform layer of particles on the bed, but, for these size ranges, this was not entirely successful. A study of initiation of motion also requires defining clearly such a transitional event. While in a channel flow, a definition can be based on a certain level of bed-load transport, this is not possible here since there is no net overall transport. Local transport can however be detected in a change in the spatial distribution of particles on the bottom. At this preliminary stage, the distinction was based purely on visual evidence. Initiation of motion was deemed to occur when relatively regular motion of particles on the bottom, by way of changing patterns in the reflected light, was observed. For the smaller size range, this proved inadequate, since particles were observed in suspension even when there seemed to be no motion on the bed. For the larger size range, visual observations indicated motion for $f > 2.7$ Hz.

An estimate of the turbulent kinetic energy, $k = (u'^2 + v'^2 + w'^2)/2$, in the near bed region can be obtained as $k \approx 0.018(fS)^2$, where the LDV results and the assumption that $u' \approx v'$ have been used. A modified Shields parameter can be defined as $\Theta_* \equiv k/g(s-1)d$, which yields a value of 0.14 for the larger particles respectively (here $d = 0.16$ mm was assumed). These may be compared with values estimated from a traditional Shields curve for the same grain size. In the constant stress layer of a channel flow, $k \approx 4u_*^2$, where u_* is the shear velocity, and so the conventional Shields parameter may be expressed as $\Theta = k/4g(s-1)d$, or $\Theta \approx \Theta_*/4$. Thus, the present result would correspond to $\Theta \approx 0.035$. If the traditional Shields curve were used, then, for these grain sizes, $\Theta \approx 0.06$. Differences in definitions of inception of sediment motion, in the structure of the turbulent flow and entrainment processes, or in inhomogeneities in the bed may account for the discrepancy. Nevertheless, the comparison suggest that values from the traditional channel-flow-based approach, appropriately transformed in terms of k, will yield reasonable values for an oscillating-grid flow.

Work is continuing on different fronts: in refining the flow system to make it even more convenient to use and to approximate more closely a zero-mean-shear zero-mean-velocity flow, in developing imaging techniques so as yield greater precision in measurements particularly for smaller particles, in extending the range of experimental variables, and to consider other flows such as an equilibrium suspension.

References

1. E, X. and Hopfinger, E. J. (1990) *Proc. 3rd Int. Symp. on Stratified Flows*, Pasadena, Calif., 1987.
2. Hannoun, I., Fernando, H. J. S., and List, E. J. (1988) *J. Fluid Mech.*, **189**, pp. 189–209.
3. McDougall, T. J. (1979). *J. Fluid Mech.*, **94**, Pt. 3, pp. 409–431.
4. Rouse, H. (1938). *Proc. 5th Int. Cong. Appl. Mech., Cambridge, Mass.* in *Hydraulics, mechanics of fluids, engineering education; selected writings of Hunter Rouse*, Dover, New York, 1981.
5. Tsai, C.–H. and Lick, W. (1986). *J. Great Lakes Res.*, **12**, No. 4, pp. 314–321.
6. Vanoni, V. A. ed. (1975) *Sedimentation Engineering*, ASCE Manuals and Reports on Engineering Practice, No. 54, ASCE, New York.

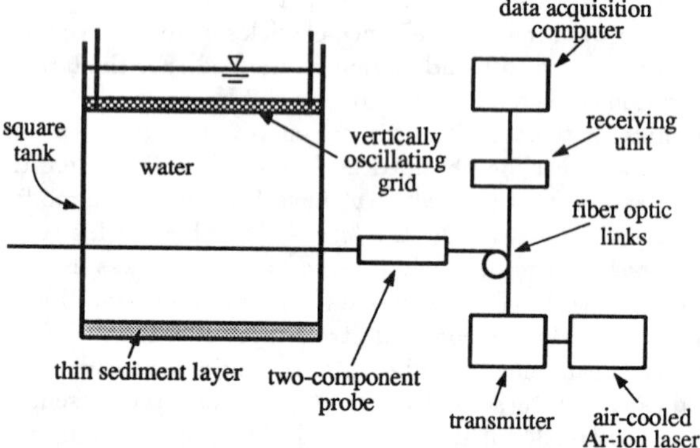

Fig. 1: Schematic diagram of flow configuration and LDV setup

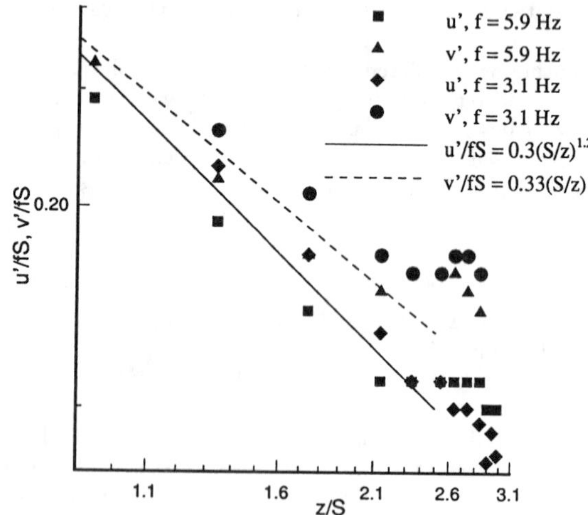

Fig. 2: Statistical characteristics of vertically oscillating grid flow from LDV measurements

FINITE ELEMENT SIMULATION OF 2–DIMENSIONAL TURBIDITY CURRENTS

Sung–Uk Choi [1] and Marcelo García[2], A.M. ASCE

Abstract

A finite element numerical model is proposed for the simulation of 2–dimensional turbidity currents. Time–dependent, layer–averaged governing equations, a hyperbolic system of partial differential equations, are employed. The Petrov–Galerkin formulation is used for the spatial discretization and a second–order finite difference scheme is used for the time integration. A deforming grid technique based on the Arbitrary Lagrangian–Eulerian description is employed to cope with the moving boundary of a propagating front. The developed numerical algorithm is applied to the simulation of laboratory observations.

Introduction

Turbidity currents are sediment–laden underflows, which occur in ocean, lakes, and reservoirs. The motion of turbidity currents results from the reduced gravitational acceleration associated with the density difference between suspended sediment and clear water. Unlike density currents, the buoyancy flux of turbidity currents changes depending upon the interaction between the current and channel bed. Turbidity currents move downslope carrying sediment through a submerged river or a submarine canyon. When a channelized turbidity current arrives to a delta or a widely open area such as a submarine fan, it starts to spread laterally and becomes a 3–dimensional flow. For non–channelized currents, both longitudinal and lateral spreading are important. The depositional pattern induced by the sediment–laden currents is of relevance to both reservoir and ocean sedimentation.

For numerical modeling of non–channelized turbidity currents, layer–averaged equations, a hyperbolic system of partial differential equations, are employed together with closure relationships. Two major difficulties are encountered in analyzing the system of equations numerically. Firstly, to solve hyperbolic partial differential equations with the finite element method requires a special numerical technique in the case of convection–dominated flows. Secondly, the governing equations of a single–layer model, as the one considered herein, constitute a moving boundary problem, the domain of which should be updated at every time step.

[1] Research Assistant, [2] Assistant Professor. Department of Civil Engineering, University of Illinois at Urbana Champaign. 205 N. Mathews, Urbana, IL 61801. USA.

For the moving boundary, both Lagrangian and Eulerian approaches are available. The Lagrangian approach can estimate the interface accurately, however, it can not allow large deformations. The Eulerian approach is good for a problem undergoing a large deformation, but it produces a blunt interface. Arbitrary Lagrangian–Eulerian (ALE) description, which is known to be superior to either method (Ramaswamy, 1990), is used herein.

The purpose of this paper is to propose a numerical technique that can be used to investigate the spreading and dilution of 2–dimensional turbidity currents.

Layer–Averaged Equations for Turbidity Currents

The 1–dimensional, unsteady, layer–averaged equations for turbidity currents in Parker et al. (1986) can be directly extended into 2–dimensional equations such as:

$$\frac{\partial h}{\partial t} + \frac{\partial uh}{\partial x} + \frac{\partial vh}{\partial y} = e_w U \tag{1}$$

$$\frac{\partial uh}{\partial t} + \frac{\partial u^2 h}{\partial x} + \frac{\partial uvh}{\partial y} = -\frac{1}{2} Rg \frac{\partial ch^2}{\partial x} + RgchS_x - u_*^2 \tag{2}$$

$$\frac{\partial vh}{\partial t} + \frac{\partial uvh}{\partial x} + \frac{\partial v^2 h}{\partial y} = -\frac{1}{2} Rg \frac{\partial ch^2}{\partial y} + RgchS_y - v_*^2 \tag{3}$$

$$\frac{\partial ch}{\partial t} + \frac{\partial uch}{\partial x} + \frac{\partial vch}{\partial y} = v_s(e_s - r_o c) \tag{4}$$

where the dependent variables are, h = the current thickness; u and v = the depth–averaged velocities in x and y directions, respectively; and c = the depth–averaged volumetric concentration. Also, e_w = water entrainment coefficient; v_s = sediment fall velocity; e_s = sediment entrainment coefficient; r_o = shape factor for near–bed concentration; u_* and v_* = shear velocities in x and y directions, respectively; S_x and S_y = slopes of the channel in x and y directions, respectively; U = the magnitude of velocity vector $[=(u^2+v^2)^{1/2}]$; and R = submerged specific gravity that usually has a value of 1.65 for quartz.

Fluid mass balance is given by eq.(1), where $e_w U$ denotes the rate of clear water entrainment from the ambient upper layer into the underflow. Conservation of momentum in x and y directions are given by eq.(2) and eq.(3), respectively. The mass balance of suspended sediment is given by eq.(4). Note that the above equations become those for conservative currents if the right hand side of eq.(4) is set equal to zero. The closure relationships for e_w, u_*, v_*, e_s, and r_o are given in Choi and García (1995).

Petrov–Galerkin Finite Element Method with Moving Mesh

The 2–dimensional, layer–averaged, governing equations, eq.(1)~eq.(4), can be recast in the following matrix form:

$$\frac{\partial \boldsymbol{Y}}{\partial t} + \boldsymbol{A}\frac{\partial \boldsymbol{Y}}{\partial x} + \boldsymbol{B}\frac{\partial \boldsymbol{Y}}{\partial y} = \boldsymbol{b} \tag{5}$$

where $\boldsymbol{Y}=\{h, q_x\ (=uh), q_y\ (=vh), p\ (=ch)\}^T$ and, $\boldsymbol{A}$ and $\boldsymbol{B}$ are convection matrices in x and y directions, respectively. Considering the mesh moving with velocities w_x and w_y in x and y directions, respectively, then eq.(5) can be written as (Akanbi and Katopodes, 1988):

$$\frac{\partial \boldsymbol{Y}}{\partial t} + \boldsymbol{A}_*\frac{\partial \boldsymbol{Y}}{\partial x} + \boldsymbol{B}_*\frac{\partial \boldsymbol{Y}}{\partial y} = \boldsymbol{b} \tag{6}$$

where $A_* = A - Iw_x$ and $B_* = B - Iw_y$ (I = identity matrix). The weighting function is defined as:

$$N_* = N + \varepsilon_x A_*^T \frac{\partial N}{\partial x} + \varepsilon_y B_*^T \frac{\partial N}{\partial y} \tag{7}$$

where the superscript T means transpose of the matrix. Notice that the weighting function N_* depends on A_* and B_*, and the level of upwinding can be controlled by the parameters, ε_x and ε_y. Using this weighting function, we can express the weighted residual form of eq.(6) as:

$$\sum_{n=1}^{n_e} \int_{\Omega} N_*^T (\frac{\partial Y}{\partial t} + A_* \frac{\partial Y}{\partial x} + B_* \frac{\partial Y}{\partial y} - b)\; dA = 0 \tag{8}$$

where n_e = total number of elements, and Ω = the area of each element. For time integration, a second–order implicit finite difference scheme is used. In order to solve eq.(8) which is nonlinear, a linearizing technique such as Newton–Raphson method can be used. The detailed procedure is described in Katopodes (1984a, b).
An analytical expression for the optimal upwind parameter ε was obtained in Choi and García (1995) for one–dimensional, homogeneous, linear equations. This can be directly extended into the 2–dimensional case as in Akanbi and Katopodes (1988),

$$\varepsilon_x = \varepsilon_y = \frac{c_\varepsilon}{\sqrt{15}} \frac{L}{U + \sqrt{Rgch}} \tag{9}$$

where c_ε is a control parameter to adjust the magnitude of upwinding and L is the length of a straight line through the centroid of the element along the flow direction. It was found from numerical experiments that a c_ε value of about 3 yields good results, free from spurious oscillations caused by the non–homogeneity and non–linearity of the equations.

The boundary conditions and initial conditions for ALE formulation are identical to those for the Eulerian and Lagrangian methods. Lagrangian description is used for the front motion at the downstream boundary, and Eulerian description is used for the upstream boundary. The rest of the nodes are translated by mesh velocities which vary between zero at the upstream boundary and the propagation velocity at the downstream boundary. For the boundaries at the two sides, zero current thicknesses are imposed. To obtain the opening angle at the downstream boundary, the velocity vector at the front edge is used. The other nodes at the front move by the way described in Akanbi and Katopodes (1988). Initial profiles of dependent variables are necessary to start the computation with the deforming grid technique. Arbitrary initial configurations, if they are reasonable compared with 1–dimensional solutions, do not have a serious effect on the convergence of the solutions.

Application Example

The developed numerical model is applied to the data from laboratory experiments by Luthi (1981). Turbidity currents were generated on an inclined (5°) channel 6 m wide and 10 m long. The initial flow variables were q_o = 116.0 cm^2/s and p_o = 0.205 cm, so that the initial Richardson number, Ri_o was 0.6. The inlet dimension was 5 cm high and 30.0 cm wide. The dense underflow was turbulent and supercritical. The use of large particles with D_s=37 μ ensured sediment deposition during the running time of the experiment.

For the computation, the domain is discretized into 196 elements, 14 for the longitudinal and 14 for the lateral direction. Bed resistance coefficient, $c_D = 0.005$, and a sediment entrainment coefficient, $e_s = 0$ are kept constant throughout the computations.

Figure 1 shows the time evolution of a 2–dimensional turbidity current after being discharged through an inlet on a sloping channel. It is interesting that the layer–averaged thickness decreases along the downstream direction because of the lateral spreading although there is a considerable amount of water entrainment from the ambient layer. A velocity field and a concentration contour at time = 70.2 s are given in *Figure 2*. This is an example of a decelerating and depositing turbidity currents. Substantial deposition occurs quite rapidly as moving downstream due to high settling velocity of the particles. In *Figure 3*, both longitudinal and lateral spreadings of a 2–dimensional turbidity current are plotted. Herein, the spreading is defined as the propagation of the maximum distance traveled in both directions. Although the measured profiles by Luthi (1981) show elliptic configurations (unlike those computed), the maximum spreadings agree well with measured data. The longitudinal spreading is no longer linear with time due to the lateral spreading and loss of buoyancy. The discrepancy shown at initial times could be caused by the shallow depth of ambient water near the inlet.

Conclusions

A numerical model for 2–dimensional turbidity currents has been proposed and applied to available experimental data. The application example demonstrates that the proposed numerical technique is capable of predicting the longitudinal and lateral spreading and dilution of turbidity currents. Coupling the bed sediment conservation equation with the preceding algorithm will provide characteristics of the sediment deposits created by the currents.

Acknowledgments

The support from the Marine Geology and Geophysics program of the U.S. Office of Naval Research (N00014–93–1–0044) is gratefully acknowledged. The writers also wish to thank A. A. Akanbi, Illinois State Water Survey, for his useful comments.

References

Akanbi, A.A. and Katopodes, N.D. (1988). "Model for flood propagation on initially dry land," *Journal of Hydraulic Engineering*, ASCE, 114(7), 689–706.

Choi, S.–U. and García, M. (1995). "Turbidity Current Modeling with Dissipative–Galerkin Finite Element Methods", submitted to Journal of Hydraulic Research.

Katopodes, N.D. (1984a). "A dissipative Galerkin scheme for open channel flow," *Journal of Hydraulic Engineering*, ASCE, 110(4), 450–466.

Katopodes, N.D. (1984b). "Two–dimensional surges and shocks in open channels," *Journal of Hydraulic Engineering*, ASCE, 110(7), 794–812.

Luthi, S. (1981). "Experiments on non–channelized turbidity currents and their deposits", *Marine Geology*, 40, M59–M68.

Parker, G., Fukushima, Y., and Pantin, H.M. (1986). "Self accelerating turbidity currents," *Journal of Fluid Mechanics*, 171, 145–181.

Ramaswamy, B. (1990). "Numerical simulation of unsteady viscous free surface flow," *Journal of Computational Physics*, 90, 396–430.

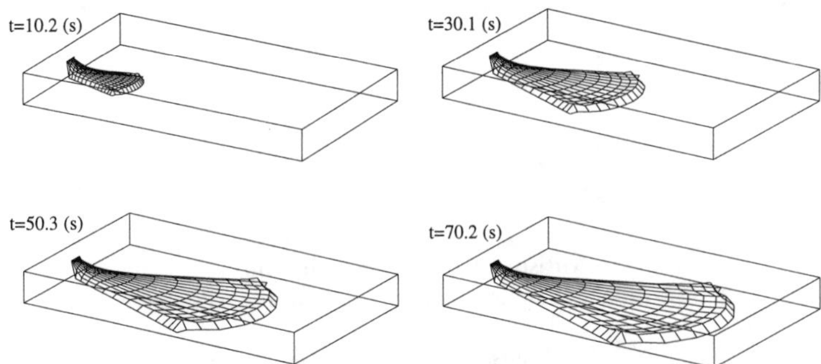

Figure 1. Time Evolution of a 2–Dimensional Turbidity Current

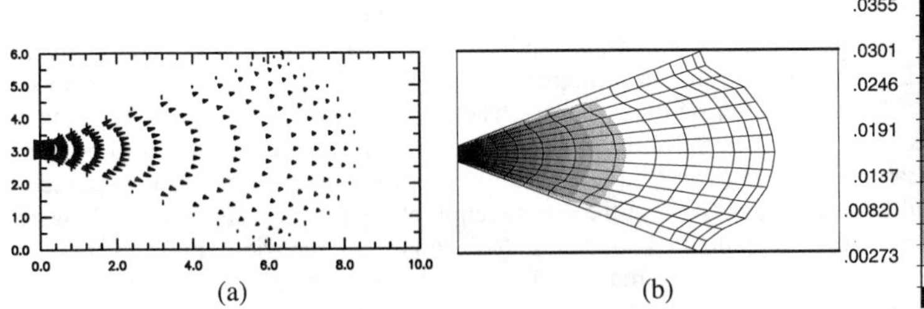

Figure 2. Velocity Field (a) and Concentration Profile (b) of a 2–Dimensional Turbidity Current at 70.2 (s)
(the right hand side table shows the values of concentration)

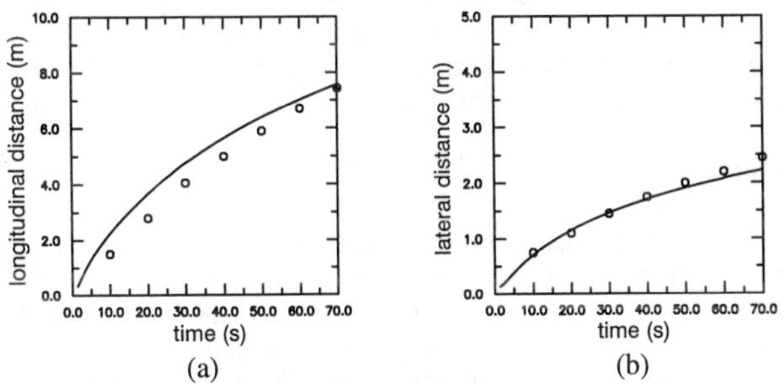

Figure 3. Longitudinal and Lateral Spreadings of a 2–Dimensional Turbidity Current

Measurements of Cross Section Integrated Sediment Transport Rate in the Nile River

Moustafa T.K. Gaweesh[1]

Abstract

Measurements of bed-load and suspended-load transport in the Nile river, in Egypt, were carried out in a cross section located downstream of the town of Beni Sweif (120 km south of Cairo). The measurements were performed using a mechanical sampler called the Delft Nile sampler which was developed to be used in sand bed rivers (Van Rijn and Gaweesh, 1992). The bed-load and suspended-load transport rates, passing the cross section of the river (width = 400 m), were measured by sampling in a number of subsections over the width of the river. The samples were taken at a representative station in each subsection. The actual sediment transport rate for the whole cross section was determined by integrating the measured sediment transport rates of each subsection. The measurements were performed during an almost constant flow discharge of 1040 m^3/s. A comparison was made between the measured sediment transport rates and the computed values using different prediction methods. It was found that the used methods for computing the bed, suspended and total load transport rates, generally, overpredict their corresponding measured values by a factor ranging between about 1 and 20 times. Future research will be aimed at measuring and analyzing the cross section integrated sediment transport rate at various cross sections along the Nile river.

Introduction

The Nile river downstream of the High Aswan Dam (HAD) has a sandy bed with median partial sizes in the range of about 300 - 600 μm. The maximum and minimum discharges downstream of HAD are about 2800 m^3/s and 700 m^3/s, respectively. The average cross section velocities are in the range from 0.5 m/s to 1 m/s. In these conditions the measurements of the bed and suspended-load transport are equally important. This paper describes the measuring procedure and

---

[1]Deputy Director, Hydraulics Research Institute, Delta Barrage 13621, Egypt.

results of sediment transport rate at a certain cross section of the river. It is the beginning of a research program aims at measuring the actual sediment transport rate at various cross sections along the Nile river, in Egypt. At that cross section, the bed and suspended-load transport were simultaneously measured using a mechanical sampler called the Delft Nile sampler (Van Rijn and Gaweesh, 1992). The sampler was extensively tested in laboratory and field conditions using different bed materials and flow velocities to determine the sampling efficiency and to study the sampling performance and statistics (Gaweesh and Van Rijn, 1994).

Field Conditions and Measuring Procedure

The field measurements of sediment transport rate were carried out in the Nile river at a cross section located downstream of the town of Beni Sweif (about 120 km south of Cairo). At that location, the cross section of the river (width of 400 m and maximum depth of 5 m) has a relatively regular profile. The bed consisted of sand with a median particle diameter of 415 μm (average of 30 bed-material samples). Sand dunes with a mean height of about 0.75 m and a mean length of about 28 m were observed from local echo sounding recordings at three longitudinal profiles conducted at 100 m, 200 m and 300 m from the left bank. The water surface slope was determined from local gauge readings and found to be 8.5 cm/km and the measured flow discharge was 1040 m^3/s. The observed water surface levels and discharge did not change during the measuring period.

Based on statistical error analysis to determine the proper number of measuring stations (Gaweesh and Van Rijn, 1994), the cross section profile was divided in 6 subsections. The measurements were conducted at a representative station in the middle of each subsection. The six measuring stations (St.1, St.2, ..., St.6) were located at 344 m, 282 m, 221 m, 179 m, 120 m and 60 m from the left bank, respectively. At each station, the measurements of the bed and suspended-load transport were performed at five locations (L_1, L_2, ..., L_5) distributed equally (7 m apart) along the bed-form length to account for the variability of the bed-load transport process (see Figure 1).

The Delft Nile sampler was used to collect bed-load samples and also suspended-load samples within a height of 0.5 m just above the bed. The bed-load sampler consists of a nozzle (entrance width = 0.096 m; entrance height = 0.055 m; length = 0.085 m) connected to a bag of Nylon material with mesh size of 250 μm. Two small propeller meters were attached to the sampler to measure the current velocities at 0.25 m and 0.5 m above the bed. An underwater video camera connected to a monitor was used to observe the sampling process. The suspended-load samples were collected at 0.1 m, 0.14 m, 0.25 m and 0.5 m above the bed by means of four intake nozzles (inner diameter = 0.003 m) that are connected to plastic hoses and operated by small pulsation pumps. A separate instrument (called Delft Fish) equipped with a propeller meter and one intake nozzle was used to measure the flow velocities and collect suspended-load samples within the height

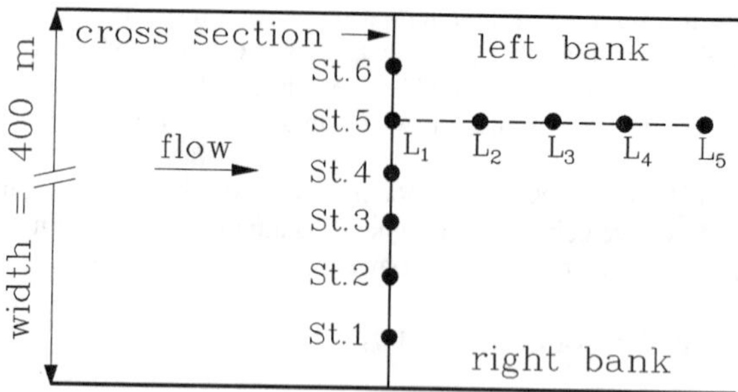

Figure 1. Layout of Measuring Stations and Locations

starting from 0.5 m above bed and up to near the water surface. A grab sampler was used to collect bed-material samples. The entire measuring procedure at each location consisted of taking the following: (A) ten instantaneous bed-load samplings; (B) eight bed-load samplings of 3 minutes each (including video recordings); (C) two velocity measurements at 0.25 m and 0.5 m above the bed during each bed-load measurement; (D) four suspended-load samplings (volume of 5 liters, each) at 0.1 m, 0.14 m, 0.25 m and 0.5 m above the bed; (E) four or five separate velocity measurements and also suspended-load samplings distributed between the water surface and 0.5 m above bed; (F) one bed material sample. The duration of this measuring procedure took about 60 - 90 minutes. The completion of measurements at each station requires a whole working day (about 7-8 hours). The bed-load samples of each station were separately dried and weighed and then put together to make a bulk sample for sieve analysis to be compared with that of the bed material samples. The individual bed-load transport rates (q_b) per unit width (in g/s/m), at each station, were determined as:

$$Q_b = \alpha \; (G-G_o) \; / \; bt \tag{1}$$

in which α = the sampling efficiency factor of the Delft Nile sampler; G = dry mass of sand catch (g); G_o = mean dry mass of instantaneous sand catches (g); b = width of the sampler nozzle (m); and t = sampling period (s). As the D_{50} of the bed material was larger than 400 μm, a value of α = 1.0 was used (Gaweesh and Van Rijn, 1994). The bed-load transport rate of each subsection was determined from the mean transport rate $(\overline{q}_b)$, per unit width, multiplied by the width of the subsection. The actual bed-load transport rate for the whole cross section (Q_b) was determined by integrating the transport rates of each subsection. The time-averaged suspended-load transport rates $(\overline{q}_s)$ per unit width (in g/s/m), at each station, were determined by depth-integration from the measurements of sand concentration and flow velocity with integration limits between the water surface level and a level equal to the height of bed-load layer. It was assumed that the bed-load transport takes place by migrating mini ripples with a height of about 0.03 m over the upper side of larger-scale dunes. Therefore, the velocity and concentration profiles were extrapolated down to a height of 0.03 m above the bed. The actual suspended-load transport rate for the whole cross section (Q_s) was determined by integrating the transport rates of each subsection. The actual total-load transport rate was obtained by summation of the bed and suspended-load transport rates.

Measured and Computed Transport Rates

The measured data were used to compute the local depth-averaged velocity and determine the effective bed roughness (k_s), (see Table 1). The measured

Table 1. Hydraulic and Bed Material Data

Station number	1	2	3	4	5	6
Median particle diameter (μm)	603	490	409	343	350	296
Mean water depth (m)	2.82	2.76	2.76	3.40	4.28	5.04
Mean velocity (m)	0.81	0.74	0.72	0.66	0.71	0.73
Effective bed roughness (m)	0.05	0.07	0.09	0.30	0.47	0.67

Table 2. Measured and Computed Transport Rates

Station number	Transport mode	Transport rate (g/s/m)							
		Measured	Frijlink (1952)	Bagnold (1966)	Mayer-P-M. (1948)	Van Rijn (1984a)	Van Rijn (1984b)	Engelund-H (1967)	Ackers-W. (1973)
1	bed	19	58	54	48	51	-	-	-
1	suspended	16	-	33	-	-	27	-	-
1	total	35	-	-	-	-	-	66	74
2	bed	12	44	47	36	35	-	-	-
2	suspended	27	-	31	-	-	26	-	-
2	total	39	-	-	-	-	-	63	61
3	bed	18	38	47	31	28	-	-	-
3	suspended	42	-	33	-	-	31	-	-
3	total	60	-	-	-	-	-	75	62
4	bed	13	30	52	25	12	-	-	-
4	suspended	42	-	40	-	-	36	-	-
4	total	55	-	-	-	-	-	97	55
5	bed	6	41	70	34	19	-	-	-
5	suspended	48	-	60	-	-	76	-	-
5	total	54	-	-	-	-	-	157	85
6	bed	4	46	84	40	19	-	-	-
6	suspended	62	-	93	-	-	189	-	-
6	total	66	-	-	-	-	-	248	127

cross section integrated bed-load, suspended-load and total-load transport rates were found to be 4.7 kg/s, 15.8 kg/s and 20.5 kg/s, respectively. The sediment transport rates, at each station, were also computed using different prediction methods. The measured and computed values are given in Table 2. The computed sediment transport rates, generally, overpredict their corresponding measured values by a factor ranging between about 1 and 20 times.

Conclusions

Field measurements of sediment transport rate were carried out at a cross section (width = 400 m) in the Nile river, in Egypt. The measurements were performed during an almost constant water levels and flow discharge of 1040 m^3/s. The cross section was divided in 6 subsections and measurements were conducted at a representative station in the middle of each subsection. The measured cross section integrated bed-load, suspended-load and total-load transport rates were found to be 4.7 kg/s, 15.8 kg/s and 20.5 kg/s, respectively. The computed sediment transport rates, using different prediction methods, overpredict their corresponding measured values by a factor ranging between about 1 and 20 times. Further measurements and analysis of sediment transport rate will be carried out at various cross sections along the Nile river, in Egypt.

Appendix. References

Ackers, P. and White W.R. (1973). "Sediment Transport: New Approach and Analysis.", Proc. Hydr. Div., ASCE, Vol. 99, No. HY11.

Bagnold, R.A. (1966). "An approach to the sediment transport problem from general physics.", U.S. Geological Survey Profl. Paper 442-1, U.S. Geological Survey, Washington, D.C.

Engelund, F., and Hansen, E. (1967). A monograph on sediment transport in alluvial streams. Teknisk Vorlag, Copenhagen, Denmark.

Frijlink, H.C. (1952). "Discussion of bed load movement formulas.", Report No. X2344/LV, Delft, The Netherlands.

Gaweesh, M.T.K., and Van Rijn, L.C. (1994). "Bed-load Sampling in Sand-Bed Rivers.", J. Hydr. Engrg., ASCE, 120 (12).

Meyer-Peter, E., and Muller, R. (1948). "Formulae for bed-load transport.", Proc. 2nd Congress IAHR, Stockholm.

Van Rijn, L.C. (1984a) "Sediment transport. Part I: bed-load transport." J. Hydr. Engrg., ASCE, 110(10).

Van Rijn, L.C. (1984b) "Sediment transport. Part II: suspended-load transport." J. Hydr. Engrg., ASCE, 110(11).

Van Rijn, L.C., and Gaweesh M.T.K. (1992) "A new total sediment load sampler." J. Hydr. Engrg., ASCE, 118 (12).

Simulation of General Scour at the US-59 Bridge Crossing of the Trinity River, Texas

Howard H. Chang[1], M. ASCE, David D. Dunn,[2] A.M. ASCE, and Jay Vose[3]

Abstract

A fluvial study of the Trinity River near US-59 in southeast Texas evaluated potential river-channel changes at the bridge crossing. The sediment-transport characteristics of the Trinity River have been altered by Lake Livingston, which retains nearly all the bed sediment. This results in a sediment deficit and erosion of the river channel downstream of Lake Livingston. Quantitative assessment of general scour in the downstream channel is based on simulation using the FLUVIAL-12 model and a 10-year record of daily mean discharges.

The simulated results show a general pattern of erosion of the river channel, with local exceptions. Simulated cross-sectional changes at the bridge crossing consist of channel-bed scour and retreat of the right channel bank for as much as 150 feet. The left bank has limited scour because of outcropping bedrock. Simulation results indicate that bank-protection structures placed on the right bank would result in as much as 16 feet of additional vertical channel degradation at the US-59 structures.

Introduction

A fluvial study of the Trinity River in southeast Texas near the two US-59 bridges evaluated the potential river-channel changes at the bridge crossing. The sediment-

1. San Diego State University, P.O. Box 9492, 6011 Avenida Alteras, Rancho Sante Fe, California 92067
2. U.S. Geological Survey, 8011 Cameron Rd., Austin, Texas 78754
3. Texas Department of Transportation, 125 E. 11th., St. Austin, Texas 78701

transport characteristics of the Trinity River have been altered by the construction of Lake Livingston, a water-storage reservoir for the Houston area. Two US-59 bridges on the Trinity River are about 12 miles downstream of Lake Livingston. The lake retains nearly all the bed sediment, which results in a sediment deficit and erosion in the downstream river channel. In the long term, the river width, depth, slope, meander pattern, and other characteristics adjust to provide a balance between the water and sediment loads from the watershed and the ability of the river to transport these loads.

Change in the river channel since construction of the US-59 bridges is characterized by erosion of the right bank (fig.1). From 1964 to December 1992, the right bank retreated about 50 feet. The channel bed has not eroded and has remained more or less stable.

The Texas Department of Transportation is interested in the potential for channel erosion at the US-59 bridges and how such erosion could affect the stability of the bridges. Piers of varying lengths support the bridges. At the bottom of each pier is a pile cap that rests on piles. The pile cap and piles at pier 6 near the right bank have been exposed by bank erosion.

A quantitative assessment of general scour based on mathematical simulation of the river channel was made using the FLUVIAL-12 model (Chang, 1988). The model simulates the hydraulics of flow, sediment transport, and fluvial processes of the river channel. The interactive effects of sediment delivery and river-channel changes are integrated in mathematical simulation. To estimate the long-term response of the channel, the simulation is based on daily mean discharge for a 10-year period (October 1, 1981 - November 30, 1991) at the U.S. Geological Survey streamflow-gaging station 08066500, Trinity River at Romayor. The largest daily mean discharge during this period is 104,000 cubic feet per second. To determine the effect of Lake Livingston, the model simulates the river channel from about river mile 40.3 to the outlet of the lake at river mile 129.1, a total length of about 89 river miles. The US-59 bridges are at river mile 117.4; the Romayor streamflow-gaging station is at river mile 96.3.

Simulation of General Scour at Bridges

Sediment load is defined as the total amount of sediment that is delivered past a given channel cross section during a specified period of time. The simulated sediment loads generally increase in a downstream direction, except for local deviations from this pattern. This pattern indicates that the river channel generally is subject to erosion, although deposition occurs in certain short river reaches. Simulated longitudinal channel-bed profiles depict channel-bed degradation along most of the channel length, with local exceptions. The general pattern of erosion is related to the sediment load retained by Lake Livingston; the primary source of sediment to the river below Lake Livingston is the channel boundary.

Simulated cross-sectional changes at the bridge crossing during a 10-year period of simulation indicate channel-bed scour and retreat of the right channel bank (fig. 2). The left bank shows limited scour based on the assumption that the outcropping bedrock resists scour. During periods of high flow in the 10-year simulation, general channel-bed scour increases with rising discharge, reaches a maximum at peak discharge, and then decreases during falling discharge. The maximum channel-bed scour averages about 2 feet at the US-59 crossing. In addition to general scour, local scour also develops at bridge piers. Local scour at the bridge piers was not analyzed during this study.

The most important channel change simulated at the US-59 bridges is widening resulting from erosion along the right bank. For the 10 years of streamflow (daily means measured from October 1, 1981, through November 30, 1991), the maximum retreat of the right bank is about 100 feet at the upstream US-59 bridge and about 150 feet at the downstream US-59 bridge.

The channel changes simulated at the US-59 bridges are consistent with observed changes from as-built conditions (from bridge plans). Bridge plans from 1964 indicate that the pile cap of pier 6 was beneath the riverbed. Currently (1995), the pile cap and piles are exposed due to bank retreat. Bank erosion as simulated in this study gradually will approach pier 7 and other piers on the right bank.

If the right bank is stabilized with bank protection, greater channel-bed scour is predicted at these crossings (fig. 3). The maximum channel-bed scour from the simulation averages about 10 feet at the upstream bridge and 16 feet at the downstream bridge. The maximum scour develops at the peak flow, and the channel bed is partially refilled during the falling flood stage.

Conclusions

On the basis of the results of the study described in this paper, the following conclusions can be made about channel conditions at the US-59 bridges:

(1) Pier 6 for the US-59 bridges has been exposed by river-channel erosion.

(2) Under the existing conditions, the river channel will widen, primarily toward the right bank. Such a development will further expose piers on the right bank outside the main channel. These piers could be protected by stabilizing the right bank; however, simulation results indicate that this will result in substantial channel-bed scour.

(3) The left channel bank has remained stable because of outcropping bedrock. The resistance of bedrock bank material to scour could be evaluated on the basis of the geotechnical characteristics of the material and the hydraulics of flow.

Reference

Chang, H. H., 1988, Fluvial processes in river engineering: New York, John Wiley & Sons, p. 330-341.

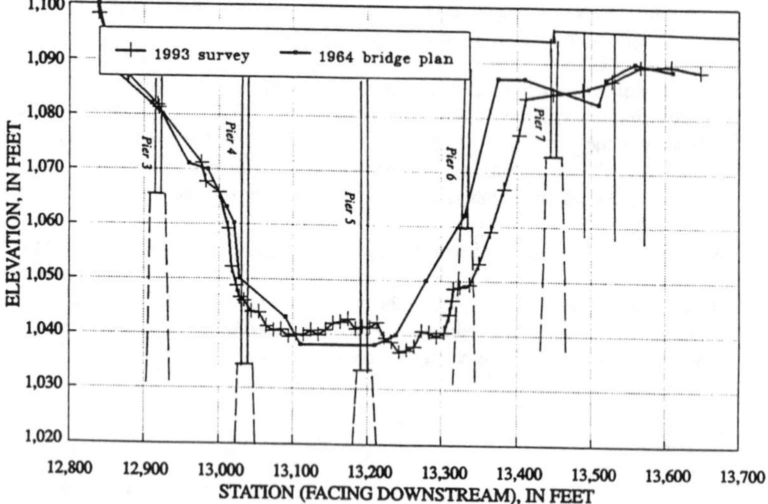

Figure 1. Cross-sectional profile of river channel at the downstream US-59 bridge

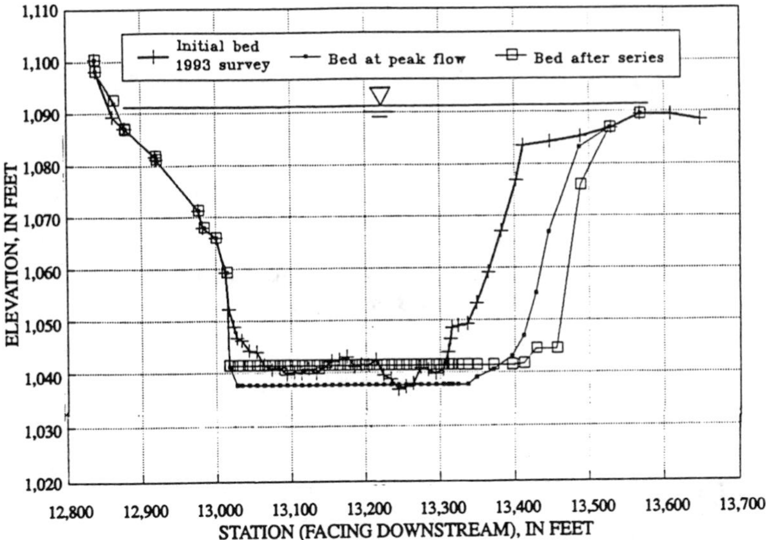

Figure 2. Simulated cross-sectional changes at the downstream US-59 bridge for existing conditions during 10-year simulation period

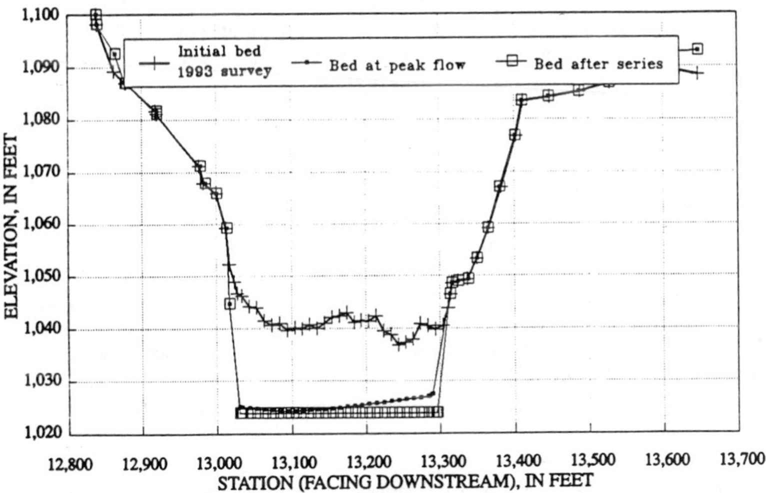

Figure 3. Simulated cross-sectional changes at the downstream US-59 bridge with bank stabilization during 10-year simulation period

Three-Dimensional Sediment Transport Modeling Using CH3D Computer Model

John J. Engel[1], Rollin H. Hotchkiss, Ph.D., P.E.[2], and Brad R. Hall[3]

Abstract

Progress is being made in three-dimensional sediment transport modeling. This paper describes the US Army Corps of Engineer's CH3D model and modifications made within the past two years to include sediment transport modeling capabilities. Unique aspects of the CH3D model include transformation of the governing equations to a boundary-fitted grid in the horizontal plane, a sigma-stretched grid in the vertical plane, and short-term uncoupling of the hydrodynamic and sediment transport operations which allows the computations to be solved by two separate modules of the program.

Hydrodynamics Module

CH3D (Curvilinear Hydrodynamics in Three Dimensions) is a three-dimensional, time-varying, hydrodynamic model developed originally for water quality studies of the Chesapeake Bay (Johnson, et. al., 1991). The basic model was developed by Sheng (1986) for Waterways Experiment Station, but has been extensively modified. It is capable of modeling physical processes such as tides, wind, density effects, inflows from tributaries, earth's rotation, and turbulence. Water surface profiles, 3-D velocity fields, and salinity and temperature variations (for density calculations) are computed at each time step. The model performs the hydrodynamic computations on a curvilinear or boundary-fitted planform grid. The

[1]Graduate Student, Department of Civil Engineering, University of Nebraska-Lincoln

[2]Associate Professor, Department of Civil Engineering, University of Nebraska-Lincoln

[3]Research Hydraulic Engineer, Hydraulics Laboratory, US Army Engineer Waterways Experiment Station

governing equations are first non-dimensionalized and then transformed from Cartesian (x, y) coordinates to the boundary-fitted (ξ, η) coordinate system. In addition to this planform transformation, the vertical grid is transformed using a sigma-stretched grid in the vertical plane (see Figure 1). The boundary-fitted grid transformation allows better representation of irregularities in grid bathymetry such as deep navigational channels and irregular bank profiles, in addition to improving computational efficiency. The sigma-stretch transformation permits better delineation of density stratifications that develop in the flow field. The effectiveness of these transformations in representing complex geometries is illustrated in its applications to the Chesapeake Bay study (Johnson, et. al., 1991) and the Old River Control study on the Mississippi River (Spasojevic and Holly, 1993).

Finite difference approximations are employed to replace derivatives in the governing equations, yielding a system of linear algebraic equations. These equations are solved in both external and internal modes. The external mode solution provides vertically averaged unit flows, U and V, as well as water surface displacements. The internal mode solution consists of the fully three-dimensional velocity field, as well as salinity and temperature values. Turbulence is modeled using the concept of eddy viscosity and diffusivity to represent the velocity and density fluctuations that arise in the time-averaging of the governing three-dimensional equations. The eddy coefficients are computed from the mean flow characteristics using a simplified second-order closure model originally developed by Donaldson (1973), and modified by Sheng (1982).

Sediment Transport Module

The sediment transport module incorporates mobile-bed modeling techniques developed at the Iowa Institute of Hydraulic Research into the general CH3D code allowing simulation of three-dimensional sediment transport and bed evolution under unsteady flow conditions. A few of the attributes of these techniques include (Spasojevic and Holly, 1993):

- Local flow conditions determine transport mode (suspended or bedload), eliminating assumption of transport mode in advance.
- Criteria and mechanisms of exchange between two modes are included.
- Ability to represent sediment mixtures through a number of size classes and model nonuniform sediment processes, such as armoring.

The relatively short time steps of the sediment computations, necessitated by their partial explicitness, enable short-term uncoupling of the water-sediment operations which are performed in separate program modules. The hydrodynamic module provides the hydrodynamic input (velocities, depths, etc.) required by the sediment module. After completing its computations, the sediment module returns bed elevations, bed-surface size distributions, and density changes (due to suspension or deposition of suspended sediment) to the hydrodynamic module for use in the solution at the next time step. Currently, the sediment transport calculations are limited to using the empirical relations presented by van Rijn (1984a, 1984b). Additional sediment transport relations including Brownlie, Toffaletti, Parker, and Laursen, are being incorporated to provide transport functions which will enable application of the model to a wider variety of study area conditions.

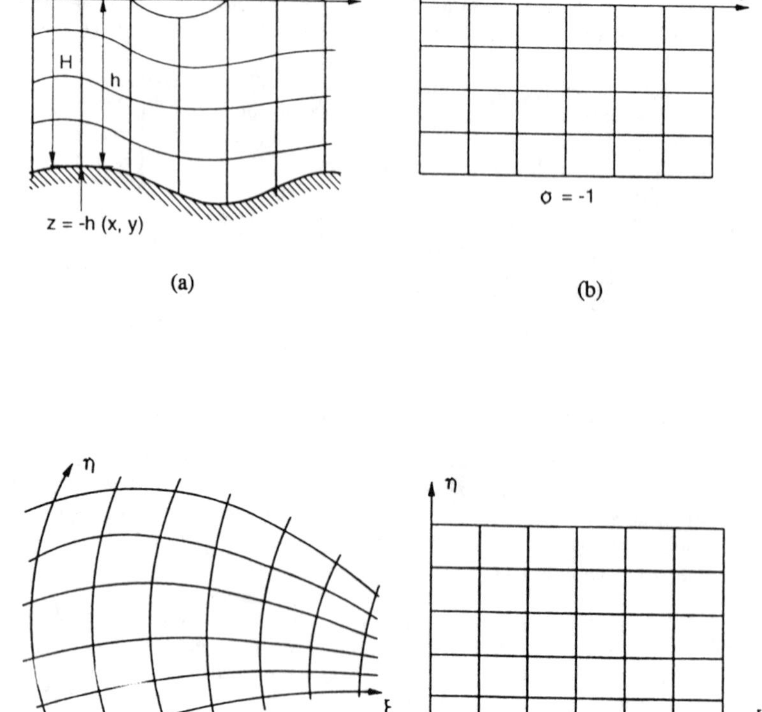

(c) (d)

Figure 1 Coordinate Transformation (from Spasojevic and Holly, 1993)

Applications

The newly incorporated transport functions will be tested initially on a relatively simple application, such as the rip-rap test facility at WES to ensure proper model performance. The model will then be further tested utilizing suspended sediment and bed material data sets collected from two areas of the Lower Mississippi River. The first site is a 12-mile stretch of the Mississippi River (RM 307 to RM 319) near the Old River Control Structure. The second site is an approximate seven mile stretch of the Mississippi River (RM 225) just south of Baton Rouge, which includes the Missouri Bend and Redeye Crossing. Each function will be evaluated based on its bed evolution prediction, densities of water-sediment mixtures, suspended-sediment concentrations, and active-layer size fractions.

Discussion

Currently, CH3D is limited to modeling non-cohesive sediment, but work is continuing on the inclusion of cohesive sediment modeling capabilities. The addition of sediment transport capabilities into the general CH3D code may highlight some of the computational concerns in using this model. The immense CPU time and memory requirements currently limit its usage to the supercomputing environment. In addition, post-processing support is in its infant state, making analysis of results tedious and time consuming.

References

Donaldson, C. 1973. "Atmospheric Turbulence and the Dispersal of Atmospheric Pollutants," AMS Workshop on Micrometeorology, D. A. Haugen, ed., Science Press, Boston, 313-390.

Johnson, B. H., Kim, K. W., Heath, R. E., Hsieh, B. B., and Butler, H. L. 1991. "Users Guide for a Three-Dimensional Numerical Hydrodynamic, Salinity, and Temperature Model of Chesapeake Bay," Technical Report HL-91-20, US Army Engineer Waterways Experiment Station, Vicksburg, MS.

Sheng, Y. P. 1982. "Hydraulic Applications of a Second-Order Closure Model of Turbulent Transport," Applying Research to Hydraulic Practice, P. Smith, ed., ASCE, 106-119.

Sheng, Y. P. 1986. "A Three-Dimensional Mathematical Model of Coastal, Estuarine and Lake Currents Using Boundary Fitted Grid," Report No. 585, A.R.A.P. Group of Titan Systems, New Jersey, Princeton, NJ.

Spasojevic, M. and Holly, F. M., Jr. 1993. "Three-Dimensional Numerical Simulation of Mobile-Bed Hydrodynamics" Iowa Institute of Hydraulic Research Technical Report No. 367, The University of Iowa, Iowa City, IA.

van Rijn, L. C. 1984a. "Sediment Transport, Part I: Bed Load Transport," Journal of Hydraulic Engineering, ASCE, Vol. 110, No. 10, pp. 1431-1456.

van Rijn, L. C. 1984b. "Sediment Transport, Part II: Suspended Load Transport," Journal of Hydraulic Engineering, ASCE, Vol. 110, No. 11, pp. 1613-1641.

CANADA-USA 1994 OIL SPILL EXERCISE ON THE GREAT LAKES - MODELLING SUPPORT AND ITS EVALUATION

Srinivasan Venkatesh (Member, ASCE)[1], Steve Clement[2], Philip Baker[2] and Bhartendu Srivastava[1]

Abstract

In this paper we examine the details of the trajectory modelling support provided during a Canada-USA oil spill exercise conducted in September of 1994 near the western end of Lake Erie. The residual currents in Lake Erie are computed from a quasi 3-dimensional hydrodynamic model that accounts for the outflow from the Detroit river. The real-time predicted winds from the Canadian Meteorological Centre Regional Finite Element Model were obtained as an integral part of the spill model system. These winds were applicable for the location and time of the spill. The model simulations are compared to the drift of the buoys that had been deployed at the spill location. With the spill source in an area of the lake affected by the Detroit river outflow, the model simulations are sensitive to the relative strength of the winds and residual water currents.

Introduction

In September 1994 a joint Canada/U.S.A. oil spill exercise was conducted near the western end of Lake Erie. The objective of the exercise was to test various aspects of emergency response to a marine spill in the area including trajectory modelling support. The scenario involved the collision, in the early morning hours of September 12, 1994, of two vessels one carrying bunker oil and the other carrying caustic soda.

The caustic soda, coming in contact with the water, sinks to the bottom of the lake and hence no trajectory calculations were made for it. Real-time simulations of the surface trajectory of the bunker oil were carried out using the

[1]Environment Canada, 4905 Dufferin Street, Downsview, Ontario, Canada M3H 5T4
[2]Environment Canada, 25 St. Clair Ave. E, Toronto, Ontario, Canada M4T 1M2

SLICK-II oil spill model of the Atmospheric Environment Service of Environment Canada (Venkatesh, 1988). The simulations were compared to the movement of a set of spill-following buoys that were deployed at the spill source location.

In addition to the buoys that were deployed on September 12, 1994, the day of the exercise, several others were also deployed at various locations in the western part of Lake Erie and their movements monitored during the period September 7-22, 1994. The drift of these spill following buoys were also simulated using the SLICK-II model.

Model Simulations

Real-time Response

The objective of the exercise was to simulate, as much as possible, a real oil spill emergency. As such no pre-preparations were done as far as trajectory modelling was concerned. Upon receiving a call from the Emergency coordinator at 7 am on September 12, 1994, efforts were initiated to produce model simulations of the trajectory of the Bunker C oil. The SLICK-II system is set up to obtain the forecast winds for the location (anywhere in the world) and time of the spill through a link to the outputs of various weather forecast models of the Canadian Meteorological Centre in Montreal. With the spill being in the Great Lakes, the winds output from the Canadian high resolution Regional Finite Element (RFE) model (Benoit et al. (1989), Mailhot et al. (1995)) can be used. While the resolution of this model is fine enough to sense the presence of Lake Erie, it is not fine enough to simulate the lake and land breezes that will be prevalent during September. If spill movement forecasts for a period longer than about two days are required then the winds from the Canadian Global Spectral Model can be used. The non wind-driven surface currents in Lake Erie result mainly from the inflow from the Detroit river at the west end of the lake and the outflow into the Niagara river at the east end. The circulation resulting from these flows is determined using a quasi 3-d model described by Tsanis and Wu (1991).

In the model simulations the oil parcels are released over a short period of time (4 hours). Using the residual water currents and RFE winds (Figure 1a), Figure 1b shows a comparison of the model simulated spill parcel positions and the corresponding motion of spill-following buoys over a period of 15 hours. The land impact location as simulated by the model is much further to the east compared to that indicated by the buoys. In order to determine the influence of errors in winds on spill motion, a hindcast of the motion of spill parcels was carried out using observed winds instead of the RFE winds. While wind observations were available from three different locations around the lake, those from South Bass Island (41°36'N, 82°49'W) near the southern shore of Lake Erie were selected. The spill source was close to this station and in line with the wind direction, thus best

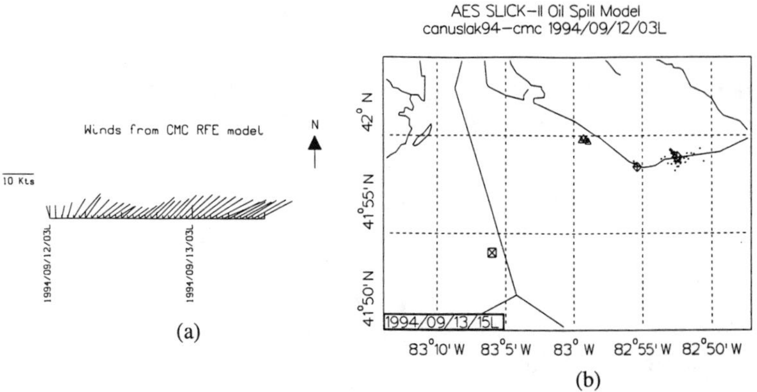

Figure 1: (a) Forecast winds for the spill location in western Lake Erie obtained from the Canadian Regional Finite Element model; (b) Model simulations of oil spill motion. Spill source is shown by □, spill parcels by (.), the land impact points by ◊. The observed locations of the buoys are shown by Δ. Valid time of the simulation is shown at bottom left.

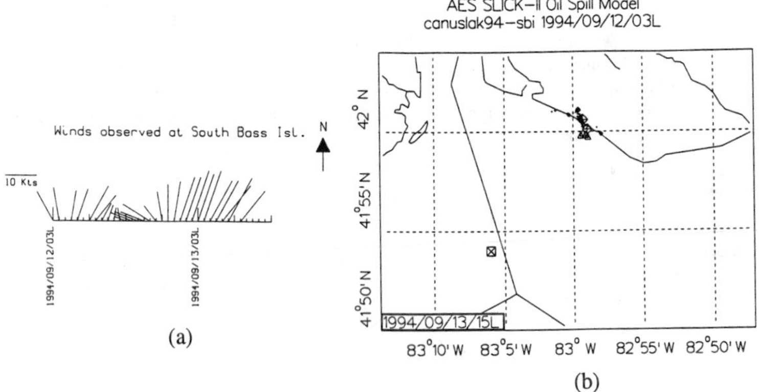

Figure 2: (a) Winds observed at South Bass Island; (b) Model simulations of oil spill motion. See Figure 1 for an explanation of symbols.

representing the wind conditions prevalent in the area of the spill. The winds observed at South Bass Island are shown in Figure 2a. Figure 2b shows the corresponding model simulations. These simulations are in very good agreement with the observed motion of the spill-following buoys.

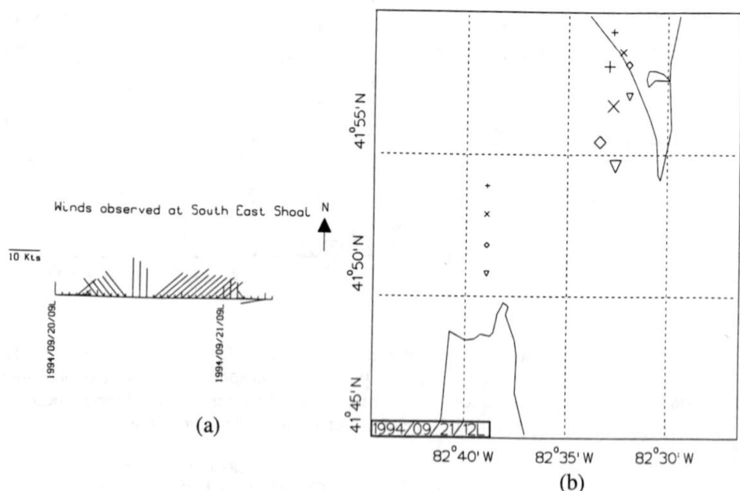

Figure 3: (a) Winds observed at South East Shoal; (b) Model simulated (includes effects of winds and residual water currents) and observed movement of four spill following buoys, each shown by a different symbol. The smallest symbols indicate initial buoy positions. Model simulated positions are shown by the medium sized symbols and the observed buoy positions by the largest symbols.

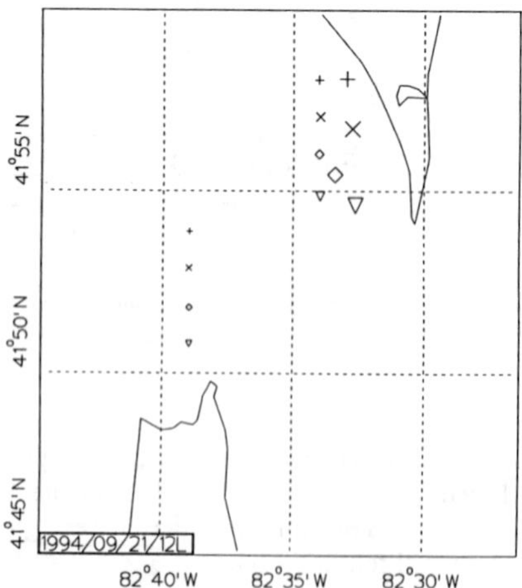

Figure 4: Same as Figure 3(b), except residual currents are set to zero.

Other Model Simulations

In addition to the real time simulations, model simulations were also carried out of the motion of buoys whose initial time and location of deployment only were given. These simulations were later compared with the observed buoy positions at the end of the simulation periods. Such simulations were carried out for a total of 10 different periods for which observations were available. The buoy deployments for these cases covered a good portion of the western part of Lake Erie. One of these ten cases is briefly discussed below.

Figure 3a shows the winds observed at South East Shoal (41°50'N, 82°28'W) beginning at 0900 local time on September 20, 1994. A series of four buoys initially deployed at this time were tracked for a period of 15 hours. Figure 3b shows the initial, final model simulated and final observed positions of the buoys. Compared to the observations, the buoy movements are slightly over-estimated by the model. The 4 km resolution of the model used to derive the residual currents is not adequate to resolve the currents around Point Pelee (41°55'N, 82°30'W) and this may have affected the model results. Figure 4 shows model simulations that are purely wind-driven. These simulations, compared to those with residual currents, are in slightly better agreement with the observed movement of the buoys. Complete details of the other simulations can be obtained from the authors.

References

Benoit, R., J. Côté and J. Mailhot. 1989. Inclusion of a TKE boundary layer parameterization in the Canadian Regional Finite-Element model. *Mon. Wea. Rev.*, Vol. 17, pp1726-1750.

Mailhot, J., R. Sarrazin, B. Bilodeau, N. Brunet, A. Méthot, G. Pellerin, C. Chouinard, L. Garand, C. Girard and R. Hougue. 1995. Changes to the Canadian regional forecast system: Description and evaluation of the 50-km version. ATMOSPHERE-OCEAN (In press).

Tsanis, I.K. and J. Wu. 1991. Study of water currents in the Canadian Great lakes. Report prepared for the Atmospheric Environment Service, Downsview, Ontario, Canada.

Venkatesh, S. 1988. The oil spill behaviour model of the Canadian Atmospheric Environment Service, Part I: Theory and model evaluation. ATMOSPHERE-OCEAN, Vol. 26, pp93-108.

Weathering Effect on the Dispersion of a Heavy Crude Oil

Hector R. Fuentes[1], Rudolf Jaffé[2], Vassilios A. Tsihrintzis[1], and Donald Boyé[3]

Abstract

In response to regulatory actions and the need to protect the ecological and economic resources of South Florida, studies are conducted to understand the fate of oil spills in coastal and inland water environments. Partial and ongoing efforts at Florida International University to develop experimental facilities and methodologies for the evaluation of crude oils and fuels are described. Emphasis has been concentrated on the characterization of physical and chemical changes as a result of weathering and their effects on potential dispersion.

Introduction

Federal and state rules and regulations address preparedness and emergency actions for accidental spills. Vital to those actions is the understanding and prediction of the fate and transport of oils and fuels in fresh and saline aquatic environments. Because of the traffic of oil shipments along the Florida coastline, laboratory experiments have become of prime concern to both regulators, users, and academicians.

A good number of papers have been published on experiments, modeling, and assessments of oil spills in inland and ocean waters (e.g., Mackay et al. 1981; Buchanan and Hurford 1988; Yapa and Shen 1994; Spaulding 1988; and Wolfe et al. 1994). The purpose of this paper is to share some experiences in the testing of a sample of a crude oil in small-scale reactors (Boyé 1994).

---

[1]Respectively, Associate and Assistant Professor of Civil and Environmental Engineering; [2]Associate Professor of Chemistry; and [3]Former Research Assistant. All researchers with the Drinking Water Research Center, Florida International University, University Park Campus, VH-160, Miami, Florida, 33199, USA.

Materials and Methods

A sample of an arabian crude was used in artificial seawater. Table 1 shows the basic characteristics of the oil. Artificial seawater was prepared to a specific gravity of about 1.022 in regular tap water.

Table 1. Sample of Crude Oil

Origin	Saudi Arabia
Source	Mobil R&D Laboratory
Sample No.	92-64371
API	27.6
Specific Gravity	0.889
Kinematic Viscosity (15°C)	55.1cS
Carbon (pct wt)	84.48
Hydrogen (pct wt)	12.35
Carbon/Hydrogen Ratio	6.84
Lighter Hydrocarbons (pct wt)	6.22

Two kinds of reactors were used: a) weathering stations (Figure 1) and b) a dispersion column (Figure 2). The weathering stations were formed by 2000-mL Erlenmeyer Flasks resting on magnetic stirring plates. Air flow inlets and samplers were passed through rubber stoppers at the mouth of the flask. Stations were set up on a roof where they were exposed to natural light. Experiments were conducted lasting from over few hours up to seven days; at the end of each experiment samples of the oil slick were retrieved to measure its density, interfacial tension, viscosity, and chemical composition of main oil constituents.

The second reactor was a 12-inch I.D. plexiglass column, 72-inch high. The reactor was used to simulate spills with the weathered oil samples in turbulence provided by a mixer located at the top of the column. Samples were collected at various ports at different depths along the column.

The following equipment was used in the determination of the physical properties and chemical composition of the weathered oil slick: Precision Glass Picnometers (density); Digital Cone Plate Viscometers (viscosity); du-Nuoy Ring Apparatus (surface tension); Gas Chromatograph/Mass Spectrophotometer (aliphatic and aromatic hydrocarbon fractions).

Results and Discussion

Results indicate that the oil slick density increased from 0.889 g/mL to 0.95 g/mL in seven days with a rapid change in the first eight hours. Viscosity also increased from 118 to 1025 cP in the same period. Figure 3 illustrates the change in dynamic viscosity. Meanwhile, the interfacial surface tension decreased from 24.8 to 18.6 dynes/cm. These changes are expected to directly impact the expected transport characteristics of the oil slick, as well as emergency response actions. The dispersion characteristics were also changed by weathering as shown by the distribution of total dispersed aromatic hydrocarbons after three hours of turbulent mixing in the column (Figure 4); in fact, dispersion decreased in time, most possibly as a result of the increase in density and viscosity of the slick.

Recommendations

Further work is undergoing to enhance the experimental facilities and procedures to better simulate natural environments. In addition, efforts are beginning to evaluate the role of various fate processes with various oils and fuels, including dissolution, photooxidation, and kinetics of sticking to solid phases, among others.

References

Boyé, D.J., Jr. (1994). The effect of weathering processes on the vertical turbulent dispersion characteristics of crude oil spilled on the sea, M.S. Degree Thesis (Major Professor: Hector R. Fuentes), Florida International University, Miami, Florida, August.

Buchanan, I. and Hurford, N. (1988). "Methods for predicting the physical changes in oil spilt at sea." *Oil and Chemical Pollution*, 4, 223-230.

Mackay, D., Paterson, S., Boehm, P.D., and Fiest, D.L. (1981). "Physico-chemical weathering of petroleum hydrocarbons from the IXTOC-1 blowout -- chemical measurements and a weathering model." *Proceedings of the 1981 Oil Spill Conference*, American Petroleum Institute, 453-460.

Spaulding, M.L. (1988). "A state-of-the-art review of oil spill trajectory and fate modeling." *Oil and Chemical Pollution*, 4, 39-55.

Yapa, P.D. and Shen, H.T. (1994). "Modelling river spills: a review." *Journal of Hydraulic Research*, IAHR, 5, 765-782.

Wolfe, D.A., Galt, J.A., Short, J., O'Claire, C., Rice, S., Michel, J., Payne, J.R., Braddock, J., Hanna, S. and Sale, D. (1994). "The Fate of the oil spilled from the Exxon Valdez." *Environmental Science & Technology*, 13, 561A-568A.

Figure 2. Dispersion Column

Figure 1. Photograph of a Weathering Station

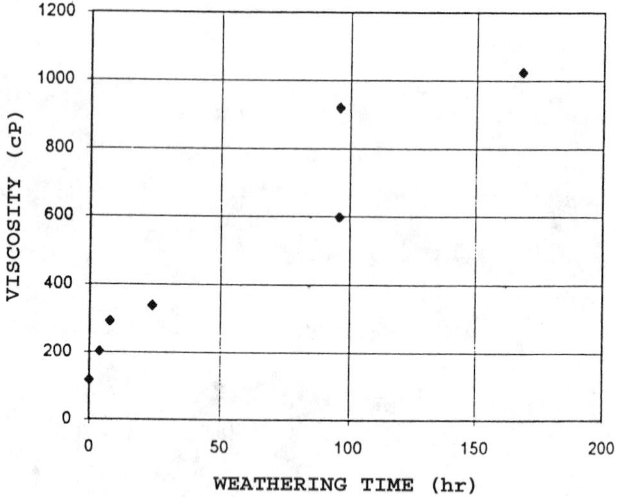

Figure 3. Viscosity Change in Time

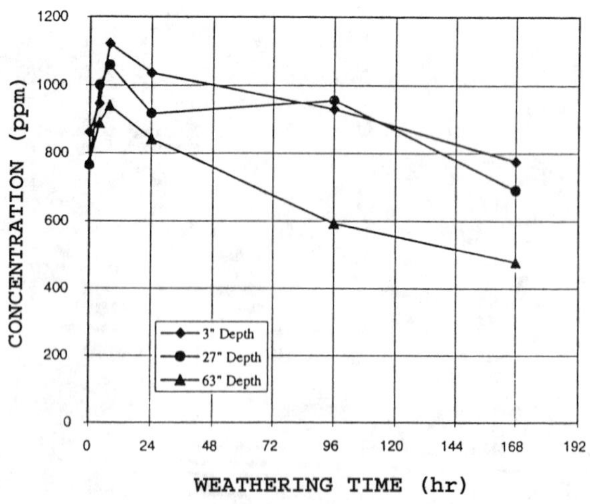

Figure 4. Dispersion of Total Hydrocarbons in the Turbulent Column

Comparison of Oil Trajectory Model with Drifting Buoys

Robert P. LaBelle and Charles F. Marshall[1]
Pearn P. Niiler[2]

Abstract

Oil-spill trajectory models are increasingly being used for both real-time spill response and longer term contingency planning and risk assessment. The Minerals Management Service (MMS) supports development of the latter type of model for use in estimating risks from offshore oil activity. A common shortcoming with all such models is a typical lack of ocean observational data to drive the models and skill assess their results.

The MMS and Scripps Institution of Oceanography funded the deployment of over 340 drifting buoys in the western Gulf of Mexico over a 1-year period during 1993-94. The buoys were satellite-tracked, Lagrangian surface drifters that floated in the top meter of the water column and reported positional information several times a day for up to several months. The data from these drifters have characterized the surface flow in the western Gulf, concentrating on shelf waters. The emerging picture of surface circulation from the drifter data is being compared with results from oil-spill trajectory model runs driven by a numerical ocean circulation model with a high-resolution curvilinear grid.

Introduction

In 1990, the MMS funded a workshop on advancing the science of oil-spill transport modeling and risk assessment (MMS, 1991). The major recommendation from the workshop was to implement a large field program of Lagrangian measurements to support and skill assess numerical ocean circulation models. Such a study was begun in 1993 through a cooperative effort between MMS and Scripps Institution of Oceanography (Niiler and Davis, 1992). Preliminary results from the Surface Current Lagrangian Program (SCULP)

----

[1]Minerals Management Service, Dept. of the Interior, Herndon, VA 22070
[2]Scripps Institution of Oceanography, La Jolla, CA 92309

study are now being used as inputs to the MMS Oil-Spill Risk Analysis (OSRA) model runs in the western Gulf of Mexico.

This paper compares the OSRA model runs using surface currents generated by a numerical ocean model with model runs using Lagrangian drifter data as a supplement to the numerical surface current field. Both model runs are also used to hindcast a recent actual oil spill in a tanker lightering zone south of Galveston, Texas.

Oil-Spill Risk Analysis Model (OSRA)

The OSRA model uses a stochastic approach to characterize the risk of spill contacts to resources over large spatial and temporal dimensions (LaBelle and Anderson, 1985). For example, the model has been used by the U.S. Coast Guard to estimate oil spill contact probabilities from various offshore zones used or proposed for tanker lightering (offloading cargo from super-tankers to smaller vessels). In estimating average contacts to coastal resources, the OSRA model uses climatological inputs from available wind and ocean data and models to drive multiple simulated trajectories from potential offshore spill sites. Figure 1 shows a simulated spill launch point and the resource area representing Texas coastal waters. The spill launch point represents the approximate location of an oil spill that occurred on February 5, 1995, from the collision of two tankers about 105 km south of Galveston Bay. Bunker C oil came ashore at the Matagorda National Wildlife Refuge and along the Padre Island National Seashore within about 3 weeks of the accident.

Oil-Spill Trajectory Simulations

The OSRA model was run using a combination of observed and numerically computed ocean currents and winds. Most of the ocean currents used were generated by a numerical model. They were supplemented with many direct observations of the currents in the western Gulf resulting from repetitive deployments of SCULP surface drifting buoys. Climatologically representative monthly mean surface currents were provided using the Princeton-Dynalysis Ocean Model (PDOM), (Dynalysis, 1994). The PDOM, an enhanced version of the Mellor-Blumberg Model, is a three-dimensional, time dependent, primitive equation model using orthogonal curvilinear coordinates in the horizontal and a topographically conformal coordinate in the vertical. The use of these coordinates allows for a realistic coastline and bottom topography, including a sloping shelf, to be represented in the model simulation. The model incorporates the Mellor-Yamada turbulence closure model to provide a parameterization of the vertical mixing process through the water column.

The PDOM model was driven by monthly climatological wind stress, heat flux, river flow, and Gulf Stream transport boundary conditions. From a 12-year simulation, a climatologically representative year was selected. Monthly averaged surface currents were then computed from that year and constitute the background ocean currents used in the oil-spill trajectory simulations.

Surface Current Lagrangian Program (SCULP)

In addition to the model surface current maps, quasi-Eulerian surface currents were derived from the positions of satellite-tracked drifting buoys (see Figure 2). Approximately 340 drifting buoys were deployed from aircraft and from three production platforms in a repeated array located southeast of Galveston, Texas. The drifters were designed to float in the top 1 meter of the water column and to report their positions several times a day for up to several months. Weekly deployments were made from mid-October 1993, running through January 1994, followed by monthly deployments from February 1994 through September 1994.

In the trajectory simulations, the drifting buoy mean velocities were substituted for the PDOM model currents in the appropriate geographical areas. In addition to the ocean currents, a wind field was employed to add the effect of the direct sea surface winds on the hypothetical spills. Note that this is a different wind field than the climatological field used by the PDOM to generate the model currents. The wind field employed was the Naval Research Lab's geostrophic sea surface winds derived from analyzed atmospheric pressure maps at 12-hour intervals for the 27-year period extending from January 1967 through December 1993 (Rhodes, Thompson, and Wallcraft, 1989). The direct wind drift on a hypothetical oil spill is modeled using an empirical "3.5% rule" with a speed-dependent wind deflection angle (Samuels, Huang, and Amstutz, 1982).

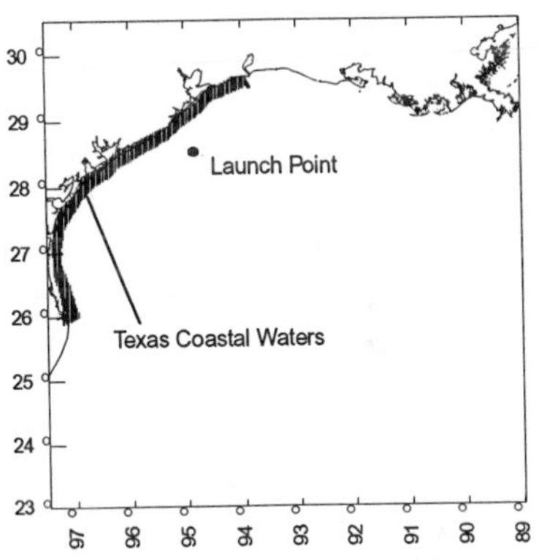

Figure 1. Launch point and Texas coastal waters

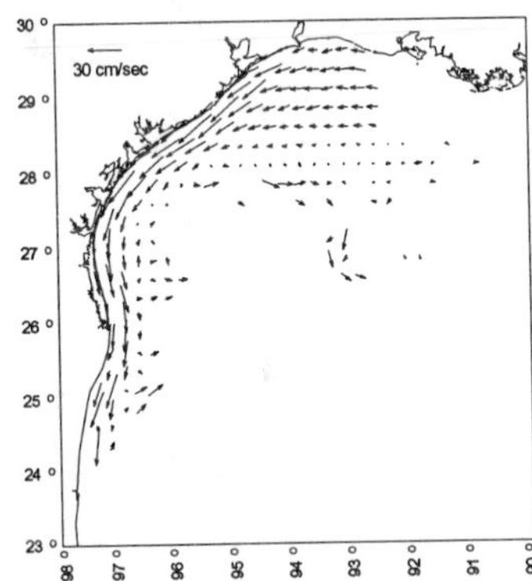

Figure 2. Mean quasi-Eulerian velocity field, October-May

Results

The primary purpose of the SCULP study was to use drifter observations directly in defining a horizontal and time-evolving surface current field to improve oil-spill trajectory analysis. The SCULP data have been averaged as seasonal mean surface currents and were input to the OSRA model.

The OSRA model was run with and without the SCULP drifter data from multiple spill launch sites in the western Gulf of Mexico. The results shown in Figure 3 are from a single launch point, the site of the tanker collision previously described (see Figure 1). Trajectories from the other launch points showed similar trends to those from the single site.

Figure 3 compares oil-spill contact probabilities when the OSRA model was run with and without the drifter data. The probabilities shown are for the winter season (January - March) for up to 30 days spill travel times from the launch point. (Five hundred trajectories were simulated for each season.) Use of the drifter data as supplement to the model run resulted in contacts more along the southwest coast than did use of the PDOM model alone. These results are

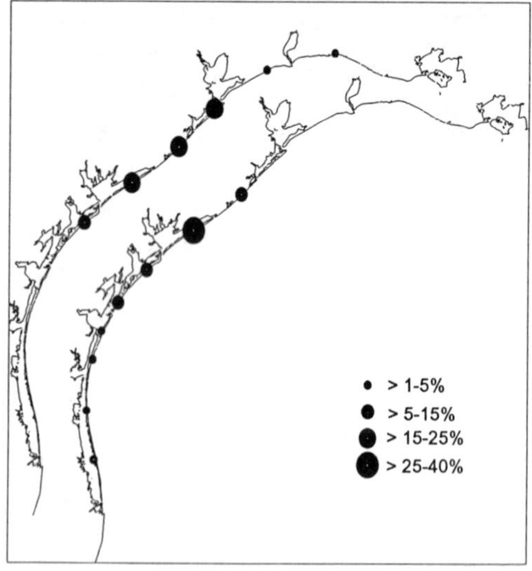

Figure 3. Comparison of chance of oil-spill contacts to the coastline using numerical model with (right coastline) and without (left coastline) SCULP drifter data.

consistent with other measurements and observations of the flow in this region, including the February 1995 oil spill.

Similar patterns were seen in the spring and fall seasons when comparing results of the two current fields. However, the drifters showed markedly different behavior in the summer (July - September), moving generally northeastward. The numerical model results were most closely matched to drifter movements in the summer.

The model hindcast of the tanker spill south of Galveston showed winter 30-day contact probabilities to the Matagorda coastline at a 15% chance when drifter data was used and 6% when not used. The Padre Island contacts ranged from 1 to 5% with drifters, zero without. However, since the simulated spills were modeled with no areal dimension, it is more realistic to record contacts to the nearshore waters (i.e., the Texas coastal waters shown in Figure 1). When the drifter data were used, contacts to this resource (30-day winter trajectories) were estimated at a 90% chance, and a 78% chance without drifters.

Although a single spill event may not reflect the characteristic nature of the surface flow; for this area, the southwestward flow has been observed and measured. Therefore, although the numerical model results are a large improvement over previous oil-spill model inputs, the judicious use of drifter data in concert with the numerical model surface flow provides a measured improvement to surface oil-spill modeling in the western Gulf of Mexico.

References

Dynalysis of Princeton, 1994. Intermediate deliverable for MMS Contract No. 14-35-0001-30631, Dynalysis of Princeton, Princeton, New Jersey.

LaBelle, Robert P. and C.M. Anderson, 1985. The Application of Oceanography to Oil Spill Modeling for the Outer Continental Shelf Oil and Gas Leasing Program, Marine Technology Society Journal 19(2):19-26.

Minerals Management Service, 1991. Offshore Oil Spill Movement and Risk Assessment Workshop, OCS Study Report MMS 91-0007, Herndon, VA.

Niiler, Pearn P. and Russ E. Davis, 1992. Surface Current and Lagrangian-Drift Program on the Louisiana-Texas Shelf, Scripps Inst., La Jolla, CA.

Rhodes, R.C., J.D. Thompson, and A.J. Wallcraft, 1989. Buoy-Calibrated Winds over the Gulf of Mexico, Jour. Atmos. and Ocean Tech., 6/89.

Samuels, W.B., N.E. Huang, and D.E. Amstutz, 1982. Oil Spill Trajectory Analysis Model with a Variable Wind Deflection Angle, Oc. Eng. 9:347-360.

TRANSPORT OF ACCIDENTAL OIL SPILLS IN TVA RESERVOIR SYSTEM

Boualem Hadjerioua[1], Vahid Alavian[2] M. ASCE, Earl G. Marcus[3]

Abstract

A study was conducted to investigate the transport of oil from accidental spills downstream of Tennessee Valley Authority (TVA) fossil power plants. The purpose of the study was to assess and test the applicability of several existing unsteady flow models in estimating advective transport of accidental oil spills in a controlled river/reservoir system. The plume characteristics of an oil spill from a plant were simulated by linking a one-dimensional unsteady flow model (ADYN) to a contaminant transport model (CONTRAN). The model results were verified by a field investigation using free-floating drogues to determine velocities in the upper 1-meter layer of the reservoirs. The results of the numerical simulation are in good agreement with the experimental data.

INTRODUCTION

The Oil Pollution Act of 1990 requires that every facility on a river develop an oil spill plan suitable for handling possible spill scenarios under a range of river flow conditions. At the request of TVA Emergency Operation Control (EOC), a study was made of the transport of accidental oil spills at Colbert, Johnsonville, Gallatin, and Allen fossil plant sites. These four plants are regulated by the U.S. Coast Guard because of the spill potential of water-borne oil transport.

It is reported that 172 spills of various types have occurred in the TVA river system since 1991, including 27 mandatory reportable oil spills. Potential worst possible spills would be associated with rupture of the combustion storage tanks. Large oil spills of such magnitude would enter the waterways almost instantaneously. High spill rates would not be of long duration because they would be detected quickly and controlled; however, slow leaks might go undetected for several hours, or more.

This study is aimed at assessing and testing existing mathematical unsteady flow models to simulate accidents and provide a tool to quickly estimate the travel time for the

[1]Civil Engineer, Tennessee Valley Authority, Engr. Lab., Norris, TN 37828

[2]Vice President, Rankin International Inc., Knoxville TN 37828

[3]Senior Spill Prevention, Control, and Countermeasure Specialist, TVA, Chatt., TN 37402

four sites. The mathematical model used to estimate travel time is the one-dimensioal unsteady flow model ADYN. This model has been used effectively throughout the TVA system and has been calibrated for each reservoir.

The oil spill travel time was estimated by using a particle tracking routine assuming that the oil particle travels with the same velocity as the water. The oil plume characteristics were simulated using a contaminant transport model CONTRAN linked to ADYN. A field drogue study was conducted to verify the mathematical model results for the Gallatin and Johnsonville sites.

FOSSIL PLANTS CONSIDERED

The plants reported on in this paper are Johnsonville and Gallatin. The location of the plants, and the upstream and downstream dams, are shown in Figure 1. Location of water supply intakes in the vicinity of potential oil spill sites is critical, and immediate measures would be required to prevent oil from getting into the water supply systems.

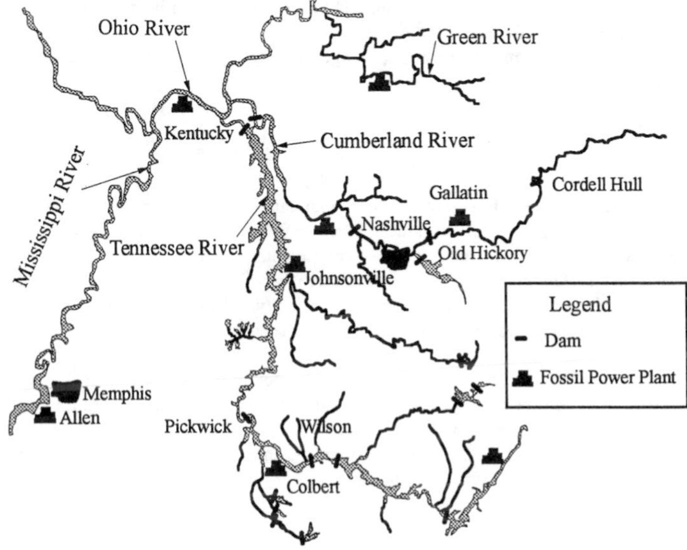

Figure 1. Location Map

FLOW MODEL

The hydrodynamic model is a one-dimensional unsteady flow model ADYN developed by Hauser and McKinnon (1987). ADYN solves the one-dimensional equations for conservation of mass and momentum and is used to study river and reservoir hydraulics. The water velocity required in determining the mean travel time of an oil spill was supplied by this hydrodynamic model. Calibration of the model was achieved by comparing computed and observed water surface elevations.

Using a one-dimensional model to evaluate the travel time of an oil spill is a reasonably cost-effective approach and met the objective of the spill planning effort. Using two- or three-dimensional models, including the physical and biochemical effects of oil transport, would have improved the predictions, but also would have increased costs exponentially.

TRANSPORT MODEL

The transport model CONTRAN used to simulate the oil plume characteristics is a numerical model which solves the mass transport equation in a two-dimensional (depth-averaged) unsteady domain (Alavian, 1986). Spill conditions are entered interactively, and the flow properties are read from an input file generated by the hydrodynamic flow model ADYN.

CONTRAN was designed for use in conditions that lend themselves to a two-dimensional (depth-averaged) formulation. Therefore, judgment should be exercised when using the model for deep reservoirs. The longitudinal dispersion coefficient was considered insignificant, a poor assumption when applied to instantaneous spills because longitudinal dispersion can be significant.

Simulations were run for a wide range of river flow conditions (Hadjerioua and Alavian, 1994).

FIELD INVESTIGATION

Simulation studies provided spill travel time from each plant to the downstream dam. Field tests included flow pattern observation and drogue tracking in the downstream reservoirs at the Johnsonville (Tennessee River) and Gallatin (Cumberland River) plants. Depth-integrating drogues were launched at predetermined positions. Each drogue consisted of a surface float for tracking and a light-weight vane vertically suspended one meter below the water surface to integrate the velocity. Drogue velocity was calculated from distance traveled and lapsed time between consecutive position measurements.

Drogues were launched in the main channel at several different longitudinal locations in order to minimize the duration of the tracking. The drogues were tracked from each plant through the downstream reservoir. Tracking was performed using a global positioning system installed on a boat. Latitude and longitude readings were converted to river miles for drogue velocity calculations.

RESULTS

Data from the field studies were compared with the corresponding simulation using the hourly flows (unsteady flow) to verify the mathematical model. Results are shown in Figures 2 and 3. Plume characteristics were simulated for an assumed oil spill of 1000 cfs for 4 hours (complete destruction of the supply tank) from Gallatin plant (Figure 4). Additional field data are needed to verify CONTRAN simulation results of the plume characteristics.

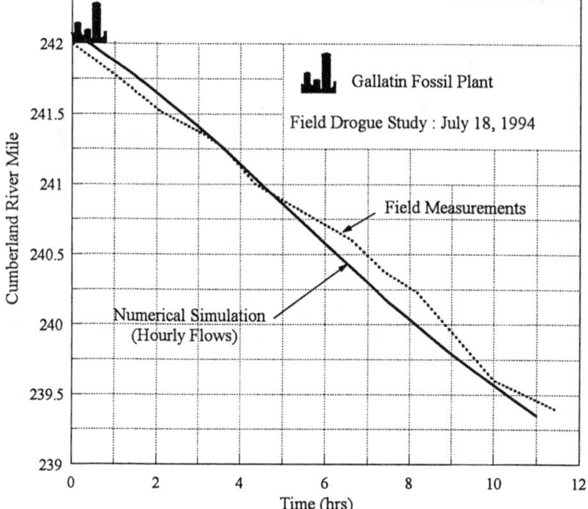

Figure 2. Water Travel Time Below Gallatin

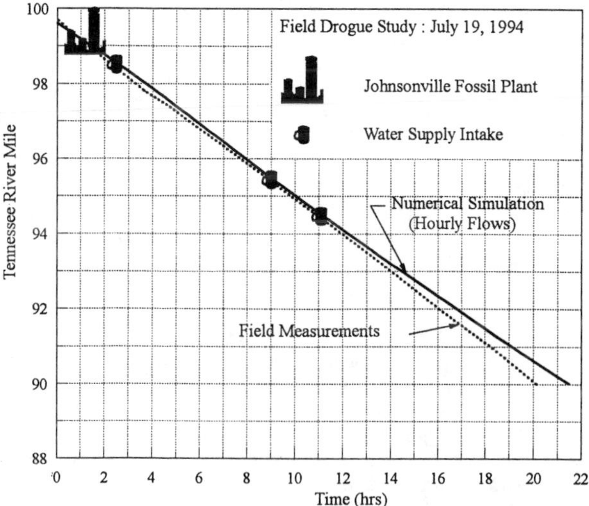

Figure 3. Water Travel Time Below Johnsonville

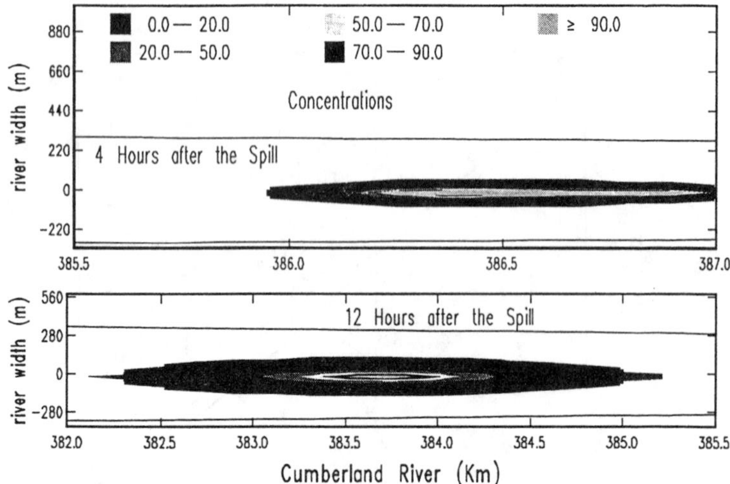

Figure 4. Plume Characteristics of a Spill of 1000 cfs/4hr, Gallatin Plant

CONCLUSIONS

Transport of an accidental oil spill from TVA fossil plants was simulated using ADYN and CONTRAN models and compared with field data. Analysis of the spill movement was limited to advective transport of oil without longitudinal dispersion. This analysis provides a conservative travel time for the spill. An actual oil spill would, in all likelihood, travel more slowly due to the effects of dispersion, decomposition of the spill, and wind effects.

Field tracking times are in good agreement with the mathematical simulations, suggesting that a one-dimensional model provides adequate results for the development of spill emergency plans for the TVA fossil plants.

ACKNOWLEDGMENTS

This study was sponsored by TVA Fossil and Hydro Environmental Affairs. The help of B. Johnson, J. Ault, and L. Cole of the TVA Engr. Lab in the field study is appreciated.

REFERENCES

1. Alavian V. "A Numerical Model for Simulation of Contaminant Dispersion and Transport in TVA Waterways," Completion Report Contract No. TV-68728A, Norris, TN, June 1986.
2. Hadjerioua B., and V. Alavian. "River Transport of Accidental Oil Spill at Selected TVA Fossil Plants," TVA, Engr. Lab Report No. WR28-1-900-268, Norris, TN, Sept. 1994.
3. Hauser, G. E., and M. K. McKinnon. "User's Manual for One-Dimensional Flow and Water Quality Model, Part I User's Manual, Version 1.31," TVA, Engr. Lab Report No. WR28-3-590-108, Norris, TN, March 1987.

MODELING OIL DEPOSITION ON SHORELINES AND RE-ENTRAINMENT

By Tomonao Kobayashi[1] and Poojitha D. Yapa, Member, ASCE[2]

ABSTRACT : *A time-dependent one-dimensional model is developed to simulate the oil-shoreline interaction. The model computes the effects of tidal level changes, wave run-up height, wave setup, and the width of the swash zone from the hydrodynamical considerations. The model also takes into account the irregular nature of waves and the wave damping due to oil. Preliminary results presented here illustrate the dynamic nature of the oil-shoreline interaction.*

INTRODUCTION

When oil from accidental spills reaches a shore, it environmentally damages the shoreline. The oil reaching a shore from a major spill (e.g. Exxon Valdez, Alaska) can be ecologically and financially disastrous. Results from models can be used in decisions related to control of oil spills, contingency planning, and post-spill damage assessment. The abilities to predict how the oil may interact with the shoreline, and to predict the potential damage to various shorelines is critical for decisions on diverting the spill, boom placement, and dispersant application.

The physics of oil deposition on shorelines and re-entrainment is a complex process that depends on the turbulence level, wave height, swash zone, tidal fluctuations, type of shoreline (rocky, tidal flat, sand, marshy land etc.), and oil characteristics. Very little research has been done on this topic. The COZOIL model developed by Read et al. (1988) takes into account some dynamic aspects of oil interaction with the shore. Other models use an exponential decay type of equation and ignore important hydrodynamical factors.

In this paper, a time-dependent one-dimensional model is developed to simulate the dynamics of oil-shoreline interaction. This model considers the effects of run-up height, the wave set-up, the tidal level, and the width of the swash zone on the oil deposition process.

[1]Research Assoc., Dept. of Civil & Envir. Engrg., Clarkson Univ., Potsdam, NY 13699–5710, on leave from Tokyo Rika Univ., Noda, Chiba 278, Japan

[2]Assoc. Prof., Dept. of Civil & Envir. Engrg., Clarkson Univ., Potsdam, NY 13699–5710

MODEL DESCRIPTION

Waves in Shoaling Zone and Surf Zone

The wave transformation approaching the shoreline from offshore, was computed based on the work by Goda (1975). The irregular wave field is accounted for by assuming that the wave height has a Rayleigh Distribution. It is important to consider the wave irregularity because the surf beats occur only in an irregular wave field. Surf beats affect the oil deposition area on the beach.

Waves around the breaking zone have a finite wave height and show nonlinearities. The present model evaluates the wave shoaling by means of the nonlinear long wave theory. The wave heights due to shoaling when computed this way, are much greater than the wave heights obtained from the small amplitude wave theory (e.g. Stiassnie and Peregrin 1980). The wave setup (or wave setdown) $\overline{\eta}$ is computed based on the radiation stress concept. The wave setup is an important factor to consider in evaluating the shoaling and swash zone, because the water depth at each point changes according to it.

The instantaneous water depth due to the wave effect varies due to surf beats. Therefore, the surf beats influence the shoaling and swash zone. Goda (1975) reported that the maximum amplitude of the surf beats at the shoreline may reach as much as 40 % of the offshore wave amplitude.

The computations using the model presented here showed a surf beat amplitude of 0.22 m for a beach slope of 1/100 when the offshore wave amplitude is 1.5 m. A sample calculation for an instantaneous run-up height of 0.56 m showed that the swash zone becomes wider when the surf beats are considered. This fact shows that for improved numerical simulation of the oil spill in the coastal zone, the effect of surf beats should be considered.

Wave Damping Due to an Oil Slick

Several studies on wave damping due to oil films on the water surface showed that the damping of capillary waves is more significant than the damping of long waves (Hühnerfuss et al. 1983; Wei and Wu 1992). The capillary waves are damped by the interfacial tension between the oil film and the water surface (Cini and Lombardini 1978). In the coastal region the wind waves and swells of more than several seconds in period are dominant. For oil reaching a shore, the wave damping due to interfacial tension is insignificant. However, the wave damping in wind waves must be considered because the deformation of the oil slick can affect the wave damping.

The wave damping due to the oil slick is formulated by assuming that oil behaves as an ideal Bingham fluid. Assumptions of a thin oil slick and small amplitude waves were applied to the Navier-Stokes Equation for a Non-Newtonian Fluid. The following equation was obtained to describe wave damping due to the oil slick:

$$\frac{\rho g}{8}\frac{\partial H^2}{\partial x} = -A\frac{d\tau_0}{\sqrt{2}}\frac{k}{2\pi}\coth kh \qquad (1)$$

where, ρ = density; H = wave height; k = wave number; h = depth; τ_0 = yield stress; d = oil slick thickness; A = a constant related to the dynamic properties of the oil.

A sample calculation for checking the wave damping (using Eq. 1) showed that the wave height was reduced before wave breaking and that the breaking point was shifted as a result of the oil slick. Therefore, the wave damping due to an the oil slick should be considered to determine both the wave height distribution and the advection of the oil slick due to the wave action.

Waves on Shore

The wave properties in the swash zone should be estimated, because the deposition on the shore is affected by them. The run-up height R_u for nonlinear waves is calculated from the relation $R_u/H_0 = \sqrt{\pi/2\beta} + \pi H_0/L_0 - \Delta$, (Le Méhauté et al., 1968). Where H_0 and L_0 = offshore wave height and period; β = the beach slope; Δ = energy dissipated due to wave breaking. The run-down wave height R_d is evaluated from the width of the swash zone ΔY; $R_d = R_u - \Delta Y$. ΔY is computed from $\Delta Y/H_{x=0} = 0.4\xi$ (Battjes, 1974), where $H_{x=0}$ = the wave height at the seaward edge of the swash zone; ξ = a surf similarity parameter ($\xi = \tan\beta\sqrt{H/L_0}$). The maximum run-up wave height and the minimum run-down wave height can be represented as $R_u + \overline{\eta}$ and $R_d - \overline{\eta}$, respectively.

The swash speed is evaluated to estimate the oil film thickness on the surface of the shore. The water particle velocity U_0 at the starting point of the run-up is evaluated from the relation $U_0 = 2.2\sqrt{gH_m}$, where H_m = the wave height at the same point as U_0, obtained by Ogawa and Shuto (1984). The run-up speed in the whole swash zone is assumed to be evaluated from the simplified equation $u_r = (1+f')\sqrt{2gz}$. f' = a coefficient that corresponds to the bottom friction, and evaluated from the run-up height R_u. The run-down speed is evaluated from an equation similar to the one for run-up speed. The thickness of the surface oil on the swash zone must be related to the run-up and run-down speed. In this model the thickness d_b is evaluated from the equation $d_b = C_A \exp(-C_B u_r)$, where C_A and C_B = coefficients related to the oil properties and the beach sediment type. For oil deposition and refloatation, the effect due to tidal level change is also taken into account.

RESULTS

Figs. 1 to 3 show the results obtained using the model described here, for the following data: beach slope = 1/10, the offshore wave height and period are 3.0 m and 10.0 s, respectively, and the tide amplitude is 1.0 m. The shoreline of the mean water level is located at x = 0 m. Oil is discharged on the sea water surface at $x = -600$ m for the first 12 hours at the rate 0.02 m^2/h. Fig. 1 shows the thickness distribution of the oil slick near the shoreline. The oil slick moves in the onshore direction due to wave action. It reaches the shoreline about 5 hours after the oil spill started. The oil thickness distribution reaches a fairly stable state at about 13 hours after the spill began. Afterwards, the oil slick reduces its mass while maintaining the shape of the thickness profile.

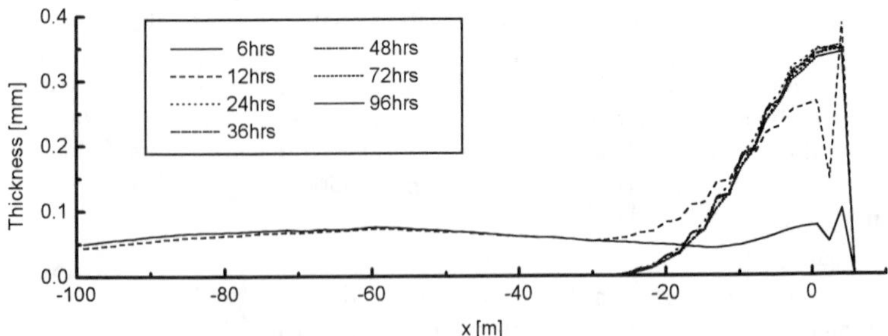

Figure 1: Oil slick thickness distribution

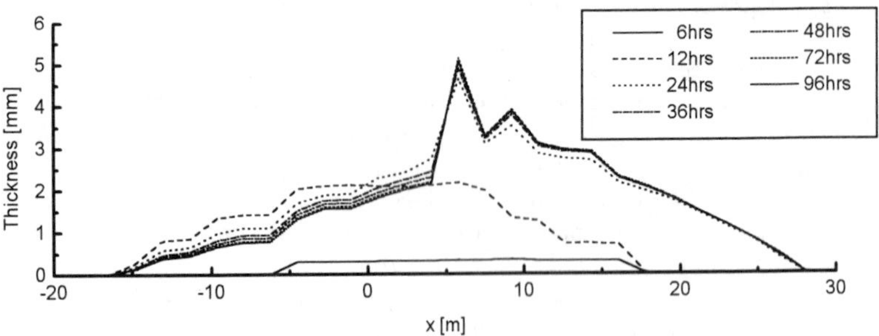

Figure 2: Oil thickness distribution on the beach

Fig. 2 shows the thickness distribution of the oil on the beach. The oil distributes around the shoreline; $x \approx 0$ m on the beach in the early stage of the spreading. About one day after the oil started leaking on the sea, the oil thickness distribution on the beach becomes stable. The oil spreads on the beach from -15 to 25 m in the horizontal direction, corresponding to the beach from -1.5 m to 2.5 m in the vertical direction, due to the effect of the wave run-up, surf beats and tidal level changes. The oil on the beach concentrates mainly above the mean water level, $x = 0$ m, due to the wave run-up action.

Fig. 3 shows the overall mass balance of the oil in the coastal zone. Most oil is on the sea surface in the early stages. As the time passes, the oil moves to and concentrates on the beach. The amount of oil moving towards the beach changes periodically, affected by the tidal level change, and reduces gradually. This periodical change of the amount of oil is caused by deposition and re-entrainment due to the tidal level change.

SUMMARY AND MODEL LIMITATIONS

A one-dimensional model to simulate the oil-shoreline interaction is developed.

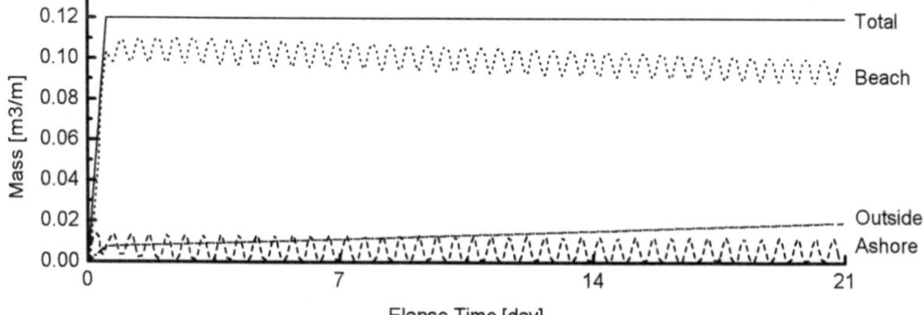

Figure 3: Overall mass balance

This model can simulate the surf beats and their effects on oil deposition. The effects due to tidal level changes and wave damping due to an oil slick are also included. The model does not consider all the complex physico-chemical processes that take place under field conditions. Therefore, this model can be considered crude, especially in its current early stages of development. The conceptual development, however, is a significant improvement over the existing models.

Future modifications are planned for the model, including laboratory experiments. The results presented here shows the time-dependent variation of oil slick thickness distribution.

REFERENCES

Battjes, J.A. (1974). "Surf similarity," *Proc. 14th Coastal Eng. Conf., ASCE*, 466-480.

Cini, R., and Lombardini, P. P. (1978). "Damping effect of monolayers on surface wave motion in a liquid ," *J. of Colloid and Interface Science*, 65(2), 387-389.

Goda, Y. (1975). "Deformation of irregular waves due to depth-controlled wave breaking ," *Report of the Port and Harbour Research Institute, Japanese Ministry of Transport*, vol. 14 No.3, 59-106 (in Japanese).

Hühnerfuss, H., Alpers, W., Garrett, W. D., Lange, P. A. and Stolte, S. (1983) "Attenuation of capillary and gravity waves at sea by monomolecular organic surface films ," *J. of Geophysical Res.*, 88(14), 9809-9816.

Le Méhauté, B., Koh, R. C. Y., and Hwang, L. S. (1968). "A synthesis of wave run-up," *Proc. ASCE*, vol. 94, No.WW1, 77-92.

Ogawa, Y., and Shuto, N. (1984). "Run-up of periodic waves on beaches of non-uniform slope ," *Proc. 19th Coastal Eng. Conf., ASCE*, 198-205.

Reed, M., Kana, T. W. and Gundlach, E. R. (1988). "Development, testing and verification of an oil spill surfzone mass transport model ," *CSE '87-88 R-12*, 1-177.

Stiassnie, M., and Peregrine, D. H. (1980). "Shoaling of finite-amplitude surface waves on water of slowly-varying depth ," *Jour. of Fluid Mech.*, vol. 97, 783-805.

Wei, Y., and Wu, J. (1992). "In situ measurements of surface tension, wave damping, and wind properties modified by natural films ," *J. of Geophysical Res.*, 97 (C4), 5307-5313.

An Australian Perspective on the Roles of Riparian Vegetation in River Restoration

Dr Cathy Wilson[1], Dr Peter Hairsine[2] and Dr Ian Rutherfurd[3]

Abstract

A framework for riparian land management is being developed through a National Australian Riparian Land Management Research Program which has recently been established and funded by the Australian Land and Water Resources R & D Corporation (LWRRDC). A major aim of the program is to quantify interactions between landscape attributes, hydrology and vegetation in riparian lands across a range of agricultural/hydrological/ecological regions in Australia. Here we present a set of knowledge gaps to be addressed, and the research activities which are planned or currently under way in 3 large (100s of km^2) Australian watersheds.

1. Introduction

There is a general belief that vegetation in the riparian zone, the interface between water and land, is endowed with unique features that will provide new opportunities for managing water quality in streams. In particular, we know that grass buffer strips can filter sediment and sorbed nutrients from overland flow before it reaches the stream's edge. We also know that riparian forests can help de-nitrify shallow groundwater as it flows from fertilised pasture into the stream. Finally, the roots of trees, shrubs and grasses have been shown to provide cohesion in soil as well as obstruct flowing water, and thus are capable of slowing stream bed and bank erosion processes.

This information suggests that the riparian zone can be managed to trap pollutants coming off hillslopes, and to reduce stream bank erosion rates. Both functions will help maintain or improve water quality. Other benefits that are likely to be derived from managing vegetation in

---

[1]CSIRO Division of Water Resources, Canberra, ACT; [2]CSIRO Division of Soils, Canberra, ACT; [3]Dept. of Civil Engineering, Monash University, Clayton, Victoria.

riparian lands include better terrestrial and aquatic habitat, reduced algae and macrophyte production and increased aesthetic value.

With so many prospective benefits accruing from riparian vegetation management, it has been given a high priority by many land and water management departments, funding agencies and community groups across Australia. A wide range of activities are on-going; these include streamside vegetation surveys, fencing and tree-planting exercises, and the development of official guidelines for riparian management in some States. Tens of millions of dollars are currently entrained in these projects throughout Australia. Riparian vegetation management is viewed by many as a first line of defence in maintaining water quality in streams.

2. An Australian Program for Research on Riparian Lands

In 1992, the Land and Water Resources R & D Corporation (LWRRDC) commissioned independent reviews to identify priorities for research and management in three areas of riparian zone function: 1) in-stream ecological processes, 2) terrestrial wildlife habitat and 3) management of diffuse pollution sources. Two review papers (Catterall, 1992; Bunn, 1992) were prepared and two workshops relating to the ecological and physical functions of riparian zones, respectively, were held in 1993 (Woodfull et al., 1993, Bunn et al., 1993).

LWRRDC then commissioned a national program of research and development on the rehabilitation and management of riparian lands. The program formally involves community groups, governing bodies, land and water management agencies and academic and research institutions in most of the states. The major objectives of this program are to: 1) identify and quantify the effects of riparian lands on channel morphology, bank stability, and the ingress of sediment and nutrients to rivers and water bodies; 2) identify the key processes by which riparian lands influence in-stream ecosystems and their functioning, and quantify major effects; and 3) demonstrate practical, cost effective and ecologically sound methods for rehabilitation and management of riparian lands.

Research and management activities within the national program are structured around these objectives. And although we recognise the importance of the ecological and management objectives above, this paper will focus mainly on the development of a conceptual framework for how vegetation in the riparian zone affects the physical and chemical aspects of water quality in streams.

3. The Role of Riparian Vegetation in Stream Water Quality

The vegetation in riparian zones can be managed for

many different purposes. In some cases a vegetation type that optimises one function, (e.g.. grass for sediment trapping) may have a detrimental effect on other functions (in-stream habitat and stream bank stability). We also believe that the impact of a given management strategy on water quality at any point along a river depends on the manner in which riparian vegetation, land use, hydrology and landscape attributes interact along the river. To predict an impact on water quality requires a quantitative framework or model which accounts for these interactions.

3.1 Conceptual Framework

Water and sediment is moved from the land through a hierarchical stream network. Channel morphology, flow dynamics and material transport processes change within this network in association with increasing drainage area and decreasing stream gradient downstream. There are also downstream changes in the geomorphology of catchments. These are characterised by low order streams occurring at higher elevations in steeper terrain, and high order streams being associated with wide gentle valleys or plains. Little work has been done to link these changes in geomorphology with changes in hydrology, ecology and riparian zone processes in catchments. Our modelling framework is based on the premise that stream order, geomorphology, and stream and catchment runoff processes are linked in a systematic manner. We can predict the functions of the riparian zone, and hence apply appropriate management techniques, at any point in the stream network if we understand these links.

3.2 Knowledge Gaps

A range of knowledge gaps relating to the mechanisms by which sediment and nutrients move through the riparian zone needs to be addressed in order to quantify these links. These knowledge gaps are summarised in the questions posed below.

- What are the mechanisms of delivery of sediment and nutrients from source areas to and through the riparian zone? How do these mechanisms depend on position in the river drainage network (headwater to floodplain), topographic convergence, and the hydraulic geometry of the river?
- What are the relative roles of subsurface and surface runoff in the transport of pollutants through riparian zones?
- How do preferred surface pathways (livestock tracks, roads, convergent topography) affect runoff rates and trapping efficiencies in riparian zones?
- Can deep rooted vegetation modify runoff pathways, water table dynamics, and the intensity of seepage forces in bank material?
- What are the efficiencies of grasses, litter, harvest

trash/slash in trapping sediment and sorbed pollutants carried in surface runoff?

- What are the residence times of sorbed pollutants in the riparian zone and what is the risk of release of nutrients and re-entrainment of sediment after deposition?
- How do soil hydrologic properties in the riparian zone depend on vegetation type, topographic setting, and amount and size of trapped sediment?
- What is the role of vegetation in providing cohesive strength to river bank material, and resistance to corrasion and erosion during floods?
- What is the effect of riparian vegetation on flood heights and duration at different points in the drainage network?

The project aims to gather existing and new experimental data to answer these questions for a range of environmental conditions. The data is currently be used to develop a suite of models which describe the hydrologic, trapping and bank stability processes occurring in the riparian zone. These models will in-turn be used to develop a set of guidelines for riparian vegetation management which account for the variations in conditions between one reach of river and the next.

4.0 Research Activities

Three types of research activities are required to develop our understanding of riparian zones; these are 1) catchment-wide surveys and rough sediment and nutrient budgets, 2) long-term monitoring of transport and deposition of sediment and nutrients through the riparian zone and small sub-catchments, and 3) short-term rainfall simulation and laboratory experiments.

Catchment-wide work needs to be aimed at identifying sources of sediment and nutrients, their production rates and their pathways to streams. With this knowledge we can target research and demonstration sites to areas where rehabilitation of riparian lands will make a difference to a range of environmental values.

Effective management also depends on the "treatability" of material delivered to the riparian zone. For instance, any place that surface runoff is concentrated as it enters the riparian zone is a place that will be hard to treat. These areas of concentration can range from cattle tracks, to natural topographic swales, to agricultural drains in floodplains. The catchment surveys will determine the extent to which runoff is concentrated as it enters waterways, the manner in which source areas are connected to the riparian zone, and how these vary as a function of position in the drainage network.

Long term reach-based and sub-catchment monitoring is required to determine the fluxes of sediment and nutrients through riparian zones, and the impacts of vegetation on material pathways over many storm events. Information on the residence times of sediment and sorbed pollutants, soil formation processes, changes in soil physical properties resulting from deposition, and rates of bank migration can also be gained in this manner.

Short-term experiments with large rainfall simulators are being performed to understand the processes by which sediment and sorbed pollutants get trapped, sorted and released from riparian zones during storm events of differing magnitudes. The inter-relationships between plant characteristics, overland flow hydraulics and soil physical properties can also be examined in this manner.

In addition, short-term experiments can be carried out in laboratory flumes and with standard soil mechanics devices to quantify the conditions under which roots provide a significant increase in bank strength and/or resistance to corrasion in river bends.

The Australian riparian research program is now developing these research activities in three catchments in Australia. We're hopeful that through collaboration between a network of research institutions, community groups and state agencies, we can provide a set of scientifically based riparian vegetation strategies in the next few years which, if adopted, will help improve water quality in Australian rivers and waterbodies.

5.0 References

Bunn, S.E., (1992) Riparian vegetation. A review of ecologicl issues, deficiencies in basic understanding and priorities relating to the influence of riparian vegetation on the structure, function and management of Australian river ecosystems. Unpublished report to LWRRDC, Canberra, 11p.

Bunn, S.E.; Pusey, B.J.; and Price, P. (Eds) (1993) Ecology and Management of Riparian Zones in Australia. Proceedings of a National Workshop on research and management needs for riparian zones in Australia, held at Marcoola, Qld., April1993. LWRRDC Occasional Paper Series No: 05/93

Catterall, C.P. (1992) Remnant vegetation scoping paper. Unpublished report to LWRRDC, Canberra, 13 pp.

Large, A.R.G and Petts, G.E . (1992) Buffer Zones for Conservation of Rivers and Bankside Habitats. Project undertaken by The International Centre for Landscape Ecology, Loughborough University of Technology. Project Record 340/5/Y (National Rivers Authority: Almondsbury)

Woodfull, J.; Finlayson, B.; and McMahon, T. (Eds.) (1993) The Role of Buffer Strips in the Management of Waterway Pollution from Diffuse Urban and Rural Sources. Proceedings of a Workshop held at the University of Melbourne, October 1992. LWRRDC Occasional Paper Series No: 01/93

FLUSHING FLOWS FOR HABITAT RESTORATION

Robert T. Milhous (M,ASCE)[1] and Russ Dodge[2]

Abstract

Water projects have caused a reduction of the quality of stream habitat for aquatic animals by deposition of fine sediment and sand. Restoration, flushing, and substrate maintenance flows are required to remove and prevent deposition of the fine sediment and sand. The maximum size of the wash load, the suspended load, and the bed load was used to determine the magnitude of these flows. An application to the Trinity River in California resulted in a calculated restoration and flushing flow of 6700 cfs and a substrate maintenance flow of 2500 cfs.

Introduction

Flushing flows are streamflows required to maintain a river in suitable conditions for the desired aquatic animals. Two types of flushing flows exist: 1) to keep most of the undesired sediment moving without significant deposition in the bed material, and 2) to remove undesired sediment, once deposited, from the substrate. In this paper the first flow is a substrate maintenance flow and the second a flushing flow. The objective of a habitat restoration flow is to remove the undesired sediment deposited over a long period of time without maintenance and flushing flows prior to starting a regime of maintenance and flushing flows. Most undesired sediment is fine material or sand. If an undesired particle is removed from the bed it may soon be re-deposited in the bed unless the flows are adequate to transport the particle through the channel as suspended

---

[1] Hydraulic Engineer. National Biological Service. 4512 McMurry Ave. Fort Collins, CO 80525-3400.

[2] Water Resources Laboratory. U.S. Bureau of Reclamation. Denver, CO

or wash load. This paper presents an approach to the calculation of the maximum size of particle that can be transported as wash or suspended load and demonstrates how the maximum size information is used to determine substrate maintenance, flushing, and habitat restoration flows. The size of the bed load is also considered in the analysis.

Calculation of Maximum Size of Wash and Suspended Load

The definitions of wash and suspended load, as used in this paper, are consistent with the general use of the terms but the specifics are different. The wash load is the material with a 100% probability the material will move either on the bed or suspended, the suspended load is the load in suspension.

The maximum size of the particles in suspension was calculated using the logic presented in Nezu and Nakagawa (1993) that the dimensionless shear stress is 0.28 for the upper limit on the suspended sediment load over a fully rough surface.

The maximum size of the wash load assumes the dimensionless shear stress for 100% probability of movement can be used to calculate the maximum size. Pazis and Graf (1977) developed a relation between the probability of movement of the sediment and the dimensionless shear stress. Using the diagrams in Pazis and Graf a dimensionless shear stress of 0.56 was selected as the upper limit on the size of the wash load.

Data from Barta, Wilcock, and Shea(1993) for the Poker Bar site on the Trinity River in Northern California was used to calculated the sizes as a function of discharge. The results are given in Figure 1.

The lower limit on the median size of the bed load is also shown on Figure 1. The equation used to calculate the lower limit on the median size of the bed load is based on the "keep sand moving" case given in Milhous (1990). The 1990 equation was rearranged to give the median size of the bed load as the dependent variable. The equation for bed load size less than or equal to the armour size is:

$$\text{d50bl} = (\text{beta}/0.066)^{2.13} * \text{d50a} \quad (\text{d50bl} < \text{d50a})$$

where beta is the dimensionless shear stress parameter calculated using the bed shear stress and the median size of the bed surface (armour) material (d50a) , d50bl is the median size of the transported material.

For a calculated bed load larger than the d50 of the armour the assumption is that the dimensionless shear stress (based on the d50 of the bed load) is 0.066.

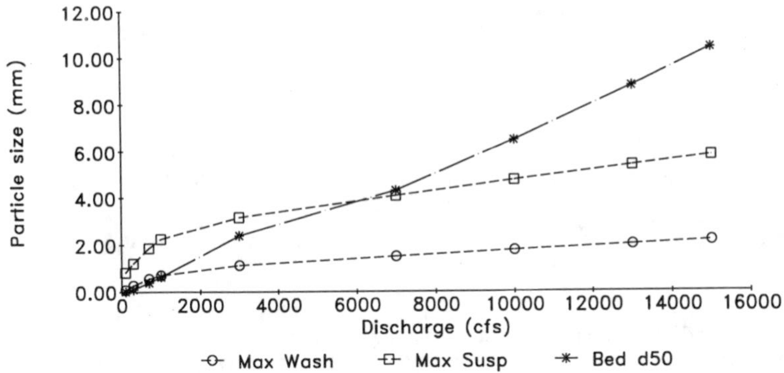

Figure 1. The maximum size of the wash load (Max Wash) and the suspended load (Max Susp), and the lower limit on the median size of the bed load (Bed d50) as a function of discharge for the Poker bar site on the Trinity River. Median size of the bed surface material (armour) used in the calculations was 30 mm.

A similar equation has been developed using the Oak Creek data for the size at which 95% are finer (d95bl). The equation is:

$$beta = 0.018\ (d95bl/d50a)^{0.32}$$

The equation was rearranged the same way as the equation for d50bl. The results of the calculation for d95bl are shown on Table 1 along with the information presented in Figure 1.

Application of Maximum Wash and Suspended Size Diagram

The habitat requirements of the aquatic animals must be determined. This is not a small task. For the purpose of this paper, the assumption is that the substrate must be free of material less than 1 mm, and material less than 4 mm must be removed at intervals from the surface (armour) and from the substrate material below the armour. There must also be a restoration flow to remove the 4 mm material.

Table 1. Maximum sizes of transported sediment and the lower limited on the median sizes of the bed load in the Trinity River at Poker Bar. The size of the bed load at which 95% is finer is used as the maximum bed load size.

Discharge (cfs)	Maximum size (mm) of Wash Load	Maximum size (mm) of Suspended Load	Maximum size (mm) of Bed Load	Median size (mm) of Bed Load
100	0.07	0.82	0.01	0.01
300	0.26	1.19	0.34	0.09
700	0.53	1.85	2.84	0.38
1000	0.69	2.25	6.03	0.63
3000	1.11	3.17	33.42	2.38
7000	1.49	4.07	44.04	4.28
10000	1.75	4.73	53.45	6.46
13000	2.00	5.37	61.68	8.76
15000	2.15	5.78	66.82	10.39

The first task is to calculate the restoration flow required to remove the 4 mm or smaller material. The assumption made is that this requires the disturbance of the surface layer and the transport of 4 mm material as suspended load. Using information in Table 1 or Figure 1, the flow required to transport the 4 mm material as suspended load is 6700 cfs. At 6700 cfs the d95 size of the bed load is 43 mm. The median (d50) size of the surface material has a range between 1.6 and 67 mm. The smaller size bed surface material probably occurs when sand is deposited on top of the gravels. The range in size of the gravel armour is probably 19 - 67 mm with a common value of 30 mm. A 6700 cfs discharge is probably adequate to restore the substrate of the Trinity River. If the fine material is cohesive the restoration flow would be larger because of the bonding of particles; the material in the Trinity River is non-cohesive which means the flushing flow required at intervals to remove the 4 mm or less material is numerically the same as the restoration flow.

The second task is to calculate the flow required to keep the 1 mm size material in transport as wash load when the tributary flows are transporting significant quantities of sediment. Again using Table 1 or Figure 1, the discharge was found to be 2500 cfs. This flow is adequate to transport 2.9 mm material as suspended load with a median size of the bed load of 1.9 mm, and a bed load d95 size of 27 mm. More than 90% of the tributary inflow is smaller than 2.9 mm which means a maintenance flow of 2500 cfs is adequate to keep most of the tributary sediment moving but some material in the undesirable size (4 mm or smaller) may be deposited in or on the stream bed.

Discussion

The thesis of this paper is that a maximum wash - suspended particle size diagram (Figure 1) is needed to calculate substrate maintenance, flushing, and restoration flows for a river. The equations used to develop Figure 1 are in a early stage of development. The criteria used to select sizes of material to be removed by flushing and restoration flows and to be prevented from deposition by the maintenance flows, while based on the needs of Chinook Salmon, must be better developed. Studies are currently (1995) underway in the office, laboratory, and field to improve and verify the various concepts and equations presented in this paper. The equations and criteria are expected to change. The concepts are expected to remain valid.

The calculation of the length of time required for the restoration, maintenance, and flushing flows is not a topic of this paper. These must be determined using some type of sediment mass balance analysis. The restoration flows, while numerically the same as the flushing flows, probably must occur for a longer period in order to be effective.

The maximum size diagram (Figure 1) was used to investigate sediment transport and flushing flow needs in the Trinity River (Milhous, 1995).

References

Barta, A.F., P.R. Wilcock, and C.C. Shea. 1993. Determination of Flushing Flows for Salmonid Spawning Gravels. Third Progress Report to Southern California Edison Company. Rosemead, CA.

Milhous R.T. 1990. The Calculation of Flushing Flows for Gravel and Cobble Bed Rivers. in H.H. Chang and J.C. Hill editors. Hydraulic Engineering. American Society of Civil Engineers. New York, NY.

Milhous, R.T. 1995. Suspended and Bed Load in Flushing Flow Analysis: Trinity River Case Study. in H.J. Morel-Seytoux editor. Proceedings of the Fifteenth Annual American Geophysical Union Hydrology Days. Hydrology Days Publications. 57 Selby Lane. Atherton, CA.

Nezu, I. and H. Nakagawa. 1993. Turbulence in Open-Channel Flows. A.A. Balkema. Rotterdam, Netherlands. 281p.

Pazis, G.C. and W.H. Graf. 1977. Erosion et Deposition; Un Concept Probabiliste. Proc. XVII Cong. Assoc. Intrern. Rech. Hydr. Baden/Baden, Vol 1.

Calculation of flow resistance for river channel restoration

Jan Zelazo[1], Zbigniew Popek[2]

Abstract

The results of investigation on flow resistance arising in natural lowland river channels were presented in the report. The impact of bottom and banks of the channel, variability of the shape and sizes of cross section and river meandering on the roughness coefficient was analyzed.

Introduction

The natural river channel is characteristic with its irregularities of the bottom and banks, alternating deep and shallow areas of different flow velocity and turbulence. According to environment conservation requirements, rivers should have such features after their restoration as well. For this reason the planned transversal profiles of river channel should be irregular, with various sizes and shapes, while from the top view the channel should be composed, at a great extent, of curvilinear sections.

In the channels of natural rivers there the flow resistance arises as the effect of the occurrence of many diverse factors. When elaborating characteristics of flow resistance one should analyze all factors and channel properties that can have a substantial impact on the size of that resistance. Such a proceeding method had been proposed, among others, by Cowan (1956), Garbrecht (1961), Morin and Favant (1975). The estimation of roughness coefficient was assessed in the most complex way in the Cowan's procedure, where the total value of the river channel roughness coefficient had been presented as the sum of partial coefficients:

----

[1]Professor, [2]Assistant professor, Department of Hydraulic Structures, Warsaw Agricultural University SGGW, ul. Nowoursynowska 166, 02-766 Warsaw, Poland.

$$n = (n_0 + n_1 + n_2 + n_3 + n_4) \cdot m_c \tag{1}$$

where: n_0 - the basic value of the roughness coefficient for a regular straight channel, n_1 - the coefficient taking into account the impact of channel surface irregularity, n_2 - the coefficient taking into account variations in shape and size of cross section, n_3 - the coefficient taking into account an impact of local obstructions, n_4 - coefficient related to the vegetation impact, and m_c - coefficient related to river meandering.

The impact of the cross section on flow resistance

The impact of the roughness of transversal profile surface is accounted in the formula (1) by the coefficients n_0 and n_1. In practice the assessment of the value of those partial coefficients is very difficult and not always necessary for calculations. For river channels formed in noncohesive grounds the state described with the n_0 coefficient is hypothetical because even at low velocities the surfaces of the bottom and banks are irregular to a certain extent (e.g. bed forms). The n_1 coefficient values are influenced by not only the irregularities caused by bed forms but also other profile deformations caused by either erosion or sedimentation. For this reason a substitution of n_0 and n_1 figures by one partial coefficient:

$$n_{st} = n_0 + n_1 \tag{2}$$

seems to be purposeful; it would characterize flow resistance in a straight line channel, caused by friction on wetted channel perimeter, characterized by a definite grain coarseness of the bottom material and irregularities on the bottom and banks. The n_{st} coefficient should moreover take into account the impact of the shape and size of cross section on flow resistance.

The relationship between the flow velocity and hydraulic parameters of the channel and the properties of water and rubble along the straight line river channel had been described by Shen (1971). In lesser rivers, where a spatial movement exists, beside the parameters mentioned by Shen, the channel width B should be taken into account too. The flow resistance could then be expressed in the form of the following function of dimensionless parameters (Zelazo, 1992):

$$\frac{V}{V_*}=\frac{C}{\sqrt{g}}=\sqrt{\frac{8}{\lambda}}=f\left(\frac{h}{d},\frac{B}{d},\frac{V_*\cdot d}{\nu},\frac{V_*}{\sqrt{g\cdot d}},\frac{h}{B},\frac{V_*}{\sqrt{g\cdot h}}\right) \quad (3)$$

Statistical analysis was applied to assess the significance of individual dimensionless parameters and to define the detailed form of the relationship (3). The study on relationship between flow resistance and channel characteristics was carried out on the basis of 670 measurements made on 23 profiles on straight line sections of 23 lowland rivers. The channels had been formed in sand grounds. The basic characteristics of profiles were as follows: profile width B = 1.60-114.0 m, mean depth of water h = 0.10-2.24 m, slope of energy line I = 0.0001-0.00225, the average diameter of bottom material d = 0.15-0.63 mm, the average flow velocity V = 0.303-1.079 m/s.

After elimination of parameters statistically insignificant and application of the product regression model, the following functional relationships were achieved for the general equation (3):

$$\frac{C_{st}}{\sqrt{g}}=0.2934\cdot\left(\frac{V_*\cdot d}{\nu}\right)^{0.1291}\cdot\left(\frac{V_*}{\sqrt{g\cdot d}}\right)^{-0.0251}\cdot\left(\frac{h}{B}\right)^{-0.1143}\cdot\left(\frac{V_*}{\sqrt{g\cdot h}}\right)^{-0.7111} \quad (4)$$

or after transformations:

$$n_{st}=\frac{I^{0.3036}\cdot h^{0.0623}}{6.1262\cdot B^{0.1143}\cdot d^{0.1416}} \quad (5)$$

Taking into account the relationship (5), the mean flow velocity in naturally shaped river channel on the straight line section will be:

$$V=6.1262\cdot h^{0.4377}\cdot I^{0.1964}\cdot B^{0.1143}\cdot d^{0.1416} \quad (6)$$

The equation (6) was tested on an independent material. A new set of data was used for that purpose, consisting of the results of 166 hydrometric measurements made in the rivers with the following parameters: B = 2 - 120 m, h = 0.26-2.66 m, V = 0.17-1.016 m/s, I = 0.00017-0.0019. The velocities measured in the nature and calculated for those conditions according to the equation (6) are presented on Fig. 1, while on Fig. 2 there the histogram of errors DV = $(V_m - V_{cal})\cdot 100\% / V_m$ ("m" - measurement in the field, "cal" - calculated value) was shown.

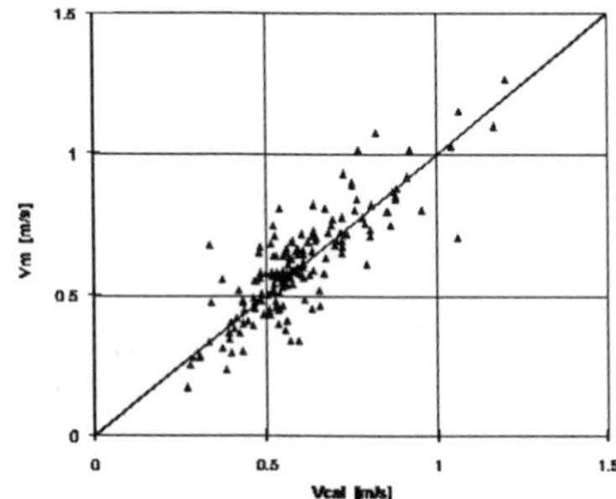

Figure 1. Comparison of velocities: V_m - measured and V_{cal} - calculated from equation (6).

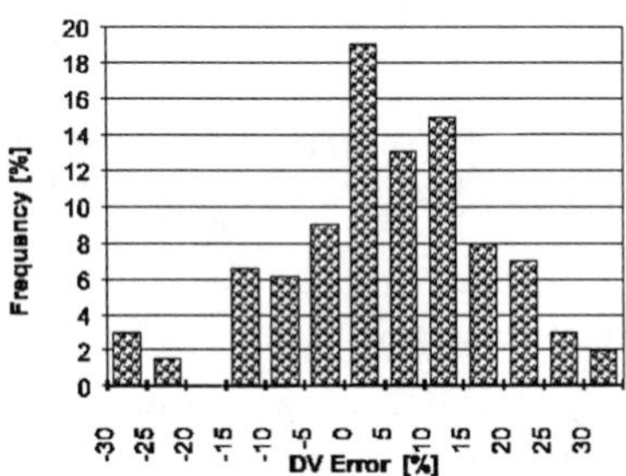

Figure 2 Histogram of *DV* errors.

<u>The impact of river meandering on flow resistance</u>

The general relationship of flow resistance and parameters of river channel on river arc is expressed in the following relation of dimensionless parameters (Zelazo 1992):

$$\frac{C}{\sqrt{g}} = f\left(\frac{h}{d}, \frac{B}{d}, \frac{V_* \cdot d}{\nu}, \frac{V_*}{\sqrt{g \cdot d}}, \frac{h}{B}, I, \frac{h}{r}, \frac{r}{B}, \frac{\alpha}{90}\right) \qquad \mathbf{(7)}$$

Because the first 6 parameters in the equation (7) express also flow resistance along the straight channel (equation 3), then the increase of resistance on the river curvature, expressed as the ratio of roughness coefficient n_c on the curvature to the n_{st} coefficient on the straight channel section can be expressed in the following dimensionless relationship:

$$\frac{n_c}{n_{st}} - 1 = f\left(\lambda_{st}, \frac{h}{r}, \frac{r}{B}, \frac{\alpha}{90}\right) \qquad \mathbf{(8)}$$

Studies on the relationship between additional losses on the river curvature and the hydraulic and geometric parameters of the channel were made for 4 lowland rivers with the following parameters: B/h = 10-40, r/B = 1-11.5, $\alpha/90^o$ = 0.3-2.0 and h/r = 0.001-0.06.

On the basis of the regression analysis carried out for measurement data from 197 natural curves of different rivers, the following functional form was established for the general relationship (8)

$$m_c = 1 + 0.0208 \cdot \left(\frac{\alpha}{90} - 0.3\right)^{0.40} \cdot \left(12 - \frac{r}{B}\right)^{1.10} \cdot \left(\frac{10 \cdot h}{r}\right)^{0.24} \cdot (\lambda_{st})^{-0.20} \quad \textbf{(9)}$$

where $m_c = n_c/n_{st}$ is a coefficient as in the equation (1).

Calculations according to the equation (8) indicate, that in the meandering rivers additional resistance on river curvature can considerably influence the value of the total roughness coefficient. For instance, for a river with the bow angle 90°, the increase of roughness coefficient for the curve radius $r = 2B$ will amount to 29.7%, for $r = 5B$ the increase will be by 16.1% and for $r = 10B$ the increase will be by about 3%.

List of symbols:

B - width of river bed, (m)
C - Chezy water velocity coefficient, ($m^{-1/2}s^{-1}$)
d - mean particle diameter, (m)
g - acceleration due to gravity, ($m\ s^{-2}$)
h - mean depth of channel, (m)
I - slope of energy line, (-)
n - Manning roughness coefficient, ($m^{-1/3}s$)
r - radius of meander curvature, (m)
V - mean velocity in the profile, ($m\ s^{-1}$)
V_*- shear stress velocity, ($m\ s^{-1}$)
α - angle of curvature, (°)
λ - Darcy-Weisbach friction factor
υ - kinematic viscosity, (m^2s^{-1}).

Literature

1. COWAN, W.L., 1956: Estimating hydraulic roughness coefficients, Agricultural Engin., 37, 7.
2. GARBRECHT G., 1961: Abflussberechnungen fuer Flusse und Kanale, Die Wasserwirtschaft, 2, 3.
3. MORIN M., FAVANT M., 1975: Protection contre les affouillements de berge dans l'empris des coudes des cours d'eaux, La Houille Blanche, 2/3.
4. SHEN H.W., 1971: River mechanics, Shen H.W. Editor, Fort Collins, USA.
5. ZELAZO J., 1992: Study of velocity and flow resistance in natural lowlands river channels, Scientific Fascicles of Warsaw Agricultural University, 149 (in Polish).

Design of Artificial Riffles using RMA-2V Two-Dimensional Hydraulic Model

Mark R. Peterson[1], M. ASCE, Lyle W. Zevenbergen[1], M. ASCE, and Jerry Blevins[2]

Abstract

This paper describes the use of two-dimensional hydraulic modeling to investigate creating an artificial riffle at the Glenn-Colusa Irrigation District (GCID) diversion on the Sacramento River. Two-dimensional modeling was selected because accurate simulation of the flow distribution in the main and side channel and flow distributions within the artificial riffle was deemed essential. The project was performed for the Sacramento District, U.S. Army Corps of Engineers (USACE).

Introduction

GCID was organized in 1920. The diversion is located on the Sacramento River at River Mile (RM) 206, 5.6 km (3.5 miles) upstream from Hamilton City, California. The pumping plant, with a capacity of 85 m^3/s (3,000 cfs) is on a side channel of the Sacramento River which forms Mongomery Island. Fish screens were installed in 1972 to prevent the diversion of endangered salmonids. The fish screens were designed for river hydraulic conditions that existed prior to a 1970 bendway cutoff centered at RM 203.5. The cutoff reduced the channel length approximately 2.4 km (1.5 miles) and reduced the low-flow water surface approximately 0.9 m (3 ft) at the GCID intake channel. The lower water surface results in reduced cross-sectional flow area and higher than design velocities at the screens. Juvenile salmon are killed by impingement on the screens and GCID has had to reduce pumping rates. An additional problem is caused by the lowered water surface and high pumping rates. Under these conditions, the pumping facility can draw water up the bypass channel necessary to return salmonids to main channel. When this occurs, it is expected that all juvenile salmon entering the diversion are lost due to predation.

[1]Manager, Hydraulic Engineering and Senior Hydraulic Engineer, respectively, Ayres Associates, 3665 JFK Parkway, Bldg. 2, Suite 300, Fort Collins, CO, 80525.
[2]Technical Project Manager, Project Engineering, Sacramento District, U.S. Army Corps of Engineers, 1325 J Street, Sacramento, CA, 95814.

In order to alleviate the problems of fish kill and reduced pumping rates, new and/or modified fish screening facilities are proposed in conjunction with a gradient restoration facility (GRF) to raise water surfaces to pre-1970 conditions. The GRF is essentially an artificial riffle that raises water levels for low flows and has little impact (drowns out) at flood flows. A key design issue for the GRF is that it not act as a barrier to fish passage. A review of fish passage literature concluded that even though there is a long history of passing fish over structures, such as dams and natural barriers, fish passage facilities have been designed primarily to pass salmonids. Very little regard has been given to passage of various species such as striped bass, American shad or white sturgeon, all of which are present in the middle Sacramento River. Data identifying acceptable hydraulic conditions for passage of other species are minimal and recommendations for use in hydraulic structure design are extremely conservative. In an attempt to overcome the obvious paucity of data, a "design riffle" concept was proposed. This approach is based on the rationale that if fish species can accommodate natural hydraulic conditions within the Sacramento River, those hydraulic conditions provide a reasonable basis for design of an artificial riffle. In order to use the design riffle concept, data were collected at natural riffles, the RMA-2V (Thomas and McAnally, 1985) hydraulic model was tested for the natural riffle conditions, and used to design an artificial riffle which poses no greater hydraulic barrier than the natural case.

Natural Riffle Model

Data were collected at three naturally occurring riffles on the Sacramento River. The following discussion focuses on a riffle located at RM 202.5, located a short distance downstream from the proposed GRF project site. The data collection procedure involved measuring velocity near the bed and at 60 percent of the flow depth relative to the water surface. When flow depths were high enough for the equipment being used, additional readings were collected at 20 and 80 percent of the depth. Near-bed velocities were measured within 60 mm (0.2 ft) of the bed. Water surface elevations were measured at the waters edge. Velocity, depth, and water surface elevation were measured at 12 cross sections (Figure 1) at the riffle site on November 1993 when the discharge was 150 m^3/s (5,300 cfs).

A detailed finite element mesh was developed for the natural riffle based on field measurements and observations. The model Manning's n of 0.035 best replicated observed water surface conditions. The observed flow distribution was 31 m^3/s (1,100 cfs) in the left branch and 110 m^3/s (4,200 cfs) in the right branch. The model predicted 37 m^3/s (1,300 cfs) and 113 m^3/s (4,000 cfs), respectively. However, at the observed shallow-flow conditions, it is unlikely that Manning's n would be constant. Therefore, the model was modified so that different element types were specified for different flow depths. Hey's (1979) flow resistance equation was used to predict Manning's n for various flow depths and a bed material (D_{84}) of 75 mm. Elements with flow depths less that 0.3 m (1 ft) were assigned an n value of 0.043, flow depths greater than 0.9 m (3 ft) were assigned an n of 0.033 and intermediate depths were assigned an n of 0.035. No further calibration was performed because the modeled water surface elevations virtually replicated observations. The predicted flow distribution improved to 34 m^3/s (1,200 cfs) and 116 m^3/s (4,100 cfs). Figure 1 shows observed and computed water surface elevations at the cross section locations.

The detailed modeling shows that shallow flow hydraulics can be simulated well with the RMA-2V model. Although a single reasonable Manning's n value yields adequate predictions of water surface an overall energy loss, varying Manning's n with depth improves velocity distribution estimates for shallow flows. Therefore, Manning's roughness was varied with depth at the design riffle for the design low-flow conditions of 113 m^3/s (4,000 cfs).

Artificial Riffle Model

A key consideration in modeling the artificial riffle (gradient restoration facility, or GRF) is accurate determination of the hydraulic roughness of the bed material. This factor is crucial in producing the desired 0.9 m (3 ft) rise in water surface elevation and in determining the distribution of depths and velocities within the GRF. It is anticipated that the GRF will be constructed with riprap to define and maintain the design configuration. The preliminary design of the GRF includes a wide variety of flow depths, including pools and riffles to mimic natural variability.

A common relation used to estimate the flow resistance of riprap is the Strickler type equation where Manning's n is proportional to the D_{50} of the riprap to the 1/6 power. The constant of proportionality is typically 0.04. Maynord (1991) reported a constant of proportionality of 0.038 using flume data. While this approach is reasonable for high relative submergences, where flow depth is high relative to the bed material size, it does not account for variability due to shallow flow. Maynord concluded there was little depth affect, although this result assumed a variable Von Karman constant. If a Von Karman constant of 0.4 is used with Maynord's data, the roughness height of riprap is, on average, $2.2D_{50}$ and approximately 80 percent of the data plotted within the range of $1.5D_{50}$ to $3.2D_{50}$. Design flows produce flow depths up to 1.5 m (5 ft) over the GRF and a riprap median size of 0.3 m (1 ft) will be used. Using $2.2D_{50}$ as the roughness height in a Keulegan (1938) type equation, Manning's n was set at 0.051 for flow depths less than 0.3 m (1 ft), 0.044 for flow depths greater than 0.9 m (3 ft) and 0.048 for intermediate depths. For flood flows, a Manning's n of 0.038 was used for all flow depths. It is interesting to note that the when Hey (1979) and Keulegan (1938) equations are put in similar form, the roughness height for natural gravels is $3.5D_{84}$ while Manord's (1991) data indicates a roughness height of $2.2D_{50}$ for riprap. Therefore, riprap produces less flow resistance than river run gravels (and cobbles) of the same size. A possible explanation is that the riprap is more uniform than natural gravels and forms a more even boundary. Also, flow through the riprap could tend to mask some of the (computational) flow resistance of this material.

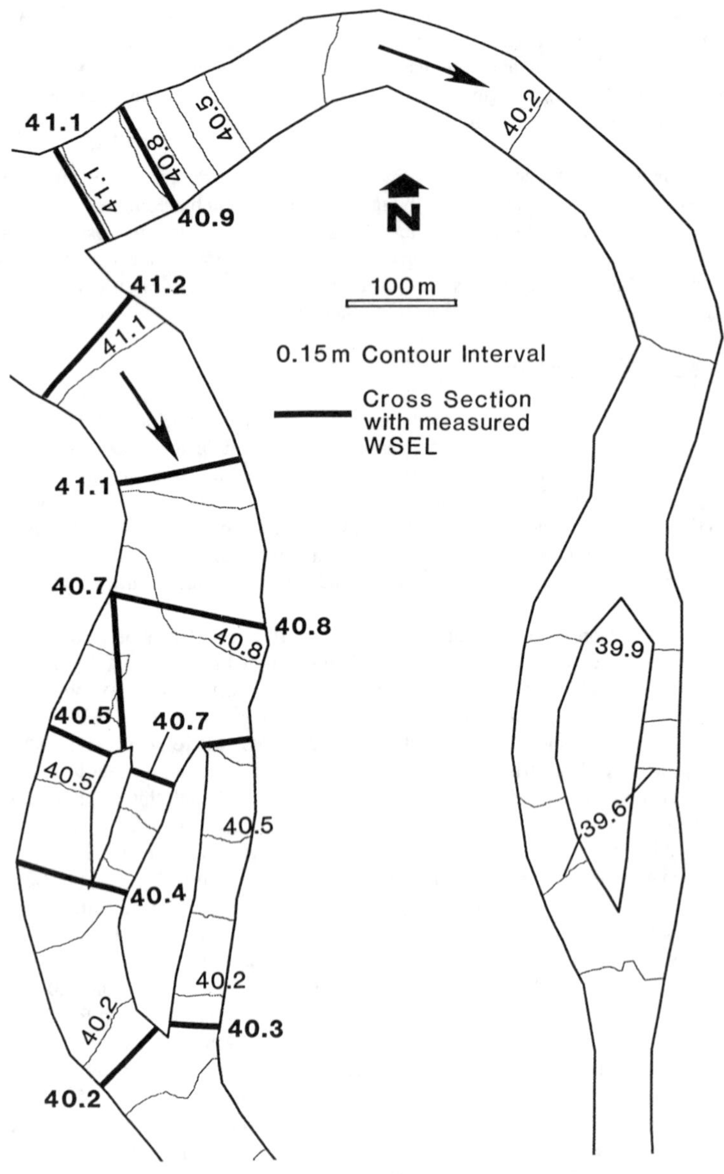

Figure 1. Cross sections and water surface elevations at natural riffle data site.

Using the depth variable Manning's n and the "allowable" velocities and depths measured at the natural riffle, preliminary hydraulic design of the GRF (artificial riffle) has been developed. The structure is to be constructed out of riprap and sheet pile cutoff walls will be driven across the structure for additional stability. The structure will divide the Sacramento River flow into two paths with a berm of riprap. Along one flow path, a pool riffle sequence will be established. The majority of the flow is conveyed along the second flow path. Normally, fish passing the natural riffle would encounter flow velocities between 0.9 and 1.5 m/s (3 and 5 ft/s) and occasionally encounter velocities outside this range. Fish passing the GRF would also normally encounter flow velocities between 0.9 and 1.5 m/s (3 and 5 ft/s) and would frequently encounter flow velocities less than 0.9 m/s (3 ft/s). Velocities greater than 1.5 m/s (5 ft/s) are not expected in the GRF. Improvements to the GRF design could include producing a more natural appearance and providing more resting and hiding places for the fish.

Conclusions

The investigation showed that shallow flow can be modeled accurately using the RMA-2V two-dimensional model and that a structure can be designed to produce hydraulic properties similar to natural riffles. Although this paper focused on hydraulic design and fish passage, other investigations have included effects of the structure on (1) flood flow water surface elevations, (2) sediment transport in the Sacramento River and GCID intake channel, (3) alternative fish screen designs, and (4) other environmental issues.

Appendix - References

Hey, R.D. (1979). "Flow Resistance in Gravel-Bed Rivers," Journal of the Hydraulics Division, ASCE, Vol. 105, No. HY4, April, 1979, pp. 365-379.

Keulegan, G.H. (1938). "Laws of Turbulent Flow in Open Channels," Journal of Research of the National Bureau of Standards, Vol. 21, Reset Paper 1151, December, 1938, pp. 707-741.

Maynord, S.T. (1991). "Flow Resistance of Riprap," Journal of Hydraulic Engineering, ASCE, Vol. 117, No. 6, June, 1991, pp. 687-696.

Thomas, W.A. and McAnally, W.H. (1985). "User's Manual for the Generalized Computer Program System: Open Channel Flow and Sedimentation, TABS-2," U.S. Army Engineer Waterways Experiment Station, Vicksburg, MS, 671 pp.

PC-Based Design of Channel Protection
Using Permanent Geosynthetic Reinforcement Mattings

David T. Williams[1], F. ASCE and Deron N. Austin[2], M. ASCE

Introduction

New erosion and sediment control legislation, coupled with enhanced public awareness toward environmental issues, has led to a rapid increase in the use of flexible geosynthetic lining systems as lining materials in inland waterways. These materials are being selected as alternatives to rigid linings because of several advantages they offer. Geosynthetic mattings:

1. Allow vegetative establishment
2. Extend performance limits of natural vegetation
3. Conform to uneven subgrades
4. Are easy to install
5. Impede water flow/capture sediment
6. Promote infiltration/groundwater recharge
7. Reduce sediment transport
8. Offer greater than 50% cost savings

This paper describes the procedures used for the hydraulic analyses and selection of a permanent geosynthetic matting as channel lining materials featured in a new computer program. EC-DESIGN™ is a complete erosion control design package recently published by Synthetic Industries, Construction Products Division. The program allows the user to select the most appropriate synthetic erosion control material for either his/her construction slope or channel application. The materials in the program consist of polypropylene roving, open weave geotextile meshes, and a wide variety of permanent geosynthetic reinforcement mattings. This paper will describe only the channel lining and design and selection process for permanent geosynthetic mats.

Review of HEC-15

Published by FHWA in 1988, HEC-15 endorses the use of flexible lining materials by providing step-by-step, well-documented design procedures (Chen and Cotton, 1988). Limiting shear stress design values for bare soil, riprap, vegetation, jute, roving, natural

[1] President, WEST Consultants, Inc., 2111 Palomar Airport Road, Suite 180 Carlsbad, CA 92009-1419
[2] Marketing Engineer, Synthetic Industries, Construction Products Division, 4019 Industry Drive, Chattanooga, TN 37416

erosion control blankets, and synthetic mats are given to assist the user in selecting the most appropriate material. However, the permissible shear stress of 9.76 kg/m^2 is a drastic underestimation of the performance of permanent geosynthetic reinforcement mattings. Dozens of well documented laboratory investigations, field studies and case histories have demonstrated performance limits of reinforced vegetation that far exceed guidelines given in HEC-15 (Hewlett, *et. al*, 1987; Carroll *et. al*, 1991). In fact, biotechnically-reinforced systems have resisted velocities and shear stresses in excess of 4.3 m/sec and 35 kg/m^2, respectively, for durations of over two days, providing twice the erosion protection of unreinforced vegetation (Carroll, *et al.*, 1991; Hewlett *et al.*, 1987; Theisen, 1991). The discrepancies in and cumbersome nature of HEC-15 helped establish the need for this computer-aided design program.

Function of Permanent Geosynthetic Mattings

Biotechnical Composites™ is a family of geosynthetic materials comprising non-degradable components that furnish temporary erosion protection, accelerate vegetative growth, and ultimately become synergistically entangled with living plant tissue to permanently extend the performance limits of natural vegetation used in channel stabilization. Exhaustive field and laboratory studies have verified that flexible, three-dimensional (3-D) geosynthetic erosion mats can create a "soft armor" channel lining system providing twice the performance of unreinforced vegetation (Theisen, 1991). Turf reinforcement is a method or system by which the natural ability of plants to prevent soil erosion is enhanced through the use of geosynthetic materials. A flexible 3-D matrix retains seeds and soil, stimulates seed germination, accelerates seedling development, synergistically meshes with developing plant roots and shoots (IECA, 1992).

Classifications and Categories

Permanent geosynthetic mattings are composed of durable synthetic materials, stabilized against ultraviolet degradation and inert to chemicals normally encountered in a natural soil environment. These mattings consist of a lofty web of mechanically or melt-bonded polymer nettings, monofilaments or fibers which are entangled to form a strong and dimensionally stable matrix. Polymers include polypropylene, polyethylene, nylon, and polyvinyl chloride (Theisen, 1991). Geosynthetic mattings generally fall into three matrices categories: Permanent Erosion and Reinforcement (PERMs), Turf Reinforcement (TRMs), or Erosion Control Revegetation (ECRMs). PERMs consist of UV stabilized polypropylene monofilament yarns woven into a dimensionally stable, uniform configuration of resilient pyramid-like projections. Woven 3-D geotextile matrices provide many times the strength of traditional geosynthetic mattings. TRMs provide sufficient thickness and void space to permit soil filling/retention and the development of vegetation within the matrix. TRMs are installed first, then seeded and filled with soil. Seeded prior to installation, ECRMs are denser, lower profile mats designed to provide long-term ground cover and erosion protection. ECRMs rely upon sediment capture for increased long-term stability. By their nature of installation, PERMs and TRMs can be expected to provide more vegetative entanglement and better long-term performance than ECRMs. However, denser ECRMs may provide superior temporary erosion protection (Theisen, 1991).

Input Parameters to Computer Program

If a geosynthetic matting is to be placed in a given channel geometry, design discharge, slope, flow resistance, and vegetative conditions are input and the program determines the flow velocities and shear stresses on the channel bottom and each side slope. To determine these values, the program requires the following input.

The duration of the design flood, in hours, must be entered. Laboratory results have shown that the critical water velocity or shear stress of the matting (*i.e.*, when it is considered "failed") decreases over time and generally reaches its lowest value after 50 hours of peak flow (Carroll, *et al.*, 1991; Hewlett *et al.*, 1987). If a hydrograph, which is the relationship of discharge over time, is available, the duration can be considered to be the time the discharge is greater than 90% of its peak discharge.

If the channel has a significant bend (bend radius of curvature divided by the bottom width is less than 10), a correction to the shear stress and velocity is made according to HEC-15 for the outside side slope of the channel as well as the bottom. The designer must specify if the channel bottom and/or side slopes are to be protected with geosynthetic mattings. This option allows for design of channels with a combination of synthetic materials and/or other channel protection methods on any channel portion.

If the channel is to be protected with a geosynthetic mat and soil filled, a Manning's roughness coefficient, "n", of 0.020 is automatically assigned for the analysis. If the mat will not be soil filled or a fully vegetative system is desired to be analyzed, Table 1 can be used to select the appropriate "n" value. The shaded boxes indicate recommended Manning's "n" for soil filled, vegetated conditions. The effective "n" value for the entire section is then determined by weighing the subsection "n" by its wetted perimeter using the Horton method (Chow, 1959).

Table 1. Recommended Manning's Roughness Coefficients* for Geosynthetic Mattings (modified after Chen and Cotton, 1988)

Geosynthetic Matting Type	Flow Depth Ranges		
	0 - 15 cm	15 - 60 cm	> 60 cm
ECRM	0.035	0.025	0.021
	0.023	0.02	0.02
TRM	0.036	0.026	0.03
	0.023	0.02	0.02
PERM	0.038	0.028	0.024
	0.023	0.022	0.021

* If vegetation exceeds 40% ground cover, species' "n" dominates for expected height

If the geosynthetic mat is to be vegetated, the vegetative class expected to be established is then entered. It is assumed that this vegetative class is applicable for all channel portions protected by the geosynthetic mat. Typical vegetative covers, their class, and "n" values based upon the hydraulic radius and slope are shown in Table 1 of HEC-15.

Geosynthetic Matting Selection Process

The computed water velocities and shear stresses are multiplied by a user specified factor of safety before determining acceptable lining materials. Factors of safety are used to adjust the design for uncertainties and should be determined based upon the magnitude of the uncertainties. Other parameters affect the safety factor, such as danger to human life if the channel fails and any extreme cost associated with repairs to the channel. The expected project life, in months, that the geosynthetic matting is to be functional must also be entered but values greater than 60 months (5 years) have no effect on selection.

Critical velocities and shear stresses of the lining material are divided by the actual values (after multiplication with a factor of safety), resulting in stability factors. If the stability factor is greater than 1, the mat is adequate for erosion protection. This analysis is performed for the channel bottom and side slopes, allowing the user to select different geosynthetic mats for each channel segment. The program takes into consideration the expected life of the project in relation to the geosynthetic performance life.

HEC-15 suggests that the shear stress is the most important parameter for stable channel design (Chen and Cotton, 1988). However, there are occasions, such as high velocities with low flow depths, where the flow velocity is the controlling factor and more appropriate to use. Because of this, the program considers both shear stress and flow velocity and allows the user to determine which selection method to use.

The maximum hydraulic properties of each product are presented in Table 2. These values are published as a result of extensive testing at a leading western university (Theisen, 1991). The thrust of this study was to develop design guidelines for mats when placed over bare soil, filled with soil, and completely vegetated. The results of the study are the basis for the geosynthetic mat selection process. The critical shear stress or velocity of the mats is at the hydraulic condition when the product has "failed", where failure is defined as any of the following conditions: 1) tractive forces tore significant chunks of sod from channel floors or from the reinforcing mats, 2) a mat was torn or its structural makeup was altered significantly due to tractive forces of the water flowing over it, or 3) the channel itself was eroded significantly in spite of the protective vegetation and/or mats placed in them.

Final Selection and Output

In the program output, the shear stress and velocity stability factors for each mat is presented for the channel bottom and each side slope if they are to be protected by geosynthetic materials. The program then lists those materials that have stability factors greater than or equal to 1.0, for both shear stress and velocity, and have an expected performance lifetime greater than or equal to the project life.

Table 2. Maximum Permissible Design Values for Geosynthetic Mattings

PRODUCT	VELOCITY (m/sec)		SHEAR STRESS (kg/m^2)	
	Short Term (1/2 hr)	Long Term (50 hrs)	Short Term (1/2 hr)	Long Term (50 hrs)
PERM, Non-Veg.	6.1	3.1	39.1	14.7
PERM, Vegetated	7.6	4.6	48.9	29.3
TRM, Non-Veg.	4.3	2.4	29.3	9.8
TRM, Vegetated	6.1	4.3	39.1	24.5
TRM, Non-Veg. w/Geotextile	3.7	2.1	24.5	9.8
TRM, Vegetated w/Geotextile	6.1	4.3	39.1	24.5
ECRM, Non-Veg.	5.5	2.4	34.2	9.8
ECRM, Vegetated	5.5	3.1	34.2	19.6

Acknowledgements

The authors would like to thank Marc Theisen for his input and review and Melissa Smith for typing this paper.

References

1. "Choosing a Softer Touch", Parts 1 and 2, IECA Spring/Summer Report, International Erosion Control Association, Steamboat Springs, CO, 1992.
2. Carroll, Jr., R.G., Rodencal, J., and Theisen, M.S., "Evaluation of Turf Reinforcement Mats and Erosion Control and Revegetation Mats under High Velocity Flows", Proceedings of the XXII Annual Conference of the International Erosion Control Association, Orlando, Florida, 1991, pp. 131-145.
3. Chen, Y.H. and Cotton, B.A., "Design of Roadside Channels with Flexible Linings", Hydraulic Engineering Circular No. 15 (HEC-15), FHWA, Publication No. FHWA-IP-87-7, USDOT/FHWA, McClean, Virginia, 1988.
4. Chow, V.T., "Open Channel Hydraulics", McGraw-Hill Book Company, New York, New York, 1959.
5. Hewlett, H.W.M., Boorman, L.A., and Bramley, M.E., "Design of Reinforced Grass Waterways", Construction Industry Research and Information Association (CIRIA), Report No. 116, London, England, 1987.
6. Theisen, M.S., "The Role of Geosynthetics in Erosion and Sediment Control: An Overview", Proceedings of the 5th GRI Seminar, Philadelphia, Pennsylvania, December 1991, pp. 188-203.

PC Reservoir Simulation System (PCRSS) - Development and Calibration of a Reservoir Simulation Model for the Salt River Project

Yvonne Reinink[1], **M. ASCE and Jon Behrens**[2], **M. ASCE**

Abstract: Salt River Project (SRP) is in the second year of a process to develop a computerized hydro-meteorological decision support system (Hydromet) that will eventually cover all aspects of its hydropower, water supply and flood-control functions. This system will be of use to all departments having an operational (as opposed to accounting) water supply or hydropower function. This means that the new system must exist on a distributed computer network having machines running different operating systems (UNIX, OS2, Windows). It is highly desirable that the same look and feel be maintained across these platforms, since SRP staff will need to use the software from different locations during flood operations.

The first year of the Hydromet system development process involved extensive studies to determine precisely what capabilities would be needed by the users and to determine where the data, which drive the system, are located. That first year study resulted in a high-level system architecture and a road map to guide the rest of the development process. This year the effort is on developing a rainfall-runoff model and a reservoir simulation model within the master architecture. To develop a new reservoir simulation, SRP hired Jon Behrens & Associates (JBA) of Longmont, Colorado. This paper describes the process of implementing and calibrating the new reservoir simulation model for the six reservoirs on the Salt and Verde rivers.

Introduction: The Salt River Project (SRP) is one of the oldest reclamation projects in the United States. With its formation in 1903, SRP's purpose was to manage and deliver surface and groundwater to agricultural users within its 250,000 acres (101,172 hectares) service area. The water demand from this service area is satisfied with surface water stored in six reservoirs and

---

[1] Senior Engineer, Salt River Project, P.O. Box 52025, Mailstop PAB203, Phoenix, Arizona 85072-2025

[2] President , Jon Behrens & Associates (JBA), 5575 Bowron Place, Longmont, Colorado 80503

groundwater pumped from about 250 wells. Surface water and groundwater are managed conjunctively.

Since the SRP's primarily function is water conservation, the spillway crests on its dams are relatively high, and large releases can only be made when the water surface rises above the spillway crest. This was not a problem when the area was mainly agricultural and little development occurred in the Salt River bed; but much of the agricultural land has urbanized in what is now the Phoenix metropolitan area. Many bridges and at-grade crossings have been built over and through the Salt River. Extensive flooding in the late 1970s and beginning 1980s resulted in most of those river crossings being closed, disrupting traffic significantly; it became clear that the reservoirs could not be operated for water conservation alone. Since the physical characteristics of the dams do not allow for early releases to create flood-control space, an extensive hydrometeorological data collection network was developed with the purpose of getting as much advance notice of weather affecting the watershed and the resulting runoff approaching the metropolitan area. Near real-time data collection and communication were the keys to warning the people in the metropolitan areas of upcoming floods. The first generation of the Hydromet system was born with data storage and programs to compute runoff and reservoir routing running on mainframe computers.

With new technology coming available, SRP started the process to develop the second generation of Hydromet in the beginning 1990s. This year's effort includes the development of a reservoir simulation model within the master plan by Jon Behrens & Associates (JBA) of Longmont, Colorado. This paper describes the process of implementing and calibrating the new reservoir simulation model for the six reservoirs on the Salt and Verde rivers.

Description of PCRSS: PCRSS, PC-based Reservoir Simulation System, is object-oriented and designed to function within the distributed SRP computing environment. The first version of the model concentrates on operational and medium- to long- range planning with a time step varying from one hour to a month. At this level, it incorporates mass balance[3] routing routines with loss calculations. The model uses rule curves, demand schedules and weighted constraints to define operating policies. Outlet works operate using head-discharge tables and tailwater-discharge tables to insure that the hydraulics are properly represented. While the initial version does not include routing of flows, the model architecture will accommodate routing and it is planned to add routing routines this year.

---

[3] Mass balance routing at short time steps works fairly well here because several SRP dams have their pools against the face of the upstream dam. In cases where this does not apply, short time step, slug flow routing requires the exercise of engineering judgment.

SRP Reservoir System: The SRP reservoir system consists of six reservoirs, two on the Verde River and four on the Salt River. The reservoirs store the runoff from a watershed of about 13,000 square miles (33,670 square kilometers), with the Salt River watershed slightly larger than the Verde River watershed. The primary function of the SRP reservoirs, with a total current capacity of 2,019,102 acre-feet ($2.490 * 10^9$ cubic meters), is water conservation. Roosevelt Dam, completed in 1911, is the oldest and largest storage reservoir of the system. Current storage capacity is 1,336,000 acre-feet ($1.648 * 10^9$ cubic meters) or about 66 percent of SRP's total reservoir capacity. The combined storage of the Verde River reservoirs is 309,613 acre-feet ($0.382 * 10^9$ cubic meters), about 15 percent of the total system capacity. Hydrogeneration is available on the Salt system, with a pumped back storage system available on the two middle dams

SRP's operating policy has generally been to maximize the releases from the Salt system up to the capacity of the generator of the most downstream dam, Stewart Mountain, and to take the remainder off the Verde system to satisfy the water demand, if necessary. For the winter months, beginning October and ending in April, water deliveries are normally taken in their entirety from the Verde system. During the winter months, SRP tries to balance the probability of spill between the reservoir systems. Safety of dams construction on both reservoir systems, and the required drawdowns to facilitate the construction activities have resulted in SRP deviating from its normal operating policy the last several years.

At the start of the 1995 runoff season the following construction constraints were in place. Roosevelt Lake elevation was not to exceed 2,110 ft. (25 percent reduction in storage capacity). No releases could be made from Roosevelt until February 1, because the contractor was working on the river outlet works, the plunge pool below the dam, access road to the power house etc. Major damage to the construction work would be incurred if early releases were required. The cofferdam in front of the right spillway would be overtopped if the lake would rise above elevation 2,136. At Bartlett Dam, the most downstream dam of the Verde system, the lake was not to exceed 1,779 ft, and maximum releases were limited to 2,600 cfs.

Besides the constraints on the reservoir system, additional constraints downstream of the dams in the urban area posed limitations on the releases. Exceeding the release constraints could cause damage to the downstream diversion dam, a recharge facility, major bridge construction, a housing development, and access to a recreation facility.

Given these constraints, important requirements of the new reservoir simulation model were the flexibility to change operational constraints easily and be able to run many alternative operational scenarios quickly.

Calibration of PCRSS: By the middle of February, the Verde and Salt River watersheds were hit hard by a series of winter storms. The reservoirs were fuller than desired, because of the release constraints due to construction, and SRP was placed immediately in a situation of emergency operations. SRP and JBA had intended to have the model of the SRP system built and calibrated before the onset of the winter/spring runoff of 1995. This did not happen because of the press of other business. At that point, we decided to use the fledgling model in an operational mode, checking the results with HEC-5 and the experience of the SRP Water Resource Operations' (WRO) staff. This decision, although nerve wracking at times, proved to be sound. Within a three-week period we got a wider variety of test scenarios and more eyes looking at the model than would have been affordable in a normal testing program. Since operations went on around the clock, we got the added benefit of excellent user training.

Communications took place by phone and by Internet, and WRO staff and JBA quickly developed a routine where a copy of each model or data modification was saved before trying to run the model. If all went as expected, the next modification simply overwrote the prior. If there was any problem, either with software performance or with the reasonableness of results, the offending model instance was downloaded over the Internet by JBA and run through a debugger. In cases where there was a bug, the repaired model was generally back in SRP hands within 3 hours. In those cases where the software was functioning properly but not as expected, both SRP and JBA would start the model running and then discuss the differences over the phone, both from the "same sheet of music".

It was these latter cases that gave us the most insight into what we were doing because we could discuss our assumptions and expectations and jointly agree how we wanted the model to treat each situation. In those cases where we agreed that the original model treatment was what we wanted, the problem became a defacto training session. When we decided that the model should behave in some other way, we either made a note so the model could be changed once things slowed down, or, if the difference was important, we made the change in the code and got the model back into service by the next morning. These latter situations really proved the value of object-oriented software. We were able to make some fairly significant code changes without worrying about having other parts of the model break. While it was a nervous time for JBA when these changes had to be made, the soundness of the original design and adherence to good software engineering practices really paid off. In no case did this practice fail to fix the problem to SRP's satisfaction and in no case did it introduce other bugs into the software.

PCRSS has now been tested during flood operations with time steps varying from one hour to one day. Communication by Internet and telephone and speed of update during runoff events proved very efficient and effective. PCRSS performed well and produced results fast. It was possible to look quickly at different inflow and release scenarios, and produce a recommended course of action.

It is obvious that in the new Hydromet operating environment, PCRSS will change the way WRO does business by allowing analysis of more alternative operating schemes, quicker presentation of output, and more informed decision making with a better handle on the uncertainty weather and watershed conditions cause in reservoir operations.

Where We Are Going From Here: SRP has always regarded PCRSS as an integral part of a larger integrated Hydromet system. Similarly, it was also recognized that the PCRSS architecture is ideally suited to modification and extension. SRP and JBA intend to extend our collaboration for at least another year. In addition to interface improvements suggested by the users, we intend to add routing of releases between reservoirs, a rule-based control language, and hydrogeneration capabilities. Also, medium- and long-term planning features of the model need to be tested against the results of the programs presently used by SRP.

An important aspect of this testing period is the realization that considerable training will be required to get the users used to and comfortable with the conventions used by the model. This may require a different way of thinking about the way WRO staff approaches reservoir operations.

Irrigation Companies Use & Prospective Of Manual 57

Vince Alberdi [1]

Abstract

The Management, Operation and Maintenance of Irrigation and Drainage Systems Manual No.57 (ASCE 1980) is a reference for the Management and Engineering practices required to operate an irrigation project.

Topics from district organization, to budgeting and personnel management are presented. Also, many aspects and problems in maintaining an irrigation distribution system are discussed and opinions to solve these problems are offered.

This manual should assist anyone who is responsible in an engineering or management role of an irrigation project.

Introduction

The interdependency of our society demands the best from each and every segment. This reliance of this interdependency in our fast paced society has been coined to mean "good customer service" some refer to it as "excellence" for our consumer. Those of us in the water distribution for irrigation purposes recognize the importance of our mission. We are charged with providing irrigation water, on a consistent reliable basis to our waterusers.

For those who aspire to manage an irrigation company, they need to become interested in further developing their personal skills to motivate the organization to provide that required level of service.

To develop the skills to cope with daily challenges of managing an irrigation district one needs a solid educational foundation in both civil engineering and business management. Along with these ingredients, experience in the practical operation of an irrigation company is needed to develop the ability to provide the direction that results in providing the service to the wateruser.

....

[1] Manager, Twin Falls Canal Company, Twin Falls, Idaho 83301-0326

With the USBR's development of most irrigation projects, management personnel evolved from key personnel during the construction phase being retained by the district as managers when the districts became an operational water distribution organization. This group of irrigation managers is very well versed on the maintenance and operation of the district and one can learn a great deal, but many years of exposure is required to gain the necessary skills. Since the Bureau has not developed new projects in recent times the "source" of experienced personnel is lacking as that generation is now rapidly retiring.

The "Manual and Reports on Engineering Practices No. 57" (ASCE 1980) does provide much information one needs to address those situations that happen with varying frequencies.

My first reading of the manual exposed me to operational, maintenance, and management procedures that we subscribe to and many procedures and suggestions of ways to address potential problems that we had not been aware of before. Indeed I found the manual educational.

My thoughts were that the Manual should be more than a reference manual that is used only when a specific issue needs to be addressed. Perhaps if the three divisions in the manual: operations, maintenance, and management were used as a standard, I could evaluate how our organization measured up. One has to be objective if you attempt the exercise, and open minded with the attitude to evaluate and make changes if necessary.

My first attempt at evaluating an organization was for my previous employer where I served as the Assistant Manager. During this evaluation, an opportunity to manage a larger canal company presented itself. I was fortunate to be selected as the Manager. Wanting to be a success at my new endeavor, I immediately began evaluating practices and procedures that were in place at my new organization and compared them to the manual. The organization had been well managed and this allowed me to make recommendations to continue practices but always with the challenge to question and improve. I have referred to the Manual to seek information to improve the following situations.

Operations

Specifically, our refinement in the operations department has focused on why waterusers supplies vary at the ends of our project. As stated in 3.1 Operation Management (ASCE 1980) the first sentence reads " The importance of delivering irrigation water to waterusers at the right time and at the proper rate and duration cannot be overemphasized." Our examination of the situation revealed many contributing factors. Small inconsistencies in deliveries, modifications to scheduling aquatic weed control efforts, and hydraulic limitations to the canal system all contributed to the problem. Reviewing the manual sparked an idea to evaluate the not so obvious hydraulic limitation. Major efforts have been undertaken to hopefully solve the situation.

Maintenance

Maintenance of the project is an ongoing responsibility. Evaluating the new structures was a pleasant experience, quality was superb. However, with further review the numbers of new small turnout structures needed was not being met. Each structure was individually specified and drafted for the specific site. Further review revealed a few standard structures could fit the majority needs and prefabricating techniques allowed personnel to pre-cast concrete structures in the summer for installation during the non-irrigation season. Productivity has been enhanced without sacrificing quality of construction.

To continue the search for the best way to measure main laterals, we reviewed many sources including the manual and concluded that the broadcrest weir fit our situation the best. The manual states the largest broadcrest can measure 1765 cfs ; and ours measures over 3700 cfs. However, the majority of our broadcrest weirs are used to measure flows of less than 200 cfs.

Safety

A safety program is something that is easy to talk about but requires discipline to initiate and continual nurturing to assure longevity of the program. Reviewing the manual reminded me that we needed to do more than just discuss. We reviewed the manual for suggestions, appointed a director, developed a committee of employees from various departments, arrived at a policy, and are pleased to report the benefits are a safer place to work and morale has improved as well. The forms we use for meetings are similar to those suggested in the manual.

Budgeting

The method and detail of the annual budget reflects on the plan for the next fiscal year. The plan needs to be well thought out and inclusive of all expenditures that are planned as well as anticipated revenues. The "Budget Guidelines" (ASCE 1980) in the manual is an excellent check to remind the participants in the budgeting process of customary expenditure categories.

Personnel

Perhaps the attitude and motivation of personnel are the most important ingredients in providing superb service. The manual has assisted me in writing job descriptions. A clear description of ones duties and management's expectations helps eliminate future problems. We evaluate our employees on an annual basis. Few actions can substitute for this one on one conversation about ones performance. In fact, we plan to have employees review the manual in their areas of responsibility, thus expanding the information to our entire staff.

District Organization and Management

During my tenure as a manager, I have been blessed by an outstanding Board of Directors that is enthusiastic, pro-active and understand their role is to set policy, budgets, and act on issues presented by the manager. However, the day may come when a new member of the Board needs to understand his role. The manual's explanation as an official source of information will be invaluable. Should a question arise as to the suggested duties of any officer the manual would serve as an official source of information as well.

Most managers are well aware of their responsibilities but the short Section, 2.6.2 (ASCE 1980) quickly reminds one of the many duties.

Conclusion

The success of an irrigation company is something like fitting the pieces of a puzzle together. Some of the puzzle pieces are: water supply, water rights, water quality, water safety, water cost, distribution system, maintenance plan, maintenance crews, operation plans, operation crews, offices, shops, equipment, and people in each respective areas from labor to Board Members working to provide the best service for the waterusers. Obviously, we don't live in a perfect world so the pieces of the puzzle may not be available or do not fit. That is where resources like the manual assist in providing the knowledge to find the missing piece or make the piece we have fit and work.

APPENDIX

References

The Committee on Operation and Maintenance of Irrigation and Drainage of the Irrigation and Drainage Division of the American Society of Civil Engineers. (1980) *Management, Operation and Maintenance of Irrigation and Drainage Systems (Manual 57)*

Designing Training Materials and Irrigation Sector Manuals for International Development

John Wilkins-Wells[1]
Tom S. Sheng[2]

Abstract

Future development of training materials for international development programs will require more emphasis on self-governance of water user associations, as well as management principles of water quality, public safety and public awareness. In addition, such training materials should emphasize the partnership relationship between government agencies and local water user groups. This partnership involves recognizing the heavy risks born by farmers in developing land and contributing fees to sustain irrigation systems, while at the same time recognizing the need for sound management of public resources.

Background

Three issues tend to arise whenever the subject is broached about the U.S. Bureau of Reclamation helping other countries develop their irrigation sector. First of all, why the Bureau? Second, should a U.S. government agency help promote foreign agricultural development in a way that could create unwanted competition in commodity production? Finally, is it presumptuous for us to think that we have answers that other nations want?

The U.S. is unique in its institutional development for water management, and this development has not been uni-dimensional. We have seen both public and private

---

[1] Research Associate, Department of Sociology, Colorado State University, Fort Collins, CO 80523

[2] President, Computer Assisted Development, Inc., 1635 Blue Spruce Dr., Fort Collins, CO 80524

sector participation in irrigation, although a middle level sector of cooperatives and quasi-governmental districts have probably played a more important role (Hutchins, et al., 1953). In many respects, we represent a laboratory of experiments and approaches. The 17 western states are littered with successes and failures, many of which unfortunately are not well documented. There is much to learn from this experience, and recent studies suggest that we may need to return to our roots, so to speak (Moore and Willey, 1991). The re-discovery of user fees and new emphasis on more local financing of natural resource conservation are only a few examples of this re-invention (Smith, 1984). In short, we feel that developing nations can profit enormously from our experience, and blend it with their own traditions to arrive at even better results.

The Bureau is a valuable source of technical assistance for developing nations. It has a long history of irrigation sector development, research and publication. There has been a tendency for the Bureau to focus on engineering at the expense of social and economic issues, but this is now changing with a new mission. Like any agency, it has its faults, but it has played an important role in assisting local organizations in developing their water resources. We feel that the relationship that has generally existed between local communities and the Bureau has been of the kind that fosters greater self-reliance and autonomy in local management than many examples of the same relationship we see outside the U.S. The relationships between public agencies and local communities have generally not been good internationally, although this is changing (Moore, 1991). We believe the Bureau has many important lessons to share on this matter.

Finally, we come to the comment we often hear from local growers. International commodity markets are very competitive, and U.S. sponsored programs designed to improve rice production in Thailand, cotton in Egypt, or fruits and vegetables in the Asian subcontinent will very likely impact farmgate prices in the U.S. for years to come. So, is it fair for a publicly funded agency like the Bureau to indirectly assist foreign governments in improving their commodity production? The short answer is probably no. However, the longer answer is that stabilizing markets through improved local organization performance, and increasing the standard of living in less developed countries, will likely improve the market for U.S. commodities in the future. Consumers in small and relatively poor villages, where most of the World lives, simply do not have the disposable income to purchase our commodities, even if they were available in local markets. We know this to be true for Nepal and Sri Lanka based on personal experience. An international development project designed to improve food production through better water user associations, thereby improving the income stream coming from commodity production, may be addressing the very thing that blocks U.S. growers from earning more income from their land. Much in the way that expanding electrical grids improves markets for small appliance producers, improving the performance of water user associations may be seen to increase disposable income for the purchase of U.S. commodities in the future.

Statement of Problem

There is need for the continued provision of international training in the administration of water resources. This training is needed for local community organizations. These may include water organizations, and a variety of conservation organizations which are developing in many countries. This trend is occurring in response to weaknesses in the public sector, and slow progress in private sector development to address the organization and financing of resource conservation of all kinds. Of particular importance is training in how to initiate a water user association, how to govern water resources effectively, how to develop a stable revenue stream to cover O & M costs, how to perform routine O & M on earthen canals, and how to administer a water supply in a way that best meets grower needs.

Past and Current Perspectives

In regard to water user association development programs, focus in the past has been on voluntary, informal organizations to manage water at the tertiary level of government-managed irrigation systems. This is often labeled the "participatory approach" but it has not been very effective. New focus is being placed on formal water user associations managing entire irrigation systems, with the government agency relegated to managing large diversions, storage facilities, etc. This newer focus is driven by the vision of a water user association as a "business house", rather than simply an informal group of farmers responsible for a field ditch serving a few farms. It is equally driven by the vision of a government agency and its personnel as facilitators providing technical assistance on demand, rather than governing the irrigation system.

Indeed this requires new legislation authorizing such organizations, adjudication of water rights and the development of water law, and public and private financial institutions to meet the needs of these organizations above the revenue they generate through fee assessments for water delivery. However, these additional needs do not prevent us from developing and providing training for foreign water agencies and community organizations. By and large, our experience is that foreign governments tend to approve such training if it is done in a thoughtful and responsible way. Furthermore, there is strong reason to believe that providing training in a way that stimulates interest in autonomous water user associations will lead to a demand in the countryside for further institutional needs in the form of legislation, water law, and local infrastructure banks and capital markets designed to finance their needs.

The Training Context

Until such time that the above needs are met, these new organizations will have a relatively weak revenue stream. In Nepal, for instance, we are participating in a program designed to initiate user fees for these organizations, and some moderate success is being achieved (Wilkins-Wells, et al.., 1994). However, income from the land is still low, although the market value of land may be quite high. In one of the areas we are working in, irrigated land values are in the range of $3000.00 per acre, and this is not uncommon for good irrigated land near commercial market centers. These high land values would seemingly fuel financial lending, but the low income levels continue to require the careful design of user fee programs and water user association self-financing activities.

Consequently, the technology employed in such organizations, and its cost of maintenance, should be relatively consistent with the level of revenue that can be generated through user fees or other forms of assessments. The type of water culture and technology that is most consistent with these new organizations is the self-managed, earthen canal systems of the western U.S. prior to World War II. The notable exception would be that electricity and the telephone are often now available in the countryside in a way that they were not in the U.S. prior to that date.

We are talking about associations that are just beginning to develop a revenue stream, generally have little experience in governing themselves, and do not exhibit sophisticated canal technologies. Admittedly, some systems have fairly substantial canal technologies in the form of regulators, headgates and access roads. However, canal lining is at a premium, and is not expected to be a factor in irrigation management for many years to come. This context sets up a framework for assessing training materials developed by the Bureau, or any other entity for that matter. What are the key needs of such organizations, and how can we best design and develop training materials to meet these needs?

Training Needs

Improving irrigation efficiencies in older irrigation systems, although highly desirable in many instances, may necessitate a thorough investigation of future O & M costs to the new association, as well as operation changes that impact current water use patterns, return flows to various areas, and irrigator labor requirements. The objective of water management in such systems is to 1) minimize maintenance problems and costs, because they reduce water control at the farm headgate and drive up the price of water, and 2) maximize water control (timing and flow rate) at the farm headgate, because it assures continued strong support for the water user association. Improved irrigation efficiencies are expected to follow in the course of time.

The basis of any good water organization is the **vote** cast by the water user and the **share** he/she pays to the operation and maintenance costs of the system. All accountability and resources flow from these units of representation and currency to governing entities, the workforce, and even the way the government agency interfaces with the organization through its technical assistance and/or extension programs.

Training materials should adequately address ways of organizing and setting up an internal government. There are many options available. Voting by shares, property qualification or weighted voting are only a few. Fees may be based on uniform rates per land unit, ad valorem assessments, benefits received, and water allotments and tolls. Designing a federation of water user groups may involve using the concept of a parent organization, with special roles and responsibilities for affiliated user groups. The steps involved in initiating a record keeping program may involve the identification of specific training needs and methods of training. Water delivery techniques that are adequate and appropriate for earthen canals, and maintenance procedures that address the needs of earthen canals and the structures in them may be more essential than focusing on the hydraulics of lined canals. Of particular importance are simple and well-explained methods of estimating canal conveyance losses in a way that allow these organizations to develop a delivery plan which ensures that tail-end farmers do not absorb the conveyance shrink in their allocated share of water. A grower's share of water should be roughly proportional to the fees he/she contributes to the organization's revenue; in cash or labor. The tighter this connection, the stronger the organization (Freeman, et al.., 1989).

There is need for continual study and expansion of knowledge in these areas. They should not be written off as "taboo" areas, but should be part of the general knowledge of water management that is communicated in such manuals.

APPENDIX 1.--REFERENCES

Freeman, David M., Bhandarkar, V., Shinn, E., Wilkins-Wells, J., and Wilkins-Wells, P. (1989). *Local Organizations for Social Development: Concepts and Cases for Irrigation Organization*. Westview Press, Boulder, Colorado.

Hutchins, W.A., Selby, H.E., and Voelker, W. (1953). *Irrigation-Enterprise Organizations*. U.S.D.A., Circular No. 934.

Moore, D., and Willey, Z. (1991). "*Water in the American West: Institutional Evolution and Environmental Restoration in the 21st Century.*" Univ. Of Colo. Law Rev., 62(4) 775-825

Moore, M. (1991). "*Rent-Seeking and Market Surrogates: The Case of Irrigation Policy*" in, Colclough, C. And Manor, J. (Eds.), States or Markets?: Neo-Liberialism and the Development Policy Debate. Clarendon Press, Oxford.

Smith, R. (1984). *Troubled Waters: Financing Water in the West*. The Council of State Planning Agencies, Wash. D.C.

Wilkins-Wells, J., Molden, D., Pradhan, P. and Rajbhandari, S.P. (1994). "Developing share systems for sustainable water user associations in Nepal." International Irrigation Management Institute, Conference on Irrigation management Transfer, Wuhan, China.

Two-Dimensional Sediment Transport Model of the Missouri - Platte River Confluence

Lyle W. Zevenbergen[1], M. ASCE, Mark R. Peterson[1], M. ASCE, and John I. Remus II[2], M. ASCE

Abstract

A study was performed under contract to the Omaha District, U.S. Army Corps of Engineers (USACE) to develop alternative confluence configurations to maintain the navigation channel nearer the outside bendway (concave bank) of the Missouri River at the Platte River confluence. In order to meet this objective, two-dimensional hydraulic and sediment transport modeling was performed. Existing conditions and project alternatives were analyzed.

Introduction

The Missouri River from Sioux City, Iowa, to the mouth has been narrowed and straightened as part of the Streambank Stabilization and Navigation Project. The project was designed to be a self-maintaining channel requiring no dredging maintenance. The thalweg was intended to remain along the outside of the bends (concave bank) and provide a reliable 2.7 m (9 ft) deep by 91 m (300 ft) wide navigation channel. This has been accomplished through a series of revetments and transverse dikes. However, at the Platte River confluence, sediment deposition has occurred within the navigation channel and shifted the thalweg abruptly to the inside of the bendway (convex bank) causing severe navigation problems.

The two-dimensional RMA-2V hydraulic and STUDH sediment transport models were used for this study. These models are part of the TABS-MD modeling system (Thomas and McAnally, 1985). An RMA-2V mesh, developed by the USACE Waterways Experiment Station (WES), was modified for use in this project. The WES model was developed using 1989 hydrographic survey data which show a

[1]Senior Hydraulic Engineer and Manager, Hydraulic Engineering, respectively, Ayres Associates, 3665 JFK Parkway, Bldg. 2, Suite 300, Fort Collins, CO, 80525.

[2]Project Engineer, River and Reservoir Engineering Section, Hydrologic Engineering Branch, Omaha District, USACE, 215 N. 17th St., Omaha, Nebraska, 68102.

bar of deposited sediment extending from the mouth of the Platte River to approximately 600 m (2,000 ft) downstream and covering as much as 75 percent of the channel width. The 1993 Missouri River flood completely removed the sand bar; however, it re-formed in 1994 and is virtually the same as the 1989 survey.

Methodology

The finite element network, developed by WES, was refined to better represent channel geometry along the Missouri River banks and at the locations of existing dikes. The model was calibrated for flow conditions of known discharges and water surfaces in the study area. The first was a July 7, 1987, flow of 991 m^3/s (35,000 cfs) on the Missouri River and 182 m^3/s (6,420 cfs) on the Platte River. The second was an October 1, 1991, flow of 890 m^3/s (31,500 cfs) on the Missouri River and 53 m^3/s (1,870 cfs) on the Platte River. The maximum difference between modeled and observed water surface elevations was 0.15 m (0.5 ft) for the 1987 event and 0.11 m (0.35 ft) for the 1991 event. The Manning's n was 0.020 for the Missouri River and 0.028 for the Platte River. The models were then run for two hydraulic conditions. The first condition was 878 m^3/s (31,000 cfs) for the Missouri River and 139 m^3/s (4,900 cfs) for the Platte River. These flows are a combination of low flow on the Missouri River (this discharge is exceeded approximately 93 percent of the time) and average flow on the Platte River (50 percent exceedance). This condition was modeled to determine whether the sediment bar could be controlled and the sailing line maintained during low flow. For these flow conditions, little sediment transport occurs on the Platte River. In order to simulate flows which cause the sediment bar, a second condition was modeled. Low flow was maintained on the Missouri River, but a discharge of 566 m^3/s (20,000 cfs) was used for the Platte River. At this discharge, significant amounts of sediment are transported to the confluence by the Platte River.

The STUDH sediment transport model was used for the sediment transport analyses. STUDH uses the same geometric data as RMA-2V uses for hydraulic calculations. Additional input includes the hydraulics calculated by RMA-2V, sediment characteristics, and inflowing sediment loads. STUDH uses the Ackers-White (1973) sediment transport formula which is appropriate for the Missouri River. Inflowing sediment loads were determined for the Missouri River using bed material grain-size data and suspended-load data provided by the Omaha District, USACE. For the Platte River, inflowing sediment loads were determined from suspended-load data from the USGS Louisville gage (approximately 30 km {20 mi} upstream of the confluence) and bed material grain-size data presented by Kircher (1993). Given the sediment supply to an area (element or node) and the computed transport capacity, STUDH computes an amount of aggradation or degradation for a selected time step. The model adjusts the velocity and depth by maintaining the original water surface elevation and unit discharge, then computes a new sediment transport capacity and makes further adjustments to the bed elevations. This procedure does not account for redistribution of the flow longitudinally or laterally because the unit discharge at each node is maintained. The hydraulics are updated periodically by running RMA-2V with the adjusted bed elevations from STUDH. For this study, the STUDH model was allowed to adjust bed elevations by 0.5 m (1.6 ft) or flow depth by 20 percent at 50 nodes prior to transferring computation back to RMA-2V to update the hydraulics. If neither of these constraints were reached, the STUDH was allowed to adjust bed elevations for a set time limit. For low-flow runs, the maximum time limit prior to revising hydraulics was set to 10 days, the total simulation time was set to 60 days, and the internal time step was set to 1 hour. For high flow runs, the time limit was set to 12 hours, the total simulation time was set to 3 days, and the internal time step was 15 minutes. At the end of each simulation, comparisons were made between initial and final bed elevations for existing conditions and between the remediation alternatives.

Alternatives

Several alternatives were considered for alleviating the sedimentation problem at the Platte River confluence. Figure 1 shows the modeled confluence area depicting the existing longitudinal dike separating the two channels and existing transverse dikes along the convex bank. Figure 1 also includes flow depth contours for low-flow conditions and clearly shows the sediment bar at the confluence. The alternatives included extending the longitudinal dike which separates the two channels, adding transverse dikes to the Platte River side of this dike to better align the Platte River with the Missouri River, extending the existing dikes A through G both above and below the navigation water surface, and adding dikes between existing dikes E and F. The existing dikes extend approximately 25 m (82 ft) from the convex bankline. The first alternative was eliminated from consideration because it would probably move the sediment bar downstream without alleviating the problem. The second alternative was excluded because dikes on the Platte River side would not affect sediment delivery to the Missouri River. Therefore, alternatives involving extending the existing Missouri River transverse dikes and/or adding dikes were analyzed in detail. These alternatives address the Missouri River's ability to convey the sediment delivered from the Platte River.

Alternative 1 consisted of extending the existing dikes (A through G) by between 27 and 55 m (90 and 180 ft) at an elevation of 1.5 m (5 ft) below the normal navigation water surface. The dike extensions were modeled perpendicular to the bankline. In addition, two dikes (H and I) were added between the existing dikes E and F. The longest dike (H) was approximately 61 m (200 ft) and extended across 33 percent of the channel width. The dike lengths were selected to constrict the channel, while maintaining a navigation channel in excess of 91 m (300 ft), and to form a smooth arc along the ends of the dikes. Alternative 2 was identical to Alternative 1 except that the dikes are entirely above water for normal navigation flows. Alternative 3 was similar to Alternative 2 and is depicted in Figure 2. Dikes A, B and G were not extended, dikes D through F were the same as Alternative 2, and dike C was extended to form a smooth transition between dikes B and D.

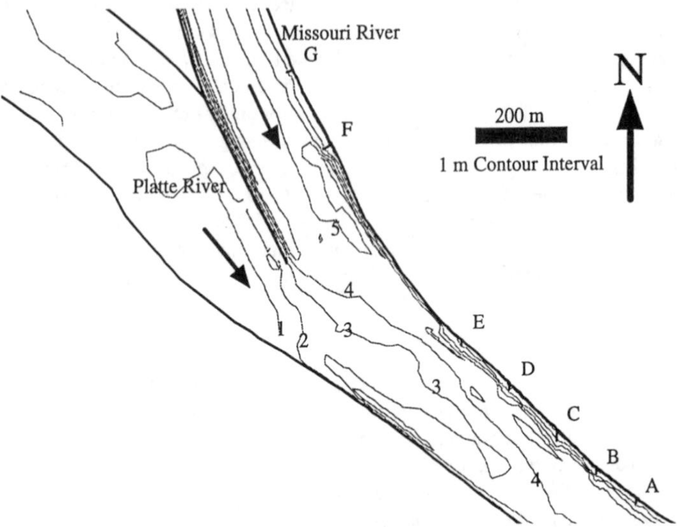

Figure 1. Existing Conditions low flow, depth contours and dike locations.

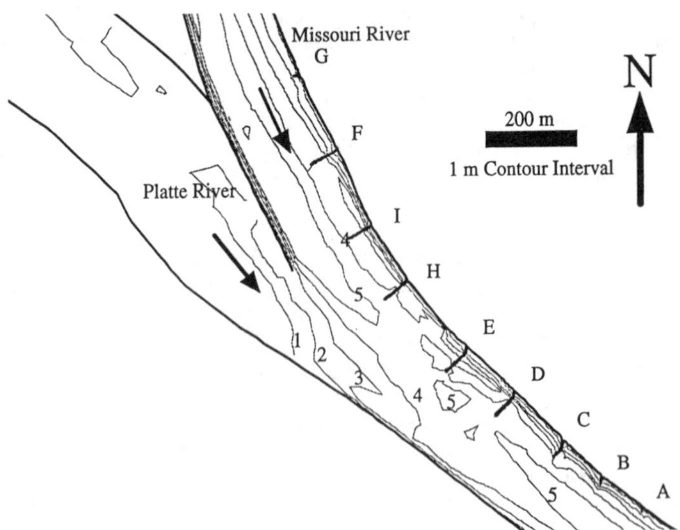

Figure 2. Alternative 3, low flow depth contours after 60 days.

Results

Prior to performing sediment transport analyses, the hydraulic model was run to determine the initial hydraulic impacts of the dikes. Alternative 1 raised water surfaces only 0.05 m (0.15 ft) while Alternatives 2 and 3 raised water surface as much as 0.18 and 0.15 m (0.6 and 0.5 ft), respectively. Even though the dikes encroach significantly into the channel, the fact that they are located on the convex bank minimizes the impact. For this reason, the dikes need to extend well into the channel to influence sediment transport properties.

From the sediment transport analyses, the existing conditions model showed that the sediment bar would be trimmed somewhat for low flow and enlarged for high flow on the Platte River. Alternative 1 runs showed little improvement over existing conditions and would not be expected to alleviate sedimentation problems at the confluence. Alternative 2 showed significant improvement over existing conditions. The sediment bar was virtually removed and the thalweg was shifted away from the convex bank to a favorable alignment for navigation traffic. Depths in excess of 3 m (10 ft) extended across the entire navigation channel. Alternative 3 produced nearly the same results as Alternative 2 and is the recommended design. The only difference was that slightly less scour occurs upstream of the confluence. After the 60-day simulation, the water surface for Alternative 3 was raised up to 0.04 m (0.13 ft). It is expected that the dikes will have negligible impact on flood elevations after the channel has adjusted. Figures 1 and 2 show 1 m (3.3 ft) flow depth contours for existing conditions and Alternative 3 after the 60-day simulation.

The mouth of the Platte River is a sensitive habitat for the endangered Pallid Sturgeon, and therefore will require a Biological Assessment before proper permits can be obtained. At the present time, the project has not been constructed. The dikes are scheduled to be constructed and modified in 1997. A pre-construction survey and at least one post-construction survey of the sand bar area will be obtained to document the accuracy of the model predictions.

Appendix - References

Ackers, P. and White, W.R. (1973). "Sediment Transport: New Approach and Analysis." Journal of the Hydraulics Division, ASCE, Vol. 99, No. HY11, Proc. Paper 10167, November, 1973, pp. 2041-2060.

Kircher, J.E. (1993). "Interpretation of Sediment Data of the South Platte River in Colorado and Nebraska, and the North Platte and Platte Rivers in Nebraska." Geological Survey Professional Paper 1277-D.

Thomas, W.A. and McAnally, W.H. (1985). "User's Manual for the Generalized Computer Program System: Open Channel Flow and Sedimentation, TABS-2." U.S. Army Engineer Waterways Experiment Station, Vicksburg, MS, 671 pp.

Movable-Bed Scale Model for the Nile River at El-Kureimat

A.F. Ahmed[1]

Abstract

A decision was made to construct a new 4 x 600 MW thermal power plant on the east bank of the Nile River near El-Kureimat City (about 100 km south of Cairo). When the construction works were started, it was observed that an efficient cooling water intake system could not be designed. This because the available flow depth in front of the selected site is too small to provide a sufficient discharge for the cooling system, (which ultimately will be equivalent to 80 m^3/s). To solve the problem a movable-bed scale model representing about 9 Km of the river including selected plant site was built. As the model was calibrated and the trend of bed configurations compares well with the prototype, three group of tests were conducted to enhance the flow depth in front of the intake structure. The tests conducted and the obtained results are the subject of this paper.

Introduction

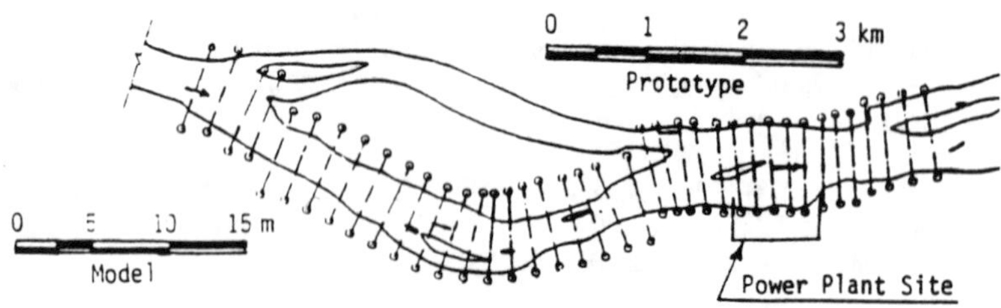

Figure 1. General Layout of the Modelled Reach

---

[1]Senior Researcher at the Hydraulics Research Institute, Delta Barrage, Egypt.

Fig.(1) shows a general layout of the reach which contains many islands. One of them is located in front of the intake structure which divides the river into two channels. The right one is forming a conversion channel at the upstream inlet, while a diversion one is created further downstream in front of the intake structure. This configuration not only caused additional resistance, but also reduced the flow depth at the downstream end at the intake structure. For this reason more than half width of the right channel, at some locations along the plant site, is above the water surface level during the winter closure period. Consequently, the required flow depth in front of the intake structure will not be sufficient to provide the required discharge for the cooling system.

Model Scaling Ratios

Tests were conducted with the dominant flow rate which produces the yearly average sediment discharge. The hydraulic and morphological parameters that represent the problem area as well as the scaling ratios are summarized as follows:

Parameter		Prototype	Scale	Model
Average width	[m]	531.0	150.0	3.54
Average depth	[m]	2.73	30.2	0.09
Mean velocity	[m/s]	0.75	2.89	0.26
Discharge	[m^3/s]	1086.0	13093.0	0.083
Slope	[-]	8.5×10^{-5}	0.055	0.00155
Chezy C	[$m^{1/2}/s$]	49.0	2.23	22.0
Grain size	[mm]	0.37	1.68	0.23

Model Construction

The model was shaped according to the field survey. To adjust the water surface level a flap gate 5.0 m wide was installed at the lower end of the model. A total number of 39 pairs of supports, located at distances ranging between 1 m and 1.5 m apart, were firmly erected along both banks of the model as shown in Fig.(1). These supports were utilized to accommodate a portable bridge to measure the bed levels at each cross-section. A mixture of water and sediment were fed into the model through a manifold. Adjustable vertical vanes were used to control the lateral variation of velocity and sediment load.

Reference Test

This test was conducted with the model representing the actual condition that was achieved by the end of the calibration stages. The purpose of the test was to acquire the necessary measurements which can be used as a reference to assess the fulfillment of any proposed solution. Using the sounding bridge, extensive measurements of bed levels along the model were acquired. The test was carried out for three days, during which the bed profile was measured at fixed intervals six times a day. Using a EMS current-meter, mid-depth velocity was measured along 12 cross sections located along the modelled reach especially in front of the power plant. The data were utilized to determine the flow distribution through the two river channels.

Testing Program

In order to improve the condition at the location assigned for the intake structure, three group of tests were carried out: The first scheme, as shown in Fig.(2), consists of several groins located at the downstream end of El-Kureimat island and along the left bank. This alignment was defined so that most of water flowing in the left branch can gently join the main channel at the southern end of the small island. The training works shown in Fig.(2) was gradually modelled and tested. For each testing step the corresponding variations in bed levels and the mid depth velocity profiles were measured. Various combinations of deflecting spur dikes were conducted but no improvement was resulted. This is because of the deep scour holes, that were created at the toe of the spur dikes, which formed a deep channel along the left side of the river.

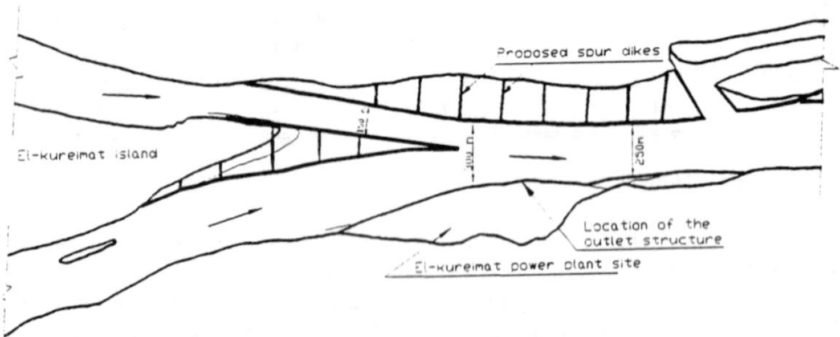

Figure 2. Alignment of the First Tested Scheme

The aim of second scheme was to reform the existing bar-pool system in such a way as to divert the thalweg-line through the right channel in front of the intake structure and consequently provides the required flow depth for the cooling system. Different lengths of spur dikes were tested. Comparison of the results obtained with respect to the reference test revealed unsatisfactory results. This is because the reformed thalweg line was confronted with the small island located in front of the intake structure.

As no improvement was obtained during the two previous test groups, different technique was applied. This technique was based on reshaping the small island in such a way as to provide a large channel width at the upstream to insure enough entering discharge and reduce the percentage of the recirculated water required for cooling system. While the converging configuration of that channel helped to cause stable deep channel at the downstream outlet where the intake structure is located. A recirculation cooling system, simulating the withdrawal and discharge of 80 m^3/s, in the prototype, was applied. Two different shapes of pre-fabricated islands were made of aluminum sheets and angles. Each one was affixed in the model and tested at various possible orientations. The design that resulted in a flow depth of about 3.0 m in front of the intake structure, which is shown in Fig.(3), was selected.

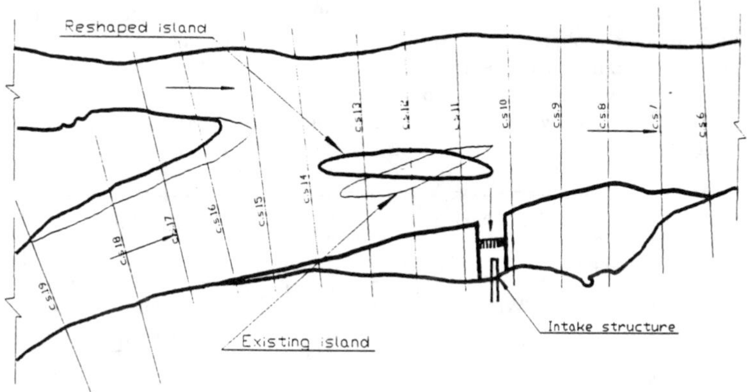

Figure 3. Alignment of the Proposed Scheme

The proposed scheme was then made of cement mortar and bricks to carry out the necessary measurements. The sounding bridge was exclusively utilized to acquire the bed level measurements along the model. The test lasted for three successive days, during which the bed profile was measured twice a day. As the steady state was established, and lasted for two days, the actual bed configuration was determined as the average of the last three acquired sounding. Moreover, mid-depth velocity was measured along the cross sections that were previously carried out during the reference test. The measurements were conducted with and without simulating the withdrawal of the cooling system.

Results of the Proposed Scheme

Comparison of average bed levels for the proposed scheme with that for reference test are presented as follows :

Condition	Average bed level (m)								
	C.S No. (8)	C.S. No. (9)	C.S. No. (10)	C.S. No. (11)		C.S. No. (12)		C.S. No. (13)	
				Left	Right	Left	Right	Left	Right
Reference test	19.06	18.85	19.11	18.96	19.08	19.00	18.61	19.07	19.17
Proposed scheme	18.20	18.28	18.63	19.20	17.67	19.02	18.17	19.02	18.88
Difference (m)	-0.86	-0.57	-0.48	+0.24	-1.41	+0.02	-0.44	-0.05	-0.29

This revealed slight scour took place at C.S. 8, 9 and 10 between the intake and outlet structures. This obviously due to the accelerated flow through the right channel as well as the additional flow from the left channel which compensated the withdrawal of the cooling system. Moreover, general enhancement occurred in the average flow depth through the right channel between C.S. (11) and (13) which reached 1.41 m at cross section (11) located in front of the intake structure. While, no substantial variation was recorded through the left channel. Moreover, the flow distributions at cross sections 11, 12, and 13, were determine as follows :

C.S. No.	Left channel			Right channel		
	Reference test	Proposed scheme		Reference test	Proposed scheme	
		Without cooling	With colling		Without cooling	With cooling
11	61.6%	54.1%	43.4%	38.4%	45.9%	56.6%
12	67.1%	55.8%	49.0%	32.9%	44.2%	51.0%
13	63.0%	53.6%	45.3%	37.0%	46.4%	54.7%
Average	63.9%	54.5%	45.9%	36.1%	45.5%	54.1%

This implied that applying the proposed alignment will enlarge the percentage of flow through the right channel and consequently the percentage of withdrawal water required for the cooling system will be substantially reduced.

Conclusions

Reshaping the small island, as shown in Fig. (3), is essential for solving the problem. This solution is not only the most economic one but also resulted general enhancement in the average flow depth in front of the intake structure while no substantial variation, for the average bed level through the left channel, was recorded. Also, applying this scheme will enlarge the percentage of flow through the right channel which will consequently reduce the percentage of withdrawal water required for the cooling system.

Rio Grande Sediment Study Supply and Transport

Elvidio Diniz[1], M.ASCE, Darrell Eidson[2], and Matthew Bourgeois[3], M.ASCE

Abstract

Very reasonable sediment transport rate comparisons for the Rio Grande near Albuquerque, New Mexico were obtained from several different computational approaches. A sediment mass transport balance was determined for each sub-reach (between USGS gages). This consisted of computing tributary and upstream sediment inflows and change in sediment storage within the subreach. These results compared very well to the total sediment transport rates determined from the measured suspended sediment data at each gage and also to the rate of sediment accumulation in Elephant Butte Reservoir at the downstream end of the study reach, as measured by reservoir re-surveys.

Introduction

The New Mexico Interstate Stream Commission (ISC) has authorized this investigation for improving water conveyance and water conservation in the Middle Rio Grande Valley from Cochiti Dam to Elephant Butte Reservoir (Figure 1). This phase of the study consisted of an analysis of the sediment contribution to the Rio Grande from its tributaries and an evaluation of the existing United States Geological Survey (USGS) sediment gage data.

Tributary Analysis

The study area sub-basins (tributary watersheds) were delineated using 1:250,000 scale maps. From a total of 171 sub-basins, eleven "representative" sub-basins were selected for detailed analysis. Their locations are shown on Figure 1. A detailed sediment analysis was then performed on each of the selected sub-basins. Finally, a regression analysis of the results was conducted to allow prediction of sediment yields from the remaining sub-basins.

Data gathered on the selected sub-basins included USGS 7½ minute quadrangle maps for detailed sub-basin boundary delineation; Soil Conservation Service (SCS) Soil Survey maps to estimate parameter values to be used in sediment yield

---

[1]Principal Engineer, Resource Technology, Inc. 2129 Osuna Rd. NE, Albuquerque, NM 87113

[2]Hydraulic Engineer, U.S.Army Corps of Engineers, P.O.B. 1580, Albuquerque, NM 87103-1580

[3]Civil Engineer, U.S.Army Corps of Engineers, Albuquerque, NM

calculations; field surveys of specific arroyo reaches for use in hydraulic modelling; and sediment samples of bed and overbank material to determine size gradations to be used in sediment transport calculations.

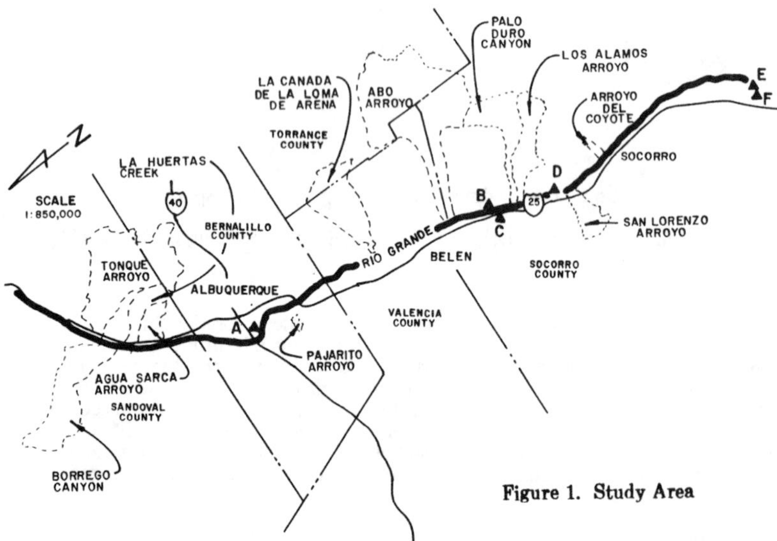

Figure 1. Study Area

Hydrologic modelling using the HEC-1, *Flood Hydrograph Package*, computer program, provided peak discharges, runoff volumes, and runoff hydrographs. The model results were calibrated using the procedure described in *Methods for Estimating Magnitude and Frequency of Floods in the Southwestern United States*, Blakemore et al, 1994. The eleven sub-basins, their drainage areas, and representative hydrologic results are shown in Table 1.

Table 1
Representative Hydrologic Results

Basin	Drainage Area (sq mi)	Peak Discharge (cfs)			Runoff Volume (Ac ft)		
		2-yr	10-yr	100-yr	2-yr	10-yr	100-yr
Agua Sarca	5.68	696	1,800	3,538	67	198	392
Palo Duro	63.5	86	4,655	11,849	23	885	2,439
Abo	290	0	9,534	28,091	0	2,663	7,954
Coyote	1.55	130	570	1,840	10	42	146
Comanche	15.0	370	1,560	5,660	81	345	1,266
Los Alamos	58.8	690	3,080	11,640	258	785	2,786
Las Huertas	29.2	490	2,110	8,260	83	359	1,441
San Lorenzo	30.5	480	2,380	8,610	117	538	1,919
Tonque	163	940	3,960	16,170	307	1,227	4,940
Borrego	75	800	3,360	12,810	296	1,183	4,482
Pajarito	0.85	80	430	1,450	3	18	78

Hydraulic modelling was performed using the HEC-2, *Water Surface Profiles*, computer program using the geometry and frictional loss values from the field surveys. Watershed sediment yield computations were performed using the Modified Universal Soil Loss Equation (MUSLE). These values were used for the wash load component of the total sediment load and for calculating the Colby correction factor, which was applied to the bed material component.

The Zeller-Fullerton (1983) Equation was used to compute the majority of bed material load, while the Meyer-Peter, Muller (1948) Equation was used for large sized material (the upper portion of the gradation curve). The sediment transport rate was then integrated over the various frequency runoff hydrographs to yield sediment volumes. These volumes were then scaled to their portion of the original gradation curve and the Colby (1964) Correction factor was applied. The wash load volume (from the MUSLE) and two bed material load volumes were added to produce total sediment volumes for the various storms for each of the eleven selected sub-basins.

A regression analysis was then performed on the resulting sub-basin sediment volumes. Total sediment volume equations were subsequently developed for the 2-, 5-, 10-, 25-, 50-, and 100-year events, as well as one for average annual conditions, as a function of drainage area. The non-linear correlation coefficients for the equations range from 0.77 to 0.86. The resulting regression equations are shown in Table 2.

Table 2

Sediment Volume Regression Equations

Return Period (yr)	Regression Coefficients* A_1	A_2	A_3	Standard Error of Estimate	Non-Linear Correlation
2	0.202	0.0469	5.91	5	0.78
5	-2.860	0.0723	26.90	16	0.81
10	-3.810	0.0386	45.60	22	0.86
25	-7.690	0.1770	83.20	47	0.86
50	-14.900	0.1790	139.00	85	0.82
100	-43.800	-0.2200	280.00	166	0.77
Average Annual	-3.250	-0.0400	24.80	12	0.80

* Q_{ts} total sediment volume (ac ft) = $A_1 + A_2$*(Drainage Area) + A_3*log(Drainage Area)

Analysis of Sediment Gage Discharge Data

The U.S. Geological Survey collects sediment data at six locations in the study area (Middle Rio Grande), as listed below. Figure 1 also shows the approximate locations of the sites, four of which are located on the main stem.

Gage Designation	Gage Number	Gage Name
A	08330000	Rio Grande at Albuquerque, NM
B	08332010	Rio Grande Floodway near Bernardo, NM
C	08353000	Rio Puerco near Bernardo, NM
D	08354900	Rio Grande Floodway at San Acacia, NM
E	08358300	Rio Grande Conveyance Chan. at San Marcial, NM
F	08358400	Rio Grande Floodway at San Marcial, NM

The construction of Cochiti Dam in 1976 on the main stem of the Rio Grande affected the sediment transport capacity of the river below the dam. Pre-1976 data are not compatible to those after the dam; and consequently, the 1976-1992 period of record was selected for the six gages used in this analysis.

The USGS measures and publishes only the daily suspended sediment and water discharges at their gage locations; whereas the total sediment discharges were required for purposes of this study. The USGS occasionally has computed the daily total sediment discharge using the daily suspended sediment discharge and a double mass regression analysis of 30 years of total sediment load data, which was previously calculated by using the Modified Einstein Equation to compute the bed material load, and then added to the suspended load to determine total sediment load.

This regression equation was applied to all the other suspended sediment load data to determine the corresponding total sediment load. The mean daily total sediment loads were then correlated to the mean daily flows at each of the USGS gages. In this case the correlation coefficients range from 0.60 to 0.89.

The developed total sediment-to-flow correlation equations for the gages are not valid for the 10 percent greatest flows as they predict total sediment loads less than the suspended load. To compensate for this, the maximum historical suspended sediment, as measured at each gage, was used to establish the endpoints of the flow duration curves (at 0.1% of time equalled or exceeded) and the total sediment was assumed to be 1.15 times this maximum suspended sediment discharge. The total sediment discharge for the 5% and 10% exceedance points were then derived graphically. The total annual sediment load was then computed as the area under the curve.

Analysis of Rio Grande Re-Survey Data

Rio Grande channel mapping from 1972 and 1992 was obtained from the U. S. Bureau of Reclamation. Cross sections along fixed range lines at approximately 500-foot intervals were developed for both mapping data sets. The area differences between corresponding cross-sections were used to compute changes in volume along the river bed between cross-sections. The computed change in sediment volume over time was used to determine the locations of aggradation and degradation.

The Rio Grande was divided into four reaches and the cumulative volume change as a result of river aggradation (positive) or degradation (negative) for each reach is presented in Table 3 below:

Table 3
Cumulative and Average Annual Volume Change (Q_{sr})
as Calculated from Rangelines for Each Reach (between Gages)

Reach	Cumulative Volume change 1972 To 1992(ac ft)	Average annual Volume change 1972-1992 Q_{sr}(ac ft/yr)
Cochiti to A	-4,671	-234
Gage A to B	-111	-6
Gage B to D	-3,463	-173
Gage D to F	19,569	978

Analysis of Elephant Butte Cross Section Data

Elephant Butte Reservoir resurveys were used to evaluate the reliability of the computed sediment data. Based on 73 years of record (1916-1988), the average annual volume loss rate is 7,800 ac-ft. per year; however, the most recent resurveys, which cover an eight-year period (1980-1988), show a current storage loss rate of 5,590 ac-ft./yr. which presumably reflects the result of upstream dam construction, including Cochiti Reservoir.

Comparison of Sediment Volumes

A sediment volume balance computation was performed at each gage location. For each gage, Table 6 shows the computed sediment discharge of the upstream tributaries (Q_{st}), the cumulative change in river sediment upstream of the gage as a result of river aggradation/degradation as measured by rangelines (Q_{sr}) and the resulting calculated sediment discharge at each gage (Computed $Q_{s\ gage}$). The sediment discharge at each gage ($Q_{s\ gage}$) computed from gage data is also provided for comparison.

Table 4
Volume Balance Using Gage Data, River Aggrad/Degrad
Data and Computed Arroyo Sediment Discharge
Gage data converted to acre-feet using 165 lbs/ft^3 unit weight

	Q_{st} Volume of Sediment from US Tributaries (ac ft/yr)	Q_{sr} Volume Change from River Aggrad/Degrad (ac ft/yr)	$Q_{s\ gage}$ Volume of Sediment Passing Gage (ac ft/yr)	$Q_{s\ gage}$ Volume of Sediment Passing Gage (ac ft/yr)
Gage A	968	-234	1,202	1,125
Gage B	2,093	-239	2,332	1,345
Gage D	4,952	-412	5,364	6,012
Gage F	6,082	566	5,516	4,403
Elephant Butte				5,590**

* Computed $Q_{s\ gage} = Q_{st} - Q_{sr}$ ** from reservoir resurveys

Conclusion

The comparison of sediment discharge rates at each gage location, using various sediment data computational procedures, indicate relatively comparable sediment transport rates. Some of the variables affecting these rates include variable unit weights of the sediment, the 1972 to 1992 period of study for the river cross-sections which includes pre and post Cochiti Dam data, no adjustments were made for the diversion of flow (and sediment) into irrigation systems, the computed sediment production rate does not exclude the sediment produced in several arroyos that do not flow directly into the Rio Grande, and the time reference for each of the four data sets is different. In spite of these variables the computed sediment transport rates appear to be very reliable.

APPENDIX - REFERENCES

Blakemore, T.E., Hjalmarson, H.W., and Waltemeyer, S.D., 1994: "Methods for Estimating Magnitude and Frequency of Floods in the Southwestern United States" USGS Open File - Report 93-419

Colby, B.R., 1964: "Practical Computations of Bed-Material Discharge," Journal of the Hydraulics Division, ASCE, v. 90, no. HY2.

Meyer-Peter, E. and R. Muller, 1948, "Formulas for Bed Load Transport," Proceedings, Third Meeting of International Association, Hydraulic Research, Stockholm, pp. 39-64.

Zeller, M.E. and W.T. Fullerton, 1983, "A Theoretically Derived Sediment Transport Equation for Sand-Bed Channels in Arid Regions," Proc. of the D.B. Simons Symposium on Erosion and Sedimentation, R.M. Li and P.F. Lagasse, eds.

Effects of Particle Packing on Bedload Motion

by

A.Papanicolaou[1], M. Balakrishnan[1], P. Diplas[2], and C.L.Dancey[3]

Abstract

The effects of particle surface packing of a bed on bedload transport are examined by running tests in a laboratory flume for different surface packing conditions. Specifically, the bedload transport rate is measured and compared for three different surface packing conditions. In addition, laser Doppler anemometer (LDA) measurements of the distribution of the instantaneous shear stress above the bed are obtained for the different bed packing conditions. The results indicate that for the same value of boundary shear stress, the bedload transport rate decreases as the packing density increases. In the case of the loosely packed layers, the phenomenon of "clustering" was also observed.

Introduction

Over the last twenty years a significant number of bedload transport studies (e.g., Fenton and Abbott, 1977; Moore and Diplas, 1994) have concentrated on the role of different parameters, namely, the size, shape and density of the particles and the relative particle protrusion on the initiation of bedload motion. Most of the existing experimental studies, that have been reported in the literature, do not examine, however, the

---

[1] Graduate Research Assistants at Civil and Mechanical Engr. Depts., respectively, Virginia Polytechnic Institute and State University (VPI & SU), Blacksburg, VA 24061.
[2] Associate Professor, Civil Engr. Dept., Virginia Polytechnic Institute and State University, Blacksburg, VA 24061.
[3] Assistant Professor, Mechanical Engineering Dept., Virginia Polytechnic Institute and State University, Blacksburg, VA 24061.

influence of the particle surface packing conditions of a bed on particle entrainment.

The aim of the present work is to determine the effects of different particle packing conditions at the surface of a flat bed on bedload motion. For this purpose three different initial surface packing conditions were studied. They include, 20%, 50%, and 100% surface coverage of the total surface area of the test section with entrainable glass beads in three separate tests. Measurements of the instantaneous kinematic Reynolds stress distributions near the top of the bed surface layer were also obtained in order to determine the local instantaneous stress applied to the bed.

Experimental Set -Up

All of the present bedload experiments were conducted in a tilting laboratory flume that is 20.1 m long, 0.60 m wide, and 0.30 m deep. This flume has a useful length of 15 m and operates in the water -recirculating mode. The slope of the flume during the present study was constant and set equal to 0.006. During the tests the flow depth was measured at seven locations to ensure flow uniformity. The measurement of the rate of transport of particles was effected by using a basket to collect the transported particles at the downstream end of the flume. The detailed point - wise velocity measurements near the bed surface were obtained with the use of a 3D - LDA, described in detail by Balakrishnan and Dancey (1994).

Methodology-Results

Glass balls of 8 mm diameter and specific gravity 2.54 were used as the bed material. This particle size was chosen so as to render the critical shear stress independent of the particle Reynolds number. The length of the test section was 3.29 m and it was sufficient to allow for the observation of the representative entrainment characteristics of the large number of balls present within the test section. The test section was located 15.5 m downstream of the flume entrance. Naturally worn gravel (d = 25 mm) was used for the first 15.5 m of the flume length. The remaining reach was covered with three layers of glass beads and topped with a fourth layer of lead balls of diameter 8 mm. The total thickness of the bed layer along the flume length was 2.88 cm. Within the test section the entrainable glass beads were placed atop the lead balls. For the 100%, 50%, and 20% particle packing conditions 30,636, 15,318, and 6,127 balls were used, respectively to cover the surface of the test section. The balls in the last two cases were distributed uniformly within the reach of the test section.

During each experimental test run the transported glass balls were collected at the exit of the flume at regularly timed intervals. Then the collected material was dried and weighed. If no significant variations in the weight of the collected material are observed within consecutive time periods of the run, it is an indication that the test has reached a steady state condition and it can be terminated. At the conclusion of each run the total weight of the transported material was attained while the water drained from the flume bed. The data collected in each run was the discharge, the flow depth, the water surface slope and the bed load transport rate. In this preliminary study 4 to 5 runs were made for each packing condition. During the runs for the 20 % and 50 % particle surface packing conditions the formation of cluster microforms were observed. This was most evident in the lowest packing, the 20 % case. The cluster structures varied with the test flow conditions. Their density altered with time, however, this change was small. During the tests a delay in particle entrainment was observed that might be attributed to the formation of clusters. A similar tendency has recently been observed, among others, by Reid et al (1992) in natural streams.

For comparison purposes the present data are plotted with those of Paintal's (1971) in figure 1. Figure 1 shows the variation of bedload transport rates with respect to the dimensionless shear stress. The results in all cases present a consistent behavior. It is evident from figure 1 that the general pattern of increasing bedload rate with increasing shear stress applies to all cases. The results for the 100% packing case compare well with Paintal's (1971) measurements for a bed of uniform material with 7.9 mm sieve diameter.

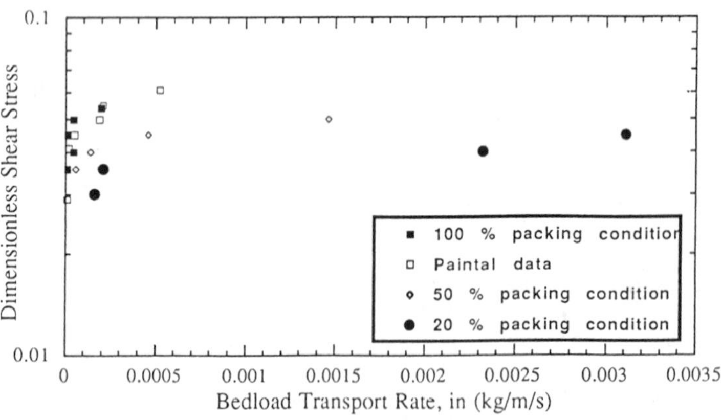

Figure 1. Relation between boundary shear stress and bedload transport rate

In figure 1 it is also illustrated that as the packing condition of the surface layer decreases the bedload transport rate increases at a certain shear stress. This increase becomes more significant when the dimensionless shear stress reaches higher values.
From the post-processing of the 3D-LDA measurements, the probability distribution functions (pdf) of the normalized Reynolds stress, uv/(u'v'), shown in figure 2, were obtained. Here u and v are the instantaneous, fluctuating streamwise and normal velocity components, respectively, and u' and v' are their corresponding standard deviations. The hyperbolas in figure 2 denote the stress events of magnitude equal to the normalized average bed shear stress. The portion of the pdf outside the hyperbola denotes the sporadically occurring stress events which have higher magnitude than the average bed shear stress. Since the particle transport was also observed to be quite sporadic it is speculated that these high stress events could cause particle entrainment.

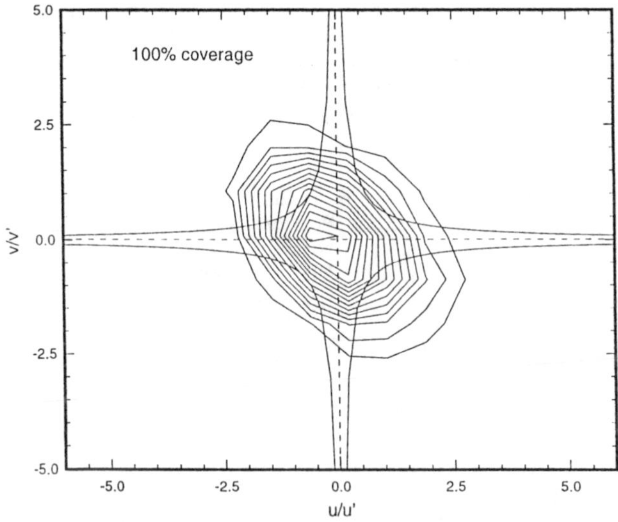

Figure 2. Probability distribution of the normalized Reynolds stress

Conclusions and Recommendations

The preliminary results of this work show that depending on how tightly packed the layer is, different bedload transport rates can be measured for the same material at a given flow condition. Under these conditions the criterion for the incipient motion will differ with the packing conditions of a layer. This may provide some explanation for the scattering of measurements of critical shear stress that is observed in existing experimental data. Alternatively, a surface layer "compactness" factor which alters the bedload transport rate to account for different packing conditions could be incorporated into current bedload transport formulas. The LDA results demonstrate the relatively large number of stress events near the bed which exceed the average bed shear stress particularly during "sweeps" and "ejections" and may contribute to instantaneous bedload motion.

Acknowledgments

This material is based upon work supported by the U.S. Geological Survey grant 1434-92-G-2271

Appendix-References

Balakrishnan, M. and Dancey, C . L, 1994. "An investigation of turbulence in open channel channel flow via three-component laser Doppler anemometry," presented at the 1994 National Symposium on Fundamentals and Advancements in Hydraulic Measurements and Experimentation, ASCE Hydraulics Division, 159-175, Buffalo, NY

Fenton, J. and Abbott, J . E., 1977. "Initial movement of grains on a stream bed: the effect of relative protrusion", *Proc.Royal Soc. London, 352:523-537.*

Moore, M . C., and Diplas, P. (1994). "Effects of particle shape on bedload transport" presented at the ASCE Hydraulic Engineering Conference, 800-805, Buffalo, NY

Paintal, A . S. (1971) " Concept of critical shear stress in loose boundary open channels", *J. Hydraulic Research.,* 9(1), 91-113

Reid, I., Frostick, L . E., and A.C. Brayshaw. (1992) "Microform roughness elements and the selective entrainment and entrapment of particles in gravel - bed rivers," Dynamics of Gravel Bed Rivers, Ch 12, 253-275, Ed. by Billi, et al, John Wiley & Sons.

CHANNEL RECTIFICATION FOR CASTOR RIVER, MISSOURI: A CASE STUDY

Roger A. Gaines[1]

Abstract

The Castor River is located along the Ozark foothills in southeastern Missouri. Channel improvements and modifications along reaches of the river have had varying effects on channel stability. Lateral stability has been a particular problem throughout one 2.09 km reach. Local instabilities created during moderate to extreme runoff events have caused extensive channel adjustments within this reach. Stabilization of this reach was critical in preventing migration of channel degradation and lateral instability to the upper reaches of the watershed. A case study of measures employed to stabilize this most dynamic reach is presented here.

Introduction

The Castor River basin originates in the foothills of the Ozark mountains in southeastern Missouri. The basin is characterized by steep slopes and rapid rise and fall of stream water levels. Ground elevations range from 86.8 meters mean sea level (MSL) in the delta region to over 152.4 meters MSL in the foothills. Part of a complex drainage system consisting of drainage canals, improved streams, and natural streams, the Castor River can be divided into three distinct regions. These regions are the upper foothills, the lower foothills and the delta.

The upper foothills region is located completely within the Ozark foothills. Stream channels in this region are generally unmodified; however, runoff from this region was diverted eastward to the Mississippi River to reduce headwater flooding in the delta region.

---

[1]Research Civil Engineer, U.S. Army Corps of Engineers, Memphis District, 167 North Mid America Mall, Room B202, Memphis, TN 38103-1894, (901) 544-3391

Below the headwater diversion, the basin continues in the foothills until reaching the St. Francis-Mississippi River delta. This segment comprises the lower foothills region. The stream channel in this region has undergone varying degrees of alterations. Improvements constructed in the stream consist of selective clearing of vegetation and channel cleanout.

The delta region is located in the St. Francis-Mississippi River delta. Only slight elevation variations occur across this region. Drainage through this region occurs primarily through drainage canals. Few stream channels are utilized for primary drainage, and those that are utilized have been straightened and enlarged.

The lower 2.09 km of the lower foothills region serve as a transition between the drainage canal and the meandering river channel. The modified outlet causes significant drawdown within this short reach. Increased velocities due to the drawdown increase sediment transport capacity. Channel instabilities result as the channel seeks a new equilibrium. Channel adjustments are both lateral and vertical. Vertical adjustment results as the channel reacts to a steeper hydraulic gradient, and baselevel adjustments brought about through an enlarged and straightened outlet. Grade control constructed at County Road 535 in 1987 reduces the tendency for significant degradation in the vicinity of County Road 535. However, because further vertical adjustment was limited, lateral adjustments began. Construction in 1992 and 1993 targeted further stabilization of the channel profile and alignment through this reach. Thalweg profiles for 1962, 1984, 1991 and the design for 1992 are shown in Figure 1.

General Design Approach

Stabilization feature design was outlined to be in accordance with acceptable engineering practices. Minimizing impacts to bank vegetation and the environment was a priority. Low-level weirs were employed to stabilize the channel grade. Secondly, the weirs restored the transition reach to an historic low-flow riffle-pool stream. Bank stability was accomplished by use of training dikes and bank protection. Training dikes and bank protection provided bank stability. Construction was conducted from within the stream channel where possible.

Environmental Considerations

Previous channel improvements in the 2.09 km transition consisted of selective vegetation removal. A link between the amount and type of bank vegetation and bank stability had been observed from this previous work. As a result, stabilization feature designs were developed to minimize disturbance of bank vegetation. Riprap placement thickness was reduced as much as possible considering all factors, including the cementing properties identified previously. Minimal riprap thickness caused less excavation and disturbance to the banks. Construction work was accomplished from within the channel where possible. Minimizing the removal of

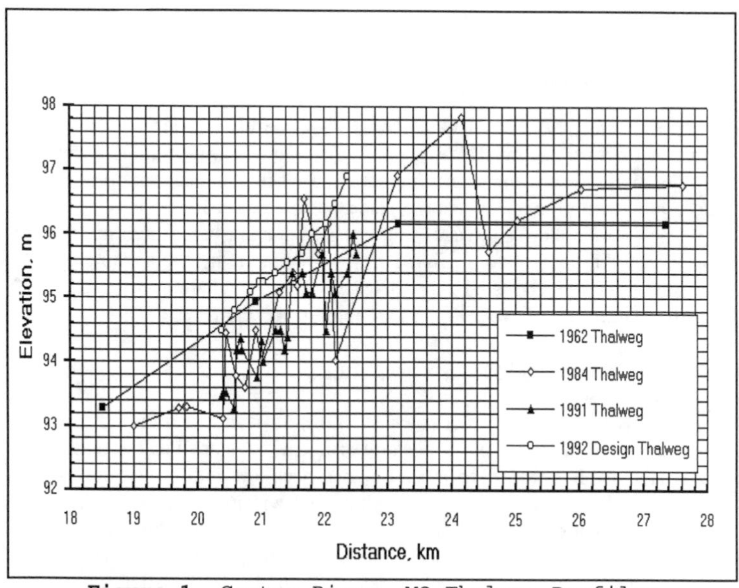

Figure 1 Castor River, MO Thalweg Profiles

bank vegetation reduced the changes to water temperature resulting from removal of canopy trees along the bank. Fishery habitat was also enhanced by reconstructing a riffle-pool sequence for low flow conditions.

Channel Stabilization Measures

Channel stabilization consisted of constructing riprap weirs, toe dikes, training dikes, and bank protection throughout the 2.09 km transition reach. Weirs were generally located at locations downstream of where natural channel bars formed historically. Crest elevations were determined from present and historic thalweg and bar elevations. In locations where large distances between bar locations existed, additional weirs were constructed to reduce the overall drop across any single weir to less than 0.3 meters. Limiting differences in weir elevations to 0.3 meters provided a riffle-pool sequence similar to historic conditions.

In addition to constructing weirs, bank protection was placed in critical areas where direct bank attack was observed and where protection was necessary to maintain channel alignment in the vicinity of weirs. Three areas were of specific concern since lateral channel adjustment had been from 9-12 meters annually. In these areas flow was diverted to the inside of the bends and closure/training dikes were constructed to form a new outer bank.

Selection of Riprap

Riprap gradation and placement thickness was determined using standard US Army Corps of Engineers Lower Mississippi Valley Division procedures. Additionally, considerations were made to incorporate experience obtained from previous riprap placed at County Road 535. Riprap placed at this site in 1987 had bonded together as a result of sediment and water properties interacting with the limestone riprap. Because of the cementing exhibited in riprap placed in 1987, riprap placement thickness was reduced slightly. Selected riprap gradation was R200, 90 kg maximum stone weight and 40 cm maximum stone diameter.

Post Construction Problems Experienced in 1992

Shortly after original construction of stabilization features in the fall of 1992, an extreme rainfall event occurred. Several problems developed in the transition reach as a result of this event. Ten areas that had been left undisturbed during original construction had undergone bank sloughing and widening. Bank widening caused trees to be downed in the channel, causing obstructions to flow. Bank widening was limited to areas immediately upstream of riprap bank protection placed the previous year. Repairs consisted of extending the bank protection. Additionally, riprap failure necessitated the repair of one tributary inlet. Repairs included a revised design for the tributary inlet.

Conclusions

Stabilization measures constructed in the 2.09 km transition reach were successful in arresting major channel instabilities during the winter of 1992 immediately following construction. Because original construction was limited to reduce disturbances to the environment, local instabilities developed in response to an extreme rainfall event that occurred shortly after construction. Repairs conducted during the summer of 1993 extended protection only where necessary to halt further bank widening. While additional construction was required to extend bank protection in 1993, limiting original construction resulted in a reduction in overall cost in dollars and in impacts to the environment from the design approach utilized for this project.

References

"Hydraulic Design of Flood Control Channels: EM 1110-2-1601", U.S. Army Corps of Engineers, CECW-EH-D, Washington, D.C., June 30, 1994.

"Report on Standardization of Riprap Gradation", U.S. Army Corps of Engineers, CELMV-ED-W, Vicksburg, MS, March, 1989 Revision.

STREAMBANK PROTECTION AND HABITAT RESTORATION

F. D. SHIELDS, JR., M. ASCE, S. S. KNIGHT, AND C. M. COOPER [1]

INTRODUCTION

Stream reaches which are candidates for habitat restoration projects often have been damaged by accelerated bank or bed erosion. Many techniques are available for streambank erosion control; selection of the best technique for a given application should be based on the dominant erosion mechanism at the site in question, economic factors, and environmental considerations. Considerable evidence suggests that intermittent structures like stone spur dikes are superior to continuous protection like revetment or stone toe protection (Shields et al., In Press). Additional work indicates approaches that involve restoration of woody vegetation on eroding banks such as the use of dormant willow posts may be beneficial to stream ecosystems. Little data or experience is available to guide the engineer in selecting the bank protection alternative that is most beneficial to warmwater stream habitat quality. Scientific information on low-gradient stream restoration in general is also scarce relative to that for less vulnerable and less frequently altered coldwater systems. This paper presents initial findings of a 3 to 5 year study of habitat response to three types of bank protection: stone toe, stone spur dikes, and dormant willow posts. A companion paper in this proceedings provides additional detail (Derrick, 1995).

STUDY SITE

Harland Creek, Mississippi, a fourth-order, meandering, sand and gravel-bed stream drains about 80 km^2 of hilly, mostly forested lands. In the study reach, bank heights varied from 1.6 to 12 m, channel widths averaged about 35 m, and bed slope averaged about 0.0012 (Neill and Johnson, 1987). Sinuosity was about 2.0. Comparison of airphotos taken in 1955 and 1986 indicated extensive lateral migration and avulsion of meanders during that interval (Neill and Johnson, 1987).

---

[1]Rsrch. Hydr. Engr., Rsrch Ecologist, and Supv. Rsrch Ecologist, respectively, U. S. Department of Agriculture, Agricultural Research Service, National Sedimentation Laboratory, P. O. Box 1157, Oxford, MS 38655-1157. Telephone (601) 232-2919, Fax (601) 232-2919. Internet shields@gis.sedlab.olemiss.edu.

Bed material was a mixture of sand and medium gravel with D_{50} of 10 samples collected at 300 m intervals in 1987 varying between 0.5 and 13 mm. Well-defined pools occurred on the outside of bends adjacent to large sandy point bars, and there were one or more riffles between bends. Water quality was adequate to support aquatic life (Slack, 1992). Suspended sediment yield measured at a station about 6 km downstream averaged 1,673 metric tons km^{-2} for 1985-1991.

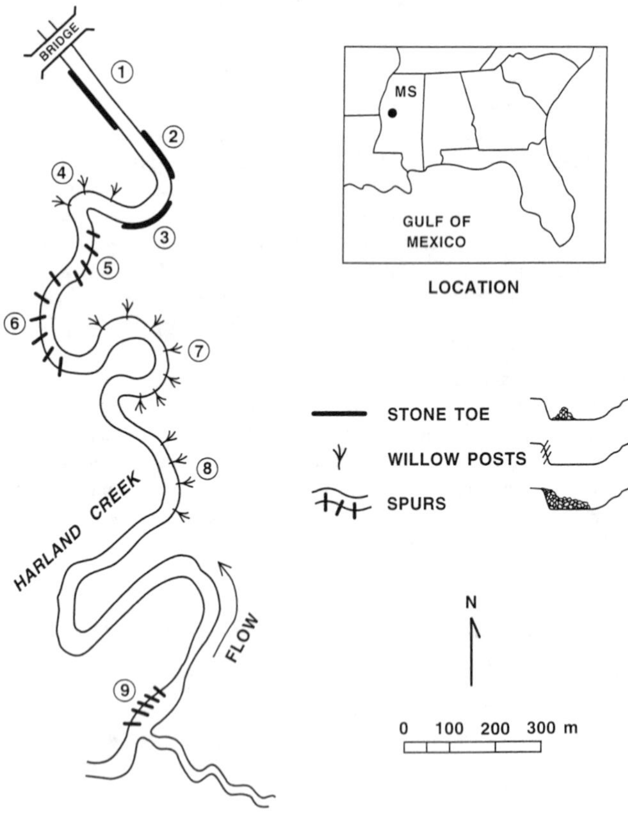

Figure 1. Map of study sites

Concave banks of bends in a 4-km reach of Harland Creek were stabilized in 1991-1994 using three techniques: stone spur dikes (also called bendway weirs), willow posts, and longitudinal stone toe. Although some bends were stabilized using combinations of these methods, sampled bends were stabilized using only one of the three at a time. Spur dikes were generally angled 10 to 30 degrees upstream with sloping crests 0.6 m wide. Dike lengths varied from about 6 to 12 m. Willow posts were dormant, living plant materials 8 to 30 cm in diameter by 3 m long and were planted 2.4 m deep using an hydraulic auger. Longitudinal stone toe was a windrow of quarried limestone riprap with triangular cross section placed at the bank toe. Stone toe dimensions were controlled by the angle of repose and a fixed application rate of 3,000 to 6,000 kg m^{-1}.

METHODS

Fish and their habitats were sampled in three reaches protected by each of the three types of protection (3 x 3 = 9 study reaches total). Sampling occurred at baseflow (0.15 to 0.16 $m^3 s^{-1}$) during June and October 1994. Each sampled reach was composed of a 100-m long bendway segment with protection applied to the concave bank. Habitat was sampled by measuring depth, velocity, and visually classifying bed material type at five points along five equidistant transects in each of the 9 reaches. Sample points were located 0.75 m from each bank, and at 0.25, 0.5, and 0.75 W from the left water's edge, where W = water surface width. Water width was measured at each transect, and notes were made regarding the number and size of woody debris formations, bank vegetation, and canopy. Bend radii were measured from contour maps. Fish were sampled using two backpack electroshockers simultaneously. Shocking crews moved through each reach in a streamwise direction, and the lower end of each reach was blocked with seines. Fish trapped in the seines were included in the sample. Riffles downstream from each sampling zone were also studied, but this aspect of the study is beyond the scope of this paper.

RESULTS

Bend radii ranged from 54 to 344 m and averaged 127 m. The largest bend radius was for a reach with spurs and was more than twice as large as the next highest value (142 m). Bend radii means and ranges for toe and posts were similar; average bend radius for spurs was elevated by the single long-radius bend (344 m) (Table 1). Reaches stabilized with stone toe tended to be shallower and narrower with higher velocity and more sand than spur or post reaches. Fall and spring habitat data were pooled and grouped by bank treatment for analysis, yielding about 150 data points (depth, velocity, bed type) per treatment. Depth, width, and velocity data were not normally distributed, so medians were compared using a one-way ANOVA on ranks. Values shown in boldface in Table 1 were not significantly different ($p < 0.05$). Results were not affected by excluding data from the long-radius bend stabilized with spurs. Because of the presence of the posts and the debris they trapped, mean woody debris densities for post reaches were 4 to 5 times greater than for other reaches.

TABLE 1. PHYSICAL HABITAT VARIABLES, HARLAND CREEK, SPRING AND FALL 1994

Type	Mean Bend Radius, m	Median Depth, cm	Median Width, m	Median Velocity, cm s^{-1}	Mean Debris Density, $m^2 km^{-2}$
Stone Toe	109	26	**7.9**	6	**24,000**
Willow Posts	99	**43**	**8.2**	**3**	114,000
Spurs	173	**38**	15.1	**2**	**20,000**

Bed type distributions differed among the three bank treatments (Chi-square, $p \leq 0.0001$). Once again, results were not affected by excluding data from the long-radius bend stabilized with spurs. Reaches with posts had no riprap, but had more organic debris on the bed (5%) than for spurs (3%) or stone toe (2%) (Figure 2). Stone toe reaches had more sand and less gravel than for the other two protection types.

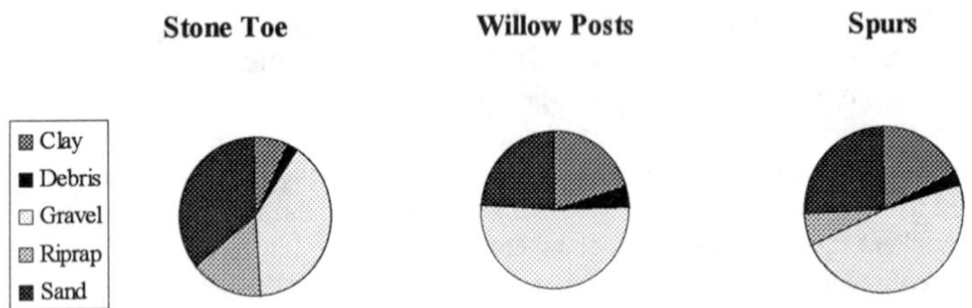

Figure 2. Bed Type Distributions for Three Types of Bank Restoration.

TABLE 2. RESULTS OF FISH SAMPLING, HARLAND CREEK, SPRING (FALL) 1994

Type	Number of fish	Average number of species per 100-m reach	Total number of species
Stone Toe	725 (435)	13 (11)	24
Willow Posts	801 (309)	10 (9)	26
Spurs	1,247 (1,122)	12 (12)	25

A total of 4,639 fishes representing 33 species was collected (Table 2). Fish numbers were higher in reaches stabilized with spurs. Species composition and relative abundance were similar for the three bank treatments. Two cyprinids comprised 65% to 72% the catch from each type of bank treatment; species composition and relative abundance were similar for all three bank treatments.

DISCUSSION AND CONCLUSIONS

Pool habitat (relatively deep areas of low velocity) is typically in short supply in warmwater streams damaged by erosion or channelization. Spur dikes and dormant willow posts used for streambank restoration on Harland Creek created more pool habitat per unit length than standard stone toe protection. Bed type distributions in reaches stabilized with posts and spurs were indicative of more desirable habitat than those in stone toe reaches. Willow posts provided significant amounts of woody debris. Unfortunately, rates of survival for posts planted along Harland Creek have been disappointingly low (~40% after one growing season, personal communication, Dr. S. R. Abt, Colorado State University). If the post plantings do not thrive, associated habitat benefits will be temporary. Fish were most numerous in reaches with spurs; effects of habitat differences on fish species composition were not detected.

ACKNOWLEDGMENTS

Restoration works on Harland Creek were designed and constructed by the U. S. Army Corps of Engineers. Terry Welch, Sam Testa, Jennifer Bowen, Rocky Smiley, Todd Randall, Shawn Yates, Robin Chipley, Belinda Garraway, Kim Damon, Monica Young, Dawn Harrington, John Wigginton, and Ken Kallies assisted with collection and analysis of data. Keith Parker collected bed material samples. David Derrick, Thomas Pokrefke, Roger Kuhnle, and Steven Abt read a draft of this paper and made many helpful comments. P. D. Mitchell drew Figure 1.

APPENDIX--REFERENCES

Derrick, D. L. (1995). Case Study: Harland Creek Bendway Weir/Willow Post Bank Stabilization Demonstration Project. In *Water Resources Engineering*, Proceedings, First International Conference on Water Resources Engineering, ASCE, New York.

Neill, C. R. and Johnson, J. P. (1987). Black Creek watershed geomorphic analyses. contract report prepared by Northwest Hydraulic Consultants, Inc. for U. S. Army Corps of Engineers, Vicksburg, Miss.

Shields, F. D., Jr., Cooper, C. M., and Testa, S. (In Press). Towards greener riprap: environmental considerations from micro- to macroscale. *Proceedings of the International Riprap Workshop: Theory, Policy and Practice of Erosion Control Using Riprap, Armour Stone and Rubble*. John Wiley & Sons, London.

Slack, L. J. (1992). "Water-quality and bottom-material-chemistry data for the Yazoo River basin demonstration erosion control project, north-central Mississippi, February 1988- September 1991." *U. S. Geological Survey open-file report 92-469*, U. S. Geological Survey, Jackson, Mississippi.

A Habitat Improvement Plan for the Big Sunflower River, Mississippi

Andrew C. Miller[1], John Harris[2], and Ken Quackenbush[3]

Introduction

State and Federal conservation agencies recently expressed concern that proposed maintenance dredging in the Big Sunflower River, Mississippi, by the U.S. Army Engineer District, Vicksburg, could adversely affect freshwater mussels (Family: Unionidae). Freshwater mussels are bivalve molluscs that filter-feed particulate organic matter from the water and live partially buried in firm gravelly sand substratum. Thick-shelled species are collected from medium-sized to large rivers and their shells used in the cultured pearl trade (Williams et al. 1992). Concern over dredging was partially due to high densities of commercially valuable species in some reaches of the river.

The Vicksburg District assemble a multidisciplinary team to develop a habitat improvement plan for the Big Sunflower River that offset some of the habitat losses caused by dredging. Although small streams are often restored, man-made habitat improvements to larger rivers are uncommon, possibly because of costs and potential for interference with navigation (Gore and Shields 1995). A successful restoration project of narrow scope was completed in 1984 in the Tombigbee River (Miller et al. 1988). Two riffles 46 m long and 24 m wide were created by capping sand fill with 2- to 80-mm diameter coarse sand and gravel. In the case of the Big Sunflower River the team was asked to consider fairly large-scale modifications over much of the area to be dredged. The intent was to modify water velocity and substratum conditions in selected reaches to maintain the existing mussel species and to improve species diversity and recruitment levels above those measured in an earlier survey (Miller and Payne 1995).

---

[1]Research Limnologist, U.S. Army Engineer Waterways Experiment Station, Vicksburg, MS 39180, (601)634-2141; [2]Adjunct Associate Professor, Department of Biology, Arkansas State University, State University, AR, 72467; [3]Fish and Wildlife Biologist, U.S. Fish and Wildlife Service, Vicksburg, MS 39180.

Study Area

The Big Sunflower River originates near Moon Lake, Coahoma County, flows south through agricultural land, and enters the Yazoo River between Sharkey and Yazoo counties (Figure 1). The gradient is low, with steep, poorly vegetated and often eroding banks. Water velocity at low flow is typically less than 15 cm/sec. With the exception of isolated gravelly sand bars in the mid reach, substratum throughout consists mainly of fine-grained silt and sand.

The most recent channel modifications were completed in the 1960s which included channel clearing (76.4 km), cleanout (34.6 km), enlargement (69.4 km), cut-offs aggregating 25.7 km, and construction of a weir at the lower end of the Holly Bluff Cutoff to control low water level. Following recent flooding it was determined that channel hydraulic conveyance had deteriorated since earlier modifications. The District proposes to hydraulically dredge sediments from much of the river channel to improve flood conveyance and reduce local flooding.

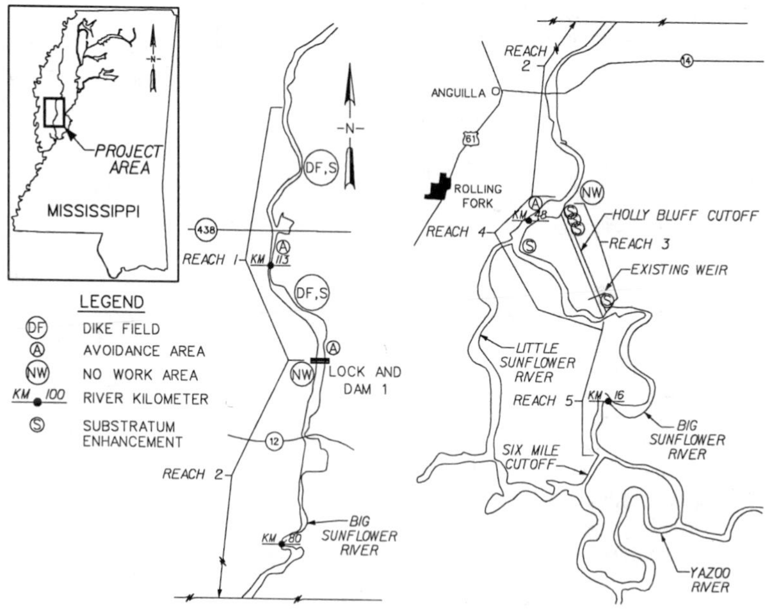

Figure 1. Map of the study reaches.

Freshwater Mussels in the Big Sunflower River

At a mussel bed immediately downriver of abandoned Lock and Dam 1, total mean density (individuals/m^2) and biomass (g/m^2) was 235.0 ±16.0 (standard error of the mean) and 52,250.1 ±3,284.8, respectively. Approximately 90% of the fauna consisted of the commercially valuable threeridge mussel *Amblema plicata plicata* that would sell for an estimated $2.20/kg ($1/lb) live weight. Regardless of localized high densities, there was virtually no recent recruitment (evidence of recent successful reproduction as noted by presence of individuals less than 30-mm total shell length), the total number of species collected was 12, and diversity (a measure of total species present and their evenness of distribution) was 0.49. In contrast, at a bed in the lower Tennessee River downriver of Kentucky Dam, 30% of the *A. p. plicata* were less than 30 mm, 23 species were identified, and species diversity ranged from 1.54 to 1.87. Mean density at 6 sites (10 samples were collected at each) ranged from 9.2 to 128.0 individuals/m^2 (Miller et al. 1992).

Poor recruitment and low species richness and diversity in the Big Sunflower River are likely the result of stress from high sedimentation, reduced dissolved oxygen, and high water temperatures during low flow in the summer. Habitat could be improved by increasing water velocity and substratum stability, and increasing fish habitat, since immature mussels must develop on the skin or gills of a host fish.

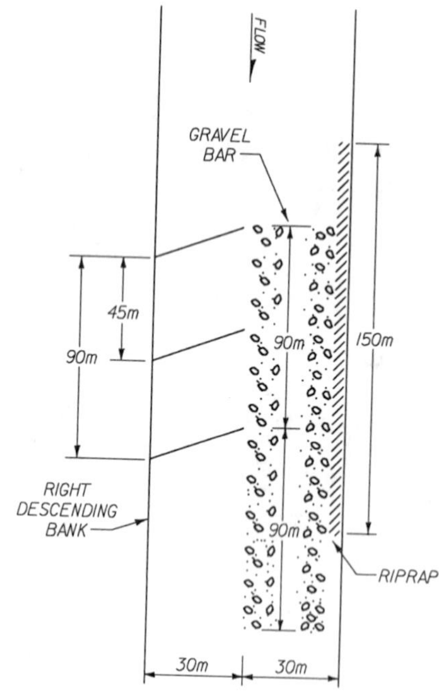

Figure 2.
Design of dike field with associated gravel substratum and bank protection.

Proposed Habitat Improvement Features

The improvement plan depended on the proper placement of a series of features to modify habitat. Dikes will be used to increase water velocity in selected reaches where sedimentation rates were high. The dikes would be placed in a field of three, constructed of riprap, and would reach halfway across the channel (Figure 2). The field was restricted to one bank, since most riverine mussels are found between the thalweg and the shore. The banks opposite the dike field would be protected from erosion with riprap and an underlying gravel filter. If needed, flood-tolerant trees (green ash, baldcypress, and Nutall oak) would be used to reduce erosion in adjacent areas. Mussel habitat would be improved by placing gravel on existing sand, mud, or clay substratum opposite and a short distance downriver of the dike field. Increased velocity (at least 40-50 cm/sec) resulting from the dikes should keep gravel sediment-free. There is a lack of natural cover that provides hiding places and food for fish. Artificial cover can be made from PVC pipe, automobile tires, logs, or brush.

Description of the Habitat Improvement Plan

The river was divided into five reaches based upon sedimentation rates, hydraulic features, and existing mussel resources. The team prepared 5 alternative plans which differed in the number of features and total cost. The District accepted the fifth plan which is described in the following paragraphs:

Two 3-dike fields with added gravel substratum, bank protection, and fish attractors will be placed in Reach 1. This reach extends approximately 35 km upriver of abandoned Lock and Dam 1 (Figure 1). Substratum consist of fine sands and silt, and the morphology is conducive to placement of physical structures to modify velocity.

Reach 2 extends from Lock and Dam 1 downriver to the upper end of the Holly Bluff Cutoff. Some sections are free of excessive sedimentation and have firmly packed sand or gravelly sand. Since there are some locally dense mussel beds in this reach, no major habitat features, with the exception of fish attractors, were considered.

The third reach includes the Holly Bluff Cutoff, a 11.9-km man-made channel. A weir is located near the downriver end that maintains flow in the excluded channel. Mussel populations are less than $1.0/m^2$ in the clay and shifting sand. Because of the hydraulic design of the cutoff, it is not possible to alter flow with dikes or weirs. Narrow strips (60 m long by 7.6 m) of gravel will be placed along one shore upstream of the existing weir, and two strips of gravel will be placed 60 and 240 m downriver of the weir.

Reach 4 consists of 24 km excluded by the Holly Bluff Cutoff. Water velocity is moderate to high in the upper section in which there is a bed where densities at 8 sites (10 samples taken at each) ranged from 10.0 ± 7.0 to 64.8 ± 9.2 individuals $/m^2$. Because flow is affected by the Holly Bluff Cutoff, dikes were not considered. Fish attractors will be put in the upper section.

Reach 5 includes 26 km of the most downriver section of the maintenance project where mussel densities are less than $1/m^2$. Fish attractors will be placed at two locations.

The estimated cost of the improvement plan is $249,808. The plan includes 2 three-dike fields, 54 fish attractors, 152.4 m of bank stabilization at 2 locations, and 670 linear m of gravel at 7 locations. In addition to the above features, the District identified two no-work reaches where there will be no dredging because of high density mussel populations. Three avoidance areas were identified where dredging will be restricted to one bank to reduce impacts to mussels.

Physical and biological effects of the features will be monitored for a minimum of 10 years. Habitat features will be inspected by divers. Sedimentation rates, water velocity, and water quality variables important to mussels (temperature, dissolved oxygen, turbidity, and total suspended solids) will be monitored regularly. Mussel recruitment rates, richness, diversity, and density will be measured annually.

Acknowledgments

Larry Banks, Tommy Shelton, Frankie Griggs, Ron Goldman, Lee Robinson, Marvin Cannon, Steve Reed, Johnny Sanders, Jack Herring, Garry Lucas, K. Jack Killgore, Alan Mueller, David Beckett, Shiao Wang, Barry Payne, Nick Chandler, and Fawn Burns provided information on the project area or assisted with plan development. Funds were provided by the U.S. Army Engineer District, Vicksburg. Permission was granted by the Chief of Engineers to publish this information.

References

Gore, J. A., and Shields, F. D. 1995. Can large rivers be restored? BioScience 45(3):142-152.

Miller, A. C., Killgore, K. J., King, R. H., and Mallory, J. 1988. Aquatic habitat development on the Tombigbee River. pp 77-89 In: Inland Waterways, Proceedings of a National Workshop on the Beneficial Uses of Dredged Material. Technical Report D-88-8, U.S. Army Engineer Waterways Experiment Station, Vicksburg, MS.

Miller, A. C., Payne, B. S., and Tippit, R. 1992. Characterization of a freshwater mussel (Unionidae) community immediately downriver of Kentucky Lock and Dam in the Tennessee River. Transactions of the Kentucky Academy of Sciences 53(3-4):154-161.

Miller, A. C., and Payne, B. S. 1995. An analysis of freshwater mussels (Unionidae) in the Big Sunflower River, Mississippi, for the Big Sunflower River Maintenance Project: 1993 studies. Technical Report, U. S. Army Engineer Waterways Experiment Station, Vicksburg, MS.

Williams, J. D., Warren, M. L., Cummings, K. S., Harris, J. L., and Neves, R. J. 1992. Conservation status of freshwater mussels of the United States and Canada. Fisheries 18(9):6-22.

DEVELOPMENT OF THE RIVERINE COMMUNITY HABITAT ASSESSMENT AND RESTORATION CONCEPT

y: S. R. Abt[1], F. ASCE, J. C. Fischenich[2], M. ASCE, C. C. Watson[3], F. ASCE, and
. B. Florentin[4], S.M. ASCE

bstract

There currently exists a need to integrate habitat enhancement concepts into the flood hannel restoration design process. It has been demonstrated that habitat quality can be directly nked to stream hydraulic properties (i.e., velocity-depth pairs). The Riverine Community labitat Assessment and Restoration Concept (RCHARC) provides a method for assessing habitat conjunction with the hydraulic properties. A field test was performed at Goose Creek near lamosa, CO and the results indicate that the methodology has potential for assessing comparison tream reaches and for predicting habitat quality replication.

ntroduction

In an attempt to provide a means of assessing habitat response to stream restoration, lestler et al. (1993a) developed the Riverine Community Habitat Assessment and Restoration oncept (RCHARC) at the U.S. Army Corps of Engineers Waterways Experiment Station. CHARC is a simulation approach for relating the effects of flow alterations on aquatic biota sing the stream system as a basis of comparison, which is a "comparison standard" for the nalysis against which the project alternatives can be evaluated. The comparison standard river ystem is considered to represent the target or ideal habitat conditions, both in terms of channel onfiguration and seasonally varying flow characteristics, for the aquatic community in the project iver system. The standard reach can be selected based upon professional consensus, physical imilarity to the project system, and/or similarity of the aquatic community to the desired standard ystem.

RCHARC does not directly use life stage-specific suitability curves as does PHABSIM Milhous et al., 1989); this is considered implicit in the methodology (Nestler et al., 1993). In sing RCHARC, it is assumed that for a given discharge, there exists a distribution of flow depths nd velocities. These distributions represent the habitat template upon which the aquatic ommunity is structured. Changes in the frequency distribution of the flow depth and velocity vill result in associated changes in the aquatic community. Depiction of a stream reach in terms f frequency distributions of depth and velocity is likely to capture the stream heterogeneity that

---

[1] Prof., Dept. of Civ. Engrg., Colo. State Univ., Fort Collins, CO 80523

[2] Res. Civ. Engrg., Env. Lab., U.S. Army Engrg., WES, Vicksburg, MS 39180

[3] Assoc. Prof., Dept. of Civ. Engrg., Colo. State Univ., Fort Collins, CO 80523

[4] Grad. Res. Asst., Dept. of Civ. Engrg., Colo. State Univ., Fort Collins, CO 80523

dictates the aquatic community composition. The holistic perception of RCHARC provides framework to evaluate the system differences between the standard reach and the project strea reach(s) by describing the pertinent physical factors that affect habitat. RCHARC compares t underlying patterns of depths and velocities in two or more comparative reaches and uses t results as the basis of the community-level impact analysis. The degree of impact is rough approximated by the degree to which the physical habitat differs between the target and t standard reaches. The analysis focuses on a comparison of bivariate velocity-depth pairs and t frequency of occurrence in the target and standard reaches.

Approach and Field Test

The RCHARC simulation provides a linkage between field observations, survey results ar an understanding of habitat diversity. Assuming habitat similarity is related to hydraul similarity, morphologic and hydraulic diversity may be compared and habitat quality correlatio drawn. To apply the concept, standard (ideal or natural) and restored comparison reaches a selected. Each site is then segmented into a minimum of ten sequential cross sections wi adjacent cross sections ranging from three to five channel widths apart. The channel geometr is measured and roughness is estimated at each cross section. Each cross section is segmented in 20 cells equally spaced across the section. The flow depth and velocity are measured in each ce resulting in a minimum of 200 velocity-depth pair measurements for each reach.

The RCHARC program consists of two primary sub programs. One sub progra processes the input files containing the field data (standard and restored). The second sub progra tabulates velocity-depth groups, velocity-depth occurrences within each group, and the percent c total occurrence for each group per discharge. The output is presented in both numeric an graphic form. A three-dimensional bivariate plot of velocity versus depth versus perce occurrence in each reach is generated. Habitat similarity is evaluated by comparing the velocity depth distributions between comparison reaches for a desired discharge. A detailed discussion c the components of the RCHARC program is presented by Nestler et al. (1993a), Nestler et a (1993b) and Peters (1994).

A field test of the RCHARC approach for evaluating habitat quality was performed fc Goose Creek, located approximately 58 miles west of Alamosa, Colorado. Goose Creek i situated in the upper plateau of the San Juan mountain range, Rio Grande National Forest; and i the upper drainage basin in the Weminuche Wilderness area. Goose Creek was selected as a stud site because (1) the stream contained natural (standard), restored and degraded reaches; (2) certa reaches are classified as gold medal trout fishing areas by the Colorado Division of Wildli (CDW); and (3) permission was granted to access the three reaches.

The Goose Creek study segment was approximately 11.5 km in length. The standard reac bordered the wilderness area and extended approximately 275 m downstream. The restored reac was centered in the stream study segment and was approximately 275 m long. The degraded reac was approximately 275 m in length and was located near the outlet of the stream study segment The channel-geometry, velocity-depth pairs (200 per reach), discharge and water-surfac elevations were determined for each section. A summary of the reach characteristics is presente in Table 1. The field data (i.e., geometry, velocity-depth pairs, etc.) were input into the RCHARC model per Nestler (1993a) and Peters (1994). An RCHARC simulation of standard, restored and degraded reaches was performed, resulting in a velocity-depth bivariate plot for the prescribed flow condition. The bivariate plots for each of the three reaches is presented in Figures 1 through 3.

Table 1 Summary of Reach Characterization

Channel Characteristic	Standard Reach	Restored Reach	Degraded Reach
Slope	0.020	0.013	0.010
Discharge (m^3/sec)	0.78	0.98	0.90
Width (m)	8.41	8.66	11.05
Temperature (°C)	19.3	18.9	16.4
Average Velocity (m/s)	0.58	0.52	0.52
Velocity Range (m/s)	0.32-1.18	0.40-0.72	0.24-0.96

Reach Comparisons

The field data and results from the RCHARC simulation were quantitatively compiled to provide a relative comparison of the three reaches. The reach slopes ranged from 0.010 to 0.020. Although there was a considerable difference in slopes, all three reaches represented steep channel conditions in which the riffle segments comprised 40 to 70% of the stream, while shallow pools comprised 20 to 50% of the stream. The compilation of alternating riffles and pools lend a similarity in reaches that enhanced aquatic habitat.

The average stream velocities ranged from 0.58 m/s in the standard reach to 0.52 m/s in the restored and degraded reaches; a nominal difference. The velocity ranges through each reach were also similar, as presented in Table 1. The stream broadened with distance downstream. However, the discharges also increased as the tributary area increased. The discharge per unit width was 0.10, 0.11 and 0.8 m/s/m for the standard, restored and degraded reaches, respectively. These results indicate that the standard and restored reaches are similar.

A comparison of the velocity-depth bivariate plots of the standard, restored and degraded reaches provides a qualitative evaluation of the flow diversity. Figure 1 presents the velocity-depth plot for the standard reach. It is observed that the flow depths and velocities span the quadrant although the distribution centroid is near a depth of 10 cms and a velocity of 0.3 m/s (30 cms/s). The velocity-depth plot for the restored reach is presented in Figure 2. The velocity-depth pair distribution for the restored reach is similar to the standard reach; however the distribution centroid is at a depth of approximately 15 cms of depth and 0.4 m/s (40 cm/s). The restored reach displays a pronounced bi-modal distribution. Figure 3 presents the velocity-depth bivariate plots of the degraded reach. It is observed that the widened reach has a more uniform depth and velocity distribution compared to the standard and restored reaches. The distribution centroid is at a depth of approximately 20 cm and a velocity of 0.2 m/s (20 cm/s).

In an attempt to quantitatively compare the hydraulic similarity of the standard, restored and degraded reaches, a Canberra metric coefficient was computed for each of the comparison reach pairs (i.e., standard vs restored, restored vs degraded, etc.). The Canberra metric coefficient, usually expressed as a dissimilarity, is the average of a series of fractions representing the inter-entity agreement of each attribute and is commonly used in ecological community comparisons. The Canberra coefficients for the standard-restored, standard-degraded, and restored-degraded comparisons are 0.85, 0.89 and 0.88, respectively. A Canberra coefficient of 0.70 indicates a reasonable similarity in comparison alternatives. In general, the standard and restored reaches exhibit greater similarity in velocity-depth distributions than the degraded reach velocity-depth distributions; however, neither the degraded reach nor the restored reach display a dissimilarity in hydraulic conditions.

Conclusions

The field test of the RCHARC methodology applied to Goose Creek, CO has potential for assessing comparison stream reaches for predicting habitat quality replication. Guidelines for determining minimum quantitative criteria for defining a reasonable similarity need to be developed.

References

Milhous, R.T., Updike, M.A., and Schneider, D.A. (1989). "Physical Habitat Simulation System Reference Manual-Version II." Instream Flow Information Paper No. 26, U.S. Fish and Wildlife Service Biological Report 89(16), September.

Nestler, J.M., Schneider, L.T. and Latka, D. (1993a). "Physical Habitat Analysis of Missouri River Main Stem Reservoir Tailwaters Using the Riverine Community Habitat Assessment and Restoration Concept (RCHARC)." U.S. Army Corps of Engineers, Waterways Experiment Station, Technical Report EL-93-22, November.

Nestler, J., Schneider, T., and Latka, D. (1993b). "RCHARC: A New Method for Physical Habitat Analysis," ASCE Symposium on Engineering Hydrology, San Francisco, CA, July 25-30, pp. 294-299.

Peters, M.R. (1994). "Low Flow Habitat Rehabilitation-Evaluation of the RCHARC Methodology." M.S. Thesis, Colorado State University, Fort Collins, CO.

Acknowledgements

Information presented herein, unless otherwise noted, was obtained from research conducted under the Environmental Impact Research Program of the U.S. Army Corps of Engineers. Permission was granted by the Chief of Engineers to publish this information.

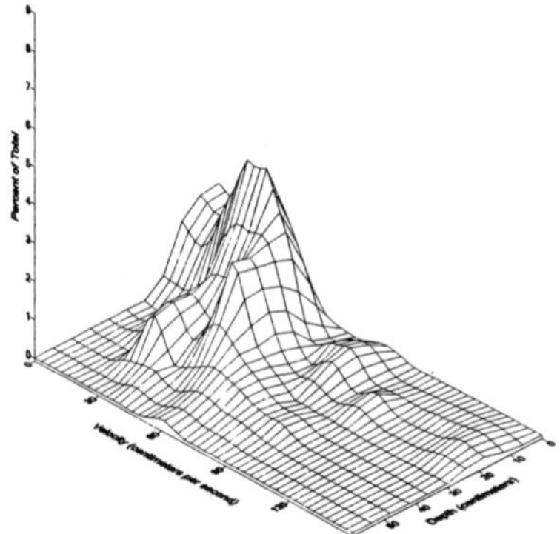

Figure 1. Goose Creek Standard Reach, Velocity-Depth Distribution

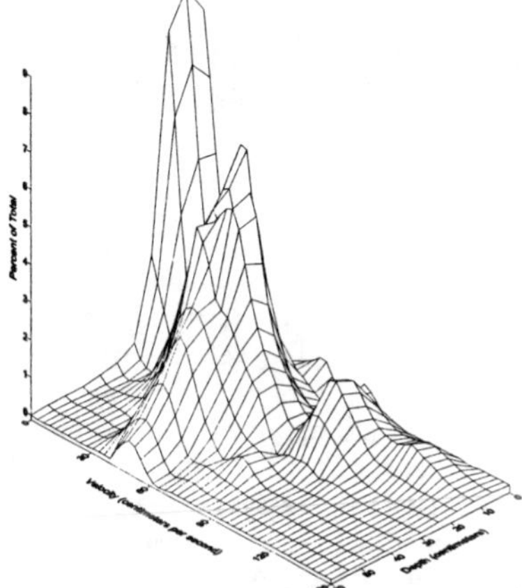

Figure 2. Goose Creek Restored Reach, Velocity-Depth Distribution

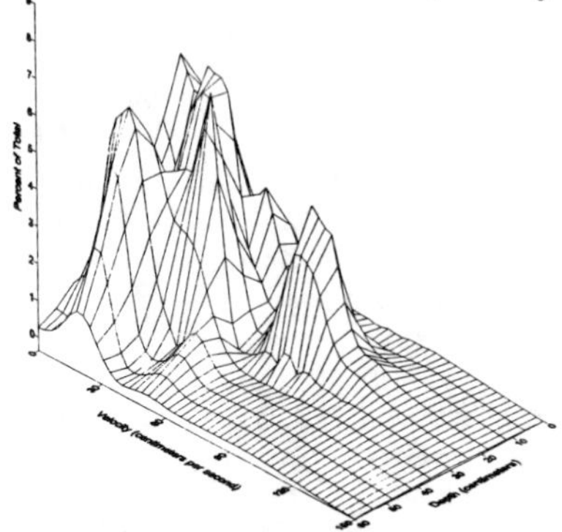

Figure 3. Goose Creek Degraded Reach, Velocity-Depth Distribution

Southern Africa Regional Water Sector Assessment for U.S. Agency for International Development

Charles A. Scheibal[1], Member, ASCE, L. Lynn Pruitt[2], and Gary W. Foster[3], Member, ASCE

Abstract

The U.S. Agency for International Development (USAID) recognizes the recurrent problem of water shortages resulting from droughts and inadequate water management practices in southern Africa. Planned and on-going projects aimed at alleviating water-related problems in the 12-country region have been cataloged. Regional approaches involving water sharing, conservation, environmental enhancement, and equitable and efficient water use management have been assessed and are among the projects being considered for funding by USAID. Potential advances in the institutional setting and legal environment affecting the southern Africa water sector complete the realm of this assessment.

Project Setting

The drought throughout southern Africa in 1991-1992 highlighted the deficiencies in many water supply systems and water resource use strategies. Some areas escaped the severeness of the drought because of relatively adequate water impoundment and well developed water infrastructure, however, many others were not as fortunate.

----

[1]SARP Regional Engineer, U.S. Agency for International Development, No. 1 Pascoe Avenue, Belgravia, Harare, ZIMBABWE
[2]Project Manager, Stanley Consultants, 225 Iowa Avenue, Muscatine, IA, USA
[3]Project Coordinator, Stanley Consultants, 225 Iowa Avenue, Muscatine, IA, USA

The threat of recurrent drought, which has been amplified by observed weather trends of reduced rainfall over the past decade, together with increased demands for water, have combined to place supply of water to the region under increasing risk. The need for mutual conservation and inter-country cooperation for utilization of scarce water resources is becoming increasingly important. Hence the need for a systematic program that identifies, analyzes and makes positive recommendations for the use, conservation and sharing of water on a regional basis.

Scope of Project

The U.S. Agency for International Development (USAID) Mission in Harare, Zimbabwe, commissioned Stanley Consultants to perform a regional assessment of the water resource sector for 12 countries in southern Africa:

Angola	Malawi	South Africa	Zaire
Botswana	Mozambique	Swaziland	Zambia
Lesotho	Namibia	Tanzama	Zimbabwe

The scope of the assessment was to evaluate the physical water resources, water laws and institutional capabilities as they presently exist with the objective of determining:

- If there is a need for additional international funding;
- Critical needs relative to increasing regional cooperation, and;
- Whether USAID's limited resources can be effectively applied to critical needs of the water sector in the region.

Under the initiative to expand regional economic cooperation and increased operational efficiency in key infrastructure areas, USAID hopes to identify and support programs that will generate substantial cost savings from coordinated investments, shared use, policy harmonization, restructuring and privatization in the priority sectors.

This assessment was not conducted within a vacuum. The investigation of all aspects of the water resources sector within the region has been the subject of intense and comprehensive study by the nations of the region themselves, bi-lateral donor countries, international organizations and non-government organizations. One of the primary goals of this study was to draw together these far-ranging information resources.

Phase I Information Collection

Key members of the Consultant's staff and the USAID Project Director initially met in May 1994 with country representatives, donors and officials of the Southern Africa Development Community Environment and Land Management Sector (SADC-ELMS) for a project briefing in Livingstone, Zambia. SADC is the organization which at that time represented 10 of the 12 countries included in the water sector assessment. (South Africa was subsequently added to SADC but Zaire is still not a member). SADC-ELMS is the specific sector unit tasked with coordinating water resources within the southern Africa region.

The primary focus of Phase I was collection of all readily available information pertaining to water resources in the 12-country study region. A two-pronged effort was initiated involving field visits to and telephone/fax contacts with the various country water authority administrators, international and bi-lateral donors, regional water agencies, consultants and other involved parties. Most of this work took place over an intense two-month period (June-July 1994) and produced over 1,100 hard copy documents and several computer databases describing ongoing and planned water resource appraisals and developments in the region.

Next came the monumental effort of managing and digesting the voluminous information to assess the current status of the water resource sector, current involvement by other donors and the need for additional outside assistance. A computer database system was considered necessary to effectively retrieve and sort the large volume of available information in a systematic manner. The FoxPro Relational Database Program became the chosen software system because of its versatility and capacity, and its "friendliness" to other database systems which may be used by organizations in the region. The overall program consists of five separate but relational databases: 1) Documents; 2) Projects; 3) Organizations; 4) Funding; and 5) Policies. While many different data sorts can be accommodated, catalogs of grouped information were generated to facilitate review and planning for the Phase II assessment:

- Water Projects by Country and Regional Projects
- Funding of Projects by United Nations, International Banks, Bilateral Agreements and Non-Governmental Organizations
- Water Resource Institutions by Country and Regional Institutions
- Water Laws by Country and Regional Treaties

Over 500 projects in various stages of completeness were identified in the Phase I database. Most projects (about 450) were considered country specific with the remainder encompassing two or more countries.

A working meeting to present the results of the information gathering effort was held in Pretoria, South Africa during September 1994. The purpose of the meeting was to review the completeness and accuracy of the information collected, and to present preliminary evaluations to USAID, SADC-ELMS, country representatives, donors and other interested parties.

Phase II Water Sector Assessment

Phase II of the USAID Southern Africa Water Sector Assessment involved a review of the collected information for planned water resource projects, financing options, status of investigations (technical, institutional, social, environmental, and financial), donor activities, opportunities for private sector involvement and investments, institutional settings and legal environment. Emphasis was placed on regional activities to promote water resource sharing between countries and an increased cooperation with technology training, free dissemination of technical database information, and concern for each nations' upstream and downstream neighbors.

One of the main activities of the Phase II investigations was the in-country assessment of regional projects, legal aspects and institutions that would benefit from USAID support. A project team meeting held in Muscatine, Iowa, in October 1994 served as a kick-off for the country visits. All 50 plus regional projects identified in the Phase I catalogs were reviewed for potential USAID support. The in-country assessments were conducted by project team members in November 1994. Personal visits to each study country, except Zaire, produced appraisals for the identified regional projects plus a few new ones. In the end, 30 projects were selected for in-depth appraisal.

Evaluation Categories and Criteria Weighting

Evaluation criteria and category weighting factors indicate the perceived importance of each relative to the overall water sector assessment.

Need
Economic Impact
Readiness to Proceed
Institutional Environment
Regional Impact
Activity Size

Sustainability
Environmental Impact
Population Impact
Public and/or Political Support
Legal Environment
Private Sector Opportunities

The "need" and "sustainability" categories have been assigned the highest weights because they are considered essential to the long-term viability of the proposed activities. Another factor that impacts the long-term success of a proposed activity is its "economic impact"; an activity with a poor economic future is unlikely to be viable. "Environmental impact" is weighted relatively high because activities that adversely impact the environment are generally considered to be socially and politically unacceptable.

Identified Water Sector Activities

In its effort to foster cooperation among southern African nations, USAID desires to sponsor projects that cross international boundaries. Thirty activities are ranked in the assessment report. It is interesting to note that none of the activities are physical construction projects, although there are feasibility studies in preparation for design. Mostly the projects are environmental assessments, planning studies, and institutional and legal support. These projects, documented in the database and library, and discussed in meetings with water officials in the governments, donor community, and regional organizations, stand out as the most important regional activities at this time. Space is not available to discuss the project details.

Given the regional context of southern Africa, inefficient water uses cannot be resolved through a single government's policy. For example, it is not in the interest of an upstream riparian to increase or maintain the flow and quality of water for downstream riparians. The end result of the lack of regional agreements may be environmental, social, and economic losses to downstream riparians that outweigh the benefits to the upstream riparian.

There are more than a dozen river basins shared by two or more countries in the southern Africa region. Coordinated management of regional river basins has not previously been considered a priority plan of action for the region. Notwithstanding the treaties that have been signed by countries to deal with specific concerns about water resources, funds have not been significantly devoted to managing these water resources. Water disputes have arisen over quantity allocation and environmental degradation.

This assessment has indicated that there is both a need and a desire among the water sector entitles in the region to move forward with a more regional emphasis on water sector management. The lacking ingredients seem to be a shortage of technical capability and lack of financing required to establish and operate basin-wide management authorities. International donors and funding agencies could provide these missing ingredients.

Protecting Water Supplies Using Waste Management

Justin D. Mahon, Jr., Member[1]

Abstract

Watershed management is of renewed interest to water purveyors because of recent events and publicity. Purveyors use strategic watershed plans to avoid the costs of expensive treatment plant upgrades. Evaluating and taking advantage of existing waste management programs is an important component of such a plan. Several recent watershed projects illustrate this concept.

Introduction

Watershed management is of renewed interest to water purveyors because of recent events and publicity questioning the sufficiency of treatment plants alone to provide high quality water to consumers. There is renewed interest in the industry in the multiple barrier approach to providing safe drinking water. This thinking is exemplified by the AWWA mission statement on Total Water Management. As it becomes widely recognized that a systematic approach to water quality management at the watershed level is needed to manage our water resources efficiently into the twenty-first century, watershed management programs prepared by water purveyors increasingly recognize that purveyors' efforts must be integrated with a wide range of other environmental regulations and programs.

Purveyors can create strategic plans to achieve the cost avoidance benefits of watershed management in a cost-effective manner. Evaluating the effectiveness of existing waste management programs is an important component of such a plan. There are many local, state, and federal waste management programs which help protect watersheds. These programs become especially important where the water purveyor controls very little of the land in the watershed and lacks the legal authority to assert control.

---

[1]Associate, Malcolm Pirnie, Inc., One International Boulevard, Mahwah, NJ 07495

Maximizing the benefits of existing waste management programs requires a water purveyor to understand the programs, communicate the purveyor's needs to their sponsors, and to work together to implement mutually satisfactory solutions. For several purveyors which have recently undertaken watershed projects, understanding was achieved by making an inventory of programs and relating key provisions to the criteria that the water industry uses to measure watershed protection (such as overall imperviousness). Subsequent education and implementation approaches have varied with the circumstances. Smaller utilities have tended toward direct action with local governments, whereas larger utilities have tended to rely on internal programs.

Planning for Watershed Management

All watershed plans share several common elements: setting goals, evaluating water quality, evaluating existing programs, identifying needed improvements, and prioritizing the needed improvements. This paper focuses on the third step – evaluating existing programs. In particular, it focuses on those programs addressing waste management. These include wastewater discharge permit programs, solid waste management programs, agricultural waste management programs, and hazardous waste management programs. There are three steps to integrating these waste management programs into watershed management plans: understanding the programs, communicating with the agencies operating the programs, and cooperating with the programs (hopefully in a win-win scenario).

Understanding Waste Management

Its difficult for many water purveyor managers not only to grasp the extent of waste management programs, but to utilize them as well. Most of their experiences with these programs are as regulated entities. Therefore it is useful for a watershed management plan to include an inventory of waste management programs pertinent to the specific jurisdiction. The inventory should identify the nature of the programs, types of data which may be accessed by the water purveyor, citations for regulations governing the programs, and names and telephone numbers of contact persons. The inventory should also express the objectives of each program in terms of criteria pertinent to watershed protection goals.

A recent impact statement for New York City's proposed watershed regulations (New York City Department of Environmental Protection, November 1993), illustrates how to relate waste management programs to watershed management needs on both the state and federal level. Subsections of the impact statement address hazardous wastes, community wastewater facilities, septic systems, storm water, and solid waste facilities. Each subsection cites the regulatory reference, discusses the regulated activities, and identifies how New York City's proposed regulations would exceed existing state and federal requirements. Regulatory agency contacts are not provided, given the nature of the document.

Communications

It is unreasonable to expect that waste management programs will always be perfectly coordinated with watershed management needs. Therefore, as watershed management programs move from planning to implementation, dialogue between water purveyor managers and waste management managers is necessary. Several avenues of dialogue exist: formal comments at the time waste management regulations are proposed, participating in watershed oriented organizations along with representatives of waste management organizations, requests for data, consultation with waste management regulators, and presentations at conferences such as this one. Flexibility and a readiness to compromise are important attributes for successful dialogue. It also helps to be able to demonstrate water quality benefits in concrete terms.

Dialogue between water purveyors and wastewater dischargers is currently occurring in New Jersey's Passaic River basin. There are approximately a dozen major wastewater treatment plants within 20 miles upstream of two major water intakes. Water purveyors have historically fought for having all discharges satisfy potable water standards, yet the purveyors' position has become less realistic as standards are promulgated for more substances. However, there has been progress over time in reducing loadings of ammonia from wastewater discharges, with the great benefit of lowering water purveyor chemical costs and reducing chlorinated organics. There have also been instances of cooperation where water and wastewater managers have jointly approached regulators to explain exactly which discharge requirements are important to both parties. Things have not yet progressed to the ideal – upstream dischargers helping to fund the downstream water treatment facilities that remove contaminants that would otherwise be removed from each of the wastewater discharges at a substantially greater cost.

Cooperation

Dialogue leads to cooperation. In fact the previous example illustrates how dialogue can evolve into cooperation. There already exists a high inclination toward cooperation between water purveyors and waste program regulatory managers. Further cooperation will come from the water purveyors understanding exactly what a waste management program can and cannot accomplish, and furthermore, asking for help in a way that the regulators can understand.

Discharge permit programs, the federal National Pollution Discharge Elimination System (NPDES), and the derivative state permit systems, are perhaps the most widely applicable waste management programs for watershed managers. These programs regulate municipal and industrial discharges, combined sewer overflows and storm water, and even large discharges to groundwater. Watershed managers can obtain lists of dischargers to help identify and prioritize water quality concerns. Information can often be obtained on electronic media which facilitates its evaluation. Discharge permit programs set effluent limits for dischargers. Discharge limits are an item a watershed manager needs to be prepared to comment upon every time a permit is issued or renewed within the watershed. In the extreme water purveyors can use the citizen suite provisions contained in most permit programs to directly seek relief from unacceptable discharges. Only about 40 percent of the watershed management programs surveyed in 1988 were taking advantage of discharge permit programs (Robbins et al. 1991). However, they are a common element of several watershed management programs reviewed in Michigan, New York, New Jersey and Ohio since then.

Hazardous waste programs established pursuant to the Resource Conservation and Recovery Act, and corresponding state statutes regulate new hazardous waste facilities, the cleanup of old hazardous waste facilities, and the transportation of hazardous waste. Watershed managers can obtain a list of regulated facilities to help identify and prioritize water quality concerns. Hazardous waste programs provide ample opportunities to make concerns known. In fact, the most common complaint with these programs is that the cleanups proceed very slowly. These programs are very important in urbanized watersheds where purveyors cannot control land uses. Only 13 percent of the watershed management programs surveyed in 1988 were controlling hazardous materials, (Robbins et al., 1991). More recent evaluations of programs for specific watersheds like the one for New York City (New York City Department of Environmental Protection, November 1993), have placed a greater emphasis in maximizing the benefits of the existing programs.

The agricultural waste of greatest concern to watershed managers is manure. Manure is receiving increased attention because it may be a source of pathogens, such as *Cryptosporidium*. While management of wastewater dischargers and hazardous waste, is regulated, agricultural waste management programs are usually voluntary and rely on incentives to encourage participation. Many manure management projects throughout the nation have been reported. Most of them involved the sponsorship of the local soil conservation district or the state agricultural extension service. It is harder to incorporate manure management into watershed management because it is voluntary. There are neither the records nor the penalties that characterize other waste management programs. In fact, especially if there are only a few farmers in the watershed it may be best to approach them individually and appeal to their own concerns about the environment. This strategy has been recommended in recent plans for smaller watersheds such as the 4 square mile Mechanicville, New York, watershed in rural Saratoga County (Malcolm Pirnie, Inc., Nov. 1994).

Appendix 1 - References

Malcolm Pirnie, Inc. "Mechanicville Reservoir Watershed Protection", City of Mechanicville New York, Nov. 1994.

New York City Department of Environmental Protection, November 1993, "Final Generic Environmental Impact Statement for the Proposed Watershed Regulations for the Protection from Contamination, Degradation, and Pollution of the New York City Water Supply and Its Sources", Volume II, Nov. 1993.

Robbins, R.W., Glicker, J.L., Bloem, D.M., and Niss, B.M., "Effective Watershed Management for Surface Water Supplies", AWWA Research Foundation, Denver CO, 1991.

Integration of Water Supplies in New Jersey

Michael S. Bennett[1] (M-ASCE), Robert J. Ulrich[2]

Abstract

The New Jersey Department of Environmental Protection (NJDEP) is in the process of revising the 1982 New Jersey Statewide Water Supply Master Plan (SWSMP). As part of the revision a consultant team recently completed a 4-year study for the NJDEP in which a comprehensive data base water balance model (DB/WBM) was developed, an interactive river-aquifer system model (IRAS) was used, and alternative strategies for meeting water supply needs through the year 2040 were evaluated. Significant water supply challenges were identified and institutional, financial, and environmental issues related to selected water supply initiatives were analyzed. The study recommends adoption of an integrated resource planning approach to accomplish protection, enhancement, conservation, and, where appropriate, expansion of water supplies.

Introduction

Compared to the 1982 New Jersey Statewide Water Supply Master Plan, the study forecasts more modest estimates of future water supply needs (CH2M HILL, 1993, 1994). Capital projects developed as a result of the 1982 SWSMP, together with those in progress, will substantially meet water supply needs through 2010 and beyond. However, there are still significant water supply challenges to be met.

Several generic water supply alternatives were developed during this study. They include: protecting the quality and yield of existing facilities, improving and expanding system interconnections, implementing appropriate water conservation and reuse measures, improving management of depletive uses, and utilizing

---

(1) Project Manager, CH2M HILL, Inc., 99 Cherry Hill Road, Suite 304, Parsippany, NJ 07054-1102, (201) 316-9300, FAX (201) 334-5847

(2) Project Engineer, Elizabethtown Water Company, 1341 North Avenue, Plainfield, NJ 07062, (908) 654-1234, FAX (908) 756-0753

innovative technologies to strategically develop new sources. These alternatives have been expressed as statewide and planning-area-based initiatives. Institutional, financial, and environmental impacts analyses were configured to study the principles defining each initiative. The results will be useful for formulating future projects using a watershed-based approach.

Opportunities and constraints that should be considered as part of developing and implementing water supply initiatives based on basin-level integrated resource planning include: (1) evaluating current operating rules and statutory minimum flows within the context of changing watershed conditions, (2) effectively relating surface water and groundwater allocations to waste discharges to track depletion of stored reserves and impacts on safe yield, (3) tracing the impacts of land-use changes on runoff and recharge quantity and quality, (4) evaluating tradeoffs of land-use control and pollution control facility upgrades for meeting water quality objectives for receiving waterways, and (5) evaluating economic and water supply impacts of proposed future water supply projects.

The first step in developing water supply initiatives was to create a comprehensive data base of water supply, demand, and transfer information. The initiatives were then formulated to address water supply deficits and surpluses forecasted through the planning period. The suitability of each initiative was then analyzed using institutional, financial, and environmental assessments. In addition, a policy-testing simulation model was applied to the Raritan and Upper Passaic watersheds to test initiatives for meeting demands forecasted by the DB/WBM.

Data Base/Water Balance Model

The NJDEP has divided the state into 23 regional watersheds called regional water resources planning areas (RWRPAs). A data base was designed so that counties and regional watersheds could be analyzed independently. The DB/WBM was developed to total the current surface water and groundwater supplies and demands and to project future demands and deficits to 2040. The collection, organization, and use of the data formed the foundation of the study.

The State of New Jersey has 21 counties, 23 major drainage basins, 567 municipalities, 333 purveyors and 7 geologic provinces. In an area totaling 7,660 square miles and with a population of approximately 7.8 million, the estimated 1995 average purveyor-supplied water demand for the state is approximately 1.1 billion gallons per day. The NJDEP provided a list of 1,708 existing surface water and groundwater allocations, which was incorporated into the DB/WBM. The model allows a number of water supply scenarios to be evaluated by changing the projection criteria and incorporates flexibility so that data can be extrapolated in a variety of ways.

The DB/WBM was developed to enable the NJDEP to routinely update the demand and water availability data as projections are replaced by operational information. It is also a useful tool for evaluating specific water supply actions identified in the study and assisting in future planning. As specific new water supply projects are implemented to address identified needs, and/or projected data are replaced by actual data, the DB/WBM can be updated and used to project and evaluate future water supply issues.

Water balances utilized were on an RWRPA basis. Analysis of the RWRPA-based deficits/surpluses led to grouping certain planning areas for analyses. However, continuous monitoring of demand and water availability should be conducted for the individual planning areas and the portions of planning areas in which special water supply concerns have been identified. The study recommends improving and extending the data base and computational tools so that more detailed analyses can be performed in the future.

Institutional Analysis

The ongoing process of developing projects that respect the needs of integrated planning, and that will implement the recommended initiatives, will require new tools and revised institutions. The institutions should continue to become the means to effective water supply development; evolving beyond site-by-site, jurisdiction-by-jurisdiction evaluations of constraints. Careful financial planning requires improved characterization and accommodation of uncertainty in water supply planning and forecasting.

Cooperation among purveyors and with the officials responsible for wastewater collection/discharge is a central premise to performing RWRPA-based analyses, and is necessary in order to implement integrated resources management. Local and county government officials who are responsible for making land-use decisions will also play a role. In addition, cooperation and coordination within the NJDEP and with other State departments will be required because management of regional water supplies will inevitably be achieved through a watershed planning and management approach. Conservation is vital to water supply planning, although it should be appropriate to the individual purveyor's financial needs and responsive to the availability of water to purveyors on a regional basis.

The study recommended that the NJDEP take the lead in setting the context and developing the data, the tools, and the organization for implementing the watershed-based approach to management of water resources (including wastewater). Within this context, county, local, and utility interests will develop institutions necessary to support their projects and meet watershed-based objectives. This includes initiatives for water resources protection, conservation, reuse, and planning. Recommendations for enhancement of the water allocations approval process include: discouraging depletive uses; encouraging reuse and aggressive water conservation; requiring analyses for conjunctive use and innovative technologies; reviewing methodologies and, where necessary, recalculating safe yields of surface water sources; and avoiding conflict with the State Development and Redevelopment Plan.

Coordination of the development of the watershed-based approach is an important part of the effort to extend and protect water supplies. In doing so, the many programs, regulations, and policies that have been implemented, at times independently, will become linked. For example, the New Jersey Water Quality Planning Act, through its requirement for development of Water Quality Management Plans, provides a regulatory framework for linking the New Jersey Pollutant Discharge Elimination System and water allocations programs. These institutional approaches could help to evaluate difficult situations, are valuable as tools in public processes, and can allow for unique solutions that may have been questionable on a regulation-by-regulation basis.

Financial Analysis

Currently, the State is faced with new project financing challenges. Although initiatives have been identified for various RWRPAs throughout the State, the available sources of funding are not as plentiful as they were in the early 1980s. Local government and private water purveyor financing are available through mechanisms such as local municipal borrowing and Economic Development Authority loans, but there are limits to the availability of funding through these sources. Although funds from the 1981 Water Supply Bond Act are available for appropriation, it is not certain that the second cycle of projects' financial needs can be met by the currently unappropriated funds and anticipated principal repayments. Consequently, there is an opportunity for new direction in the State's activist role in planning for and assuring implementation of needed water supply facilities throughout the State.

Environmental Analysis

In certain instances, the aggregate of environmental regulations, combined with the numerous other rules that affect water supply projects, may severely limit or distort implementation, to the point where less-than-optimal solutions would result. Given sufficient levels of study to fully demonstrate needs, address concerns, and develop necessary mitigation measures, regulations can be streamlined to ultimately allow effective implementation. Integration of water supplies in New Jersey may require this type of approach.

Conjunctive use can be environmentally sound, but must be based on quantitative analyses to avoid double-counting of water resources. Enhanced interconnections and additional surface water withdrawals are faced with regulatory constraints associated with depletive-use restrictions and passing-flow requirements. These issues must also be resolved using sound quantitative procedures. Reservoir construction has a range of potential environmental impacts and may be the most difficult to implement, although several projects are under development in New Jersey or have been extensively studied. Additional groundwater withdrawals are possible in certain areas, but will require additional interconnections, with associated impacts and limitations, and their implementation will be dependent on inter- or intra-basin cooperation. Aquifer storage and recovery generally has benign environmental impacts, although appropriate planning and design studies are necessary. Reuse of treated water will require public education and approval.

Interactive River-Aquifer System Model (IRAS)

In order to test various scenarios for meeting demands forecasted by the DB/WBM, IRAS was adapted to the Raritan and Upper Passaic river basins, 2 of the 23 RWRPAs. A primary task of the IRAS modeling study was to graphically illustrate water resource relationships in a simple and interactive way. This approach was shown to be particularly suited to encouraging participation by decision makers early in project planning, supporting risk-based decision making, engaging the public, and developing effective policies (Bennett, 1994).

Policies tested by using water balance models, such as IRAS, will allow affected purveyors, property owners, jurisdictions, and other stakeholders to observe the net benefits of basin-level management. Subsequently, detailed studies should be conducted in regions where modeling reveals critical problems. The application of IRAS to the Raritan and Passaic basins was illustrative and has demonstrated the value of conceptually simple, but flexible and easy to use, simulation tools.

Summary

The master plan study resulted in a document that presents institutional mechanisms, financial opportunities, and environmental considerations necessary for implementation of selected water supply initiatives. The analysis of the statewide and RWRPA initiatives resulted in confirmation of the value of integrated management of water resources on a watershed basis.

References

CH2M HILL, Inc., Metcalf & Eddy, Inc., New Jersey First, Inc., *New Jersey Statewide Water Supply Master Plan, Task 3 Report: Development and Projection of Water Demands and Comparison to Net Available Water*, Parsippany, NJ, May 1993.

CH2M HILL, Inc., Metcalf & Eddy, Inc., New Jersey First, Inc., *New Jersey Statewide Water Supply Master Plan, Task 5 Report: Institutional, Financial and Environmental Impact Analyses*, Parsippany, NJ, April 1994.

Bennett, M.S., M.R. Taylor, C. Wan, D.P. Loucks, *Simulating Water Supply Systems of the Raritan Basin*, ASCE WRPM Division, Denver CO, May 1994.

Travel Times in Water Distribution Systems

Lin Wu[1], Steven G. Buchberger[2], Greg J. Wells[3]

Abstract

Under the assumption that residential water demands occur as a Poisson rectangular pulse process, expressions are presented for the cumulative distribution function of travel times along dead-end stems in a municipal water supply system. An example illustrates application of the travel time concept.

Introduction

Water quality in a municipal distribution system can deteriorate between the points of treatment and consumption (LeChevallier, et al., 1987). The Safe Drinking Water Act will soon force utilities to meet drinking water standards at the consumer's tap rather than at the treatment plant. It is likely that network models, now indispensable in sizing and operating a distribution system to satisfy conventional hydraulic criteria, will become essential tools for predicting the system's ability to comply with evolving water quality standards. Modeling quality in water supply systems is complicated because flows fluctuate over time and across space in response to variable demands imposed by consumers dispersed throughout the network.

---

[1] Graduate Research Assistant, Department of Civil and Environmental Engineering, PO Box 210071, University of Cincinnati, Cincinnati, Ohio 45221-0071.

[2] Assistant Professor, Department of Civil and Environmental Engineering, PO Box 210071, University of Cincinnati, Cincinnati, Ohio 45221-0071.

[3] Engineer, Willis Engineering, Charlotte, North Carolina; Formerly Graduate Research Assistant, Department of Civil and Environmental Engineering, University of Cincinnati, Cincinnati, Ohio 45221-0071.

From a practical point of view, the randomness of individual water demands can probably be safely ignored in the main arteries of the distribution system. Here the aggregate demand from many downstream consumers is usually sufficient to maintain high velocity flows. Travel times through mainlines tend to be short. In peripheral regions, however, random demands of individual consumers measurably influence the local flow regime (Buchberger and Wu, 1995). For this case, it may be important to have fine temporal and spatial resolution of the local hydraulics in order to accurately predict the fate and transport of substances traveling to remote points of consumption. In this paper, we present a stochastic model of residential water demand and use this model to estimate the probability distribution of travel times through dead ends of a municipal distribution system.

Travel Time Distribution

Water use at a typical residence is an intermittent process. Wells (1994) has shown that residential water demands can be modeled as rectangular pulses. In what follows, we assume that water use pulses at a single residence originate according to a Poisson process with rate λ_1. When a pulse occurs, it has a random volume X_k. The total volume of water use at the single residence up to time t $\{V_1(t), t \geq 0\}$ can be described by a compound Poisson process (Ross, 1993)

$$V_1(t) = \sum_{k=1}^{N_1(t)} X_k; \qquad t \geq 0 \tag{1}$$

where $\{N_1(t), t \geq 0\}$ is the random number of pulses generated by the residence up to time t. The probability mass function of $N_1(t)$ is given by

$$P[N_1(t) = k] = \frac{(\lambda_1 t)^k e^{-\lambda_1 t}}{k!}; \qquad k = 0,1,\ldots \tag{2}$$

Pulse volumes X_k are assumed to be independent and identically distributed, as well as independent of $\{N_1(t), t \geq 0\}$. The cumulative distribution function (cdf) of total water demand $V_1(t)$ at a single residence can be obtained by conditioning on the number of pulses that have occurred until time t

$$P[V_1(t) \leq v] = \sum_{k=0}^{\infty} P[N_1(t) = k] \cdot P[X_1 + \ldots + X_k \leq v] \tag{3}$$

For multiple residences, it is assumed that pulses at each home follow a Poisson process and that water demands at all residences are mutually independent. From the additive property of the Poisson distribution, the total number of pulses produced by a block of m residences $N_m(t)$ is also Poisson with rate λ_m given by

$$\lambda_m = \sum_{i=1}^{m} \lambda_i \tag{4}$$

where λ_i is the pulse arrival rate at residence i in the block of m residences. Pulse volumes at different residences may have different characteristics. As a first approximation, for a given number of block pulses, moments of total water volume consumed across the entire block are estimated from a weighted average of pulse moments from the individual homes. Then the cdf of total water demand from a block of m residences $V_m(t)$ is calculated with (3) where $N_m(t)$ is used in place of $N_1(t)$ and the pulse moments represent demands from the entire neighborhood block.

Having the cdf of residential water demands, now the cdf of water travel time through a dead-end pipe can be found. Imagine that at time t=0, a parcel of water enters the upstream end of link m shown in Figure 1. The parcel is pulled deeper into link m whenever downstream residences exert water demands. The travel time of this parcel through link m corresponds to the time at which the downstream water demands sum to the link volume. Hence, the cdf of travel time through link m, T_m, can be expressed in terms of the cdf of $V_m(t)$

$$P[T_m \le t] = 1 - P[V_m(t) \le L_m] \tag{5}$$

where L_m is the volume of link m. Equation (5) implies that the travel time through link m must exceed t if the downstream demand at time t is less than the link volume. Clearly water demands diminish and travel times increase as the parcel moves deeper into the dead-end branch.

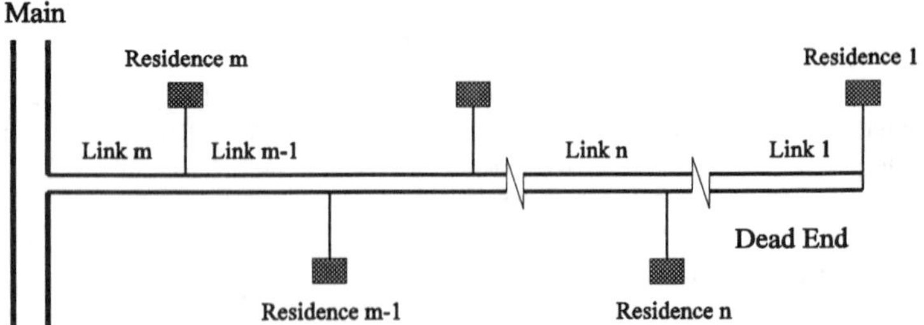

Figure 1. Dead-end branch is modeled as a series of individual links.

The travel time between any two residences m and n is the sum of travel times through all links between these two homes

$$T_{m,n} = T_{m-1} + T_{m-2} + \cdots + T_n \tag{6}$$

Here m is the upstream home and n is the downstream home and $T_{m,n}$ represents the travel time between the two. The travel time distribution for links m-1 and m-2 are calculated first. By assuming T_{m-1} and T_{m-2} are independent, the cdf of the travel time between residences m and m-2 can be written as a convolution integral

$$P[T_{m,m-2} \leq t] = \int_0^\infty P[T_{m-2} \leq t-\tau] f_{T_{m-1}}(\tau)\, d\tau \tag{7}$$

where $f_{T_{m-1}}(t)$ is the probability density function of travel time through link m-1. Once the cdf of travel time between residences m and m-2 is found from (7), it is used to get the cdf of travel time between residences m and m-3. The above step is repeated until residence n is reached. In such a way, the cdf of travel time between any two residences in the deadend can be obtained.

Example Application

These concepts are illustrated with a hypothetical deadend composed of five links each of length 24.4 m (80 ft) and inside diameter 15 cm (6 in). Pulse volumes at each of the five residences are assumed to be exponentially distributed. Table 1 summarizes selected input parameters (arrival rates and mean demands) and computed travel times (mean and standard deviation) for each link. Figure 2 shows the cdf of travel times through link 1, link 5, and the entire dead-end.

Deadend Link	Residence Values		Link Values		Travel Times	
	arrival rate (hr^{-1})	mean demand (L)	arrival rate (hr^{-1})	mean demand (L)	mean (hr)	std dev (hr)
1	2.94	10.0	2.94	10.0	15.47	3.23
2	3.90	8.5	6.84	9.14	7.26	1.45
3	3.42	11.5	10.3	9.93	4.46	0.93
4	3.54	10.5	13.8	10.1	3.27	0.68
5	3.60	7.0	17.4	9.44	2.77	0.56
1-5	--	--	--	--	33.23	3.77

Table 1. Travel times through hypothetical deadend branch with five homes.

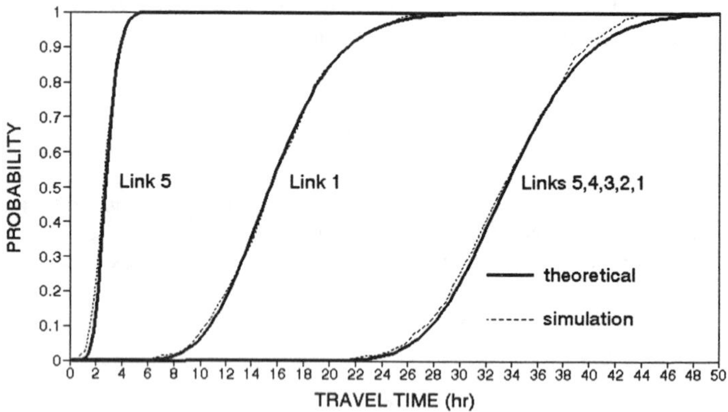

Figure 2. Cumulative distribution function of travel time through dead-end links.

Theoretical results are numerical solution of the travel time cdf given in (7); simulation results are estimates from a program which tracks water movement through a deadend subject to residential demands that occur as a Poisson process. The mean travel time is less than 3 hours through link 5 at the head of the deadend. In contrast, the mean travel time exceeds 15 hours through link 1 at the terminal end of the branch. Travel times along the entire deadend stem often exceed one day.

Conclusions

A simple example illustrates that water travel times through dead-end stems can be quite long and highly variable even under normal operating conditions. Methods presented here can be coupled with water quality models to predict the probability distribution of disinfectant concentration at various points of consumption along dead-end stems of the water supply network.

References

Buchberger, S.G., and L. Wu, (1995). "A model for instantaneous residential water demands." *J. of Hydraulic Engineering*, ASCE, 121(3):232-246.

LeChevallier, M.W., T.M. Babcock, and R.G. Lee (1987) "Examination and characterization of distribution system biofilms", *Applied & Environmental Microbiology*, 53(12):2714-2724.

Ross, S.M. (1993). *Introduction to Probability Models*, Academic Press, Inc., Orlando, Florida.

Wells, G.J. (1994) *Statistical Characteristics of Residential Water Demands*, MS Thesis, University of Cincinnati, Cincinnati, Ohio, 82 pages.

Effects of Karst Environments on Streamflow in South Georgia

Reggina Garza[1]

Abstract

This paper presents an overview of the hydrologic responses and interaction between the Flint River and the Upper Floridan aquifer during the flood produced by Tropical Storm Alberto in July 1994. There also is a discussion on the incorporation of this process as part of the National Weather Service River Forecast System model (NWSRFS) used by the Southeast River Forecast Center (SERFC).

The flood of July 1994 was the largest ever recorded in southwestern Georgia. Analyses of streamflow and ground-water-level data suggest that part of the streamflow measured at Albany moved as groundwater, past Newton.

1.Introduction

Karstification involves multiple geologic processes that dissolve rocks, thus forming complex morphological features, increased porosity, and altered hydrologic environment. Sinkholes are common in karst terrain but solution openings also might be formed in soluble rocks producing preferential flow paths.

Streamflow forecasting requires an accurate representation of hydrologic processes. Where streams incise an aquifer, knowledge of the interaction between the aquifer and the stream is essential to adequately address flow changes (gain or loss) in a channel reach. Although flow in a homogeneous aquifer can be reasonably evaluated, in karst terrain the development of secondary porosity adds complexity to the system. Karst terrain is predominant in south Georgia and adjacent parts of Florida. Sinkholes, solution openings, and fractures in the aquifer are well

----

[1] Hydrologist, National Weather Service, Southeast River Forecast Center 4 Falcon Drive, Peachtree City, Georgia, 30269

documented in south Georgia, particularly in the Albany (Hicks and others, 1987) and Valdosta (Krause, 1979) areas.

The Flint River is the principal surface-water drain in the Albany-Newton area (fig. 1) and incises the Upper Floridan aquifer. The Upper Floridan aquifer, composed of the Ocala Limestone is typical of a karst area. Where the aquifer is homogeneous, ground-water flow is laminar. However, because of the presence of cavities, sinkholes, and solution openings, ground-water flow can be turbulent.

Knowledge of stream-aquifer relations in a karst terrain is important to understanding gains or losses in streams. The incorporation of such relations could improve forecasts of streamflow, not only for floods, but also for low-flow conditions.

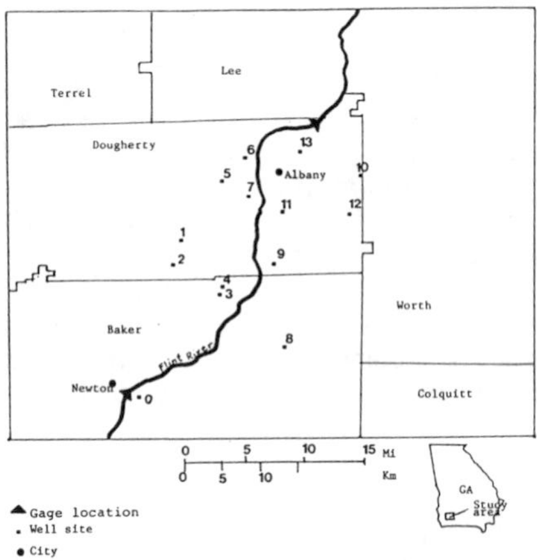

Figure 1. Study area showing wells and stream-gage locations

2.Hydrologic effects during the flood of 1994

Rainfall is the primary source of recharge to the Upper Floridan aquifer in the Albany area and its effect on ground-water levels may be significant. Most of the rainfall occurs from December through March, and the aquifer receives maximum recharge then.

During the flood of July 1994, many of the streams in the Flint River Basin, including the Flint River at Albany and at Newton, experienced record floods (table 1). Flow-mass curves at Albany and Newton (fig. 2), for July-August 1994, suggest that the volume of water that passed through Albany was greater than that at Newton. Even 30 days after peak streamflow had occurred at Albany, the volume of water that had flowed past Newton was less than that through Albany during the same period. It is hypothesized that stream-aquifer relations at flood flows might be anomalous because of the occurrence of karst terrain.

Table 1. Peak stage of Flint River at Albany and Newton, in meters

Station	Flood Stage	January 1925 Peak Stage	Day	July 1994 Peak Stage	Day
Albany	6.1	11.5	21	13.1	11
Newton	7.3	12.6	21	13.8	13

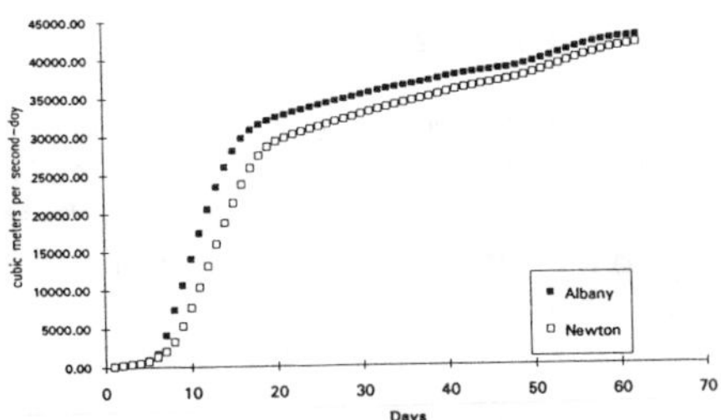

Figure 2. Flow-mass curve for Albany and Newton, July-August 1994

The hypothesis of flood flow through karst features was formulated after examining ground-water-level data from several wells (fig. 1) plus river stages at Albany and Newton. Most of the ground-water-level data analyzed for the flood period indicate that ground-water levels peaked later than the Flint River and that the period that they remained high, varied from well to well. Because of the characteristics of the karst terrain in the area, differing altitude of water-bearing zones in wells, and head gradients between the aquifer and the Flint River, ground-water-level response varied from comparatively fast to comparatively slow. The quick response suggests that in a flood of this magnitude, in addition to the effects of direct recharge from rainfall, sediment in sinkholes, solution openings, and cavities may be flushed out by the flood waters. If this occurs, the volume of water loss from the river to the aquifer may increase, thereby decreasing streamflow. In addition, streamflow hydrographs for the Flint River at Albany and at Newton, show a greater volume of water going into aquifer storage than that returning from storage, for the period analyzed. Therefore, this flood water either continued flowing as groundwater and diverted in various directions, or remained as ground-water storage for a longer period. Further analysis of these data would be needed to determine the validity of this hypothesis on the movement of the flow wave.

3. Stream-aquifer relations for period of record.

To determine the relation between streamflow at Albany and at Newton under other flow conditions, data were plotted for a variety of discharges. Flow-mass curves were analyzed for a range of peak discharges at Albany between 45 (low-flow conditions) and 3400 cubic meters per second (flood of July 1994). Results indicate that during low-flow conditions, the average streamflow for the period analyzed is higher at Newton than at Albany. This was expected considering the connection between the river and the aquifer (Hicks and others, 1987) and the contribution of tributary streams. For the high flow conditions examined, the volume of water passing through Newton equalled and later exceeded that of Albany, except for July 1994.

4. Streamflow forecast in karst terrain.

The National Weather Service River Forecast System (NWSRFS) is a comprehensive set of hydrologic techniques used by the National Weather Service River Forecast Centers to perform streamflow forecasting. This is a flexible system, governed by operations that constitute the components of the simulation. This model is used for operational purposes and run on a daily basis.

The incorporation of an operation to account for gain or losses in a channel reach has long been discussed by hydrologists. Presently there exists an operation, CHANLOSS, designed to account for losses or gains of water through the channel bottom and evaporation from the stream surface. This operation is very useful when the losses or gains are constant. However, in the case of karst terrain, where losses vary with time and space, the operation may need modification.

The SERFC is in the process of developing an operation similar to CHANLOSS, but that would modify flows at a flow-point by adjusting the instantaneous discharges for the losses, then route the streamflow. The ideal adjustment to the discharge would be done through a loss to the aquifer that would depend on the degree of saturation of the conduits in the karst terrain, and the head difference between the river and the aquifer. Both of these factors are difficult to determine; thus, the loss might have to be determined based on historical records and computations of change in storage.

5. Summary and Conclusions

The flood of July 1994 in its area of occurrence in Georgia was the largest ever recorded. Such large volumes of water could have had subsurface effects not observable nor occurring under normal conditions.

The hypothesis, after analyzing streamflow and ground-water-level data in the Albany-Newton area, Georgia, is that part of the volume of water that passed through Albany moved downstream as groundwater, by-passing the Newton gaging station. The aquifer acted as a reservoir and some of the flow might have slowly returned to the river later than the period analyzed or even downstream of Newton. If this had occurred, then a flatter peak and one extended in time would be experienced downstream of Newton. Further analysis of the discharges, in conjunction with data on ground-water levels and precipitation, could be used to validate the hypothesis of water loss during the flood of 1994.

After the flood of 1994, sinkholes have developed in the flooded area and their effect on streamflow are yet to be evaluated. Existing models and others being developed incorporate groundwater and streamflow. These might become the tools to quantitatively determine the effect of karst terrain on streamflow. If it is determined that the effects of karst terrain on streamflow are significant, these need to be accounted for in the forecasting model. In such case, the inclusion of a karst operation in the NWSRFS would be useful.

Acknowledgements

The author thanks Richard E. Krause, U. S. Geological Survey, for his technical input and review of this manuscript, and Charles Joiner and Roger McFarlane, U.S. Geological Survey, for providing data. Thanks are also extended to the SERFC for their support during the development of this paper.

References

1. Hicks D.W., Gill, H.E., and Longsworth, S. A., 1987, Hydrogeology, chemical quality, and availability of ground-water in the Upper Floridan aquifer, Albany area, Georgia. U.S. Geological Survey WRIR 87-4145

2. Krause, R.E., 1979, Geohydrology of Brooks, Lowndes, and Western Echols Counties, Georgia. WRI 78-117

3. U.S. Geological Survey Water-Data Report. (Georgia, various years)

Water Management Aspects of Southeastern Flood of 1994

Memphis Vaughan, Jr.[1]

Abstract

Tropical Storm Alberto brought extensive flooding to areas of Georgia, Florida and Alabama as it moved inland from its landfall in northwest Florida on July 4, 1994. Heavy convective rains of 250 mm (10 in) or greater fell across a large area of central Georgia and southeastern Alabama as the storm moved northward into central Georgia, stalled, and moved southward back toward southeastern Alabama and northwestern Florida, before finally breaking up. Rainfall amounts exceeding 500 mm (20 in) fell in parts of the Flint, Ocmulgee, and Choctawhatchee Rivers in Georgia and Alabama. Record or near record flooding occurred at numerous sites along these rivers. This event was difficult to predict due to the fact that it had been many years since any comparable floods had occurred, the time of year of the occurrence, and the distribution of the rainfall. This paper will describe the actions taken to forecast the flood and manage the floodfight and some aspects of the hydrometeorological setting of the flood. Water control actions will be discussed and how these actions affected flood levels.

Introduction

Tropical Storm Alberto formed in the southeastern Gulf of Mexico on June 30, 1994 between the Yucatan Peninsula and the western tip of Cuba. The storm drifted westward for the first 18 hours and then began a more northwestward course until July 2 when it began to turn more northerly. Alberto neared hurricane strength as it made landfall near Fort Walton Beach, Florida on the morning of Sunday, July 3. A maximum wind gust of 127 km/h (79 mph) was reported but the maximum sustained winds only reached 96 km/h (60 mph).

----

[1]Hydraulic Engineer, Water Management Section, U.S. Army Corps of Engineers, Mobile District, P.O. Box 2288, Mobile, AL 36628

Upon reaching land the storm moved slowly north-northeastward over the next several days. Although Alberto was a fairly weak tropical storm, it proved to be a greater than normal producer of rain. As it drifted toward the northeast, it produced from 80 to 200 mm (3 to 8 in) of rain over an area ranging from the central Florida panhandle to just south of Atlanta, Georgia on July 4.

Lakes at the Corps' projects on the Apalachicola-Chattahoochee-Flint (ACF) River System, which were the focus of much of the flooding, were typically near their summer levels and were being operated to maintain full pool elevations as much as possible due to a drier than normal May and a near normal June. Antecedent conditions in the basin were normal in most areas prior to the storm. Soils in the area were not completely saturated and some of the rivers were only carrying their base flow.

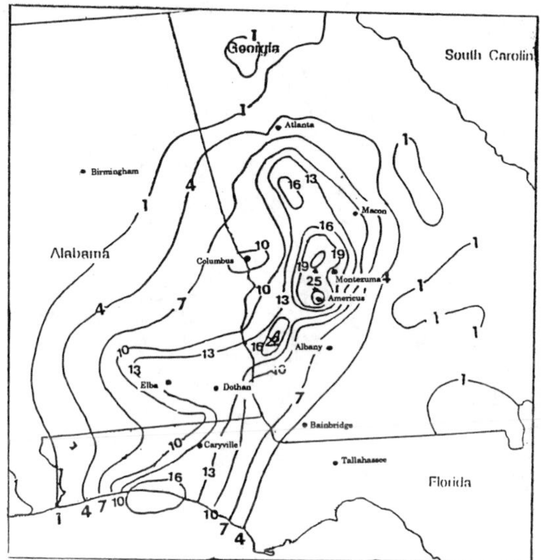

Figure 1. 5-Day Storm Total Precipitation 7/2/94 to 7/7/94 (in inches).

Tropical Storm Alberto produced rainfall amounts that resulted in one of the largest floods in Alabama, northwest Florida and Georgia. Figure 1 shows the storm totals for the period of July 2-7, 1994. By July 5, the storm system had moved from the central Florida panhandle northeasterly towards central Georgia and stalled in an area just south of the Atlanta airport. The system began to drift back to the south and produced the heaviest rainfall over the Flint River Basin around the city of Americus, Georgia. Americus received over 530 mm (21 in) of rainfall in a 24-hour period ending on July 6 and sustained heavy flooding. The nearby town of Plains, Georgia received over 400 mm (16 in) in the same period. Flooding occurred on the Ocmulgee and Altamaha River Basins in central Georgia as well as on the Flint River Basin. Montezuma, Albany, and Newton, Georgia, along the Flint River, received heavy flooding and all set new record flood levels. Figure 2 shows the location of these cities. The city of Bainbridge also experienced flooding but did not surpass its record flood stage set in 1929. Levees were overtopped at Albany and Montezuma. The Crisp County Dam located near Cordele, Georgia was breached but it did not cause any significant impacts since the headwater and tailwater were both almost at an equal level at the time.

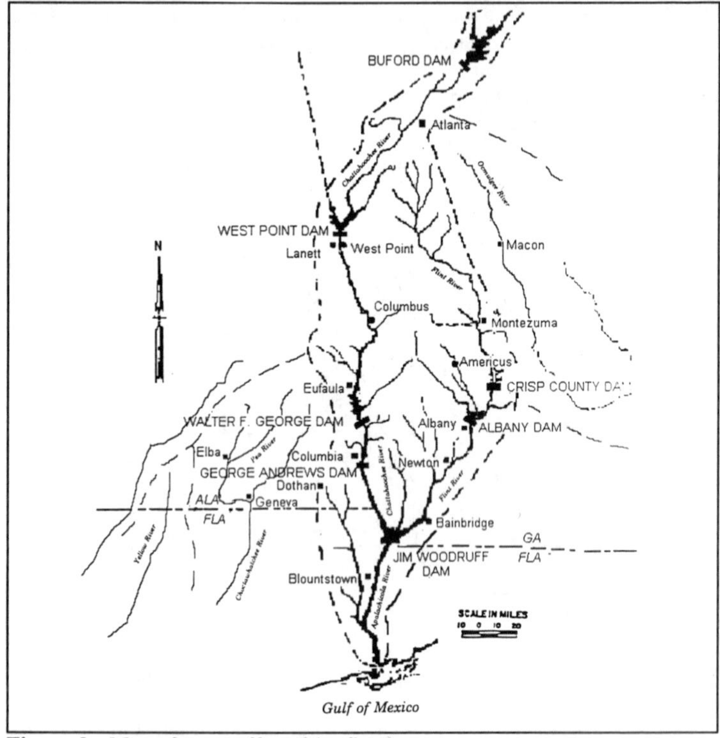

Figure 2. Map of areas affected by flood.

The storm system continued its southwesterly trek back across many of the areas that it had originally passed. By July 7, it was raining heavily over southeastern Alabama and produced as much as 230 mm (9 in) of precipitation in some areas of the Pea, Choctawhatchee, Yellow, and lower Chattahoochee River Basins, causing flooding in areas such as Geneva, Elba, and Columbia, Alabama and Caryville, Florida. By July 8, only a remnant of the storm system of Alberto remained over central Alabama.

The basins above the Corps of Engineers projects at Lakes Lanier or Allatoona did not receive significant rainfall and were not impacted by the tropical system. Releases from both projects were reduced for a few days to avoid adding to high river stages downstream. However, the reservoirs at West Point, Walter F. George, George W. Andrews and Jim Woodruff Dams all experienced significant increases in their lake levels. See Figure 2 for the location of these projects.

West Point Dam, which is authorized to operate for flood control, prevented significant flooding in the downstream towns of West Point, Georgia and Lanett and Valley, Alabama. West Point Dam reduced the peak flood wave by approximately 40% as the flood wave moved downstream. Walter F. George, although not a flood control project, had some reduction in the peak flood wave as it moved through the project. Releases from George during the July 1994 Flood was significantly lower than the releases made during the March 1990 Flood, yet due to the heavier rainfall that occurred between George and Andrews Dams in July, the elevation at Andrews surpassed the record set in 1990 and established a new record elevation. Flooding occurred in the Andrews Dam area, inundating the lock and dam and parts of the town of Columbia, Alabama. The lock operators at Andrews were forced to evacuate the lock and river level readings became unavailable for that segment of the river. However, communication between the Corps and the Alabama Power's Farley Nuclear Plant, several miles downstream from the lock, provided hourly river level readings at Farley that were correlated to the Andrews Lock elevations.

Jim Woodruff Dam, which is located at the confluence of the Chattahoochee and Flint Rivers, is a run-of-the-river dam and has very limited storage capacity. Flows from the Flint and Chattahoochee combined for record inflows into Woodruff. All of its functioning gates were opened by July 9 and the lake seemed poised to overtop the auxiliary fixed crest spillway for the first time in its history. However, it reached a maximum level of 23.81 m (78.05 ft) on July 10, which was 0.29 m (0.95 ft) below the fixed crest elevation and lower than the record pool level set in April 1960. The tailwater set a new record of 23.27 m (76.29 ft) on July 10. Records were also set on average daily inflow and average daily outflow of approximately 6,524 cms (233,000 cfs) and 6,272 cms (224,000 cfs) respectively.

One of the primary concerns during the flood was the peak level of flooding at various cities, towns, and points along the rivers. Since the heaviest and most damaging flooding occurred on the Flint River, most of the concern centered around the flood crests at Montezuma, Albany, Newton, and Bainbridge. The official river stage forecasts for the region is provided by the Southeast River Forecast Center of the National Weather Service near Atlanta, Georgia. The Corps obtains forecast information via the NWS AFOS computer and is disseminated to various entities such as the Emergency Management Offices of the affected states and counties. The Water Management Office received hundreds of calls from the public during the floods requesting stage and flood crest information and the NWS information was provided to the Corps' Public Affairs Office for distribution to the news media in the affected areas.

The river crested on the upper Flint River as early as July 6. As the flood wave progressed downriver and each town surpassed its record stage, there was increased anxiety at Bainbridge, Georgia and other downstream points concerning the river stage forecast that was significantly higher than the record stage. On July 14, the river crested at Bainbridge and due to a number of factors, it did not exceed its record stage.

The residents downstream of Jim Woodruff Dam, especially in the Blountstown area, were also extremely concerned about the forecasts on the Flint River. They felt that a huge flood wave would eventually reach their area in addition to the water that was approaching record levels at Blountstown. Fortunately, because of the slower travel time on the Flint River and the fact that the Chattahoochee part of the ACF basin did not receive the same magnitude of flows as the Flint, Blountstown's stage peaked on July 10 about four days before the Flint peaked at Bainbridge. Flooding occurred in the Blountstown area but the river did not surpass its record stage set in 1929. The Apalachicola River remained out of its banks from July 6 to August 4.

The flooding caused the ACF navigation channels to become unusable due to insufficient clearance under several bridges as a result of the high river stages. At Woodruff Dam, the generators were shutdown from July 6 to July 20 due to insufficient head to produce hydroelectric energy.

Overall, the Corps' flood management efforts were handled very effectively with close coordination among key participants during the storm event. Information exchange between the responsible agencies such as the Corps, the NWS, the state and local emergency management offices and FEMA was performed in a timely and professional manner. It is through the unified efforts of all involved agencies and personnel that the Corps of Engineers can continue to perform a major role in managing flood events of this magnitude.

Flooding in Southeastern United States from Tropical Storm Alberto, July 1994

Timothy W. Hale[1/] and Timothy C. Stamey[2/]

Introduction

In July 1994, parts of central and southwestern Georgia, southeastern Alabama, and the western panhandle of Florida were devastated by floods resulting from rainfall produced by Tropical Storm Alberto. President Clinton declared 78 counties as Federal disaster areas: 55 in Georgia, 10 in Alabama, and 13 in Florida.

Whole communities were inundated by floodwaters as numerous streams reached peak stages and discharges far higher than previous floods in the Flint, Ocmulgee, and Choctawhatchee River basins. The flooding resulted in 33 human deaths in towns and small communities along or near swollen streams (Federal Emergency Management Agency (FEMA), 1994a,b,c). The towns of Montezuma and Newton, Ga., were almost entirely encompassed by floodwaters from the Flint River. In Macon, Ga., the municipal water-system operations were flooded, leaving about 150,000 people without a water supply for three weeks (FEMA, 1994a).

Travel was disrupted as railroad and highway bridges and culverts were overtopped and, in many cases, washed out. Numerous small earthen dams failed after being overtopped and water from small recreational lakes and farm ponds quickly emptied into local streams. Sinkholes formed in the Albany, Ga., area, that is underlain by cavernous limestone formations. Total flood damages to public and private property were estimated at nearly $1 billion dollars (FEMA, 1994a,b,c).

This paper highlights severe stream flooding resulting from Alberto and describes U.S. Geological Survey (USGS) efforts to document this event and provide information to those who need it.

Tropical Storm Alberto

Tropical Storm Alberto grew from a tropical depression which formed in the Gulf of Mexico on June 30, 1994. Alberto first came over land on the morning of July 3, near Fort Walton Beach, Fla., rapidly lost energy, and was downgraded to a tropical depression by mid-afternoon of the same day. Remnants of Alberto drifted north to near southwest Atlanta early on July 5, changed course, and moved in a southwesterly direction before dissipating about two days later. Slow movement of Alberto and abundant tropical moisture combined to produce historic rainfalls.

[1/]Supervisory Hydrologist, U.S. Geological Survey, 3039 Amwiler Road, Suite 130, Peachtree Business Center, Atlanta, GA 30360-2824.

[2/]Hydrologist, U.S. Geological Survey, Atlanta, GA.

Storm-rainfall totals of more than 13 inches (in.) commonly were recorded in the areas of the heaviest rainfall throughout central and southwestern Georgia, southeastern Alabama, and the western Florida panhandle (fig. 1). The highest total rainfall of 27.6 in. (July 3-7) and the highest 24-hour rainfall of 21.1 in. (24-hour period ending at 7 a.m. on July 6) were recorded in Americus, Ga. The 24-hour rainfall was nearly 2.5 times greater than the area's estimated 100-year recurrence interval for 24-hour rainfall (U.S. Department of Commerce, 1961). The maximum 5-day total rainfall recorded in Alabama was 15.0 in. at Elba (U.S. Department of Commerce, National Weather Service, written commun., November 1994).

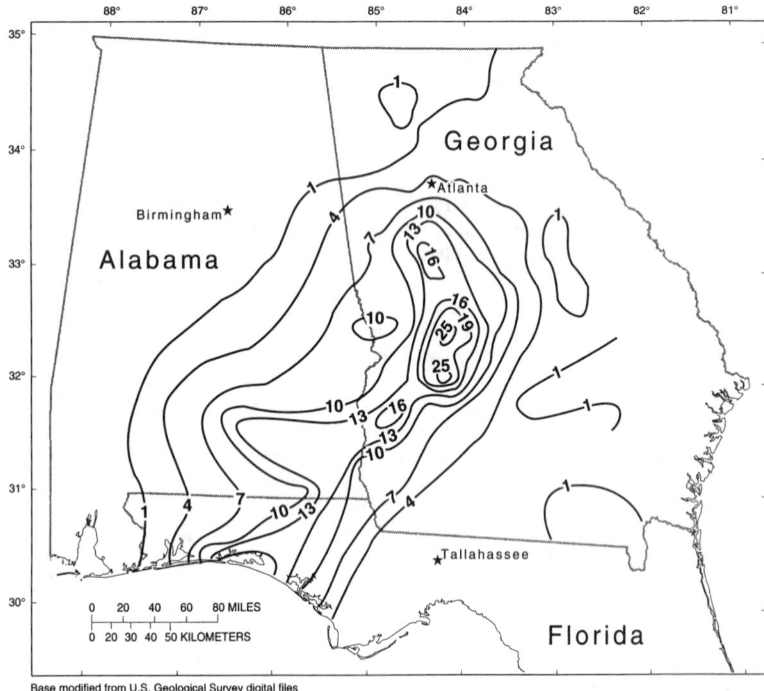

Figure 1. Lines of equal precipitation for tropical storm Alberto, July 3-7, 1994. Modified from U.S. Department of Commerce, National Weather Service, 1994.

Stream Flooding

Stream flooding resulting from Tropical Storm Alberto was as extreme as the rainfall that caused the flooding. The most significant flooding in Georgia occurred along the Flint and Ocmulgee Rivers and their tributaries (fig. 2). Major flooding in Alabama primarily was in the Choctawhatchee River basin and, to a lesser extent, in the Yellow River basin. The Florida flooding principally was along the Apalachicola River and tributaries.

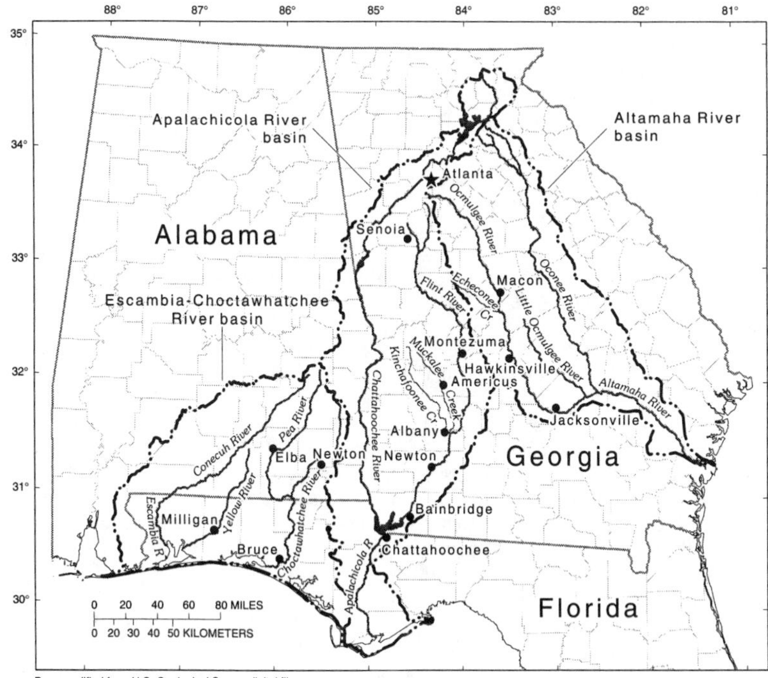

Figure 2. Selected streams in Alabama, Florida, and Georgia, affected by the July 1994 flooding.

Tributary flooding

Damaging flash floods occurred from the southern suburbs of Atlanta to Macon, Ga., on the night of July 4 and morning of July 5. The Line Creek near Senoia, Ga., gaging station recorded a peak discharge of about 28,400 ft^3/s, which was about 2.4 times the 100-year flood (1-percent chance) discharge (Stamey and Hess, 1993). The maximum stage at the Senoia gaging station was about 5.2 feet higher than any other recorded flood during 30 years of operation.

As the heaviest rains continued to move south, more destructive flash flooding occurred in the Americus area on the night of July 5 and morning of July 6. Muckalee Creek at Americus peaked on July 6 at a discharge of about 33,500 ft^3/s (about 4 times greater than the 100-year flood discharge) that was likely affected by numerous local dam failures. On July 6, Echeconnee Creek near Macon peaked at an estimated discharge of 64,700 ft^3/s, which is 3.2 times the 100-year flood. At the gaging station on the Kinchafoonee Creek near Dawson, Ga., a peak discharge of about 29,500 ft^3/s, was recorded on July 7, which is 1.7 times greater than the 100-year flood discharge. Floodwaters at this gaging station were greater than the 100-year flood discharge for about 48 hours.

Mainstem flooding

Tributary floodwaters combined and moved downstream in the Flint and Ocmulgee Rivers contributing to mainstem flooding. Peak discharges greater than the 100-year flood discharge were recorded at Flint River gaging stations (fig. 2) from about 20 miles south of Atlanta to Bainbridge, Ga. (fig. 2). At Montezuma, the Flint River peaked on July 8 at a stage about 6.7 ft higher than the 1929 flood, the highest stage previously recorded at this gaging station. At Albany, the Flint peaked on July 11 at a stage of about 43 ft, about 5 ft higher than the 1925 flood, which was the previous maximum flood.

Peak discharges on the Ocmulgee River exceeded the 100-year flood discharge from near Juliette, about 25 miles north of Macon, to Jacksonville, Ga. At Macon, the Ocmulgee peaked on July 6 at a stage of about 35.4 ft, which was about 5.5 ft higher than the 1990 flood, and the highest stage since 1887. The peak discharge of the Ocmulgee at Macon was about 107,000 ft^3/s, the highest of record which exceeded the previous high at 83,500 ft^3/s in 1948 by 23,500 ft^3/s. At U.S. Highway 341 at Hawkinsville, the Ocmulgee peaked at a stage of about 40.9 ft, which was about 4.4 ft higher than previous high in 1925 and probably was the highest since 1841.

Peak discharges greater than those of the 100-year flood (Olin, 1984) were experienced on the Choctawhatchee River at Newton, Ala., on July 7 and all along the mainstem to Bruce, Fla., on July 11. Peak discharges equivalent to the 40-year flood (Bridges, 1982) were experienced on the Apalachicola River at Chattahoochee, Fla., and the Yellow River at Milligan, Fla. Peak discharges on July 10 on the Apalachicola River, which is formed by the confluence of the Flint and Chattahoochee River at the Georgia-Florida state line were caused by peak flooding on the lower Chattahoochee River, July 7-8. The lower Flint River peak flooding did not occur until July 14.

Computations of the estimated 100-year flood discharges are based on recent USGS flood-frequency studies. The estimated 100-year flood discharges presented here do not include data from the July 1994 flood event. However, stages given for the 100-year flood discharges are based on current stage-discharge relations that do reflect any changes in the these relations resulting from the July 1994 flood.

Data Acquisition And Dissemination

USGS personnel monitored and reported flood information to other Federal, State, and local agencies from the onset of the storm until floodwaters receded. Stage and discharge data from many streams were collected and reported to the U.S. Army Corps of Engineers (COE); the U.S. Department of Commerce, National Weather Service (NWS); Federal Emergency Management Agency; Federal Highway Administration; various State natural resource and highway departments; electrical power companies; and numerous county and city officials as these groups worked to minimize loss of life and property. Using flood data provided by USGS and other information, the NWS provided updated flood warnings to people in affected areas. Flooding was so severe and widespread that 18 USGS gaging stations were severely damaged or destroyed, requiring much of the data to be collected manually and reported by cellular telephone to agencies involved in public safety. At the height of the flooding, almost 50 USGS personnel were working in the field to collect and provide hydrologic information vital to protecting life and property.

Despite the extraordinary effort to collect and document needed hydrological and flood information as it occurred, it was impossible to visit every site where data were needed. In some instances, bridges and roadways were inundated and floodwaters were too dangerous to work from boats. In other cases, personnel could not travel to points of interest before floodwaters receded. Therefore, immediately following the flood, field crews were dispatched to flag and document highwater marks along the Flint, Ocmulgee, and Choctawhatchee Rivers and many of their tributaries. Data collection and documentation will serve as a basis for determining flood-elevation profiles and indirect computations of peak discharges at several key locations. By October 1994, 18 USGS gaging stations damaged or destroyed by floodwaters were repaired or temporarily restored to operation; and by February 1995, indirect peak-discharge measurements were computed for 32 sites.

Flood Documentation

Hippe and others (1994), as part of the National Water-Quality Assessment Program, presented preliminary water-quality data in the Flint and Ocmulgee River basins during the July 1994 flood. The report compares the types and concentrations of selected pesticides in surface waters and presents preliminary information on the occurrence of nitrates and commonly used pesticides in shallow ground water.

Additional field- and office-work activities are continuing by USGS, COE, and Georgia Department of Transportation to further document information pertinent to the July 1994 flood. Flood-elevation profiles are being determined for the Flint and Ocmulgee Rivers and their major tributaries. A report documenting the magnitude and frequency of this flood is planned by USGS, when flood discharges are finalized. Flood-frequency analyses will be updated to include the July 1994 flood data.

The death and suffering caused by this storm serves to emphasize once again the high cost imposed upon life and property by flood disasters; and thus, the importance of preparing for, monitoring, and documenting such occurrences. Hydrologic data collected and analyzed by the USGS is extremely valuable to agencies responsible for future land-use activities and minimizing potential flood damages. As in the past, flood-related data collected, analyzed, and documented will be available to Federal, State, local agencies, and to the general public.

References

Bridges, W.C., 1982, Techniques for estimating magnitude and frequency of floods on natural-flow streams in Florida: U.S. Geological Survey Water-Resources Investigations Report 82-4012, 49 p.

Federal Emergency Management Agency, 1994a, Interagency hazard mitigation team report, Tropical Storm Alberto, July 1994—Georgia: FEMA 1033-DR-GA, 57 p.

____ 1994b, Interagency hazard mitigation team report, Tropical Storm Alberto, July 1994—Alabama: FEMA 1034-DR-AL, 21 p.

____ 1994c, Interagency hazard mitigation team report, Tropical Storm Alberto, July 1994—Florida: FEMA 1035-DR-FL, 34 p.

Hippe, D.J., Wangsness, D.J., Frick, E.A., and Garrett, J.W., 1994, Water quality of the Apalachicola-Chattahoochee-Flint and Ocmulgee River basins related to the flooding from Tropical Storm Alberto; pesticides in urban and agricultural watersheds; and nitrate and pesticides in ground water: U.S. Geological Survey Water-Resources Investigations Report 94-4183, 36 p.

Olin, D.A., 1984, Magnitude and frequency of floods in Alabama: U.S. Geological Survey Water-Resources Investigations Report 84-4191, 105 p.

Stamey, T.C., and Hess, G.W., 1993, Techniques for estimating magnitude and frequency of floods in rural basins in Georgia: U.S. Geological Survey Water-Resources Investigations Report 93-4016, 75 p.

U.S. Department of Commerce, Weather Bureau, 1961, Rainfall frequency atlas of the United States: Washington, D.C., Technical Paper 40, 61 p.

Rating Tidal Bridges for Vulnerability to Scour Damage

Stanley R. Davis
Member[1]

Abstract

The Maryland State Highway Administration (SHA) is presently rating tidal bridges for vulnerability to scour damage. Structures are classified using a modification of a system developed by C. R. Neill (Richardson et al. 1993, Neill, 1973). This modification involves estimation of storm tide characteristics and peak discharges from flood runoff in order to estimate the strength of the tidal currents. Special studies are required for structures spanning passageways between the mainland and island or between two islands. Scour estimates are developed using the design procedures developed by the Federal Highway Administration (Richardson et al. 1993), and the structures are field checked using procedures developed by the U.S. Geological Survey and SHA. A screening process is used to minimize or eliminate scour calculations for low risk bridges.

Introduction

The Maryland State Highway Administration (SHA) is rating its 251 on-system tidal bridges for vulnerability to scour damage. This presentation outlines the methods used to analyze flow conditions and to estimate worst-case scour depths. The following observations are made, based on the SHA's experience in evaluating tidal bridges:

- The SHA concurs with the observations of C.R. Neill (Neill, 1973) that "rigorous analysis of tidal crossings is difficult and is probably unwarranted in most cases, but in important cases consideration should be given to enlisting a specialist in tidal hydraulics".

[1]Hydraulic Engineer, Maryland State Highway Administration 707 North Calvert St. Baltimore, Md. 21202

- Structures over tidal waters are normally designed to span the tidal channel and adjacent wetlands. Such designs do not significantly constrict the tidal flow. Since the currents of storm tides in unconstricted channels are usually low (0.3 to 1 m/s), contraction scour generally is not a problem. Tidal Bridges are almost always founded on piles, and abutments are generally protected with riprap or concrete slope protection. For these reasons, tidal bridges crossing unconstricted channels are normally rated as low risk. Concerns about scour and erosion for such bridges normally focus on (1) the extent of local pier scour and (2) the protection of abutments and approach roads from wave action.

- The Federal Highway Administration's (FHWA) equation for pier scour (Richardson et al. 1993) can be expected to over-estimate the extent of local pier scour at tidal bridges with wide piers and low velocities of flow.

- The tidal bridges most likely to experience significant scour are those built in constricted channels or those which create constrictions in tidal channels

Most of the tidal bridges in Maryland are located on the Chesapeake Bay or on estuaries or inlets tributary to the Bay. Previous studies commissioned by the Federal Emergency Management Agency (Virginia Institute of Marine Science, 1978) have defined the elevation of the 100-year and 500-year storm tide elevations throughout the bay area. Studies by the SHA have identified a storm tide period of 24 hours, based on measured historic storm tides on the bay. Using this information, and the results of hydrologic studies of flood runoff from upland drainage areas, the SHA conducts hydraulic studies of most tidal bridges following Neill's method (Richardson et al. 1993, Neill, 1973). Special cases where this method does not apply are addressed later in this presentation.

Evaluation Process

SHA uses a screening process for rating bridges for Item 113, Scour Critical Bridges. The basic tool used in the screen is a classification system. Following the guidance presented by Neill (Richardson et al. 1993, Neill, 1973), tidal bridges are categorized by the geometric configurations of the bays and estuaries and the flow patterns at the bridges being rated:

I. bridges in enclosed bays or lagoons,

II. bridges in estuaries, and
III. bridges across islands or an island and mainland.

These categories are further subdivided to take into account whether:

- there is a single inlet or multiple inlets,
- there is a channel constriction at the bridge crossing,
- river flow or tidal flow predominates for the anticipated worst-case condition for scour, and
- tidal flow or wind establishes the anticipated worst-case condition for scour for Category III bridge crossings.

<u>Categories I and II - Bridges in enclosed bays, lagoons or estuaries.</u>

Upland runoff is normally quite limited for Category I bridges, so the discharge through the bridge consists primarily of tidal flow. For an enclosed bay with only one inlet, the tidal flow must enter and exit through the inlet, and the hydraulic analysis is relatively straightforward. If there are multiple inlets to the bay, special studies must be made to determine the portion of the total flow through each inlet for the design conditions.

Flow in estuaries consists of a combination of riverine (flood runoff) and tidal flow. The ratio of the magnitude of these flows depends on many factors including the size and character of the upland drainage area, the surface area of the tidal estuary, the elevation and period of the storm tide and the rainfall and wind characteristics associated with the event creating the storm tide.

If a highway structure constricts a tidal waterway, there may be a significant energy loss at the structure which needs to be taken into account in the hydraulic analysis. SHA has developed design procedures for evaluating the effect of the constriction on the flow through the bridge (Maryland State Highway Administration, 1995). The method involves routing the storm tide along with any upland runoff in order to calculate the differential head across the structure and the resulting depth and velocity of flow through the structure. The routing can be accomplished with either a spread sheet or a computer program prepared for in-house use that is called Tiderout.

Category III. Bridges connecting two islands or an island and the mainland.

The hydraulic analysis of each bridge in this category is unique, and no general guidelines have been developed for such locations. The effect of wind may become a primary factor to be considered in such locations. The analysis of such tidal problems should be undertaken by Engineers knowledgeable about tidal hydraulics.

An example of a Category III crossing is the Route 33 bridge spanning Knapps Narrows. It connects Tilghman Island with the mainland of the eastern shore of Maryland. Flow through Knapps Narrows is essentially a function of the intensity, direction and duration of the wind. In order to evaluate the anticipated worst-case scour condition for the replacement bridge at this location, the upwind setup and wave setup and the downwind setdown were determined for design conditions. These factors were used to determine the head differential between the two sides of the island. Once this information was determined, flow quantities,depths and velocities could be calculated for estimating the extent of scour at the bridge.

The Screening Process.

The SHA uses the following process to rate tidal bridges for Item 113, Scour Critical Bridges:

1. The location of each bridge is plotted on USGS topographic maps or NOAA navigation charts. Preliminary information is collected on the tidal waterway, upland drainage basin the highway crossing using a Tidal Bridge Data and Analysis Worksheet.

2. A preliminary estimate is made of the depths and velocities of storm tides, taking into account the expected contribution to the flow of flood runoff from the upland drainage basin. (This is based on a variation of Neill's method).

3. The bridge is categorized in accordance with the criteria discussed above.

4. Bridge plans, files and inspection reports are reviewed along with the Channel Stability Reports prepared by the U.S.Geological Survey for the SHA. This step may or may not include another bridge site inspection by the hydraulic engineers or the interdisciplinary team.

5. The structure is rated for Item 113 by the SHA Interdisciplinary Team. Structures on piles with no history of scour normally will be rated as low risk when the preliminary hydraulic analysis indicates that the velocities of flow and anticipated scour depths are low. In those locations where estimated velocities are high, additional studies will be undertaken to evaluate the risk of damage.

Judgment is required when estimating local scour at wide piers in tidal waters, since the FHWA's pier scour equation (Richardson et al. 1993) is believed to over-estimate scour depths for such conditions. This observation has been substantiated by preliminary laboratory tests performed by Torrico (Torrico, 1993).

At this time, about half of the on-system tidal bridges have been rated. It is estimated that the cost of rating a typical on-system tidal bridge for Item 113, Scour Critical Bridges, will be under $2000. Information regarding detailed procedures developed by SHA for evaluating tidal bridges can be obtained from the Office of Bridge Development.

Acknowledgments

The author wishes to acknowledge the contributions made by Raja Veeramachaneni, Dr. Fred Chang, Len Podell and Prasad Inmula in the development of Maryland's procedures for evaluating scour at tidal bridges.

References

1. Maryland State Highway Administration, Manual of Hydrologic and Hydraulic Design, Final Draft, 1995.

2. C. R. Neill, Guide to Bridge Hydraulics, Roads and Transportation Association of Canada (RTAC), 1973,

3. Richardson, E.V. et al Hydraulic Engineering Circular No. 18, Evaluating Scour at Bridges, Second Edition, Federal Highway Administration, April, 1993.

4. Torrico, E.F. "The Effect of Wide Piers in Shallow Water on Scour Around Circular Piers", Submitted to the University of Maryland in partial fulfillment of the requirements for the Master of Science, 1993

5. Virginia Institute of Marine Science, A Storm Surge Model Study, Volumes 1 and 2, Glouster Point, Virginia, 1978.

Time Rate of Local Scour

Mark S. Gosselin[1] and D. Max Sheppard[2], M. ASCE

Abstract

Compared with predicting equilibrium scour, the determination of the time varying evolution of local scour has received relatively little attention. A general review of the current understanding of time dependent scour is presented. Also, the main mechanisms of local scour are discussed. Finally, the authors' research on this topic is outlined.

Introduction

An important criterion for the design of bridge piers is the depth of scour. Current practice dictates that the design scour depth be the maximum equilibrium scour depth for steady flow achieved for design flow conditions. In many cases, however, the time it takes for the scour hole to reach this equilibrium depth is longer than the length of the design event producing these conditions. There have been a number of studies performed on determination of equilibrium scour depths for both clear-water (e.g. Raudkivi and Ettema (1983), Jain (1981)) and live bed (e.g. Sumer et al. (1992), Jain and Fischer (1979)) scour. These studies relate the maximum scour depth under steady flow conditions to the hydrodynamic and sediment parameters. Empirical equations based on these results are used in the design of bridge piers. For a riverine system the use of equilibrium scour depths is reasonable since in many cases, even though the flow is unsteady during storm events, high velocities can persist for long periods of time. In tidal (coastal) waters both normal and extreme event flows are much more unsteady and are of shorter duration. Viewed in this light, the temporal variation of scour depth becomes an important aspect of accurate prediction of design scour depths.

Description of Mechanisms

In order to better understand the scouring process, it is first advantageous to examine the individual mechanisms at work. These processes have been described in

[1] Graduate Assistant, Coastal and Oceanographic Engineering., University of Florida, Gainesville FL 32611.

[2] Professor, Coastal and Oceanographic Engineering, University of Florida, Gainesville FL 32611.

detail in a number of papers (e.g. Shen et al. (1966)). To begin, take the situation where a single vertical cylindrical pile is placed in a wide stream. Viewing the cylinder from above, the presence of the cylinder causes the flow to accelerate around the sides of the pile reaching a maximum value at 90 degrees to the flow direction. In the wake of the cylinder, the flow field consists of vertical (wake) vortices. The shedding frequency and intensity of these vortices is dependent upon the Reynolds number based on the cylinder diameter ($R = \frac{UD}{\nu}$).

Viewing the cylinder from the side, the approach flow has a velocity profile that varies with depth and is logarithmic in shape. This velocity profile causes a pressure gradient to develop on the leading edge of the cylinder. Flow is thus directed downward, inducing the formation of a vortex at the base of the cylinder. The ends of the vortex are swept downstream by the surrounding flow bending the vortex into a horseshoe shape. This flow feature is the horseshoe vortex.

Scouring occurs when these hydrodynamics are coupled with the fact that the bed is erodible. Sediment is transported when the shear stress at the bed exceeds the critical value. During initiation of scour, sediment is removed from the sides (at 90°) of the pile where the flow has been accelerated. From here, the scour hole works its way to the front of the pile where the horseshoe vortex has been established. As the horseshoe vortex removes sediment the scour hole begins to deepen. As more sediment is removed, the hole expands both in depth and in width. It resembles an inverted cone whose axis corresponds with that of the pile. The angle that the slope of the hole makes with the vertical is equal to the angle of repose of the sediment. As the hole deepens, the horseshoe vortex grows in diameter and descends into the hole. When the vortex scours out sand from the bottom of the hole, an "avalanche" of sediment can occur from farther up the side of the hole causing the hole to widen. Once a sediment particle is put into suspension by the horseshoe vortex, it is swept around the side of the cylinder. Here, a number of things can happen to it. If it interacts with the wake vortices it can be brought higher in the water column due to the upflow associated with the center of the vortex and then brought downstream where it is eventually deposited. If it is swept directly behind the cylinder, it experiences the relatively calm wake region in between the vortex shedding. Here it will settle out to form the characteristic mound behind the pile. During certain phases of the scour process, sediment from the wake mound has been observed to reenter the scour hole in the form of bedload.

The increase in size of the horseshoe vortex is accompanied by a decrease in bed shear stress. For the case of clear-water scour, when the shear stress drops below the critical value, the equilibrium scour depth has been reached. For the case of live bed scour, equilibrium occurs when the flow of sediment into the hole is equivalent to the flow of sediment out of the hole.

Survey of Time Dependent Scour Literature

Due to the complexity of the scouring process, the majority of the literature has concentrated on determining the maximum equilibrium scour depth for given flow and sediment conditions while relatively few researchers have delved into the dependence of this process on time. Ettema (1980) observed that under steady flow conditions the rate at which local scour occurred in his experiments changed rather abruptly as they proceeded toward equilibrium. In general the initial rate is quite large, followed by a more moderate rate that decreases with time. For clear-water scour, it slowly approaches an asymptotic value, which is called the equilibrium scour depth. For the case of live bed scour, the depth of scour oscillates around the equilibrium value with an amplitude equal to the height of the sand waves migrating in and out of the scour hole. While this much is known, there are relatively few papers on the rates at which these processes occur and the conditions under which the sudden change in rate takes place. Six papers that address the scour rate problem are discussed below.

One of the first instances where this subject is broached is in the study by Shen et al. (1965). They concluded that the Froude number was the most significant parameter in determining the scour depth, and provide empirical curves of scour depth as a function of time. This study was limited by the fact that only a single sediment size and pile diameter were used.

In a later study, Shen et al. (1966) define t_{75} as the time it takes the depth of scour to reach 75% of its equilibrium value. They found that t_{75} tends to increase as the sediment size increases and that it decreases as the upstream velocity increases. They also concluded that t_{75} was independent of the pile diameter for the range of the values they studied. Their tests spanned both live bed and clear-water regimes.

In his Ph.D. thesis, Ettema (1980) presented a more thorough analysis. He noted that when the depth of scour was plotted versus $\log_{10}$(time) there were three distinct phases of the scour process characterized by three straight line segments on this plot. He termed the three phases the initial phase, the erosion phase, and the equilibrium phase. In the initial phase, rapid scouring occurs by the down flow on the pile impinging on the planar bed. This phase is characterized by a steeply sloping line on the graph. During the erosion phase, the scour hole develops as the horseshoe vortex grows in both size and strength. The slope of this line on the graph is considerably less than the previous stage. In the final stage, the equilibrium depth has been reached and hence no further scour occurs. At this point the graph attains a constant value with a zero slope. Ettema, however, does nothing to relate the slopes of these lines (the rate of scour) to the flow and sediment conditions and also does not give a method for predicting when one phase ends and another begins.

A few researchers have attempted to model the rate of scour using semiempirical math and computer models. Yanmaz and Altinbilek (1991) make use of an empirical sediment pickup function to relate the rate of sediment transport over a flat bed to that around bridge piers. To do this they define a proportionality coefficient that is found through regression analysis of existing scour experiments. Then, by making certain assumptions about the scour hole geometry, they relate the rate of sediment volume lost around the pile to the flow and sediment conditions as a function of time. The depth of scour as a function of time can then be found by numerically integrating this expression. This method is very sensitive to the sediment number ($Ns = u/\left[(S-1)gD_{50}\right]^{0.5}$) and does not work particularly well outside the range of their experiments (u = velocity, S = ratio of specific gravity of sediment to that of water, D_{50} = median grain diameter).

Sumer et al. (1992) take a simpler approach to this problem. They assume that the time varying depth of scour (ds_t) normalized by the equilibrium value (ds_e) can be represented by the formula $\frac{ds_t}{ds_e} = 1 - \exp\left(-\frac{t}{T}\right)$, where t is time and T is a function of the pile diameter, gravity, specific gravity of the sediment, mean sediment diameter, Shield's parameter and water depth. T was determined using a regression analysis of experimental data. Their experiments were performed under live bed conditions, and therefore, only apply in this regime.

In a study by Kothyari et al. (1992), the depth of scour time history is determined using an iterative approach. The researchers begin by making a few assumptions about the scouring mechanisms; i.e. that the horseshoe vortex is the primary mechanism and that it grows in direct proportion to the depth of scour. In their model, they relate bed shear stress to the size of the vortex. Using a stochastic model for bed load transport, they associate this shear stress to a certain time interval. Over this change in time they assume that the depth of scour is incremented an amount equal to the diameter of the sediment. The model then returns to the initial step of finding the size of the vortex and repeats the process until the shear stress drops below the critical value for the sediment.

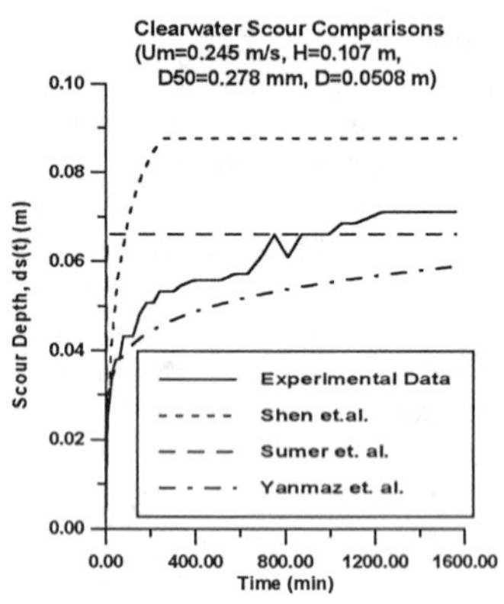

Figure 1. Clear-water Scour Comparisons.

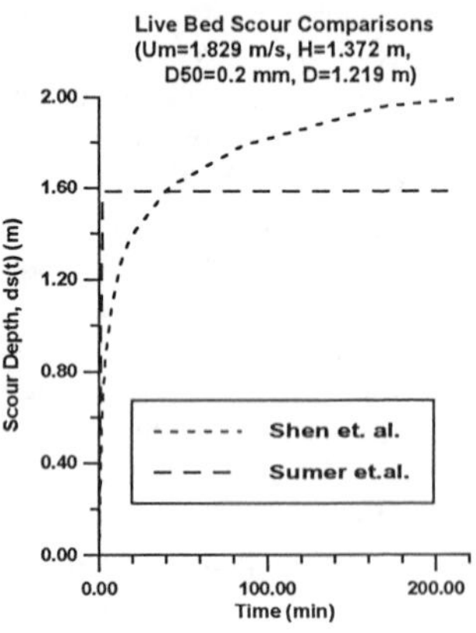

Figure 2. Live Bed Scour Comparisons.

A comparison of some of the techniques outlined above was performed for both clear-water and live bed conditions. The results are contained in Figures 1 and 2. For the case of clear-water, laboratory scale scour (see Figure 1), the error from experimental values ranged from 7% to 23%. In fact, the method by Yanmaz and Altinbilek does a fair job at predicting the rate of scour. For the case of live bed, prototype scale scour, neither method seems to give reasonable results. As demonstrated, more work is certainly needed in this area.

Time Dependent Scour Research at the University of Florida

At the University of Florida, a phenomenological approach to time dependent scour research is being pursued. Investigations are currently being carried out on the main mechanisms believed to transport sediment in the immediate vicinity of the pile. These mechanisms are 1) the scouring action of the horseshoe vortex, 2) the "avalanching" of sediment down the sides of the scour hole and 3) the formation of the characteristic mound behind the cylinder.

To examine the hydrodynamics, a three dimensional flow solver is being used (FLOW3D from Flow Science, Inc.). Through flow simulations it is possible to determine the shear stress distribution as a function of the flow parameters and also as a function of scour depth. To examine the avalanching phenomena and the sediment return from the downstream mound, an experimental device is being constructed that will allow the rate of sediment influx to be determined as a function of the rate of sediment removal from the base of the hole.

The crux of this research is to produce a time dependent model for clear-water scour. This will not only produce more accurate predictions of scour by factoring in the time of a design event, but it will eventually lead to a model that will accept time dependent flows and water levels as an input. This would be particularly useful for the design of coastal structures.

References

Ettema, R. (1980) "Scour at Bridge Piers." Ph.D. Thesis, Auckland University, Auckland, Report No. 216.

Jain, S. C. (1981) "Maximum Clear-Water Scour Around Circular Piers." JHD, Proc. A.S.C.E., V.107, HY5, 611-626.

Jain, S. C. and E. E. Fischer (1979) "Scour Around Bridge Piers at High Froude Numbers." Report No. FHWA-RD-79-104 Federal Highway Administration.

Kothyari, U.C., R.J. Garde, and K.G. Ranga Raju (1992) "Temporal Variation of Scour Around Circular Bridge Piers." J. of Hyd. Engrg, V.118, No.8, 1091-1106.

Shen, H.W., Y. Ogawa and S.S. Karaki (1965) "Time Variation of Bed Deformation Near Bridge Piers." Proc. IAHR, 11th Cong., 3.14.

Shen, H.W., V.R. Schneider and S.S. Karaki (1966) "Mechanics of Local Scour." C.S.U., Civil Engineering Dept., Pub. No.CER66HWS22, 56p.

Sumer, B.M., J. Fredsoe and N. Christiansen (1992) "Scour Around Vertical Pile in Waves." J. Waterway, Port, Coastal, and Ocean Eng., V. 118, No. 1, 15-31.

Raudkivi, A. J. and R. Ettema (1983) "Clear-Water Scour at Cylindrical Piers." J. of Hyd. Engrg, V. 109, No. 3, 338-350.

Yanmaz, A. M. and D. H. Altinbilek (1991) "Study of Time-Dependent Local Scour Around Bridge Piers." J. of Hyd. Engrg, V. 117, No. 10, 1247-1268

Computer Models for Tidal Hydraulic Analysis at Highway Structures

Lyle W. Zevenbergen[1], M. ASCE, Everett V. Richardson[1], F. ASCE, and Billy L. Edge[2], M. ASCE

Abstract

This paper presents the findings of an investigation on simulating hydraulics of tidal waterways. Accurate hydraulic information is necessary for calculating scour at bridge crossings, assessing channel stability, and designing bridge foundations and countermeasures. This study identifies and evaluates existing models and methods for hydraulic engineers to analyze highway encroachments in tidal waterways, and recommends improvements to selected models.

Introduction

Tidal waters are subjected to dynamic flow conditions caused by astronomical tides, ocean currents, storm surges, and upland runoff. Highway encroachments are subjected to stream instability and foundation scour resulting from these dynamic flow conditions. Computer modeling is the most accurate method for determining the hydraulic conditions for extreme storm events that cause scour at tidally affected bridge crossings. A consortium of seven East Coast State Departments of Transportation (Connecticut, Georgia, Maine, New Jersey, New York, North Carolina, and South Carolina) sponsored this study. The study focused on three tasks: (1) compile a database of literature on tidal processes and computer models, (2) evaluate sources and methodologies for determining ocean tides and storm surge characteristics, and (3) evaluate which computer models are best suited for use by bridge engineers for tidal hydrodynamic and investigations. Task 2 includes determining the storm surge hydrograph which consists of the storm surge height and the duration of the rise and fall. The storm surge hydrograph may be superimposed on normal astronomical tides. Task 3 includes accurate hydraulic representation of bridge, culvert, and embankment overtopping hydraulics.

[1]Senior Hydraulic Engineer and Senior Associate, respectively, Ayres Associates, 3665 JFK Parkway, Bldg. 2, Suite 300, Fort Collins, CO, 80525.

[2]President, Edge and Assoc., Inc., 4911 Bay Oaks Court, College Station, TX, 77845.

Literature Database

References were obtained by conducting computer searches of bibliographic databases and investigating other sources. The information was incorporated into a database which contains over 600 citations on tidal processes and analytical procedures. The citations were assigned keywords (from a collection of 51) under the general topics of modeling (physical, computer, statistical), study topic (hydraulics, hydrology, water quality, sediment transport, geomorphology, etc.), coastal features (bays, estuaries, inlets, etc.), tides (astronomical, storm surge, etc.), data reports (hydrologic, tidal, currents, etc.), and locations (South Carolina and other eastern seaboard states). Many of the citations include abstracts.

Storm Surge Evaluation

Tidal hydraulic studies require estimates of tides and storm surge stage hydrographs as boundary conditions. In many cases, upstream flood hydrographs must also be included in the hydraulic modeling. Flow velocities through inlets and bridges are controlled by the storm surge elevation and the rate of rise and fall of the surge as well as energy losses due to flow resistance and channel constrictions.

The Federal Emergency Management Agency (FEMA) and National Oceanographic and Atmospheric Administration (NOAA) publish peak storm surge elevations related to the frequency of occurrence or hurricane severity. The FEMA elevations were determined using the SURGE two-dimensional finite difference model (FEMA, 1988). Because FEMA's focus is on flooding potential, maximum surge elevations are reported and the individual storm hydrographs are not retained. NOAA developed the SLOSH (National Weather Service, 1992) two-dimensional finite difference model to determine surge elevations for each class of hurricane. The peak elevations are reported as maps of maximum water surface contours for use by emergency managers. Although the NOAA data provide an alternative to the elevations reported by FEMA, storm surge hydrographs are also not retained.

It appears that the most appropriate method for predicting storm surge hydrographs along the East and Gulf Coast States has been developed by Cialone et al. (1993). Using a computer program, SSEL, this methodology combines peak stage information from existing sources (FEMA or NOAA) with storm properties (storm forward speed and radius of maximum winds) to estimate storm surge hydrographs. Representative storm properties as a function of location and recurrence interval are also included in the Cialone et al. (1993) report. Local average astronomical tides can be included to develop a set of eight hypothetical surge hydrographs.

Since both the SURGE and SLOSH models were designed to estimate only peak storm surge elevations, their application is limited for hydrodynamic modeling of estuaries. The Cialone et al. (1993) approach to developing surge hydrographs can utilize the data generated from SURGE, SLOSH, or other peak surge elevation data to generate a surge hydrograph. The hydrograph generated by the SSEL program is produced with readily available data and is suitable for hydrodynamic tidal waterway modeling.

Computer Model Evaluation

Twenty-one models were reviewed to determine their applicability to tidal hydraulic and scour studies. It is unlikely that a single hydraulic model will be capable of efficiently modeling all of the tidal hydraulic situations which are likely to be encountered in practice. One-, two- and three-dimensional hydraulic models were included in the evaluation. The evaluation was performed in two steps: (1) developing necessary and desirable model characteristics, and determining which models best met these criteria, and (2) subjecting the retained models to more detailed evaluation, including testing.

The necessary hydraulic modeling capabilities included (1) applicability to a range of bridge types, sizes, and geometries ranging from culverts to large multi-span bridges, (2) applicability to a range of tidal conditions, including bays, inlets, estuaries, and tidally affected inland river crossings, and (3) ability to simulate multiple inlets and branching channels. The necessary physical characteristics of the models were: (1) PC based, (2) able to operate under Windows™, (3) public domain, (4) well documented, (5) modular and efficient, (6) maintained by an agency, (7) equipped with both a pre- and postprocessor with graphical interface, (8) operable in both English and SI units, and (9) minimal user uncertainty in selection and use of modeling parameters. The desirable modeling capabilities included (1) simulation of wind effects, (2) ability to simulate overtopping of land barriers, (3) ability to predict effects of highway embankments and openings on flows in tidal marsh lands, and (4) simulate or link to models for transport sediment and chemical pollutants. Although no single model met all the necessary and desirable characteristics, four models met most of them and were retained for detailed evaluations.

The four models subjected to detailed evaluation included two one-dimensional and two two-dimensional models. The one-dimensional models were DYNLET1 (Cialone and Amein, 1993) and UNET (Barkau, 1993). The two-dimensional models were FESWMS (Froehlich, 1989) and RMA-2V (Thomas and McAnally, 1985). Each of these models was tested using data collected at the Indian River Inlet and Bay, and Rehoboth Bay on the coast of Delaware. The inlet is over 370 m (1,200 ft) long and the narrowest part is approximately 150 m (500 ft) wide. The inlet depth is 12 m (40 ft) on average, but can exceed 24 m (80 ft) locally. The connected bays each have a surface area of approximately 36 km^2 (14 mi^2) and are approximately 2.1 m (7 ft) deep. Hydrographic survey and tide data were obtained in 1988. The tide gages included four in the bays, one in the inlet and one offshore. The offshore gage data were collected for a 63-hour period and used to test the models. The tidal range of the offshore gage was 1.7 m (5.5 ft) which was dampened to only 0.55 m (1.8 ft) at the most remote gage in Rehoboth Bay.

Each of the four tested models performed well for the Indian River tests. The models well replicated observed tide gage readings, generally within 0.12 m (0.4 ft). The one-dimensional models were easier to set up and ran much faster than the two-dimensional models, and calibrated Manning's n values for the inlet and bay areas were similar for all the tested models. For the 63-hour simulation, DYNLET1 required 13 minutes and UNET required 40 seconds on a 33 MHZ 486 PC, and FESWMS required 12 hours and RMA-2V required 3 hours on a 90 MHZ Pentium PC. For this case, the one-dimensional models produced similar results to the two-dimensional models, although it is anticipated that many complex hydraulic situations would require two-dimensional modeling.

Because analyzing the hydraulics and scour potential at highway structures are a focus of this study, tests were performed of flow through culverts and bridges and over embankments. The FHWA WSPRO and HY-8 models, and the Corps of Engineers HEC-2 model were used as a basis for comparison for the structure hydraulics tests. RMA-2V contains limited structure analysis capabilities which consist of specifying various types of rating curves at structure locations. Since the specific characteristics of a structure are not included as input, RMA-2V was not included in the structure hydraulic tests. The other models use various methods for computing structure hydraulics, and their performance varied significantly. UNET provided the best structure hydraulic computations, although questionable results were obtained for submerged weir flow. FESWMS performed well for culvert and embankment overtopping flows, but did not give reasonable results for bridge pressure flow, although this feature may have been corrected in the most recent version of the program. Of the three models tested, it appeared that DYNLET1 gave the least acceptable structure hydraulic analysis.

Based on the results of the hydraulic tests, UNET, FESWMS and RMA-2V will be retained for further development. UNET was selected because it accurately simulates tidal and structure hydraulics. In comparison to the other models, UNET is most capable of modeling numerous cross sections and very long river reaches, including branched and looped channel networks. DYNLET1 performed well on tidal hydraulics, but was not as powerful as UNET and did not simulate structure hydraulics as well. FESWMS was selected because it accurately simulates tidal and structure hydraulics, and is well suited for simulating complex flow conditions. FESWMS would benefit greatly from enhanced pre- and postprocessing software, which is currently being developed. RMA-2V was selected because it is currently available with excellent pre- and postprocessing, and includes numerous other features which warrant further development.

Conclusions and Recommendations

The research identifies existing methods and models which can currently be used for hydraulic analysis of encroachments on tidal waterways. A second phase of this project will include the following five tasks: Task 1: The current models and methods will be presented in a workshop to the participating states. The states will use the models on current projects to identify deficiencies and desired changes to the models. Task 2: Enhancements will be made to the three selected models (UNET, FESWMS and RMA-2V). Enhancements include correcting known bugs and deficiencies, adding new capabilities, allowing output of information necessary for scour calculations, and improving performance of the models. Task 3: Enhancement and refinement of methods for determining storm surge hydrographs will be performed and design events will be developed for the participating states. Task 4: The developed methods and enhanced models will be tested with various real applications from the participating states. Task 5: A users manual and training course will be developed for the participating states. The users manual will include case studies from Task 4.

Appendix - References

Barkau, Robert L. (1993). "UNET - One Dimensional Unsteady Flow Through a Full Network of Open Channels," (CPD-66), U.S. Army Corps of Engineers, Hydrologic Engineering Center, Davis, CA, 240 pp.

Cialone, M.A. and Amein, M. (1993). "DYNLET1 Model Formulation and User's Guide," CERC-93-3, Coastal Engineering Research Center, U.S. Army Corps of Engineers Waterways Experiment Station, Vicksburg, MS, 62 pp.

Cialone, M.A., Butler, L., and Amein, M. (1993). "DYNLET1 Application to Federal Highway Administration Projects," CERC-93-6, Coastal Engineering Research Center, U.S. Army Corps of Engineers Waterways Experiment Station, Vicksburg, MS, 93 pp.

Federal Emergency Management Agency (1988). "Coastal Flooding Hurricane Storm Surge Model," Vol. 3, Office of Risk Assessment, Federal Insurance Administration, Washington, D.C.

Froehlich, D.C. (1989). "Finite Element Surface-Water Modeling System: Two Dimensional Flow in a Horizontal Plane, Users Manual," No. FHWA-RD-88-177, FHWA, Office of Research, Development and Technology, McLean, VA, 285 pp.

National Weather Service (1992). "SLOSH: Sea, Lake, and Overland Surges from Hurricanes," NOAA Technical Report NWS 48, Silver Spring, MD.

Thomas, W.A. and McAnally, W.H. (1985). "User's Manual for the Generalized Computer Program System: Open Channel Flow and Sedimentation, TABS-2," U.S. Army Engineer Waterways Experiment Station, Vicksburg, MS, 671 pp.

Sensitivity of Bridge Scour Producing Currents to Storm Surge Parameters

Christopher W. Reed[1], Susan Harr[2] and D. Max Sheppard[3], M. ASCE

Abstract

A study has been conducted to determine the sensitivity of storm surge induced currents in a tidal system to variations in the storm surge parameters. Numerous storm surge hydrographs have been developed which are representative of surges predicted for the southeast coast of Florida. A depth averaged, finite-element hydrodynamic model (RMA2) has been used to calculate the velocities within a shallow water estuary system for systematic variations in the surge hydrograph. Results indicate significant sensitivity to surge parameters such as the peak surge elevation, duration, and rate-of-rise for the tidal system studied.

Introduction

Design scour computations in tidal inlets, bays, estuaries and rivers use one in one-hundred and one in five-hundred year return interval storm conditions. Storm surge hydrographs, predicted at the coastline for each storm, are used as the basis for calculating the associated surge velocities in the inlets and adjacent tidal waters. Comparisons of predicted and measured storm surges on the open coast indicate that the predicted peak water elevations are relatively close to the measured value, but other features of the hydrograph, such as the rate of rise or fall, can differ greatly. The propagation of meteorological tides (storm surges) through a tidal inlet or river mouth into a bay-estuary-river system can be extremely complex. The complexity is enhanced when there is flooding of barrier islands and other low lying subaerial lands. It is by no means obvious how the "errors" in the predicted coastal hydrograph influence the currents and water elevations at various locations in the bay-estuary-river system. Variations in design currents and water elevation can have a major impact on design scour depth predictions and therefore it is important to

1 Post Doctoral Fellow in the Coastal and Oceanographic Engineering Department, University of Florida, Gainesville, FL 32611.

2 Graduate Assistant in the Coastal and Oceanographic Engineering Department, University of Florida, Gainesville FL 32611.

3 Professor of Coastal and Oceanographic Engineering, University of Florida, Gainesville FL 32611.

know how sensitive these quantities are to variations in the storm surge parameters. In order to address these issues, a study has been conducted to determine the sensitivity of design scour producing currents in a tidal system to variations in the storm surge parameters. A calibrated depth averaged, two-dimensional, finite-element hydrodynamic model has been used to provide the velocity predictions for each variation in the surge hydrographs.

It should be noted that surge models, such as SLOSH (Jarvinen and Gebert, 1987), were developed primarily for obtaining peak elevations for use in flood evacuation planning for severe storms events. To this end the models have been relatively successful. For instance, a comparison of SLOSH predicted surge elevations with measured water elevation data for Hurricane Gloria's landfall over Long Island (Jarvinen and Gebert, 1987) and for Hurricane Hugo's landfall near Charleston, South Carolina (Garcia et al., 1990) show generally good agreement. In these cases, the variations between the predicted and measured hydrographs may be insignificant since the peak water elevations are usually well predicted. However, for the purposes of design scour calculations, it is necessary to accurately predict the storm driven currents as well as water elevations. The effects of hydrograph variations on storm currents is not well documented. Furthermore, many of the other hydrograph features, such as rate of rise and fall, duration of the peak, etc., are not well predicted. Since hydrographs are typically used in design storm velocity predictions, it is important to quantify the sensitivity of the predictions to velocity uncertainties in the hydrographs.

Sensitivity Analysis

The approach for determining the sensitivities is based on the application of a two-dimensional, depth averaged hydrodynamic model to a southern Florida tidal system, namely the St. Lucie Estuary and portions of the Indian River Lagoon. The hydrodynamic model used is RMA2 (Norton and McAnally, 1973; Thomas and McAnally, 1991) with the FASTTABS pre and post processor (Fasttabs, 1992).

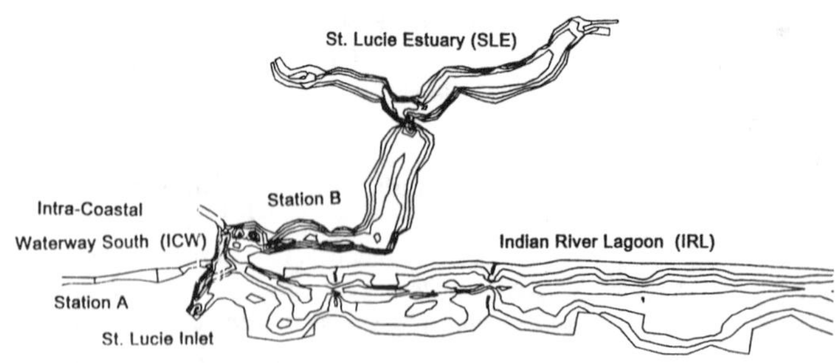

Figure 1. Contour and Station Map of Estuary System

RMA2 is a depth averaged two-dimensional model employing finite-element solution methods to solve the shallow water wave equations. The tidal system is characterized by relatively shallow water, generally 2 to 6 ft deep at MLW, throughout most of the system with maximum depths of 10 ft in narrow maintained channels. A portion of the modeled area is shown in Figure 1 which shows bathymetric contours on 2.5 ft intervals. The Intracoastal Waterway extends southward to Jupiter Inlet. A section of the Indian River Lagoon is represented, including Fort Pierce Inlet and portions of the lagoon to the north.

Calibration of the model to normal tidal conditions was completed prior to beginning the sensitivity analysis. A number of modeling studies have been conducted for portions of the estuary system, (Williams, 1985; Morris, 1987; Sheng et al., 1990; Smith, 1990) and were used to investigate estuary hydrodynamics, salinity transport and water quality. Data obtained for these studies as well as information from the NOAA Tide Tables was used for calibration of the St. Lucie Estuary model.

The sensitivity analysis was conducted for the tidal system by driving the flows at the inlets with a typical surge hydrograph superimposed on a normal tide. The surge parameters such as amplitude, duration and shape were varied and then the calculated flows for each surge within the tidal system were compared. In each case the surge parameter variations were scaled such that each of the surges had the same energy. This was necessary to eliminate effects of the surge size on the results, and focus only on the effects due to "shape" parameters. The energy associated with each surge was calculated by considering the surge as a solitary wave propagating with speed $\sqrt{gh}$ where g is gravity and h is the water depth. This definition allows one to transform the surge hydrograph (i.e. time series) into a wave profile, from which the total kinetic and potential energy could be calculated. Typical perturbations used are

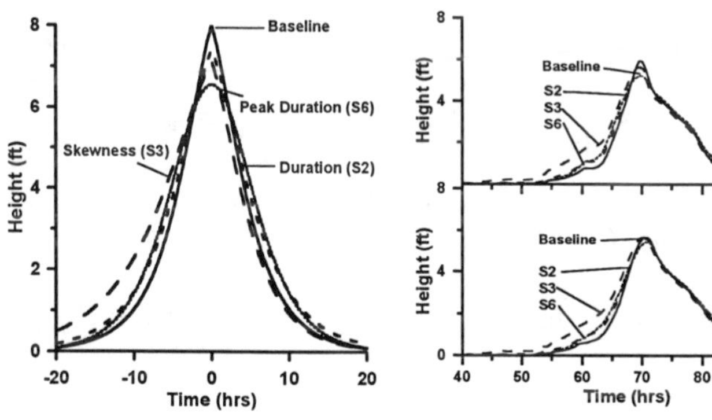

Figure 2. Surge Variations

Figure 3. Computed Hydrographs

shown in Figure 2, representing changes in height and duration, rate-of-rise, skewness and peak duration. The variations in height and duration of these surges averaged about 15% of the baseline surge values.

Results

The water elevation time series plots shown in Figure 3 (corresponding to points A and B in Figure 1), are representative of results from the hydrodynamic model. Note that the tidal influence associated with the astronomical tide has been removed from the curves. Comparison of the three curves representing perturbations to the baseline indicate that the variations in the surge height and width are of similar magnitude (averaging 15%) to those of the input hydrographs. The discharge time series corresponding to the water elevation plots in Figure 3 are shown in Figure 4. The differences in the maximum discharge (and subsequently the maximum velocity) at these stations vary greatly with the type of surge perturbation and can exceed 50%. Using changes in maximum discharge as an indicator of sensitivity, plots of the sensitivity for points along the St. Lucie Estuary, Intracoastal Waterway (South) and the Indian River Lagoon can be developed. These plots are shown in Figure 5. The results indicate that the highest sensitivity and largest range of sensitivities are in the Intracoastal Waterway (South), and the least occur in the Indian River Lagoon. Note that the Intracoastal Waterway and the Indian River Lagoon represent the smallest and largest water volume respectively within the system.

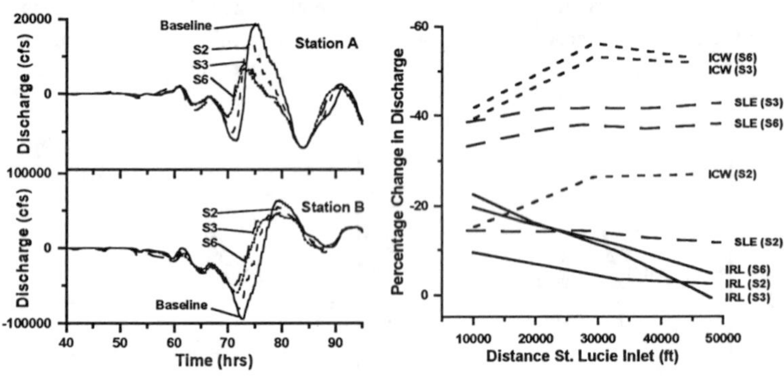

Figure 4. Computed Discharges

Figure 5. Sensitivities of Discharge

Conclusions

Equilibrium sediment scour depths and the rates at which these depths are reached depend on, among other quantities, the local depth average velocity. Just how dependent scour is on velocity again depends on several quantities, including the magnitude of the velocity, but under certain circumstances it can be sensitive to changes in velocity. Thus, the relatively large dependence of velocity on the storm surge parameters found in this study can translate into an even greater dependence of scour on these parameters. The results of this study provide some guidelines for determining the locations within a tidal system and the conditions under which the currents are most sensitive to variations in storm surge parameters. Improvements in storm surge predictions are needed. Meanwhile, for points of interest within the tidal system that are in sensitive areas, a range of the critical design storm surge hydrograph parameters (around the predicted values) should be investigated.

Acknowledgments

The authors would like to thank the Florida Department of Transportation for supporting this research. The authors also acknowledge the valuable technical contributions to this work made by Shawn McLemore and Luis Maldonado of FDOT in Tallahassee.

References

Fasttabs User's Manual (1992) Boss Corporation and Brigham Young University.

Garcia, A.W., B.R. Jarvinen and R.E. Schuck-Kolben (1990) "Storm Surge Observations and Model Hindcast Comparison for Hurricane Hugo," in Shore and Beach, October, Vol.58-59, p.15.

Jarvinen, B.R. and J. Gebert (1987) "Observed Versus SLOSH Model Storm Surge for Connecticut, New York and Upper New Jersey in Hurricane Gloria, September 1985," NOAA Technical Memorandum, NWS NHC 36, August.

Morris, F.W. (1987) "Modeling of Hydrodynamics and Salinity in the St. Lucie Estuary," SFWMD Technical Publication 87-1, January.

NOAA Tide Tables, Department of Commerce.

Norton, W.A. and W.H. McAnally (1973) "A Finite Element Model for Lower Granite Reservoir," Water Resources Engineers, Inc., Walnut Creek California.

Sheng, Y.P., S. Peene and Y.M. Liu (1990) "Numerical Modeling of Tidal Hydrodynamics and Salinity Transport in the Indian River Lagoon," Florida Scientist, Vol.53, No.3, p.147.

Smith, N.P. (1990) "Longitudnal Transport in a Coastal Lagoon," Estuarine, Coastal and Shelf Science, Vol.31, pp.835-849.

Thomas, W.A. and W.H. McAnally (1991) "User's Manual for the Generalized Computer Program System: Open-Channel Flow and Sedimentation, TABS-2," US Army Engineer Waterways Experiment Station, Vicksburg, Mississippi.

Williams, J.L. (1985) "Computer Simulation of the Hydrodynamics of the Indian River Lagoon," M.S. Thesis, Florida Institute of Technology, Melbourne.

Effect of Geomorphic Hazards on Bridge Reliability

George K. Cotton, PE[1]

Abstract

A method for addressing the effect of river morphology on hydrologic reliability at bridges. Geomorphic hazards are part of a set of hydrologic hazards at bridges. In general, we can identify four types of hydrologic hazards that can cause bridge failure. We can identify two general groups of geomorphic hazards: natural river instability due or river instability created by man-caused actions. We can classify the remaining hydrologic hazards as at-bridge hazards that are the result hydraulic conditions created by the obstruction of the river valley by the bridge during a flood.

Introduction

Reliability analysis provides a general framework for quantitative assessment of geomorphic risk at bridges. This approach is useful since we can integrate it into a quantitative assessment of hydrologic reliability of bridges.

$$\text{Reliability} = (1 - \text{Probability}_{\text{failure}} * \text{HWF}) * \text{Loss Foregone} \quad \text{(1)}$$

where HWF is a hazard weighting factor.

River instability, either lateral migration or scour, is the prevailing geomorphic hazard at bridges. The proposed procedure uses the stream classification method given in HEC No. 20 and observations of local stream stability near the bridge. Important observations for lateral and vertical stability include: alignment of the stream at the bridge, the proximity of channel bends, the amount of erosion at the bridge abutments, and the history of scour problems at the bridge. Finally, we note the type and condition of appurtenant countermeasures.

Methodology

The methodology consists of three steps:

1. Identify the hydrologic hazards at the bridge;
2. Determine the probability of bridge survival; and
3. Assess the benefits of bridge survival (i.e. the losses foregone).

[1] Senior Water Resources Engineer, Carter & Burgess, 216 Sixteenth Street Mall, Suite 1700, Denver Colorado 80202

Hazard Identification. We identify four groups of hydrologic hazards at a bridge. Hazard groups 1 and 2 are external fluvial geomorphic processes that characterize a particular stream environment. These processes create uncertainty over the long-term that river instability may result due to natural events or man-caused actions. Hazard groups 3 and 4 result from hydraulic conditions occurring at the bridge foundation and superstructure during a flood.

HAZARD GROUP 1.	Existing river instability	External
HAZARD GROUP 2.	Potential for induced instability	Factors
HAZARD GROUP 3.	Structure foundation vulnerability	At-bridge
HAZARD GROUP 4.	Structure collapse vulnerability	Factors

We deal with external factors, in a different manner than internal factors. First, we assume that given a long enough span and proper safeguards, any stream channel can be crossed, regardless of the channel pattern. Second, we recognize that bridges spans over certain channel forms, require countermeasures to maintain acceptable reliability. Likewise, if additional natural or man-caused hazards exist, then additional safeguards are required. The effect of external hazards is based on an assessment of channel form, channel contraction ratio, and countermeasures. The probability of a collapse is then adjusted using a hazard weighting factor.

Hazard Weighting Factor - Existing River Instability

Hazard Group 1 consists of hazards associated with river instability. First we make a classification of stream stability using the method given in HEC No. 20 (see Figure 1). Next, we observe stream alignment at the bridge, the proximity of channel

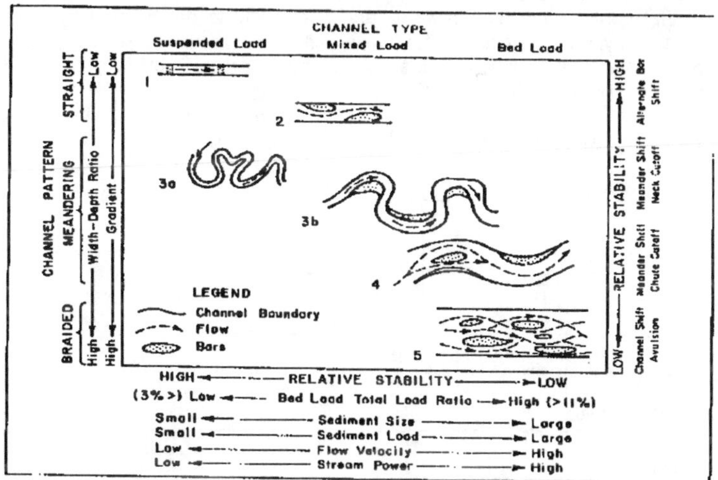

Figure 20. Channel classification and relative stability as hydraulic factors are varied (After [5]).

Figure 1. Channel classification and relative stability

bends, the amount of erosion at the bridge abutments, and the history of scour problems at the bridge. Finally, we note the type and condition of appurtenant countermeasures.

The external hazard weighting factor (XHWF) is determined by comparing weighting factors associated with lateral (WF_l)and vertical channel stability (WF_v), where:

$$WF_l = f_c * f_l * f_{cm} \quad \text{and } WF_l \geq 1.0; \qquad (2.a)$$

$$WF_v = f_c * f_v * f_{cm} \quad \text{and } WF_v \geq 1.0; \text{ and} \qquad (2.b)$$

where XHWF = larger of WF_l or WF_v , and, f_c is the stream stability weighting factor; f_l is the lateral stability weighting factor; f_v is the vertical stability weighting factor; and, f_{cm} is countermeasure weighting factor.

The stream stability weighting factor is a function of channel classification, and the channel contraction ratio, σ, where:

$$f_c = \left(\frac{1}{\sigma}\right)^b \qquad \text{where } \sigma \leq 1 \qquad (3)$$

The factor f_c increases as the contraction ratio decreases or as the stability of the stream channel decreases. This relationship shows that long-span bridges with little contraction of stream flows are not affected by stream channel stability, while bridges that contract the flow severely are more likely to be affected by stream channel stability. The combination of severe flow contraction and poor channel stability will significantly increase the risk to the bridge. Table 1 provides the recommended exponent values for (2).

Table 1. Stream Stability Weighting Factor

Channel Classification[(a)]	Exponent, b
Type 1	0.08
Type 2	0.34
Type 3a	0.68
Type 3b	0.77
Type 4	0.86
Type 5	1.10

Table 2. Lateral Stability Weighting Factor

Waterway Condition	Factor, f_l		
	Low	Moderate	High
Abutment erosion	1.00	1.65	2.80
	Good	Fair	Poor
Channel alignment[(b)]	1.00	1.75	3.00
	None	Moderate	Severe
Channel bend[(c)]	1.00	1.75	3.00

Table 3. Vertical Stability Weighting Factor

Waterway Condition	Factor, f_v		
Degradation within waterway	Low	Moderate	High
	1.00	1.65	2.80
Degradation at abutment	Limited	Moderate	Severe
	1.00	1.75	3.00

Table 4. Existing Counter Measure Weighting Factor

Structure Type	Factor, f_{cm} (d): Counter Measure Effectiveness Good	Fair	Poor
Spurs(e)	0.45	0.75	1.00
Guide Bank(e)	0.25	0.55	1.00
Check Dam(f)	0.20	0.50	1.00
Revetment(e)	0.25	0.55	1.00

Notes on Table Parameters:

(a) The stream channel classification types are according to Figure 20 in HEC No. 20.
(b) Channel alignment: Good 0- to 15-degree angle of attack
Fair 15- to 30-degree angle of attack
Poor >30-degree angle of attack
(c) Channel bend: None, bend radius is greater than 10 time the channel width
Moderate, bend radius is 5 to 10 times channel width
Severe, bend radius is less than 5 times channel width
(d) If multiple counter measures exist, the total weighting factor is the product of the weighting factors for each counter measure.
(e) Use for lateral stability only.
(f) Use for vertical stability only.

We determine the overall effect of geomorphic effects by a hazard weighting factor (HWF) that compares weighting factors associated with lateral (WF_l) and vertical channel stability (WF_v), where:

$$WF_l = f_c * f_l * f_{cm} \quad \text{and } WF_l \geq 1.0; \tag{2.a}$$

$$WF_v = f_c * f_v * f_{cm} \quad \text{and } WF_v \geq 1.0; \text{ and} \tag{2.b}$$

where HWF = larger of WF_l or WF_v, and
f_c is the stream stability weighting factor,
f_l is the lateral stability weighting factor,
f_v is the vertical stability weighting factor, and
f_{cm} is countermeasure weighting factor.

Summary

The goal in adding geomorphic analysis to hydrologic reliability analysis is to develop a method that consistently ranks the scour susceptibility of bridges. A ranking of most scour reliable to least scour reliable provides an overview of the bridge structure population in a way that resources can be prudently allocated. Also because the ranking is in economic terms, transportation agencies can make decisions based on the cost of scour hazard mitigation versus the benefit derived. The ranking can also serve as an indicator for non-economic risks to the traveling public.

References:

AASHTO, "A Manual on User Benefit Analysis of Highway and Bus Transit Improvements", 1977

Annandale, G.W. "Risk Analysis of River Bridge Failure", from Hydraulic Engineering '93, edited by Hsieh Wen Shen, S.T. Su and Feng Wen, proceedings of the 1993 ASCE Hydraulics Division Conference.

Benjamin, J.R., Cornell, C.A., Probability, Statistics and Decisions for Engineers, McGraw-Hill Book Company, 1970.

Callander, Stephen J., "Fluvial Processes Occurring at Bridge Sites", Masters Thesis, Colorado State University, 1980.

Elias, V., "Strategies for Managing Unknown Bridge Foundations", Federal Highway Administration, FHWA-RD-92-030, February, 1994.

FHWA, "Recording and Coding Guide for the Structure Inventory and Appraisal of the Nation's Bridges" Bridge Division, 1988.

Lagasse, P.F., Schall, J.D., Johnson, F., Richardson, E.V., Richardson, J.R., and Chang, F., "Stream Stability at Highway Structures", Hydraulic Engineering Circular No. 20, FHWA-IP-90-014.

Li, R.M, Cotton, G.K., Zeller, M.E., Simons, D.B., Deschamps, P.Q., "Effects of In-Stream Mining on Channel Stability" Arizona Department of Transportation, Report HPR-PL-1-31, 1988.

Rhodes, J., Trent, R., "Economics of Floods, Scour, and Bridge Failures", from Hydraulic Engineering '93, edited by Hsieh Wen Shen, S.T. Su and Feng Wen, proceedings of the 1993 ASCE Hydraulics Division Conference.

Richardson, E.V., Harrison, L.J., and Davis, S.R., "Evaluating Scour at Bridges", Hydraulic Engineering Circular No. 18, FHWA-IP-90-017.

Sabol, G.V., Nordin, C.F., and Richardson, E.V., "Scour at Bridge Structures and Channel Aggradation and Degradation Field Measurements", Arizona Department of Transportation, FHWA-AZ90-814, 1990.

Smith, P.N., Vose, W.O., "Texas Secondary Evaluation and Analysis for Scour (TSEAS)" Division of Bridges and Structures, Hydraulics Section, September, 1993.

Young, K., Stein, S.M., Trent, R., "Risk Cost for Scour at Unknown Foundations", from Hydraulic Engineering '92, edited by Marshall Jennings and Nani G. Bhowmik, proceeding of the 1992 ASCE Hydraulics Division Conference.

Soil Bioengineering for Stream Restoration

Robbin B. Sotir[1]
Nelson R. Nunnally[2]

Abstract

Soil bioengineering, sometimes referred to as biotechnical stabilization, is an effective technology for reducing streambank erosion and restoring degraded aquatic and riparian ecosystems. Soil bioengineering systems that are most effective for these purposes employ native, woody pioneer species that provide immediate bank protection and cover. These pioneer systems evolve through natural invasion and plant succession into diverse ecosystems capable of supporting a rich abundance of riparian and aquatic species. Three case studies are presented to illustrate the value of soil bioengineering in protecting streambanks and restoring stream channels in different environmental settings.

Introduction

Except for water, woody vegetation has more effect on channel stability and aquatic habitat than any other single factor. The stabilizing influence of woody vegetation on streambanks is well documented, and its loss can result in bank erosion that leads to wider channels and shallow flow. Removal of streamside vegetation results in the loss of shade and organic debris, which can result in higher water temperatures, reduced cover, and lower aquatic productivity. Clearly, then, restoration of streamside vegetation is an important aspect of stream restoration.

Soil bioengineering systems that employ native, woody pioneer vegetation can be effective in stabilizing streambanks and restoring degraded aquatic and riparian ecosystems. When properly designed and installed, soil bioengineering systems provide immediate bank protection and cover. Over time they evolve through invasion and plant succession into healthy diverse plant communities. Three case

[1] Principal, Robbin B. Sotir & Associates, 434 Villa Rica Road, Marietta, Georgia 30064
[2] Fluvial Geomorphologist, Robbin B. Sotir & Associates, 434 Villa Rica Road, Marietta, Georgia 30064

studies that follow are presented to illustrate the value of soil bioengineering for protecting streambanks and restoring riparian vegetation.

Crow Creek

Crow Creek, in Marshall County, Illinois, drains a 220 km^2 rural watershed that is heavily used for agriculture. The stream was channelized and straightened in the late 1930's and early 1940's. Following a series of large flood events in 1969 and 1970 the steepest reach of the channelized stream developed accelerated bank erosion. Local officials of the Marshall-Putnam Conservation District estimated that in 1986 alone more than 5000 m^3 of sediment was eroded from this site. The top width of the eroding reach had widened from about 25 m to about 100 m and vast quantities of sediment were deposited in the downstream reaches.

Vegetated dikes, called live booms, were designed to move the thalwag away from the bank toe. Live siltation constructions were installed between the structures to help trap sediment. Upper bank areas were treated with live fascines and coir. Photo 1, taken during the first growing season following installation provides an overall view of the project. The new soil bioengineering installation survived a record flood of 140 m^3 s^{-1} per second five months after installation just as the vegetation was beginning to establish, without damage. The flood uprooted trees, destroyed highway bridges, and caused extensive damage upstream and downstream of the project site.

Photo 1. Upstream view of Crow Creek six months after installation.

Longfellow Creek Bypass Channel

Longfellow Creek, in Seattle, Washington, is a small stream that drains a residential watershed. A bypass channel was constructed to alleviate flooding and stream erosion problems in the lower reaches of Longfellow. The constructed bypass channel was a straight, trapezoidal shaped drainage ditch that was to be lined with gabion wire baskets filled with rock, but residents objected to the sterile appearance of the gabions and the lack of fish habitat. A cost effective soil bioengineering design was selected that would stabilize the steep banks (as steep as 0.5:H to 1:V) while providing the desired habitat and aesthetic benefits.

Channel banks were constructed from vegetated geogrids, consisting of soil lifts wrapped in coir (a woven geotextile made from coconut husk) with layers of live branches between each lift. The geogrids were constructed on a foundation of rock wrapped with a polymer mesh that was installed below the channel invert. A special gravel substrate was used for the channel bottom, and large boulders were placed in the channel to provide cover for fish. Photo. 2 shows the channel two years after construction in the second season of growth.

Photo 2. Longfellow Creek bypass channel after two years.

Johnson Creek

Johnson Creek drains an urbanized watershed in the Portland, Oregon metropolitan area. Johnson Creek is a third order stream with a 100-year flood discharge at the project site of about 140 m^3s^{-1}. Land uses in the watershed range from low density residential to heavy industry. Because of increased public concern about the condition of the stream, which has significantly degraded water quality and aquatic habitat, and an interest in restoring an andromous fishery, a citizens' committee was appointed to monitor developments within the stream corridor. Largely due to the recommendations of this committee, Oregon Department of Transportation (ODOT) elected to use soil bioengineering for bank stabilization and habitat restoration along a section of Johnson Creek relocated for construction of a new bridge and interchange.

The cross-section of the realigned channel was redesigned as a compound channel with a low flow channel to improve stability and deeper water during low flows. Soil bioengineering systems were designed to stabilize the streambanks and to provide improved cover and riparian habitat. Soil bioengineering systems installed along Johnson Creek include live fascines, live siltation constructions, vegetated geogrids, and brushmattress. (See Photo 3).

Photo 3. Downstream view of Johnson Creek in the first spring.

The soil bioengineering systems were installed during the winter of 1993 and spring of 1994, and the first evaluation and monitoring report was completed during the fall of 1994. The systems were undamaged by spring floods and were providing excellent bank protection and habitat benefits. Based on observations made during the evaluation visits, there appears to have been no decline in the use of the stream by waterfowl, herons, kingfishers, and other birds attracted to the stream. Over the next five (5) years as the number of plant species and structural diversities increase the habitat value and species diversity of the site for birds, small animals, and reptiles will also increase. Photo 4 shows the excellent cover provided by the live siltation constructions in the first year's growth.

Photo 4. Live siltation constructions on Johnson Creek after one summer's growth.

The projects range in age from one (1) year (Johnson Creek) to five (5) years (Longfellow bypass channel and Crow Creek). All have survived major flood events shortly after construction and have been flooded several times since without experiencing significant damage. Environmentally, they have shown to improve habitat value for a variety of species. The success of these projects attests to the value of soil bioengineering in stream restoration efforts.

Vegetation and stream stability: a scale analysis

Ian D. Rutherfurd[1], Kathryn Jerie, and Michael Wright

Abstract

Vegetation is increasingly being used as a first-line management technique for river bank erosion. This paper argues that the influence of native vegetation on erosion rates and stream hydraulics will vary through the stream network. This is because the size and shape of the river, the erosion processes, and the suite of vegetation species, all change through the stream network. The interaction of vegetation and erosion is demonstrated using a 'scale analysis' of the Latrobe River in SE Australia. A scale analysis matches the stream erosion processes to the vegetation characteristics. This analysis shows that vegetation has little effect on hydraulic resistance below the upper 10% of the catchment area. Similarly, the size of root balls becomes too small to influence slumping or undercutting processes through most of the floodplain tract. Finally, the period of inundation of riparian vegetation varies dramatically along the river. Considering all of these variables we define the river reaches in which vegetation will be most effective for erosion control.

Introduction

Vegetation is increasingly being used as a first-line management technique for river bank erosion. The role of vegetation in slope stability (particularly rotational failures) has been well studied (Gray and Leiser 1982). Much less is known about the role of vegetation in river bank erosion processes (Thorne 1990). It is widely believed that vegetation canopies reduce erosion by reducing flow velocity, and by protecting the bank face from scour, whilst roots increase the resistance of boundary material. On the strength of this belief, many millions of dollars are being spent in Australia every year in revegetating stream banks, particularly with native vegetation species. This paper argues that the influence of native vegetation on erosion rates will vary through the stream network. This is because the size and shape of the river, the erosion processes, and the suite of vegetation species, all change through the stream network. The interaction of vegetation and erosion is demonstrated using a 'scale analysis' of the Latrobe River in SE Australia. A scale analysis matches the stream erosion processes to the vegetation characteristics so that managers can plant vegetation in the reaches of the river where it will be most effective. This work is a preliminary part of a national research project investigating physical and chemical processes in Australian riparian zones.

The Latrobe River is located in Eastern Victoria (Lat. 38°10', Long 146°15'). The river is 242 km long, and drains a catchment of 5,200 km^2. The river is alluvial for almost its full length, with few bedrock reaches. The lower Latrobe has a meandering, single-thread channel that is about 35 m wide and 5-6 m deep, with silt-clay banks and a sand bed. The lower 30 km of the Latrobe is in the backwater of Lake Wellington. Rainfall ranges from 600 mm to 1600 mm. Much of the headwaters remain forested, whilst the floodplain is cleared for cattle grazing.

---

[1]Research Fellow, Cooperative Research Centre for Catchment Hydrology, Department of Civil Engineering, Monash University, Clayton, Victoria, 3168,

We surveyed eleven sites from the headwaters to the mouth of the Latrobe River (six of the sites are at stream gauges) (Figure 1), describing boundary sediments, erosion processes, and vegetation at each site. For the purposes of the study, we considered vegetation on the edge of the floodplain, on the stream banks, and in the channel. Vegetation in the headwaters is wet, closed canopy *Eucalyptus regnans* forest, whilst on the lower floodplain the riparian vegetation was originally an open forest of wattles (*Acacia dealbata*) and redgum (*E. camaldulensis*). Today the lower channel is lined with basket willows, with wattles often lining the bank face (Figure 1). We will discuss the role of vegetation in increasing flow resistance (and so reducing velocity and erosion) on the Latrobe, after which we will consider the role of vegetation in the erosion processes along the stream.

Vegetation and channel hydraulics

Flow velocity is affected by live vegetation projecting into the channel area, but also by dead vegetation (large woody debris, LWD) in the channel bed. The hydraulic effects of vegetation in flow are complex (Kouwen 1988), so we assume here that the hydraulic effect is proportional to the area that the vegetation projects into the bankfull flow (the blockage ratio). This area is estimated for the natural suite of vegetation that would have lined the banks. Below Willowgrove, LWD in the Latrobe has been artificially removed up to four times, so the LWD projected area is estimated from measurements made by Gippel et al. (1992).in the nearby lower Thomson River which has not been snagged. The median LWD projected area on the lower Thomson (similar size to the Latrobe at about Thoms Bridge) is about 1 m^2/m. Figure 2 shows that more of the channel is blocked by LWD than by live vegetation, and both live vegetation and LWD occupy a progressively smaller proportion of the channel downstream. Gippel et al. (1992) suggest that LWD will have little influence on velocity below 10% blockage ratio. In the upper reaches the vegetation and LWD occupy up to 80% of the cross-section, having a large impact on flow resistance. The combined blockage ratio falls rapidly to 20% at the Ada River Junction (L3 on Figure 1), and to only 2% at the mouth of the river. Thus, for over 80% of the river's length, vegetation probably has little influence on flow velocity.

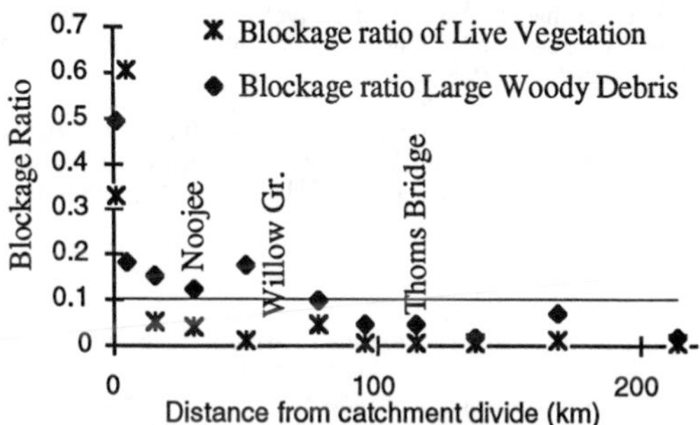

Figure 2. Proportion of cross-section area (blockage ratio) occupied by vegetation

Erosion processes and vegetation

The role of riparian vegetation in channel erosion processes can best be understood by dividing the Latrobe River into six reaches, with each having a specific relationship between bank vegetation and erosion (refer to Figure 1 for the location of the reaches).

Reach 1: Streams within about 10 km of the drainage divide (1st to 4th order streams) are narrower than the dominant floodplain vegetation (<10 m) (Figure 1) so that fallen trees span, and choke the channel. The main erosion process is bank undercutting below the 0.3 m - 0.5 m root-zone of vegetation. Banks can be undercut by 0.5 m. The floodplain is narrow (less than 10 m on each side) and permanently saturated. Importantly, it is only in this reach that riparian vegetation will buffer the direct runoff from hillsides. Further downstream the floodplain becomes too wide (>40 m each side).

Reach 2: From 10 - 40 km from the divide (300 km^2 catchment, ie. 6% of catchment area) the channels are still narrower than the height of the dominant *Eucalyptus* trees. The channel banks are vertical and too low to sustain trees on the bank face. Undercutting below the root-zone is still the major erosion process.

Reach 3: Three major changes occur in this transition zone from the confined floodplain reaches to the broad unconfined alluvial floodplain (40 km - 60 km from the divide). First, the banks become high enough to sustain trees on the bank face and as a result, the root-zone of trees on the bank face extends below the low water line. Second, the channel becomes wider than the dominant riparian tree (>15 m), and fallen trees tend to be swept against the bank at an angle of about 30°. At this angle, the LWD does not divert flow into the banks. The third change in this reach is that the channel begins to meander so that undercutting is concentrated on one bank.

Reach 4: In this upper floodplain reach (60 km - 100 km, up to 600 km^2 catchment area) the slope of the floodplain is away from the channel as levees develop. Clearly the riparian vegetation can play little role in buffering runoff. The channel begins to meander in this reach, and erosion is concentrated on the concave bank. Bank slumping is the dominant erosion process. The average bank slump is 2 m - 2.5 m long and 0.5 m - 1.5 m wide. These bank-slump blocks are smaller than the root-balls of the dominant wattle trees, thus the roots from the trees extend through the slump blocks and increase the shear strength of the bank.

Reach 5: (100 km - 200 km) The river reaches its largest dimensions in this reach, being up to 50 m wide and 7 m deep (almost twice as deep as reach 4). Erosion, by slumping, is most pronounced on the concave bank where the banks are vertical. Hence, the roots from trees on top of the concave banks do not reach the mean water-level of the river. Where there are trees on top of the outside banks they are often undercut by up to two metres. Importantly, the slump blocks in this reach are over twice as large as those in reach 4 (3 m - 5 m wide, 10 m long, and 1.5 m - 2 m deep). This means that the failure blocks can be larger than the root ball of the typical *Acacia* vegetation. Therefore, the roots from trees on the banks would provide little shear strength to the bank, and the trees could even increase block failure by surcharging the banks.

Reach 6: In the backwater of Lake Wellington (200 km - 230 km) the channel widens, banks are vertical, and the stage varies by only 0.5 m. Water-logged sediments limit the root depth of vegetation to about 1 m deep, and as in reaches 1 to 3, erosion is by undercutting below this root zone. There is little slumping.

Flow duration

In relation to all interactions between vegetation and channel processes it is important to consider flow duration. In terms of flow resistance and direct protection of the banks from scour, vegetation can only influence channel processes if it is in contact with the flow. When stage duration is plotted relative to bankfull depth (Figure 3) it is clear that, on the Latrobe, the shape and hydrology of different reaches means that different portions of the bank are underwater for different amounts of time. For example, planting vegetation on the upper-half of the bank at Hawthorn Bridge, where flow only occupies the top three fifths of the bank for less than 2% of the time, may be less effective than at Noojee, where flows occupy the top 3/5 for 97% of the time. Similarly, revegetating the top metre of the bank at Rosedale would provide direct protection for more than 10% of the time, but less than 1% of the time at Willowgrove or Thoms Bridge.

Conclusions

Replanting stream banks with native vegetation is becoming the most common stream management technique in Australia. However, the effectiveness of riparian vegetation in improving stream stability, altering stream hydraulics, or buffering runoff varies dramatically down a stream system. The qualitative assessment of the Latrobe River described in this study demonstrates this

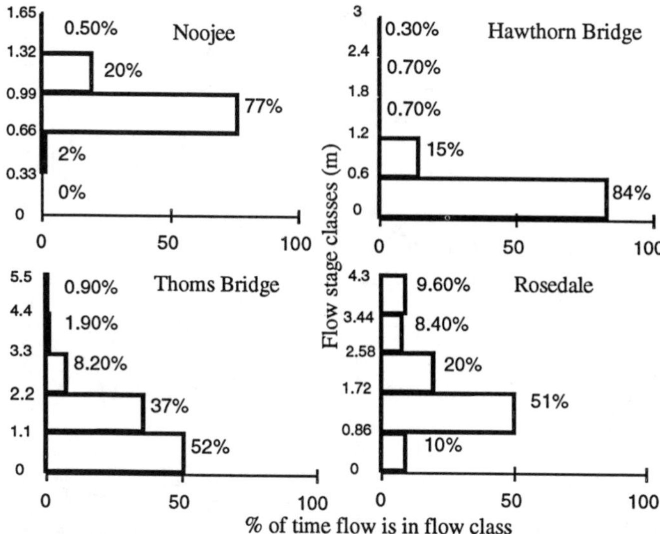

Figure 3. Bankfull depth at four stream gauges on the Latrobe River is divided into 5 stage classes. The graphs show the percentage of time that the flow stage lies within each class.

variation. Under natural conditions (which managers are often hoping to imitate) native vegetation has a diminishing effect downstream as the channel increased in size. Flow resistance by both live vegetation and large woody debris are probably negligible in the floodplain tract (below the upper 10% of the catchment area on the Latrobe). Similarly, vegetation slows the erosion process in the upper reaches (in reach 3 and above) where the roots reduce undercutting. But by reach 5 the roots of riparian vegetation do not extend to the mean water level, and slump blocks (the major erosion mechanism) are larger than the root ball.

For erosion control on the Latrobe River, bank revegetation will be most effective in reach 4 where the root ball of the dominant vegetation extends below the low water mark, and matches the size of the slump blocks. Finally, the effectiveness of vegetation is also influenced by the period of time vegetation is inundated. Vegetation planted on the top 40% of the banks on the Latrobe could be inundated for as much as 20% of the time, or as little as 1% of the time. To conclude, it is imperative that river managers match the scale of the process that they are wishing to influence, to the scale of the suite of vegetation that they plant along rivers. The majority of revegetation work in Australia is now carried-out in the larger channels where it is likely to have least effect.

References

Gippel, C.J., Ian O'Neill, and B. Finlayson (1992) The hydraulic basis of snag management. Melbourne University, Report to the Land and Water Resources Research and Development Corporation, Canberra, Australia.

Gray, D.H. and A.T. Leiser (1982) Biotechnical slope protection and erosion control. Van Norstrand Reinhold. New York .

Kouwen, N. (1988) "Field estimation of the biomechanical properties of grass." Journal of Hydraulic Research 26(5), 559-568.

Thorne, C.R. (1990) "Effects of vegetation on riverbank erosion and stability." In Thornes, J.B., (ed).) Vegetation and Erosion. 125-144. John Wiley and Sons Ltd..

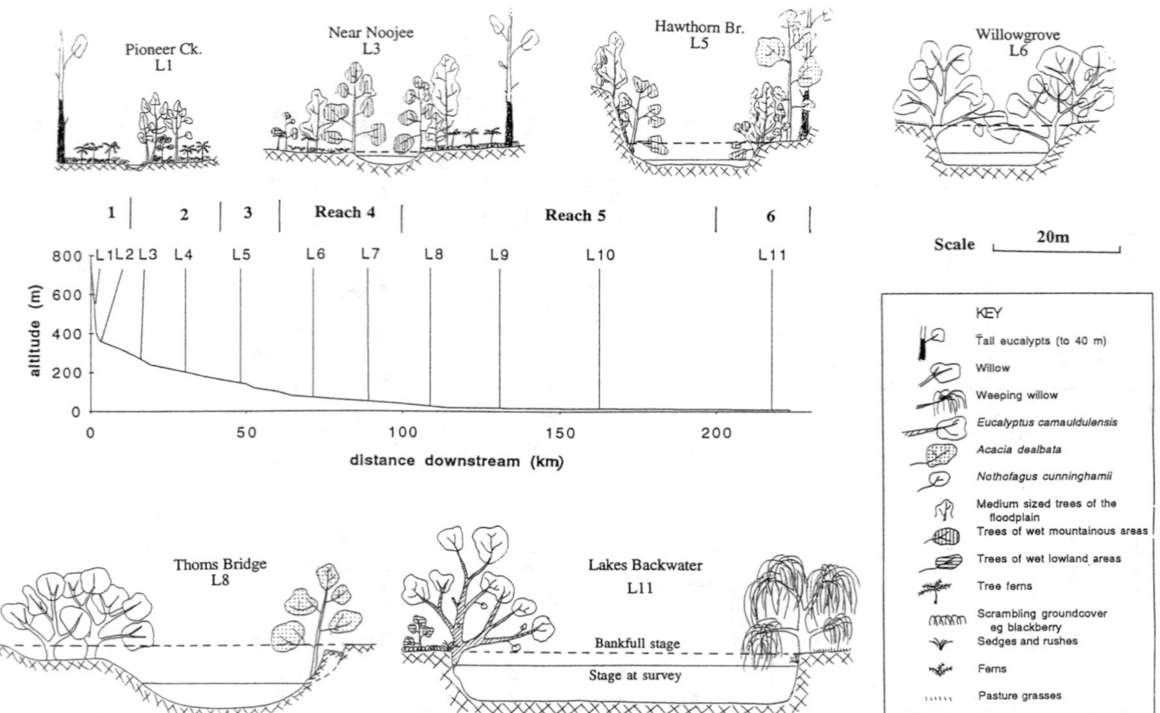

Figure 1. Long-profile of Latrobe R. with reaches, cross-section locations, and some typical cross-sections and vegetation.

ESTIMATING FLOW RESISTANCE IN VEGETATED CHANNELS

J. Craig Fischenich (M ASCE)[1] Steven R. Abt (F ASCE)[2]

Abstract

The authors are conducting research to identify procedures for the estimation of resistance coefficients for flows in densely vegetated floodways. Existing methods are presented and evaluated by comparing resistance coefficients obtained by the application of each existing method to coefficients computed from measured friction slopes for several river reaches containing vegetation in the floodway. Results of the comparison are presented, each method is discussed, and the attributes of an appropriate resistance relation are proposed.

Introduction

Hydraulic engineers have become increasingly involved in channel restoration projects and modifications to existing flood control projects as the conventional "flood control" ideology is replaced with a new "flood management" philosophy. The incorporation of vegetation into these projects is often mandated. Healthy riparian vegetation tends to stabilize stream banks, provides shade that prevents excessive water temperature fluctuations, performs a vital role in nutrient cycling and water quality, improves aesthetic and recreational benefits of a site, and is immensely productive as wildlife habitat.

Concurrent with these benefits are impacts to the channel and floodway conveyance with subsequent sedimentation and stability impacts. Proper evaluation of these impacts requires an analysis of flow depths and velocities in the channel and floodplain. Although more complicated analyses are often warranted and sometimes performed, engineers typically evaluate river systems using one-dimensional, steady, gradually-varied flow models. Depending upon the model used, engineers can account for the effects of dense vegetation by adjusting one or more of the following parameters: channel cross section, velocity coefficients, momentum coefficients, flow area subdivision, or resistance coefficients.

Virtually no guidance exists for the adjustment of the first four parameters and only limited guidance is available for the selection of resistance coefficients in vegetated floodways. In the U.S., it is customary to express the flow resistance in terms of the resistance coefficient

1. Research Civil Engineer, Environmental Laboratory, US Army Engineer Waterways Experiment Station, Vicksburg, MS 39180

2. Director, Hydraulics Laboratory, Department of Civil Engineering, Colorado State University, Fort Collins, CO 80523

from Manning's Monomial Equation, n. Procedures for the computation or estimation of Manning's n in vegetated channels can be grouped into four general categories: direct measurement, analytical approaches, and handbook methods. Direct measurement, although important for model calibration and verification, is of little practical use in prediction and is not presented.

Existing Techniques

Analytical Approaches: Cowan (1956) proposed a procedure for estimating Manning's n that takes into account the contributions of various factors, including vegetation, to total flow resistance. The procedure, assumes that the resistances induced by various contributing factors can be summed to establish total resistance as follows:

$$n=(n_b+n_1+n_2+n_3+n_4)m \quad (1)$$

where m = ratio for meandering, n_b = base n value, n_1 = addition for surface irregularities, n_2 = addition for variation in channel cross section, n_3 = addition for obstructions, and n_4 = addition for vegetation. Arcement and Schneider (1989) summarize estimates for each of the factors as presented by several previous investigators. Adjustment factors for vegetation vary from 0.002 to 0.100 and, as with the other factors, selection of an appropriate adjustment is based upon interpretation of qualitative descriptions of channel conditions.

Petryk and Bosmajian (1975) proposed a modification to Cowan's approach that more quantitatively addresses the effect of vegetation. By summing the forces in the longitudinal direction and substituting into the Manning formula, they developed the following equation:

$$n=n_0\sqrt{1+\left(\frac{C_*\Sigma A_i}{2gAL}\right)\left(\frac{k_n}{n_0}\right)^2 R^{4/3}} \quad (2)$$

where n_0 = Manning's boundary-roughness coefficient excluding the effect of the vegetation, C_* = the effective-drag coefficient for the vegetation in the direction of flow, ΣA_i = the total frontal area of trees in the flow section, g = the gravitational constant, A = the cross-sectional area of flow, L = the length of channel reach, and other parameters are as previously described. The total boundary roughness n_0 is determined from a modified form of Cowan's equation:

$$n_0=n_b+n_1+n_2+n_3+n_4' \quad (3)$$

The drag coefficient can be approximated from the relation $C_*=22-(3.75)R$. The roughness factors n_b and n_1 through n_3 are the same as those for Cowan's method. The n_4' factor is for vegetation such as shrubs, brush and grass, which was not accounted for in the vegetation density computations, and ranges from 0.001 to 0.025.

Ree and Palmer (1949) summarized research of flow in vegetated channels conducted by several SCS researchers from 1935 to 1943 at Spartansburg, GA and Stillwater, GA. The investigators found that most flow data for a particular grass, when plotted with n as a function of the product of velocity and hydraulic radius, would fall approximately along a single line. The most frequently reproduced graph from these experiments summarizes the n - VR curves for five "classes" of vegetation, each class considered to have similar properties. Use of this graph is referred to as the SCS method. Kouwen and Unny (1973) proposed an improvement to the SCS method by suggesting that vegetation be classified on the basis of its flexural rigidity, defined by the product MEI. Kouwen, Li, and Simons (1981) present tables of MEI values for several species of vegetation.

Handbook Methods: Establishment of flow resistance with procedures that do not rely on direct measurement or numerical analysis are referred to as "handbook methods". Included are tables of roughness values and compilations of photographs with descriptive information and calculated n values. Chow (1959) presented tables of minimum, normal, and maximum values of Manning's n for conduits, lined canals, and natural channels. This work is probably the most prevalent reference for the selection of n values. Other tables of roughness values have been compiled, but offer little additional insight. Of the 111 channel and floodplain types listed in Chow's table, 27 include vegetation, and for these Chow suggests values of n ranging from 0.022 to 0.200. Selection of an appropriate value is accomplished by interpreting brief anecdotal descriptions of channel conditions.

Barnes (1967) presents color photographs and descriptive data for 50 stream channels, nearly all of which have vegetation on the banks. Hicks and Mason (1991) presented similar information for 78 rivers reaches in New Zealand encompassing a broad range of conditions. Unlike Barnes, Hicks and Mason use multiple photographs for each reach, present bed material gradations, summarize all pertinent hydraulic data, compute values for both Manning's n and Chezy's C, and provide an estimate of computation error. They also evaluated multiple discharges for each reach. Like Barnes, Hicks and Mason avoided computation of flow resistance in floodplains, although their work does provide some insight into the contribution of bank vegetation to channel roughness.

Arcement and Schneider (1989) presented photographs for 15 densely vegetated floodplains for which roughness coefficients have been verified. Using Cowan's procedure and the method proposed by Petryk and Bosmajian, they used measured vegetation density in the floodplain and an effective drag coefficient to calculate the contribution of vegetation to the total roughness. Values for Manning's n ranged from 0.10 to 0.20. The contribution due to vegetation ranged from 0.065 to 0.145 which constituted 64 to 81 percent of the total value.

Evaluation of Techniques

Fifteen channel reaches with measured n values were used to assess the prediction methods described above. Four of the reaches were excavated canals, one was a laboratory flume and the remainder were natural channels. Six of the channels were evaluated for a discharge contained within the banks and seven for overbank flows. Three general cases of vegetal retardence were represented; 1) dense vegetation on the streambanks, 2) submerged or partially submerged aquatic vegetation, and 3) dense vegetation on the floodplains. Limitations in the length of this paper prevent a complete description of each reach. However, comparisons of the estimated and actual n values are summarized in the following figure and the accompanying discussion.

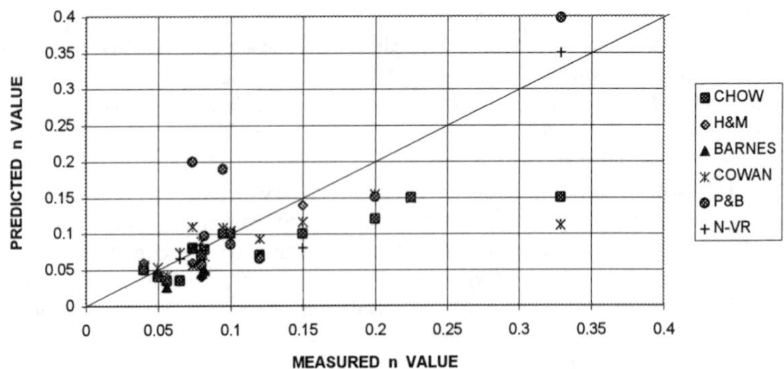

In general, prediction error increased with increasing degree of resistance. Coefficients of variation for each method were: n-VR - 0.177, Chow - 0.247, Cowan - 0.249, Barnes - 0.326, Hicks and Mason - 0.330, and Petryk and Bosmajian - 0.471. While estimates could be made for all 15 reaches using the methods by Chow and Cowan, the others proved far less applicable. Reasonable matches in Hicks and Mason's handbook could be found for only eight cases, and Barnes' handbook could be used for only six. Measurements of vegetation density were available for only four sites, but were estimated for an additional four, allowing the method by Petryk and Bosmajian to be used on eight reaches. The n-VR method was applicable to only three of the fifteen cases investigated, and these were somewhat outside the range of data for which the method was developed.

Conclusions

Most engineers rely upon experience and judgement, along with a standard reference such as Chow's tables, to select resistance coefficients for hydraulic and channel stability analyses. The authors have speculated that a more quantitative approach would improve accuracy and prove more useful for investigations where dense vegetation is present in the floodway. A predictive method incorporating variables such as flow depth, percentage of the wetted perimeter covered by vegetation, vegetation density, vegetation response to flow, and vegetation alignment would be expected to outperform standard methods. No such predictor has been proposed, but the methods evaluated in this investigation include some of these variables. This investigation revealed several faults in the application of these methods.

Petryk and Bosmajian's vegetation density technique is hindered by the additional data collection requirements and is only applicable to densely vegetated floodplains. The n-VR method is limited by the range of slopes over which data were collected (greater than three percent), the types of vegetation used (grasses only), and difficulties in selecting an appropriate curve. The MEI analyses proposed by Kouwen et. al. explicitly includes the geometric and physical properties of the vegetation in the analysis of flow resistance. Though not substantially verified, this method appears to provide a sound process for the assessment of resistance due resistance due to some types of vegetation. Unfortunately, the method could not be applied to any of the reaches in this investigation because of a lack of data.

Handbook methods relying upon pictoral descriptions of channels are convenient and offer the advantage of implicitly compositing channel resistance, but they offer little utility in cases where vegetation is present or where floodplain flows are anticipated. Only a dozen such circumstances are addressed in the collective works of the handbooks cited. The authors found that reliance upon these handbooks can actually be misleading, resulting in greater errors than simply using qualitative channel and floodplain descriptions.

Though Chow's handbook tables and Cowan's procedure rely upon interpretation of qualitative descriptions for the assignment of resistance values, they proved to be the most applicable techniques and had lower prediction errors than all but the n-VR method (which only applied to three cases). Additional work on the development of resistance relationships for vegetated floodways should focus on a procedure that exhibits the following characteristics: resistance should be related to readily defined, measurable characteristics of the channel, vegetation, and flow; resistance should be described as a continuous function of the independent variables involved; and the resistance function should be dimensionally homogeneous. None of the methods evaluated possess these qualities.

References

Arcement, G. J., Jr., and Schneider, V. R. (1989). Guide for selecting Manning's roughness coefficients for natural channels and flood plains, *U.S. Geological Survey Water-Supply Paper 2339*, Denver, CO.

Barnes, H.H. (1967) Roughness characteristics of natural channels. US Geological Survey, Water-supply Paper No. 1849, 214 pp. Washington, DC.

Chow, V.T. (1959) *Open-Channel Hydraulics*. McGraw-Hill Book Co., New York.

Cowan, W.L. (1956) Estimating hydraulic roughness coefficients. *Agricultural Engineering 37*, 473-475.

Hicks, D.M. and Mason, P.D. (1991) Roughness Characteristics of New Zealand Rivers. *Water Resources Survey*, 1-9.

Kouwen, N. and Unny, T.E. (1973). Flexible Roughness in Open Channels. *Journal of the Hydraulics Division, Proceedings of the American Society of Civil Engineers 99*, 713-728.

Kouwen, N., Li, Ruh-Ming, and Simons, D.B. (1981). Flow Resistance in Vegetated Waterways. *Transactions of the American Society of Agricultural Engineers 24*, 684-690, 698.

Petryk, S. and Bosmajian, G., III. (1975). Analysis of Flow Through Vegetation. *Journal of the Hydraulics Division, Proceedings of the American Society of Civil Engineers 101*, 871-885.

Ree, W.0. and Palmer, V.J. (1949) Flow of water in channels protected by vegetative linings. *US Department of Agriculture. Soil Conservation Service Technical Bulletin No. 967*, 115 pp.

Assigning Weights to Precipitation Stations

Thomas T. Burke, Jr.[1] and A.R. Rao[1]

Abstract

In rainfall-runoff analysis the areal average precipitation over a basin is estimated by methods such as arithmetic-mean, Thiessen polygon, isohyetal and others. The greatest drawback in these procedures is that the weights associated with rainfall stations do not accurately reflect their contribution to runoff.

In order to better delineate the relationship between $Q(t)$, the measured runoff at a station, and the rainfall $P_i(t)$ measured at the i-th precipitation station, Sugawara (1992) has proposed the linear model $\hat{Q}(t) = \sum_i w_i \hat{P}_i(t)$ in which w_i are the weights, $\hat{Q}(t)$ is the runoff estimated by using the model and $\hat{P}_i(t)$ is the runoff contributed by each precipitation station. The weights w_i are estimated by using principal axis transformation and factor analysis. The weights w_i define the contribution of different precipitation stations to the total runoff $Q(t)$.

In the present study, Sugawara's method is used to analyze the contribution of different rainfall stations to the flow in the Upper Wabash River basin in Indiana. The choice of input sequence $\hat{P}_i(t)$ is discussed. The relationship between the weights and the cross correlation between inputs and outputs are also discussed.

Introduction and Statement of Problem

Estimating the areal average precipitation over a basin is a common hydrologic problem. There are several methods which have been developed to determine the average rainfall depth. Three of the most common methods are the *arithmetic-average, Thiessen* and *ishohyetal* methods. Each of these has its benefits and drawbacks. The simplest of these methods is the arithmetic-average

[1] School of Civil Engineering, Purdue University, W. Lafayette, IN 47907

method in which point rainfall values in an area are arithmetically averaged to find the average depth. This method is unsatisfactory if the gages are not distributed uniformly and if the measurements at different raingages vary significantly from the mean. The Thiessen averages account for the distribution of raingages over an area. However, a new set of polygons must be drawn each time the location of gaging stations changes. Also, the method assumes linear variation of precipitation between stations and assigns each segment of area to the nearest station. The isohyetal method is the most accurate method of averaging precipitation over an area. The isohyetal method represents an accurate map of the rainfall pattern. However, an extensive gage network is needed to accurately draw the isohyets. A new set of isohyets must be drawn for each storm.

In order to improve the accuracy of estimation of rainfall over an area, Sugawara (1992) proposed that the weights associated with precipitation stations be determined by their contribution to runoff. In Sugawara's method the weights associated with precipitation stations are found by minimizing the difference between the observed discharge and the discharge computed by using the precipitation measured at each station. All the stations are initially assigned equal weights, and a runoff model is used to calculate the contribution to runoff from the precipitation measured at each recording station.

In Sugawara'a method a linear relationship is assumed between the contribution of runoff from each of the rainfall stations to the runoff from the watershed. The weights are derived by using this relationship. In essence, the problem is reduced to a form of the unconstrained inverse problem in system identification, in which the parameters describing a system are estimated by using the input and output data.

The theory and method to obtain the optimum set of weights for precipitation stations has been described in Sugawara's paper. However, the method does not appear to have been tested by using observed rainfall-runoff data. It is the purpose of this paper to apply the techniques presented by Sugawara to the rainfall-runoff data from Upper Wabash River watershed in Indiana to examine the characteristics of the method.

Theory

The principal axis transformation and factor analysis are the techniques used by Sugawara (1992). Details of the method are found in Sugawara's paper and hence it is only briefly discussed here. In this method, precipitation from different stations are multiplied by weights to get the observed runoff. The weights which yield an optimum estimate of observed runoff is computed by this method. Monthly data are used in the following discussion.

Let the number of stations from which the data are available be k. The monthly average runoff is calculated from the input at the ith station in the lth month; $x_i(l)$ $(i = 1, \ldots, k;\ l = 1, \ldots, n)$, where n is the length of the time series. Similarly, the monthly observed runoff (the output) in the lth month is denoted by $y(l)$ $(l = 1, \ldots, n)$.

The vectors $\boldsymbol{x}_i$ and $\boldsymbol{y}$ of input and output are normalized by dividing them by their root mean square values to get $\boldsymbol{X}_i$ and $\boldsymbol{Y}$.

$$X_i = \frac{x_i}{s_i} \ ; \ s_i = \left(\sum_{l=1}^{n} \frac{[x_i(l)]^2}{n} \right)^{1/2} \qquad Y = \frac{y}{s} \ ; \ s = \left(\sum_{l=1}^{n} \frac{[y(l)]^2}{n} \right)^{1/2}$$

The problem, as posed by Sugawara, is to find the coefficients c_i which make the norm of the difference $\boldsymbol{d}$ in eq. 1 a minimum.

$$\boldsymbol{d} = Y - \sum_{i=1}^{k} c_i X_i \tag{1}$$

The problem of finding c_i in eq. 1 is ill posed and the solution is unstable because of the correlation in the rainfall data, especially if the stations are close. Hence, Sugawara uses the method of principle axis transformation to estimate the coefficients c_i.

Using the input data vectors $\boldsymbol{X}_i$, several new matricies are defined as

$$A = \frac{1}{n} X^T X \ , \quad B = \frac{1}{n} X^T Y \ , \quad C = [c_1, c_2, \ldots c_k]^T \tag{2}$$

The eigenvalues and eigenvectors of **A** are computed. The method of factor analysis is then used to develop a new vector system $\boldsymbol{Z}_i$, from the characteristic eigenvectors and precipitation matrix. Since the new vector system $(\boldsymbol{Z}_1, \boldsymbol{Z}_2, \ldots \boldsymbol{Z}_k)$ is an orthogonal system, the runoff can be determined as shown in eq. 3.

$$Y = \sum_{i=1}^{k} a_i Z_i + d \tag{3}$$

The weights $\boldsymbol{a}_i$ are computed by using the transformed variables **Z**. The weights a_i computed by using eq. 3 are for normalized variables. They are transformed back to the sequence $\boldsymbol{W}$ which corresponds to the observed data. If only k' eigenvalues and eigenvectors are retained for final computation, we will have k' weight values, where k' is the number of significant eigenvalues.

Data

The monthly rainfall and runoff data from the Upper Wabash River watershed in Indiana were used in the study. There are 14 stations inside the basin that have complete monthly precipitation data for the time period from 1970 through 1990. Stations with incomplete data were not considered. The data were obtained from the NWS records.

The measured runoff from the Wabash River at Lafayette was also used.

This station is at the downstream end and at the western tip of the Upper Wabash basin.

Results

Several different approaches were used to determine the weights for the precipitation stations. The input sequence X_i can be total or effective rainfall or some other function of rainfall. For this reason, four different approaches were used. In the first approach precipitation recorded at the fourteen stations inside the watershed were used as the inputs X_i. In the second approach a smaller part of the Upper Wabash River basin and the data from the rainfall stations in the smaller part of the watershed were used. In a small watershed the relationship between rainfall at individual stations and the runoff is stronger. Consequently, the weights associated with rainfall should also be higher than the weights computed by using data from a larger watershed. The second experiment was conducted to test whether Sugawara's method would reflect this commonly observed phenomenon.

In the third approach the precipitation values were converted to runoff by using runoff coefficients for each station and analyzed. In this case the relationship between rainfall from each station and runoff would obey the law of conservation of mass. The weights associated with rainfall stations in this approach would be analogous to the unit hydrograph.

Runoff from tributaries to the Wabash River are considered as X_i values in the last approach. Runoff from each of the tributaries reflects the integrated effect of rainfall within that basin. Consequently, in this case the weights would quantify the contribution of rainfall within each of the subbasins as they are reflected through the runoff from them.

The results from the EOF analysis showed that there was only one significant eigenvector in all these cases. However, the analysis was continued with all the eigenvectors because there was little significance in reducing the calculations to a single eigenvector.

The results from the first trial with 14 precipitation stations showed that only 3 stations had significant weights. The rainfall from these stations also had the highest cross-correlations with the Wabash River runoff. The lag zero cross-correlations showed that, in general, there is a strong relationship between the stations with high cross-correlation coefficients and significant weights. Stations with higher lag zero cross correlation coefficients have higher weights and stations with lower cross correlation coefficients have low or insignificant weights.

For the second (a smaller watershed) and third (precipitation was converted to runoff) cases, again the station with the highest weight corresponds to the station with the highest cross-correlation coefficient.

For the fourth case, flow data from gaging stations were used to compute weights. All gaging stations within the Upper Wabash River Basin boundary that has a complete set of data for the 20 years in consideration were used. The total number of gaging stations was 14. These were compared to the observed runoff from the Wabash at Lafayette. The results were as expected. The highest weights were associated with the gaging stations closest to Lafayette and encompassing a large area. Data from rainfall stations around Logansport, which is in the center of the watershed can be assigned higher weights. Complete details of these results are presented in Burke and Rao (1995).

Conclusions

The following conclusions may be presented on the basis of results presented herein.

(1) Sugawara's method may be used to assign weights to precipitation values measured at different locations in a watershed.

(2) The choice of inputs, whether it is total rainfall or some other variant of it needs study for individual basins.

References

Burke, Thomas T. and Rao, A.R., "Assigning Weights to Precipitation Stations", Tech. Report No. CE-EHE-95-1, School of Civil Engineering, Purdue University, W. Lafayette, IN 47907, March, 1995, pp. 31.

Sugawara, M. "On the weights of precipitation stations", *Advances in Theoretical Hydrology, A Tribute to James Dooge*, European Geophysical Society Series on Hydrologic Sciences, 1, Elsevier, 1992, p. 59-74.

Drainage Network Simulation using Digital Elevation Models

A.R. Rao[1] and A.S. Al-Wagdany[2]

Abstract

Digital elevation models (DEMS) are used to delineate geomorphologic features such as stream channels and watershed divide. Numerous DEMs have been developed. The most common method used in DEMs to extract a watershed boundary and stream network is the eight neighbors method.

In the present study ten drainage networks in watersheds in Indiana are analyzed manually and by using DEMs and the results are presented. The elevation data have a 73 x 94 meter resolution. They are stored in a raster format of the GRASS GIS system developed by the U.S. Army Corps of Engineers. The results produced by the DEMs are compared to manually delineated watershed boundary and drainage networks. The results from DEMs are very sensitive to input data. If the digital elevation data are accurate, then the DEMs give reliable results. Otherwise the results can be very erroneous.

Introduction

Digital elevation models (DEMs) are algorithms which are used to delineate stream channels and watershed divides. Considerable research has been conducted in this field by hydrologists and geologists. The most common method used in DEMs to extract a watershed boundary stream network is the eight neighbors method.

The accuracy and details of automatically extracted stream networks depend on the quality and resolution of the digital elevation maps. Watershed data extracted from DEMs with fine resolutions are expected to be closer to those of the actual basins than those extracted from DEM with course resolution.

[1] School of Civil Engineering, Purdue University, W. Lafayette, IN 47907

[2] Dept. of Hydrology, King Abdul-Aziz University, Jeddah, Saudi Arabia

However, to check the validity of an automatically extracted drainage network, it should be compared with a drainage network manually extracted from topographic maps.

In this paper, two DEMs are used to derive the drainage networks which were also manually delineated and presented. The drainage basins and stream networks resulting from these algorithms are compared to the corresponding manually delineated basins and networks.

Delineation of Basins Using Digital Elevation Models

Data from ten watersheds in Indiana were used in this study. The watershed areas ranged from 6.8 to 242.5 mi^2. The elevation data used in this study have a 73x94 meter resolution. The data is stored in a raster format of a Geographic Information System (GIS) [GRASS, (1991)]. GRASS is an integrated set of programs which has the capability of image processing and map production.

Two DEMs are available on GRASS. The first program is called r.drain.pt. It uses a raster elevation map to determine boundaries of watersheds. It has one input parameter which is the location of the basin outlet. Then it determines the drainage boundaries of the area draining to the given location. The second program is called r.watershed. It divides the input raster elevation map into a number of subbasins depending on an input parameter of the model called the threshold. Threshold is the minimum size of an exterior watershed basin (in cells or area units). Threshold is relevant only for those basins that have no basins draining into them (exterior basins), [GRASS, (1991)]. The interior drainage basins can be of any size, and it has the area that flows into an interior stream segment.

Both the DEMs were used to generate drainage networks of the study basins. Given the locations of the basin outlet (gauge stations), the r.drain.pt program was used to determine the drainage boundaries of the basins. Then, program r.watershed was used to delineate the stream networks of the basins.

The model r.watershed was applied to each basin using different threshold values. Drainage areas, orders and magnitudes (number of exterior links) of the basins resulting from the DEMs and those delineated manually are given in table 1. In table 1, A_T is the basin drainage area, the highest stream order or the order of the basin is Ω and Λ is the basin magnitude.

Drainage basins extracted by the DEMs were compared to those delineated manually. These results indicate that the automatically extracted drainage network of five of the study basins are completely different from the corresponding actual drainage network. They have different drainage areas, boundaries, stream orders, magnitudes and orientations from those delineated manually. These are the Brush

Creek, Buck Creek, Little River, Salamonie River and Sand Creek basins. The automatically derived drainage area of Little River basin is 245% larger than its actual area. For Salamonie River basin the derived area is only 16% of the actual drainage area.

The only factor which is common to all of these basins with erroneous results is that they are all located in the eastern part of Indiana. This result is consistent with the uneven quality of elevation data used for the DEMs. The quality of elevation data for the southern and western parts of Indiana is much better than that for the eastern part.

For the other five basins, located in western and southern Indiana, geomorphologic properties of automatically extracted and manually delineated drainage basins are similar. The error in estimating catchment area of these basins is less than 5%. When the proper threshold size is used, the resulting basins have the same Strahler's order as those extracted manually. For these basins, the error in the resulting basin magnitude is less than 15%.

Table 1. Geomorphologic Properties of Basins Delineated Manually and by DEMs.

Basin	Method of Delineation					
	Manual			DEM		
	A_T	Ω	Λ	A_T	Ω	Λ
Bear Creek	6.8	3	22	6.6	3	20
Brush Creek	12.2	3	35	7.2	2	11
Buck Creek	34.9	4	69	46.2	4	73
Carpenter Creek	42.3	4	55	49.2	4	63
Youngs Creek	98.1	4	116	98.6	4	117
Busseron Creek	225.6	5	357	231.5	5	338
Indian Creek	127.7	5	180	125.1	5	165
Little River	242.5	5	269	659.4	4	107
Salamonie River	81.1	5	168	27.0	4	35
Sand Creek	142.1	5	184	140.7	4	80

The automatic derivation of basin drainage area significantly reduces the time required to estimate it, compared to the manual method. A main drawback of the r.watershed program is that it does not identify the stream links and the corresponding areas as defined by Strahler (1957) or Shreve (1967). There are two procedures which can be used to estimate the area draining to each stream of the basin drainage maps obtained from the DEMs. In the first procedure, these areas are determined manually. Drainage boundaries of each stream can be approximately determined by assuming that each area in the basin drainage to the stream nearest to it. Then a planimeter or an electronic digitizer may be used to measure these areas. The second procedure is to use an output map of the DEM r.watershed to estimate these areas. As mentioned above, the r.watershed program divides the input raster elevation map into a large number of subbasins. These subbasins may not correspond to the extracted streams. The number of subbasins is much greater than the number of streams.

The two basins which have the smallest drainage areas among the five basins located in western Indiana are the Bear Creek and Carpenter Creek basins. They are selected to determine their stream numbers N_i, mean stream length of each stream order $\bar{L}_i$ and mean area draining to streams of each order $\bar{A}_i$. Lengths of each stream of the two basins were measured by using a measuring wheel. Area draining to each stream is estimated by using the second procedure explained above. A comparison between values of N_i, $\bar{L}_i$ and $\bar{A}_i$ measured from the manually delineated maps and those determined from the automatically derived maps is presented in table 2.

Table 2 Number of Streams

Basin	Method of Delineation							
	Manual				DEMs			
	Number of Streams							
	N_1	N_2	N_3	N_4	N_1	N_2	N_3	N_4
Bear Creek	22	5	1		20	6	1	
Carpenter Creek	55	13	2	1	63	7	2	1
	Length of Streams							
	$\bar{L}_1$	$\bar{L}_2$	$\bar{L}_3$	$\bar{L}_4$	$\bar{L}_1$	$\bar{L}_2$	$\bar{L}_3$	$\bar{L}_4$
Bear Creek	0.58	0.88	2.69		0.43	0.8	2.81	
Carpenter Creek	0.84	1.09	8.26	3.90	0.56	1.69	6.03	9.58
	Areas Draining to Streams							
Bear Creek	0.20	0.29	0.88		0.19	0.28	1.19	
Carpenter Creek	0.46	0.65	4.82	1.55	0.37	1.33	3.84	8.60

For the Bear Creek basin, values of N_i, $\bar{L}_i$ and $\bar{A}_i$ determined from the automatically derived drainage map are very close to those directly measured from the actual drainage map of the basin. For the Carpenter Creek basin, first order streams resulting from the DEM are shorter and drain smaller areas compared to those of the actual drainage map. The automatically extracted drainage map of

Carpenter Creek shows only 7 second order streams compared to 13 second order streams shown on the manually delineated map. The drainage map obtained from the DEM shows that the fourth order stream of Carpenter Creek basin originated at the upstream part of the basin. In the manually extracted drainage map of the basin, the fourth order stream starts at the down stream portion of the basin about 3.9 miles from the basin outlet. Therefore, similar values of $\bar{L}_3$, $\bar{L}_4$, $\bar{A}_3$ and $\bar{A}_4$ are not expected from the two drainage maps, as shown in table 2. Details of these results are found in Al-Wagdani and Rao (1994).

Conclusions

The following conclusions may be presented based on the results from this study.

1. The quality of the drainage networks derived from the digital elevation data depends on the quality of input data. If the data are poor, the results are also unreliable.

2. It is preferable to compare the results from DEMs with those derived manually.

References

Al-Wagdani, A.S. and A.R. Rao, 1994, "Geomorphologic Characteristics and Instantaneous Unit Hydrographs of Indiana Watersheds", Tech. Rept. CE-EHE-94-1, School of Civil Engineering, Purdue University, W. Lafayette, IN 47907, pp. 215.

GRASS, User's Reference Manual, Version 4.0, U.S.A. CERL, 1991.

Shreve, R.L., 1967, "Infinite Topologically Random Channel Networks", J. Geo., 75, pp. 17-37.

Strahler, A.N., 1957, "Quantitative Analysis of Watershed Geomorphology", Trans. Am. Geophys. Union, 38, pp. 913-920.

A Framework for Efficient Watershed Management

Kenneth R. Wapman, Nathan Y. Chan, and Donald S. Wilson[1]
Robert A. Goldstein[2]

I. Introduction

Environmental regulatory agencies are increasingly moving toward watershed-wide management of surface water resources. There are several advantages of this approach. It offers a single "system-wide" vision for addressing interrelated problems, and it explicitly considers all important activities, including both point and non-point sources, pathways of exposure, and ecological and human health impacts. In addition, synergistic and antagonistic effects can be captured, including explicit tradeoffs among pollutant sources, and cost-effective strategies may be identified that may not be apparent when considering each discharge separately.

Despite intuitive appeal and considerable advantages, the management of watersheds on a basin-wide basis has been difficult and costly to effect. Decisions related to water resources typically affect many different groups and multiple regulatory jurisdictions, often with conflicting goals. The diversity, complexity, and uncertainty associated with the processes that occur within a watershed are very difficult to characterize and model. To date, efforts to model key watersheds have created customized, detailed models—at great cost. In the future, it is unlikely that such resources will be available to manage most watersheds. Without more cost-effective tools, few watersheds will benefit from the current management programs.

This paper proposes an alternative, and more efficient, approach for managing water resources on a watershed-wide basis. The proposed conceptual framework includes capabilities to:

- efficiently develop and calibrate an engineering representation of a watershed,
- simulate the flow of water through a watershed to characterize water quality, and
- identify preferred options for enhancing water quality.

The information developed would be used to help stakeholders and decision-makers identify the critical factors and processes that affect water quality; and

[1]Decision Focus Incorporated, 650 Castro Street, Suite 300, Mountain View, CA 94041

[2]Electric Power Research Institute, 3412 Hillview Avenue, Palo Alto, CA 94304

identify, evaluate, and compare options for enhancing the watershed. Information about stakeholders' preferences and the impacts associated with each option would be evaluated, facilitating the identification of preferred options.

II. Critical Difficulties and Obstacles

Watershed management allows a broader view of management options and tradeoffs, and supports development of coordinated and consistent policies. However, watershed-wide management is difficult for several reasons:

- **Complexity.** Much of the complexity associated with managing water resources on a watershed-wide basis arises from the large number and diversity of point and nonpoint pollutant sources. Many of the underlying natural and anthropogenic influences that determine the quality of water in a watershed are extremely complex and poorly understood.
- **Data.** Because of this complexity, large amounts of data and information are required to support decisions. While much of this information is available for point sources, data for nonpoint sources and for nontraditional water quality criteria is sparse. Collection of data for complex site-specific watershed models is extremely time consuming and costly.
- **Uncertainties.** High levels of uncertainties are associated with meteorological conditions, discharges, and the natural processes that determine water quality. Such uncertainties compound the complexity of watershed management decisions, and shift the focus from a search for the "right" answer to options which are robust over a broad range of likely scenarios. Hence, it is important to explicitly consider uncertainties, including average conditions and extreme events.
- **Stakeholders.** Typically, multiple stakeholders are involved, including industries, environmental interest groups, and regulatory agencies. The parties involved often have conflicting goals and priorities, and different types and magnitudes of resources. Complex tradeoffs must often be made among multiple water quality criteria and competing economic and environmental objectives.

III. Conceptual Framework

Given the trends toward integrated watershed management, in light of the difficulties discussed above, how can policy options and decisions be appropriately and efficiently evaluated? The primary goal of the framework proposed here is to help stakeholders identify preferred options for enhancing the water quality of a watershed. To accomplish this goal, the framework must:

- use sound engineering and scientific methods,
- allow users to efficiently characterize a watershed,
- quickly and consistently evaluate options for enhancing water quality,
- explicitly consider all important inherent uncertainties,
- characterize the statistical water quality, economic, political and other attributes associated with each option, and
- provide impartial results.

The proposed framework includes three linked components, as shown in Figure 1:

- linked, deterministic engineering models to develop a representation of a watershed,
- a Monte Carlo sampling and results processor to address uncertainties, and
- capabilities to consider multiple attributes and evaluate tradeoffs among options.

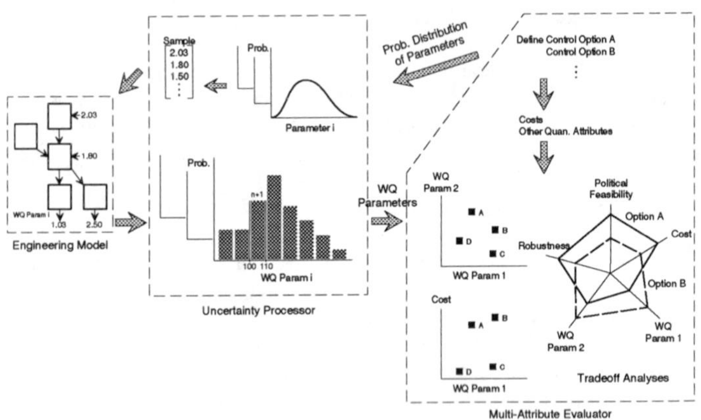

Figure 1. Framework for Efficient Watershed Management

As an overview, the framework would be used as follows: The watershed configuration (or "base system") would first be characterized using the engineering models. The user would identify the attributes of interest, including water quality parameters, cost, and others. Then, the Monte Carlo sampling module would help the user statistically characterize each component, including its (perhaps uncertain) impacts to water quality and other attributes, under normal conditions and extreme events. Next, the base system would be calibrated, and modified as required. Subsequently, the user would define options for enhancing water quality and modify the base system accordingly. Each option would be evaluated with the Monte Carlo processor, and the resulting water quality parameters and other attributes would be recorded for comparison with other options. After all options have been evaluated, decision analytic techniques would be used to identify preferred options and evaluate tradeoffs.

The framework includes reduced-form engineering models (i.e., simplified, parametric models that could be readily calibrated to each watershed) to capture the critical characteristics of important watershed sources, sinks, and processes, and to estimate water quality parameters of interest. Well understood physical processes would be modeled using simple functional forms; for processes that are less well understood, flexible relations, designed to capture the important processes and interactions without going into excessive detail, would be used. This approach balances the need for physical and engineering accuracy with the need to efficiently evaluate options for enhancing a watershed.

The framework also addresses the issue of data requirements. The framework proposed here can be readily tailored to the data available for each watershed, then use decision analytic techniques, facilitated by the Monte Carlo sampling capabilities, to identify the key relationships and areas where additional data would be of greatest value. In this way, the framework can focus the data gathering efforts on areas that most affect the decisions under consideration.

Explicit treatment of uncertainties is one of the most powerful features of the approach. Rather than a deterministic "answer," the framework includes Monte Carlo sampling capabilities to provide probabilistic representations of the water quality attributes of interest. In this way, the user can develop an understanding of exceedances, extreme events and other low-probability occurrences, as well as the underlying natural variations.

The decision analytic component of the framework uses the information developed by the two components previously described to address the issues associated with multiple stakeholders and potentially competing objectives. In addition to water quality attributes, the framework can include other quantitative and qualitative factors that might affect a decision, such as the cost associated with an option, equity among stakeholders, and political feasibility. The framework can provide valuable technical support in communication of stakeholder preferences, the identification of attractive options, and in tradeoffs and negotiation.

The decision analytic component would include several graphs and tables that summarize each option, compare multiple options, and summarize tradeoffs. Figure 2 illustrates the trading off of one water quality parameter against another, showing how each control option performs.

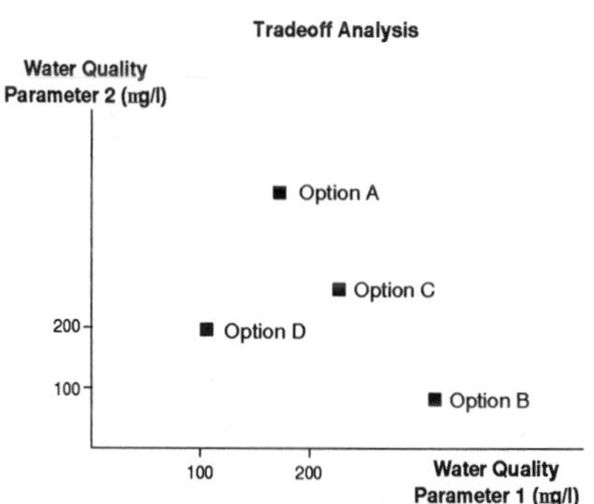

Figure 2. Tradeoff of Two Water Quality Parameters

Figure 3 presents a Kiviat diagram with each attribute represented as a separate axis. Preferred outcomes (e.g., lower cost, greater political feasibility) are plotted further from the intersection of the axes. Each control option would be represented as a polygon defined by its "performance" with respect to each attribute. Other graphical representations, including simpler bar and pie charts, would also be available. Collectively, the graphs and tables would facilitate explicit consideration of the multiple quantitative and qualitative attributes associated with each option. If desired, multi-attribute utility functions could be used to develop a single metric, or "score," for each option.

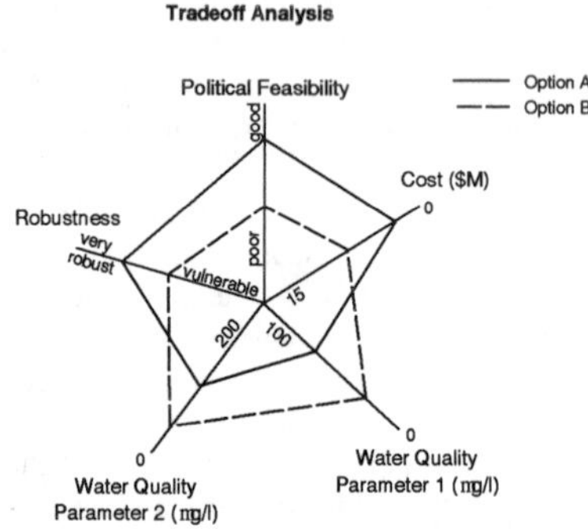

Figure 3. Tradeoffs of Multiple Qualitative and Quantitative Attributes

IV. Summary

The growing emphasis on integrated watershed management and its clear advantages necessitate planning and management techniques and tools consistent with the basin-wide philosophy. Historically, watershed management models have been large, complex, and costly. More efficient tools are needed to provide relevant analyses of policies and decisions affecting the watershed. These decisions must consider the multiple stakeholders involved, and the uncertainties associated with natural processes and anthropogenic sources.

This paper discussed a conceptual framework to efficiently analyze and evaluate decisions related to the management of watersheds. The framework contains three components: 1) reduced-form engineering models to quickly model and simulate a watershed and options for its enhancement, 2) a Monte Carlo processor to explicitly address uncertainties, and 3) a decision analytic component to compare options by assessing tradeoffs among multiple criteria, and to facilitate communication and negotiation among various stakeholders.

The framework would provide quantitative and qualitative information that characterizes the performance, costs and tradeoffs associated with each control option. The framework would facilitate significantly more extensive implementation of the watershed-wide planning approach than is currently feasible.

TWO DIMENSIONAL DETENTION POND ROUTING MODEL

Laura J. Howard[1] and A. Osman Akan[2] M. ASCE

Abstract

A two-dimensional non-inertial wave model is presented for routing flood hydrographs through detention ponds. The model is applied to several hypothetical detention basins to study the effects of pond shape on outflow hydrographs and on detention time.

Introduction

It has long been recognized that detention ponds with large length to width ratios perform better from the standpoint of pollutant removal from stormwater runoff. However, the conventional pond routing methods, such as the storage-indication method, can not reflect this fact. The reason is that the conventional methods employ a lumped stage-area-discharge relationship without considering the exact shape of the pond and the positions of inlet and outlet structures. Also, because the velocities along different flow paths are not calculated, a precise detention time can not be defined and determined using conventional methods.

The mathematical model presented herein accounts for the exact shape of the pond and the positions of the inlet and outlet structures. Also, flow velocities and volumes along different paths are calculated, enabling the determination of minimum and average detention times.

The governing equation for the two-dimensional non-inertial wave model is

$$\frac{\partial}{\partial x}\left(\frac{Y^{5/3}}{n} S_{fx}^{1/2}\right) + \frac{\partial}{\partial y}\left(\frac{Y^{5/3}}{n} S_{fy}^{1/2}\right) + \frac{\partial Y}{\partial t} = 0$$

[1] Research Assistant

[2] Professor, Old Dominion University, Department of Civil and Environmental Engineering, Norfolk, Virginia 23529

with

$$S_{fx} = \frac{\partial h / \partial x}{\sqrt{|\partial h / \partial x|}}$$

$$S_{fy} = \frac{\partial h / \partial y}{\sqrt{|\partial h / \partial y|}}$$

where

Y = flow depth,
n = Manning roughness factor,
h = water surface elevation,
S_{fx} = friction slope in x-direction,
S_{fy} = friction slope in y-direction,
x,y = spatial coordinates,
t = time

The flow velocities in the x- and y- directions are calculated, respectively, using

$$u_x = \frac{1}{n} Y^{5/3} S_{fx}$$

$$u_y = \frac{1}{n} Y^{5/3} S_{fy}$$

At pond boundaries u_x= 0.0 or u_y= 0.0 as appropriate.

To solve Eq. 1 using a finite-difference scheme, a rectangular grid is superposed over the pond dividing it into a number of rectangular cells. Using a fully-implicit central difference scheme, Eq. 1 is written in finite difference form for each cell. This results in a set of N nonlinear simultaneous equations in N unknowns for each time step of computation, where N= total number of cells. The unknowns are the flow depths in N cells. A generalized Newton's iterative scheme is employed along with an upper triangulation matrix inversion routine to solve these simultaneous equations.

A user-input source term is included in the finite difference equations for the inflow cells to describe the inflow rates. The equations for outflow cells include a sink term evaluated as some function of the flow depth in the cell depending upon the size and the type of the outlet structure.

Model Applications

The mathematical model is applied to three hypothetical detention basins to study the effects of pond shape on outflow hydrographs and detention time. All three ponds are rectangular in shape with vertical side walls and a constant surface area of 6400.0 m^2 for simplicity. The outlet structure is a rectangular weir with a crest 1.0 m above the pond bottom, a crest length of 0.5 m, and a discharge coefficient of 0.50. The ponds receive inflow at the midpoint of the upstream wall, and the outlet weir is placed at the middle of the downstream wall. The length to width ratios are 1:3, 1:1, and 3:1 for ponds 1,2, and 3, respectively.

The inflow hydrograph shown in Fig. 1 is routed through the three detention ponds using an initial water surface that is at the same elevation as the weir crest. The outflow hydrographs calculated for the three ponds are nearly identical, indicating that the pond shape has negligible effect on outflow rates. The hydrograph displayed in Fig. 1 represents the individual outflow rates from all three ponds.

Different flow paths are identified in each pond to determine the detention time. For example, pond 1 is represented by a grid system consisting of five columns and five rows. This results in five different flow paths, the shortest one being the path straight from the inflow cell to the outflow cell along the axis of symmetry. The longest two flow paths are those furthest away from the axis of symmetry.

The average velocity along each flow path is calculated for every time step. At the end of the simulation a time average of these velocities is found for each path. Also, the total volume of runoff reaching the outlet structure from each of the different paths is found. The detention time along each path is found by dividing the length of the path by the average path velocity. The detention time for the shortest path is the "minimum detention time." The "average detention time" is the average of path detention times weighted with respect to the runoff volumes along the paths. The calculated minimum and average detention time values are listed in Table 1.

An inspection of Table 1 shows that the shape of a detention pond has a significant effect on flow detention times. The detention time increases with increasing length to width ratio, and this explains the superior performance of longer ponds for pollutant removal.

Concluding Remarks:

A two-dimensional non-inertial wave flood routing model can account for the shape of a detention pond and the positions of inflow and outflow structures. The model also enables determination of detention times.

Although further study is needed to generalize the results of this paper, the general conclusions of the paper are that

(a) The shape of a pond has negligible effect on outflow rates as long as the surface area is specified.

(b) The length to width ratio has significant effects on detention times.

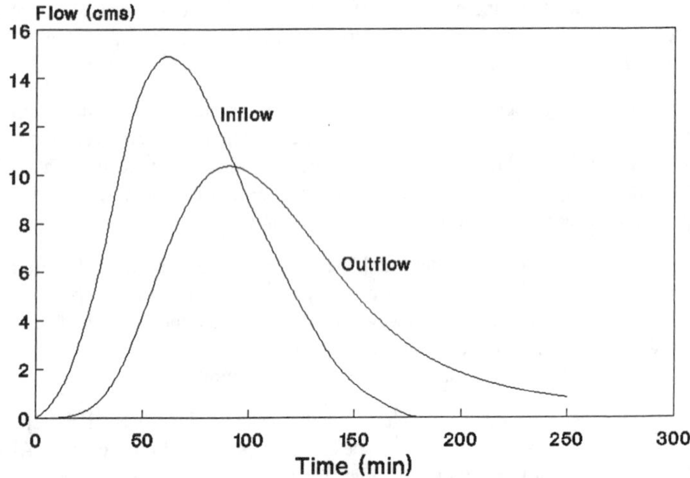

Figure 1. Inflow and Outflow Hydrographs

Table 1. Comparison of Detention Times

Pond	Length to Width Ratio	Number of Flow Paths	Volume Fraction from Shortest Path	Average Detention Time (min)	Minimum Detention Time (min)
1	1:3	9	34.9%	76.2	27.2
2	1:1	5	36.2%	86.0	59.7
3	3:1	3	42.8%	98.3	79.9

A COMPREHENSIVE ENVIRONMENT FOR WATERSHED MODELING AND HYDROLOGIC ANALYSIS

E. James Nelson[1], Norman L. Jones[2], and Jeffrey D. Jorgeson[3]

ABSTRACT

A comprehensive graphical Watershed Modeling System (WMS) has been developed to address the needs of surface water hydrologic computer simulations. WMS, developed at the Engineering Computer Graphics Laboratory (ECGL) at Brigham Young University, is part of a multi-year project funded in part through the U.S. Army Corps of Engineers, Waterways Experiment Station. WMS is a graphically based software tool providing functionality for modeling all aspects of the rainfall/runoff process. Facilities include triangulated irregular network (TIN) generation from scattered and digital elevation model data sources, automated watershed and sub-basin delineation from TINs, an interface to drive the HEC-1 surface runoff modeling program, and an interface to a two-dimensional, grid based, distributed hydrologic model CASC2D. An interface to both a lumped parameter model like HEC-1 and a distributed model like CASC2D in a single comprehensive environment provides a system which can be used comfortably by engineers as they transition from traditional to more state of the art hydrologic modeling.

INTRODUCTION

While many advances in distributed watershed modeling have been made over the past several years, lumped parameter models such as HEC-1, TR-20, SWMM, and others which were developed in the late 60's and early 70's, continue to be the accepted standard for most regulatory agencies. While the knowledge base required to perform more comprehensive and sophisticated distributed modeling has

---

[1] Research Associate, Dept. of Civil Eng., 300 CB, Brigham Young University, Provo, Utah, 84602.

2 Assistant Professor, Dept. of Civil Eng., Brigham Young University.

[3] Research Hydraulic Engineer, U.S. Army Engineer Waterways Experiment Station, 3909 Halls Ferry Rd., Vicksburg, Mississippi, 39180

existed for some time, the data required to drive these models has been overwhelming for potential users. With the emergence of more user-friendly software, and particularly Geographic Information Systems (GIS), the data collection and processing problem for spatially distributed runoff models can be solved.

In recent years, a comprehensive watershed modeling system named WMS (formerly known as GeoShed) has been developed. WMS incorporates digital terrain modeling, GIS data, and analytical hydrologic models in a single environment. It can be used to automatically delineate watershed and sub basin boundaries from TINs. Once boundaries have been computed, geometric parameters such as area, slope, and runoff distances can be computed for each basin. WMS includes a direct interface to HEC-1 (HEC, 1990) for performing rainfall/runoff analysis. All input can be defined through a series of user-friendly dialogs. Results of one or more analyses can be viewed simultaneously to allow for calibration of basin parameters. The WMS environment is shown in Figure 1.

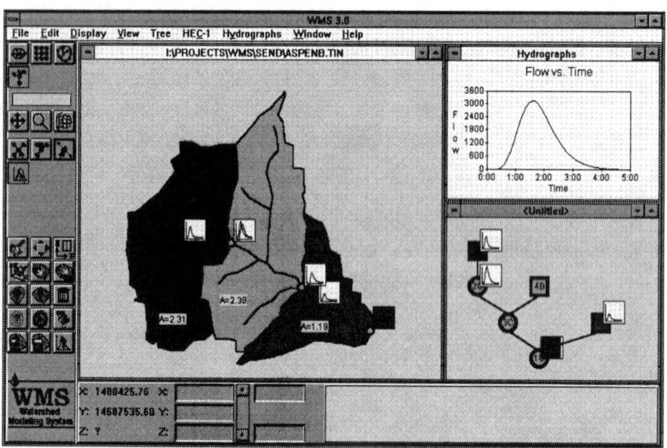

Figure 1. Watershed Modeling System.

In addition to providing a direct link to an established model like HEC-1, WMS can also be used to drive the CASC2D model being developed through the U.S. Army Corps of Engineers (Julien et. al., 1995). CASC2D is a physically based rainfall/runoff model which uses rectangular grid cells to represent the distributed watershed and rainfall domains. The model uses a two-dimensional diffusive wave equation to simulate overland flow and a one-dimensional diffusive wave equation to simulate channel flow. The model also includes an advanced soil moisture accounting procedure, primarily based on the Green-Ampt infiltration model. WMS can create a finite difference grid and provide all necessary input, such as stream locations and properties, lakes, soil properties, precipitation, etc. Spatial data can be prepared using a standard GIS and then imported into WMS. Results of real time analysis can then be displayed with animations of soil moisture, runoff depth, precipitation and other time-varying quantities.

WMS is divided into five basic components, each designed to perform specific tasks. These modules include: DEMs, Feature Lines, TINs and Drainage, Topologic Trees, and Grids.

DEMS

Digital Elevation models are the most common source of elevation data. The USGS has compiled 1:250000 scale, 3 arc-second digital elevation models for the entire United States. These models are available via the internet, and while their resolution is limited, they are adequate for most large-scale models. In addition to the 3 arc-second DEMS, WMS can also be used import USGS 7.5 minute quadrangle DEMS and Arc/Info grids. While WMS is a TIN based modeling system, DEMs provide a convenient background elevation source when creating a TIN. DEM elevations can be smoothed to eliminate errors associated with roundoff (Nelson and Jones, 1995) and contoured using a variety of options.

FEATURE LINES

Feature lines are used in WMS to represent important terrain features such as streams, ridges, and boundaries. Typically a bounding polygon (at least as large as the watershed being analyzed) and a stream network of feature lines are created. These lines can be generated inside of WMS by using simple point and click techniques with a contoured image of a DEM (or other background elevation map) as a guide. As an alternative, points digitized from maps or other sources such as digital line graphs (DLGs) can be imported and used to define feature lines.

TINS AND DRAINAGE

TINs can be created using a combination of a bounding polygon and stream and ridge feature lines. The bounding polygon is used to define the extent of the resulting TIN and triangle edges are made to conform to all interior stream and ridge feature lines. The point density of the TIN is determined by the spacing of points along the feature lines, forming smooth transitions from the bounding polygon to the stream network. Elevations for TIN points are interpolated from the background elevation map.

Typically DEMs are used as background elevation maps, but WMS can also be used to triangulate a scattered set of data points to create a TIN. TINs created from scattered data can be used directly for performing a drainage analysis. However, because triangle edges will conform to streams, and ridges, it is preferable to define a set of feature lines, and create a new TIN using the original TIN as a background elevation map.

The TIN module in WMS contains many operations for delineating watershed and sub-basin boundaries from TINs. Once a TIN has been created, stream networks can be created (this is done automatically if stream feature lines are used to create a TIN) and the watershed boundary delineated (Nelson et. al., 1994). Any number of outlet points can be placed along the stream network so that sub-basins of the watershed can be created as shown in Figure 1. Once basins boundaries have been delineated geometric attributes such as area, slope, and runoff distances can be calculated for each basin.

TOPOLOGIC TREES

As streams and basins are created on a TIN, a topologic tree for the watershed is automatically generated (see the lower right portion of Figure 1). Nodes or icons for each component, such as outlet points (confluences), basins, diversions, and reservoirs are linked together in an identical manner to the underlying stream network of the watershed.

Once the tree diagram for a watershed is established, all necessary data to run an HEC-1 simulation can be defined using a series of user-friendly dialogs. Any of the different precipitation, loss, unit hydrograph, routing, etc. methods available in HEC-1 can be defined. Potential errors such as undefined basin/outlet data can be detected and corrected prior to running HEC-1 using a sophisticated model checker. WMS automatically creates a properly formatted HEC-1 input file and then launches the HEC-1 program. After successfully running HEC-1, hydrograph results can be viewed (Upper right portion of Figure 1) in the same user-friendly environment. Several different hydrograph sets, resulting from different runs of HEC-1 with different parameter definitions, can be viewed simultaneously inside of WMS to aid in model calibration.

GRIDS

In addition to providing a link to an industry standard lumped parameter model such as HEC-1, WMS can be used for pre- and post processing of a two-dimensional, spatially distributed model named CASC2D. CASC2D models are based on finite difference grids, which can be constructed in WMS by specifying the spacing and number of cells in the two primary coordinate directions. Since a grid occupies a rectangular region, cells outside the watershed boundary are inactivated. Watershed boundaries, and feature lines, can be used to automatically map boundaries, and streams on the grid. Figure 2 shows a grid created with WMS.

Several post processing capabilities for visualizing results of a 2-D analysis are provided in WMS, including contour plots, hydrographs at any point, and a variety of animation tools. Like the commands for grid generation, post processing utilities have been implemented in a generic fashion so that additional 2-D models could be supported without having to rewrite significant parts of the program.

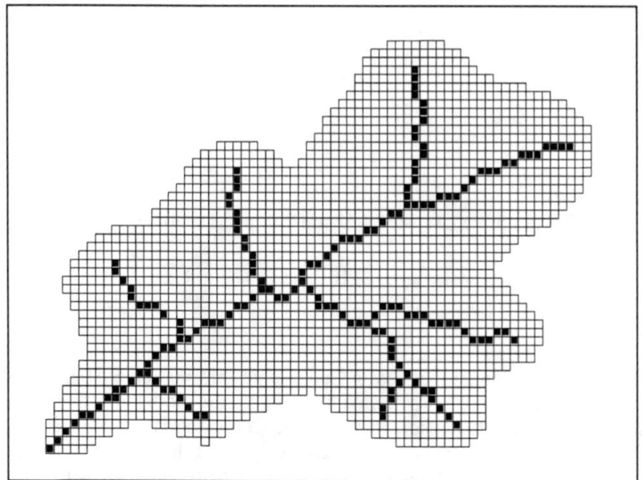

Figure 2. Finite Difference Grid for a Watershed.

CONCLUSIONS

The development of WMS provides a comprehensive hydrologic modeling environment allowing for computations to be done using both traditional lumped parameter methods as well as state of the art spatially distributed models.

ACKNOWLEDGMENTS

Funding and support provided by the US Army Corps. of Engineers Waterways Experiment Station is gratefully acknowledged.

APPENDIX. REFERENCES

Hydrologic Engineering Center, 1990, "HEC-1 Flood Hydrograph Package User's Manual," US Army Corps of Engineers Hydrologic Engineering Center, Davis, Ca.

Julien, P.Y., B. Saghafian, F.L. Ogden, "Raster-based hydrologic modeling of spatially-varied surface runoff", AWRA Water Resource Bulletin (in press).

Nelson, E.J., and N.L. Jones, "Reducing elevation roundoff errors in digital terrain models", Journal of Hydrology, Elsevier Editorial Services, (in press).

Nelson, E.J., N.L. Jones, A.W. Miller, "Algorithm for precise drainage-basin delineation", ASCE Journal of Hydraulic Engineering, Vol. 120, No. 3, Mar. 1994.

Crop Consumptive Use Model Using GIS and Database

Garcia, L.A., Manguerra, H.B.[1] and Bennett, R.[2]

Abstract

A crop consumptive use model that uses a Geographic Information System (GIS) and Database has been developed. It was tested by applying the model to estimate the consumptive use (CU) for the Gunnison River Basin, Colorado. The model is fully interactive and supported by a graphical user interface (GUI). It has the capability to browse map sets which will eventually allow users to graphically select areas for crop consumptive use calculations. The model also uses GIS to determine the types of soils in the area of interest and create combinations of soil type and crop type for CU calculations. Physical (crop and soil characteristics) and time series data (weather) are stored and extracted from a relational database (Informix). The model estimates crop consumptive use by three methods: 1) Blaney-Criddle; 2) an enhanced Blaney-Criddle that includes a soil moisture budget; and 3) Penman-Monteith method with a soil moisture budget.

Introduction

The US Bureau of Reclamation (USBR) is mandated to prepare every successive five years a report of the annual consumptive uses and losses of water in the Colorado River Basin. It involves the estimation of crop evapotranspiration (ET) from irrigated areas, evaporation from reservoirs and stockponds; transmountain diversions; and water use for livestock, municipal, and industrial purposes.

About 64-69 percent of the estimated consumptive use and losses in the Upper Basin tributaries of the Colorado River Basin is agricultural-related crop ET from irrigated areas (USBR, 1986). Monthly estimates are normally obtained by using an ET estimation method based on the SCS Blaney-Criddle method (USDA, 1970). The USBR has been using this method to estimate ET as part of its five-year consumptive uses and losses report.

The estimation of consumptive use and losses is just one of the many decisions involved in the operation and management of the Colorado river. The need to analyze

1. Assistant Professor and Graduate Research Assistant, respectively, Department of Chemical and Bioresource Engineering, Colorado State University, Fort Collins, CO 80523.
2. Senior Professional Engineer, Colorado Division of Water Resources, 1313 Sherman Street, Denver, CO 80203

the best available information with the most appropriate analytical tools has paved the way for the development of a Colorado River Decision Support System (CRDSS). In addition to estimating consumptive use, CRDSS includes various other water resource planning and operation modeling.

The consumptive use component of CRDSS (CRDSS-CU) is a stand-alone model developed and implemented in the UNIX environment. The large amount of geographic data involved in ET estimation are handled through the use of a relational database (Informix) and Geographic Information System (Arc/Info and Grass).

This paper presents the development and application of CRDSS-CU. The subsequent discussions are limited to the estimation of CU from crop evapotranspiration in irrigated areas. Its application is demonstrated using the Gunnison River Basin as a test application. The results presented are for demonstration purposes only and have not been adopted by the State of Colorado or the USBR.

Model Description

The estimation of crop evapotranspiration in CRDSS-CU is based on a monthly or daily estimation method. The monthly estimation uses the SCS-Blaney Criddle method expressed as follows:

$$u = k_t k_c \frac{tp}{100} \tag{1}$$

where u = monthly consumptive use (in); kc = crop coefficient reflecting the growth stage of the crop; kt = climatic coefficient related to mean temperature; t = mean monthly temperature (degree F); and p = monthly percentage of daylight hours.

The daily estimation uses the Penman-Monteith method which is expressed as follows:

$$\lambda ET = \frac{\Delta}{\Delta + \gamma^*}(R_n - G) + \frac{\gamma}{\Delta + \gamma^*} K_1 \frac{0.622\lambda\rho}{P} \frac{1}{r_a}(e_z^o - e_z) \tag{2}$$

$$\gamma^* = \gamma(1 + r_c/r_a) \tag{3}$$

where ET = rate at which water, if readily available, would be removed from the soil and plant surface; Δ = the slope of the saturation vapor pressure; γ = psychrometric constant; Rn = net radiation; G = soil heat flux; ρ = air density; e_z^o and e_z = saturation and actual vapor pressures at the z level above the surface; r_a = aerodynamic resistance to sensible heat and vapor transfer; r_c = surface resistance to vapor transfer; K_1 = dimensionless constant; P = atmospheric pressure. A comprehensive discussion of this method is provided by Jensen et al. (1990). The maximum crop evapotranspiration is obtained by multiplying ET with appropriate crop coefficients calibrated using either a grass-based or alfalfa-based reference crop.

Daily and monthly soil moisture budgets may be incorporated in both techniques. They allow water supply information to be accounted for in ET estimation. If water supply is not adequate, water short areas can be calculated. On the other hand, when water supply is at all times adequate, the consumptive use values correspond to estimates of irrigation water requirements.

The above computational component of the model is written in the FORTRAN language. The basic routines required for the implementation of SCS Blaney-Criddle ET estimation method were taken from the USBR XCONS2 program. A few routines of the Penman-Monteith method were taken from SMB program (Wheeler and Assoc, 1994).

The current version of the model is written to run on an SGI workstation. The Graphical User Interface (GUI) is a combination of pop-up windows, pull-down menus, and icon selections. It is developed using the C programming language combined with OSF/Motif and X intrinsic libraries. Figure 1 shows the main interface window of CRDSS-CU. It shows the map of the Western Slope of Colorado as divided by river basins namely: the Yampa, Colorado, Gunnison, Dolores and San Juan River Basins (read from top to bottom). The Gunnison River Basin, which is discussed in the application part of this paper, is shown in the darkest shade.

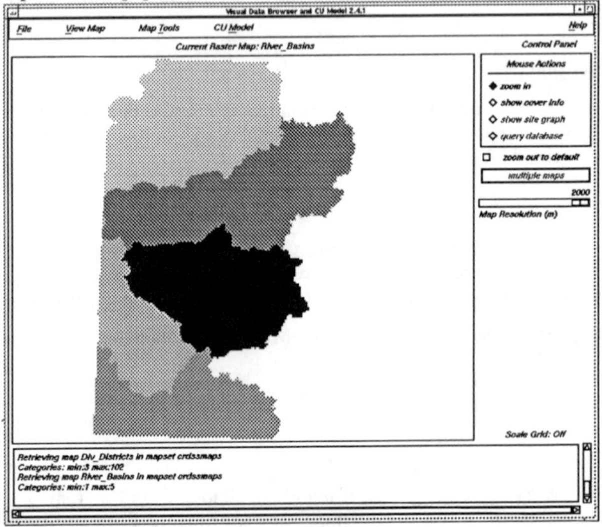

Figure 1. Main GUI interface of CRDSS-CU.

CRDSS-CU offers three main pull-down menus namely: View Map, Map Tools, and CU Model. The first two menus are part of the Visual Data Browser (VDB) capability of CRDSS-CU. They provide browsing capabilities to the several vector and raster map sets including land use, crop acreage, elevation, contour map, irrigation structures and boundary information of river basin, county and hydrologic units. They also provide standard browsing tools such as zoom capabilities and access to the GRASS shell. The CU model menu option includes capabilities such as creating, editing, running consumptive use scenarios and viewing the results. The Create Scenario menu allows the user to query the Informix database and extract a baseline scenario which in turn can be modified through the Edit Scenario menu. The consumptive use is calculated using the Run Scenario Menu, and the results viewed using the View Scenario menu.

Test Application to the Gunnison River Basin

To demonstrate the functionality of CRDSS-CU, a test application to the Gunnison River Basin was performed. The basin was subdivided into 15 subareas that are obtained by intersecting the counties and USGS hydrologic units. These include the subareas: (1) Delta county in hydrologic unit (HU) 14020002, (2) Delta in HU14020004, (3) Delta in HU14020005, (4) Delta in HU14020006, (5) Gunnison in HU14020001, (6) Gunnison in HU14020002, (7) Gunnison in HU14020003, (8) Gunnison in HU14020004, (9) Hinsdale in HU14020002, (10) Mesa in HU14020005, (11) Montrose in HU14020002, (12) Montrose in HU14020005, (13) Montrose in HU14020006, (14) Ouray in HU14020006, and (15) Saguache in HU14020003. The subarea in county Delta and HU 14020002 will be referred to by the county and the last digit of the HU, for example, Delta2. This convention is used for the rest of the paper.

The irrigated acreage, including the type of crop cover for each subarea was obtained through GIS using a 1993 irrigated acreage map developed by the Colorado Division of Water Resources and the USBR. Based on the review of annual agricultural statistics (Colorado Agricultural Statistics Service, 1985-90) and local knowledge, the 1993 acreage data was used directly for years 1985-90. The total irrigated area in the Gunnison River Basin was estimated to be 102,658 ha (Table 1). About 69 percent of the irrigated area is devoted to pasture, and 12 percent to alfalfa. The remaining 19 percent is devoted to corn, beans, small grains, orchard, vegetable and tree farms, in decreasing order.

Table 1: Irrigated area (ha) by crop and by subarea in the Gunnison River Basin.

Sub Area	Al-falfa	Pas-ture	Corn	Small Grain	Bean	Or-chard	Tree Farm	Veg.	Total
Delta2	450	2599	0	96	0	4	0	0	3149
Delta4	1175	10388	79	295	0	1002	19	1	12959
Delta5	2514	6865	2517	801	401	1161	73	428	14760
Delta6	253	270	655	58	427	22	0	260	1945
Gunnison1	0	4752	0	0	0	0	0	0	4752
Gunnison2	0	11609	0	0	0	0	0	0	11609
Gunnison3	0	5883	0	0	0	0	0	0	5883
Gunnison4	27	2725	0	0	0	0	0	0	2752
Hinsdale2	0	523	0	0	0	0	0	0	523
Mesa5	490	1615	26	22	0	3	0	3	2159
Montrose2	511	4349	0	84	62	0	0	0	5006
Montrose5	206	570	338	57	205	0	0	188	1564
Montrose6	6221	7177	5858	1255	2530	138	6	553	23738
Ouray6	314	5708	20	3	0	0	0	0	6045
Saguache3	0	5814	0	0	0	0	0	0	5814
Total	12161	70847	9493	2671	3625	2330	98	1433	102658

The data set was prepared to support the input requirements for estimating evapotranspiration by SCS Blaney-Criddle. The ET using the Penman-Monteith method cannot be calculated due to lack of representative daily weather data for the Gunnison River Basin.

Figure 2 shows the total estimated crop CU from 1985-90 for the Gunnison River Basin. This preliminary calculation does not account for the availability of water supply or water short areas. The six-year average annual use is 49,443 ha-m (400,730 acre-ft). The six-year average annual consumptive use in terms of unit depth is 47.8 cm (18.8 in).

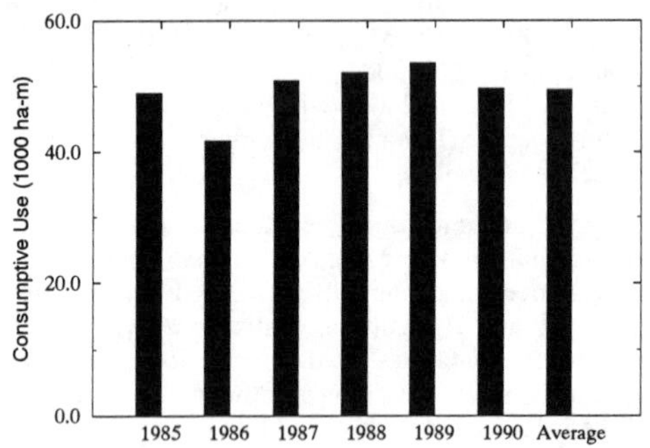

Figure 2. Estimated consumptive use for the Gunnison River Basin, 1985-90.

Summary

A crop consumptive use model that uses GIS and Database has been developed. The ET methods that are available include 1) Blaney-Criddle; 2) an enhanced Blaney-Criddle that includes a soil moisture budget; and 3) Penman Monteith method with a soil moisture budget. The model was applied in Gunnison River Basin, Colorado as a test case to demonstrate the model's functionality.

References

Colorado Agricultural Statistics Service. (1985-90). "Colorado Agricultural Statistics", Lakewood, Colorado.

Jensen, M.E., R.D. Burman, and R.G. Allen. (1990). "Evapotranspiration and Irrigation Water Requirements." ASCE Manuals and Reports on Engineering Practice No. 70. American Society of Civil Engineers, New York, NY.

USBR (1986). "Colorado River Systems Consumptive Uses and Losses Report (1981-1985) Technical Appendix Volume 1.

USDA., Soil Conservation Service. (1970) "Irrigation water requirements. Tech Release No.21," (rev.), 92 pp.

W.W. Wheeler and Associates, Inc. (1994). "SMB and Supporting Programs." Water Resources Engineers, Englewood, Colorado.

Use of GIS to Predict Erosion in Construction

David G. Parker[1], Member, ASCE, Sandra C. Parker[2], and Thomas N. Stader[3]

Abstract

A graphic software system is designed and implemented to allow for the analysis of erosion potential on proposed highway construction sites. The system is based on Geographic Information System technology and allows for the consideration of erosion prevention products such as straw and other mulches as well as other types of cover products designed to prevent or minimize erosion from construction practices. The use of this system will allow for effective decisions concerning erosion control before construction has begun and erosion damage has already occurred.

Introduction

Topographic surface modeling using a Geographic Information System (GIS) can be useful for the prediction of soil erosion resulting from highway construction projects. The assumption is that terrain, along with other parameters, will influence the potential for soil erosion in a given area. Disturbance of the surface in highway construction will result in soil erosion and deposition, a source of pollution for streams and lakes. Modeling these various parameters with a GIS can provide an analysis tool for determining the potential for erosion while construction is in the planning stage, thereby minimizing the deleterious effects of construction on water quality. The authors have designed and implemented a GIS based system which allows for predictive modeling of erosion potential and effectiveness of erosion control products for proposed highway construction projects.

Geographic Information Systems

A Geographic Information System is an information system that is designed to

----

[1]Prof., Civil Engrg. Dept., University of Arkansas, Fayetteville, AR 72701.
[2]Prof., Indus. Engrg. Dept., University of Arkansas, Fayetteville, AR 72701
[3]Project Manager, Intel Corporation, Albuquerque, NM 87125

work with data referenced by spatial coordinates (Antenucci et al. 1991). A GIS provides an automated manner of collecting, storing, manipulating, combining, and displaying this data. A significant aspect of such systems is that they incorporate both a database for 'layers' of spatially referenced data, each representative of a spatial parameter of interest for analysis, as well as a set of operations for manipulating the data layers. Soil type, topography, and streams, for example, are spatial attributes that would be essential in a GIS database to be used for environmental analysis.

The data layers in a GIS are generally handled in one of two ways, either by a raster or a vector method. Raster data are represented by uniform grid cells of specified resolution, and data are stored as a matrix of cells. Vector data layers are handled as lines between points. Generally, operations involving these two types of data are primarily oriented to either raster structures or vectors, although a GIS usually incorporates algorithms that convert these structures from one to the other form depending on the actual system being used.

Geographic Resources Analysis Support System

The Geographic Resources Analysis Support System (GRASS) is the GIS that has been used in designing this system. GRASS is a public domain, general purpose, grid-cell based geographical modeling and analysis software package developed at the U.S. Army Corps of Engineers Research Laboratory. GRASS databases consist of three major forms, site or point, vector or line, and raster or grid. While the users of GRASS can model and conduct operations with vector data, it is primarily oriented to raster data.

Soil Erosion Models and the Universal Soil Loss Equation (USLE)

A variety of mathematical models to predict soil erosion have been proposed, and each is an attempt to represent the actual erosion process over a given time horizon. The USLE is one such model. It was originally developed for calculating field soil erosion losses for agricultural lands. The USLE is widely known and understood and is compatible with a raster-based GIS. Additionally, large databases for the application of this model, including such parameters as soil type and slope, were available for this research, therefore, the predictive system is based on the USLE.

The USLE is a mathematical model that is used to compute the longtime average soil losses from sheet and rill erosion under specified conditions and can be used for construction sites and other non-agricultural conditions. As a result of the unpredictable short-time changes in the levels of influential variables, the USLE is less accurate for prediction of specific events than for prediction of longtime averages. However, since the primary purpose of this research is dedicated to construction projects that take a large amount of time to complete, for example six months to a few years, specific events are not as great a factor, therefore the USLE is suitable for evaluating soil erosion for the present purpose.

Model and Components

The Universal Soil Loss Equation is given as follows (Wischmeier and Smith 1978):

$$A = R \times K \times L \times S \times C \times P$$

where:

A is the computed soil loss per unit area usually expressed in tons per acre per year,

R is the rainfall and runoff factor which quantifies the raindrop impact effect and provides relative information on the amount and rate of runoff likely to be associated with the rain. The map layer for this factor is a secondary map produced from an isoerodent map for the state (Wischmeier 1959),

K is the soil erodibility factor which is a secondary map derived from the primary data layer of soil type,

L and S are combined into a secondary map layer called slope length. This factor reflects the steepness and length of the land from which the entire slope length drains into a particular spot and is derived from another secondary map, slope, which is derived from a topographic map of the study area,

C is the cover and management factor which comes from land use and land cover from such things as vegetative canopies and trees,

and P is the support practice factor used primarily in croplands. In general, the value for P will usually equal 1.0 for construction sites because the erosion-reducing effects of shortening slopes or reducing slope gradients are accounted for through the LS factor (Wischmeier and Smith 1978).

System Design for Erosion Prediction

The erosion prediction software system is based on GRASS as the underlying GIS and uses the USLE to calculate erosion potential. It runs on a SUN work station under the X Windows System which is a network-based graphics window system. The software developed to predict erosion takes into account construction practices and erosion prevention measures and uses the existing GRASS databases as the source of the primary data layers and some secondary data layers, such as slope. Other secondary data layers are derived from primary data layers, such as slope length from elevation (McKimmey 1994).

The system uses both primary data layers and secondary data layers to conduct an erosion estimate for a proposed construction path or area as defined by the user for a session with the software. The initial erosion estimate is based on the present conditions of the proposed path. The system provides numerical data in the form of tons per acre per year as well as providing graphic representation of this data on the screen. The graphic output is a color coded raster map of the area of interest, along with a color legend which allows the user to identify portions of the area that have the same potential for erosion. Each color is assigned to a class of erosion potential, arranged from the lowest to the highest. Highway construction procedures, such as clearing and grubbing, may be defined for the proposed area, and the system will respond with a new calculation of erosion potential considering the procedure and a new color raster map indicating the erosion potential for each raster in the area, as well as numeric data which indicates the new predicted erosion potential.

After a proposed area is defined for construction procedures, i.e. clearing and grubbing, the user of the system can conduct analyses concerning possible use of several different erosion control products. Those products that may be used for ground cover and are presently built into the system are mulches of straw, crushed stone or wood chips. The user may choose one of three different straw mulches (1, 1.5, or 2 tons per acre), or two different crushed stone mulches (135 or 240 tons per acre), or one of three different wood chip mulches (7, 12, or 25 tons per acre). After the selection of a particular mulch type and amount, the system recomputes the USLE by incorporating the cover factor (C, as defined above) for the mulch instead of the cover factor for the actual ground cover at present. The cover factors for these mulches have been computed and are given in (Wischmeier and Smith 1978). By doing such analyses, the user can decide on an effective ground cover for reducing or preventing erosion in the proposed construction area.

Conclusions

The use of such a system based on GIS can be a valuable tool in evaluating erosion potential and recommending appropriate ground covers. As cover factors for new erosion prevention products become available, they can be used with the present system by virtue of its being designed for the input of user-defined erosion control products, such as soil conditioners, stabilizing emulsions, erosion control blankets, and other such products, provided that there is access to effectiveness data for these various products. Such analyses allow for effective decisions regarding construction paths before work has begun and damage from erosion has already occurred.

Appendix I. References

Antenucci, J. C., Brown, K., Croswell, P. L., and Kevany, M. J. (1991). Geographic Information Systems: A Guide To The Technology, Van Nostrand Reinhold, NY.

McKimmey, J. M. (1994). *Prediction and Management of Sediment Load and Phosphorous in the Beaver Reservoir Watershed Using A Geographic Information System*, Masters Thesis, University of Arkansas, Fayetteville, AR.

Wischmeier, W. H. (1959). "A Rainfall Erosion Index for a Universal Soil Loss Equation". *Soil Science Society of America Proceedings* 23.

Wischmeier, W. H. (1978). *Predicting Rainfall Erosion Losses*, U.S. Department of Agriculture, Agriculture Handbook 537.

ADVANCES IN AUTOMATED LANDSCAPE ANALYSIS

Jurgen Garbrecht[1], M. ASCE, and Lawrence W. Martz[2]

Abstract

Automated watershed segmentation from raster Digital Elevation Models (DEM) is weak with respect to (1) drainage definition within sinks and over flat areas, and (2) identification of network topology. Advances in these areas are presented for DEM processing models based on the D-8 method. The proposed treatment of sinks and flat areas leads to a better representation of actual drainage conditions. And, the channel and subcatchment indexing establishes an important link to distributed hydrologic models and enables an automated control of channel flow routing in channel networks.

Introduction

The automated extraction of landscape features directly from Digital Elevation Models (DEM) has gained increasing attention as DEM coverages for many areas of the United States are becoming available. In the field of water resources and hydrology, automated landscape evaluation has focused on watershed segmentation and definition of drainage divides and channel networks. However, the procedures used to extract such information from DEMs have limitations and require a number of assumptions which sometimes are simplistic and can lead to inaccuracies. This paper presents advances in the treatment of sinks, flat areas and network indexing.

---

[1] Hydraulic Research Engineer, US Dept. of Agriculture, Agricultural Research Service, P.O. Box 1430, Durant, OK 74702

[2] Professor, Dept. of Geography, 9 Campus Drive, University of Saskatchewan, Saskatoon, Canada, S7N0W0

Treatment of sinks in DEMs

A common DEM processing method is the D-8 method (Fairchild and Leymarie, 1991) in which drainage patterns are identified using downslope flow concepts (Martz and Garbrecht, 1992). The steepest downslope flow direction is determined at each raster cell and used to trace the flow paths. The primary difficulty in this approach is the treatment of sinks. Sinks are groups of raster cells which are completely surrounded by other cells at higher elevation. Thus, they are features that have no downslope flow path that lead out of them, which is why the approach based on downslope flow direction fails.

Sinks are commonly found in raster DEM and are usually considered to be spurious features which arise from interpolation and truncation of interpolated values. The traditional approach to the treatment of sinks is: (1) to fill each sink to the elevation of its lowest outlet, and (2) subsequently directing flow across these flat areas from adjacent areas of higher elevation to those of lower elevation. There are two assumptions implicit to this approach. The first is that sinks do not represent true landscape features, but are spurious features that arise through interpolation errors or insufficient precision in elevation values. The second is that all sinks arise through the underestimate of elevations and should be filled. However, if DEM interpolation errors are considered random, elevations are as likely to be overestimated as underestimated. It follows, therefore, that at least some sinks arise from the obstruction of flow paths by overestimated elevations. In such cases, breaching the obstruction is more appropriate than filling the sink it creates. The following procedure simulates the breaching of narrow obstructions along flow paths.

The spatial extent of each sink is delineated and its lowest available outlet is identified and evaluated for possible lowering to simulate breaching. The number of cells at the outlet that may be lowered is termed the breaching length. To restrict breaching to relatively narrow obstructions, the breaching length is limited to a maximum of two cells.

All cells inside the sink and at the same elevation as the outlet are examined to determine if they are (1) adjacent to a cell outside the sink and at a lower elevation than the outlet, and (2) within the breaching length of a cell inside the sink and at a lower elevation than the outlet. Cells meeting these criteria are potential breaching sites. If no such cells exist, outlet breaching is not possible and the sink is simply filled to the outlet elevation. If more than one potential breaching site exists, the one with the greatest breaching depth (primary criterion), and the shortest breaching length (secondary criterion) is selected. If these criteria are met at more than one site, one of the sites is selected arbitrarily. The elevation of the cell at the selected breaching site is lowered to the elevation of the lesser of the cell outside or inside the sink. This changes the elevation of the outlet and effectively breaches the obstruction responsible for the sink. Regardless of whether a breach is effected, the elevations of all cells inside the sink and at a lower elevation than the outlet are then changed to the elevation of the outlet. This fills any remaining cells in the sink to the outlet elevation and produces a continuous flat area.

The purpose of the proposed outlet-breaching procedure is simply to assign part of the apparent elevation data error responsible for sinks to elevation overestimation. By imposing a relatively short breaching length, breaching is restricted to narrow obstructions on defined flow paths. This approach reduces or eliminates the filling of sinks created by narrow obstructions, and is particularly effective for low relief landscapes.

Treatment of flat areas in DEMs

When landscape elevations are interpolated into a DEM, areas of low relief are often translated into a sequence of small, perfectly flat surfaces. This is the result of too low of a vertical and/or horizontal DEM resolution to represent the landscape. Flat surfaces are also generated in the DEMs when sinks are filled to the level of their lowest outlet (see previous section). Whatever their origin, flat surfaces are problematic because flow direction on a perfectly flat surface is indeterminate.

The problem is particularly difficult for raster processing schemes that rely on the D-8 method because the method defines landscape properties based only on the DEM cells at the point of interest and its immediately surrounding 8 neighboring cells. Since all DEM cells in a flat surface have the same elevation value, a unique flow direction cannot be assigned.

Traditionally, flow direction over flat surfaces in DEMs are defined using a variety of methods ranging from landscape smoothing to arbitrary flow direction assignment. For a short review of existing methods the reader is referred to Tribe (1992). In the following, a numerical scheme for use with the D-8 method is presented that allows the identification of flow direction over flat surfaces as a function of the non-local rising and falling topography surrounding the flat surface.

Since the D-8 method cannot operate on landscape information beyond its immediate vicinity, the elevation values of flat surfaces must be amended ahead of time in a manner to subsequently produce, by means of the D-8 method, the expected flow direction. The necessary amendments to elevations in flat surfaces is based on the recognition that natural landscapes generally drain towards falling terrain while simultaneously draining away from rising terrain. This effect, as well as flow merging on flat areas, is incorporated into DEMs by incrementing elevations of flat surfaces to produce two gradients: one which forces flow away from rising terrain, and a second which draws flow towards the nearest downslope outlet. It is emphasized that incrementation is applied only to DEM cells belonging to flat surfaces. Furthermore, the incremental elevation height is selected arbitrarily small (say 1 mm) because it only serves to numerically identify flow direction. Therefore, the addition of incremental elevations much smaller than the vertical DEM resolution has little bearing on the digital landscape elevations. Yet, it provides the necessary information to identify a realistic flow direction that is consistent with the surrounding rising topography and that also provides for flow concentration.

Network and subcatchment indexing

For use in hydrologic modelling, the automatically extracted channel network links and their direct drainage areas must be explicitly identified by code numbers and be associated with topological information on upstream and downstream connections. Such identification is often possible in vector GIS, but usually not in raster GIS. The following network and subcatchment indexing system for rasters is proposed.

The channel network is first delineated as all cells with a drainage area greater than a specified threshold. This yields a raster representation of a uni-directional, full-connected network. In the first phase of network analysis, the flow direction grid is used to move through the network from upstream to downstream to determine the Strahler order (Strahler, 1957) of each link and to find the location of all source and junction nodes. In the second phase, the flow direction grid is used to simulate a walk along the left bank of the channel network beginning and ending at the watershed outlet. Each node is assigned a sequential number the first time it is passed during the walk.

In the third phase, the subcatchment area of each network link is determined. The subcatchment areas of each link are subdivided into left-bank, right-bank and, in the case of exterior links, source node contributing areas. Subcatchment are assigned unique identification codes based on previously assigned node numbers. For all subcatchments, a base identification number is assigned which is the node number (NN) at the upstream end of the link to which the subcatchment drains multiplied by 10 (NN*10). For source node subcatchments a value of 1 is added to (NN*10), for left-bank subcatchments a value of 2 is added, and for right-bank subcatchments a value of 3 is added. Channel cells (which are not considered to be part of the right-bank, left-bank or source node subcatchments) are assigned an identification code of (NN*10) plus 4. These identification codes are based on the initial node number scheme and provide a basis on which network links, nodes and subcatchments can be associated with one another. Most importantly, the node numbering scheme on which the identification system is based can be used to determine the optimal routing sequence to be used in modelling streamflow through large and complex networks (Garbrecht, 1988).

Conclusions

The above advances in drainage pattern identification reduce the level of arbitrary decision-making in the presence of poorly defined topography due to limited DEM resolution, and are believed to better represent actual drainage patterns. The channel and subcatchment indexing is important for a direct coupling of DEM based watershed segmentation and hydrologic modelling. These improvements have been incorporated into the topographic parameterization model TOPAZ which automatically segments and parameterizes watersheds from DEMs for water resources, hydraulic and hydrologic applications.

References

Fairchild, J. and P. Leymarie. 1991. Drainage Networks from Grid Digital Elevation Models. WRR 27(4):29-61.

Garbrecht, J. 1988. Determination of the Execution Sequence of Channel Flow for Cascade Routing in a Drainage Network. Hydrosoft 1(3):129-138.

Martz, L.W. and J. Garbrecht. 1992. Numerical Definition of Drainage Networks and Subcatchment Areas from Digital Elevation Models. Computers & Geosciences 18(6):747-761.

Strahler, A.N. 1957. Quantitative Analysis of Watershed Geomorphology. Trans. AGU, 38(6):913-920.

Tribe, A. 1992. Automated Recognition of Valley Lines and Drainage Networks from Grid Digital Elevation Models: a Review and a New Method. J. of Hydrology, 139:263-293.

HISTORICAL FLOODS ON THE MIDDLE MISSISSIPPI RIVER

Claude N. Strauser[1], Fellow, ASCE and Donald M. Coleman[2]

Introduction

A review of historical floods is one of the key elements in understanding the many parameters relating to the present day Mississippi River. This review of the past will be a journey of learning and will help to reveal the cherished secrets of the river. Scientific investigators need to have an understanding of the past in order to appreciate present day conditions and to also plan for the future. This paper will focus on the time period from the flood of 1543 to the flood of 1903 (360 years).

Background

The period of record of hydrologic data in the United States is often found to be insufficient to accomplish detailed engineering investigations. European and Asian countries often have periods of record which are more extensive and comprehensive. These periods of record sometimes reach into several centuries of comprehensive documentation of hydrologic data. This is a disadvantage to investigators in the United States, both from an engineering point of view and from an ability to have a proper perspective of past events. Even the data that are available for the limited period of record in the United States is not homogeneous. Data were collected using various types of equipment, methods and procedures. Comparisons with contemporary data often will lead the unsuspecting investigator to misleading conclusions.

----

[1] Chief, Potamology Section, U.S. Army Corps of Engineers, St. Louis District, 1222 Spruce St., St. Louis, MO 63103

[2] Chief, Water Control Management Unit, U.S. Army Corps of Engineers, St. Louis District, 1222 Spruce St., St. Louis, MO 63103

Stage Data

The stage data collected from historical events and narrative accounts can, for the most part, be determined with a high level of confidence. Usually high water marks on trees, buildings, etc. are reliable and can be correlated with the contemporary data without too much suspicion. The stage data for the events being discussed in this paper are as follows:

STAGE DATA AT ST. LOUIS

Date	Gage Reading
1543	unknown
1724	unknown
1745	unknown
1772	unknown
1785	38.9
1811	less than 1785
1823	unknown
1826	33.81
1844	41.3
1851	36.5
1855	unknown
1858	37.1
1860	unknown
1862	31.45
1876	32.0
1881	33.65
1882	32.4
1883	34.8
1892	36.0
1893	34.8
1903	38.0

The elevation equal to zero on the St. Louis gage is 379.94 NGVD.

Flow Data

A discussion relating to the volume of flow for the known river stages is more complicated and has confused many contemporary investigators. The first issue to be discussed is the measurement of flow on the Mississippi River, especially in the St. Louis reach of the river. The data are definitely not homogeneous. Several different types of equipment, methods and procedures have been used to measure the volume of flow.

The first discharge measurement made at St. Louis was done by Mr. Homer, City Engineer in 1866. The first discharge measurement made by the Corps of Engineers was made at St. Louis in May of 1872. The equipment used was varied and continued to evolve over the years. For example, during the months of February and March of 1881, seven discharge measurements were made at St. Louis using ice cakes as the surface velocity measuring vehicle. These are the only discharge measurements recorded which involved the use of surface floats. Rod floats were used to measure discharge at St. Louis during the period of 14 October 1881 to 20 September 1919. During this extended period, 400 discharge measurements were made utilizing rod floats. Double floats were used at St. Louis from 14 October 1881 until 31 December 1930. During this period, 102 discharge measurements were made utilizing double floats. There were 32 discharge measurements made at St. Louis with meters prior to January 1882. Therefore, these, and possibly others, were not made with Price current meters. During the years 1872 until 1933, discharge measurements in the St. Louis District were made by personnel from the Corps of Engineers. From 1933 to the present, the majority of discharge measurements were made by the U. S. Geological Survey (USGS) for the Corps of Engineers.

Besides the different types of equipment used, different methods and procedures were used. The Corps of Engineers made most of their discharge measurements from floating platforms (barges or small vessels). Most of the USGS measurements were made from fixed structures (bridges). In addition, some measurements were made using the .6 depth method, while other measurements were used using the .2 and .8 depth method.

This lack of homogeneity has created a problem in analyzing the period of record and has obscured any easy possibility of determining long term trends in a hydrologic cycle (if one exists).

FLOW DATA

Date	Volume
1543	Unknown
1724	Unknown
1745	Unknown
1772	Unknown
1785	Unknown
1811	Unknown
1823	Unknown
1826	Unknown
1844	not measured, est. 36,790 cms (1,300,000 cfs)
1851	not measured, est. 28,920 cms (1,022,000 cfs)

FLOW DATA (Continued)

Date	Volume
1855	not measured, est. 29,710 cms (1,050,000 cfs)
1858	not measured, est. 29,820 cms (1,054,000 cfs)
1860	not measured
1862	not measured, est. 20,150 cms (712,000 cfs)
1876	not measured, est. 20,970 cms (741,000 cfs)
1881	25,350 cms (896,000 cfs), rod float measurement
1882	not measured, est. 20,910 cms (739,000 cfs)
1883	not measured, est. 24,420 cms (863,000 cfs)
1892	measurement discarded, est. 26,200 cms (926,000 cfs)
1893	not measured, est. 19,810 cms (700,000 cfs)
1903	not measured, est. 29,430 cms (1,040,000 cfs)

DEFINITION OF ABBREVIATIONS:

cms = cubic meters per second
cfs = cubic feet per second

It is believed that the estimates for flows made before discharge measurements were actually made came from rating curves developed from the period 1900 to 1904. Discharges determined from the extrapolation of a rating curve are subject to large errors. Discharges determined from rating curves developed subsequent to the time of occurrence of the flow are questionable and are susceptible to considerable error. As the time interval increases, the probability of the discrepancy is enhanced.

Present day understanding of rating curves certainly confirm the aforementioned conclusions. Rating curves are single valued functions, implying that the cross sectional area, water surface slope, and etc., are constant for a given stage. Everyone knows this is not true, even in a report prepared in 1861 by Humphreys and Abbot this was understood.

"It is evident that the condition of the river, whether

rising or falling, makes a great difference in discharge at any given stand; but it is equally evident that a mean line between these two extremes can be drawn that shall form the basis of a table by which the annual discharge can be deduced from the recorded gauge readings. For any given day, its indication will be erroneous, but for the entire year, which includes both rising and falling branches of the curve, it will be sufficiently accurate."

Discharges determined from manufactured rating curves should be treated as historical data and used only to determine the rank of various hydrologic events. Subsequent studies performed for the St. Louis District at the Waterways Experiment Station (Mississippi Basin Model) have shown the published flood discharge measurements along the Middle Mississippi River prior to the mid 1930's are over estimated. The major floods of 1844 and 1903 may be significantly over estimated by as much as 33% and 23% respectively.

Summary

Amateur hydrologists and uninformed investigators often make the mistake of using published hydrologic data without completely understanding the relative value of the data. This has led to faulty and totally misleading conclusions which can result in a disservice to the general public. A comprehensive and fully informed researcher can assign relative weights to hydrologic data and evaluate which information can be scientifically used and which can be used only in a general and/or historical context.

This paper represents the findings and opinions of the authors and are not necessarily those of the Corps of Engineers.

River Forecasting During the Great Flood of 1993

"Knowing When to go Beyond Computer Models"

Dave Busse, P.E.* and Robert Rapp, P.E.*

Abstract

A discussion of the difficulty in choosing the most accurate and effective river forecasting methods is given. The computer models which have been developed for forecasting are examined and are compared to other methods. During the Flood of 1993, computer models were used. Their accuracy and timeliness in this situation are evaluated and are used to discuss their limitations. Conclusions about forecasting for future events with computer models and alternative methods are discussed.

Introduction

Floods are a function of the location, intensity, volume, and duration of rainfall. Land cover and the capacity of a rivers' floodplain to convey or store water is also important. These factors must be considered while forecasting river levels. This was highlighted during the Great Flood of 1993. The U.S. Army Corps of Engineers, St. Louis District Water Control Office was tasked by the District Commander to provide real-time forecasts to the District flood fight teams. These forecasts had to be provided to the teams as soon as possible so increases in flood protection could be achieved in a timely manner.

The Basin

The part of the Mississippi Basin above the confluence of the Ohio River is known as the Upper Mississippi. It is this sub-basin that experienced what is known as "The Great Flood of 1993." Draining all or part of 13

----

*Hydraulic Engineer, U.S. Army Corps of Engineers, St. Louis District, 1222 Spruce St., St. Louis, MO 63103

states, the basin encompasses about 1,849,000 km^2 (714,000 square miles). From its source at Lake Itaska, Minnesota, to its confluence with the Ohio River it is 2,198 km (1,366 miles long). The principle tributaries are the Missouri and Illinois Rivers. The Missouri River enters the Mississippi 24 km (15 miles) upstream of St. Louis and drains 1,371,000 km^2 (529,300 square miles). The Illinois River enters the Mississippi 61 km (38 miles) upstream of St. Louis and drains 74,866 km^2 (28,906 square miles). This area is one of the most complex hydrologic areas in the country to forecast.

The Flood Event

The Great Flood of 1993 was a hydrometeorological event without precedent in modern times. Precipitation amounts, river levels, duration, and areas inundated surpassed all previous floods in the basin. From June through September, record and near record precipitation occurred. This fell on soil saturated from previous precipitation events that began occurring in late 1992. This resulted in flooding along major rivers and tributaries in the Upper Mississippi River Basin. The flood was extensive in magnitude, and also in duration. Many locations remained above flood stage for months. For example, the Mississippi River at the confluence of the Illinois River (Grafton, Illinois) was above flood stage for 203 days in 1993. Downstream at St. Louis, the river was above flood stage for 148 days.

The Role of the Water Control Office

The National Weather Service (NWS) is the official river forecasting agency of the United States Government. This may lead some to question why the Corps of Engineers Water Control Office (WCM) does its own river forecasts. Forecasts are a byproduct of project regulation of the locks and dams, and reservoirs operated by the District. Changes in regulation of these projects significantly alter the forecasts.

The District has flood fight responsibilities that require stage forecasts for locations that the NWS does not forecast. The WCM update their forecasts as often as the District Commander requires. The District also needs at it disposal a series of "what if" forecasts. These "what if" forecasts are for various possible future precipitation events. These forecasts are necessary to ensure the flood fight teams have the necessary resources(sand, sandbags, manpower, equipment, etc.) necessary to provide additional protection if required.

Challenges

The magnitude and duration of this flood presented challenges to the WCM that their existing computer forecast models were not equipped to handle. The rivers rose higher than existing rating tables. Levees overtopped or failed which had major impacts on the flood hydrograph. Some of these levees had never failed or been overtopped since they were first constructed. Others, mainly private agricultural levees fail during most large floods, and provide some attenuation to the flow. Existing computer models(calibrated to past events), took this attenuation into account and therefore produced forecast that were later proven to be low. In mid July the WCM determined that the existing computer models must be ignored if the flood fight teams were to have accurate stage forecasts. The majority of the levee systems on the Missouri River between Kansas City and St. Louis Missouri had already been breached or overtopped. Therefore any future increase in flows would not attenuate as they had in previous events.

Solutions

The WCM requested the United States Geological Survey (USGS) to take discharge measurements. USGS crews from six different states took discharge measurements at key locations. Measurements were collected along the Missouri River from St. Joseph, Missouri to St. Charles, Missouri. On the Mississippi River, discharge measurements were taken from Hannibal, Missouri to Thebes, Illinois. On the Illinois River, discharge measurements were taken at Valley City, Illinois. Also, measurements were taken on the Grand River, a major tributary of the Missouri River. The data collected by the USGS proved to be invaluable.

The WCM believed that the traditional attenuation would not occur. The majority of levee systems had failed or overtopped due the extreme stage and duration of the flood. Satellite pictures revealed that the majority of the levees upstream of St. Louis were overtopped by mid-July. This would prove critical as the area behind the levees (traditional off-channel storage) would provide little benefit in reducing flows. On July 24, the WCM provided the flood fight teams a forecast for the St. Louis gage to crest at 14.9 m (49+ feet) on August 1 or 2. This forecast considered both the water on the ground and the rain that was forecast that day. As USGS data came in from the various locations it helped confirm that St. Louis would reach a stage more than 14.9 m (49.0 feet) (the highest stage ever recorded).

Discharge data were also being collected by the USGS at St. Louis on almost a daily basis for the entire month of July. This data was used to get a reliable stage-discharge relationship for the St. Louis gage. This data also confirmed that with the magnitude of the measured flows upstream of St. Louis, that 14.9 m (49+ feet) at St. Louis was very likely.

Flow measurements had never been taken for such an extreme event. Significant overbank flows provided a very difficult measuring environment. However, it was quickly revealed that each successive measurement made by different crews confirmed the validity of each upstream measurement. They also confirmed that very little attenuation was occurring. The USGS did an outstanding job in measuring the flow during this epic event. Table 1 shows the USGS measurements, and clearly illustrates how the flows tracked during the week leading to the crest at St. Louis.

Results

The forecast provided by the WCM proved to be both timely and accurate. The forecast that was given on July 24 (a week ahead of the actual crest) of more than 14.9 m (49 feet) at St. Louis gave the flood fight teams a head start that would not have been possible using existing computer models. The computer models would have continually raised the forecast until right before the crest. This would have been disastrous for the flood fight teams. They would not have been able to add the required height and width to the temporary flood protection structures. Members of the flood fight team credit the WCM forecast with saving many areas of St. Louis as well as other communities along the Mississippi River.

Conclusions

The most important lesson to be learned is that the forecaster must know the limitations and reliability of existing computer models. When stages and flows are eclipsing previous records, the results must be viewed with careful scrutiny. Computer models are calibrated from previous events. For record events, results from computer models must be closely analyzed. Conditions for these record events will most likely be far different from those previously experienced. For the Great Flood of 93, the rivers were virtually running bluff to bluff before the last major rain event. In this case, the conditions were more similar to a flume, than to a typical river system with significant off-channel storage. Except for minimal channel attenuation, what went in at the upper end of the flume came out at the lower end. To forecast this event, it was a matter of determining what was entering the various flumes(Missouri, Upper Mississippi, and Illinois River flows, taken by USGS measurements), and how they would combine at St. Louis.

For this event, a sophisticated computer model was not necessary. This is not to say computer models are not necessary. If the existing off-channel storage areas had been available, significant attenuation would have occurred. This would have definitely required a computer model to simulate the changing conditions.

This paper illustrates that the forecaster must examine all available data and resources, and determine if the results being generated by computer models are accurate and dependable. The most reliable forecast, be it computer generated or paper generated, must be passed along to flood fight teams as soon as possible. To provide additional flood protection, time can be the flood fight teams biggest obstacle. An accurate forecast provided hours before a crest is of little help to flood fight teams. To be helpful, forecasts during a flood must be both **accurate and timely.**

The aforementioned discussion represents the views of the authors and are not necessarily the views of the Corps of Engineers.

EVOLUTION OF USACE/USCG RADIOBEACON NETWORK FOR WESTERN/INLAND RIVERS

Robert D. Mesko[1], ACSM, and LCDR George Privon[2]

Abstract

Efforts within the Lower Mississippi Valley Division (LMVD) hydrographic surveying work group to implement a Division wide DGPS system lead to joint testing and development efforts with USACE and the U.S. Coast Guard (USCG) during the summer of 1993 resulting in the combining of resources, technology, and expertise to implement a marine radiobeacon based DGPS service on select inland rivers. The evolution and implementation of the system to date is discussed as well as challenges for the future and obstacles yet to be overcome.

Introduction

Demonstrations of Differential Global Positioning System (DGPS) technology by the Technical Engineering Center (TEC) at Savannah, Georgia during Sept. 1990 and by many vendors, contractors, and manufactures have made it apparent that DGPS technology would be replacing many aging microwave hydrographic surveying systems throughout the Corps of Engineers. DGPS can achieve meter level (1-3 meters) accuracy in real time using Coarse Acquisition (C/A) or Precise (P) code of the GPS signal if the reference station position is known. The pseudorange corrections are transmitted from the reference station across a radio data link to the user. The user receives the differential pseudorange corrections and applies them to measurement at the remote site. This technique can achieve accuracies of 1-3 meters (2drms) which meets or exceeds class one

----

[1] Cartographer, U.S. Army Corps of Engineers, St. Louis District, 1222 Spruce St., St. Louis, MO 63103

[2] Chief, Radionavigation Planning Branch, U.S. Coast Guard Headquarters, 2100 2nd Street, S.W., Washington, D.C. 20593

hydrographic surveying standards as stated in Engineer Manual EM 1110-2-1003, "Hydrographic Surveying".

LMVD Hydrographic Survey Work Group

Within the LMVD, individual districts were beginning to implement DGPS in various ways, ie. raw range systems, Radio Technical Commission for Maritime Services (RTCM) formats of different versions, commercial Wide Area DGPS (WADGPS), and combinations of differing frequencies of UHF and VHF data transmission formats.

The Districts within LMVD have a hydrographic surveying mission with a focus that is perhaps unique from other Districts, involving the surveying of river channels for revetments, mat-laying, dikes, dredging, monitoring structures, and bendway weirs. With the central focus on the Lower Mississippi River, a hydrographic survey committee was set up to include the four districts, New Orleans, Vicksburg, Memphis, and St. Louis. The hydrographic committee included work groups to coordinate CADD, Electronic Charting (ECDIS), DGPS, Articulated Concrete Mattress (ACM), and Data Management (REGIS) with a dual chairperson oversight manager from Engineering and Construction Divisions. The group's goal is to standardize data being sent to division for review and to allow division offices to utilize CADD technology and electronic transferring of data.

The DGPS work group considered the Civil Works Guide Specification for procurement of real-time DGPS as well as other systems being developed (Coast Guard, Wide Area Networks, SERCEL, and others) and then began formulating plans for an LMVD network. Some approvals for UHF radio frequencies had already been obtained for DGPS by two LMVD Districts. The group then decided to stay with the UHF band because of previous approval in that part of the radio spectrum and the update rates being transmitted were achieving 1-3 meter accuracy.

Tower sites were being selected for the system and GPS reference station units were being procured. In addition TEC had prepared a report with recommendations on the network, and frequency authorization had been requested. The frequency application procedure is:

a. District Frequency Manager
b. HQUSACE Frequency Manager
c. Communications Electronics Services
d. Electromagnetic Compatibility Analysis Center
e. National Telecommunications and Information Administration

The LMVD DGPS network was progressing on schedule until three major developments changed its direction.

1. Frequency approval would only be for VHF range (162 Mhz thru 174 Mhz)

2. USCG announced an interest to expand their DGPS service to select inland rivers.

3. An OMB auditor requested that coordination be made with the USCG to avoid duplication of efforts.

A joint test was then performed in June 1993 utilizing the Coast Guard Radiobeacon based DGPS and the New Orleans DGPS (UHF) system. This testing demonstrated the USCG DGPS system could meet a majority of USACE hydrographic surveying requirements at very low cost.

Memorandum of Agreement

Subsequent coordination meetings and testing of the USCG DGPS service with HQUSACE, TEC, USCG, LMVD, and participating districts led to a broad Memorandum of Agreement between the Corps and Coast Guard effective 23 February 1994. Major points of the MOA are:

(1) Provides for inter-agency DGPS activities
(2) Establishes overall policies and relationships
(3) Provides for the maximum use of existing infrastructure and resources
(4) Provides for USCG coordination of frequency approval
(5) Allows for supplemental agreements between divisions, districts, etc.

LMVD districts began implementation of the system by performing site surveys for the location of the Radiobeacon antennas, acquiring the necessary real estate, ordering and installing towers and antennas, providing power and communication to the sites, building equipment shelters, installing the ground planes and MF transmitters, surveying the location of the GPS antenna locations, and turning on and testing the system. The Coast Guard's coverage analysis of the system indicates that many Western/Inland Rivers could be adequately covered with additional stations in Corps Districts. The LMVD DGPS system is currently providing accuracies of 1-3 meters and will be upgraded with standard USCG DGPS hardware and connected into the USCG DGPS control system by 1996.

The first supplement under this MOA was signed Oct. 1994 to allow the siting of Coast Guard DGPS broadcast sites on ACOE property at Detroit, MI, Saginaw Bay, MI and Sault Ste Marie, MI.

A second supplement to this MOA for the first three (3) sites constructed in the LMVD at Vicksburg, MS, Memphis, TN, and St. Louis, MO was signed in April 1995. Table 1 presents a summary of the supplement as it pertains to these LMVD sites.

SUPPLEMENTAL AGREEMENTS

	AGENCY		
TASK	USCG	USACE	NGS
Training	X		
Site Construction		X	
DGPS Equipment Funding		X	
DGPS Equipment Procurement	X		
DGPS Equipment Installation		X	
Site Maintenance		X	
Geodetic Monumentation			X
DGPS Equipment Depot	X		
Control Station Operation	X		

TABLE 1

The DGPS Project Manager at Coast Guard Headquarters has requested 1996 funding to extend the Coast Guard DGPS network into the inland rivers. If fully funded, this request would provide for the establishment of thirteen (13) new sites at or near the following locations:

- -St. Paul, WI
- -Rock Island, IL
- -Omaha, NE
- -Kansas City, MO
- -Huntington, WV
- -Louisville, KY
- -Chattanooga, TN
- -Pittsburgh, PA
- -Andrews L&D, GA
- -Miller Ferry, AL
- -Gunthersvile,AL
- -Tulsa, OK
- -Demopolis, AL

Predicted DGPS coverage is for the navigable portions of the following rivers:

- Alabama
- Arkansas
- Black Warrior
- Chattahooche
- Tennessee & Tombigbee Waterway
- Illinois
- Ohio
- Mobile
- Tennessee

Whether this second phase of the USCG DGPS program will be funded by Congress in 1996 is unknown at this time. This budget request was based on the following assumptions:

- The Coast Guard will fund all DGPS hardware, antennas, and equipment shelters.

- The USACE will provide suitable Corps owned or leased property for each site in addition to constructing the site.

- Coast Guard will install all DGPS hardware and provide depot level support for the same.

- The USACE will perform all site maintenance as the Coast Guard does not have an infrastructure in these areas.

- The Coast Guard will maintain the responsibility for operation of the overall navigation service.

Conclusion

The coastal USCG DGPS project consisting of approximately 50 broadcast sites remains on schedule with an operational date of January 1996. The reference station and integrity monitor contracts have been awarded and equipment delivery is underway. Although this service was primarily designed for marine navigation, it will provide significant inland coverage in many areas. This coverage will lead to numerous applications such as terrestrial surveying and mapping. The USCG DGPS service with an estimated accuracy approaching 1.5 meters (2 drms) can meet all but the most stringent USACE positioning requirements.

Disclaimer

The information and opinions in this paper are attributable solely to the authors and are not necessarily representations of the U.S. Army Corps of Engineers or the United States Coast Guard.

Analysis of Negative Effects of Major Floods on Man..

by Glenn D. Lloyd JR[1]

Abstract

The Flood of 1993 in the upper Mississippi and Missouri river systems has given new impetus to analyze the interaction of man and nature. The flood has caused us as a nation to again ask the age-old question of the optimum mix of man and his environment. This research looks at the economic/engineer interface and points to some non-engineer reasons for the continued problems and proposes ways to better utilize our limited economic and natural resources. Additionally, a look at the possible beneficial effects of more hydropower facilities in the system has been undertaken.

Introduction

As long as man exists on this planet, two things are certain, there will be rain events that cause floods and man kind will want to reside in proximity to the water sources that are the absolute necessity of life. The river courses that carry water are in a constant state of flux, however the change is measured in a way much different manner than the way man does. We can construct and build structures in months, while nature changes the rivers over decades. Regardless of who is doing the changing, flood events will occur and cause river modifications as well as damage, and the goal of mankind must be to minimize the damages by working with, rather than against, the forces of nature. Like a well-run business, we needed to dissect the various components of the damage costs and then analyze them to see which can and cannot be decreased in a cost effective manner. In addition to damage reduction, man must also look for opportunities to gain synergistic benefits, such as hydropower, where possible.

----

[1] Student Member, Teaching Fellow, Department of Civil Engineering, University of Missouri-Rolla, Rolla, MO 65401

Flooding/Investment

Excess rainfall events result in runoff that gathers and eventually become the swollen rivers which result in floods of various magnitudes depending on the rainfall event. Mitigation of the peak and volume of runoff can be accomplished naturally with features such as wetlands and natural lakes or thru man made enhancements such as reservoirs both with and without hydropower facilities, channel improvements that increase flow capacity for the same cross sectional area, levees that increase the cross sectional flow area, and diversion channels that take the water around the effected area. The estimates are that the nation has invested approximately $25 billion in man made flood protection systems and just in the 93 Flood alone, these features prevented an estimated $19 billion in damages which can be added to the savings already totaled from all previous floods that occurred since the system has been in place. A respectable portion of this savings occurred because of reservoirs, some of which have hydropower facilities. The case can easily be made that the cost of the system has more that paid for itself, but that does not address the question of how to further reduce the damages that did occur and if hydropower facilities can play a role.

Cost of 93 Flood

The estimated damages for the flood range from $15 to $20 billion dollars with the most reliable number being $15.6 billion. Closer inspection of this number shows that it can be further broken down into sub categories and that when this is done over half the damages are related to agricultural costs. Other areas of damage expense include communications systems, utilities, urban areas and damage to flood protection facilities(levees). The cost reimbursements are not yet final because they have not all been quantified, but, the most recent subtotal is $10.8 and still rising. While much has been said about the negative effects of river improvements, much of the agricultural damage was due to upland flooding away from the river, which was not affected by the rivers. Because all the dollar amounts must be obtained from various agencies and repair work etc. is still on going, the final cost comparisons are not yet available, however, there is a good indication that a large portion of the cost of this flood event is NOT related to the channel improvements, reservoir effects and levees that have been built. What is yet to be done, is breaking down of the costs by the categories of transportation systems utilities, large cities, small cities, rural areas and open land (agricultural and natural), and comparing these to the reimbursements using the same categories. Once done, we can begin to assess if in fact flood control measures such as flood zoning, etc. have worked to reduce costs to man occupied regions and our problem is one of flooding in other than urban areas. Another aspect that is also being analyzed is the damage, by category, and the benefits returned to the effected people. No one agency really sums all the inputs of funds put back into a region as a result of floods, and there is a theory that the net cost vs regional benefits may be a net positive which rather than cause an economic hardship creates an economic boom. If this is the case, then this net increase of revenue eliminates the incentive to change the way we do business. The issue of flood relief benefits exceeding costs is a key area that no one appears to have tackled. The costs of the flood appear to be reasonably well documented, but the relief benefits are not summarized by any one agency, to include FEMA. Issues like payments of workman unemployment compensation, contributions from a host of volunteer agencies, federal and state highway dollars, farm subsidies, SBA loans, free FEMA provided housing for up to 18 months and the like are not rolled up into one package for future use and comparison.

Other Systems

Of interest also, is a look at river systems where a single agency has been able to create a comprehensive flood prevention plan by using all the tools available. Systems like the Tennessee River Valley and the Columbia River have had little damage since their systems have been put in place as well as enjoying benefits from power production. By contrast, the upper Mississippi and Missouri Rivers systems are not under the control or oversight of any one agency and no comprehensive plan has ever been developed for this system for flood control or power production. Even an inexperienced hydrologic engineer can see system gaps in this region. Additionally, the myth that many levees failed must be corrected. Only a few levees failed, most worked exactly as designed and were over topped when the design capacity was exceeded, just as any structural facility will fail if the allowable load is exceeded.

Conclusions

The systematic or holistic approach is being taken in review of this flood and comparison made of all the costs and benefits. The task under way is the assembly of all cost data, engineer, human, agricultural etc., and the breaking of it down by category. Against this, the benefits received to the region and the people residing there are being assembled in like categories. The necessary cost indexing of this event against other events, by category will begin to show where the real costs are rising and/or falling by area affected. Clearly some measure such as hardening key facilities, like water treatment plants, need to be done as well as building in failure points on levees to minimize damage. There is also a need to compare this area to other more sophisticated river systems and their damage costs over time. Here, the benefits of hydropower facilities on both flooding and environmentally clean power production are being analyzed. Finally, the question of the proper balance of activities such as cities, towns, agriculture, man made water-resource facilities and natural land will be dictated by the economic judgement based on sound engineer practice.

References

FEMA After action Report, Kansas City Office, The Great Flood of '93, dated 11 December 1993, Jefferson City Field Office

U.S. Army Corps Of Engineers, North Central Division, The Great Flood of 1993 Post-Flood Report, September 1994.

U.S. Department of Commerce, NOAA Natural Disaster Survey Report, The Great Flood of 1993, February 1994.

U.S. Army Corps Of Engineers, Kansas City Office, Annual Report of Reservoir Regulation activities, Summary for 1993 and 1994

Report of the Interagency Floodplain Management Review Committee, Sharing the Challenge: Floodplain Management into the 21st Century, June 1994 Washington D.C.

ASCE Magazine, January 1994 VOL. 64 Number 1, Page 38, " When the Levee Breaks" by James Denning

Evaluating Stormwater Model Performance in Northern Virginia

Aaron B. Small, David F. Kibler, and R. Fernando Pasquel[1]

Abstract

The purpose of the research underlying this paper was to evaluate desk-top methods and numerically complex models on their ability to reproduce observed runoff events from developing watersheds in northern Virginia. The models included EPA SWMM, PSRM-QUAL, TR-20, HEC-1, TR-55, variations of the rational method, three unit hydrograph procedures, the USGS regression equations, and the Anderson method. Simulation performance was evaluated with respect to peak flow, time to peak flow, time base, volume, and overall shape using statistics which measure the modeled hydrograph against a baseline reference hydrograph. Simulations were performed on eight watersheds in northern Virginia. A total of nineteen models and methods were applied to these test watersheds and compared to gaged runoff events as well as to calibrated storm events. The purpose of this paper is to provide a brief summary of principal findings from the larger research effort at Virginia Tech (Small, 1993).

Models and Methods Tested

Table 1 lists the models and methods tested in this study. These models and methods were chosen because of their applicability to the developing watershed and their widespread use in the design of urban drainage systems. There are many ways in which engineers design drainage and storm water management structures. The choice of hydrologic model often depends on the type of structure and the risk associated with it. Some models, such as the USGS regression equations, provide only peak flows for design rainfall events, while others, such as EPA SWMM,

[1]Authors are, respectively, Project Engineer, Maguire Associates, Inc., Virginia Beach, VA; Professor, Department of Civil Engineering, Virginia Tech, Blacksburg, VA; and Chief, Watershed Management Division, Prince William County Department of Public Works, Prince William, VA.

provide full hydrographs for any rainfall event. Table 1 classifies the urban drainage models used in this study by the type of input required and output received. Many of the inputs for the models are similar. All models require certain basic watershed physical characteristics, such as basin slope and drainage area. Precipitation, infiltration, and other losses play an important role in all of the models studied.

Test Watersheds

The watersheds chosen for the model evaluation are located in northern Virginia. The northern Virginia area consists of many land uses ranging from the urban streets immediately surrounding Washington, D.C. to the farmland of rural Loudoun County. Table 2 describes the watersheds used in the simulations. Six of them are located in Fairfax County, Virginia; the other two are located in Prince William County, Virginia and Loudoun County, Virginia.

Table 1. Models and Methods Tested

Full Hydrograph Simulation Models:

- EPA Storm Water Management Model, SWMM 4.05 (1991)
- SCS Computer Program for Project Formulation-Hydrology, TR-20 (1983)
- US Army Corps of Engineers Flood Hydrograph Package, HEC-1 (1990)
- Penn State Runoff Quality Model, PSRM-QUAL (1993)

Unit Hydrograph Methods:

- SCS Curvilinear Unit Hydrograph (1949), applied in HEC-1 (1990)
- Snyder Unit Hydrograph (1938), applied in HEC-1 (1990)
- Clark Unit Hydrograph (1945), applied in HEC-1 (1990)

Design Rainfall, Hydrograph Methods:

- SCS Urban Hydrology for Small Watersheds, TR-55 (1986)
- Rational Method (1850)
- Modified Rational Method (1974)
- Universal Rational Method

Design Rainfall, Peak-Only Methods:

- USGS Virginia Regression Equations (1978)
- The Anderson Method (1968)

Method of Analysis

The methods chosen for this study rely on many types of parameters, but an important input not yet discussed is the rainfall distribution. Since a gaged storm is not usually available at the project site, a design storm approach must be taken. Two basic types of design rainfall generation methods are employed in order to compare all of the models equally. The first is the widely used SCS 24-hr rainfall. Another is the method by Yarnell which is used here to generate 1-hr and 2-hr central peaking rainfall events from regional IDF charts. Design rainfall events were generated for the 2, 5, 10, 25, 50, and 100 year return periods.

Table 2. Watershed Characteristics

Name	Area Acres	Imperv. %	Slope %	No. Sub-Basins	No. of Channels	Land Use Conditions
Broad Run Tributary	505	4.6	1.7	5	10	Woods
South Fork Broad Run	2592	6.6	2.1	13	30	Farmland
Snakeden Branch	526	42.9	2.5	6	8	Suburban
Stave Run	50	56.4	3.3	1	1	Commercial/Ind.
Smilax Branch	201	22.6	3.3	5	9	Commercial/Ind.
Holmes Run I	4115	32.8	2.7	22	52	Suburban/Com.
Holmes Run II	3163	33.2	2.7	14	38	Suburban/Com.
Holmes Run IV	1669	33.2	2.7	5	15	Suburban/Com.

Direct comparison between a hydrograph computed by a design rainfall method and one recorded by a stream gage is not possible because of obvious differences in rainfall and antecedent moisture. In order to make a comparison between all of the models, the authors developed a calibrated PSRM/QUAL model for each of the watersheds. The calibrated model is based on the rainfall and runoff recorded at the site and then used to simulate a design rainfall or runoff event. The latter hydrograph is taken as the base-line hydrograph resulting from the design rainfall event and is used as a benchmark. Approximately 2600 different hydrographs were entered into the evaluations of model performance.

Model Performance Measures

The model performance evaluation was based on the following hydrograph parameters: peak flow; time to peak; total time base; volume; and shape. The systematic error (bias) and the standard error (accuracy), and the standard deviations of these measures were computed for each hydrograph parameter and averaged across watersheds by method. The accuracy and bias statistics have been combined in a numerical performance indicator which is summarized in Table 3.

Table 3. Summary of Model Performance Indicators

Scale

Worst 1 — Best 100

0 = Not Applicable

Model	Peak Flow Estimation	Volume Estimation	Time to Peak Estimation	Time Base Estimation	Hydrograph Shape Prediction	Urban Area Simulation	Partially Developed Area Simulation	Rural Area Simulation	Small Basins (< 2 sq. miles)	Large Basin (> 2 sq. miles)	Gaged Storm Simulation	Short (1-2 hr) Design Storms	Long (24 hr) Design Storms	Overall Performance*
EPA SWMM (4.05)	72	87	92	84	40	1	83	86	83	12	63	76	85	65
PSRM-QUAL	76	87	89	83	43	78	86	87	86	81	74	73	82	79
TR-20	79	87	93	86	53	85	87	86	86	86	80	79	80	82
HEC-1 Kinematic Wave	1	82	87	84	1	48	84	79	83	55	52	1	72	44
HEC-1 SCS UH-Whole Basin	84	84	86	81	57	83	83	86	83	84	77	80	76	80
HEC-1 SCS UH-Sub-areas	80	87	89	84	55	83	85	87	85	85	77	80	81	81
HEC-1 Snyder UH-Whole Basin	86	86	88	85	63	88	84	87	84	88	76	82	86	83
HEC-1 Snyder UH-Sub-areas	82	87	88	84	57	83	84	88	84	85	76	80	83	82
HEC-1 Clark IUH-Whole Basin	86	86	88	85	62	88	83	87	84	88	76	82	84	83
HEC-1 Clark IUH-Sub-areas	80	87	88	84	55	83	83	87	83	85	75	80	81	81
TR-55	89	99	100	98	88	83	91	89	89	86	0	0	79	76
Rational-Whole Basin	88	0	0	0	0	95	98	96	97	96	0	95	0	51
Rational-Sub-areas	83	93	93	90	56	77	82	84	81	80	0	66	0	68
Modified Rational-Whole Basin	89	0	0	0	0	96	98	96	97	96	0	96	0	51
Modified Rational-Sub-areas	65	94	94	90	1	63	79	83	78	70	0	36	0	58
Universial Rational	91	96	87	89	71	83	85	76	84	80	0	84	80	77
USGS Regression-3 parameter	87	0	0	0	0	93	97	97	96	94	0	95	0	51
USGS Regression-7 parameter	86	0	0	0	0	92	96	97	96	94	0	94	0	50
Anderson Method	70	0	0	0	0	84	90	96	91	88	0	88	0	47

*Overall Perofrmance = Arithmetic Average of Other Perofrmance Indicators

Summary of Model Performance

Table 3 is a summary of the model performance indicators calculated for each model or method for a given class of storm, watershed size, and level of urban development. The performance indicator is computed for each model and each hydrograph attribute as the weighted sum of various statistical error measures, including the bias, standard deviation of the bias, the average standard error, and the standard deviation of the standard error. The performance indicator shown in Table 3 has been normalized so that it falls in the range 1 - 100. In certain instances, the performance indicator fell below zero and out of the normal range. This was assigned a nominal value of 1. A zero in Table 3 indicates that the model or method has no applicability to the particular hydrograph attribute in question.

Because Table 3 is based solely on the results of this study, it is important that the table be restricted to the range of watershed conditions and storms encountered in northern Virginia. With this caveat in mind, there are several significant patterns in Table 3. The first is that the more complex models do not necessarily provide more accurate hydrograph or peak discharge estimates than the simple desk-top methods. The second observation is that discretizing or sub-dividing the watershed does not improve model results significantly. A third concern is the very weak performance of EPA SWMM in this study. The zero rating under urban area simulation capability in Table 3 must be regarded as an anomaly. It is attributed to the lack of a highly developed drainage system in any of the test watersheds. Even the three urban watersheds had only limited storm sewering. The consequence is that overland flow paths tend to be longer than in the typical urban catchment, a condition that is not favorable to EPA SWMM hydrograph computation.

Another aspect of the summary in Table 3 is the highly localized nature of the test watersheds. Because all eight test watersheds lie within 30 miles of each other, there is no opportunity to isolate the regional effects of similar land use, topography and climate. Furthermore, one must consider that the data set available in this study includes only two rural, three suburban, and three urban watersheds, making assessment of model application for particular development conditions very difficult. It is anticipated that future expansion of the watershed data base to areas outside the limits of this investigation will help to clarify the notable patterns evident in Table 3.

References

Small, A. B., A Comparative Evaluation of Surface Runoff Models and Methods on Small Developing Watersheds in Northern Virginia, Master of Science Thesis, Department of Civil Engineering, Virginia Tech, Blacksburg, VA, September 1993.

HOLISTIC APPROACH TO HYDROLOGICAL MODELLING AND HYDROLOGIC MONITORING SYSTEMS DESIGN

A.G. Capodaglio[1], M. ASCE, L. Natale[1], M. ASCE, L. Ubertini[2], M. ASCE

Abstract

Hydrology has developed as an independent specialty discipline only in relatively recent times: this status was however achieved mostly due to the influence and demands of other applied sciences than by virtue of its own theoretical and experimental foundations. Hydrologic phenomena constitute the driving force for most of the environmental processes of interest to engineers and scientists: to name just a few, the interaction between surface runoff and interflow on natural slopes, the relationship between pollutant discharges and water quality in natural water bodies, the generation of contaminant loads from urban and rural diffuse sources, and many others.

In formulating mathematical models, new interpretative schemes should be employed to properly apply fluid mechanics equations, since hydrologic phenomena are only partly represented by the classical equations; in designing monitoring systems it would be advisable to include the measurement of state parameters, in order to help formulating or operating a mathematical model; in applications requiring real-time capabilities, attention should be paid to all relevant aspects of the problem, and the ideal model may be found by combining distributed and lumped parameter models.

Introduction

Hydrology has been defined as the branch of science that deals with the occurrence, distribution, circulation and properties of water. It has developed as an independent

[1] Dept. of Hydraulic & Environmental Engrg., University of Pavia. I-27100 PAVIA ITALY

[2] Inst. of Hydraulics, College of Engineering, Univ. of Perugia, I-00128 PERUGIA, ITALY

specialty discipline only in recent times; however, this status was achieved mostly due to the influence and demands of other applied sciences than by virtue of its own theoretical and experimental foundations.

Historically, man has related abundance and scarcity of water to abundance and scarcity of crops since the earliest civilizations, thus implying the need of knowledge about water and its management: water gages were installed on the River Nile since 3000 B.C., but methods for measuring streamflow based on cross-sectional areas and velocity were not developed until the XVII century. Quantitative hydrology was "baptized" only in the second half of that century, when Perrault studied the relationship between rainfall and streamflow at the Seine River springs catchment. It is only in the XIX century that the field of hydrology shows a rapid increase in knowledge, with the development of empirical discharge formulas, the estimation of flood flow with the "rational formula" and the definition of groundwater flow laws (Biswas, 1972). Hydrology was since considered a practical application area serving the "more scientific" field of hydraulics until well after the end of the II World War; its justification lied in the need to forecast the availability of water for crop irrigation and other beneficial uses, and the possible hazard situations resulting from sudden, excess availability. Its domain was limited to the description of the acting processes at the earth surface, which were de-coupled from the phenomena occurring in the atmosphere.

Evolution in the very same ideas underlying the purposes and boundaries of hydrology, and in the technological tools available to engineers and scientists are in the process of transforming it from a specific engineering specialty into a multi-disciplinary geoscience that encompasses large-scale geophysical interactions. This transformation was also prompted by the recognition of the need for international cooperation in the use of water resources at the global scale, which occurred in the late 1960s (Eagleson, 1994). In very recent years, the recognition of the role of hydrological forcing functions in the generation and effects of pollutant loads on the water environment and resources has further accentuated the need for multidisciplinariety in the hydrological sciences.

Hydrological modelling: status and purposes

Models used in hydrology can be classified, in extremely rough terms, in two main categories: input-output models and mechanistic models. Either class can be used at an almost arbitrary scale of interest, if sufficient information is available.

Input-output models are only applicable to observed phenomena: the model acts in fact as an "engine" that translates the observed inputs to the observed outputs. Once the "engine" is "tuned", models can be used to generate forecasted outputs from measured inputs or even scenaria where to hypothetical inputs are associated probable outputs.

The tendency to use input-output type hydrologic modelling was at its apex in the

1960's and 70's, and was mainly a consequence of the popularity of system analysis and time series analysis theories, of which these models were ideal applications. These were also justified by the characteristics of the computing hardware available at the time, and by the necessity of obtaining results rapidly for practical purposes.

Mechanistic models are based on a strict physical representation of the underlying mechanisms of the studied phenomena (universal laws), and are those that must necessarily be applied to unobservable phenomena. Phenomena may be unobservable for different reasons: for example because they have not yet occurred, or because the system under study is being modified in time, thus making present observations unapplicable to future conditions. It follows that no empiricism is allowed in mechanistic model building; an interesting discussion on this topic was presented by Konikow and Bredehoeft (1992).

Either set of models requires some special attention with respect to its conditions of application. It is clear, for example, that input-output models must undergo a calibration and validation process before their ability to describe the system under study can be accepted. In the so-called "black-box" models (e.g. stochastic models, neural nets), the input-output relationships are determined by the very same data introduced in an "identification algorithm". This is referred to as "learning", rather than calibration process. Although black-box models are the prototype of the input-output class of models, the latter include also other examples. These are represented by any kind of relationship (whether deterministic, statistical or stochastic) that needs to be "fitted" or "calibrated" to the observed data. Once the learning or calibration has occurred, the model needs to be verified with a set of data different from the one previously used. The verification is deemed necessary to reduce the risk of using a non-representative model, and in general to increase the degree of confidence in its representation capabilities. Problems with input-output model application arise when one of the validating conditions is violated, for example by extrapolating the model identified for a specific system or condition to another situation, or by the onset of natural or man-induced changes into the same system.

On the other hand, mechanistic models require in-depth knowledge of the system under study, as the response rendered is strictly dependent on the discretization of real system description and of the physical mechanisms underlying the acting processes. For example, in a model describing the erosion by overland flow in addition to the governing laws of motion the grain size of soil particles must be described with the maximum possible accuracy; in a groundwater flow model, the geometry and properties of the medium must also be specified. The resulting models are usually very complex, and both the uncertainty about the geometric parametrization of the system and related scale effects affect the model response. Investigations to determine model parameters can thus become extremely burdensome; in addition it may ultimately be impossible to state whether the results obtained are significant, and to which degree. To this effect it has been pointed out (de Marsily, 1994) that the so-called "physically-based" (mechanistic) models may, after, all, turn out to be nothing more than very sophisticated "black-box" models if parametrization is carried out with inappropriate "space-and-time-averaging" criteria.

Hydrology and information

From the previous exposition, it appears that a "correct" (i.e. spatially accurate and scale-appropriate) description of the physical system is essential for the application of mechanistic models. However, experimental investigation of field quantities was hardly developed until recently, mainly due to technological and economical factors; furthermore, even when state-of-the-art techniques are employed, measurement systems are usually still geared towards the acquisition and handling of input-output type quantities, hence paying only a marginal attention to the hydrologic system status. Undoubtedly, input-output type observations are of more immediate implementation, and their interpretation is not dependent on the scale of the basin or the phenomenon observed.

Emerging technologies (satellite imaging, geographic information systems, etc.) can dramatically extend data gathering capabilities even at small resolution and extended geographical scale, however, some physical quantities must still be measured point-by-point on the field. Once these data points are introduced in, for example, a geographic information system, the temptation to "generalize" such observations to a wider domain can be dangerously appealing. Research is being conducted on "geostatistical" (Rossi, 1992) and other sophisticated approaches, such as stochastic and fractal characterization, or artificial media generation (Mackay and Riley, 1994) in order to extend point measurements to full scale applications. Although these results seem promising, their full validity is yet to be demonstrated.

Concerns about environmental pollution, water resources management and natural hazards forecasts (not necessarily in that order) are often encouraging, in Italy, as well as in Europe and around the world, to make substantial investments to implement monitoring systems that can provide information about the current hydrological and physical status of the system (basin, subbasin or hydrologic compartment). The design of monitoring systems is therefore becoming an issue of considerable economic and practical importance, as investments required are relatively large and the accuracy of the gathered information will reflect on all its subsequent uses. To date, at the national (Italy) or supernational (Europe) level, there are no accepted guidelines for the standardization of such monitoring systems. Too often the design of a monitoring system is carried out by a different subject from the one that is actually going to use the observations obtained: situations may arise where available knowledge and predictive requirements are not compatible or suboptimal.

Modelling tools and design of the monitoring system should be jointly evaluated to globally reduce errors (or uncertainties) and their propagation in the simulated system. The required accuracy or detail of measurements should be consistent with the detail of knowledge and the schematization (and thus forecastability) of the process considered. An appropriately balanced conjunctive use of black-box and mechanistic models may, in this respect, yield satisfactory and cost-effective results in practical applications (Cunge et al., 1990).

Discussion

The evolution of hydrology from empirical discipline to multi-disciplinary geoscience encompassing the occurrence, distribution and properties of water resources has placed on it increasing demands with respect to the study and forecasting, among others, of environmental pollution, water resources management and natural hazards occurrence. The development and application of "models" is essential to meet such demands of knowledge and predictability of hydrologic phenomena, however, results must never be accepted acritically. It is especially important that the descriptive information be commensurate to the type of model used and the category of problem being investigated. Conjunctive use of different types of models may provide satisfactory and cost-effective answers to complex problems. It is in any case desirable that modelling tools and monitoring and data acquisition systems be jointly evaluated to obtain a global error minimization and gather consistent information.

Acknowledgement

This work was partly supported by C.N.R.- G.N.D.C.I.

References

Biswas A.K. (1972) *History of Hydrology*. North-Holland Publishing Co. Amsterdam

Cunge J.A., J.L. Rahuel, E. Todini and R. Vignoli (1990) The Hydrological Forecasting System (HFS)- A comprehensive system for real-time flood forecasting. *Proceedings, NATO ARW "Computer-Aided Support Systems for Water Resources Research and Management"*. Ericeira, Portugal, 22-28 Sept.

de Marsily, G. (1994) Quelques réflexions sur l'utilization des modèles en hydrologie. *Revue des Sciences de l'Eau,* **7**:219-234

Eagleson, P.S. (1994) The evolution of modern hydrology (from watershed to continent in 30 years). *Adv. Wat. Res.*, **17**:3-18

Konikow, L.F. and J.D. Bredehoeft (1992) Ground-water models cannot be validated. *Adv. Wat. Res.*, **15**:75-83

Mackay R. and M. Riley (1994) A method for geological media generation. *J. of Contaminant Hydrology*, **16**

Rossi, M.E. (1992) Assessment of uncertainty using geostatistics. *Environmetrics,* **3**(1):71-80

Kinematic Wave Application to Rangeland Watershed

Daniel H. Hoggan[1] M. ASCE and Dean D. Taylor[2]

Abstract

The kinematic wave method of computing rainfall-runoff was applied along with the more commonly used unit hydrograph method to four ungaged rangeland watersheds in Southeastern Idaho. The results from the two methods were very close, indicating that the kinematic wave method can be useful in non-urban watershed analysis provided the assumptions of the method are reasonably satisfied.

Introduction

The kinematic wave method has been widely used in urban hydrology studies because of its capability to model basin characteristics in a more spatially distributive manner than the unit hydrograph method. It also provides for nonlinear response, and in ungaged basin analysis offers the advantage of having model parameters that are related to measurable watershed characteristics.

Criteria for applying kinematic wave theory to catchment analysis relate primarily to channel bottom or land surface slope. One such criterion for stream channel flow indicates the theory to be 95% accurate, if the following inequality is satisfied:

$$\frac{t_r S_o V_o}{d_o} \geq 85 \qquad \textbf{(1)}$$

---

[1]Professor, Utah Water Research Laboratory and Civil and Environmental Engineering Department, Utah State University, Logan, UT 84322-8200

[2]Senior Engineering Specialist, Lockheed Idaho Technologies Company, Idaho Falls, ID 83415

where t_r is the time of rise of the inflow hydrograph, S_o is the bottom slope, V_o is the average velocity, and d_o is the average water depth. From the above condition it is apparent that the theory holds best for steeper slopes, generally greater than .01 (Ponce, 1989). Provided that the criteria for applicability are met, there is little reason to presume that the kinematic wave method cannot be applied to non-urban watersheds, although the practice is relatively uncommon. The advantages of using the method in an ungaged non-urban setting are similar to those of an urban setting. The major difference in application comes in defining the representative basin characteristics used to determine model parameters. For example, how are the shapes, lengths, and slopes associated with runoff in a rangeland or other non-urban watershed defined as compared with those related to roofs, parking lots, streets, etc. in an urban setting?

The INEL Waste Processing Facility Site Investigation

In studies of flooding potential at four of six candidate sites for a waste processing facility at the Idaho National Engineering Laboratory (INEL) desert site west of Idaho Falls, Idaho, runoff was computed with two methods, the SCS unit hydrograph method and the kinematic wave method. No gaged flow data was available at any of the sites, so the two different methods were used and compared to provide greater confidence in the results. The major objectives of the studies were to determine 100-yr and 500-yr flood elevations and areas of inundation at each of the sites due to localized flooding.

The watersheds for the four sites all consist of undeveloped rangeland covered with sagebrush and wild grass; however there is considerable diversity in relief among the four locations. One of the sites which is fairly flat is on a lava flow with surface elevation varying between 4843 ft(1476 m) and 4864 ft(1482 m), and includes several natural ponding areas. Two of the sites with gentle slopes and no natural ponding areas vary in elevation from about 5006 ft(1526 m) to 5076 ft(1547 m). One site is located on an alluvial fan next to high mountains where the elevation varies from about 5000 ft(1524 m) at the site to 10000 ft(3048 m) at the high point of the drainage.

The HEC-1 flood hydrograph program was used to model precipitation-runoff at each site. For the SCS unit hydrograph method, the lag time parameter was determined using the Kirpich equation for time of concentration. The length and slope parameters for the equation were determined from 2 ft(0.61 m) or 4 ft(1.22 m) contour maps, depending on the degree of vertical relief. Channel routing between subbasins was done with the Muskingum-Cunge and kinematic wave routing methods.

Taking a conservative approach, ripe snow and frozen ground conditions were assumed in all of the models at elevations where runoff was generated, and consequently all snowmelt and rainfall losses and loss rates were set equal to zero. Base flow was considered negligible under conditions of frozen ground and the short time frame for generating runoff (24-hour storm duration). The 100-yr event consisted of a 2-yr, 24-hr winter rain storm together with a 50-yr, 24-hr snowmelt. The 500-yr event consisted of 10-yr, 24-hr winter rain storm together with a 50-yr, 24-hr snowmelt.

Application of Kinematic Wave Method

The kinematic wave method required the definition of the parameters (basin characteristics) so as to be consistent with the open range conditions of the sites being investigated. A typical overland flow plane of 300-ft(91.4-m) length was used to model flow into the collector channel system. The slope of the plane was taken as the area-averaged slope of the subbasin. A roughness coefficient of 0.3 was selected to be typical of rangeland conditions at the sites. The subcollector channel was determined by dividing the average width of the subbasin by two, assuming that the main channel followed the longitudinal center of the area. The slope for the subcollectors was assumed to be the same as the basin area-averaged slope. Manning's "n" for the channels was assumed to be 0.05. The channel cross section was assumed to be triangular with side slopes of 2%, based on an analysis of several measured cross sections. The surface area drained by a single representative subcollector channel was determined by multiplying the subcollector channel length by 600 ft(182.9 m), assuming that 300 ft(91.4 m) overland flow planes drained into the channel from each side. The length of the main channel was taken as the length of the ephemeral channel shown on the quad sheets. The slope was estimated from this length and the elevation differences indicated by the contours crossed. The cross section and "n" values were assumed to be the same as for the subcollector. The area associated with the main channel was considered to be the total area of the subbasin.

The results of the simulations using the two methods, the SCS unit hydrograph method and the kinematic wave method, were found to be very close by comparison in every case. The peak discharges varied by a difference of 0.1% to 4.0%, the largest being at the site with the smallest relief, and the smallest at the site with the greatest relief. In all cases, the time to peak discharge found by the kinematic wave method differed by less than 60 minutes from the time to peak discharge found by the other method, or within roughly 4% of the 24-hr duration of the assumed rainfall event.

Site 5 Results

The results for one of the sites in the study, site 5, are presented to show how close the computations of the two methods compare. Other sites produced results in the same order of closeness as site 5. A plan of the site is shown in Figure 1, and a tabulation of results for selected subbasins is shown in Table 1.

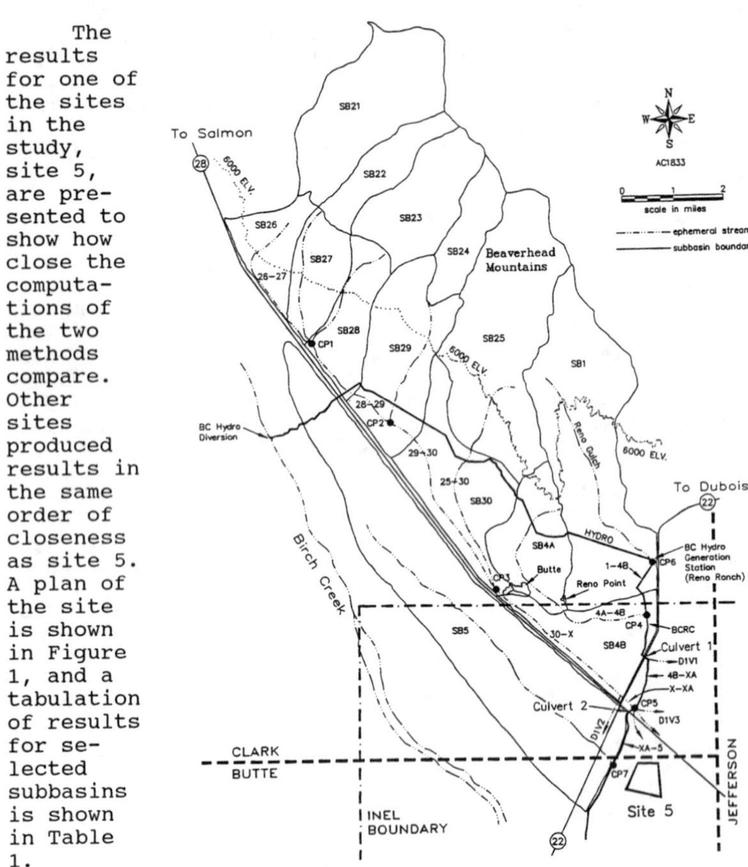

Figure 1 Site 5 Watershed

Although the Beaverhead Mountains within the watershed reach 10000-ft(3048-m) elevation, colder temperatures and unripened snow conditions at elevations above 6000 ft(1829 m) generally prevent runoff from these higher elevations during a flood event. Thus, a 100-yr(or 500-yr) rain-on-snow event was applied to the contributing part of the watershed below 6000 ft(1829 m) to produce a 100-yr(or 500-yr) flood discharge at site 5. A dotted line in Figure 1 indicates the 6000-ft(1829-m) contour, and subbasin areas in Table 1 reflect only areas below 6000 ft(1829 m).

Subbasin No.	Area (km^2)	Area - Ave Slope (m/m)	Main Channel		Sub-collector Channel Length (m)	Peak Discharge (m^3/sec)	
			Length (m)	Slope (m/m)		UH	KM
SB 25	4.27	0.18	1974	0.06	531	10.48	10.48
SB 26	2.33	0.04	1078	0.05	607	5.75	5.66
SB 29	8.16	0.03	4524	0.03	582	19.48	19.60
SB 1	13.51	0.08	6422	0.04	1277	32.20	32.45

Table 1 Subbasin 5 characteristics and 500-yr peak flows

The 500-yr discharge hydrographs computed by both methods for CP5 (including runoff and routing from eleven subbasins are plotted in Figure 2.

Conclusions

These results suggest that the kinematic wave method can be useful for computing runoff from non-urban as well as urban watersheds, particularly in the case of ungaged conditions, provided the assumptions of the method are reasonably satisfied. Whether such is the case can be checked in a straightforward, *ad hoc* fashion by evaluating the dimensionless expression in Eq (1) against the suggested threshold value of 85, using the results of the kinematic wave analysis to estimate the required parameters in the expression.

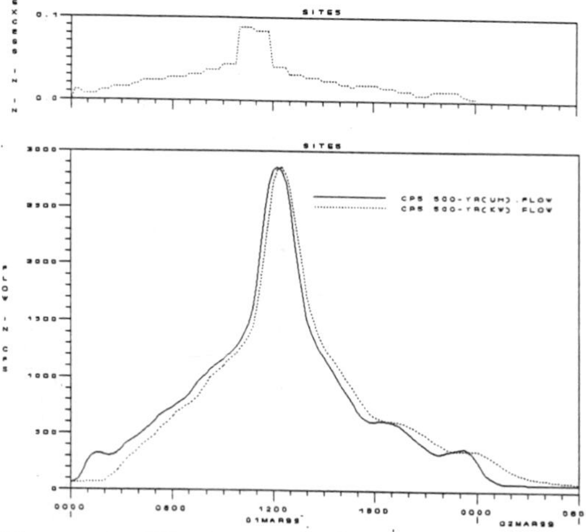

Figure 2 CP5 500-yr hydrographs

Literature Cited

Ponce, V. M., *1989, Engineering Hydrology*. Prentice Hall, Engelwood Cliffs, New Jersey

PLANNING OF MULTIUSE HYDROLOGIC SYSTEMS IN THE CENTRAL VALLEY BY IMPLEMENTING INTEGRATED RESOURCE MANAGEMENT

Andrew E. Cullen, Associate Member[1]

The Central Valley, situated in the fertile area between the Sierra Nevada and Costal Range Mountains, provides a majority of the 15 billion dollar a year agricultural output of California. With a mean annual precipitation of approximately 10 inches per year, the Central Valley must depend on irrigation water from sources other than rainfall. Beginning in the early 1900's, agricultural interests have been consuming more and more of the state's water and currently control approximately 80 percent of the developed water supplies of California.

Long thought to have an inexhaustible supply of water or the means to achieve one, the farmers of the Central Valley are coming under considerable scrutiny as urban and environmental demands for water increase. Because of the scarcity of water in California, "conjunctive use" (multiuse) projects are being embraced as a way to balance these interests. Municipalities including city and county agencies that implement multiuse projects are seeking to satisfy several goals including: flood control, wildlife habitat protection and restoration, groundwater recharge, preservation of agricultural lands, addition of recreational areas, and preservation of the cultural heritage in the Central Valley. However, one or even several projects may not be able to meet all these goals. A compromise to maximize the amount of project goals must be obtained. Integrated Resource Management (IRM) seeks to achieve this compromise.

The IRM technology evaluates project goals within a watershed or watersheds. Technologies including reapportioning, conservation, recycling, or other methods are examined and their effectiveness in meeting these goals is measured. Many terminologies exist for IRM and it is important to note that is it is basically the development of a regional watershed management scheme. However, the

[1]Camp Dresser & McKee Inc., 280 Granite Run Drive, Lancaster, PA 17601

distinguishing feature of IRM can be seen in its multidisciplanary approach to satisfy the project goals.

City, county and private agencies in the Central Valley are in tune to the agricultural, urban, and environmental needs of the particular watersheds they encompass. These entities identify the specific areas that need to be addressed (i.e. flood control, groundwater recharge, etc...) IRM utilizes technical and managerial skills necessary to analyze the situation, apply the correct technologies, and determine cost effective and feasible alternatives to achieve the project goals. IRM can be broken into the following tasks:

- Feasibility Study
- Environmental Assessment
- Prioritization Plan
- Implementation Plan

The Feasibility Study is a generalized data collection and analysis stage that identifies information gaps and potential technologies and sites to apply them to. The outcome of the Feasibility Study is an answer to the question "can it be done?". A scope of work particular to the project goals is developed and encompasses the remaining three phases.

Once the Feasibility Study determines if further analysis is warranted, an environmental assessment is made. This assessment is a more detailed study that actually analyzes the potential sites by modifying National Environmental Policy Act (NEPA) guidelines, (if Federal funding is available) or California Environmental Quality Act (CEQA) guidelines (if private or state funding is available) within IRM. The deliverable at this stage is an Environmental Impact Statement (EIS) or Environmental Impact Report (EIR). A more detailed explanation of alternatives, their subsequent impacts on the environment, and mitigation measures are analyzed. Alternative identification includes detailing costs of construction and operation and maintenance, adaptability to future change, reliability, and conformance with local and state regulations. Advantages and disadvantages with respect to the beforemented aspects are listed (but not prioritized) with respect to project goals.

The Prioritization Plan is a ranking of the sites based on how they meet the specific project goals identified by the municipalities. The prioritization task of IRM includes meeting with the potential funding parties to select alternatives identified and screened in the environmental assessment. The key to the prioritization is to measure the effectiveness of these alternatives in meeting the project goals.

The Implementation Plan includes the securing of funds, pre-design, design, construction, and evaluation portions of IRM. Funding is available from a number of sources including federal, state, and local agencies, as well as non-profit and private interests. Once designed and constructed, the effectiveness and compliance of the selected alternative is monitored throughout the life of the project. Status reports, data collection, and overall performance parameters are collected and documented. This information will provide valuable information for the next application of Integrated Resource Management.

IRM, as outlined above, has many different aspects. The interests of agriculture, urban, and environmental entities are the driving force to implement this approach. Various low and high technologies can be used to achieve the project goals. However, sole responsibility for making it work lies with the agricultural and urban interests. It is their duty to protect the environment because they are the ones who are reshaping it.

EMBANKMENT DAM BREACH PARAMETERS REVISITED

by David C. Froehlich[1], Member, ASCE

INTRODUCTION

The extent of flooding and the travel time of a flood wave that would result from failure of a dam because of breaching needs to be predicted to establish spillway capacities and to prepare emergency action plans. Accuracy of numerical simulation of the outflow from a breached dam and the movement of the flood wave downstream can have a significant effect on the cost of a proposed dam or decisions regarding needed modifications to an existing structure. Because the predicted severity of flooding downstream of a dam and estimates of flood arrival times depend largely on the size and shape of the breach that forms in the dam and the development of the breach with time, it is important to model the formation of the breach as accurately as possible.

The United States Committee on Large Dams (USCOLD) estimates that about 80 percent of all major dams in operation in the United States are embankment dams [*Lessons from dam incidents, USA* (1975)]. Earthfill and rockfill embankment dams are the most common and are classified on the basis of the predominant composition of embankment material. Building on an earlier study [Froehlich (1987)], empirical models of breach formation in embankment dams are reviewed, and data from 63 embankment dam failures are analyzed to find methods for estimating breach formation model parameters including the height, average width, side slope ratio, and formation time of the ultimate breach that would develop in an embankment dam during a failure.

EMBANKMENT DAM FAILURE DATA

Data for 63 embankment dam failures were assembled from a variety of sources and are summarized in table 1. These data include a description of the dam, the sources of information, the failure mode, reservoir characteristics at the time of failure, and the measured dimensions and formation time of the breach that ultimately formed in the dam.

For most of the dam failures ultimate breach dimensions were found from a cross section taken through the most narrow part of the breach or from topographic maps of the breached section of the embankment. For other dams, breach dimensions were ultimate breach shape at a dam was approximated as a trapezoidal opening having a horizontal bottom. Breach height was found as the difference between the elevation of the top of the dam at the location of the breach and the elevation of the bottom of the trapezoidal approximation. Average breach width is one-half the sum of the trapezoid

[1]Consulting Engineer, 4612 Thornwood Circle, Lexington, Kentucky 40515-6129.

TABLE 2. Characteristics of Breached Embankment Dams and Ultimate Breach Dimensions.

Dam name and location	Failure mode[a]	Average width of embankment, $\overline{W}$, in meters	Volume of water above breach bottom, V_w, in m^3 $\times 10^6$	Height of water above breach bottom, H_w, in meters	Height of breach, H, in meters	Average breach width, in meters	Average breach side slope ratio, z	Breach formation time, t_f, in hours
(1)	(2)	(3)	(4)	(5)	(6)	(7)	(8)	(9)
Apishapa, Colo.	P	82.4	22.8	28.0	31.1	93.0	0.44	0.750
Baldwin Hills, Calif.	P	59.6	0.910	12.2	21.3	25.0	0.31	0.333
Bearwallow Lake, N.C.	S	17.1	0.0493	5.79	6.40	12.2	1.43	--[b]
Buckhaven No. 2, Tenn.	O	13.4	0.0247	6.10[c]	6.10	4.72	0.73	--
Bullock Draw Dike, Utah	P	18.6	0.740	3.05	5.79	12.5	0.21	--
Butler, Ariz.	O	9.63	2.38	7.16	7.16	62.5	0.85	--
Castlewood, Colo.	O	47.4	6.17	21.6	21.3	44.2	0.50	0.500
Caulk Lake, Ky.	P or S	32.0	0.698	11.1	12.2	35.1	1.38	--
Coedty, England	O	--	0.311	11.0[c]	11.0	42.7	2.22	0.250
Cougar Creek, Alberta, Canada	O	21.7	0.0298	11.1	10.4	--	--	0.050
East Fork Pond River, Ky.	P	38.9	1.87	9.80	11.4	17.2	0.44	--
Elk City, Okla.	O	50.4	1.18	9.44	9.14	36.6	1.00	--
Emery, Calif.	P	22.2	0.425	6.55	8.23	10.8	0.35	--
Fogelman, Tenn.	P	21.3	0.493	11.1	12.6	7.62	0.36	--
French Landing, Mich.	P	34.3	3.87	8.53	14.2	27.4	0.97	0.583
Frenchman Creek, Mont.	P	37.3	16.0	10.8	12.5	54.6	0.50	--
Grand Rapids, Mich.	O	14.8	0.0255	6.40[c]	6.40	19.0	2.26	--
Haas Pond, Conn.	P	16.7	0.0234	2.99	3.96	10.7	0.38	--
Hart, Mich.	P	31.1	6.35	10.7	10.8	73.9	3.03	--
Hatchtown, Utah	P	44.8	14.8	16.8	18.3	151	2.42	1.00
Hell Hole, Calif.	P	103.2	30.6	35.1	56.4	121	0.96	0.750
Herrin, Ill.	O	28.8	--	10.7[c]	10.7	47.2	1.14	--
Horse Creek, Colo.	P	26.8	12.8	7.01	12.8	73.1	0.83	--
Iowa Beef Processors, Wash.	P	--	0.333	4.42	4.57	16.8	0.33	--
Ireland No. 5, Colo.	P	18.0	0.160	3.81	5.18	13.5	0.38	0.500
Jacobs Creek, Penn.	P	--	0.423	20.1	21.3	17.5	0.61	--
Johnston City, Ill.	P	21.5	0.575	3.05	5.18	8.23	1.00	--
Kelly Barnes, Ga.	P	19.4	0.777	11.3	12.8	27.3	0.85	--
La Fruta, Tex.	P	40.0	78.9	7.90	14.0	58.8	0.30	--
Lake Avalon, N. Mex.	P	42.7	31.5	13.7	14.6	130	0.52	--
Lake Francis, Calif.	P	47.4	0.789	14.0	17.1	18.9	0.65	--
Lake Genevieve, Ky.	P	19.8	0.680	6.71	7.92	16.8	1.54	--
Lake Latonka, Penn.	P	28.0	4.09	6.25	8.69	39.2	1.18	--
Lambert Lake, Tenn.	P	53.9	0.296	12.8	14.3	7.62	0.21	--
Laurel Run, Penn.	O	40.5	0.555	14.1	13.7	35.1	2.40	--
Lawn Lake, Colo.	P	14.2	0.798	6.71	7.62	22.2	0.96	--
Lily Lake, Colo.	W	--	0.0925	3.35	3.66	10.8	0.13	--
Little Deer Creek, Utah	P	63.1	1.36	22.9	27.1	29.6	0.75	0.333
Long Branch Canyon, Calif.	P	11.3	0.284	3.17	3.66	9.14	0.40	--
Lower Latham, Colo.	P	25.7	7.08	5.79	7.01	79.2	6.3	1.50
Lower Otay, Calif.	O	53.3	49.3	39.6	39.6	133	1.00	1.00
Lower Two Medicine, Mont.	P	--	29.6	11.3	11.3	67.0	1.50	--
Lynde Brook, Mass.	P	41.8	2.88	11.6	12.5	30.5	1.22	--
Melville, Utah	P	25.1	24.7	7.92	9.75	32.8	0.70	--
Otter Lake, Tenn.	P	20.6	0.109	5.00	6.10	9.30	1.28	--
Oros, Brazil	O	110	660	35.8	35.5	--	--	8.50
Pierce Reservoir, Wyo.	P	--	4.07	8.08	8.69	30.5	0.77	1.00
Potato Hill, N.C.	O	23.5	0.105	7.77[c]	7.77	16.5	1.25	--
Prospect, Colo.	P	13.1	3.54	1.68	4.42	88.4	0.69	2.50
Puddingstone, Calif.	O	--	0.617	15.2[c]	15.2	--	--	0.250
Quail Creek, Utah	P	56.6	30.8	16.7	21.3	70.0	0.10	1.00
Rainbow Lake, Mich.	O	28.2	6.78	10.0	9.54	38.9	2.52	--
Renegade Resort Lake, Tenn.	O	11.0	0.0139	3.66[c]	3.66	2.29	0.63	--
Rito Manzanares, N. Mex.	P	13.3	0.0247	4.57	7.32	13.3	0.77	--
Schaeffer, Colo.	O	80.8	4.44	30.5[c]	30.5	137	2.25	0.500
Scott Farm Dam No. 2, Alberta	P	39.3	0.086	10.4	11.9	15.0	0.00	--
South Fork, Penn.	O	64.0	18.9	24.6	24.4	94.5	1.38	0.750
Teton, Idaho	P	250	310	77.4	86.9	151	1.00	1.25
Trial Lake, Utah	P	7.62	1.48	5.18	5.18	21.0	0.82	--
Trout Lake, N.C.	O	21.6	0.493	8.53[c]	8.53	26.2	1.79	--
Upper Pond, Conn	O	--	0.222	5.18[c]	5.18	16.5	1.71	--
Wheatland No. 1, Wyo.	P	--	11.6	12.2	13.7	35.4	0.75	1.50
Winston, N.C.	O	7.76	0.662	6.40	6.10	19.8	0.20	--

[a]O = overtopping, P = piping, S = sliding.
[b]Information not available.
[c]Height of water was assumed equal to height of dam breach but might have been greater.

top-width and bottom-width. Breach formation time is considered to be the time from the beginning of rapid growth of a breach to the time when significant lateral erosion of the embankment had stopped. Estimates of breach formation time were made from eyewitness accounts of the failure, from photographs, and from recorded stage and discharge measurements that helped to establish the time when the breach began to form.

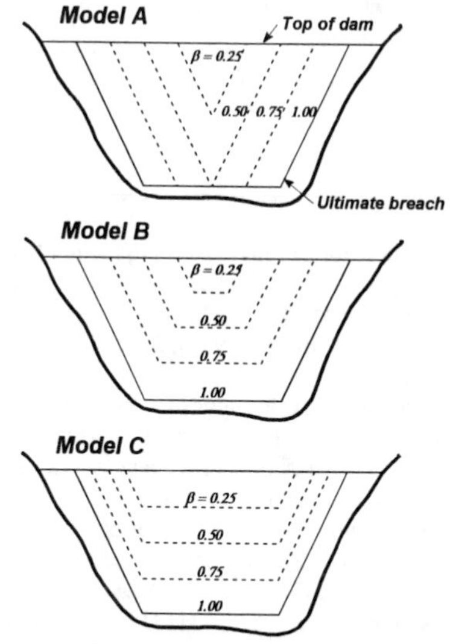

FIGURE 1. Schematic Representation of Three Empirical Breach Formation Models.

EMPIRICAL BREACH FORMATION MODELS

The way a breach forms in an embankment dam depends on numerous factors including the embankment geometry, embankment materials and composition, construction methods, reservoir dimensions, inflow to the reservoir during failure, and the mode of failure. Both causal and empirical models have been used to simulate formation of a breach in an embankment dam. Causal breach formation models are based on physical laws and empirical relations governing the flow of water and erosion of embankment materials. Empirical models, on the other hand, are not as rigorously based on physical principles as causal models. In the empirical approach, breaches are allowed to form in a predetermined manner that is controlled by specified model parameters. Empirical models are easier to apply than causal models, needing fewer data and coefficients. Three empirical models of breach formation that are in common use are illustrated in figure 1. Each of the three breach formation models is based on the assumption that a breach begins to form at the top of a dam and enlarges with time in the shape of a triangle or a trapezoid. Breach formation is assumed to begin when the reservoir water-surface reaches a predetermined elevation, Y_f.

BREACH FORMATION MODEL PARAMETERS

Breach formation model parameters that need to be estimated consist of the reservoir water-surface elevation at which breach formation begins Y_f (or, equivalently, the critical overtopping depth d_c), the height H, average width $\overline{B}$, and average side slope ratio z of the ultimate trapezoidal breach as shown in figure 2, and the

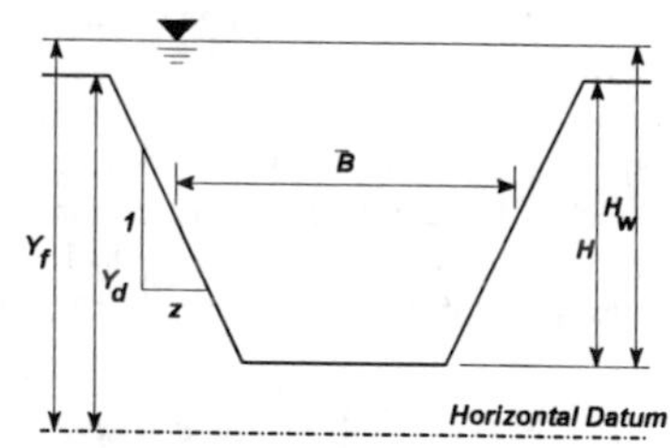

FIGURE 2. Dimensions of the Ultimate Trapezoidal Dam Breach and Related Measures.

TABLE 2. Critical Overtopping Depth for Embankment Dams of Varying Condition ["Risk-based" (1987)].

Condition of dam	Description of dam	Critical over-topping depth d_c, in meters
Good	Practically no seepage, no noticeable settlement, and both upstream and downstream slopes in good condition.	0.6
Fair	Moderate seepage, some settlement of crest, and some erosion on embankment slopes.	0.3
Poor	Excessive seepage, significant slump of crest, cracks in embankment, and erosion of embankment slopes.	0.0

breach formation time t_f (that is, the time needed for complete development of the ultimate breach). The analysis to follow builds on previous studies of embankment dam breach formation by Johnson and Illes (1976), McMahon (1981), the U.S. Army Corps of Engineers ["User's manual" (1981)], Ponce (1982), Fread (1984), MacDonald and Langridge-Monopolis (1984), and Froehlich (1987).

Critical depths of overtopping d_c for embankment dams suggested by the Federal Emergency Management Agency ["Risk-based approach to dam safety assessment" (1987)] are given in table 2. The ultimate bottom of a breach usually is the dam foundation, which is more resistant to erosion than the embankment material. However, the final height of a breach might be limited by the volume of water in the reservoir at the time of failure, or by the presence of a layer of erosion-resistant material located in the embankment. It is assumed that the ultimate height of a breach can be presupposed.

Prediction equations for the parameters $\overline{B}$, t_f, and z were obtained from multiple regression analysis of the assembled data. Logarithmic transformation of all other variables was found to provide the best linear relations. The best prediction equation for average breach width based on the assembled data was found to be

$$\overline{B} = 15\, k_o\, V_w^{0.32}\, H^{0.19} \tag{1}$$

where

$$k_o = \begin{cases} 1.4, & \textit{if the failure mode was overtopping} \\ 1.0, & \textit{otherwise} \end{cases} \tag{2}$$

V_w = reservoir volume at the time of failure, in millions of cubic meters, and H = height of the final breach, in meters. The coefficient of determination of equation 1 is 0.752. Equation 1 suggests that an overtopping failure mode produces a breach that is 40 percent wider than other causes of failure. The factor k_o also might indirectly account for the effect of large inflows to a reservoir during an overtopping failure, which are not included in the estimate of V_w and which would increase the ultimate breach width. A rough estimate of the side slope ratio obtained from the data is given by

$$z = \begin{cases} 1.4, & \textit{if the failure mode is overtopping} \\ 0.9, & \textit{otherwise} \end{cases} \tag{3}$$

which is in general agreement with the findings of others. The average side slope ratio of all the measured breaches is nearly one. Breach formation time was found to be closely related to the reservoir volume and embankment height. Regression analysis gives breach formation time, in hours, as

$$t_f = 3.84\, V_w^{0.53}\, H^{-0.90} \tag{4}$$

The coefficient of determination of the relation for the logarithm of t_f is 0.922. Breach formation time increases with the volume of water impounded by the dam at the time of failure, and decreases with the ultimate breach height.

SUMMARY AND CONCLUSIONS

The extent of flooding and travel time of a flood wave that would result from the failure of a dam need to be predicted to establish needed spillway capacities and to prepare emergency action plans that would be used in the event of an actual dam failure. Data from 63 embankment dam failures were assembled to evaluate breach formation model parameters. From these data, empirical models of breach formation were evaluated and prediction methods for the height, average width, side slope ratio, and formation time of the ultimate breach were developed. Results of the study will help improve the accuracy of numerical simulations of dam-break flood waves.

APPENDIX I. — REFERENCES

Fread, D. L. (1984). "DAMBRK -- the NWS dam-break flood forecasting model." *National Weather Service Hydrological Technical Note No. 4*, Hydrologic Research laboratory, Silver Spring, Maryland.

Froehlich, D. C. (1987). "Embankment-dam breach parameters." *Proceedings of the 1987 National Conference on Hydraulic Engineering*, Williamsburg, Virginia, August 3 to 7, 1987, 570-575.

"HEC-1 flood hydrograph package -- user's manual." (1990). U.S. Army Corps of Engineers, Hydrologic Engineering Center, Davis, California.

Johnson, F. A., and Illes, P. (1976). "A classification of dam failures." *Water Power and Dam Construction*, 28(12), 43-45.

Lessons from dam incidents, USA (1975). U.S. Committee on Large Dams, American Society of Civil Engineers, New York.

MacDonald, T. C., and Langridge-Monopolis, Jennifer (1984). "Breaching characteristics of dam failures." *Journal of Hydraulic Engineering*, 110(5), 567-586.

McMahon, G. F. (1981). "Developing dam-break flood zone ordinance." *Journal of the Water Resources Planning and Management Division, ASCE*, 107(WR2), 461-476.

Ponce, V. M. (1982). "Documented cases of earth dam breaches." *Civil Engineering Series No. 82149*, San Diego State University, San Diego, California.

"Risk-based approach to dam safety assessment." (1987). *Federal Emergency Management Agency/Stanford University Workshop Notes*, Denver, Colorado, November 4-5, 1987.

"User's manual, DAMBRK, the NWS dam-break flood forecasting model." (1981). U.S. Army Corps of Engineers, Hydrologic Engineering Center, Davis, California.

Hydrologic Modeling Within GRASS- r.hydro.CASC2D

F.L. Ogden[1], A.M. ASCE and B. Saghafian[2]

Abstract

We have developed a distributed hydrologic modeling component within the raster-based Geographic Resources Analysis Support System (GRASS) Geographic Information System (GIS). This component, called **r.hydro.CASC2D**, simulates the hydrologic response of a watershed subject to a series of given rainfall fields. Sources of rainfall input may include isohyetal patterns, raingage, or weather-radar estimates. Main components of the model include interception and retention storage, infiltration, surface runoff and channel routing. Interception is calculated using empirical methods while retention storage is user specified as a raster map layer. A modified form of the Green and Ampt equation with six parameters is applied to model infiltration. For continuous soil moisture accounting, redistribution of soil moisture is calculated whenever the non-intercepted rainfall intensity falls below the saturated hydraulic conductivity of the soil. Excess rainfall is routed as overland flow using a two-dimensional explicit finite-difference technique. Channel flow is routed using either an explicit 1-D diffusive wave approach or an implicit 1-D dynamic wave method. The performance, limitations, and applicability of **r.hydro.CASC2D** is discussed.

Introduction

CASC2D originated as a two-dimensional overland flow routing algorithm developed in APL by Prof. P.Y. Julien at Colorado State University. The overland flow routing module was converted from APL to FORTRAN by Bahram Saghafian, then at Colorado State University, with the addition of Green & Ampt infiltration and explicit channel routing (Julien et al. 1995, Julien and Saghafian, 1991, and Saghafian, 1992). The FORTRAN version was reformulated, significantly enhanced, and re-written in the C programming language by Bahram Saghafian at the U.S. Army Construction Engineering Research Laboratories. Implicit channel routing code was developed and added to **r.hydro.CASC2D** by Fred L. Ogden (Ogden 1994). This version became known as **r.hydro.CASC2D**, part of the GRASS GIS for hydrologic simulations (Saghafian, 1993).

[1] Assistant Professor, Department of Civil and Environmental Engineering U-37, University of Connecticut, Storrs, CT 06279.

[2] Post-Doctoral Research Associate, U.S. Army Construction Engineering Research Laboratories, Champaign, Illinois.

r.hydro.CASC2D is a GRASS raster command to execute fully integrated distributed hydrologic modeling within the GRASS command shell. **r.hydro.CASC2D** simulates hydrologic response of a watershed subject to a given rainfall field. Input rainfall is allowed to vary in space and time. Major components of the model include interception, infiltration, and surface runoff routing. Interception is a process whereby rainfall is being retained by vegetation and is determined by a two parameter empirical relation. Infiltration calculations determine the rate at which rainfall or surface water passes into the permeable soil. The Green and Ampt equation with four parameters is applied to model event-based infiltration. For continuous soil moisture accounting, redistribution of soil moisture can is simulated whenever the non-intercepted rainfall intensity falls below the saturated hydraulic conductivity of the soil. The redistribution option requires two additional soil hydraulic parameters. Excess rainfall becomes surface runoff and is routed as overland flow and channel flow. The overland flow routing formulation is based on a two-dimensional explicit finite difference (FD) technique, while channel flow may be routed using either an explicit FD or implicit FD method. Through a step function, a depression depth may be specified, below which no overland flow is routed.

Options

The command **r.hydro.CASC2D** invokes the physically-based hydrologic model CASC2D with a variety of required and optional arguments. These arguments control the detail and complexity of a given simulation. The list of options is given below. Optional arguments are listed in square brackets, while required arguments have no brackets. The *name* requirement in the arguments below is the name of a GRASS raster map.

```
SYNOPSIS: r.hydro.CASC2D, r.hydro.CASC2D help,
r.hydro.CASC2D [-toepiaduq] elevation=name
time_step=value outlet_east&north&slope=value,value,value
rain_duration=value tot_time=value discharge=name
[watershed_mask=name] [initial_depth=name]
[storage_capacity=name] [interception_coefficient=name]
[roughness_map=name] [conductivity=name] [capillary=name]
[porosity=name] [moisture=name] [pore_index=name]
[residual_sat=name] [lake_map=name] [lake_elev=name]
[radar_intensity_map=name] [links_map=name]
[nodes_map=name] [channel_input=name] [dis_profile=name]
[wat_surf_profile=name] [hyd_location=name]
[r_gage_file=name] [Manning_n=value]
[unif_rain_int=value] [num_of_raingages=value]
[gage_time_step=value] [radar_time_step=value]
[write_time_step=value] [unit_el_conv=value]
[unit_lake=value] [unit_space=value] [d_thresh=value]
[dis_hyd_location=name] [depth_map=name]
[inf_depth_map=name] [surf_moist_map=name]
[rate_of_infil_map=name] [dis_rain_map=name]
```

Note that in certain cases (e.g. surface roughness coefficient), at least one of the optional arguments in the above list must be specified.

The required arguments to run **r.hydro.CASC2D** are the name of an elevation map, computational time step, discharge outlet location, rainfall duration and rate, overland flow Manning n, total simulation time and name of discharge hydrograph file. This minimum number of arguments will perform a simulation assuming uniform rainfall intensity in space and time for a specified duration, without infiltration, retention storage, interception losses, or channel routing.

The optional arguments are extensive, and allow the user to tailor the simulation to the quantity of and uncertainty in available data. These optional arguments include the names of maps which contain soil parameter values for infiltration calculations, retention storage values, interception relation parameters, radar-estimated rainfall rates, and initial surface and subsurface moisture conditions. An important optional argument is the name of a watershed mask map. The presence of a mask map greatly speeds execution by limiting calculations to raster elements contained within the watershed.

FLAGS: There are several flags whose utility is driven by data availability and/or user's choice. The most important flags are summarized here:

-t interpolates raingage rainfall intensities using Thiessen polygon method. The default technique is the inverse-distance-squared method for spatial interpolation of rainfall intensity.

-e performs one-dimensional explicit finite difference channel routing. May be suitable for low- to medium-intensity rainstorms over semi-arid regions with no to very low base flow discharge. Use of this option often limits the computational time step to small values. Channel bed smoothing is recommended. No hydraulic structures, except reservoir spillways, can be simulated.

-i performs Preissmann double sweep implicit channel routing. Particularly suitable for watersheds with some base flow to avoid dry-bed condition. Supercritical slopes cannot be simulated; a warning message is printed if supercritical flow is encountered.

-a Alternative power function approximation (APFA) for channel cross sections composite with floodplain (requires -i option). Suitable for irregular channel cross sections. With this option the floodplain associated with the channel cross section should also be included, particularly when overbank flows are expected. For a detailed description of APFA cross-section types see Garbrecht (1990).

Data Requirements

An elevation map in the form of Digital Elevation Model (DEM) is undoubtedly one of the most important inputs for terrain modeling. The quality of the DEM plays a major role in success of distributed hydrologic simulations. DEMs almost always contain errors. Depending on the source of the DEM these errors may be attributed to various causes. Large flat areas in the DEM may be due to the limited vertical resolution of elevation data. Routing over such flat areas sometimes causes problems for the numerical techniques used in distributed physically-based models. Unreal pits in the DEM may be artifacts of interpolation scheme used to rasterize digitized contours. As a rule of thumb, the user must cross check the DEM with topographic maps of the area. Errors in the DEM may be discovered by running **r.hydro.CASC2D** with no infiltration and uniform rainfall and observing the time evolution of the surface depth maps.

Green and Ampt infiltration calculations require four maps: soil porosity, saturated hydraulic conductivity, capillary head, and initial water content. If the user desires soil-moisture redistribution during rainfall hiatus, then two additional maps, namely pore distribution index, and residual saturation are required. Soil moisture redistribution during rainfall hiatus is calculated using a method similar to Smith et al. (1993).

Estimation of rainfall interception requires two parameters, namely: storage capacity and interception coefficient. The storage capacity and interception coefficient maps are usually a reclassification of a vegetation or land cover maps. Typical values of these parameters are given in Gray (1973). Surface runoff retention storage may be specified through a retention depth map. Overland flow roughness coefficient may be spatially-uniform or spatially-varied depending upon the optional arguments.

Rainfall rates may be input to the model as raingage data from gauges located at specified coordinates, or as maps of precipitation such as those provided by NEXRAD weather radars. Raingage data require an ascii file containing the number of gauges, temporal resolution of the data, and the rainfall rates recorded at each gauge.

Channel routing is another option. At present, the options are implicit and explicit 1-D routing. Inclusion of channel routing requires maps of stream channel locations, and a file containing cross section geometry, and talweg elevations. The implicit channel routing is based on Preissmann's double-sweep method. Base flow is required for this implicit method to prevent dry-bed conditions. The implicit method offers improved numerical stability and the ability to accommodate a wide variety of internal boundary conditions such as: reservoirs, weirs, and culverts.

Model Output

Discharge hydrographs may be printed to a file for any location in the catchment. Coordinates of desired hydrograph locations are entered through the hyd_location option. The default location is the specified outlet. Hydrographs can be printed for any number of internal points along the channel network at a frequency determined by the parameter write_time_step.

Raster maps of surface water depth, cumulative infiltrated depth, soil surface water content, infiltration rate, or rainfall rate may be saved at user specified time intervals. These output maps are GRASS raster maps and may be displayed using GRASS. The time series of maps are given sequential numbers to denote the time during the simulation when they were written.

Applicability and Limitations

The GRASS module **r.hydro.CASC2D** is currently under evaluation and continued development. At present, the algorithms used are applicable for Hortonian runoff. Channel routing routines used in **r.hydro.CASC2D** are suitable only for one-dimensional sub-critical flow.

While using any hydrologic model, the user must be aware of the limitations of the model, and assumptions used in its development. The development of **r.hydro.CASC2D** is advanced compared to the state of knowledge in the field of distributed hydrologic modeling. Pertinent questions regarding appropriate grid-sizes, applicability of fundamental equations over those grid-sizes and others require further research. Applications of **r.hydro.CASC2D** thus far appear promising (Johnson et al., 1993) over watersheds ranging from a few km^2 to several hundred km^2.

Acknowledgements:

The development of **r.hydro.CASC2D** has been funded by the U.S. Army Construction Engineering Research Laboratories, and the U.S. Army Corps of Engineers Waterways Experiment Station under contract DACW39-94-K0007.

References:

Garbrecht, J., 1990, Analytical representation of cross-section hydraulic properties, *J. Hydrol.* Vol. 119, pp.43-56.

Gray, D.M., ed. 1973, *Handbook on Principles of Hydrology*, Port Washington, N.Y.: Water Information Center.

Johnson, B.E., N.K. Raphelt and J.C. Willis, 1993, Verification of hydrologic modeling systems, Proc. Federal Water Agency Workshop on Hydrologic Modeling Demands for the 90's, USGS Water Resources Investigations Report 93-4018, June 6-9, Sec. 8.9-20.

Julien, P. Y., and B. Saghafian, 1991, CASC2D users manual - A two dimensional watershed rainfall-runoff model, Civil Engr. Report, CER90-91PYJ-BS-12, Colorado State University, Fort Collins, CO.

Julien, P.Y., B. Saghafian, and F.L. Ogden, 1995, Raster-based hydrologic modeling of spatially-varied runoff, *Water Resources Bull.*, AWRA (in press).

Ogden, F.L., 1994, de-St Venant channel routing in distributed hydrologic modeling., Proc. Hydraulic Engineering '94, ASCE Hydraulics Specialty Conference, G.V. Cotroneo and R.R. Rumer, eds., Vol. 1, pp. 492-496.

Saghafian, B., 1992, Hydrologic analysis of watershed response to spatially varied infiltration, Ph.D. Dissertation, Civil Engr. Dept., Colorado State University, Fort Collins, CO.

Saghafian, B., 1993, Implementation of a distributed hydrologic model within Geographic Resources Analysis Support System (GRASS), Proceedings of the Second International Conference on Integrating Environmental Models and GIS, Breckenridge, CO.

Smith, R. E., Corradini, C., and F. Melone, 1993, Modeling infiltration for multistorm runoff events, *Water Resources Research*, 29(1), pp. 133-144.

WATERSHED RESPONSE FROM DIGITAL ELEVATION MODELS

Mark Michelini [1] and Rafael G. Quimpo [2]

Abstract

Public-domain and commercial software are interfaced with project developed computer programs to enable the automatic generation of the instantaneous unit hydrograph directly from the digital elevation model. Initial DEM processing is done using public domain (USGS) software. Image display and manipulation is carried out using commercial software. Starting from an ASCII file for the DEM, the sequence of steps consists of depression filling, determining flow directions, calculation of flow accumulations, identification of watershed boundary, drainage network delineation, determination of the time of concentration, development of time-area diagram and instantaneous unit hydrograph estimation. The IUH model makes use of Clark's approach, a common option in hydrologic simulation packages such as the U.S.Army Corps of Engineers HEC-l Hydrograph package. The SCS method may also be used for hydrograph prediction. The DEM for two watersheds of different shapes were used to test the software.

Introduction

The prediction of runoff from a watershed is one of the major tasks of a hydrologist. The initial steps that the hydrologist takes to carry out this task have been tedious and time-consuming. Manual extraction of watershed information from topographic maps is subjective and prone to errors. The advent of digital elevation models (DEM) and geographic information systems (GIS) and the increasing availability of data and software for processing spatial information have alleviated this problem. Public-domain and proprietary

---

[1]Civil Engineer, Fay, Spofford & Thorndike, Inc., Burlington, MA 01803

[2]Professor, Dept. of Civil Engineering, Univ. of Pittsburgh, PA 15261

software are now available to delineate watershed boundaries and identify the drainage network from digital elevation models. The development of automatic techniques to predict watershed response would be a logical consequence of these advances. This would involve the incorporation of hydrologic theory to the basic techniques of spatial data processing. This study takes the automatization one step further by developing a technique wherein the instantaneous unit hydrograph (IUH) for a watershed may be automatically identified. This method relieves the hydrologist of the tedious task of hand processing topographic maps. It also reduces errors due to subjectiveness of manual map feature extraction.

Software Overview, GIS and Image Display

The starting point in this approach is the digital elevation model of the watershed. For the continental United States this may be obtained from the US Geological Survey. It may also be accessed on the Internet by e-mail and downloaded using a file transfer protocol. The DEM file is downloaded to a disk and converted to raster format. After file conversion, the following operations enable the user to obtain a unit hydrograph: (1) depression filling, (2) determining flow directions, (3) calculating flow accumulations, (4) drainage network generation, (5) determining the time of concentration, (6) calculation of time-area diagram, and (7) IUH and unit hydrograph estimation. These steps are explained below.

DEM files typically contain *depressions*. These are cells or groups of cells with no adjacent cell having an elevation lower than the lowest elevation in the group. A depression is artificially filled by raising its elevation to that of its lowest neighbor. This results in a flat surface through which flow may be traced without being trapped. An alternative filling procedure is to provide the filled area with a slight slope. After all depressions are filled, one can calculate the flow *directions* across each cell by following the steepest downstream slope. For a drainage area which is gridded, each point in the grid may therefore be thought of a being downstream of a cell or group of cells above it. The *flow accumulations* at each cell may be calculated as equivalent to the number of cells above it. The calculated flow directions will also define the *drainage network*. Drainage may consist of overland and channelized flow. To distinguish between the two, a threshold value corresponding to the smallest area that can support channel development must be selected. From the drainage network thus generated, the outlet of a watershed and its corresponding drainage area may be identified. Having identified the drainage area of the watershed, the *time of concentration* may be calculated using suitable assumptions for calculating the velocities for both overland and channel flow. In calculating the travel time over channels, one may divide stream lengths by velocities based on slopes or other suitable method for estimating velocity. The calculated values will then allow one to define isochrones of travel time which are used to develop the *time-area* diagram. Using Clark's method the *time-area diagram* is routed through a conceptual reservoir to obtain the *instantaneous unit hydrograph* from which any *finite duration unit hydrograph* may be derived. If the Soil Conservation Method is used, the unit hydrograph parameters are derived from the *time of concentration.*

The steps in the above procedure are implemented through a set of public-domain and commercial programs interfaced by routines which were written specifically for the problem. The package consists of ten separate routines. The first six were obtained from the USGS and are used for the initial processing of the DEM. The remaining programs were written specifically for this study. These programs can be run on either a DOS or UNIX operating system. To increase processing speed and reduce disk space requirements, the programs are designed to run on binary data.

The DEM, which is an ASCII text file, must first be converted to the binary code for the particular system that is to be used, since DOS uses a different binary code than UNIX. Some of the programs have two versions for use with either integers or real numbers, depending on the type of processing needed. The use of integer data is faster. Depressions may be filled level or given some slight slope. If a slope is preferred, real variables are recommended so that elevations in adjacent cells can be raised by small increments to provide a slope to the outlet. The process starts with depression filling. This is done with FILLSNGL for single cell depression and PPUPDATE for depressions with multiple cells. After the depressions are filled, the next step is to determine flow directions. DIRECT carries out this assignment by computing slopes between adjacent cells with flow occurring along the steepest slope. Once the flow directions are determined, RCOUNT is run to determine flow accumulations. The next step is to mark the stream network using CHANNEL. Then, the location of the outlet where the hydrograph is required is marked and the watershed delineated using WTRSHED. If the Clark method is to be used, the unit hydrograph is developed by first running TIMAREA to determine the time-area curve, and then HYDRO which computes the unit hydrograph ordinates. This may be input into HEC-1 and used for event simulation. If the SCS unit hydrograph is desired, TIMAREA is used to determine the time of concentration, which can also be input in HEC-l to develop the direct runoff hydrograph.

For image display and manipulation, the commercial softwares, IDRISI and TRANSFORM were used. Images that can be displayed include the stream network, elevation contours, and marked watersheds. These programs also have hard-copy capabilities. Data management was carried out using IDRISI. This is a grid-based geographic information system (GIS) package developed at

Clark University (1992). The main element of this GIS is the database, which consists of two parts: (1) a spatial database describing the shape and position of earth surface features, and (2) an attribute database describing the characteristics of these features. IDRISI contains several specialized modules which perform specific functions. These modules may be accessed separately through a menu. The display system enables the user to view the database as a map, either on the screen or from a hard copy. The output from IDRISI is rather basic and must, in some cases, be used with an image enhancement software for map generation. Even though IDRISI can produce color images, all color images in this study were produced with Spyglass Transform. Transform is a UNIX-based program which is capable of producing standard images as well as contours and 3-dimensional views. Because it operates on UNIX, the image data sets processed by IDRISI must be converted to ASCII format. Transform is menu driven and most commands are carried out with the use of a mouse. Once the image data file menu has been accessed, Transform prompts the user to select the type of image desired and then displays it on the screen. To print the image, the file must first be saved in TIFF format and then processed with an image editor.

Implementation - Case Study

This software package was tried on two watersheds identified from a DEM - that for "Short Run, Ellisburg Project Pennsylvania", a 7.5-minute quadrangle located in Potter County. Two watersheds, chosen for their size and shape (Figure 1) were used. The smaller watershed has an area of 1.46 square miles and a triangular shape with all three sides roughly the same size. The larger watershed has an area of 3.67 square miles and is elongated in shape. For the SCS method, a curve number of CN = 60 was arbitrarily chosen to compute the time of concentration. Since only one curve number can be used for each area, if soil and land use vary greatly, the watershed must be divided into subwatersheds. In developing the time area diagram, the choice of the time interval is important. Time increments which are too small could "diffuse" the time-area information. This information is used by HYDRO for developing the unit hydrograph. Figure 2 shows the time area diagram using a 10-minute time interval. Figure 3 shows the Clark 10-minute unit hydrograph. The value of K is estimated using a method in Linsley et al (1975). Other methods are possible.

Conclusions - The study has demonstrated that indeed, one can determine the watershed response automatically from a digital elevation model (DEM) of the watershed and data on land use and soils which may be stored in a GIS. It is concluded that the procedure gives consistent results, is less time consuming and hence, efficient. Refinements in the steps as well as tests for the sensitivity of results with different time scales and resolution may also be easily carried out.

References.

1. Hydrologic Engineering Center, Flood Hydrograph Package User's Manual, U.S. Army Corps of Engineers, Davis, CA. 1990
2. Eastman, J. R. IDRISI User's Guide, Clark Univ., Worcester, MA, 1992
3. Linsley, Kohler and Paulhus, Hydrology for Engineers, McGrawHill, 1975.
4. Jenson , S., Application of Hydrologic Information Automatically Extracted from Digital Elevation Models, Hydrologic Process, (1991) 5:31-44

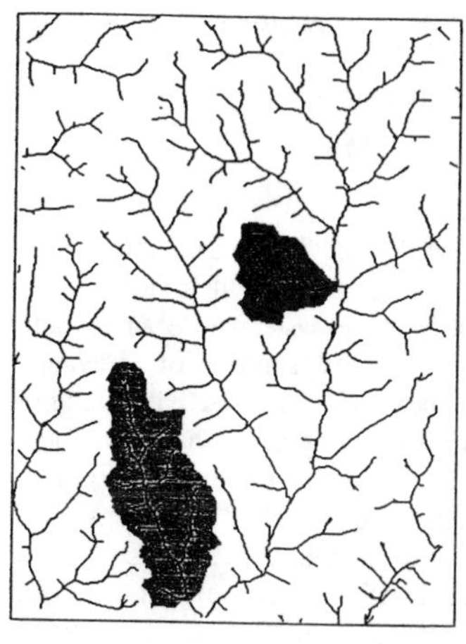

Figure 1

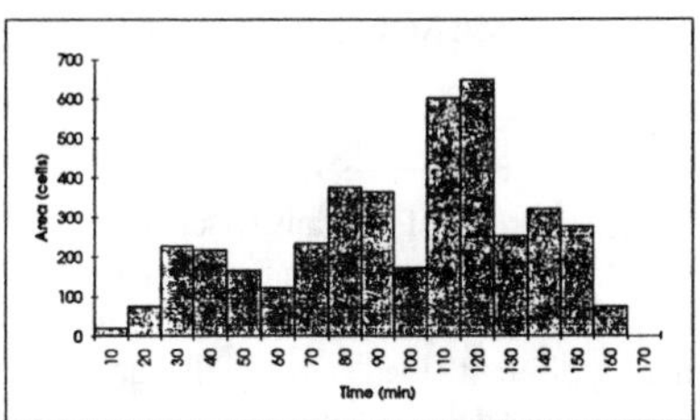

Figure 2

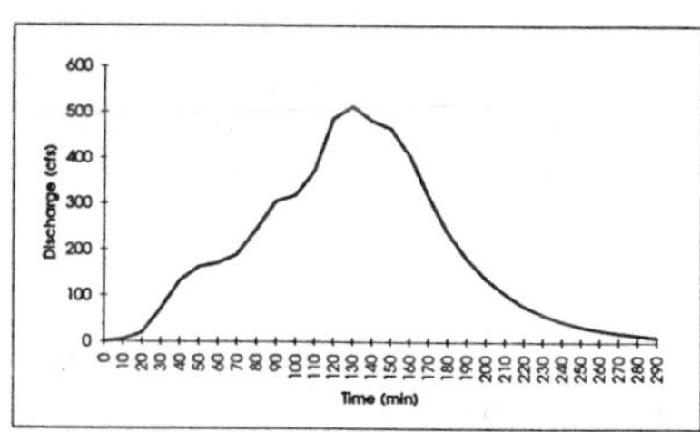

Figure 3

Figure 1. Watersheds Delineated from DEM

Figure 2. Time-Area Diagram

Figure 3. 10-minute Unit Hydrograph Derived From Time-Area Diagram

GIS MODELING FOR STORM WATER PERMIT APPLICATION

Zhida Song-James, Assoc. M.[1] **and Michael P. Sullivan**[2]

ABSTRACT

The National Pollutant Discharge Elimination System (NPDES) requires large municipalities to address water quality as well as water quantity in their storm water discharge permit applications. Municipalities must estimate system-wide annual and seasonal pollutant loads from all major outfalls; identify sources of individual pollutants; and evaluate the performance of control measures. These activities require accurate information on location of outfalls, their contributing sewersheds, type and location of best management practices (BMPs), and potential pollutant sources to outfalls. A Geological Information System (GIS) and related modeling was developed for the District of Columbia to support its NPDES application. The system was applied to address three issues: to estimate system-wide storm water volume and pollutant loads; to conduct the cost-effectiveness evaluation of BMPs; and to track sources of chemicals discharged into the storm water system. Our experience revealed that GIS has strong potential to be used as a storm water management tool, and civil engineers' involvement is a key to its successful application.

INTRODUCTION

The US EPA established NPDES permit requirements for storm water discharges (EPA, 1992). Under these regulations, municipal separate storm water sewer systems must be covered by NPDES permits. The permit process requires large municipalities (with population over 100,000) to identify sources of pollutants, to link sources of pollutants in runoff to specific water quality impacts, to identify those activities or physical factors which have most significant impacts on water quality, and to define control measures that would yield improvement in storm water quality. In addition, city engineers and planners need a tool to evaluate alternatives to

[1] Project Engineer , LTI-Limno-Tech, Inc. 1155 Connecticut Avenue, N.W. Suite 910, Washington DC 20036.

[2] Regional Manager, LTI-Limno-Tech, Inc.

meet the above requirements, and to implement cost effective measures to obtain environmental benefits through imposing storm water control.

GIS has been recognized as a powerful tool to compile, manage and demonstrate information required by the permit application, such as landuse, waterbody, watershed, and soil type. Since the initial work of data collection and map delineation are time consuming, past experience also showed that GIS may not be an efficient tool in some case (Kilgore, 1991), especially if it is developed only for a single application. Our intention is to develop a GIS for assistance with the storm water permit application and for ongoing assistance with implementation of the storm water management program. The GIS has been applied to watershed modeling for load estimation, cost-effectiveness analysis of storm water BMPs for pollutant removal, and potential pollutant source tracking.

STUDY AREA

As the Nation's Capital, the District of Columbia (DC) is a highly developed urban area of 179 km^2 (69 mi^2). The storm water drainage system covers 74.1 km^2 (28.6 mi^2), about 41% of total area. Combined sewers cover 51.2 km^2 (19.2 mi^2). The storm water drainage discharges into three major waterbodies: Potomac River in the west, its tributaries Anacostia River in the east and Rock Creek cutting through the center of DC. Among landuses, parks cover about 12% of the drainage area; low density residential area is about 28%; and the rest is industrial, commercial and mid/high density residential use as well as for federal facilities. About one third of total drainage area is impervious.

GIS DEVELOPMENT

A GIS (ARC/INFO) was developed for data management. Coverages can be divided into three general categories. General descriptive information consists of DC boundary, census tracts and blocks, streets, waterbodies, watersheds, soil types, and landuses. Storm water management information created for this study includes floodplain coverages for 100- and 500 year events, sewershed (defined as a drainage area for a single outfall), outfall, and BMP coverages. Pollutant sources information includes industrial facility and underground storage tank coverages.

APPLICATION 1: POLLUTANT LOAD ESTIMATION

The NPDES permit application requires estimation of annual and seasonal pollutant loads for each outfall and for the entire storm water drainage system. The "Simple Method" using precipitation (annual, seasonal or event), runoff coefficient and event mean concentration of pollutant as key parameters was selected to simulate load generated from each sewershed:

$$L_i = P * CF * R_i * C * A_i * UC \qquad (1)$$

Where: L_i = Pollutant load from loading area i

P = Precipitation
CF = Correction factor to adjust for storms where no runoff occurs
R_i = Runoff coefficient of loading area i
C = Event mean concentration of pollutant (from monitoring data)
A_i = Loading area
UC = Unit Conversion factor

A total 39 of landuse categories were identified. Most sewersheds are composed of several landuses, each with its own imperviousness and runoff coefficient. Using the GIS, loading areas with unique imperviousness and runoff coefficient were obtained by intersecting landuse and sewershed coverages. Pollutant loads were calculated for loading areas and then were summed as total loads for a sewershed. Watershed-wide and system-wide loads were calculated using the same procedure. Five major landuse groups and four pollutant loads are presented in Table 1 for each watershed. The result can be presented by GIS to assist in identifying critical loading areas and evaluating loading variation related to landuse changes.

TABLE 1. Estimated Pollutant Loads from Landuse Groups

	Area	Landuse (Grouped by Imperviousness)					Pollutant Loads (Ton.)			
Watershed	km^2	Park	<=30	31-50	51-75	>75	COD	TSS	TN	TP
Potomac	20.23	2.66	7.19	6.04	3.32	0.99	723.4	430.2	35.6	3.7
Anacostia	38.21	7.25	9.96	0.89	17.50	2.84	1357.4	807.3	6.7	6.9
Rock Creek	15.58	1.60	10.99	2.96	2.59	0.03	464.8	276.4	22.9	12.4
Total	74.02	11.51	28.14	9.90	23.41	3.86	2545.7	1514.0	65.2	12.9

APPLICATION 2: COST-EFFECTIVENESS ANALYSIS OF BMPS FOR POLLUTANT REMOVAL

The NPDES permit application requires evaluation of the performance of storm water controls. Although more than 200 structural BMPs have been constructed or are under construction in DC, very limited work has been done to evaluate their performance in respect of both pollutant load reduction and cost effectiveness. The GIS was applied to this evaluation. First, load reduction attributed to BMPs for each sewershed was calculated. By intersecting the BMP coverage with sewershed and land use coverages, we were able to locate each BMP and to identify its service area. The amount of pollutant loads removed by a given BMP was estimated as:

$$RL_j = P * CF * R * C * SA_j * RE * UC \qquad (2)$$

Where: RL_j = Pollutant load removed by BMP j
R = Runoff Coefficient of the BMP service area
SA_j = Service area of the BMP j
RE = Pollutant removal ratio of the BMP j, %
P, CF, R, C and UC are as same as in Eq. (1)

Total load reduction can be summed for a sewershed, a watershed, or for the entire storm water drainage system. Second, costs of using BMPs to remove pollutants were estimated. The most commonly used BMPs operated in DC consist of detention tanks, oil grit chambers, sand filters and infiltration trenches. The use of wet or dry ponds is limited by the lack of available space in the city. Assuming that the average life of a BMP is 5 years and using the discount rate of 0.08, the annual cost (construction cost only) of per pound pollutant load removed by each type of BMPs was calculated (Table 2). Maintenance cost was not available for this study.

TABLE 2. Unit Costs of Pollutant Removal by BMPs ($/kg)

	Oil Grit Chamber	Sand Filter	Detention Tank	Infiltration Trench	Dry Pond	Wet Pond
COD	262.8	25.3	459.8	18.7	33.4	8.1
TSS	441.9	54.2	1283.1	31.5	112.4	20.5
TN	5939.0	1027.9	10536.4	412.0	776.1	283.4
TP	51881.3	6992.1	126731.3	3431.2	10558.6	2400.8

Results indicated that BMPs have limited ability to reduce pollutant load and are costly. Wet pond is the most cost effective BMP but it requires large space. Searching on GIS maps showed that the suitable space is generally not available in DC. Infiltration trenches have low construction cost but they need intensive maintenance. Other BMPs are not cost-effective. Consequently, reliance on structural BMPs for developed urban areas can be a costly and ineffective way to achieve pollutant reduction. Recognition of these limitations is useful for the management of programs where pollution reduction is important. It leads managers to evaluate other key factors in load generation (Eq.1). The runoff coefficient (R_i) can be affected by reducing imperviousness, and event mean concentration (C) can be reduced through non-structural means. In DC, pollutant reduction might be achieved on a larger scale across sewersheds by placing greater emphasis on existing programs for pollution prevention, source control and public education.

APPLICATION 3: POLLUTANT SOURCE TRACKING

Although there is very little heavy industry within DC, there are over 3,000 small industrial facilities including printing houses, photo and film services and animal hospitals. The amount of pollutants entering the storm water system from those potential sources is largely unknown. DC has developed an inventory listing these sources and is establishing a monitoring program. To link chemicals detected at an outfall to their potential sources, the inventory must be organized by sewershed and watershed, and be able to identify source locations and potential chemical discharges associated with each source. The GIS was applied to establish a dynamic linkage between source inventory and water quality monitoring. Three parts are involved in the source tracking. The first is a GIS point coverage with the industrial source's name, address, SIC code, and permit number (if existing) in its attribute table. Similar to the loading calculation, this coverage is to be intersected with sewershed and outfall coverages to connect industries with sewersheds. The second part is a pollutant database containing chemicals regulated for each industry identified by its SIC. If industry A discharges the chemical X, the corresponding value in the database would be one; otherwise it is zero. The third one is a tracking procedure written by ARC/INFO's macro programming language, designed for situations when a chemical is detected from an outfall and potential sources need to be identified. The tracking consists of the following steps:

1. Select the outfall by its user ID. The corresponding sewershed shares the same user ID and it would be identified automatically;
2. List all industries within the sewershed;
3. Link industries with the pollutant database by their SIC;
4. Select the chemical being tracked; and
5. List all industries discharging the chemical.

The tracking procedure is still being developed. Other features may be added including to extend source tracking for buffer zones of a sewershed and to use tracking results to guide monitoring in "problem" sewersheds.

CONCLUSIONS

A GIS system was developed to support the NPDES permit application for the District Columbia. It was clear from the beginning that the system would be designed for multiple applications and would be a management tool for the city. The system was developed by civil engineers and emphasized problem solving. Three applications, pollutant loading estimation, cost-effectiveness evaluation of BMPs and pollutant source tracking, were reported. The system has much potential for other applications.

GIS application to water resources and water environment projects has been growing in recent years. Establishing a GIS requires intensive time and efforts in data collection and management. In many cases, systems have not been used after their initial development and application. Often the reason is that the system was developed by GIS experts without enough involvement of civil engineers. As a result, the GIS simply became a tool to generate "pretty" maps. We believe, and our experience showed, that civil engineers' active involvement is key to improvement of GIS applications in our field.

APPENDIX. REFERENCES

Guidance manual for the preparation of part 2 of the NPDES permit applications for discharges from municipal separate storm sewer systems. (1992). *EPA833-B-92-00.* EPA, Washington, DC

Kilgore, R.T. and Zatz, M. N. (1991) "Is GIS in water resources better or different," *Proc., 18th Conf. Water Resour. Plan. Mgmt & Urban Water Resour.*, ASCE, New York, Ny., 893-897.

Distributed Modeling of Snowmelt and Influence of Wind

Stephen Harelson[1] and Lynn E. Johnson [2]

Abstract

A distributed snowmelt model was developed for the North Fork of the St. Vrain Creek in Colorado using a geographic information system linked with computational modules to compute the energy balances. The GRASS GIS was used to develop spatial datasets for terrain and aspect, soils and snow cover, and to identify hydrologic response units. The Modular Modeling System package developed by the U.S.G.S. was used to simulate energy balance of snowmelt processes, and extensions of these computations were made to represent the influence of wind across the basin. Accounting for wind influence showed significant differences in the amount and timing of snowmelt runoff.

Introduction

The most important contributor to the water supply in Colorado and similar regions throughout the world is the runoff that results from melting snow. The recent advent of Geographic Information Systems (GIS) technology has provided the hydrologist with a new tool to catalogue the wide array of factors that influence snowmelt. This tool can be used to more effectively model the phenomenon of melting snow and its influence on hydrology.

In order for snow to melt, heat energy must be transferred to the snowpack. There are three major types of heat transfer; radiation, convection, and conduction. In addition to the three modes of heat transfer; energy can be transferred to and from the snowpack via the mass transfer mechanisms of precipitation, condensation and evaporation. Liquid rain heats the snowpack because its temperature is greater than the 0° C snowpack, while condensation and evaporation introduce the influence of latent heat of vaporization.

----

[1]Project Engineer, Carroll and Lange, Inc., Lakewood, Colorado, and [2]Associate Professor, Dept. Civil Engineering, CB 113, Univ. of Colorado at Denver, P.O. Box 173364, Denver, CO 80217-3364

The heat balance for snowpack is given by the relation: (U.S. Army, 1956)

$$H_m = H_{rs} + H_{rl} + H_c + H_e + H_g + H_p + H_q$$

Where: H_m is heat of melt

H_{rs} is short wave (solar) radiation

H_{rl} is long wave radiation

H_c is convective heat transfer

H_e is latent heat from condensation

H_g is conduction from ground

H_p is heat content of rain

H_q is the change in the heat content of the snow.

Using the basic principles of heat and mass transfer, one can identify the necessary variables required to create a snowmelt model. In the field of snow hydrology, atmospheric data is difficult to gather, and can vary greatly throughout a watershed. To apply the classic heat and mass transfer relations to the problem of snowmelt, a relatively small number of data points must be distributed, interpolated, and extrapolated over often large and rugged expanses of space and time. The collection and cataloguing of data are as important as the design of the snowmelt model that uses these data.

In many previous basinwide snowmelt models, the effect of wind speed in convective heat transfer has been simplified, because there is not often a record of wind speed data in the area of interest. Most models have concluded that radiation is the dominant mode of heat transfer in the melting of snowpack. The effects of high speed, adiabatically warmed Chinook winds on both the melting and evaporation of the snowpack have not been thoroughly studied.

Study Site

The University of Colorado has operated a climatological research center on the slopes of Niwot Ridge in Boulder County, Colorado since the early 1950s. The Niwot Ridge Long Term Environmental Research (L.T.E.R.) Center has recorded temperature, precipitation, solar radiation, wind and other environmental factors at various elevations throughout several periods since 1952. The records provide a rare picture of the meteorological conditions of the alpine environment on the eastern slope of the Colorado Rockies.

Approximately 15 kilometers north of Niwot Ridge lie the headwaters of North St. Vrain Creek. Near the town of Allenspark, the United States Geological Survey operates a streamflow gage on this creek. This gage has recorded the average daily flows of the creek since 1986. The watershed flowing into this gage contains an area of 85 square kilometers, and varies in elevation from 2525 meters to 4344 meters above sea level. It is wilderness, covered by pine forests at its lower reaches and alpine tundra at its upper elevations.

Model Construction

The model consists of two parts; the Geographical Information System (GIS) portion where the spatial analysis of the watershed was performed, and the Hydrologic portion where the energy and water balance analysis were performed. The GIS analysis of the model was done using GRASS 4.1, a raster based GIS written by the United States Army Construction Engineering Research Laboratory. Using GRASS, the influences of elevation, slope, aspect and land use were analyzed and catalogued for use by the hydrologic portion of the model.

The hydrologic analysis was performed using a model written within the Modular Modeling System, or MMS. MMS is a system that allows linking of small hydrologic submodules into a whole model. MMS also includes features that aid the hydrologist in testing and calibrating models and in sensitivity analysis and streamflow prediction. MMS was developed using modules first developed for the Precipitation Runoff Modeling System or PRMS (Leavesley, et al 1983.) Using MMS, these modules can be linked in different ways, modified, and linked with other modules developed independently of PRMS.

In this study, the PRMS model was modified to calculate differently the convective heat transfer involved in snowmelt. In PRMS, the convective heat transfer is calculated using a monthly index. The energy transferred between the air and the snowpack via convection are a function of this monthly index multiplied by the daily air temperature. In this study, the PRMS snowmelt module was modified to make the convection energy transfer a function of daily wind speed, humidity temperature and barometric pressure. Using MMS, it is possible to insert this change into PRMS and run both PRMS and the Modified PRMS model on the same data set, and review the influence of the modification.

The PRMS module "snowcomp.f" computes the rate of convection heat transfer based on a monthly index, CEC. This monthly index is multiplied at each 12 hour time step of the energy balance model by the average temperature during this time step. The product of these two factors, CECSUB, is the estimate of latent and sensible heat flux for the time step, and has units of calories.

The complete equation by the U.S. Army (1956) includes terms for wind speed, vapor pressure and atmospheric pressure. Snowcomp.f does not directly account for these effects because they are seldom measured in real watersheds, and they are difficult to extrapolate in space and time. To account for wind effects, snowcomp.f assumes areas within timber receive one half of the magnitude of CECSUB. To account for vapor pressure, CECSUB is only calculated on days of rainfall or when the ratio of observed to potential shortwave radiation is less than or equal to 0.33.

In this study, CECSUB was calculated using the relation from the U.S. Army (1956):

$$CECSUB=0.3174[(T_a-T_s)(P_a/P_s)+15.46(e_a-e_s)]V_b$$

Where: T_a is Air Temperature, Degrees Celsius
T_s is Snow Temperature, 0° Celsius
P_a is Atmospheric Pressure, millibars
P_s is Standard Atmos. Pressure, 1013 millibars
e_a is vapor pressure of air, millibars
e_s is vapor pressure of snow, 6.11 millibars
V_b is wind speed, meters per second

CECSUB was first calculated using average monthly values derived from the L.T.E.R. data. These values were divided by the average monthly temperature to obtain CEC, the monthly convective-condensation coefficient to be used in the PRMS model.

The snowmelt.f module from PRMS was then modified to calculate CECSUB using daily values of temperature, humidity and wind speed from the L.T.E.R. data. The modified and un-modified models were run to demonstrate the influence of the averaging of the convective-condensation coefficient. Results indicate that accounting for the winds has significant influence on the magnitude and timing of snowmelt runoff.

Conclusions

The effect of convection is of considerable importance, high temperatures and moderate wind speed can make its magnitude considerably higher than the monthly average. Downsloping Chinook winds on the Front Range of Colorado can create these conditions, and appreciable snowmelt results from these winds. In areas where Chinook winds are common, the design of snowmelt models should take into account their effect.

Acknowledgments

The authors would like to acknowledge the assistance of George Leavesley and Steve Markstrom at the U.S.G.S. Water Resources Division in obtaining, compiling and running the Modular Modeling System used in this project; and Bob Jarrett and Bob Ugland also at the U.S.G.S. Water Resources Division for assistance in obtaining streamflow data. Logistical support and/or data were provided by the Niwot Ridge Long Term Ecological Research Project (NSF DEB9211776 and the Mountain Research Station (BIR 9115097). Rick Ingersoll at the L.T.E.R. was helpful in providing the data in a useful form. This work would not have been possible without the original investigators at L.T.E.R.; Mark Losleben and Jim Halfpenny.

References

Frankoski, L (1994) Effect of spatial resolution on hydrologic model results. Masters Thesis, University of Colorado, Boulder Colorado

Leavesley, G.H., Lichty, R.W., Troutman, B.M. & Saindon, L.G. (1983) Precipitation Runoff Modeling System: Users Manual. Water-Resources Investigations 83-4238 United States Geological Survey, Denver, Colorado, U.S.A.

U.S. Army (1956) Snow Hydrology. U.S. Army Corps of Engineers, Portland, Oregon, USA.

MANNING'S N FROM AN EXTENDED RAINFALL-RUNOFF DATA SET ON A CONCRETE SURFACE

Joseph R. Reed[1], Randy T. Watts[2], John C. Warner[2], Richard S. Huebner[3]

Abstract

Three equations of Manning's n for shallow flows on Portland Cement Concrete (PCC) surfaces are presented . In the equations, Manning's n is a non-linear function of Reynold's number. The equations were based upon experimental data and extend work by Reed and Stong (1992) for Reynold's numbers up to 1000.

Introduction

Gallaway, et al, (1971) and (1979), conducted a series of experiments using artificial rainfall on ten different pavements and developed a regression equation for water film thickness (WFT) above the pavement asperities based upon mean texture depth (MTD), rainfall intensity (I), drainage path length (L), and slope (S). The single equation for all ten surfaces ($R^2 = 0.83$) did not include a fundamental hydraulic resistance variable. Subsequently, work by Reed and Kibler (1983), and Reed, et al, (1983), suggested a hydraulic resistance variable for individual types of pavement surfaces. Reed and Stong (1992), developed an equation for Manning's n on PCC roadway surfaces based on experimental results on three different PCC surfaces. The equation was bounded by a maximum R_e of 145 and based on a regression equation with an R^2 of 0.90. Reed, Warner, and Huebner (1994) developed a similar expression for dense graded asphaltic concrete surfaces.

[1] Professor, Department of Civil and Environmental Engineering, The Pennsylvania State University, University Park, PA 16802, Member ASCE.

[2] Graduate Assistant, The Pennsylvania Transportation Institute, The Pennsylvania State University, University Park, PA 16802.

[3] Assistant Professor, School of Science, Engineering and Technology, The Pennsylvania State University at Harrisburg, Middletown, PA 17507, Member ASCE.

Approach

The purpose of this research was to extend the set of data which was collected by Reed and Stong (1992) beyond a Reynold's number of 145. To do this, flow was introduced at the upper end of a channel to simulate that from a preceding plane. In this manner, 552 additional flow depth values were collected and merged with Reed and Stong's data.

The kinematic wave equation solved for Manning's roughness coefficient, n, and shown in Reed and Stong (1992), among others, is :

$$n = 36.1 \frac{S^{0.500}\, y^{1.667}}{I\,L} \tag{1}$$

where:

y	=	hydraulic flow depth (mm)
S	=	drainage surface slope (m/m)
I	=	rainfall intensity (mm/hr)
L	=	drainage path length (m)

The kinematic wave approximation is valid when certain criteria are met (Reed and Kibler, 1983). One criteria requires that the slope of the energy grade line be approximately equal to the slope of the flow plane. Observations were eliminated if the difference between the friction slope and plane slope was greater than 5 percent. This reduced the total number of observations to 1,124.

Analysis and Results

Manning's n for Reynold's Numbers Less Than 1,000

The remaining points had Reynold's numbers less than 1,000. The best regression equation ($R^2 = 0.926$) which included all significant experimental parameters used in testing was:

$$y = \frac{122.428\, q^{0.308}\, MTD^{0.0316}}{S^{0.286}} \tag{2}$$

where:

y	=	flow depth, m
q	=	flow rate (= L I / 43200), $m^3/s/m$
MTD	=	mean texture depth, mm

The analysis was simplified by removing MTD from the relationship which resulted in:

$$y = \frac{126.66\, q^{0.312}}{S^{0.285}} \tag{3}$$

The R^2 value of equation (3) was 0.94. However, MTD was retained in the equation since hydraulic flow depth, y, is the sum of the water film thickness and mean texture depth. The relationship developed as equation (3), an unbiased regression equation for hydraulic flow depth, was algebraically combined with equation (1), the kinematic wave equation. The average of the smallest and largest slope raised to the 0.025 power was calculated and placed in the resulting equation. The maximum error introduced by replacing $S^{0.025}$ with a constant was 2.1%. This yielded:

$$n = \frac{0.319}{R_e^{0.480}} \tag{4}$$

Figure 1 is a log-log plot of equation (4) and the calculated Manning's n values from 1,124 experimental data points. Because of a high degree of scatter of data points around equation (4) especially as the Reynold's number approaches 1,000, an improved equation was sought.

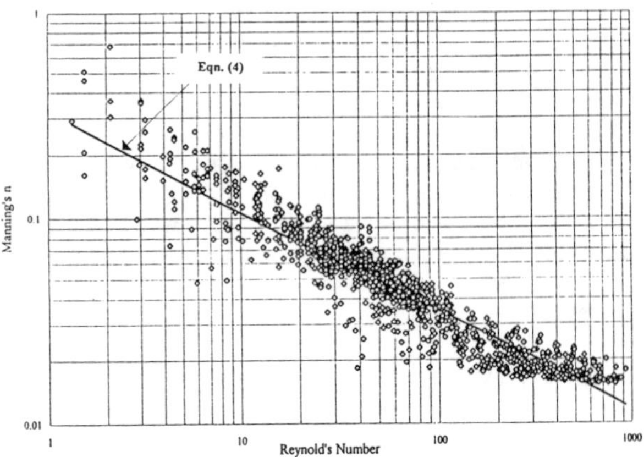

Figure 1. Manning's n vs. Reynolds Number (R_e < 1000)

Manning's n for Reynold's Numbers Less Than 500

The accepted transitional region for open channel flow is from a Reynold's number of 500 to 1,000. A Reynold's number of 500 was of particular interest because of this range. The authors decided to perform a similar analysis on the data for Reynold's numbers less than 500, the traditional laminar flow range which, in this case, was disturbed or made turbulent by the pelting rain.

A total of 1,070 of 1,124 points had Reynold's numbers less than 500 and were used in the subsequent analysis which was identical to the one implemented above for Reynold's numbers less than 1000. The best regression equation ($R^2 = 0.916$) was:

$$y = \frac{105.81\, q^{0.295}\, MTD^{0.032}}{S^{0.289}} \tag{5}$$

Once again, MTD was removed from the regression ($R^2 = 0.914$). The hydraulic flow depth equation was algebraically merged with equation (1) and, using an average slope, resulted in equation (6):

$$n = \frac{0.345}{R_e^{0.502}} \tag{6}$$

Equation (6) appeared to overestimate Manning's n as the Reynold's number approached 500. Also of concern was the manner in which the data points appeared to level off at a Reynold's number of approximately 240 as shown in Figure 1.

Manning's n for Reynold's Numbers Less Than 240

The entire data set was reduced to a data set of 940 data points with Reynold's numbers less than 240. The regression equation for hydraulic flow depth (y) was ($R^2 =$ 0.889):

$$y = \frac{83.74\, q^{0.275}\, MTD^{0.031}}{S^{0.295}} \tag{7}$$

Removing MTD from the regression of equation (7) only reduced R^2 by 0.2%. Algebraically combining the equation with the kinematic wave equation (1), and using the average slope, yielded:

$$n = \frac{0.388}{R_e^{0.535}} \tag{8}$$

When plotted, the experimental data points appeared to be consistently scattered around equation (8). Also, equation (8) did not appear to overestimate Manning's n as the Reynold's number approached 240.

Conclusions

Three unbiased fundamental equations for predicting hydraulic flow resistance on PCC surfaces based upon experimental results are:

$$n = \frac{0.319}{R_e^{0.480}} \qquad (R_e < 1000) \tag{9}$$

$$n = \frac{0.345}{R_e^{0.502}} \qquad (R_e < 500) \tag{10}$$

$$n = \frac{0.388}{R_e^{0.535}} \qquad (R_e < 240) \tag{11}$$

Acknowledgments

The authors wish to thank the National Cooperative Highway Research Program for their support of this study which is part of a larger study administered by the Pennsylvania Transportation Institute at The Pennsylvania State University.

References

Gallaway, B. M., R. E. Schiller, and J. G. Rose (1971). The Effects of Rainfall Intensity, Pavement Cross Slope, Surface Texture, and Drainage Length on Pavement Water Depths. Federal Highway Administration Research Report No. 138-5, Texas Transportation Institute, College Station, TX: Texas A&M University, May.

Gallaway, B. M., D. L. Ivey, G. Hayes, W. B. Ledbetter, R. M. Olson, D. L. Woods, and R. F. Schiller, Jr. (1979). Pavement and Geometric Design Criteria for Minimizing Hydroplaning. Federal Highway Administration Research Report No. FHWA-RD-79-31, Texas Transportation Institute, College Station, TX: Texas A&M University.

Huebner, R. S., J. R. Reed, and J. J. Henry (1986). Criteria for Predicting Hydroplaning Potential. Journal of Transportation Engineering, ASCE, 112(5):549-553, September.

Reed, J. R., Kibler, D. F., and Proctor, M. L. (1983). Analytical and Experimental Study of Grooved Pavement Runoff, Final Report. Federal Aviation Administration Report No. DOT/FAA/PM-83/84 65 pp., August.

Reed, J. R. and D. F. Kibler (1983). Hydraulic Resistance of Pavement Surfaces. Journal of Transportation Engineering, ASCE, 109(TE2):286-296, March.

Reed, J. R., and J. B. Stong (1993). Manning's n From Experiments With Rainfall-Runoff on Portland Cement Concrete Surfaces. Proceedings, International Symposium on Engineering Hydrology, ASCE, pp. 995-1000, San Francisco, CA, July.

Reed, J. R., Warner, J. C., & Huebner, R. S. (1994). Sheet Flow Resistance of Asphaltic Pavements. Proceedings. ASCE Hydraulics Division Conference, Buffalo, NY. pp. 1100-1104.

Effects of Pulsating Flow on Current Meter Performance

Janice M. Fulford,[1] Associate Member, ASCE

Abstract

Summarized are laboratory tests for current meter response to pulsating flows. Included are results for mechanical and electromagnetic water-current meters that are commonly used for stream gaging. Most of the vertical-axis and horizontal-axis types of mechanical meters that were tested significantly underregistered the mean flow velocity when the magnitude of the pulsating portion of the flow velocity was greater than half the mean velocity but less than the mean velocity. Errors for all meters tested were largest at the lowest mean flow velocity, 0.076 m/s.

Introduction

Current-meter measurements are the foundation for many studies of hydraulic and hydrologic phenomena. Ideally a current meter, whether it employs mechanical or electromagnetic principles, should respond instantly and consistently to changes in water velocity. Meters, however, are not perfect instruments and may not accurately register velocity in all measurement conditions encountered. Turbulent or pulsating flows can cause registration errors in meters. Previous studies have contradicted each other, finding that meters either overregistered or underregistered in unsteady flows (Jepson, 1967; Yarnell & Nagler, 1929)

This study presents data for one electromagnetic and thirteen mechanical current meters. The mechanical meters include six vertical-axis and seven horizontal-axis meters. For mechanical meters, the inertia of a meter's moving parts and the efficiency with which the meter translates linear velocity into angular velocity affects the ability of a meter to measure accurately in a pulsating flow. For electromagnetic meters, the response time of the circuitry and the probe shape affects the ability of the meter to measure accurately in pulsating flow. Tested meters herein measure one vector component of flow for a small flow volume and are listed in table 1 with rotor descriptions where applicable.

---

[1]Hydrologist, U.S. Geological Survey, Bldg. 2101, Stennis Space Center, MS 39529.

Table 1. Description of Meters Tested (includes the rotor properties and switch type used to signal revolutions: v, vertical-axis; h, horizontal-axis; e, electromagnetic; alum., aluminum; c/w, cat-whisker; n/a, not applicable; m, meters; gm, grams).

Meter	Type	Switch	Rotor Properties			
			diameter (m)	pitch or cup (m)	weight (gm)	material
Price type-AA	v	c/w	0.127	0.051	174.8	brass
optic Price type-AA	v	optic	.127	.051	174.8	brass
Price pygmy	v	c/w	.051	.020	20.9	brass
winter Price type-AA,m	v	c/w	.127	.051	174.8	brass
winter Price type-AA,p	v	c/w	.127	.051	226.8	plastic
modified Price type-AA	v	c/w	.127	.051	104.2	plastic
Swoffer 2100[2]	h	optic	.051	.100	6.9	plastic
Valeport BFM002	h	reed	.050	.100	25.1	plastic
Valeport BFM001	h	reed	.125	.27	506.3	plastic
Ott C-31, metal	h	reed	.125	.25	277.8	brass
Ott C-31, plastic	h	reed	.125	.25	458.7	plastic
Ott C-31, A	h	reed	.100	.125	241.4	brass
Ott C-31, R	h	reed	.100	.25	214.3	alum.
Marsh McBirney 2000	e	n/a	n/a	n/a	n/a	plastic

[2]Brand names used in the report are for identification purposes only and do not constitute endorsement by the U.S. Geological Survey.

Test procedure

Testing was conducted in the tow tank at the U.S. Geological Survey Hydraulics Laboratory, Stennis Space Center, Miss. The tow cart provided the mean velocity, Uo. The sinusoidal, pulsating flow component was superimposed in the direction of Uo by a device attached to the tow cart. The device consists of a variable-speed d-c motor and motion controller attached to a drive wheel that oscillates an attached carriage back and forth on rails. The meter is attached to the carriage by a rigid rod. Both the amplitude, d, or half stroke length of the meter in the water and the frequency, p, or speed at which the drive wheel turns are adjustable. The maximum pulsating velocity component for a test is $U'=2d\pi p$. Tests for Uo of 0.076, 0.457, and 0.914 m/s at various combinations of frequency (0.03, 0.05 0.1 and 0.2 cycles/sec) and amplitude (0.0127, 0.0635, and 0.2540 m) are presented (figs 1. and 2). Tested frequencies are considerably lower than Jepson's (1967) tests and more representative of large scale turbulence.

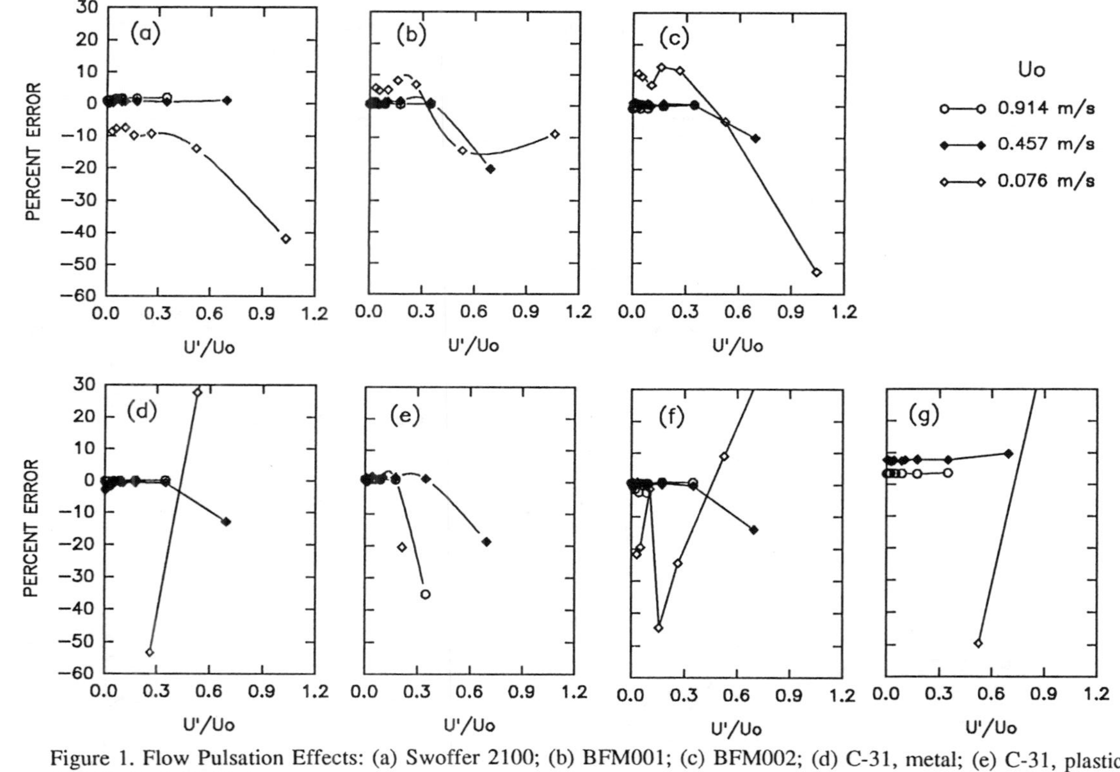

Figure 1. Flow Pulsation Effects: (a) Swoffer 2100; (b) BFM001; (c) BFM002; (d) C-31, metal; (e) C-31, plastic; (f)C-31, R; (g)C-31,A

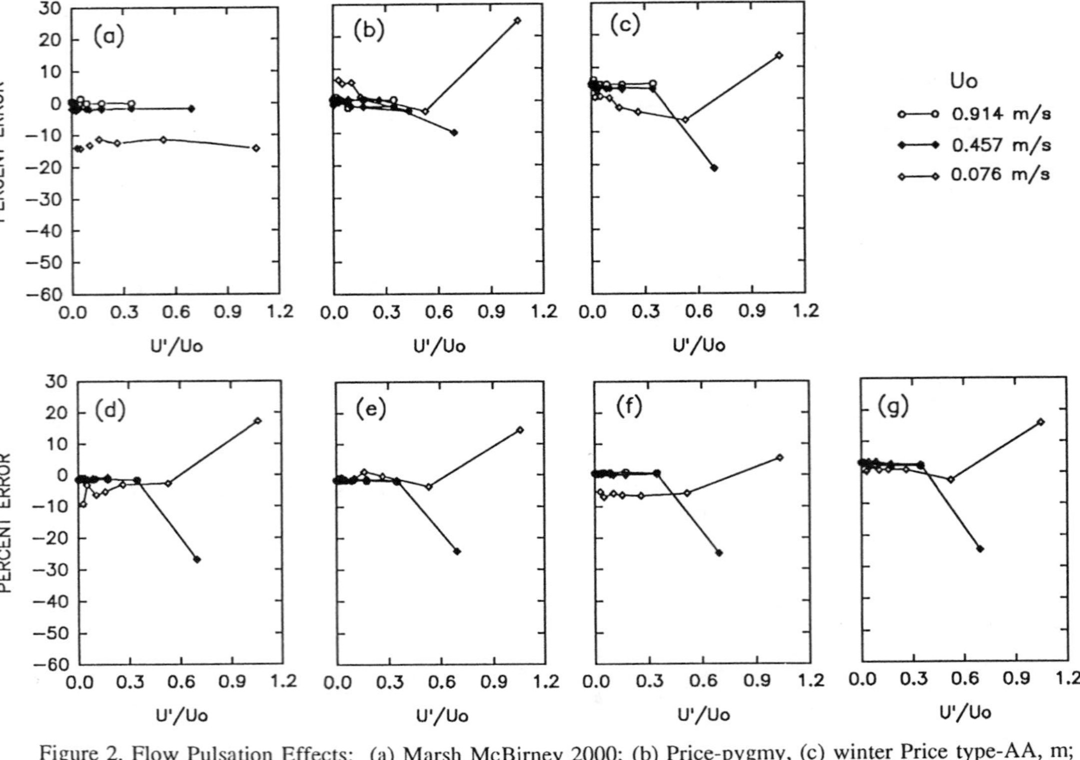

Figure 2. Flow Pulsation Effects: (a) Marsh McBirney 2000; (b) Price-pygmy, (c) winter Price type-AA, m; (d) Price type-AA; (e) optic Price type-AA; (f) modified Price type-AA; (g) winter Price type-AA, p

Test Results

Error magnitudes tended to decrease with increasing Uo for most meters tested (figs. 1 and 2). Mechanical meter error magnitudes decreased with decreasing ratios of maximum pulsating velocity component to mean velocity, U'/Uo, for Uo=0.457 m/s and Uo=0.914 m/s. The percent error, ε, in velocity registered by the mechanical meters is estimated by $\varepsilon=100(r_m-r_{est})/r_{est}$ where r_m is the meter revolutions per second (rps) measured in the sinusoidally pulsating flow and r_{est} is the rps estimated from the meter rating equation for Uo. For the electromagnetic meter, percent error is estimated by ε=100(Um-Uo)/Uo where Um is the meter registered velocity.

The poorest performance of most tested meters performed was at the lowest mean velocity, 0.076 m/s. Horizontal-axis meters, except for the BFM001, had average absolute errors >15% at Uo=0.076m/s for all U' tested. The C-31 metal and A did not register any velocity for U'<0.040m/s and the C-31 plastic registered only when U'/Uo>1. Vertical-axis meters average absolute errors are between 3% and 7% for Uo=0.076 m/s. The Marsh McBirney 2000 (MMB2000) error is 13% for Uo=0.076 m/s and may have been affected by noise from possible poor electrical grounding.

For the remaining mean velocities, Uo=0.457 m/s and Uo=0.914 m/s, average absolute meter errors are <10% for mechanical meters when U'/Uo<0.3 and <2% for the MMB2000 and Swoffer 2100 for all tested U'/Uo. All meters except the C-31 plastic and A have average absolute errors of <5% for Uo>0.076 m/s and U'/Uo<0.5. However, the C-31 A and Swoffer 2100 were relatively insensitive to change in U'/Uo for Uo>0.076 m/s. For U'/Uo>0.5, error magnitudes are larger for the vertical-axis than for the horizontal-axis meters. The MMB2000 was relatively insensitive to changes in U'/Uo for all Uo tested.

Conclusions

Registration errors for mechanical meters increased as the mean velocity decreased and as U'/Uo approached 1. Except for the electromagnetic meter (MMB2000), error was large for all meters when U'/Uo>1 and the meters were subjected to a flow reversal. The electromagnetic meter was insensitive to changes in U'/Uo. Registration errors decreased for all meters with increasing Uo.

References

Jepson, P. (1967), "Current meters errors under pulsating flow conditions." Journal of Mechanical Engineering Science, v. 9, no. 1, p. 45-54.

Yarnell, David L. and Nagler, Floyd A. (1931), "Effect of turbulence on the registration of current meters." Trans. of the ASCE, p. 766-860.

DETERMINATION OF THE FRICTION FACTOR

FOR A TURN OF THE CENTURY AQUEDUCT

E. J. Gemperline[1], R. G. Sabri[2], W. J. Marold[2], J. Iannuzzi[3]

Abstract

The New Croton Aqueduct (Aqueduct) is a brick-lined, 13.5 ft (4.1 m), horseshoe shaped tunnel, constructed between 1885 and 1891, that provides 10% of the water supply for New York City. Flow tests were undertaken in conjunction with the City Department of Environmental Protection's (DEP) long range planning for rehabilitating the Aqueduct to increase its capacity. The friction factor ("n" in Manning's equation) indicated by the flow tests was determined to be 0.014, a low value for such a conduit. The flow tests indicate that areas of the Aqueduct which were recently washed to remove slime appear to have a lower friction factor of 0.013.

Background

The Aqueduct is a 30.8 mile long conduit (49.6 km) constructed in the 1880's which delivers water from Croton Lake, in Westchester County, to New York City (City). It is the oldest of the three operating Aqueducts that serve the City and provides approximately 10% of the average daily flow, primarily to the Borough of Manhattan. Flows are controlled at a gatehouse at Croton Lake, constructed in the 1980's to replace the original structure. The gatehouse contains multiple level intakes, mechanical screens, regulating devices and Venturi-type flow meters for measuring the Aqueduct discharge. Chemical feed facilities are provided to pre-treat the water. Water

---

[1] Member, ASCE, Sr. Hydraulic Engineer, Harza Engineering Company, Sears Tower, 233 South Wacker Drive, Chicago, Il. 60606-6392

[2] Project Engineer, Harza Engineering Company

[3] Acting Chief, Planning and Programs, New York City Department of Environmental Protection, Bureau of Water Supply and Wastewater Collection, LeFrak City, New York.

is withdrawn from the reservoir through the East, Central and West intake bays and through a 2.3 mile long connection to the Old Croton Aqueduct which delivers water from a point closer to New Croton Dam which impounds Croton Lake. The Aqueduct flows open channel except for a short section at Gould's Swamp.

The Aqueduct alignment is nearly straight with a few long radius curves. There are more than 30 vertical shafts and structures. At some shafts, the Aqueduct widens to a rectangular chamber that contains gate handling or "blow off" facilities. At other shafts, local municipalities have pump intakes. These communities have withdrawn approximately 1.5 million gallons daily (mgd) to 7.0 mgd annually over the last ten years.

The Aqueduct was designed to be unpressurized with a capacity of 300 mgd. After a few years operation the capacity was noted to have declined to approximately 275 mgd (Brown, 1965) and has remained nearly constant since that time. Until recently Aqueduct flows were measured near the downstream end using flow meters or a template. Flows are now measured by Venturi meters at the new Croton Lake Gate House. DEP information indicates that annual flow rates have averaged between 80 mgd and 260 mgd over the last ten years.

Figures 1 and 2 are a cross section and profile of the Aqueduct, respectively. Between Croton Lake and Jerome Park Reservoir, approximately 25 miles (40 km), the Aqueduct is a brick lined tunnel with a 13.5 ft (4.1 m) horseshoe section set at a grade of 0.7 ft/mile (.00133%). The invert at the intake is at El. 140 ft msl (42.7 m). Approximately 12.1 miles (19.5 km) downstream of the intake the Aqueduct changes to a 900 ft long (274 m) pressurized, 14.25 ft (4.34 m) diameter circular inverted siphon section, approximately 60 ft (18.3 m) below normal grade under Gould's Swamp. At Gatehouse No. 1, approximately 24 miles (38.6 km) downstream of the intake, the Aqueduct divides into two branches. The main branch is a pressure tunnel approximately 12 miles (19 km) long, 115 ft (35 m) deeper than the grade tunnel, terminating at 135th St. in Manhattan. The Jerome Park Branch (the Branch) is approximately 2 miles (3.2 km) long, with the same grade and cross section as the upstream Aqueduct terminating at Jerome Park Reservoir (the Reservoir). Water from the Branch enters the Reservoir at Gatehouses No. 5 and 7. The Aqueduct invert is at El. 122.1 ft msl (37.2 m) at Gatehouse No. 5. The Reservoir is normally maintained at approximately El. 134.5 ft msl (41.0 m) and currently serves as the receiving basin for the Aqueduct. Distribution and regulation of flow to the service area are from this location.

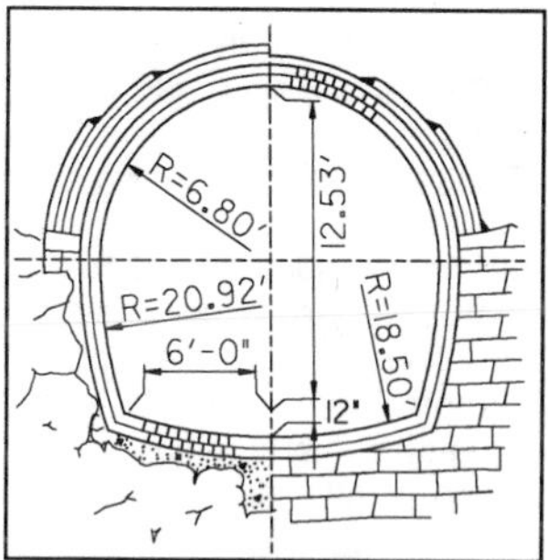

Figure 1 Typical Aqueduct cross section

The quality of the City's water supply is excellent and treatment currently consists of chlorination and fluoridation. However, the Croton basin has been subject to

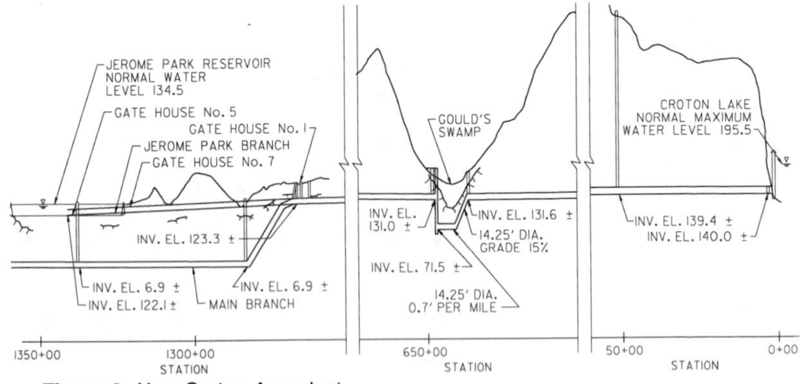

Figure 2 New Croton Aqueduct

continuing development, and this has recently affected water quality. A filtration plant is planned for Jerome Park Reservoir. Additionally, the City is examining the option of pressurizing the Aqueduct to increase its capacity and increase flexibility in utilizing its sources to meet demands during droughts. Harza Engineering Company (Harza) was retained by the DEP to evaluate the costs, benefits, and feasibility of pressurizing the Aqueduct. Accurate estimation of the Aqueduct capacity is necessary in order to assess the potential benefits of pressurization and to coordinate the capacities of the treatment plant and Aqueduct. Harza's services included a review of previous flow tests, inspection of the current condition of the Aqueduct, and a new flow test.

Previous Flow Tests

Results of four previous flow tests were evaluated. Tests made in 1963 and 1983 by the DEP using the color velocity method were the most well documented and indicated "n-values" of 0.016. However, they were for a section of the Aqueduct downstream of Jerome Park. Friction factors were expected to be higher than indicated by the Croton Aqueduct test based on published information (USBR). A new flow test was recommended to resolve the differences and to consider the Aqueduct upstream of Jerome Park Reservoir.

Aqueduct Inspection

A visual inspection of the entire Aqueduct was made during a dewatering in 1993. The brick lined tunnel was covered with a "slime" approximately 1/4 inch (.6 cm) thick. When the slime was removed in a program that included washing approximately 6 miles (9.7 km) of the Aqueduct walls, the underlying surface was found to be remarkably smooth. Standard size bricks, similar to current brick, had been used. The brickwork was uniform and the mortar joints were recessed from the brick face by up to 1/4 inch (.6 cm). The Aqueduct cross-section was found to be uniform throughout and of nominal dimensions.

Flow Test of 1994

A flow test program that included three different flow rates between 150 mgd and 300 mgd was developed and conducted in November and December of 1994. It was hoped to use the highest flow rate to pressurize the upstream portion of the Aqueduct to increase the hydraulic gradient and improve the accuracy of the roughness determination. However, this rate exceeds the combined demand and spill capacity at Jerome Park Reservoir and could not be used - the reservoir water level could not be stabilized during the test. Two test flow rates were selected - 150 mgd (232 cfs -6.6 m^3/s) and 200 mgd (309 cfs - 8.8 m^3/s). The first was approximately equal to the average demand on the Croton system at the time of the test and required few system adjustments. The second test required special considerations. Jerome Park water levels were lowered prior to the test so that, during the test, they would not exceed the spillway elevation, potentially wasting water. Additionally, the water level was monitored to verify that fluctuations during the test were minor. For the first test, flow conditions were stabilized more than 24 hours prior to water level measurements. For the second test, water levels were measured on two days - the first and second days after Aqueduct flows were stabilized.

Water levels were measured at six locations. Measurements commenced at the most downstream location and proceeded upstream to the new intake gatehouse where pressure gage readings, flow rates and other information were recorded. The procedure was then repeated in the opposite direction with the final measurement taken at the most downstream shaft. The procedure required an entire day. Water levels were measured downward from pre-set datums using a steel reinforced polyethylene flat tape water level probe with 0.01 ft gradations. The probe was lowered slowly until a steady audio tone was obtained. Multiple readings were taken to verify repeatability. Datum elevations had been established during the 1993 Aqueduct dewatering when the same tape and probe were used to measure the distance between the Aqueduct inverts and the datums.

Considerable coordination was required among DEP, upstate communities, and flow test personnel during the tests. Normal water supply operations were interrupted. Aqueduct flows were stabilized between one and two days prior to the tests. Water levels in Jerome Park Reservoir, which normally provides some daily flow regulation, were also stabilized. Upstate communities were asked not to withdraw water during the tests - all had alternative sources. Several communities gave permission for access to their facilities so that water level measurements could be made.

Results

Table 1 shows prevailing conditions during the two flow tests. Results of the flow tests including roughness coefficients are shown in Table 2. Roughness ("n") values were computed for Manning's equation using the standard step backwater method.

TABLE 1
Prevailing Conditions During Flow Tests

Condition	Flow Test 1	Flow Test 2
Average Daily Flow Rate, mgd	148	201
Water Level in Croton Lake, ft msl	195.95	196.31
Water Level in Jerome Park Reservoir, ft msl	133.3-133.4	132.4-132.6

TABLE 2
Flow Test Results

				Water Depths, ft		Roughness "n"[4]	
Location	Station	Invert Elev.	Minor Loss Coeff.[5]	Flow Test No.1	Flow Test No.2	Flow Test No.1	Flow Test No.2
u/s end	2+30	130.50	0.6	16.56	18.4[6]		
Shaft 3	176+97	137.72	-	6.51	8.15	.015	.016
Shaft 6	373+95	135.11	-	6.53	8.17	.014	.014
Shaft 9	495+30	133.50	2	6.81	8.37	.013	.013
Shaft 12	724+68	130.46	-	7.22	8.44	.014	.014
Shaft 18	1121+25	125.20	-	10.05	10.38	.013	.013

The lower friction factors were determined for areas of the Aqueduct that had been washed - Stations 373+95 to 495+30, 724+68 to 826+25 and 1021+25 to 1121+25. Minor loss coefficients were computed from standard texts. No minor losses were added for shafts and gate facilities. The apparent friction factor between Shaft 3 and the upstream end of the Aqueduct is higher because the Aqueduct in this area undergoes considerable change of section and alignment.

References

Brown, Abraham, "Report on Proposed Rehabilitation of Croton Water Supply Distribution System," prepared for New York City Board of Water Supply, 1965.

New York City Department of Environmental Protection, Bureau of Water Supply, "Sources of Supply and Consumption," Water Years 1982 - 1992.

U. S. Bureau of Reclamation, "Friction Factors for Large Conduits Flowing Full."

---

[4] "n" values shown are between indicated shaft and next uptream location.

[5] Coefficient of velocity head in Aqueduct

[6] Average of readings of 18.87 on day 1 of test and 17.95 on day 2 of test.

Irrigation/Drainage for Water Quality Management

J.C. Guitjens,[1] Member, ASCE, J.E. Ayars,[2] M.E Grismer,[3] Member, ASCE, and L.S. Willardson,[4] Member, ASCE

Abstract

Simplified equations used in drainage design for water table control must be replaced by numerical hydrodynamic models to delineate the flowpaths. Similarly, chemical transport must be modeled as a mechanistic process involving the chemical changes in the vadose and saturated zones. Hydrodynamic and chemical transport modeling can be used to explain the chemical changes of marginal quality waters used for irrigation and the influence of irrigation management on drainwater quality.

Introduction

Canal seepage and excessive irrigation have caused water logging and soil salinization in many irrigated areas. Drainage systems brought relief by removing excess soil water and salts leached from the root zone. Conceptually, designs dealt with water table control for predetermined rates of deep percolation.

When drainwater disposal became a problem, first in California (SJVDP 1990) and then elsewhere in western irrigated agriculture, emerging irrigation technologies and management practices were introduced to reduce

[1]Prof. of Irrig. Engrg., Dept. of Environmental and Resour. Sci., Univ. of Nevada, Reno, Reno, NV 89557.

[2]USDA-ARS, 2021 S. Peach Ave, Fresno, CA 93727.

[3]Assoc. Prof. Depts. of Agric. Engrg. and Land, Air and Water Resour. --Hydrol. Sci., Univ. of California, Davis, CA 95616.

[4]Prof. Emeritus, Dept. of Biological and Irrig. Engrg., Utah State Univ., Logan, UT 84322-4165.

drainage. The link between irrigation and drainage became obvious when irrigation scheduling on the basis of modeled evapotranspiration combined with improved irrigation efficiency through more uniform and controlled water application substantially reduced drainage flows. Salt mass discharge also decreased in rough proportion to decreasing drainwater volume (Ayars and Meek 1994).

The risks posed by drainwater to the ecology of receiving waters, such as wetlands, has already started to change the responsibilities assumed by irrigation enterprises. In some areas of California, farmers must dispose of drainwater through evaporation from collection ponds, while facing an uncertain future over the fate of accumulating salts precipitating as water evaporates (Tanji and Grismer 1987).

Our objective is to review traditional drainage design and to propose water quality related criteria for future drainage design.

Hydrodynamic Modeling for Drainage Design

Drainage design typically considered the effect of drain spacing and depth on controlling a water table that rose too close to the ground surface of irrigated fields. Assuming mostly horizontal flowpaths simplified the drainage equation. Steady state flow offered a simplification over transient flow (Workman et al. 1990, Bureau of Reclamation 1993). Hydrodynamic modeling considers accurate flow paths. Solute transport couples salt and water movement. Spatial heterogeneity of soils, soil water movement and chemical behavior still present many challenges to drainage design.

MODFLOW (McDonald and Harbaugh 1988) includes a drainage module that solves the equation for saturated flow using the finite difference method. Solution of the governing partial differential equation of flow eliminates the need to assume horizontal flow.

Pohll and Guitjens (1994) used MODFLOW to model transient flow to 15 parallel drains and to delineate the region of local flow to the drains and a deeper region exhibiting regional flow. As time after an irrigation event increased, the region of local flow to the drains first became deeper and then diminished in depth. Flowpaths indicated how drains collected displaced waters that may have resided in the aquifer for many years and have their origin in areas beyond the boundaries of a specific field. Fio and Deverel (1991) used the MODFLOW module to develop the flowpaths to drains at depths of 1.8 m and 2.7 m below the land surface and show that flowpath depth was increased as the drain depth increased. Water displacement resulted in long residence time and poor quality drainwater. Fio (1994) demonstrates the significance of regional flow in identifying contributing sources to drain flow.

Transport Modeling

Dudley et al. (this issue) consider leaching fraction and leaching requirement to be conceptual models whereas process-oriented models describe realistic mechanisms. Ideally, mechanistic models should give reasonable estimates of the mass of water, salt and salt components involved in deep percolation. When the water flows from the vadose zone to the water table and becomes groundwater, the saturated conditions change the chemical processes (Jurinak and Tanji 1994). Even process-oriented models of the vadose and groundwater zones are based on simplifying assumptions such as chemical equilibrium. Knowledge of vadose zone and aquifer minerals, pH and redox status, and the heterogeneous nature of the medium is difficult to acquire. To determine pathways, particle velocity and mass movement of water, hydrodynamic models are needed; but in combination with the chemical processes along the pathways for the duration that water is in residence, solute transport can also be modeled.

Version 1.21 of SWMS_2D (Šimůnek et al. 1994) is a comprehensive 2-D flow and solute transport model that documents a range of published research experience in parameter definition. Richard's equation applied to saturated and unsaturated flow is the governing flow equation. A two-dimensional chemical transport equation considers solution concentration, adsorbed concentration, volumetric flux, rate constants, a sink term for water and salt, and dispersion.

Irrigation Management

Collection of drainwater and the use of drainwater for irrigation offers an opportunity to conserve water (Guitjens 1993; Rhoades 1989). If drainage from field irrigation equals 35% of the applied water and if the drainwater is collected and used for irrigation at the same drainage proportion of 35%, the net drainage will drop to 12%. The steady-state concept (Hoffmann 1990) applied to 35% leaching would increase the salt concentration to 2.86 times the initial salt concentration of the irrigation water. Similarly, the 12% net drainage would elevate the salt concentration by a factor of 8.33. However, the drainwater is a combination of water displaced from the aquifer into the drains and deep percolation, and will reflect the relative combination from each source.

Research has shown that a net drainage reduction will decrease the mass output of salts (Ayars and Meek 1994). An increase in salt concentration has a much smaller effect on mass output than the influence of volume reduction. If the aquifer contributes excessive amounts of salts or a trace element like B, the tolerance of sensitive crops could become a problem if drainwater is used for irrigation.

Summary

Simplifying equations used in conventional drainage design for water table control have limited value in drainage design for the control of drainage quality. Instead, hydrodynamic models that display pathways and calculate particle velocity and mass movement are needed together with mechanistic chemical models that give reasonable estimates of salt transport in the vadose and saturated zones. Mechanistic models will also show the effect of irrigation management alternatives.

Appendix. References

Ayars, J.E. and Meek, D.W. 1994. "Drainage load-flow relationships in arid irrigated areas." Transactions of the ASCE, 37:431-437.

Bureau of Reclamation. 1993. "Drainage manual." U.S. Dept. Interior, Bureau of Reclamation.

Dudley, L.M., Grismer, M.E., Suarez, D.L. and Willardson, L.S. 1995. "Hydrodynamics and chemical transport in the root zone and shallow groundwater system: modeling."

Deverel, S.J. and Fio, J.L. 1991. "Groundwater flow and solute movement to drain laterals, western San Joaquin Valley, California .1. Geochemical Assessment." Water Resour. Res., 27(9), 2233-2246.

Fio, J.L. 1994. "Calculation of a water budget and delineation of contributing sources to drainflows in the western San Joaquin Valley, California." USGS Open-File report 94-45.

Fio, J.L., and Deverel, S.J. 1991. "Groundwater flow and solute movement to drain laterals, western San Joaquin Valley, California, 2. Quantitative hydrologic assessment." Water Resour. Res., 27(9), 2247-2257.

Guitjens, J.C. 1993. "Using drainage effluent for irrigation." Trans. ICID, v. 1-D, Q44, R 115:1447-1459.

Hoffman, G.J. 1990. "Leaching fraction and root zone salinity control." IN: Tanji, K.K. (ed.), Agricultural salinity assessment and management. ASCE Manuals and Reports on Engr. Practice No. 71, p. 237-261.

Jurinak, J.J. and Tanji, K.K. 1994. "Geochemical factors affecting trace element mobility." J. Irrig. Drain. Engr., ASCE, 119(5), 848-867.

McDonald, M.G., and Harbaugh, A.W. 1988. "A modular three-dimensional finite difference ground-water flow model." Techniques of water-resource investigations of the United States Geological Survey, Book 6, modeling techniques, U.S. Geological Survey.

Pohll, G.M., and Guitjens, J.C. 1994. "Modeling regional flow and flow to drains." J. Irrig. Drain. Engrg., ASCE, 120(5), 925-93

Rhoades, J.D. 1989. "Intercepting, isolating and reusing drainage waters for irrigation to conserve and protect water quality." Agric. Water Mgt. 16: 37-52.

San Joaquin Valley Drainage Program. 1990. "The problem." p. 183 IN U.S. Department of Interior and California Resources Agency (ed.) A Management Plan for Agricultural Subsurface Drainage and Related Problems on the westside San Joaquin Valley. California Department of Water Resources, Sacramento, California.

Šimůnek, J., Vogel, T., and van Genuchten, M. Th. 1994. "The SWMS_2D code for simulating water flow and solute transport in two-dimensional variably saturated media." Research Report No. 132, U.S. Salinity Laboratory, Riverside, CA.

Tanji, K.K. and Grismer, M.E. 1987. "Evaporation ponds for the disposal of agricultural water. California State Water Resources Control Board." Contract No. 3-190-150-0. Quarterly Prog. Report, July-Sept. 1987. 35 pp.

Workman, S.R., Skaggs, R.W., Parsons, J.E., and Rice, J. 1990. "DRAINMOD." North Carolina State U., Raleigh, NC.

WATER QUALITY AS A DESIGN CRITERIA IN IRRIGATION AND DRAINAGE WATER MANAGEMENT SYSTEMS

J.E. Ayars,[1] M.E. Grismer and J.C. Guitjens

Abstract

As water quality criteria become more restrictive in streams and rivers receiving drainage flows, it will be necessary to include drainage water quality in the design of new drainage systems. The existing drainage design criteria and methods are reviewed and changes are suggested to account for water quality. Proposed changes include reducing recommended depth to mid-point water table, reducing drain depth and spacing, and including crop water use in design.

Introduction

With a population of nearly 33 million people there is a constant, often conflicting, demand in California between agricultural, municipal, industrial, and environmental uses for water. Approximately 85% of the developed water supply in California is used primarily in irrigated agriculture and drainage is a necessary consequence of this usage.

The State Water Resources Control Board is currently revising the standards for water quality in the San Joaquin River to protect wildlife habitat and the public. Part of this effort is the reduction of permissible concentrations of boron, toxic trace elements and nitrate in the river. The net effect on irrigated agriculture will be to reduce the load of salt and other inorganic elements being discharged into the river via drainage sloughs and canals.

The water quality impact of drainage water discharge into the San Joaquin River will be determined, not by the absolute concentration of an element, but by the total load of salt or the element being discharged. This will have to be

---

[1]Acting Research Leader, USDA-ARS Water Management Research Laboratory, 2021 S. Peach Ave., Fresno, CA 93727

considered in the context of the existing flow and concentration in the river at the time of discharge. Concentration of salt and other elements in the river can be managed by storing drainage water in reservoirs and pumping it into the river during high stream flows. An alternative would be to reduce net drainage flows to a level not requiring summer storage. The Drainage Program Report (San Joaquin Valley Drainage Program, 1990) recommended improved irrigation water management to reduce deep percolation and drainage as the first step in solving the drain water disposal problem. Little consideration was given to developing new criteria for the design and management of subsurface drainage systems since they are no longer an option on the west side of the San Joaquin Valley. Drainage is an option for other areas and new designs should include criteria which consider the load-flow relationship of salt and elements contained in drainage water (Ayars and Meek, 1994).

The objective of this paper is to review existing design methods and criteria for drainage systems in arid and semi-arid irrigated areas and to suggest new design methods and criteria for subsurface drains that include drain water quality and load considerations.

Review of Current Design Practices (USBR)

The U.S. Bureau of Reclamation (USBR) is the principal U.S. government agency involved with the design and installation of drainage systems in arid irrigated areas. Their manual details the design process from preliminary field investigation to actual installation (U.S. Department of Interior, 1993). A few of the salient features of the process will be reviewed in this paper but for a complete treatment of the subject the reader is referred to the Drainage Manual (U.S. Department of Interior, 1993).

Some of the data required for design of a subsurface drainage system include soil hydraulic properties, soil layering, depths to restricting layers, cropping patterns, irrigation schedules, type of irrigation system, irrigation efficiency, climatic data, depth to water table, sources of water other than deep percolation, and salinity status of soil and groundwater. Lateral spacings in the design are determined using the transient design procedure pioneered by the USBR.

Soils data collection and analysis is common to any design procedure. Current soils investigations go to the depth of potential drain placement (2-4 m) with selected deeper investigations to determine the presence of a restricting layer. The salinity status in the soil profile above the drains is noted as part of the investigation. A representative crop rotation is selected and an irrigation schedule is established for the rotation based on the water holding capacity of the soil, an allowable depletion, and existing climatic data. After a representative irrigation efficiency is selected and the leaching requirement is established based on the crop salt tolerance, the data are combined to calculate a deep percolation value for the transient design.

There are no specific design criteria lists for all to use. There are many implicit considerations in the process which have worked well over the years, e.g., the mid-point water table height. The USBR recommendation is to keep the water table from 1.1 to 1.5 m below the soil surface at the mid-point, depending on the rooting depth of the crop, which should result in achieving at least 90% of maximum production. The transient design starts when the water table is nearest the soil surface and ends when the final irrigation results in a buildup of the water table to the prescribed design depth. The lateral spacing is adjusted until the mid-point criteria is met.

The recommendation to install the drains at a depth of 2.4 meters, if possible, provides a balance between the cost and spacing. Deep placement results in a wide drain spacing, lowering the system cost relative to the shallow, more narrowly spaced drains. The leaching requirement is calculated using the soil salinity, the crop salt tolerance and the irrigation water quality.

The shallow groundwater quality is noted in the investigation phase, as is any available water quality data from existing drainage systems. These data can then be used to apply for any required discharge permits, but as of yet, are not used to modify the lateral depth and spacing.

Proposed Changes in Design Practices

The proposed changes in design practices are based on findings that deep, widely spaced drains tend to remove more water than is needed to prevent waterlogging (Doering et al., 1982) and also creates deep flow paths that potentially drain poorer quality groundwater than is found at the drain installation depth (Deverel and Fio, 1990). The first proposed design change is to relax the required depth to the water table to approximately 0.9 m for all situations. Secondly, drain installation depths should be reduced resulting in a reduction in the lateral spacing. By reducing the drain depth and spacing, the depth of the flow lines is reduced, and where the water quality gets poorer with depth, less water will be extracted from this zone (Grismer, 1990). The reduction in drain depth will also lead to reduced volumes of water requiring disposal. In order to maintain a reasonable drain spacing it will be necessary to relax the requirement for the mid-point water table depth.

Ayars and McWhorter (1985) demonstrated that by including crop water use from shallow groundwater, the lateral spacing increases for a given drain depth. By including the crop contribution to groundwater removal, the drain flow is reduced, the depth to water table is increased, and the volume of irrigation water is reduced. This is only true if the shallow groundwater quality is suitable for use by the crop.

Implicit in the proposed changes is the need to revise the crop tolerance data with regard to the management of salinity in the root zone. Research has found that the plant responds to the average root zone salinity. Rhoades (1984) demonstrated that plants are more tolerant than previously thought and as a result, higher levels of soil salinity can be tolerated before yield losses were incurred.

The proposed changes relate to positioning of the drains, both the depth and lateral spacing, and the effect of irrigation management. Determining the depth and spacing through numerical modelling may require exhaustive amounts of data not routinely collected during the preliminary investigation phase. This procedure is possible on a very limited scope and not practical for a wide range of applications. Grismer (1990) has demonstrated the possibilities using both steady state and transient analysis. Garcia et al. (1994) have a model which includes many of the proposed changes in the design approach, providing a capability of evaluating the impact of these proposed changes. If water quality is to be included in the design, it is important to establish what parameters are to be considered. There is evidence suggesting that the concentration of salt and other elements in shallow groundwater, and consequently drainage water quality, is relatively invariant with time (Ayars and Meek, 1994). If this is true for the area under investigation, controlling the salt or trace element output load is accomplished by controlling the drainage discharge. Implementation of the suggested changes in the design will result in reduced discharge and probably changes in the timing of the discharge. Water quality parameters will be included in the lateral depth and spacing based on estimates of total salt load in the drain water and timing of the discharge.

After completing the design for several drain depths and spacings it would be possible to do a flow net analysis and determine the depth of the stream tubes and the probable contribution to flow and load from each layer of soil (Grismer, 1990). The calculated salt load being discharged into the receiving stream and the existing salt load in the stream will be used to calculate the final concentration which will be compared to the receiving water quality standards.

Summary

Existing drainage design criteria were reviewed to identify areas that might be modified to include water quality in the design process. Modifiable criteria include the mid-point water table height and the depth and spacing of the drain laterals. These parameters are inter-connected, so it is not feasible to change one without changing the others. Water quality impacts will have to be accounted for by designing for reduced load. This is possible by reducing the drain volume and improving the quality of the drain water discharge.

References

Ayars, J.E., and D.B. McWhorter. 1985. Incorporating crop water use in drainage design in arid areas. pp. 380-389 *IN*: C.G. Keyes and T.J. Ward (ed.) Proc. Specialty Conference, Irrigation and Drainage Division, ASCE. American Society of Civil Engineers, New York, NY.

Ayars, J.E., and D.W. Meek. 1994. Drainage load-flow relationships in arid irrigated Areas. Transactions of the ASAE, 37:431-437.

Deverel, S.J., and J.L. Fio. 1990. Ground-water flow and solute movement to drain laterals, western San Joaquin Valley, California: I. Geochemical assessment. Sacramento, CA, U.S. Geological Survey, Report 90-136.

Doering, E.J., L.C. Benz, and G.A. Reichman. 1982. Shallow-water-table concept for drainage design in semiarid and subhumid regions. pp. 34-41 *IN*: American Society of Agricultural Engineers (ed.) Advances in Drainage, Proc. Fourth National Drainage Symposium. ASAE Publication 12-82, American Society of Agricultural Engineers, St. Joseph, MI.

Garcia, L.A., K.M. Strzepek, and T.H. Podmore. 1994. Design of agricultural drainage with adaptive irrigation management. ASCE Journal of Irrigation and Drainage Engineering, 120:179-194.

Grismer, M.E. 1990. Subsurface drainage system design and drain water quality. ASCE Journal of Irrigation and Drainage Engineering, 119:537-543.

Rhoades, J.D. 1984. New strategy for using saline waters for irrigation. pp. 231-236 *IN*: J.A. Replogle and K.G. Renard (ed.) Proc. ASCE Irrig. and Drainage Specialty Conference, American Society of Civil Engineers, New York, NY.

San Joaquin Valley Drainage Program. 1990. The problem. p. 183, *IN*: U.S. Dept. of Interior and California Resources Agency (ed.) A Management Plan for Agricultural Subsurface Drainage and Related Problems on the Westside San Joaquin Valley. California Department of Water Resources, Sacramento, CA.

U.S. Department of Interior. 1993. Drainage manual. U.S. Department of Interior, Denver, CO. 321 pp.

Irrigation with Marginal Quality Waters: Issues

K.K. Tanji, M. ASCE

Abstract

Due to increasing regulations on the discharge of marginal quality waters as well as decreasing availability of freshwater resources, there is a need to consider the expanded use and reuse of marginal quality waters. This paper addresses the potential reuse of treated municipal wastewaters, food processing wastewaters, confined animal lagoon waters, and irrigation drainage waters on croplands and agroforestry systems. The efficacy of irrigation with marginal quality waters is dependent on several factors such as water quality characteristics and site-specific crop, soil and climatic conditions. Opportunities exist to beneficially use marginal quality waters. However, site-specific BMPs are required to manage potentially adverse impacts.

Introduction

There is increasing competition for water quantity and water quality from various sectors of society. Since agricultural water use is large in comparison to other water usages, it is likely that some shift in water from agriculture to other uses will occur. Moreover, with increasing concern for the environment, additional water is expected to be transferred from agriculture to fish and wildlife. Therefore, it is essential that wise management and judicial use of water for crop and animal production be promulgated. When freshwater resources become limiting, there is a need for expanded use of marginal quality waters. This paper addresses the potential reuse of treated municipal wastewaters, food processing wastewaters, confined animal lagoon waters, and irrigation drainage waters on croplands and agroforestry systems.

Municipal Wastewaters

The use of reclaimed sewage effluents on croplands and landscapes is widely but not extensively practiced in the U.S. and abroad (Pettygrove and Asano, 1985; EPA, 1992). Although irrigation with sewage effluents is, in itself, an effective form of land treatment, there is a need for preapplication treatment (Asano, 1994). Preapplication treatment is practiced to protect public health from pathogens and prevent nuisance conditions such as

----

Prof., Hydrologic Science Program, Dept. of Land, Air and Water Resources, University of California, Davis, CA

odors during storage and application. Guidelines for the interpretation of water quality for irrigation purposes are available, particularly for arid regions (e.g., Ayers and Tanji, 1981; Pratt and Suarez in Tanji, 1990). These guidelines stress the importance of the effects of water salinity on plants and the need to maintain a favorable salt balance in the root zone, water sodicity (SAR) on water intake rates and soil permeability, and specific ions toxic to sensitive plants These guidelines apply not only to wastewaters but also to freshwaters to evaluate their suitability for irrigation.

Particular constituents of concern in the use of municipal wastewaters are the pathogens that can potentially affect human health, and nitrates and trace organics that can potentially contaminate ground waters. Disinfection of wastewaters by chlorination and other oxidants reduces pathogen concentrations but not complete elimination. Waterborne diseases can be transmitted to humans by either direct contact (body contact, ingestion or inhalation of infectious agents) or indirect contact with objects previously contaminated by the wastewater (Asano, 1994).

Organic nitrogen and ammonia are oxidized to nitrate in the treatment of sewage effluents and in land application. Although nitrates are readily assimilated by plants, they are subject to leaching into ground water basins because the mass loading of nitrates may exceed N requirements of plants (Bouwer and Idelovitch, 1987). The hydraulic loading rates of reclaimed waters are commonly based on soil hydraulic properties, climate and consumptive use of water, and rarely on the basis of potential contaminant leaching losses such as nitrates.

Trace organics in wastewaters, especially those listed in EPA's priority pollutants, are of concern. As in pathogens, treatment processes may reduce the concentration levels of trace organics but not complete elimination. The fate of trace organics in soils is affected by sorption, biodegradation and volatilization. Because of these mechanisms, trace organics are strongly attenuated in soils and their concentrations in percolating waters are expected to be insignificant (Frankenberger in Pettygrove and Asano, 1985). However, ground water contamination may occur if the sewage effluent contains unusually high concentrations of refractory organics (e.g., chlorinated hydrocarbons) from industrial contributions (Bouwer and Idelovitch, 1987).

Processing Wastewaters

There is increasing reuse of wastewaters generated from food processing plants such as canneries, sugar mills and wineries due to increasingly stringent wastewater discharge requirements and increasing costs to dispose into municipal wastewater treatment facilities. Wastewaters from food processing facilities contain high levels of BOD and COD, organic SS (e.g., fruit and vegetable pulp), inorganic SS (e.g., soil), and residues of chemicals used in processing (e.g., lye used in the peeling of fruits) (Eckenfelder, 1970; SCS, 1992).

In the land application of food processing wastewaters by the slow rate method, BOD and SS are effectively removed by filtration and bacterial action as the wastewater percolates through the soil. BOD and SS are typically reduced to concentrations of less than 2 mg/L and 1 mg/L, respectively, and their loading rates are not normally of concern (EPA, 1981). The principal concern is prevention of runoffs.

Confined Animal Lagoon Waters

Liquid and slurry manures are impounded into waste storage ponds known as lagoons In confined animal facilities such as dairies. The physical and chemical quality characteristics of lagoon waters are highly variable within any given confined animal operation (dairy cattle, beef cattle, swine, or poultry) and are also dependent on the treatment process (SCS, 1992). Anaerobic lagoons are widely practiced for the treatment of animal wastes. The anaerobic treatment reduces BOD and COD by microbial and chemical degradation, N by ammonia volatilization, and odors to some extent. In contrast, aerobic lagoons are used if reduction of odors is the major objective.

Since the impounded lagoon waters may become quite concentrated in dissolved solids (DS), total solids (TS) and nutrients (N, P, K), lagoon water is frequently diluted with the fresh water and then applied to croplands. The principal constituents of concern in applied lagoon waters are salinity and nutrients. In arid regions, salinity is a major detriment to crop production while in humid regions it may not be of so much concern. About 50% of the N and over 75% of the K are in the urine while 80% of the P are found in animal feces (SCS, 1992). Phosphorus is a comparatively insoluble element while K tends to be fixed by soils. Consequently, the highly mobile nitrate is the nutrient of most concern in regard to ground water pollution and P to surface waters. Other concerns about lagoon waters are water borne pathogens such E. Coli, Giardia and Cryptosporidium.

Irrigation Drainage waters

In terms of volumes reused, the intentional as well as well as nonintentional uses of irrigation drainage waters far exceed other marginal quality water uses. The nonintentional uses of wastewaters include irrigation diversion of streams receiving upstream discharge of return flows are agriculture, municipalities and industries. As pointed out earlier, the suitability of any water for irrigation is evaluated with water quality guidelines.

Under aridic conditions, irrigation drainage waters are saline from the evapoconcentration of applied waters and naturally occurring sources of salts in the soil and subsurface (Tanji, 1990). In some regions geochemical sources of trace elements are of concern, including boron, selenium, arsenic, molybdenum, uranium and vanadium. Elevated concentrations of boron, sodium and chloride in waters are toxic to many plants.

A number of different strategies for reusing saline drainage waters have been proposed (Grattan and Rhoades in Tanji, 1990). The blending strategy involves mixing two waters of different qualities to obtain water suitable for irrigation. Blending freshwater into saline drain waters is not recommended, although commonly practiced. When a high quality water is blended, its full use by salt- sensitive plants is impaired because of the admixture of salts. The cyclic strategy uses multiple sources of water with one of them of high quality. This cyclic strategy requires a crop rotation of moderately salt sensitive to salt tolerant crops. The low salt water would be used to germinate salt tolerant crops and all irrigations of the moderately salt tolerant crops. The saline waters would be used to irrigate salt tolerant crops after they have reached a salt tolerant growth stage. Low salt water may be used at later irrigations of the salt tolerant crop to leach salts from the upper root zone prior to planting the more salt-sensitive succeeding crop.

Saline drainage waters may be used to irrigate salt tolerant trees and halophytes to reduce the drainage volume and concentrate the soluble salts for easier management and disposal (Tanji and Karajeh, 1973). A eucalyptus plantation successfully used 10 dS/m drainwater containing 12 mg/L B for five years. With a Leaching Fraction of about 15% for this marginal quality water, there was however a buildup of salts and boron in the root zone that reduced ET of the eucalyptus trees.

Discussion and Summary

Opportunities exist to beneficially use waters of marginal qualities. The use of marginal quality waters can conserve higher quality waters for other beneficial uses of water. Wastewaters such as sewage effluents and irrigation drainage waters should be considered as part of the total water resources, especially in water-short regions. The uses of marginal quality waters as irrigation water will help reduce pollution of surface and ground waters. Such uses of marginal quality waters require more intensive management and monitoring than use of higher quality waters. Successful long-term usage of marginal quality waters requires a combination of site-specific water, soil and cropping practices. The mass loadings of salts, nutrients, trace elements, pathogens and trace organics need to be considered so as not to exceed the soil's capacity to assimilate or degrade them. The upper bounds of wastewater usage are not fully known and cumulative impacts on the environment and public health require additional multidisplinary research.

References

Asano, T. 1994. Irrigation with Treated Sewage Effluents. Chapter 9, pp 199-228, IN Management of Water Use in Agriculture, K.K. Tanji and B. Yaron, ed., Springer-Verlag.

Ayers, R.S., and K.K. Tanji. 1981. Agronomic Aspects of Crop Irrigation with Wastewater. ASCE Water Forum '81, Vol I: 578-586.

Bouwer, H, and E. Idelovitch. 1987. Quality Requirements for Irrigation with Sewage waters. ASCE Journal of Irrigation and Drainage Engineering 113(4):516-535.

Eckenfelder, Jr., W.W. 1970. Water Quality Engineering for Practicing Engineers. Barnes and Noble, Inc. 328 p.

EPA. 1981. Process Design Manual for Land Treatment of Municipal Wastewater. EPA 625/1-81-013, Wash., D.C. pp 1-1 to 9-29 and app A-1 to G-3.

EPA. 1992. Guidelines for Water Reuse. Manual. EPA/625/R-92/004. U.S. Environmental Protection Agency and U.S. Agency for International Development, Wash., D.C. 247 p.

Pettygrove, G.S., and T. Asano. 1985. Irrigation with Reclaimed Municipal Wastewater- A Guidance Manual. Lewis Publishers, Inc., Chelsea, MI, pp 1-1 to 15-20 and app A-1 to I-1.

Soil Conservation Service. 1992. Agricultural Waste Management Field Handbook. A National Engineering Handbook. U.S. Soil Conservation Service.

Tanji, K.K., Editor. 1990. Agricultural Salinity Assessment and Management. ASCE Manual No. 71, American Society of Civil Engineers. 619 p.

Tanji, K.K., and F.F. Karajeh. 1973. Saline Drainwater Reuse in Agroforestry Systems. ASCE Journal of Irrigation and Drainage Engineering 119(1):170-180.

Options for Saline Water Disposal - Irrigation Case Studies

Guy Fipps[1]
Member ASCE

Abstract

Two large-scale projects are reviewed in which saline wastewater was disposed of by irrigating crops and forage. One project is located in West Texas and disposes municipal effluent. The other projected receives runoff collected at a 45,000 head cattle feedlot located on the Texas Southern High Plains. Information is provided on the quality of the wastewater and changes in soil salinity after several years of irrigation. Lack of adequate leaching fraction resulted in a serious increase in soil salinity in both projects.

Introduction

In semi-arid and arid regions, wastewater is viewed as an important resource that can be used to provide or supplement irrigation water for crop and forage production. Most design procedures and regulations are based on the nitrogen content of effluent. Thus, permitted application rates and amounts are designed to prevent excessive leaching and runoff of nitrogen, not for salinity management.

In West Texas, wastewater often contains high levels of dissolved solids and, in many areas, high concentrations of sodium. For sustainability, detailed salinity and water management plans must be formulated and implemented, including adequate irrigation to allow for leaching. Here, two irrigation projects are reviewed. In both cases, no excess irrigation was provided for leaching. As a result, soil salinity levels rose significantly after only a few years.

---

1 Associate Professor, Department of Agricultural Engineering, Texas A&M University, College Station 77843-2117.

Case 1: Municipal Effluent Irrigation

A city in West Texas established an irrigation project to dispose of secondary effluent. The project involved application of the effluent on 1920 acres with 22 center pivot machines. The city generates from 4 to 8 mgd (million of gallons per day) of secondary effluent. Eleven of the pivots covered from 110 to 120 acres each. The remaining area was composed of 3 full-circles and 8 half-circles ranging from 36 to 95 acres each. The site was leased to an individual who had total control of maintenance of the equipment and management of the irrigation systems. After five years, soil salinity problems had developed to such an extent that the city requested assistance in evaluating the project by the Texas Agricultural Extension Service.

Effluent Quality

A detailed salinity analysis of the effluent is not available. However, the TDS (total dissolved solids) of the effluent ranges from 1925 to 2225 mg/l and the EC ranges from 3.6 to 5.3 mmhos/cm (Table 1). Contributing to the salinity problem was the travel time of the effluent though 8 storage ponds. In certain months, EC increased by as munch as 22% and total dissolved solids by as much as 25 % (outflow effluent compared to inflow effluent) by the time it reached the terminal storage pond.

Table 1. Analysis of effluent quality in the municipal effluent irrigation project during one year.

month	TDS mg/l	EC micromhos/cm	Chlorides mg/l	Sulfates mg/l
JAN	2225	4050	737	413
FEB	2085	3600	800	562
MAR	1965	3900	675	637
APR	2075	4900	825	512
MAY	2090	4200	837	562
JUN	2170	4300	850	600
AUG	2335	4100	800	612
SEP	1925	5300	762	512
OCT	2330	5000	700	537
NOV	1960	4250	725	38
DEC	2175	3650	-	-

Irrigation and Soil Salinity

Land use at the site was originally planned as a cotton and forage rotation. However, cotton was abandoned after one or two seasons by the farmer. Saline areas formed under parts of 18 of the 22 circles. Table 2 gives the irrigation amounts, crops and soil types of three of the areas which are representative of the entire site. Table 3 gives the approximate size of the salt-affected areas and the soil analysis of these affected areas. A decrease in soil infiltration rate was also noted in the saline areas, falling from a range of 1.6 cm/hr to 3.7 cm/hr, to a range of 0.3 to 0.4 cm/hr. Most of the pivots were designed to discharge water at rates above 2.5 cm/hr.

Table 2. The area, soil type, crops and amount of irrigation under three circles during the fifth year of the municipal irrigation project.

circle #	total area acres	soil type	crops	feet of irrigation
10-4	119	sandy loam	forage/millet	3.7
11-2	120	loamy sand	forage	3.7
11-4	120	loamy sand	forage	1.6

Table 3. The total salt-affected areas and soil analysis for the three circles given in Table 2.

circle #	salt-affected acres	EC mmhos/cm	sodium mg/l	SAR	SSP
10-4	20	24	2431	15	50
11-2	55	36	3923	21	57
11.4	10	23	2699	19	61

Recommendations

A team of engineers and scientists from the Texas Agricultural Extension Service studied the effluent irrigation project. Among their recommendations were the following:

1. A high managerial team capability should be formed to properly managing the project.

2. The project should be viewed as effluent disposal, not as a profit making operation site. Annual operating expenses of the intensive farm must be met by the city, estimated at $200,000 to $300,000.

3. Permanent forage crops should be established (capital requirements estimated at $200,000 to $400,000). Recommended crops include Bermuda grass and Jose tall wheat grass.

4. A soil moisture monitoring program should be implemented to assess irrigation water penetration and crop uptake of water, and to assist in water management by the farmer.

Case 2: Irrigation with Feedlot Runoff

Sweeten, et. al (1994) reported on the results of a four-year study involving irrigation of winter wheat with collected runoff from a 45,000 head cattle feedlot on the Southern High Plains of Texas. Four irrigation application rates were examined: 0, 4.0, 6.7 and 9.2 inches annually. Rainfall averaged about 15 inches during the first three years of the study.

Wastewater Quality and Soil Salinity

The effluent used for irrigation had an average EC of 12.1 mmhos/cm, an SAR of 11.5, and an SSP of 31.8. The soil salinity levels increased significantly during the first three years of the project as shown in Table 4.

Following the third year of the study, the irrigated field was deep chiseled. Fortunately, the area sequentially received nearly 37 inches of rainfall (during the fourth year). The deep chiseling and large rainfall combined to produce sufficient leaching to lower soil salinity by a factor of about 2 (Table 5). However, even with this large amount of rainfall, soil salinity still remained much higher than in the unirrigated plot. The EC remained at least 3 times higher and the SAR ranged from 8 to 20 times higher.

Table 5. Soil salinity analysis under four levels of irrigation after three years in the feedlot runoff project.

parameter	annual irrigation amount (inches)			
	0	4.0	6.7	9.2
EC, mmhos/cm	0.5	4.0	3.7	5.5
Na, ppm	12	276	273	402
Ca, ppm	79	503	459	586
Mg, ppm	9	55	53	69
Cl, ppm	499	3193	2970	4607
SAR	0.3	3.2	3.2	4.3

Table 6. Soil salinity analysis under four levels of irrigation after year four following 36 inches of rainfall in the feedlot runoff project.

parameter	annual irrigation amount (inches)		
	4.0	6.7	9.2
EC, mmhos/cm	1.4	1.8	1.3
Na, ppm	155	142	206
Ca, ppm	211	241	162
Mg, ppm	35	31	22
Cl, ppm	577	784	352
SAR	8.5	3.3	4.6

Reference

Sweeten, J.M., G.L. Sokora, R.M. Seymour, M.G. Hickey, and S.M. Young. *Irrigation of Cattle Feedlot Runoff on Winter Wheat*. Proceedings of the Great Plains Animal Waste Conference on Confined Animal Production and Water Quality, Oct. 19-21. Great Plains Agricultural Council Publication Number 151. 1994.

SUBJECT INDEX

Page number refers to first page of paper

AUTHOR INDEX

Page number refers to first page of paper